21st Century Nanoscience –

21st Century Nanoscience – A Handbook

Advanced Analytic Methods and Instrumentation (Volume Three)

Edited by

Klaus D. Sattler

CRC Press
Taylor & Francis Group
Boca Raton London New York

CRC Press is an imprint of the
Taylor & Francis Group, an **informa** business

CRC Press
Taylor & Francis Group
6000 Broken Sound Parkway NW, Suite 300
Boca Raton, FL 33487-2742

First issued in paperback 2022

© 2020 by Taylor & Francis Group, LLC
CRC Press is an imprint of Taylor & Francis Group, an Informa business

No claim to original U.S. Government works

ISBN-13: 978-0-815-38473-1 (hbk)
ISBN-13: 978-1-03-233649-7 (pbk)
DOI: 10.1201/9780429340420

Library of Congress Cataloging-in-Publication Data

Names: Sattler, Klaus D., editor.
Title: 21st century nanoscience : a handbook / edited by Klaus D. Sattler.
Description: Boca Raton, Florida : CRC Press, [2020] | Includes bibliographical references and index. | Contents: volume 1. Nanophysics sourcebook—volume 2. Design strategies for synthesis and fabrication—volume 3. Advanced analytic methods and instrumentation—volume 5. Exotic nanostructures and quantum systems—volume 6. Nanophotonics, nanoelectronics, and nanoplasmonics—volume 7. Bioinspired systems and methods. | Summary: "This 21st Century Nanoscience Handbook will be the most comprehensive, up-to-date large reference work for the field of nanoscience. Handbook of Nanophysics, by the same editor, published in the fall of 2010, was embraced as the first comprehensive reference to consider both fundamental and applied aspects of nanophysics. This follow-up project has been conceived as a necessary expansion and full update that considers the significant advances made in the field since 2010. It goes well beyond the physics as warranted by recent developments in the field"—Provided by publisher.
Identifiers: LCCN 2019024160 (print) | LCCN 2019024161 (ebook) | ISBN 9780815384434 (v. 1 ; hardback) | ISBN 9780815392330 (v. 2 ; hardback) | ISBN 9780815384731 (v. 3 ; hardback) | ISBN 9780815355281 (v. 4 ; hardback) | ISBN 9780815356264 (v. 5 ; hardback) | ISBN 9780815356417 (v. 6 ; hardback) | ISBN 9780815357032 (v. 7 ; hardback) | ISBN 9780815357070 (v. 8 ; hardback) | ISBN 9780815357087 (v. 9 ; hardback) | ISBN 9780815357094 (v. 10 ; hardback) | ISBN 9780367333003 (v. 1 ; ebook) | ISBN 9780367341558 (v. 2 ; ebook) | ISBN 9780429340420 (v. 3 ; ebook) | ISBN 9780429347290 (v. 4 ; ebook) | ISBN 9780429347313 (v. 5 ; ebook) | ISBN 9780429351617 (v. 6 ; ebook) | ISBN 9780429351525 (v. 7 ; ebook) | ISBN 9780429351587 (v. 8 ; ebook) | ISBN 9780429351594 (v. 9 ; ebook) | ISBN 9780429351631 (v. 10 ; ebook)
Subjects: LCSH: Nanoscience—Handbooks, manuals, etc.
Classification: LCC QC176.8.N35 A22 2020 (print) | LCC QC176.8.N35 (ebook) | DDC 500—dc23
LC record available at https://lccn.loc.gov/2019024160
LC ebook record available at https://lccn.loc.gov/2019024161

Visit the Taylor & Francis Web site at
http://www.taylorandfrancis.com

and the CRC Press Web site at
http://www.crcpress.com

Contents

Editor

Klaus D. Sattler pursued his undergraduate and master's courses at the University of Karlsruhe in Germany. He earned his PhD under the guidance of Professors G. Busch and H.C. Siegmann at the Swiss Federal Institute of Technology (ETH) in Zurich. For three years he was a Heisenberg fellow at the University of California, Berkeley, where he initiated the first studies with a scanning tunneling microscope of atomic clusters on surfaces. Dr. Sattler accepted a position as professor of physics at the University of Hawaii, Honolulu, in 1988. In 1994, his group produced the first carbon nanocones. His current work focuses on novel nanomaterials and solar photocatalysis with nanoparticles for the purification of water. He is the editor of the sister references, *Carbon Nanomaterials Sourcebook* (2016) and *Silicon Nanomaterials Sourcebook* (2017), as well as *Fundamentals of Picoscience* (2014). Among his many other accomplishments, Dr. Sattler was awarded the prestigious Walter Schottky Prize from the German Physical Society in 1983. At the University of Hawaii, he teaches courses in general physics, solid state physics, and quantum mechanics.

Contributors

Nasser Mohieddin Abukhdeir
Department of Chemical Engineering
University of Waterloo
Waterloo, Ontario, Canada

P. R. Bandaru
Program in Materials Science
Department of Mechanical
    Engineering
and
Department of Electrical Engineering
University of California, San Diego
La Jolla, California

Alex Belianinov
Center for Nanophase Materials
    Sciences
Oak Ridge National Laboratory
Oak Ridge, Tennessee
and
Institute for Functional Imaging of
    Materials
Oak Ridge National Laboratory
Oak Ridge, Tennessee

Eduardo Bolea
Group of Analytical Spectroscopy and
    Sensors (GEAS)
Institute of Environmental Sciences
    (IUCA)
University of Zaragoza (SPAIN)
Zaragoza, Spain

Alice Chinghsuan Chang
Department of Materials Science and
    Engineering
National Cheng Kung University
Taiwan, China
and
National Institute for Materials
    Science
Tsukuba, Japan

Cristina Cocchiara
Laboratorio di Chimica Fisica
    Applicata
Dipartimento dell'Innovazione
    Industriale e Digitale
Ingegneria Chimica Gestionale
    Informatica Meccanica
Università di Palermo
Palermo, Italy

Ludovic Duponchel
Université de Lille, CNRS, LASIR
Villeneuve d'Ascq, France

E. V. Egorov
IMT RAS
Moscow, Russia
and
RUDN
Moscow, Russia

V. K. Egorov
IMT RAS
Moscow, Russia

Mario El Kazzi
Electrochemistry Laboratory
Paul Scherrer Institut
Villigen, Switzerland

Johannes A. A. W. Elemans
Institute for Molecules and Materials
Radboud University
Nijmegen, The Netherlands

Xufei Fang
Technische Universität Darmstadt
Darmstadt, Germany

Xue Feng
Department of Engineering Mechanics
Tsinghua University
Beijing, China

Edward S. Fry
Physics and Astronomy
Texas A&M University
College Station, Texas

Fabrizio Ganci
Laboratorio di Chimica Fisica
    Applicata
Dipartimento dell'Innovazione
    Industriale e Digitale
Ingegneria Chimica Gestionale
    Informatica Meccanica
Università di Palermo
Palermo, Italy

Seren Hamsici
Institute of Physical Chemistry
Göttingen University
Göttingen, Germany

Chuan He
Shaanxi Joint Laboratory of Graphene
State Key Laboratory for Incubation
    Base of Photoelectric Technology
    and Functional Materials
Institute of Photonics &
    Photon-Technology
Northwest University
Xi'an, China

Patrick J. Herbert
Department of Chemistry
The Pennsylvania State University
University Park, Pennsylvania

Yuanyuan Huang
Shaanxi Joint Laboratory of Graphene
State Key Laboratory for Incubation
    Base of Photoelectric Technology
    and Functional Materials
Institute of Photonics &
    Photon-Technology
Northwest University
Xi'an, China

Rosalinda Inguanta
Laboratorio di Chimica Fisica
    Applicata
Dipartimento dell'Innovazione
    Industriale e Digitale
Ingegneria Chimica Gestionale
    Informatica Meccanica
Università di Palermo
Palermo, Italy

Maria Grazia Insinga
Laboratorio di Chimica Fisica
    Applicata
Dipartimento dell'Innovazione
    Industriale e Digitale
Ingegneria Chimica Gestionale
    Informatica Meccanica
Università di Palermo
Palermo, Italy

Maria S. Jimenez
Group of Analytical Spectroscopy and
    Sensors (GEAS)
Institute of Environmental Sciences
    (IUCA)
University of Zaragoza (SPAIN)
Zaragoza, Spain

Yanping Jin
Shaanxi Joint Laboratory of Graphene
State Key Laboratory for Incubation
    Base of Photoelectric Technology
    and Functional Materials
Institute of Photonics &
    Photon-Technology
Northwest University
Xi'an, China

Kateřina Kůsová
Institute of Physics of AS CR
Prague, Czech Republic

Alexander Khmaladze
Physics Department
SUNY University at Albany
Albany, New York

Armin Kleibert
Swiss Light Source
Paul Scherrer Institut
Villigen, Switzerland

Kenneth L. Knappenberger Jr.
Department of Chemistry
The Pennsylvania State University
University Park, Pennsylvania

Christine Kranz
Institute of Analytical and
    Bioanalytical Chemistry
Ulm University
Ulm, Germany

Sebastian Kruss
Institute of Physical Chemistry
Göttingen University
Göttingen, Germany
and
Center for Nanoscale Microscopy and
    Molecular Physiology of the Brain
    (CNMPB)
Göttingen, Germany

Francisco Laborda
Group of Analytical Spectroscopy and
    Sensors (GEAS)
Institute of Environmental Sciences
    (IUCA)
University of Zaragoza (SPAIN)
Zaragoza, Spain

Yan Li
Department of Engineering Mechanics
Tsinghua University
Beijing, China

Nguyen Thi Phuong Linh
Department of Materials Science and
    Engineering
National Cheng Kung University
Taiwan, China

Bernard Haochih Liu
Department of Materials Science and
    Engineering
National Cheng Kung University
Taiwan, China

Changji Liu
Shaanxi Joint Laboratory of Graphene
State Key Laboratory for Incubation
    Base of Photoelectric Technology
    and Functional Materials
Institute of Photonics &
    Photon-Technology
Northwest University
Xi'an, China

Kyle T. Mahady
Materials Science and Engineering
    Department
University of Tennessee
Knoxville, Tennessee

Luigi Malavolti
Max-Planck Institute for Solid State
    Research
Stuttgart, Germany

Florian A. Mann
Institute of Physical Chemistry
Göttingen University
Göttingen, Germany

Matteo Mannini
Department of Chemistry "Ugo
    Schiff" and INSTM RU
University of Florence
Sesto Fiorentino, Italy

John Mason
Physics and Astronomy
Texas A&M University
College Station, Texas

Pierre-Emmanuel Mazeran
Roberval Laboratory
University of Technology of
    Compiegne
Compiegne, France

Martha R. McCartney
Department of Physics
Arizona State University
Tempe, Arizona

Daniel Meyer
Institute of Physical Chemistry
Göttingen University
Göttingen, Germany

Robert Nißler
Institute of Physical Chemistry
Göttingen University
Göttingen, Germany

Olivier Noel
Institute of Molecules and Materials
    of Le Mans
Le Mans University
Le Mans, France

Marc Offroy
Université de Lorraine, CNRS, LIEC
Nancy, France

Olga S. Ovchinnikova
Center for Nanophase Materials
    Sciences
Oak Ridge National Laboratory
Oak Ridge, Tennessee
and
Institute for Functional Imaging of
    Materials
Oak Ridge National Laboratory
Oak Ridge, Tennessee

Bernardo Patella
Laboratorio di Chimica Fisica
    Applicata
Dipartimento dell'Innovazione
    Industriale e Digitale
Ingegneria Chimica Gestionale
    Informatica Meccanica
Università di Palermo
Palermo, Italy

Salvatore Piazza
Laboratorio di Chimica Fisica
    Applicata
Dipartimento dell'Innovazione
    Industriale e Digitale
Ingegneria Chimica Gestionale
    Informatica Meccanica
Università di Palermo
Palermo, Italy

Philip D. Rack
Center for Nanophase Materials
    Sciences
Oak Ridge National Laboratory
Oak Ridge, Tennessee
and
Materials Science and Engineering
    Department
University of Tennessee
Knoxville, Tennessee

Daniel Rolles
J.R. Macdonald Laboratory
Department of Physics
Kansas State University
Manhattan, Kansas

Giulia Serrano
Department of Chemistry
    "Ugo Schiff" and INSTM RU
University of Florence
Sesto Fiorentino, Italy

Michele Serri
Graphene Labs
Istituto Italiano di Tecnologia (IIT)
Genova, Italy

Hidemi Shigekawa
Faculty of Pure and Applied Sciences
University of Tsukuba
Tsukuba, Japan

David J. Smith
Department of Physics
Arizona State University
Tempe, Arizona

Carmelo Sunseri
Laboratorio di Chimica Fisica
    Applicata
Dipartimento dell'Innovazione
    Industriale e Digitale
Ingegneria Chimica Gestionale
    Informatica Meccanica
Università di Palermo
Palermo, Italy

Artem A. Trofimov
Center for Nanophase Materials
    Sciences
Oak Ridge National Laboratory
Oak Ridge, Tennessee
and
Institute for Functional Imaging of
    Materials
Oak Ridge National Laboratory
Oak Ridge, Tennessee

C. A. F. Vaz
Swiss Light Source
Paul Scherrer Institut
Villigen, Switzerland

Przemysław W. Wachulak
Institute of Optoelectronics
Military University of Technology
Warsaw, Poland

Xinlong Xu
Shaanxi Joint Laboratory of Graphene
State Key Laboratory for Incubation
    Base of Photoelectric Technology
    and Functional Materials
Institute of Photonics &
    Photon-Technology
Northwest University
Xi'an, China

Zehan Yao
Shaanxi Joint Laboratory of Graphene
State Key Laboratory for Incubation
    Base of Photoelectric Technology
    and Functional Materials
Institute of Photonics &
    Photon-Technology
Northwest University
Xi'an, China

Shoji Yoshida
Faculty of Pure and Applied Sciences
University of Tsukuba
Tsukuba, Japan

Mengkun Yue
Department of Engineering Mechanics
Tsinghua University
Beijing, China

Longhui Zhang
Shaanxi Joint Laboratory of Graphene
State Key Laboratory for Incubation
    Base of Photoelectric Technology
    and Functional Materials
Institute of Photonics &
    Photon-Technology
Northwest University
Xi'an, China

Lipeng Zhu
Shaanxi Joint Laboratory of Graphene
State Key Laboratory for Incubation
    Base of Photoelectric Technology
    and Functional Materials
Institute of Photonics &
    Photon-Technology
Northwest University
Xi'an, China

Nguyen Anh Dung
Institute of Molecules and Materials
    of Le Mans
Le Mans University
Le Mans, France

High-Temperature Scanning Probe Microscopy

Yan Li, Mengkun Yue, and
Xue Feng
Tsinghua University

Xufei Fang
Technische Universität Darmstadt

1.1 Introduction

With the rapid development of nanotechnology and advanced instruments, investigations of materials/structures towards the nanoscale or even to the atomic scale have been widely carried out. For instance, ultrathin two-dimensional (2D) nanomaterial has drawn significant attention in recent years due to its unique advances on physical, electronic, chemical and optical properties [1]. A direct and accurate characterization of such materials is of particular importance for understanding the relation between structural characteristics and functional capabilities, hence to provide guidance for rational design of materials with desired properties. Multiple advanced characterization techniques offer a direct access to material at micro to atomic scales, which greatly helps unravel the essential mechanism associated with properties of materials/structures. The wide range of experimental characterization methods (from near surface to in-depth examination) generally include, for instance, optical microscopy, scanning probe microscopy (SPM), scanning electron microscopy (SEM), transmission electron microscopy (TEM), scanning transmission electron microscopy (STEM) and so on.

Optical microscopy is one of the most traditional and widely used instruments for magnifying imaging, which can be roughly classified according to different application purposes. There are four common types of optical microscopes: biological microscopes [2], stereoscopic microscopes, metallographic microscopes and polarizing microscopes. Nowadays, with the advancement of computer technology and tools, and the continuous improvement of the theory

and methods of optical design, the imaging quality of optical microscopes is close to the limit of diffraction [3].

As an effective tool for microstructure analysis, SEM can perform various forms of surface observation and analysis on various samples with microscopic geometry [4]. The main features of SEM are: (i) the sample can be directly observed; (ii) the image resolution obtained is high; (iii) there is small damage to the sample and (iv) the energy dispersive spectrometry analysis can be implemented to test the composition of the sample. Limitations of SEM include image distortion and determination of measurement boundary.

TEM techniques include bright-field image (BF), dark-field image (DF), electron diffraction (ED), high-resolution image (HRTEM) and chemical analysis methods. It is widely used in the analysis of material defects, structural changes and component changes. The resolution of HRTEM is less than 0.2 nm, and the microstructure of materials can be observed at the atomic scale. Meanwhile, with the improvement of observation requirements, in situ TEM has gradually developed in recent years. On the other hand, STEM is made by scanning the sample with a concentrated electron beam [5]. Therefore, each point scanned on the sample corresponds to the image point generated. A STEM is formed after a continuous scan of a sample area. Currently, along with the development of TEM and its related parts, field emission gun STEM equipped with spherical aberration corrector, monochromator and high-energy resolution filter can be used in the sub-angstrom scales in order to fundamentally understand the relationship between macro properties and microstructure.

Unlike the abovementioned techniques that require light or electron signal for image acquisition, SPM is an alternative tool of using a probe (or tip) driven by a piezoelectric actuator to physically scan the surface of the specimen and to investigate the morphological evolution of materials from several 100 μm to sub-nanometer [6–9]. In 1982, Binning and Rohrer successfully developed the first member of the SPM family—scanning tunneling microscope (STM), and showed images of silicon surface with atomic resolution [10]. However, STM uses the tunneling current between the probe and the surface of the sample as the detection signal, so it is only suitable for observing conductors and semiconductor samples, which greatly limits the application of STM. In order to solve the problems encountered in the application of STM, Binning, Quate and Gerber invented the first atomic force microscope (AFM, one type of SPM) in 1986 [11], which has achieved rapid development and extensive application subsequently. Nowadays, SPM has grown to a big family with various types [12] and has been a powerful tool in exploring the world of nanotechnology.

While conducting SPM tests, a sharp probe tip is used to scan over the target surface, at the same time, a feedback mechanism constantly adjusts the tip height by approaching or retracting the tip to a fixed tip-sample distance, and thus, properties of the surface are sensed at the nanoscale or atomic scale. The high precision is mainly attributed to accurate control of the motion at such a small scale by the piezoelectric actuators on electronic command. By properly choosing the shape, size and material of the scanning probe according to the specific object to be investigated, SPM can be applied to study a broad range of materials such as metals and alloys, polymers and biomaterials for a variety of phenomena, including growth and fabrication of crystals, thin films and nanoparticles [13–17]. A typical application of SPM tests is the measurement of the surface morphology and its changes of materials under micro-/nanoscale [18]. Meanwhile, it is worth noticing that, since SPM can also be used as a loading tool at the micro-/nanoscale, some researchers also use the nonuniform stress generated by the probe to conduct research on ferroelectric polarization [19,20]. This is mainly because the tip is small in size and its contact can produce a sufficiently large stress gradient. In addition, as a member of the SPM family, piezoresponse force microscopy (PFM) has emerged as a powerful tool to characterize piezoelectricity and ferroelectricity at the nanoscale over the past two decades, yielding considerable insight into domains, defects, nucleation and switching of ferroelectric materials [21].

In recent years, the material performance at high temperature has been drawing increasing attention. Materials, including those used in electronic circuits to those in structural components of aircrafts could undergo harsh service environment at elevated temperatures. Understanding the failure of the materials/structures requires a thorough investigation down to nanoscale, where the crack or failure initiates. Due to its unique capability of mapping the real-time topography at the micro-/nanoscale even at high temperatures, SPM has also been adopted for high-temperature study [14,22]. High-temperature SPM

(HTSPM) is a relatively new experimental method in studying the temperature-dependent mechanical properties as well as the surface evolution materials that are subjected to high-temperature oxidation [14,22,23]. Quantitative analysis of in situ SPM images provides new insights into the mechanisms underpinning the complicated oxidation processes. In this chapter, we present several detailed examples on high-temperature oxidation mechanism of materials/structures using HTSPM. Because of the limitation of chapter length, some technical instrumental issues such as piezoelectric effects, actuator system and electronic design are not discussed here. More technical aspects can be found elsewhere [24].

1.2   Application

1.2.1   Surface Characterization and Surface Topography Evolution during Oxidation

By using HTSPM, the in situ surface evolution of sample at nanoscale can be recorded real time at high temperature. Based on the continuous surface characterization that is made possible by HTSPM, it greatly helps to reveal the mechanism at micro-/nanoscale.

Here we take investigations on niobium-based alloy conducted by Li et al. [14] as an example. HTSPM tests were conducted by employing a specified probe in heating chamber based on TI 950 (Hysitron Inc., USA). A Berkovich indenter with a radius 150 nm used as the SPM probe is brazed to a Macro-shaft with very low thermal conductivity for reducing the thermal conduction from the sample surface to the transducer. The heating stage integrated in the equipment is specially designed to reduce the thermal radiation in order to keep the constant temperature on the sample surface [25]. The heating stage consists of two heating plates, between which the sample is properly clamped and is subsequently heated to the pre-set temperature, as shown schematically in Figure 1.1. Detailed description of this heating stage can be found elsewhere [25]. During high temperature test, the SPM probe is first brought in contact with the sample with a very small contact load (2 μN), and this contact is maintained until the temperature of the sample becomes thermally stable. This ensures a uniform distribution of temperature on the sample surface before

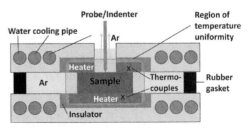

FIGURE 1.1   Schematic drawing of the heat stage for HTSPM. (Reproduced from [25], with the permission of Elsevier.)

an SPM scan is performed [28]. Nitrogen gas was used to provide protection during the whole test to reduce the oxidation effect.

The surface of the specimen was firstly indented with a load of 7,250 μN at room temperature. A residual indent (indentation marker) was created as shown in Figure 1.2 ($T = 20°C$). Then the temperature was raised from room temperature up to 800°C with a temperature interval of 100°C. When the expectant temperature was reached and the temperature remained steady with a tiny fluctuation, the in situ SPM imaging was conducted to obtain the surface topography of the indentation marker and the results are shown in Figure 1.2.

The SPM results in Figure 1.2 demonstrate clearly the variation of the indentation marker at different temperatures. When the temperature is below 400°C, the profile of the indentation is identifiable. However, when the temperature is above 400°C, the surface morphology in each profile (4 μm × 4 μm) in Figure 1.2 starts to change sharply. The surface roughness is evidently increased especially at temperature higher than 400°C compared to the smooth

surface at the initial stage (room temperature) and lower temperature (below 400°C).

Further analysis reveals that the surface evolution is mainly caused by oxidation. By choosing the indentation pit (the marker) as a representative area, the thickness of the oxide film can be calculated from the different depths and geometry of the indentation obtained by SPM [14]. In Figure 1.3, it shows the calculated oxide film thickness at different temperatures by using the aforementioned methods. As can be seen, the oxide thickness remains stable under temperature 400°C, while it accumulates quickly when the temperature is increased above 400°C. The thickness of the oxide film calculated shows a consistent correlation with the surface evolution.

By using HTSPM to record real-time the surface evolution of the indentation pit on Nb-based alloy at nanoscale from room temperature to 800°C, the oxide scales were observed and carefully calculated through geometrical method focusing on the deliberately pre-indented indentation marker. Surface morphology indicates that the oxidation behavior of Nb-based alloy has a transition temperature

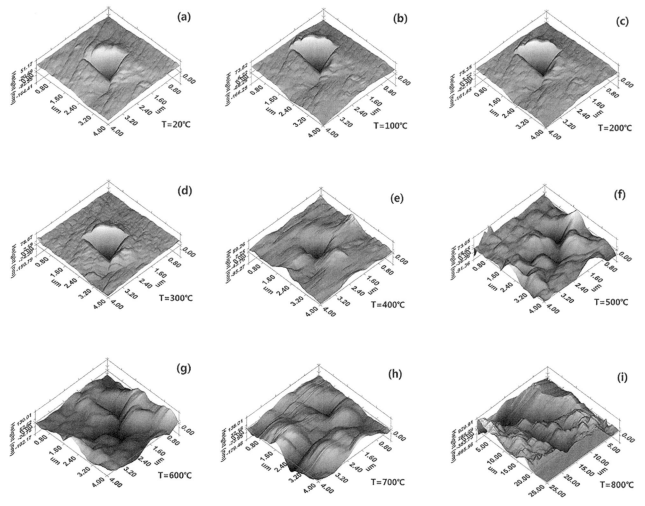

FIGURE 1.2 In situ SPM images for the surface topography at different temperatures. Panels (a–i) correspond to temperatures at 20°C, 100°C, 200°C, 300°C, 400°C, 500°C, 600°C, 700°C, and 800°C, respectively. (Reproduced from [14], with the permission of Elsevier.)

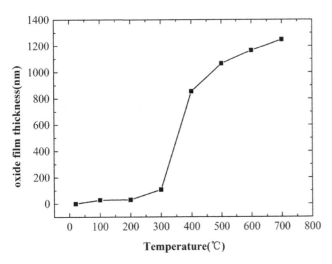

FIGURE 1.3 Calculated oxide film thickness at different temperatures. (Reproduced from [14], with the permission of Elsevier.)

at 400°C. In situ HTSPM image combining with nanoindentation technique can be properly used to study the oxidation at elevated temperature. The experimental technique and method can serve as a state-of-the-art method to study oxidation bridging the atomic scale and meso-/macro scale.

1.2.2 Oxidation Mechanism

Oxidation of materials is a classic research topic in many fields. Traditional studying methods include mass gain method, gas consumption method and accelerated life test. However, these methods are mainly postmortem characterization since none of these methods is capable of measuring the real-time growth of the oxide film and its morphology evolution. Although great advance in understanding the oxidation mechanisms has been made based on these methods, however, important information during the oxidation process is missing. Furthermore, in most cases, it is of great difficulty using these postmortem methods to characterize the oxide product quantitatively and separately (e.g. for multilayer oxidation) due to the complex structure of the oxide composition and structure at the end of the oxidation process.

There has been some pioneering and profound work concerning the oxidation at the earliest stage and atomic structure evolution at an atomic scale [26–28] using TEM, which revealed the oxidation at the early stage with a thickness at magnitude of nanometer. However still, these researches are limited to the earliest stages of the oxidation and they are unable to show the oxidation evolution once after the oxide film (which exceeds the atomic scale) is piled up on the substrate for a longer oxidizing time. Neither are these methods available to show how the oxidized layer would evolve during the oxidation process. Thus, it is of great importance to study the oxidation mechanism at a micro/nanoscale bridging the atomic scale and meso/macro scale to quantitatively obtain and characterize the process of the oxidation.

Competing Mechanism between Local Curvature and Stress during Oxidation

Many efforts have been focused on the surface curvature effect at micro/nanoscales in various research aspects in recent years. For instance, Lv et al. [29] studied the movement of liquid bead at the surface of a tapered structure at nanoscale using in situ method and proposed that the curvature as well as the curvature gradient act as the driving force for the unexpectedly higher moving speed of the liquid bead compared to those on nominally flat surface. Such a curvature and curvature gradient effect is also confirmed by Yin et al. [30], who pointed out that at micro/nanoscale, the highly curved surface could induce a driving force for the movement of the particles that are outside the curved surface.

As for an expedient tool to obtain the outline of the sample surface, HTSPM is appropriate to study local curvature effect and stress effect during oxidation. We investigate here the process of surface evolution of a chemically etched stepped structure at nanoscale during oxidation at 600°C using HTSPM, as shown in Figure 1.4.

As shown in Figures 1.5 and 1.6, experimental results obtained by HTSPM reveal that this curved stepped structure becomes flat after being oxidized for a short period of time. However, after a longer time of oxidation, it is observed that the originally flat surface near the stepped structure becomes rough again (Figure 1.7).

The mechanisms for the phenomenon that uneven surface becomes firstly flattened while the originally smooth surface roughens for a longer oxidation time was given by Fang et al. [23]. The analysis is briefly summarized here.

As shown in Figure 1.8b, the interaction between a semi-infinite plane and an outside particle is first illustrated. For

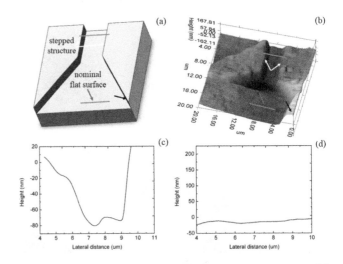

FIGURE 1.4 Surface topography at room temperature: (a) a pictorial illustration of the stepped structure and flat surface nearby; (b) SPM image; (c) cross-sectional view of the stepped structure corresponds to the upper line in (b) and (d) cross-sectional view of the nominally flat surface nearby corresponding to the lower line in (b). (Reproduced from [23], with the permission of AIP Publishing.)

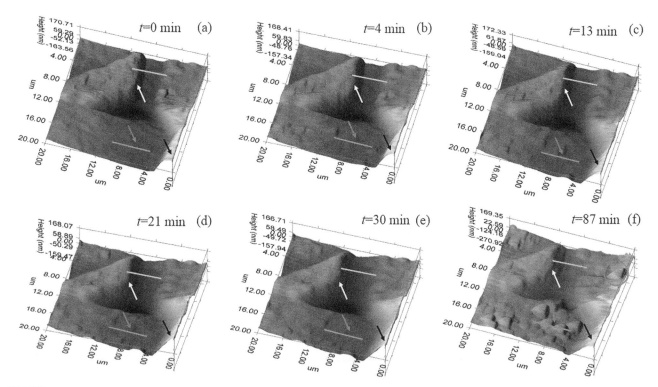

FIGURE 1.5 Evolution of surface topography at 600°C measured by HTSPM: (a) Incipient oxidation at 600°C, so the oxidation time $t = 0$ min is set here as a reference state; (b–f) oxidation times are defined with respect to the reference time $t = 0$ min in (a). (Reproduced from [23], with the permission of AIP Publishing.)

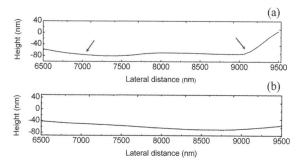

FIGURE 1.6 Example for the cross-sectional views (a) before and (b) after oxidation at 600°C. (Reproduced from [23], with the permission of AIP Publishing.)

convex and concave curved surface (cross-section of correspondingly curved body), it is assumed that the particle P has coordinates $(0, 0, -h)$ and $(0, 0, h)$, respectively. Assume that the pair-potential between particles is $u(r) = C/r^n$, where r is the distance and C is the constant, then the expression of the potential for the curved body with outside particle can be derived:

$$U_n = \overline{U}_n \left(1 + \frac{3}{4}\widetilde{c}_1\right) \qquad (1.1)$$

where $\overline{U}_n$ is a constant, and for gravitational potential, $\overline{U}_n$ is negative. $\widetilde{c}_1 \neq 0$ is the principal curvature. As for the experimental process (concave curved surface interacting with the outside particle), at the very beginning of the

oxidation, the curvature radius at the cornered area (take the stepped structure as an example) is the smallest, indicating that the value of $\widetilde{c}_1$ is the largest. Thus, at the initial stage, the potential at the corner of the stepped structure is the lowest, indicating that the oxygen particles in the vicinity of the curved surface move towards the corner. This results in a higher oxygen concentration at the corner than other locations. Based on oxidation kinetics, such a higher oxygen concentration would result in a higher oxidation rate at the corner, and thus, the corner would be flattened and $\widetilde{c}_1$ would decrease.

The interaction between a curved surface and an outside single particle is described above. The limitation is, however, it only explains one single particle interacting with the surface. Therefore, the following analysis focuses on the effect of surface curvature or radius on the chemical potential, which gives a thermodynamic and statistical effect of the particles' movement near/on the surface. For surface diffusion process, consider the following diffusion equation [31]:

$$J_s = -\frac{D_s v}{k_B T} \nabla_s \mu = -\frac{D_s v}{k_B T} \nabla_s (\gamma \kappa \Omega_0) = -\frac{D_s \gamma v \Omega_0}{k_B T} \nabla_s \kappa$$
$$(1.2)$$

where J_s is the diffusion flux, D_s is the surface self-diffusion coefficient, k_B is the Boltzmann constant, T is the temperature, and v is the number of atoms per unit area during diffusion. Based on Eq. 1.2, it shows clearly that the gradient of mean curvature acts as the driving force for surface configuration evolution (provided that the strain effect is negligible

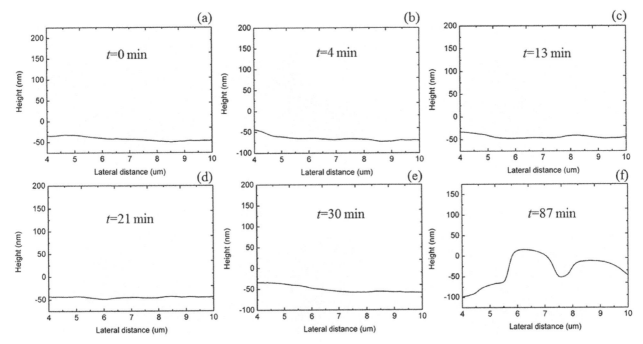

FIGURE 1.7 Roughening effect of the original nominal flat surface near the stepped structure as the oxidation time increases. Panels (a–f) correspond to oxidation time at 0, 4, 13, 21, 30, and 87 min, respectively. (Reproduced from [23], with the permission of AIP Publishing.)

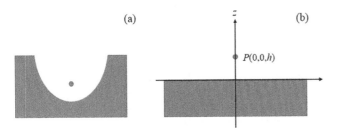

FIGURE 1.8 (a) Schematic of the interaction between concave surface (cross section of a curved body) and an outside particle; and (b) the interaction between a semi-infinite plane body and an outside particle. (Reproduced from [23], with the permission of AIP Publishing.)

at the incipient stage of oxidation). In fact, for isotropic surface, Eq. 1.2 determines that the surface evolves in the direction to meet the requirement of $\nabla_s \kappa$, corresponding to a flattening effect of the surface. This analysis agrees well with the experimental results.

The transition of flat surface to rough surface is explained by the stress effect. It has been reported that the stress and the strain energy in the thin film plays a vital role in affecting the instability of surface evolution during the film growth. The surface chemical potential is expressed by adopting Asaro–Tiller equation [32–34]:

$$U = U_0 - \gamma \kappa \Omega_0 + w \Omega_0 \qquad (1.3)$$

where U_0 is the chemical potential for a smooth and stress-free surface, γ is the surface energy, κ is the mean curvature (positive in sign for concave surface), w is strain energy density along the surface and Ω_0 is atomic volume. $\gamma \kappa \Omega_0$

stands for the chemical potential related to curvature, and $w\Omega_0$ stands for the chemical potential related to the stress effect. The opposite sign for these two items indicates a competing effect between them, i.e., the surface energy tends to flatten the surface while the strain energy tends to destabilize and roughen a flat surface [34]. In order to better illustrate such a competition mechanism, here a dimensionless factor is defined

$$\Lambda = \gamma \kappa \Omega_0 / w \Omega_0 = \gamma \kappa / w \qquad (1.4)$$

Comparison of values as well as magnitude of the symbols shows that, for curvature at the scale of several nanometers, the curvature related surface energy is much larger than that of strain energy (i.e., $\Lambda \gg 1$) at the early stage of oxidation, indicating that the flattening effect on the topography change of the nanoscale stepped structure is the dominating mechanism. However, after a longer exposure time to the oxidizing environment, the oxide films thicken and the compressive stress within the oxide film increases [35]. This increase in compressive stress during oxidation results in an increase of the strain energy. Meanwhile, as the originally curved structure flattens, the curvature reduces gradually to zero. Both the increase of the strain energy and decrease of the curvature will lead to a decrease of Λ, and the stress-driving surface roughening effect would dominate and result in a roughened surface.

To sum up, analysis shows that such a surface evolution is attributed to the competition between the nanoscale curvature effect (related to surface energy) and the stress developed in the oxide film during oxidation (related to strain energy). It is demonstrated that both the surface energy and strain energy can modify the surface chemical potential, which acts as the driving force of the surface diffusion of oxygen and further affects the oxide formation on the surface. This example demonstrates that with the advantages of HTSPM, such as in situ, real time and high resolution, it can be used to reveal the relationship between surface curvature, stress and oxidation process.

Surface Feature Fabrication: Pillar Fabrication as Surface Markers for Oxidation Study

As is known, the oxidation of Ni-based alloys has attracted considerable attention due to its common occurrence under extreme service conditions and an urgent need to improve the oxidation resistance of those alloys [25,36,37]. The oxide formation of Ni-based alloys is known to be greatly influenced by the nature of the starting surface, material purity and many other factors [23,38–40]. Several conventional experimental methods have been used to study the oxidation of Ni-based alloys, including mass gain/loss method, oxygen uptake measurement and accelerated life test [37,38,41–45]. However, few of them have the ability of providing the important information of spatial variation of growing oxides, since the results are usually obtained by smearing out the regional difference of oxidation evolution.

By combining with other technique, HTSPM will be able to quantitatively measure the full-field oxide growth rate of any arbitrary point on the oxide surface. In the following example, in situ SPM was used to monitor the full-field profiles of oxides on the (111) surface of single crystal Ni-based alloy during oxidation at temperatures ranging from 300°C to 800°C [46]. An array of SiO_2 micro-pillars was fabricated as "markers" to facilitate the mapping of surface oxide evolution at the micro- to nanoscale. A continuous increase of thickness of the oxide film in comparison to the nonoxidizing SiO_2 micro-pillar was recorded with increasing temperature.

Figure 1.9 shows the SPM images of oxidized surface obtained at different temperatures and times, with the SiO_2 micro-pillar serving as a reference. When the oxidation experiment started at 300°C, the SiO_2 micro-pillar was clearly seen in Figure 1.9a. As the time increased to 80 min and the temperature remained at 300°C, no obvious surface oxide was detected (Figure 1.9b). When the temperature reached 500°C, small oxide islands emerged on the surface (Figure 1.9c). The number of the islands increased as the temperature was raised to 700°C (Figure 1.9d), while the size of the islands remained similar to those at 500°C. With a continuous increase of temperature, the islands grew bigger and began to merge with each other, such that the surface became considerably rough, as can be seen in Figure 1.9e at 700°C and Figure 1.9f at 800°C. Meanwhile, the micro-pillar became less distinguishable from the surrounding oxide with increasing temperature. As the temperature reached 700°C –800°C in Figure 1.9e–f, the micro-pillar was almost indistinguishable from the surrounding oxide. Evidently, compared to the smooth surface at the beginning of oxidation experiment (Figure 1.9a), the roughness of the sample surface progressively increased, especially at temperatures higher than 500°C.

The full-field profile of the oxide scale around a SiO_2 micro-pillar can be quantitatively characterized from the time series of SPM images by choosing the SiO_2 micro-pillar as a reference marker. Figure 1.10 shows the calculated full field of oxide film thickness, thus revealing a nonuniform oxidation process. Each map of oxide thickness in Figure 1.10a–f was obtained based on the corresponding SPM image in Figure 1.10a–f. Notice that the big round spot in each sub-figure of Figure 1.10 corresponds to the SiO_2 micro-pillar. The uniform map in Figure 1.10a and b corresponds to the case at $T = 300°C$ when no oxidation occurred. As seen from Figure 1.10c, tiny spots with light color appeared, indicating the thickness of the surface oxide film increases slightly as the temperature was raised

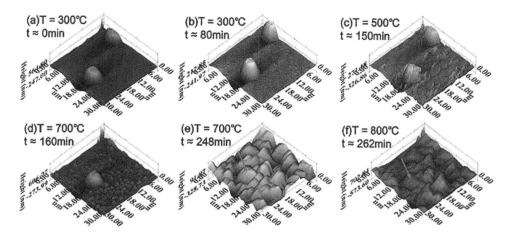

FIGURE 1.9 In situ SPM images showing the progressive surface oxidation of single crystal Ni-based alloy around the pre-fabricated SiO_2 micro-pillar at different temperatures and oxidation times: (a) $T = 300°C$ and $t = 0$ min; (b) $T = 300°C$ and $t = 80$ min; (c) $T = 500°C$ and $t = 150$ min; (d) $T = 700°C$ and $t = 160$ min; (e) $T = 700°C$ and $t = 248$min; (f) $T = 800°C$ and $t = 262$ min. For the ease of comparison among sub-figures, the micro-pillar is marked by an arrow in (e) and (f). (Reproduced from [46].)

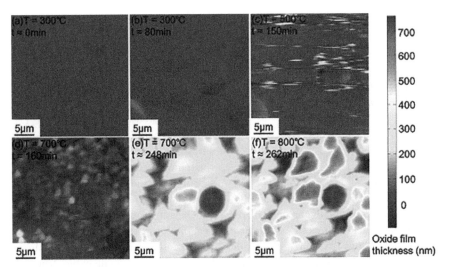

FIGURE 1.10 Full-field mapping of the oxide film thickness at different temperatures and times using the SiO_2 micro-pillar as a reference marker. Panel labels (a–f) correspond to that in Figure 1.9. (Reproduced from [46].)

from 300°C to 500°C. More island-shaped oxides appeared in Figure 1.10d–f during oxidation at 700°C and 800°C. In particular, the dark areas in Figure 1.10f indicate the occurrence of local severe oxidation.

The 3D SPM images in Figure 1.9 and the calculated maps of oxide film thickness in Figure 1.10 clearly reveal the oxide islands formed on the sample surface, indicating a nonuniform growth of the oxide film.

Meanwhile, Figure 1.11a reveals the oxide film thickness distributions for different times at the temperature $T = 700$°C. The numbers 1–9 indicate the sequence of SPM scanning. In addition, one curve obtained at $T = 800$°C is added in Figure 1.11a for comparison. It is seen from Figure 1.11a that, as the oxidation process proceeds, the maximum oxide film thickness increases. For the first two curves ($T = 700$°C-1 and 2), the thinnest oxide film accounts for the highest distribution probability while the thickest oxide film accounts for the lowest. For the third to seventh curves ($T = 700$°C-3 to 7), the distribution of the oxide film thickness varies significantly, where the largest portion of the oxide film thickness falls in between the thinnest and the thickest oxide film. Moreover, the distribution peak (i.e. the highest probability) of the oxide film thickness shifts to a larger thickness. The shifting direction

of the distribution peak is indicated by the arrow. For the last two curves at 700°C ($T = 700$°C-8 and 9) and the curve at 800°C, the oxide film thickness corresponding to the peak distribution remains almost unchanged, while the maximum distribution decreases gradually at 700°C and increases slightly at 800°C.

Figure 1.11b shows a more direct view of the evolving morphology of the oxide islands by extracting the cross-sectional profiles of the oxide islands. The oxide islands nucleate and grow at $T = 700$°C-1 to 2. During this stage, the distribution of the oxide film thickness is monotonic (see Figure 1.11a), meaning that the islands have not yet merged with each other (Figure 1.11b). Thus, these two curves represent the process of nucleation of the oxide islands undergoing independent growth. As the oxidation proceeds, the oxide islands grow bigger and the neighboring oxide islands coalesce, as indicated by the right arrow in Figure 1.11b. Once the coalescence process continues (see the curve at $T = 700$°C-3 to 8), the distribution of the oxide thickness is no longer monotonic in Figure 1.11a. Hence, the non-monotonic distribution of oxide film thickness in Figure 1.11a reflects the coalescence process of the growing oxide islands, as indicted by the left arrow in Figure 1.11b.

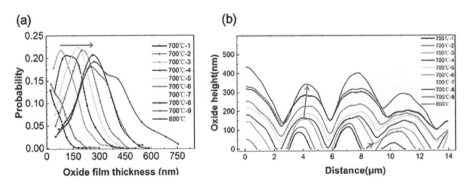

FIGURE 1.11 (a) Distribution probability of the nonuniform oxide film thickness and (b) cross-section view of the oxide islands as the oxidation evolves with time and temperature. (Reproduced from [46].)

Furthermore, the average oxide film thickness versus oxidation time is calculated using the conventional assumption of uniform growth of oxide scale. The result in Figure 1.12 shows an initial linear increase of average film thickness with time, followed by a parabolic growth. The initial linear growth and subsequent parabolic growth of the oxide film are consistent with the previous reports by other researchers [41,43–45]. The initial linear law refers to reactions whose rate is controlled by a surface-reaction step, corresponding to the first-stage oxide island nucleation and growth. The followed parabolic law suggests a rate-determining process by the diffusion of Ni and Co through the scale, facilitating the second-stage oxide island growth. This result suggests that some conventional methods of evaluating the oxide growth may not accurately reflect the non-uniform formation and growth of oxide islands, particularly during the early stage of oxidation.

By comparing the TEM images of oxide cross sections with the SPM maps of surface profiles, we find that the oxide islands in Figures 1.10 and 1.11 contain several to tens of small oxide grains. In Figure 1.13a, the dashed line represents one large oxide island in the SPM map of Figure 1.11b. This island is composed of several small oxide grains. In contrast, Figure 1.13b shows the TEM image from an independent oxidation experiment of a well-polished (110) single crystal of Ni-base alloy. It is seen that small oxide islands with diameters around 80 nm form on the surface after oxidization at $T = 600°C$ for 30 min.

The combined results in Figure 1.13a and b indicate that much smaller oxide islands form on the surface at a shorter time (30 min) and lower temperature ($T = 600°C$). This is the first stage of oxidation involving the nucleation of the single grain oxide islands and their coalescence, which has been experimentally observed on the Cu surface during oxidation [27]. At longer oxidation time and higher temperature, the second stage of nucleation and coalescence of poly-grain oxides occur on the surface. Similar phenomenon has been investigated during the oxidation process of 99.95 pct Ni [47]. The two stages of oxidation are schematically illustrated in Figure 1.14.

To depict a more complete picture of oxidation in our experiment, we note that the oxidation process of single crystal Ni at the very early stage presumably involves the chemisorption, nucleation and lateral growth of extremely small oxide islands of approximately several nanometers in thickness, as discussed in detail in an early review article [48]. These small oxide islands continue to grow and coalesce, similar to Figure 1.12b. During this stage of oxidation, Ni and Co atoms diffuse through both the SiO_2 layer and the oxide layer to react with the oxygen at the oxide-gas interface and form the oxide islands.

In order to further understand the mechanism for the occurrence of poly-grain oxide islands at the second stage, we compare the formation of the second-stage oxide islands with the surface roughening or undulation on the originally flat surface. A nominally flat but in reality slightly undulated surface at small scale is configurationally unstable during its growth or thickening [32,34], which is attributed to the stress-induced surface instability [49]. During the second stage of oxide islands' growth, the oxide islands are in conjunction with each other, the lateral growth results in compressive stress. Such compressive stress has been extensively reported in literature [35,50–52] and could be responsible for the surface instability. A critical wavelength, $\lambda_{cr} = \pi U_s E/\sigma_m^2$, is proposed to estimate and characterize the evolution of surface configuration [32,51]. Here, λ_{cr} is the critical wavelength, U_s is surface energy of the oxide thin film and can be treated as a material constant, E is the

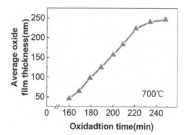

FIGURE 1.12 Average thickness of oxide film versus oxidation time at $T = 700°C$. (Reproduced from [46].)

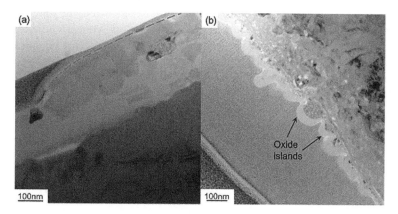

FIGURE 1.13 TEM images showing (a) an oxide island consisting of poly-grains and (b) small oxide islands after 30 min oxidation at $T = 600°C$. (Reproduced from [46].)

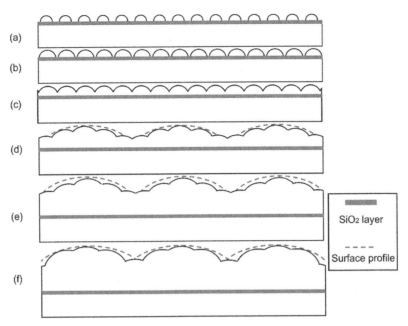

FIGURE 1.14 Schematic illustration of the two-stage oxidation. (a–c) The first stage of nucleation and coalescence of single grain oxide islands. (d–f) The second stage of nucleation and coalescence of poly-grain oxide islands. (Reproduced from [46].)

modulus and σ_m is the stress within the film. An intrinsic wavelength λ_{in} is assigned to a given surface which is nominally flat but slightly undulated. The surface is stable during the growth of the thin film and does not become rougher if $\lambda_{in} < \lambda_{cr}$. When $\lambda_{in} > \lambda_{cr}$, the surface is unstable and roughens during the film growth [32,51].

Assume that for a given pre-polished specimen, the value of λ_{in} is a constant determined by the surface morphology, while the increasing value of compressive stress during the growth of oxide film reduces λ_{cr} according to $\lambda_{cr} = \pi U_s E / \sigma_m^2$. When the critical condition of $\lambda_{in} > \lambda_{cr}$ is reached, further surface undulation is promoted, as observed in the present experiment as large secondary island formation and illustrated in Figure 1.14. Here we take $U_s = 1$ N/m, $E = 175$ GPa for the oxide film, the compressive stress was taken to be $\sigma_{ox} \approx -0.8$ GPa [52], it yields a critical wavelength of $\lambda_{cr} = \pi U_s E / \sigma_m^2 = 859$ nm. The second-stage oxide island has a length scale of 1.5–2 μm, which is larger than the critical wavelength and agrees with the above argument.

Thus, the in situ HTSPM enables the real-time quantification of the full-field oxide profile on material surfaces at the micro-/nanoscale, bridging the gap between the atomic- and macro-scale characterization of oxidation. This novel method by combining SPM and nonreactive SiO_2 pillar on the sample surface as reference marker yields quantitative results and can be applied for various materials systems besides Ni-base alloy.

1.2.3 Other Applications: Stress and Strain Measurement

As mentioned above, oxidation is a process featured by stress-chemical coupling effect. However, stress induced by

oxidation is difficult to be measured in situ with SPM, which poses a great challenge to the experimental study of the stress-oxidation coupling behavior. It is well known that stress can be calculated by deformation and the constitutive relation of the material. Therefore, provided that the strain and deformation information of the material surface could be obtained using SPM, then the stress on the surface can be calculated by combining with the constitutive relation mentioned above. For example, by measuring the undulated surface wavelength based on the surface morphology obtained by SPM, the critical stress in the oxide film can be inferred from the theoretical film/substrate model [51]. Meanwhile, digital image correlation (DIC) method has been widely accepted and commonly used for the surface deformation measurement based on surface feature/speckle detection [53]. The idea and the algorithm can be readily considered in combination with the SPM images, although the image generation mechanisms are different for these two methods. It is possible to first obtain the surface texture data by SPM scanning and then convert it into grayscale image. Subsequently, the full-field deformation before and after oxidation can be calculated based on the algorithms used in DIC method. This possible application is not limited to oxidation study but also to general cases in high-temperature environment in vacuum. It should be noted that when used for oxidation study, changes in surface topography caused by oxidation will also adversely affect the surface feature/speckle matching, which needs further study.

1.3 Challenges in HTSPM

SPM has a wide range of application fields—both in basic disciplines, such as physics, chemistry, biology and medicine, and in applied disciplines, such as materials and microelectronics. However, there are some deficiencies in the current SPM application that need to be treated with care [54–56]. First of all, the limitation of scanning speed leads to low measurement efficiency. Secondly, unlike the electron microscopes, it is difficult to achieve a wide range of continuous zoom especially at elevated temperature, making it difficult to locate and find feature structures. Thirdly, the most widely used tubular piezoelectric scanner in the SPM has a limited vertical expansion range, which leads to higher requirements on the surface roughness of samples. In addition, the system is to infer the surface topography of the sample by detecting the motion track of the probe during scanning, so the geometrical width, the radius of curvature and the anisotropy of the probe could cause undesirable imaging distortion, etc. When applied in high-temperature environment, the in situ HTSPM measurement is faced with the two most challenging subjects in the field of experimental mechanics, namely, micro-/nanoscale and high-temperature environment, which undoubtedly poses a higher challenge to HTSPM.

It is worthwhile to bear in mind the advantages/shortcomings of the technique, so that for specific problem to be investigated, a better choice can be made to avoid certain errors to be caused during the experiments. There would always be issues and conditions that are less optimal than desired. For example, the high temperature exerts demand on the materials in proximity to the hot zone, such as the probe used for HTSPM, as well as the entire assembly of the circuits. Meanwhile, circulating cooling water driven by a pump may induce noise during the image acquisition. We discuss here the following issues that need serious consideration or improvement in the future.

1.3.1 Probes Used for HTSPM

Probes for in situ HTSPM must be physically, chemically and mechanically stable against the high temperature and the corrosive atmosphere. Note that the thermal expansion of the probe material is an important factor that can influence image quality at high temperatures. From the aspect of system assembling, the temperature could also influence the fixing of the probe due to different thermal expansion coefficients between the probes and the probe holder. Measures such as using a long macor shaft to mount the tip have been applied [57,58]. The low thermal conductivity of the macor material reduces the heat conducted to the tip end and the holder, minimizing the thermal heat effect on the system assembling. Obviously, it is also necessary to avoid the adhesion of the probes to the sample or prevent the probes from undergoing chemical reactions at high temperatures. For instance, diamond tips are greatly discouraged to be used for Fe-based alloy, since the carbon could serve as an alloying element at high temperature (e.g., 400°C) and result in severe degradation of the tip. This undoubtedly limits the types of available materials and raises the requirements for processing techniques. Meanwhile, due to the inevitable

thermal degradation of the tips, new tips are frequently required to maintain the high accuracy of measurements. This frequent update of tips can be costly as well.

1.3.2 Effects of Different Test Environments

For most tests using HTSPM, the common operating conditions include two types, namely, atmosphere and vacuum. In vacuum system, the vibration of vacuum pump could affect the test accuracy. In atmospheric tests, it is also practical to, for instance, introduce inert environment in the confined tested zone to minimize the oxidation at elevated temperature. Nonetheless, the influence of protective atmosphere on HTSPM should be taken into consideration. For example, air disturbance caused by the gas flow will affect the positioning accuracy, probe contact force, etc., in the atmospheric environment. Such disturbance can be amplified especially at high temperature.

1.3.3 Thermal Drift

Sample drift can always interfere the image quality for scanning probe techniques, especially at high temperature, which contributes significantly to this drift as the different parts of the system are heated at different rates. After the pre-set temperature for experiments is reached, allowing some time for system stabilization and thermal equilibrium is necessary to minimize such drift. Although this helps to reduce the effect of heat bleaching to some extent, in general, thermal drift cannot be completely avoidable. For smaller scan areas and longer scan times, it will be more problematic. Note that the larger thermal drift will cause great deviation on the obtained experimental data, hence a compromise between the experimental time, scan area and the target problem should be evaluated.

1.3.4 Thermal Gradients

Thermal gradients through and laterally across the sample, or between the tip and sample, are difficult to avoid in the current available equipment, whether the sample is heated from one side or both sides. Finite element calculations and experiments have shown that large temperature gradients are present in the vicinity of the contact points between the probe's tip and the sample [59]. This means that during surface scanning, the effective temperature at the contact point will depend on the tip-sample contact area as well as the scan rate.

1.3.5 Near-Field Thermal Radiation

The radiative heat transfer at nanoscale has become one of the hottest spots in the relevant field in recent years. Radiative heat transfer in the far field where gap sizes are larger than Wien's wavelength (~10 μm at room temperature) has been well established. However, for near-field radiative heat transfer where the gap sizes are smaller than Wien's wavelength, studies are still under way [60]. The thermal radiation at such a small scale is of great importance to a variety of technologies, such as heat-assisted magnetic recording [61], near-field thermophotovoltaics [62] and lithography [63]. It has been demonstrated that radiative heat transfer between objects separated by nanoscale gaps considerably exceeds the predictions of far-field radiation theories [60,64,65]. This phenomenon certainly reminds us of the SPM tests, especially at high temperatures, where the probe is brought in contact with the surface of the sample. With the near-field thermal radiation effect, it may well lead to local thermal heating at the probe tip, resulting in possible material adhesion between the probe and the sample. Hence, a deviation of the load value perceived by the probe could be induced. This issue is of great importance for the measurement precision and is worth further study.

1.4   Outlook

The increasing application of structures operating down to micro-/nanoscale or up to macroscale components in space/aircrafts exerts great demand on understanding the material failure starting at micro-/nanoscale. With its unique advantages for characterizing material evolution at elevated temperatures, including capturing in situ and real time surface evolution/degradation of material surface, HTSPM has exhibited a promising potential for fundamental studies. HTSPM is able to provide more quantitative measurements on the surface topography at elevated temperatures, which is essential to quantifying the evolution process and exploring the mechanism at the micro-/nanoscale. Beyond the study of surface topography, HTSPM can be used to study the electrical and magnetic properties of materials operating at high temperatures due to its ability of accurate in situ loading, which is of great significance for electronic devices, etc.

In addition, by applying HTSPM to study the multiple chemo-mechanical coupling phenomenon and interface reaction, it opens a revenue with abundant new mechanical problems worth exploring. Meanwhile, multiple environment is also the focus of future research on mechanics and materials science. With its strong environmental adaptability and high-speed feedback equipment, HTSPM is expected to apply to different temperatures, pressures and even liquid environments in the near future. Bearing in mind the aforementioned advantages and by solving the shortcomings of HTSPM step by step, it will contribute greatly to the investigations of the mechanism at micro-/nanoscale.

1.5   Summary

We briefly reviewed the development history, applications, challenges and prospect of HTSPM technology, with a special focus on high-temperature oxidation at micro-/nanoscale. HTSPM is a rising experimental method

that is being used for studying the temperature-dependent mechanical properties as well as the surface oxidation of materials. Applications such as surface characterization and surface topography evolution, mechanism study on small-scale oxidation as well as stress/strain measurement were exploited and summarized. The results display vast potential for this technique in micro-/nanoscale study at elevated temperatures.

Acknowledgment

We acknowledge the support from the National Basic Research Program of China and National Natural Science Foundation of China.

References

1. Tan, C. et al., Recent advances in ultrathin two-dimensional nanomaterials. *Chemical Reviews*, 2017. **117**(9): pp. 6225–6331.

2. Sun, T. et al., Dual-interference-channel Fresnel biprism phase microscope and its application to quantitative phase imaging for transparent biological cell. *Japanese Journal of Applied Physics*, 2018. **57**, 090308: pp.1–4.

3. Molle, J. et al., Towards structural biology with super-resolution microscopy. *Nanoscale*, 2018. **10**(35): pp. 16416–16424.

4. Che, S.N. et al., Direct observation of 3D mesoporous structure by scanning electron microscopy (SEM): SBA-15 silica and CMK-5 carbon. *Angewandte Chemie-International Edition*, 2003. **42**(19): pp. 2182–2185.

5. Browning, N.D., M.F. Chisholm, and S.J. Pennycook, Atomic-resolution chemical-analysis using a scanning-transmission electron-microscope. *Nature*, 1993. **366**(6451): pp. 143–146.

6. Kawakatsu, H. et al., Dual optical levers for atomic force microscopy. *Japanese Journal of Applied Physics*, 1995. **34**(6S): p. 3400.

7. Das, S. et al., Controlled manipulation of oxygen vacancies using nanoscale flexoelectricity. *Nature Communications*, 2017. **8**(1): p. 615.

8. Meyer, E., H.J. Hug, and R. Bennewitz, *Scanning Probe Microscopy: The Lab on a Tip*. 2013. Springer Science & Business Media, Berlin Heidelberg.

9. Hapala, P. et al., Origin of high-resolution IETS-STM images of organic molecules with functionalized tips. *Physical Review Letters*, 2014. **113**(22): p. 226101.

10. Binnig, G. and H. Rohrer, Scanning tunneling microscopy. *Helvetica Physica Acta*, 1982. **55**(6): pp. 726–735.

11. Binnig, G., C.F. Quate, and C. Gerber, Atomic force microscope. *Physical Review Letters*, 1986. **56**(9): pp. 930–933.

12. https://en.wikipedia.org/wiki/Scanning_probe_microscopy.

13. Salapaka, S.M. and M.V. Salapaka, Scanning probe microscopy. *IEEE Control Systems*, 2008. **28**(2): pp. 65–83.

14. Li, Y. et al., In situ measurement of oxidation evolution at elevated temperature by nanoindentation. *Scripta Materialia*, 2015. **103**: pp. 61–64.

15. Hegner, M., P. Wagner, and G. Semenza, Ultralarge atomically flat template-stripped Au surfaces for scanning probe microscopy. *Surface Science*, 1993. **291**(1–2): pp. 39–46.

16. Li, J.B., V. Chawla, and B.M. Clemens, Investigating the role of grain boundaries in CZTS and CZTSSe thin film solar cells with scanning probe microscopy. *Advanced Materials*, 2012. **24**(6): pp. 720–723.

17. Kucsko, G. et al., Nanometre-scale thermometry in a living cell. *Nature*, 2013. **500**(7460): p. 54.

18. Fang, X., Y. Li, and X. Feng, Curvature effect on the surface topography evolution during oxidation at small scale. *Journal of Applied Physics*, 2017. **121**, 125301: pp.1–6.

19. Lu, H. et al., Mechanical writing of ferroelectric polarization. *Science*, 2012. **336**(6077): pp. 59–61.

20. Park, S.M. et al., Selective control of multiple ferroelectric switching pathways using a trailing flexoelectric field. *Nature Nanotechnology*, 2018. **13**(5): pp. 366–370.

21. Yu, J. et al., Quardratic electromechanical strain in silicon investigated by scanning probe microscopy. *Journal of Applied Physics*, 2018. **123**, 155104: pp.1–6.

22. Fischer, H., H. Stadler, and N. Erina, Quantitative temperature-depending mapping of mechanical properties of bitumen at the nanoscale using the AFM operated with PeakForce TappingTM mode. *Journal of Microscopy*, 2013. **250**(3): pp. 210–217.

23. Fang, X. et al., Surface evolution at nanoscale during oxidation: A competing mechanism between local curvature effect and stress effect. *Journal of Applied Physics*, 2016. **119**, 155302: pp. 1–9

24. Voigtländer, B., *Scanning Probe Microscopy*. 2016. Springer-Verlag Berlin Heidelberg.

25. Li, Y. et al., Effects of creep and oxidation on reduced modulus in high-temperature nanoindentation. *Materials Science and Engineering: A*, 2016. **678**: pp. 65–71.

26. Zhou, G. and J.C. Yang, Reduction of Cu_2O islands grown on a Cu (100) surface through vacuum annealing. *Physical Review Letters*, 2004. **93**(22): p. 226101.

27. Zhou, G., W.S. Slaughter, and J.C. Yang, Terraced hollow oxide pyramids. *Physical Review Letters*, 2005. **94**(24): p. 246101.

28. Kresse, G. et al., Structure of the ultrathin aluminum oxide film on NiAl (110). *Science*, 2005. **308**(5727): pp. 1440–1442.

29. Lv, C. et al., Substrate curvature gradient drives rapid droplet motion. *Physical Review Letters*, 2014. **113**, 026101: pp. 1–5.

30. Wang, D. et al., Interaction potential between parabolic rotator and an outside particle. *Journal of Nanomaterials*, 2014, doi: 10.1155/2014/464925.

31. Nichols, F.A. and W.W. Mullins, Morphological changes of a surface of revolution due to capillarity-induced surface diffusion. *Journal of Applied Physics*, 1965. **36**(6): pp. 1826–1835.

32. Freund, L.B. and S. Suresh, *Thin Film Materials:Stress, Defect Formation, and Surface Evolution*. Cambridge University Press, New York.

33. Gao, H.J. and W.D. Nix, Surface roughening of heteroepitaxial thin films. *Annual Review of Materials Science*, 1999. **29**: pp. 173–209.

34. Kim, K.S., J.A. Hurtado, and H. Tan, Evolution of a surface-roughness spectrum caused by stress in nanometer-scale chemical etching. *Physical Review Letters*, 1999. **83**(19): pp. 3872–3875.

35. Dong, X. et al., Diffusion and stress coupling effect during oxidation at high temperature. *Journal of the American Ceramic Society*, 2013. **96**(1): pp. 44–46.

36. Peters, K., D. Whittle, and J. Stringer, Oxidation and hot corrosion of nickel-based alloys containing molybdenum. *Corrosion Science*, 1976. **16**(11): pp. 791–804.

37. Bian, H. et al., Regulating the coarsening of the γ' phase in superalloys. *NPG Asia Materials*, 2015. **7**(8): p. e212.

38. Graham, M. and M. Cohen, On the mechanism of low-temperature oxidation (23°–450°C) of polycrystalline nickel. *Journal of the Electrochemical Society*, 1972. **119**(7): pp. 879–882.

39. Veal, B., A. Paulikas, and R. Birtcher, Mechanisms and control of phase transition in thermally grown aluminas. *Applied Physics Letters*, 2006. **89**(16): p. 161916.

40. Qin, H. et al., Oxidation-driven surface dynamics on NiAl (100). *Proceedings of the National Academy of Sciences*, 2015. **112**(2): pp. E103–E109.

41. Mrowec, S. and Z. Grzesik, Oxidation of nickel and transport properties of nickel oxide. *Journal of Physics and Chemistry of Solids*, 2004. **65**(10): pp. 1651–1657.

42. Chyrkin, A. et al., Modelling compositional changes in nickel base alloy 602 CA during high temperature oxidation. *Materials at High Temperatures*, 2015. **32**(1–2): pp. 102–112.

43. Smeltzer, W. and D.J. Young, Oxidation properties of transition metals. *Progress in Solid State Chemistry*, 1975. **10**: pp. 17–54.

44. Stathis, G. et al., Oxidation and resulting mechanical properties of $Ni/8Y_2O_3$-stabilized zirconia anode substrate for solid-oxide fuel cells. *Journal of Materials Research*, 2002. **17**(5): pp. 951–958.

45. Peraldi, R., D. Monceau, and B. Pieraggi, Correlations between growth kinetics and microstructure for scales formed by high-temperature oxidation of pure nickel. II. Growth kinetics. *Oxidation of Metals*, 2002. **58**(3–4): pp. 275–295.

46. Li, Y. et al., In situ full-field measurement of surface oxidation on Ni-based alloy using high temperature scanning probe microscopy. *Scientific Reports*, 2018. **8**(1): p. 6684.

47. Rhines, F.N. and J.S. Wolf, The role of oxide microstructure and growth stresses in the high-temperature scaling of nickel. *Metallurgical Transactions*, 1970. **1**(6): pp. 1701–1710.

48. Atkinson, A., Transport processes during the growth of oxide-films at elevated-temperature. *Reviews of Modern Physics*, 1985. **57**(2): pp. 437–470.

49. Clarke, D.R., Stress generation during high-temperature oxidation of metallic alloys. *Current Opinion in Solid State & Materials Science*, 2002. **6**(3): pp. 237–244.

50. Clarke, D.R., The lateral growth strain accompanying the formation of a thermally grown oxide. *Acta Materialia*, 2003. **51**(5): pp. 1393–1407.

51. Fang, X. et al., Transition of oxide film configuration and the critical stress inferred by scanning probe microscopy at nanoscale. *Chemical Physics Letters*, 2016. **660**: pp. 33–36.

52. Tolpygo, V.K., J.R. Dryden, and D.R. Clarke, Determination of the growth stress and strain in α-Al$_2$O$_3$ scales during the oxidation of Fe–22Cr–4.8Al–0.3Y alloy. *Acta Materialia*, 1998. **46**(3): pp. 927–937.

53. Pan, B. et al., Two-dimensional digital image correlation for in-plane displacement and strain measurement: a review. *Measurement Science and Technology*, 2009. **20**, 062001: pp.1–17.

54. Hansen, K.V., Controlled atmosphere high-temperature scanning probe microscopy (CAHT-SPM). *Metal Oxide-Based Thin Film Structures*, 2018: pp. 203–223.

55. Ng, C.S. et al., Scanning probe-based in-situ high temperature electrical and electrochemical measurements at atmospheric pressure. *ECS Transactions*, 2016. **72**(28): pp. 11–20.

56. Hansen, K.V. et al., Controlled Atmosphere High Temperature SPM for electrochemical measurements, *Journal of Physics: Conference Series*, 2007. **61**, pp. 389–393.

57. Trenkle, J.C., C.E. Packard, and C.A. Schuh, Hot nanoindentation in inert environments. *Review of Scientific Instruments*, 2010. **81**(7): p. 073901.

58. Li, Y. et al., Microstructure evolution of FeNiCr alloy induced by stress-oxidation coupling using high temperature nanoindentation. *Corrosion Science*, 2018. **135**: pp. 192–196.

59. Huber, T.M. et al., Temperature gradients in micro-electrode measurements: Relevance and solutions for studies of SOFC electrode materials. *Solid State Ionics*, 2014. **268**: pp. 82–93.

60. Kim, K. et al., Radiative heat transfer in the extreme near field. *Nature*, 2015. **528**(7582): pp. 387–391.

61. Challener, W.A. et al., Heat-assisted magnetic recording by a near-field transducer with efficient optical energy transfer. *Nature Photonics*, 2009. **3**(4): pp. 220–224.

62. Basu, S., Z.M. Zhang, and C.J. Fu, Review of near-field thermal radiation and its application to energy conversion. *International Journal of Energy Research*, 2009. **33**(13): pp. 1203–1232.

63. Pendry, J.B., Radiative exchange of heat between nanostructures. *Journal of Physics-Condensed Matter*, 1999. **11**(35): pp. 6621–6633.

64. Rousseau, E. et al., Radiative heat transfer at the nanoscale. *Nature Photonics*, 2009. **3**(9): pp. 514–517.

65. Song, B. et al., Radiative heat conductances between dielectric and metallic parallel plates with nanoscale gaps. *Nature Nanotechnology*, 2016. **11**(6): pp. 509–514.

2

Low Temperature Investigation of Magnetic Molecules by Scanning Probe Microscopies

Giulia Serrano and
Matteo Mannini
*Laboratory for Molecular Magnetism,
University of Florence and INSTM Research
Unit of Florence*

Luigi Malavolti
Max-Planck Institute for Solid State Research

Michele Serri
*Istituto Italiano di Tecnologia—IIT Graphene
Labs*

2.1 Introduction

Since the introduction of the "molecular device" idea (Ratner 2013; Launay & Verdaguer 2013), extensive efforts have been dedicated to the development of novel electronic components exploiting the molecular properties and phenomena with the aim of increasing the performances of the traditional inorganic devices. Indeed, molecules offer a cornucopia of properties that can be also fine-tuned by modifying their chemical structure. Their use has already enabled the fabrication of low-cost and flexible devices such as displays and solar cells. One of the most fascinating aspects is however related to the possibility of *scaling down* the devices' dimensions to a *single* molecule, thus entering natively the *quantum realm* achieving functionalities that go beyond traditional devices. An increasing interest, for instance, has been recently devoted to the study of single molecules as quantum bit (qubit) candidates for quantum technology.

This approach requires tools and techniques capable of operating at the molecular scale. In particular, the evaluation of the molecular when isolated on a surface, *i.e.* the preservation of the structure as well as the chemistry (*e.g.* the stoichiometry, the oxidation state of the different elements, etc.) and the functional properties, is of paramount importance. Furthermore the possible effects of molecule-surface interactions must be taken into account, being these crucial when aiming at developing a molecular-based device. In this respect, a leading role is played by scanning probe microscopies (SPM) allowing a spatial resolution well beyond the molecular scale and offering multiple spectroscopic tools to detect functional properties, *i.e.* their electronic or magnetic characteristics. Although in many cases SPM-based data need to be supported by other measurements and theory to gain a more comprehensive picture, here we want to highlight the potentialities of local probe techniques and focus on their use for the study of *magnetic molecules*.

In Section 2.1.1, we will introduce some relevant classes of magnetic molecules and their specific properties. Sections 2.2 and 2.3 are focused on three different SPM techniques that are the scanning tunneling microscopy (STM), the atomic and magnetic force microscopy (AFM and MFM), respectively. We will give a short description of their working principles, and we will report on key results dealing with imaging of magnetic molecules as well as with the local characterization of their electronic and magnetic properties using the local spectroscopies.

2.1.1 Magnetic Molecules and Molecular Magnetism

The quest for molecular analogues of classical **magnetic units** for data storage, magneto and magneto-optical switches, magnetic field sensors and classical or quantum logical units motivated and legitimated the efforts of the "molecular magnetism" (Kahn 1993; Gatteschi et al. 2006) community in the last 30 years. This multidisciplinary community of chemists and physicists have synthetized and studied several classes of molecules, whose structure was designed *ad hoc* to obtain specific magnetic properties. These molecules have been obtained by carefully combining magnetic atoms (transition metal ions, rare earth ions, as well as light elements bearing unpaired electrons) with organic ligands, leading to molecular systems with peculiar behaviors suitable for technological applications. For instance, even in a simple coordination compound, the fine selection of the coordinating ligands alterates the magnetic anisotropy parameters of the ion. In more complex architectures, these ligands can act as linkers between different magnetic units, promoting exchange, super-exchange and dipolar interactions within the molecule. Additionally, the introduction of specific functional groups on the molecular backbone can serve to limit the intermolecular interactions or to chemically anchor molecules on surfaces. The almost infinite combinations of different molecular structures, functionalizations and magnetic properties of single molecular units and their assemblies make the molecular magnetism a field plenty of opportunities.

An interesting class of molecules is represented by single-molecule magnets (**SMMs**). This fascinating family of molecular complexes features, at low temperature, magnetic bistability that is characteristic of the individual molecule and manifests as magnetic hysteresis loops (Gatteschi et al. 2006). The SMM properties usually originate from a giant spin, resulting from the presence of (super-)exchange-coupled magnetic ions, and an easy axis of magnetic anisotropy defining two preferable orientations of the magnetic moment. Recently, a similar behavior has also been obtained using single ion systems achieving the manifestation of SMM behavior at unprecedented temperatures (Ishikawa 2007; Goodwin et al. 2017). In all SMMs, reversal of the magnetization can be thermally activated, overcoming the barrier between the two lower energy states, or, in specific resonance conditions, by a quantum tunneling process (Quantum Tunneling of the Magnetization, QTM). The former mechanism is predominant above the so-called *blocking temperature* of the system, while the latter is responsible for the relaxation of the magnetization also observed at low temperature, for example when, in zero magnetic field, the two low-energy states are degenerate. By applying a magnetic field along the easy axis of magnetization, the spin levels are shifted according to the Zeeman interaction and different tunneling paths are activated or made inactive. This leads to the observation of a stepped magnetization hysteresis loop below the blocking temperature, as it was demonstrated in ensembles of SMMs such as single crystals (Sangregorio et al. 1997) or monolayers of well-ordered molecules (Mannini et al. 2010). Early experiments addressing individual molecules have opened fascinating perspectives for developing devices that profit from both magnetic bistability and quantum features (Urdampilleta et al. 2011; Thiele et al. 2014).

Furthermore, other classes of molecules behave as bistable systems, featuring metastable states that can be interconverted by altering the external conditions (temperature or pressure) or via external stimuli such as light irradiation,

application of strong electric field or magnetic field. Such a behavior has been observed in Spin Crossover (**SCO**) complexes and in Redox Isomers (**RI**). In the former, a conversion between two magnetic states of the molecules, specifically a low-spin (LS) and a high-spin (HS) state, is observed. This mechanism relies usually on the presence of a transition metal ion, such as the d^5 iron (II), in an octahedral environment (Gutlich et al. 1999). Specific ligands determine an electronic configuration where the LS state is favored enthalpically (i.e. the S=0 state of the Fe^{2+} ion in an octahedral geometry) and the HS state entropically (S=2 in the same ion). A conversion between the HS and the LS state can be induced by lowering the temperature or by increasing the pressure. Indeed, at high pressure, the contraction of the bond length enhances the crystal field stabilizing the LS state. At low temperature, an HS trapped state is accessible by irradiating the system in the LS state with light (Létard et al. 1999). Typically, the SCO transition is highly influenced by the molecular organization, with cooperative effects playing a crucial role in altering the SCO transition temperature (Félix et al. 2014). On the contrary, these factors are less important for the switching of RI molecules. Here, temperature and entropy control the equilibrium between the two molecular states varying the oxidation state of the metallic ion and its magnetic moment via electron transfer with the ligand. This is for example the case of the cobalt catecholate complexes in which the Co^{3+}/Co^{2+} redox couple exchanges one electron with a catecholate ligand that can be converted in its semiquinone form transferring an electron to the Co^{3+} ion (Caneschi et al. 2001). At low temperatures, these two states can be mutually interconverted via light irradiation resonant with the electronic transitions (Beni et al. 2006). Another example of RI systems is constituted by the Prussian blue analogues (PBAs), widely studied as bulk and 2D materials, or in the form of nanoparticles and isolated molecules (Aguilà et al. 2016; Catala & Mallah 2017). PBA molecules are probably among the most studied RI systems because of their electronic, magnetic and magneto-optical properties (Ferlay et al. 1995; Bleuzen et al. 2000). In PBAs, the coupling obtained by bridging two transition metal ions with cyanide ligands promotes a metal-to-metal charge transfer that can be thermally and optically activated and is responsible for the switching of the magnetic properties.

The molecular magnetism community is also operating with simpler molecular structures such as purely organic radicals that were initially studied as ferromagnets (Palacio et al. 1997) or simple coordination complexes from which the whole know-how of this discipline originated (Kahn 1993). This sort of *revival*, has been catalyzed by the perspectives of using them in novel devices for ***molecular spintronics*** and ***quantum computation*** (Bogani & Wernsdorfer 2008; Ganzhorn & Wernsdorfer 2014; Coronado & Epstein 2009; Bonizzoni et al. 2017). These molecules, in fact, can be used not only as simple bistable individual objects but also for influencing spin transport effects (injection diffusion and detection) in spintronic devices and as quantum bit units exploiting their molecular spin (Affronte 2009; Aromí et al. 2012; Tesi et al. 2016; Zadrozny et al. 2015). In this perspective, the interaction between the *core* of the molecules with the substrates, which acts as "connection" between molecular and inorganic components of these devices, is crucial. Recent studies (Barraud et al. 2010; Lodi Rizzini et al. 2011) demonstrated the key role of these interactions in altering, influencing and tuning the magnetic properties of the magnetic objects. This provides the opportunity for the realization of innovative hybrid devices in which the properties of the molecular and the inorganic components benefit from their mutual influence. Indeed, at the single molecular scale, inorganic layers can act in synergy with the molecular counterparts to show enhanced or even generate novel properties. This is, for instance, the case of metal phthalocyanines assembled on topological insulators, where the presence of a monolayer of interacting molecules allows to control Dirac Fermions in topologically protected states of these substrates, as observed by Angle Resolved Photoelectron Spectroscopy (ARPES) experiments (Caputo et al. 2016; Sk et al. 2018). On the contrary, decoupling the magnetic molecules from the conductive components (Wäckerlin et al. 2016) can be desirable to maintain or even enhance the individual molecular properties. For instance, this is the case of the terbium(III) bis-phthalocyaninato ($TbPc_2$) molecule, a well-known SMM complex, that once assembled on graphene or MgO decoupling layers shows a slower magnetization dynamics than in the bulk (Serrano et al. 2018; Wäckerlin et al. 2016).

In this context, a growing interest is devoted to the observation of the properties of molecules on surfaces and to understand how the interaction with the surface may alter them exploiting the capabilities of SPM techniques. Sections 2.2 and 2.3 are focused on the two principal and consolidated scanning probe techniques that are the STM and the scanning force microscopy, respectively. These paragraphs will provide an overview of the most significant results that have been reported by these techniques on the investigation of magnetic molecules at low temperature where advanced resolution and spectroscopic capabilities allow the observation of key molecular properties on surfaces. Furthermore, Section 2.4 gives a brief overview of the last developments and perspectives of scanning probe techniques for the active excitation and detection of magnetic transitions at the molecular scale that we believe might be of high interest for the study of magnetic molecules.

2.2 Scanning Tunneling Microscopy

Among the scanning probe techniques, STM has played a major role in the characterization and identification of magnetic molecules on surfaces. For this reason, we dedicated large part of this chapter to review its basic principles and its use in the study of magnetic molecules.

Since its invention (Binnig et al. 1982a, b), the STM technique has become a leading and widespread method for the investigation of the morphology of molecules and the study of their local electronic and spin properties on surfaces. Due to its ultimate spatial resolution, nowadays STM represents a well-established tool for the characterization of molecules on surface. The implementation of advanced functionalities, ranging from the observation of molecular spin excitations and fast time-resolved experiments, has made the STM an appealing tool for the investigation of magnetic nanomaterials and quantum materials.

Given the vast number of scientific reports concerning these topics, it is worth to specify that this paragraph does not represent a comprehensive review, but it rather aims to provide a general view of the most representative STM investigations on magnetic molecules. It is divided in two parts. In the first part, the theory and working principles of the STM technique are presented (Section 2.2.1), while the second part is focused on reviewing its application in the study of magnetic molecules (Section 2.2.2–2.2.4). In particular, Section 2.2.2 focuses on the characterization of the structural properties of magnetic molecules on surfaces, while Section 2.2.3 is dedicated to the study of their electronic properties detected through the *Scanning Tunneling Spectroscopy (STS)* technique. Section 2.2.4 shows further spectroscopic potentialities of the STM for the investigation of spin in molecules exploiting the *Inelastic Electron Tunneling Spectroscopy (IETS)*.

2.2.1 Theory and Working Principles of STM

In the early 1980s, Binnig and Röhrer developed the first scanning tunneling microscope (abbreviated commonly to STM) (Binnig et al. 1982a, b), receiving the Nobel Prize only a few years later, in 1986. However, *quantum tunneling* had been a well-known phenomenon observed in multiple experimental studies prior to the introduction of the STM. In quantum mechanics, a particle, such as an electron, has a certain probability of entering a classically forbidden region, such as a potential barrier, that divides two classically allowed regions. This phenomenon is referred to as *quantum tunneling*. Experimentally, forbidden regions for an electron can be insulating materials or a vacuum space positioned between two metals. Tunneling phenomena were first observed by Giavier et al. in metal-insulator-metal (MIM) junctions of superconductors (Giaever & Megerle 1961; Giaever 1960a, b) and later (1971) in metal-vacuum-metal junction by Young et al. (1971). The latter also developed the *topografiner* (Young et al. 1972), an instrument consisting in a field emission microscope. Only in 1981, Binnig et al. reported the first successful and well-controlled combination of a vacuum tunneling barrier separating a metal probe from a conducting surface. Such an STM allowed, for the first time, to achieve real-space atomic resolution (Binnig et al. 1982a, b).

Here, we report the elementary theory of the tunneling phenomenon. Let us consider a one-dimensional potential barrier $U(z)$ and an electron, represented in the quantum-mechanical treatment by a wave function $\psi(z)$.

The Schrödinger equation that describes the system is (Chen 1993):

$$-\frac{\hbar^2}{2m}\frac{d^2}{dz^2}\psi(z) + U(z)\psi(z) = E\psi(z) \qquad (2.1)$$

with E, the electron energy and m, the mass of the electron. For the classically allowed regions where $E>U(z)$, the electron is free to move either in a positive or in a negative direction with a constant velocity. In a classical picture, the electron cannot overcome a potential barrier higher than its energy $(E<U(z))$. Conversely, in quantum mechanics, the solution to the above equation for $E<U(z)$ is (Chen 1993):

$$\psi(z) = \psi(0)e^{-kz} \qquad (2.2)$$

where k is the decay constant $k = \frac{\sqrt{2m(E-U)}}{\hbar}$ and the probability that the electron penetrates the barrier takes the form of $|\psi(z)|^2 = |\psi(0)|^2\, e^{-2kz}$. It is important to point out that this strong nonlinear dependence of the tunneling events on the barrier width (z) is the key to achieve the ultimate STM spatial resolution. A scheme of the tunneling event is given in Figure 2.1a, which will be explained in detail later in the text.

In an STM, a metallic tip is brought in close range (tens of Angstrom) with a conductive surface. The tip (usually a W or Pt-Ir wire) and the sample form two electrodes separated by a few Angstroms insulating barrier. By applying an electric potential difference between the two electrodes, referred as bias potential, V_b, $(eV_b < U(z)$ in the tunneling regime) a net tunneling current will flow between them according to the Fowler–Nordheim expression (Fowler & Nordheim 1928):

$$J_T \propto V_b \exp(-A\sqrt{\phi}d) \qquad (2.3)$$

where J_T is the tunnel current density, A is a constant $(A \approx 1)$, d is the distance between tip and surface and

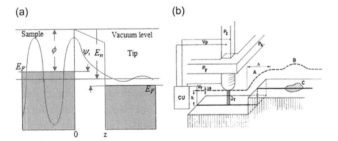

FIGURE 2.1 (a) Scheme of a one-dimensional metal-vacuum-metal tunneling junction. (b) Scheme of the working principle of an STM. A voltage is applied by the control unit (CU) to the piezodrive P_z to keep the tunnel current (J_T) constant at the tunnel voltage V_T (or V_b) (constant current mode). The piezodrives P_x and P_y scan the surface along the x and y direction. In the constant height mode, the tip-surface distance s is held constant to a fixed value and the J_T is measured. The z displacements are representative of the surface topography being sensitive to the presence of surface steps (A) or contaminants (B). (Reprinted with permission from Binnig, G. et al., *Physical Review Letters* 1982, 49, 57–61. Copyright (1982) American Physical Society.)

$\phi = \frac{\phi_1+\phi_2}{2}$, that is the average between the electrodes *work functions* ϕ, *i.e.* the minimum energy required to remove an electron from the bulk to the vacuum level.

In order to acquire an STM image, the tip is moved with respect to the surface in a raster fashion while the tip-surface distance is recorded keeping the tunneling current constant (*constant current mode*) or the current is recorded keeping the tip-surface distance constant *(constant height mode).* In the first case, a feedback circuit changes the z position of the tip (perpendicularly to the sample surface) during the scanning to maintain the tunneling current at the set value. If the tunneling current becomes higher than the reference value, the tip is retracted; conversely, if it becomes smaller, an approaching movement is performed. In the second mode, a certain z position of the tip is fixed and the tunneling current variation is recorded during the scan. In a two-dimensional image, the Z position of the tip or the current is typically represented by a color scale. Usually, the first mode is used to record STM images since it reduces the chances of crashing the tip on the surface during the scan of rough areas. A scheme of the STM functioning is presented in Figure 2.1b.

The tip movement is realized by means of piezoelectric crystals. They are capable of controlling very fine movements allowing to exploit the STM spatial resolution: 0.05 Å along the z direction and less than 1 Å in the x and y directions (Stroscio & Kaiser 1993).

STM images, however, do not simply return the sample topography and their understanding requires a more careful treatment, which considers the density of filled and empty states on the tip and the sample. Assuming for simplicity two electrodes (here the tip and the sample) with the same work function ϕ, at a distance d, then electrons from the tip can tunnel into the sample and *vice versa*. The direction of a net tunneling current depends upon the applied bias V_b, hereafter V. Assuming that the density of states in the tip is constant with energy, when electrons tunnel from the tip into the sample, the samples' *empty states* are probed, while the opposite flow direction probes the samples' *filled states.* Let us consider, for example, the case in which electrons tunnel from the sample to the tip as depicted in Figure 2.1a. The tunneling probability for an applied voltage $eV << \phi$ is proportional to $|\psi_n(0)|^2\, e^{-2kd}$, with $k = \frac{\sqrt{2m\phi}}{\hbar}$ and ψ_n being the sample's states near the Fermi level, E_F, *i.e. the electrochemical equilibrium potential.* The tunneling current, I is given by summing up the contributions of all the electrons coming from the sample states with E close to the E_F and ranging between E_F and $E_F - eV$:

$$I \propto \sum_{E_n=E_F-eV}^{E_F} |\psi_n(0)|^2\, e^{-2kd}. \qquad (2.4)$$

In a rigorous approach, one should consider in the tunneling expression also the states of the tip. In the approach proposed by Bardeen (Bardeen 1961), in the limit of small bias voltage and small $k_B T$ values, Eq. 2.4 takes the form (Chen 1993):

$$I \propto V \cdot \sum_{S,T} |M_{S,T}|^2 \cdot \delta(E_S - E_F)\delta(E_T - E_F) \quad (2.5)$$

where $M_{S,T}$ is the tunneling matrix element between the states of the sample and the tip, E_S and E_T are the energy of the sample and tip states, respectively.

The exponential decay of the tunneling current with respect to the distance allows to consider that the electron tunneling process mainly involves the last atom of the tip providing the ultimate STM resolution. Therefore, it is possible to assume a spherical shape of the tip apex. Under this assumption, the model proposed by Tersoff and Hamann (1983, 1985) simplifies the above expression including only the matrix element for S-wave tip wave function:

$$I \propto \sum_n |\psi_n(\mathbf{r_0})|^2 \delta(E_n - E_F) \quad (2.6)$$

This represents the local density of states (LDOS) of the sample at the Fermi level below the tip position $\mathbf{r_0}$. STM images are then more properly representative of the local surface density of states than a real topography. The imaging of charge density waves (Carpinelli et al. 1996) and quantum corrals (Stroscio et al. 1991) are typical examples. This property can be exploited to gain additional information about the sample. By means of the STS, *I vs. V* curves are obtained on a specific point of the surface by changing the bias and measuring the tunneling current while keeping the z position of the tip constant (see also Section 2.2.3). The dependence of the tunneling current on the bias voltage is not trivial, and in general, the interpretation of the tunneling spectra is not straightforward. In the derivation from the Tersoff and Hamann model by Selloni et al. (1985), the current assumes the form:

$$I(V) \propto \int_{E_F}^{E_F+V} \rho_S(E) T(E, V) dE \quad (2.7)$$

where $T(E, V)$ is the transmission barrier coefficient. As such, in a first rough approximation, the quantity *dI/dV(V)* is proportional to the LDOS of the sample at $E = E_F + V$. Increasing the bias voltage, the approximation of negligible bias with respect to the metal work function (usually around 4–5 eV) does not hold. The transmission coefficient T strongly depends on the voltage, introducing a distortion in the spectra (Stroscio & Kaiser 1993). For such voltages, it is convenient to use the normalized quantity *(dI/dV(V))/(I/V)*, which has the advantage of cancelling out the transmission coefficient contribution, as suggested by Stroscio et al. (Stroscio & Kaiser 1993).

2.2.2 Structural Characterization of Magnetic Molecules by STM

The STM technique has been widely used to carry out a morphological characterization of the molecular assemblies on surface. In particular, for magnetic molecules, it has been used to verify molecular integrity upon deposition on substrates, the molecular coverage and geometry

of the molecular assembly; these aspects were all found to strongly affect molecular magnetic properties and have been therefore the subject of careful investigations. Thanks to the STM and STS technique (see details in Section 2.2.3), substrate–molecule and molecule–molecule interactions were found to be the driving force in the formation of different molecular assemblies on surfaces, whose properties strongly depend on the substrate and molecular coverage. However, in order to obtain a high STM resolution, it is crucial to carefully control the substrate preparation and work in a controlled atmosphere, as for instance ultra-high vacuum (UHV) conditions, to ensure the surface cleanliness.

Phthalocyanines and porphyrins, in single- or multi-layer architectures (double- or triple-deckers) including a magnetic core in their structure, can be considered as the archetypal magnetic molecules for STM investigation. Their simple and almost flat structure, enabling a high-lateral sub-molecular resolution, is well recognizable in the STM images, often allowing a straightforward comparison with the molecular DOS predicted by calculations. Due also to the relative facility of depositing them by means of sublimation in UHV, they are one of the most commonly characterized systems by means of STM. Figure 2.2a shows an STM image of the TbPc$_2$ SMM, adsorbed on the Au(111) surface (Komeda et al. 2011). TbPc$_2$ is composed of a Tb ion sandwiched between two phthalocyanine (Pc) ligands staggered by 45 degree (see structure depicted in Figure 2.2c). A variation of the angle can be observed by a change of apparent height of the molecule within the assembly (Figure 2.2d) (Komeda et al. 2011). As one may observe, the STM image is not directly comparable to the molecular structure but more properly to the density of states of the molecular orbitals calculated by density functional theory (DFT) methods (Komeda et al. 2011). The STM-simulated image shows an

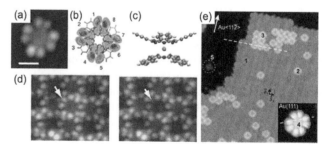

FIGURE 2.2 TbPc$_2$ molecules on Au(111): (a) STM image of an isolated molecule and (b) the simulated STM image by DFT methods; (c) schematic model of the double-decker molecule, side view. (d) TbPc$_2$ contrast change after modification of the angle between the two phthalocyanine ligands. (Panels a–d; adapted with permission from Komeda, T. et al., *Nature Communications* 2011, 2, 217 under Creative Commons Attribution 3.0 International License.); (e) TbPc$_2$ molecules in first (1) and second (2 and 3) layer. A single TbPc$_2$ (4) and phthalocyanine (5) molecules on the Au(111) are also shown. (Adapted with permission from Serrano, G. et al., *The Journal of Physical Chemistry* C 2016, 120, 13581–13586 under Creative Commons Attribution 4.0 International License.)

eight-lobe structure corresponding to the upper Pc ligand molecular orbital and well matching the observed $TbPc_2$ geometry in the STM experiments (Figure 2.2b). This structure is generally observed for double-decker molecules on different metals, such as Cu(111) (Vitali et al. 2008; Stepanow et al. 2010), Cu(100) (Deng et al. 2015), Au(111) (Ara et al. 2016; Katoh et al. 2009; Komeda et al. 2014; Serrano et al. 2016; Pan et al. 2017; Fu et al. 2012; Zhang et al. 2010, 2015, 2018; Komeda et al. 2011), Ag(111) (Pan et al. 2017; Ara et al. 2016) or metallic thin films (Malavolti et al. 2013; Pan et al. 2017), as well as non-metallic substrates (Smykalla et al. 2012; Zhang et al. 2018; Pan et al. 2017; Wäckerlin et al. 2016). STM studies have also evidenced the formation of assemblies or multi-layer structures of double-deckers at increasing molecular coverage (Zhang et al. 2009; Deng et al. 2015; Serrano et al. 2016). Figure 2.2d shows a well-ordered cluster of $TbPc_2$ molecules on the Au(111) surface where molecules are arranged in a square lattice having unit cell dimensions of about 1.4×1.4 nm^2; the second layer molecules lie on top of the first layer reproducing the squared lattice arrangement but with an azimuthal orientation rotated by 45°. Increasing the number of layers induces a deviation from the flat-lying geometry with respect to the surface. For example, tilting angles between 4° and 7° have been observed in second and third layer of $LuPc_2$ on highly oriented pyrolytic graphite (HOPG) (Smykalla et al. 2012) or in CoPc films on metals where a gradual molecular reoganization was also found (Takada & Tada 2004; Chen et al. 2008). Another exemplary class of SMMs is represented by Fe_4 complexes, which show slow relaxation of the magnetization in the sub-Kelvin temperature range and preserve this magnetic behavior when adsorbed as a monolayer on surfaces by chemical methods (Mannini et al. 2009, 2010). The robustness of their magnetic properties has been confirmed after UHV thermal deposition and the molecular integrity was checked by STM. The almost spherical conformation of Fe_4 molecules is visible at the STM inspection with the appearance of round objects ascribed to single Fe_4 molecules, whose detailed sub-molecular structure is not resolved (Malavolti et al. 2015; Lanzilotto et al. 2016; Luo et al. 2015; Gragnaniello et al. 2017; Erler et al. 2015). On the Au(111) surface, the $[Fe_4(L)_2(dpm)_6]$ (Fe_4Ph) derivative forms large assemblies where a short-range hexagonal order was observed (Malavolti et al. 2015). By means of an STM study, it was established that a fraction of Fe_4 molecules is fragmented by the UHV sublimation process, resulting in the concomitant adsorption of molecular fragments on the surface (Lanzilotto et al. 2016). The deposition of Fe_4H SMM molecules by electrospray results in highly ordered layers with an hexagonal pattern and prevents molecular fragmentation (Gragnaniello et al. 2017; Erler et al. 2015). In this case, a higher sub-molecular STM resolution was achieved, favored by a reduced complexity of the molecular structure, lacking the phenyl ligand and inducing a flat-lying molecular orientation with respect to the substrate.

STM has also been used to resolve nontrivial geometries of magnetic molecules. For Er(trensal) SMMs, it was shown that a layer of graphene (Gr) on the Ir(111) and Ru(111) surfaces induces a higher order in the molecular arrangement when compared to the adsorption on the bare substrates (Dreiser et al. 2016), as later observed also for $TbPc_2$ SMMs on Gr/SiC(0001) (Serrano et al. 2018). High-resolution STM images show Er(trensal) molecules arranged in an alternating fashion with phenyl rings pointing up and down inside the lattice unit cell (Dreiser et al. 2016). Similarly, STM imaging of VOPc molecules with sub-molecular resolution was used to identify two opposite configuration geometries having the vanadyl group pointing up or down the surface plane and producing a bright and dark inner core respectively. Both configurations were observed on the Au(111) (Niu et al. 2012, 2014), Ni(111) (Adler et al. 2015), Cu(111) (Niu et al. 2014) and Ag(111) (Niu et al. 2014) surfaces; on the contrary, a preferential oxygen up configuration was observed on a more inert substrate such as graphite (Zhang et al. 2014; Niu et al. 2014), thus evidencing the importance of surface-molecule interactions in the molecular assemblies.

The high lateral and vertical resolution of the STM was fully exploited to detect small conformational variations in SCO complexes on surfaces (Bairagi et al. 2016; Jasper-Toennies et al. 2017; Miyamachi et al. 2012; Gopakumar et al. 2012). Spin transition between HS and LS states was observed for $[Fe(II)((3, 5-(CH_3)_2Pz)_3BH)_2](1)(Pz = $ pyrazolyl) sub-monolayers on Au(111) (Bairagi et al. 2016). The transition was achieved by light illumination at 405 nm at low temperature (4.6 K) and observed through the variation of the molecular height in the STM contrast. Miyamachi et al. induced the HS–LS transition of the $[Fe(phen)_2(NCS)_2]$ SCO complex by locally ramping the electric field through a bias sweep (Miyamachi et al. 2012). The procedure resulted in a reversible transition that was observed by a variation in the differential conductance ($dI/dV(V)$) signal and a structural change of the molecules (Figure 2.3a); it was realized that the presence of an insulating layer (Cu_2N) on the copper substrate, serving as a decoupling layer, was essential to observe the molecular switching. A similar technique has been used to switch the $[Fe(pap)_2]^+$ (pap = N-2-pyridylmethylidene-2-hydroxyphenylaminato) molecules on Cu_2N/Cu(100) in a controlled manner that was further exploited to build an example of a two-state molecule memory (Jasper-Toennies et al. 2017).

A particular STM contrast is obtained by scanning with a magnetic tip (Spin-Polarized STM, SP-STM) (Brede et al. 2010; Iacovita et al. 2008; Schwöbel et al. 2012). This technique allows, for instance, to resolve the spin structure of magnetic molecules in the real-space. In this way, Schwöbel et al. (2012) reported the possibility to image the spin state of a $TbPc_2$ molecule on a magnetic Co nano-island template grown on Ir(111); the spin sensitivity is achieved by tunneling with a ferromagnetic (iron-coated) tip with an out-of-plane magnetization parallel or antiparallel

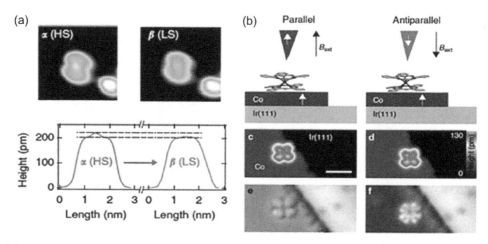

FIGURE 2.3 (a) STM images of the HS and LS state of the Fe-phen molecules on the CuN/Cu(100). The different STM contrast between the two states is evidenced from the line profiles at the bottom. (Reprinted with permission from Miyamachi, T. et al., *Nature Communications* 2012, 3, 938 under Creative Commons Attribution 3.0 International License.) (b) Example of a SP-STM measurement: parallel and antiparallel directions of the tip magnetization with respect to that of the Co nano-island (scheme at the top) are used to record SP-STM topographic images (images at the center) and spin-resolved maps of differential conductance of a TbPc$_2$ molecule. (Adapted with permission from Schwöbel, J. et al., *Nature Communications* 2012, 3, 953 under Creative Commons Attribution 4.0 International License.)

with respect to that of Co. The magnetization in the tip is reversed upon application of a magnetic field of ± 1 *T*, while it remains fixed in the Co nano-islands behaving as hard magnets. A marked difference between the parallel and antiparallel magnetization can be observed in the differential conductance images ($dI/dV(V)$ signal) of the magnetic molecule (Figure 2.3b). Here, TbPc$_2$ appears as a four-lobe structure on a bright Co-island in the first case and having eight lobes on a dark area in the latter (Schwöbel et al. 2012). Similar spin-related contrast was also achieved for Co-phthalocyanine molecules (Brede et al. 2010; Iacovita et al. 2008).

Besides the imaging capabilities, STM can be used as a tool to locally interact with the molecular system to get insights into its magnetic properties on surfaces or to manipulate them. The manipulation procedure can be carried out by applying voltage pulses, or exploiting attractive or repulsive forces between tip and molecules to drag them on the surface. On assemblies of TbPc$_2$ on Au(111), this technique was used to selectively alter the spin of a single molecule inside the molecular assembly (Komeda et al. 2011). Such procedure induced a change in the apparent height of the manipulated molecule (see Figure 2.2d) and the reversible appearance and disappearance of a Kondo resonance in the $dI/dV(V)$ spectra (further details on this phenomenon are given in Section 2.2.3). Dehydrogenation of a CoPc on Au(111), by application of multiple bias pulses, was found to produce a similar effect. A strong substrate-molecule interaction is induced by the dehydrogenation process resulting in the alteration of the molecular electronic structure and the appearance of a Kondo resonance (Zhao et al. 2005). The four-lobe feature of the CoPc was strongly affected as well, turning into a spherical shape for the dehydrogenated molecule, likely due

to a change of the molecular conformation on the surface (see Figure 2.4a). Spin alteration of YPc$_2$ complexes was obtained by the use of dopants introduced by STM manipulation (Robles et al. 2012). STM manipulation was used to artificially fabricate TbPc$_2$ assemblies (Figure 2.4b) and demonstrate the relevance of the molecule–molecule interaction for the molecular spin characteristics (Amokrane et al. 2017). The STM tip interaction was also exploited to detach nitric oxide (NO) molecules dosed on Co-TPP (TPP = tetraphenylporphyrin) molecules. In this process, the appearance of the Co-TPP is restored along with its spin configuration, as confirmed by the reappearance of a Kondo resonance (Kim et al. 2013). Other groups reported the possibility to use voltage pulses to switch molecular chirality (Fu et al. 2012) or to graft single magnetic molecules on surfaces (Luo et al. 2015).

2.2.3 Scanning Tunneling Spectroscopy (STS)

The capability of the STM of accessing the LDOS has been already introduced in the Section 2.2.1. In particular, it was pointed out that the derivative of the current with respect to applied bias ($dI/dV(V)$) is proportional to the LDOS. By recording the current during a bias sweep while keeping the tip position constant, the I *vs.* V curve can be obtained. The numerical derivative of such a curve, thus, provides the $dI/dV(V)$ spectrum. On the other hand, $dI/dV(V)$ can be directly recorded operating a lock-in measurement (Ametek n.d.) where, on top of the bias sweep, a modulation bias (ac bias) with a tunable amplitude and frequency is applied. The lock-in device monitoring the current signal at the ac frequency allows to directly obtain the $dI/dV(V)$ signal (Ametek n.d.). By using high harmonics of the signal, higher

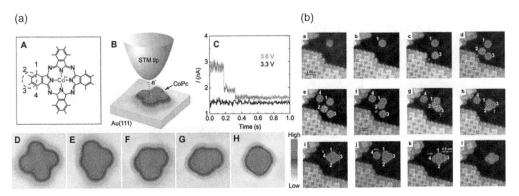

FIGURE 2.4 Examples of STM manipulation: (a) Dehydrogenation of a CoPc molecule on Au(111), the evidence of the STM topography change during multiple manipulation steps is shown at the bottom. (Adapted with permission from Zhao, A. et al., *Science* 2005, 309, 1542. Copyright (2005) Science.); (b) Fabrication of an artificial nano-architechture by STM manipulation of TbPc$_2$ molecules on the Au(111) surface. (Adapted with permission from Amokrane, A. et al., *ACS Nano* 2017, 11, 10750–10760 under Creative Commons Attribution 4.0 International License.)

derivatives are obtained, allowing to resolve small conductance variations. This method usually offers the advantage of improving the signal-to-noise ratio by choosing the lock-in frequency in a range where the electrical and vibrational noise of the STM is minimum.

The energy position of the highest occupied molecular orbitals (HOMO, HOMO-1, etc.), the lowest unoccupied molecular orbitals (LUMO, LUMO+1, etc.) and, in case the molecule possesses it, the singly occupied molecular orbital (SOMO) can be investigated by recording the *dI/dV(V)* spectrum of a molecule (energy range of few eV) (Binnig, G. et al. 1982a). When the applied bias allows the electrons to tunnel into the molecular orbitals, an increase of the conductance is observed. The identification of a LUMO orbital by STS is schematically reported in Figure 2.5. The STS provides a direct way to obtain the HOMO and LUMO position with respect to the Fermi level of the substrate and the HOMO–LUMO gap, which are otherwise difficult to determine by means of ab initio calculations. This information is relevant for the understanding of the molecular electrical conduction and the implementation of molecules in spintronic and quantum devices. Moreover, STS provides crucial information that can integrate the data obtained using different techniques such as photoelectron spectroscopies. A remarkable example of the role played by the orbital energy alignment is the magnetic field-dependent

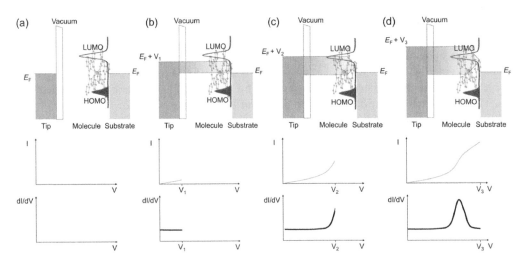

FIGURE 2.5 A simplified model of a bias spectroscopy is presented in the top panels. A molecule is in contact with the substrate surface and is investigated by the STM tip. For the sake of simplicity, the LDOS of the sample is dominated by the HOMO and LUMO while the DOS of the tip and the substrate are considered flat. The electron states of the tip and the substrate are filled up to the respective Fermi energy (E_F), the temperature broadening is not considered here. From left to right (a–d), by increasing the applied bias (V), a tunneling current, represented by a light gray curve, flows from the tip to the surface. By plotting the current history during the bias sweep, the curves presented in the central panels are obtained. In the bottom panels are reported the *dI/dV(V)* spectra obtained by numerical differentiation of the current plot or by employing a lock-in measurement. The *dI/dV(V)* curves reproduce the LDOS of the sample, and in this case, the LUMO. In a similar fashion, changing the polarity of the applied bias, the HOMO feature can be observed.

negative differential conductance (NDR) (Warner et al. 2015) features on iron phthalocyanine molecules deposited on copper nitride layer (Cu_2N) (Warner et al. 2015). The NDR shows a dependence on the magnetic field about two orders of magnitude higher than the one expected by a Zeeman splitting mechanism. As such, this phenomenon was attributed to a bias-dependent alignment of the molecular orbital with the Fermi energy of the substrate intimately related to the presence of the Cu_2N insulating layer.

The combination of the spectroscopic technique with the atomic resolution of the STM allows also to monitor the spatial dependence of specific molecular states. By mapping the $dI/dV(V)$ signal in space using a technique referred as current imaging tunneling spectroscopy (CITS) (Hamers et al. 1986), it is indeed possible to visualize specific molecular orbitals (Repp et al. 2005; Dey et al. 2007). In case of large magnetic molecules hosting many different metallic centers, this technique allows to visualize their positions. Indeed, if the metallic centers exhibit specific electronic features that can be energetically resolved from those of the ligands, a $dI/dV(V)$ map reveals their position even in the presence of a complex topographic image (Dey et al. 2007). A clear example is reported in Figure 2.6. The correlation of spectroscopic results with the spatial information allows to identify the molecule-surface interaction dependence on the adsorption site. This was observed, for instance, by the different spectroscopic behavior of $TbPc_2$ molecules (see also Section 2.2.2) evaporated on Au(111) in large molecular islands compared to isolated molecules. The reduced molecule-surface interaction within the molecular film translates into the emergence of a strong resonance in the $dI/dV(V)$ spectrum associated to the observation of the $TbPc_2$ SOMO (Komeda et al. 2014).

Additional information can be extracted by employing a SP tip (Wiesendanger 2009). As reported by Schwöbel et al. (2012), the spin polarization of the molecular orbital of a $TbPc_2$ molecule deposited on a cobalt film grown on the Ir(111) surface can be resolved and imaged (see Section 2.2.2). Another interesting example of SP-STS was reported by Garnica et al. (2013) in which the authors were able to individuate the presence of magnetic order

in tetracyanoquinodimethane (TCNQ) molecular domains deposited on graphene/Ru(0001). The nonmagnetic TCNQ molecules develop a magnetic moment due to a charge transfer from the graphene layer.

STS allows also to access electronic features related to the presence of a molecular spin on the surface. A relevant example is the observation of Kondo resonances, as previously mentioned in Section 2.2.2 (Kondo 1964; Scott & Natelson 2010; Hewson 1993). The interaction of magnetic impurities with the electron bath of the surface can lead to the formation of a multiparticle ground state (Hewson 1993). This is observed in the $dI/dV(V)$ spectrum as a Fano line shape resonance near the Fermi energy (see Figure 2.7b) (Scott & Natelson 2010; Hewson 1993). Depending on the impurity-electron bath coupling, the Kondo resonance can be observed in a temperature range spanning from sub-Kelvin to hundreds of Kelvin (Scott & Natelson 2010).

Due to the impracticability to provide here a comprehensive review about Kondo resonances in molecular systems, the observation made on the terbium double-decker system is presented as an instructive example. In Section 2.2.2, we already reported the molecular structure of the $TbPc_2$ neutral complex (De Cian et al. 1985). It is however important to recall some additional information about its magnetic properties. The $TbPc_2$ behaves as SMM due to the presence of the Tb^{3+} ion. This is characterized by a $4f^8$ electronic configuration with total angular momentum $J = L + S = 6$, which is split by the crystal field. This results in an energy separation between the ground state doublet $J_z = \pm 6$ and the first excited state of the order of a few hundred Kelvin ($k_B T$) (Ishikawa et al. 2003; Takamatsu et al. 2007). The molecule is also characterized by a second spin system due to the presence of an unpaired electron delocalized on the π orbitals of the ligands (Ishikawa et al. 2004). By means of STS spectroscopy, Komeda et al. (2011) identified the presence of a strong Kondo resonance (Kondo temperature around 250 K) studying $TbPc_2$ molecules deposited on gold (see Figure 2.7). They also found that the coupling of the magnetic impurity can also be switched by applying a

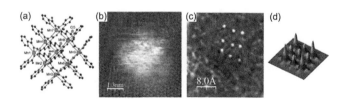

FIGURE 2.6 The X-ray structure of a grid-like hexacation $[Mn_9^{II}(L^2)_6]^{6+}$ is reported in (a). (b) Topographic STM image of a single molecule (right), showing clearly single C atoms of the HOPG substrate; (c) the experimental CITS map at -0.73 V. (d) DFT electron density map of the Mn $[3 \times 3]$ grid molecule, integrated between EF and -0.9 eV. The scale is 1.7 nm $\times$ 1.7 nm. (Adapted with permission from Dey, S.K. et al., *Inorganic Chemistry* 2007, 46, 7767–7781. Copyright 2007 American Chemical Society.)

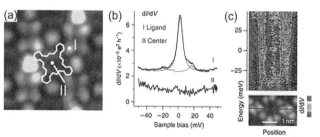

FIGURE 2.7 STS detection positions on the film together with a model of the upper Pc plane. (b) $dI/dV(V)$ spectra at I and II in (a). (c) Change in the Kondo peak height when the tip position was moved along the line in the STM topography panel. The palette on the right correspond to the $dI/dV(V)$ intensity, described below the figure. (Reprinted with permission from Komeda, T. et al., *Nature Communications* 2011, 2, 217 under Creative Commons Attribution 3.0 International License.)

bias pulse (Komeda et al. 2011). Interestingly, the magnetic impurity responsible for the Kondo resonance was identified as the spare electron hosted on the ligands whereas no direct information of the magnetism of the Tb(III) ion was obtained. Although possible ways to access the magnetism of electrons in the f shell orbitals were reported (Wiberg et al. 2009), this task is generally difficult and presents issues that affect also the inelastic spectroscopy technique that will be introduced in the next paragraph.

2.2.4 Inelastic Electron Tunneling Spectroscopy (IETS) and Spin Excitation

The first examples of IETS were performed by Jaklevic and Lambe (1966) characterizing molecular vibrations in planar tunneling junctions. The vibrational inelastic spectroscopy was implemented in STM junctions only much later (Persson & Baratoff 1987). The possibility to observe specific molecular vibrational and rotational states in an STM junction has allowed the characterization of these excitations at the single molecule scale and opened the way to the so-called single molecule chemistry (Ho 2002). Moreover, Heinrich et al. (2004) demonstrated that the IETS technique can be exploited for the study of quantum magnetism allowing the detection of spin excitations of magnetic atoms (Heinrich et al. 2004) and molecules (Chen et al. 2008).

The general concept of electron tunneling, along with the STM and STS techniques, has been introduced in Section 2.2.1. Here the reader will be guided into a specific application of the tunneling spectroscopy, *i.e.* the IETS, which allows the study of excitation processes. So far, the possibility for the tunneling electrons to exchange energy with the inspected system has not been taken into consideration.

However, electron tunneling events, can occur in an elastic or inelastic fashion (Loth, Lutz, et al. 2010). In the first case, the electrons tunnel between the two electrodes without exchanging energy with the system under investigation. On the contrary, in the inelastic processes, the electrons exchange energy with the system promoting it to an excited state. The basic principle of an inelastic process is illustrated in Figure 2.8. For the sake of simplicity, let us assume a two-level system inside the tunneling junction. By applying a bias voltage between the two electrodes, a net tunneling current starts to flow. If the energy of the tunneling electrons is below the excitation threshold of the system, only elastic tunneling processes are allowed. As soon as the tunneling electrons are provided with the ΔE necessary to excite the system, the inelastic process can take place. This opens a new conduction channel that manifests as an increase of the junction conductance (Loth, Lutz, et al. 2010). It is important to point out that the inelastic tunneling is a non-resonant process. In principle, all the electrons possessing enough energy to promote the system in an excited state can pass through this channel. For this reason, the opening of an inelastic channel usually results in a step in the conductance at the relative transition energy (Figure 2.8b) (Loth, Lutz, et al. 2010). Sometimes, and especially for vibrational spectroscopy, a second derivative of the current with respect to the bias is acquired $(d^2I/dV^2(V))$ to better evidence the inelastic threshold. In a real system, characterized by many excited states such as atoms and molecules, this technique provides information concerning vibrational or spin excited states.

In order to better illustrate the great potential offered by spin excitation measurements of magnetic molecules, the model used for the simulation and the fit of the spectra is briefly presented here. The spin system of the investigated

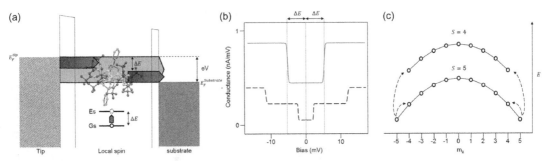

FIGURE 2.8 The scheme in panel (a) illustrates the tunneling processes through a molecule. For the sake of simplicity, we consider only two levels of the many possible: the ground state (Gs) and an excited state (Es) separated by an energy of ΔE. The pale gray arrow represents the elastic tunneling channel accessible by every electron that possesses enough energy to end in an empty state of the substrate. Dark gray arrows depict the inelastic tunneling process. The electrons interact with the molecule and only those that can exchange the ΔE energy ending in an empty state of the substrate can access this channel. If at the beginning the ground state was populated, at the end of the inelastic process, the populated state is the excited one. By measuring the conductance of the system, at a bias (mV) corresponding to the ΔE (meV), a jump in conductance is recorded due to the opening of the inelastic channel (black line in (b)). In case of two distinct excitations, as in the case represented in (c) dotted line, two steps are expected. The level diagram of an $S = 5$ spin system with strong uniaxial anisotropy is reported in panel (c) where only the $S = 5$ and $S = 4$ states are considered. Assuming that at the beginning of the experiment the two ground states are the only populated levels, two kind of spin excitations that fulfill the selection rules are expected (dotted arrows in (c)).

molecule is usually described using a spin Hamiltonian of the form:

$$H = -g\mu_B \vec{S} \cdot \vec{B} + D \cdot \hat{S}_z^2 + E\left(\hat{S}_y - \hat{S}_x\right)^2 \qquad (2.8)$$

where μ_B is the Bohr constant, g is the g-factor and $\vec{S}$, D and E are respectively the spin vector operator (with components $\hat{S}_i$), the uniaxial and transverse anisotropy parameters. Additional terms can also be added to the Hamiltonian such as, for example, exchange interactions of the form $J\hat{S}_i\hat{S}_j$. By diagonalization of the Hamiltonian, the energies and eigenstates $|\Phi_i\rangle$ of the local spin can be obtained. These states, along with the spin states of the tunneling electron $|\sigma_i\rangle$, are used as a basis set of the form $|\Phi_i\rangle \cdot |\sigma_i\rangle$. If a tunneling electron interacts with the local spin, it can exchange energy and promote a spin excitation. This interaction must fulfill the angular momentum conservation rule. Therefore, the spin excitations obey the following selection rules $\Delta m = \{+1, 0, -1\}$, where m is the expectation value of the $\hat{S}_z$ operator (see Figure 2.8c). The interaction between the tunneling electron and the local spin can be accurately modeled considering an exchange interaction of the form $\vec{S} \cdot \vec{\sigma} + u$ (Loth, Lutz, et al. 2010) where $\vec{\sigma}$ is the spin vector operator of the tunneling electron and u is a real constant accounting for spin-independent components. The possibility to describe the interaction allows to calculate the states' population of the local system as a function of the applied bias during an IETS experiment (Rolf-Pissarczyk et al. 2017). From this, the $I(V)$ curve can be calculated, and by numerical differentiation, the $dI/dV(V)$ spectrum is obtained (Loth, Lutz, et al. 2010; Rolf-Pissarczyk et al. 2017). The complete procedure is reported elsewhere (Rolf-Pissarczyk et al. 2017). This model allows, for example, to obtain the parameters of the spin Hamiltonian by fitting the experimental data (see Figure 2.9a and b) (Rolf-Pissarczyk et al. 2017). As such, the spin excitation technique enables the characterization of the magnetic behavior of individual molecules and atoms. A comparison with the data obtained by X-ray magnetic circular dichroism (XMCD) and theory, such as ab initio calculation, is however useful, if not necessary, to have a complete picture of the magnetic behavior of molecules and atoms on surfaces. It is also worth to mention some further developments of the spin excitation technique such as the employment of a SP tip (Loth, Lutz, et al. 2010; Wiesendanger 2009) and the implementation of all-electron pump-probe spectroscopy. The latter is used to measure the relaxation time of the spin excited state of atoms and nanostructures (Loth, Etzkorn, et al. 2010). To the best of our knowledge, however, no pump-probe measurement has been successfully carried out on a molecular system yet.

Considering the capabilities offered by the IETS, it is natural to ask whether there are some limitations to perform spin excitation measurements on molecules. Since the energies involved are usually of the order of few meV or less, a fundamental requirement for the STM is the capability to operate at low temperature ($T \leq 4$ K) (Loth, Lutz, et al. 2010). Moreover, the possibility to apply and tune a magnetic field is a valuable option to distinguish between low-energy vibrational excitations and spin excitations. These requirements, together with the necessity of an UHV

FIGURE 2.9 (a) $dI/dV(V)$ spectrum recorded on a Fe$_4$ molecule at $0\,T$ magnetic field (dark gray line, initial current $I_o = 75\,\text{pA}$ at $V_o = 10$ mV). A background spectrum recorded with the same tip on bare Cu$_2$N is shown in gray. (b) Spectra acquired on a single molecule without magnetic field (dark gray line, $I_o = 25$ pA, $V_o = 10$ mV) and under a 9 T out-of-plane field (gray line on top, $I_o = 50\,\text{pA}$, $V_o = 15$ mV), background spectrum (BG). The fits yield exchange coupling, $J = 2.93\,\text{meV}$ ($23.6\,\text{cm}^{-1}$), and magnetic anisotropy, $D = -52\,\mu\text{eV}$ ($-0.42\,\text{cm}^{-1}$) for the molecule in (a) and $J = 2.89\,\text{meV}$ ($23.3\,\text{cm}^{-1}$), $D = -26\,\mu\text{eV}$ ($-0.21\,\text{cm}^{-1}$) with a g factor of 2 for the molecule in (b). (c) Fe$_4$ molecule on the Cu$_2$N surface. Adsorption geometry and molecular structure are computed by DFT. (d) Calculated top view image of relaxed Fe4 on Cu$_2$N showing the spatial distribution of the density of states integrated between 0 and +3 eV in energy. (e) STM image of an Fe$_4$ molecule. It appears as a spheroid of $\sim$2 nm diameter with a multi-lobed substructure consistent with the calculated image in (d). (f) Overview STM image of the Cu$_2$N surface after deposition of molecules (scale bar, 8 nm). (Reprinted with permission from Burgess, J.A.J. et al., *Nature Communications* 2015, 6, 8216 under Creative Commons Attribution 4.0 International License.)

environment, make for a quite complex and high-priced experimental apparatus. Although in many laboratories it is possible to operate in such extreme conditions, there are also few intrinsic properties of the sample that have to be taken into consideration. These concern the investigated molecule and the coupling between the spin center and the electron bath of the substrate and, in some cases, of the tip. The tunneling junction set point used for recording spin excitation spectra usually requires a small distance between the tip and the conductive surface. The presence of the tip and its interaction with the local spin can modify the response of the molecule (Burgess et al. 2015) and, in extreme cases, promote its fragmentation. For such a reason, very stable and planar molecules (Chen et al. 2008) are usually preferred over complex molecular structures. Nevertheless, molecules with an elaborate tridimensional structure like the Mn_{12} (Kahle et al. 2012) or the Fe_4 (Burgess et al. 2015) SMMs were successfully characterized (see also Figure 2.9). Although spin excitations can be observed on bare metal substrates (Khajetoorians et al. 2013), a nonconductive decoupling layer such as alumina (Heinrich et al. 2004), boron nitride (Kahle et al. 2012), copper nitride (Hirjibehedin et al. 2006), magnesium oxide, (Rau et al. 2014) or molecular layers (Chen et al. 2008) is usually employed. The accurate design of the molecules (Hatter et al. 2015) or their local modification by dehydrogenation (Dubout et al. 2015) could also be exploited to tune the coupling of the spin center with the electron bath. The influence of the coupling between the spin center and the electron bath of the substrate or the tip on the $dI/dV(V)$ spectrum is à very fascinating topic. The overall effect of a strong coupling is to reduce the lifetime and the intensity of the observed spin excitation (Ternes 2015). Moreover, as already presented in the case of the $TbPc_2$ molecule on gold (Komeda et al. 2011), a strong coupling of the spin $\frac{1}{2}$ with the electron bath results in the appearance of a Kondo resonance. This feature is the result of a many-body state due to the scattering of the substrate electrons with the local spin (Kondo 1964). Decoupling layers, acting as an insulating barrier for the electrons of the substrate, reduce the scattering probability and the Kondo temperature of the system. Another interesting approach to decouple the spin center from the substrate is the use of a type-I superconducting surface (Heinrich et al. 2018). Below the superconducting temperature, the formation of Cooper pairs creates a depletion of the electron density around the Fermi level. This results in a decoupling action (Heinrich et al. 2013, 2018). However, the coupling between the local spin with the Cooper pairs of a superconductor gives origin to a complex competition between Kondo, spin excitation and the emergence of Yu-Shiba-Rusinov states (Franke et al. 2011).

Despite these requirements, the spin excitation spectroscopy stands as one of the few techniques able to combine the ultimate spatial resolution of the STM with the capability to evaluate the intimate magnetic nature of the inspected objects.

2.3 Atomic Force Microscopy and Magnetic Force Microscopy

AFM has become one of the most important tools for the investigation of the structure and properties of nanomaterials. Contrary to STM, AFM does not require conducting samples and it is applicable to a wider range of materials. Since its invention in 1985 by Binning, Gerber and Quate (Binnig & Quate 1986), the AFM has expanded into a large and growing family of force microscopy techniques that allow imaging of the topography with atomic resolution, precise mapping of surface forces of chemical and electromagnetic type and manipulation of atoms, just to mention a few possibilities. One of these techniques, MFM, was developed to characterize ferromagnetic materials and magnetic recording media. Recently, MFM has been successfully applied to molecular systems and therefore will be treated in detail in this chapter. This section is divided into four paragraphs: Section 2.3.1 introduces the fundamental principles of AFM and MFM; Section 2.3.2 describes the main modes of operation of an AFM/MFM instrument; in Section 2.3.3, we present applications of these techniques to magnetic molecular systems.

2.3.1 Theory and Working Principles of AFM and MFM

Forces between Tip and Surface

A force microscope relies on the measurement and control of the force acting between a sharp probe (the tip) and the sample. This force is the resultant of several possible interactions that act on quite different length scales. At very short distance, of the order of the atomic radius, the steric repulsion between the tip apex and the sample surface constitutes the dominant contribution. This interaction decays quickly with the tip-surface separation z and at distances larger than a few angstrom, attractive van der Waals (vdW) interactions become relevant.

A simple model describing the energy of the interaction between a pair of atoms or molecules is the Lennard-Jones potential (illustrated in Figure 2.10), with a repulsive term proportional to z^{-12} and the attractive vdW term proportional to z^{-6}:

$$U(z) = \frac{A}{z^{12}} - \frac{B}{z^6} \quad (2.9)$$

where A and B are fitting constants. The force acting between the two particles is oppositely equal to the gradient of the potential (or its derivative in a one-dimensional case):

$$-U'(z) = F(z) = -\frac{12A}{z^{13}} + \frac{6B}{z^7} \quad (2.10)$$

The energy minimum of the potential is the position at which the net force is zero and marks an ideal separation between the regime of contact interaction, where the force is repulsive, from that of noncontact interaction, where it is attractive.

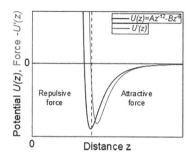

FIGURE 2.10 Lennard-Jones potential $U(z)$ (black) and its first derivative $U'(z)$ (gray), corresponding to the force; regions corresponding to net repulsive and attractive forces are divided by the dashed line.

In practice, the interaction depends on the geometry of the probe and the topography of the surface and a more realistic potential can be calculated with a so-called Hamaker integration (Israelachvili 2012). The results are qualitatively similar but show a more complex energy-distance dependence. Measurements performed in ambient may present the additional problem of the presence of a water layer on the surface, which can exert strong capillary forces on the tip and complicate the interaction.

Electrostatic and magnetic interactions are longer range compared to the vdW interaction since the force between pairs of electrostatic charges decays as z^{-2}, whereas the interaction between static magnetic dipoles decays as z^{-4}, allowing their detection at distances of tens of nanometers. The magnetic force can be measured using special tips covered by ferromagnetic material, which senses the gradient of the magnetic stray field $\vec{H}$ in proximity of the sample. The probe is, therefore, only sensitive to the inhomogeneous part of the field. Usually, the probe is most sensitive to the z component of the force, which is equal to:

$$F_z = \mu_0 \int_{V_{\text{tip}}} \vec{M}_{\text{tip}} \cdot \frac{\partial}{\partial z} \vec{H} \, dV \qquad (2.11)$$

where μ_0 is the vacuum permeability and $\vec{M}_{\text{tip}}$ is the magnetization distribution in the tip, corresponding to the microscopic local average of the magnetic moment, and integration is performed over the tip volume. Since an accurate description of $\vec{M}_{\text{tip}}$ is elusive, especially in commercial tips, the interpretation of magnetic force measurements usually relies on simplified models where the tip is modeled with an effective magnetic dipole (m_{tip}):

$$F_z = \mu_0 m_{\text{tip}} \frac{\partial}{\partial z} \vec{H} \qquad (2.12)$$

where m_{tip} is aligned in the z direction. Specially fabricated tips, with a high aspect ratio of the magnetic part, can behave as bar magnets. The prevailing interaction is, in this case, with the magnetic pole closer to the surface (Hug et al. 1998). In this case, the expression of the magnetic force becomes proportional to the stray field:

$$F_z = \mu_0 q_{\text{tip}} H_z \qquad (2.13)$$

where q_{tip} is the magnetic "charge" of the pole. The spatial dependence of $\vec{H}$ is determined by the magnetization $\vec{M}$ of the sample, which may be due to ferromagnetic ordering or, in paramagnetic materials, polarized by an external magnetic field. Magnetic structures smaller than the tip radius and the tip-sample distance, as for instance isolated molecules or nanometric particles, can be treated as point dipoles with a field distribution:

$$\vec{H}(\vec{r}) = \frac{1}{4\pi} \left(3 \frac{\vec{r}(\vec{m}_1 \cdot \vec{r})}{r^5} - \frac{\vec{m}_1}{r^3} \right) \qquad (2.14)$$

where $\vec{m}_1$ is the net magnetic moment of the molecule or the particle and $\vec{r}$ is the distance vector from it. This is also valid in the case of uniformly magnetized spherical particles, regardless of their size, since in this special case, the field distribution is the same of a magnetic dipole located at the center of the particle. In a general case, the stray field is the resultant of the dipolar fields of the distributed magnetic moments and must be evaluated numerically.

Electrostatic forces may be strong and overshadow the magnetic responses of interest, creating instability or spurious contributions in the measurement. In case of conductive substrates and probes, connecting them to the electric ground may be effective at dissipating charges, however, leaves a residual contact potential V_{CP} equal to the work function difference between them. This effect can be compensated by applying a voltage bias V_B between the tip and the substrate, as shown by the expression for the electrostatic force between two conductors:

$$F_e(z) = -\frac{1}{2} \frac{dC(z)}{dz} (V_{CP} - V_B)^2 \qquad (2.15)$$

where $C(z)$ is the capacitance of the capacitor formed by the conducting tip and the sample at a distance z. The dependence of C and its derivative on the distance is determined by shape of the tip and the morphology of the surface and can be calculated for some ideal geometries (Hao et al. 1991; Olsson et al. 1998). In the discussion above, we considered an average contact potential; however, the work function may vary on the surface of the sample. For example, this can be due to inhomogeneous molecular coverages generating local interactions that should be taken into account for a correct interpretation of noncontact experiments, including the magnetic force measurement. Scanning Kelvin probe microscopy is an SPM technique able to map the surface potential and can be used, in combination with MFM, to characterize samples presenting problematic electrostatic contributions (Jaafar et al. 2011).

The following sections will explain how force microscopy experiments detect either forces or their gradients along the z direction, depending on the measurement method. Most frequently, long-range electromagnetic forces are detected in gradient mode, which can achieve better sensitivity and spatial resolution than the direct force measurement.

Probes and Their Response to Forces

AFM/MFM instruments commercially available use high aspect ratio tips mounted on microfabricated cantilevers made of doped silicon or silicon nitride. The cantilever responds to the tip-surface force F_{TS} at the position z by opposing an elastic force F_s

$$F_{TS}(z) = -F_s = k(z - z_0) \qquad (2.16)$$

where $z - z_0$ is the displacement of the tip from the non-deflected position (z_0) and k is the elastic constant, which is determined by the Young modulus of the material of the cantilever and its geometry. A measurement of the deflection of the cantilever provides a value of the force applied by the tip. This requires sensitive mechanisms for the detection of the cantilever movement. Most commercial instruments measure the deviation of a laser beam reflected on the metalized backside of the cantilever, exploiting the amplification effect of an optical lever. When a higher precision is needed, as in the noncontact measurements (see the next paragraph), this motion can be measured with a fiber-optic interferometer that achieves a sensitivity of the order of 10 pm (Rugar et al. 1989).

The most sensitive noncontact force methods exploit the dynamic behavior of the cantilever oscillating close to its natural oscillation frequency f_0. The following relation links f_0 to the elastic constant k

$$f_0 = \frac{1}{2\pi}\sqrt{\frac{k}{m_0}} \qquad (2.17)$$

where m_0 is the effective mass load of the cantilever spring.

Dynamic measurement modes are sensitive to gradients of the tip-surface force along the oscillation direction, which cause a shift of the resonance frequency as demonstrated hereafter. The cantilever is oscillated perpendicular to the sample surface, along the z direction. For small oscillation amplitudes, the tip-surface force $F_{TS}(z)$ can be approximated by a first-order series expansion around $z = z_0$ and [2.16] becomes:

$$F_{TS}(z_0) + F_{TS}'(z_0)(z - z_0) \cong k(z - z_0) \qquad (2.18)$$

After rearranging

$$F_{TS}(z_0) \cong (k - F_{TS}'(z_0))(z - z_0) \qquad (2.19)$$

showing that the cantilever responds to the force $F_{TS}(z_0)$ as if its elastic constant was modified by the force gradient $F_{TS}'(z_0)$, which introduces an effective elastic constant $k_{\text{eff}} = k - F_{TS}'(z_0)$. This results in a Δf shift of the resonance frequency, which is proportional to F_{TS}', since for $k \gg F_{TS}'$:

$$\Delta f = f - f_0 = \frac{1}{2\pi}\sqrt{\frac{k_{\text{eff}}}{m_0}} - \frac{1}{2\pi}\sqrt{\frac{k}{m_0}} \simeq$$

$$-\frac{1}{4\pi\sqrt{m_0 k}}F_{TS}'(z_0) = -\frac{f_0}{2k}F_{TS}'(z_0) \qquad (2.20)$$

Attractive forces decaying with the distance have positive gradients and produce negative Δf, while the opposite is

true for repulsive forces. In close proximity of the surface, the force gradients may vary by orders of magnitude within the oscillation range of the tip and a quantitative prediction is more complex (Morita et al. 2009), although the same relation between the sign of Δf and that of the gradient is expected.

When the cantilever is sinusoidally excited with amplitude A_{exc} at frequency f_{exc}, for example by using a piezoelectric transducer, its tip oscillates with an amplitude A given by:

$$A = \frac{A_{\text{exc}}}{\sqrt{\left(\frac{f_{\text{exc}}}{Qf}\right)^2 + \left(1 - \frac{f_{\text{exc}}^2}{f^2}\right)^2}} \qquad (2.21)$$

where Q is the resonance quality factor, an adimensional parameter that is inversely proportional to the damping ratio and determines how broad the resonance peak is ($Q \approx \text{FWHM}/f_0$). The cantilever oscillates at the same frequency of the excitation (f_{exc}) but with a phase shift φ between 0 and π with respect to it, such that

$$\tan\varphi = \frac{f_{\text{exc}}}{Qf\left(1 - \frac{f_{\text{exc}}^2}{f^2}\right)}. \qquad (2.22)$$

Phase and amplitude of the oscillation are a function of the ratio between the frequency of the applied excitation f_{exc} and the resonant frequency of the cantilever. For small relative frequency shifts $\Delta f/f_0$, the effect is to translate the amplitude and phase curves in frequency (see Figure 2.11). The force gradient can be therefore detected by monitoring the shifts of the amplitude and phase signals at fixed f_{exc}, with a sensitivity that improves proportionally to the Q factor (Morita et al. 2009). However, high Q values may limit the bandwidth of the measurement, i.e. the scan speed, since the transient response time of the cantilever after a force change also increases with the quality factor. Some detection schemes continuously tune the excitation to the resonant frequency of the cantilever and allow a direct measurement of the frequency shift with bandwidths not limited by the Q factor (Albrecht et al. 1991). This method, called frequency modulation (FM), is very advantageous in UHV, where Q can easily reach values larger than 10^4, allowing measurements with both high sensitivity and speed. Another advantage of FM is that Δf, being independent of Q, is easier to compare between different experiments than a phase shift.

Working Modes of AFM/MFM

An AFM can work according to one of three main tip-surface interaction modes, which are the contact, the noncontact and the tapping (or intermittent contact) modes. As discussed below, MFM uses a combination of noncontact and tapping methods to measure both topography and magnetic forces.

The first method to be successfully implemented in AFM was the contact mode, in which the probe is approached to the surface until repulsive forces cause a certain level

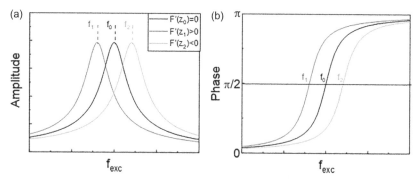

FIGURE 2.11 Amplitude (a) and phase (b) of a cantilever with natural frequency of oscillation f_0 as a function of the excitation frequency f_{exc}. Black lines show the response in the absence of an external force gradient; a positive tip-surface force gradient decreases the resonance frequency to f_1 (dark gray), while a negative gradient increases the frequency to f_2 (pale gray).

of upward deflection of the cantilever and the desired force is obtained. A feedback loop controls either the tip or the sample position along the perpendicular z direction in order to keep the deflection at the desired (set point) value while scanning the tip along the x and y directions in a raster pattern, producing a topographic image $z(x,y)$ of the surface at constant force. During contact mode scanning, wear of the tip and damage of the sample surface may occur, especially on soft materials. To reduce this problem and to increase force sensitivity, soft cantilevers, with elastic constants k of 0.3 N/m or less, are used.

Noncontact AFM, which was developed soon after, is considered nondestructive for the probe and the surface being measured, since strong interactions are avoided by keeping the tip at short distance (up to a few nm) away from the surface, where attractive vdW forces are dominating the interaction. Non contact (NC)-AFM is usually performed in dynamic mode, oscillating the tip at a small amplitude (also up to few nm), and can achieve outstanding performances in UHV environment, where it is able to image many crystalline surfaces or individual molecules with atomic lateral resolution. The force gradient detection is usually based on FM. The positive gradient of the attractive vdW force determines a negative shift Δf of the frequency as the tip is approached to the surface. The tip-sample distance is therefore stabilized by a feedback loop that aims to keep Δf equal to a set point value.

The tapping mode combines aspects of the previously described AFM modes and was developed to study delicate or soft samples (Zhong et al. 1993) in ambient conditions. In this mode, the cantilever is oscillated to a sufficient amplitude to move the tip across the repulsive and the attractive regimes of the force curve achieving an intermittent contact between the tip and the surface, hence reducing friction, damage and wear associated with a contact measurement. Furthermore, the large tip oscillation amplitude (≈ 20–100 nm) can provide sufficient energy to overcome capillary forces due to adsorbed water in experiments performed in ambient conditions. The measurement is based on amplitude modulation (AM) at constant oscillation frequency, in which the tip-sample distance is controlled by a feedback

loop that aims at keeping the oscillation amplitude constant to the set point value, allowing the mapping of the surface topography.

MFM is a dynamic noncontact measurement that uses a magnetic probe to detect the magnetic force gradient over the surface of a sample (Hartmann 1999). Depending on the capability of the instrument, FM or phase detection mode is used, expressing the signal as a frequency (Δf) or phase ($\Delta \varphi$) shift, respectively. Compared to NC-AFM, the tip-surface distance is increased to larger values, typically between 10 and 100 nm, in order to reduce the influence of vdW forces. Knowledge of the surface topography is needed for establishing the correct tip distance. A common measurement scheme is the dual pass (or two pass) method, which consists in measuring each line of the raster scan twice. First, the topography profile is acquired by tapping the surface with the magnetic probe, and then, the magnetic signal is obtained by scanning in noncontact mode, moving along the measured profile vertically translated by the desired lift height. On relatively flat surfaces, a simple measurement of the tilt of the sample plane performed along the edges of the region of interest can be a valuable alternative to replace the first scan of the surface. This latter approach has the advantage of being faster and less perturbative to the magnetic state of the sample since it largely avoids contacts between the tip and the surface.

MFM probes commercially available are AFM probes coated with a magnetic layer (thin film probes). The material of the coating, its thickness and the tip geometry determine the magnetic characteristics of the tip, first of all its coercive field and magnetic moment. Basic MFM experiments can be performed with ordinary AFM instruments; however, standard setups present several drawbacks due to ambient operation and limitations on the use of external magnetic field and temperature control. It was estimated that the minimum magnetic moment detectable by a single MFM measurement in ambient conditions is of the order of 10^{-18} Am^2, as found in magnetite nanoparticles with a diameter of about 15 nm (Raşa et al. 2002; Schreiber et al. 2008; Sievers et al. 2012), while in the case of molecular materials, the minimum size would be larger due to the lower density of magnetic moment and their paramagnetic nature. Vacuum conditions and low temperature allow significant improvements in sensitivity because of the increased resonance Q factor and reduced noise (Albrecht et al. 1991). The interpretation of weak magnetic signals, such as those of paramagnetic molecules, is complicated by the interference of electrostatic interactions or van der Waals forces. In these cases, the availability of strong external magnetic fields is extremely useful, if not necessary, to assert the magnetic origin of the image contrast. Also, the use of demagnetization protocols can provide a way to decouple different contributions to the measured signal, by cancelling the magnetic force and allowing the evaluation of nonmagnetic signals which may be eventually subtracted from the MFM data (Angeloni et al. 2016).

Experiments performed at cryogenic temperatures are especially important for molecular systems since paramagnetic molecules reach much higher polarization of the magnetization, and the most interesting phenomena, such as magnetic hysteresis, can only be observed at low temperatures.

Qualitative interpretation of MFM measurements rely on point dipole approximations of the tip magnetization, according to which the force gradient signal is proportional to the second derivative of the stray field $\vec{H}$ along the z direction. Slightly more accurate models may also include an effective magnetic monopole on the tip, introducing an additional contribution proportional to the first derivative of $\vec{H}$ (Hartmann 1999). Obtaining an accurate description of the magnetization distribution in the tip is a challenging task that requires detailed characterizations (*e.g.* electron holography) coupled with numerical micromagnetic simulations (Hartmann 1999). Unfortunately, even when the characteristics of the probe are known with precision, electromagnetic theory does not allow to derive the magnetization distribution in the sample from measurements of the stray field without ambiguity (Vellekoop et al. 1998). In general, $\vec{M}$ must be obtained by comparing the experimental data with the results of numerical simulations of the sample based on the micromagnetic properties of the material. Nevertheless, with the help of calibration measurements (Lohau et al. 1999; Jaafar et al. 2008; van Schendel et al. 2000), it is possible to determine the values of the effective parameters (position and magnitude) of the point dipole and monopole model and, in some cases, to obtain absolute values of the magnetization of the sample (Sievers et al. 2012).

2.3.2 Applications to Magnetic Molecular Systems

Magnetic Force Microscopy Studies

Thanks to its nanometric resolution, relatively high magnetic sensitivity and low sample preparation requirements, MFM has been applied to the characterization of magnetic molecular composites and nanostructures. The group of Prof. H. Nishide was the first in 2001 to claim the use of an MFM to characterize paramagnetic organic molecules on surfaces (Nishide et al. 2001; Michinobu et al. 2003, 2010; Miyasaka et al. 2003; Tanaka et al. 2006; Fukuzaki & Nishide 2006). His group synthesized a family of organic polymers containing controlled amounts of phenoxyl, nitroxyl or aminium cation-type radicals which were deposited from diluted solutions on flat surfaces (HOPG, mica) and identified by MFM. Their measurements were performed in phase detection mode in air at room temperature. By working at small lift heights and averaging a large number of scans, the authors were able to detect attractive interactions with the paramagnetic molecules polarized by the magnetic stray field of the tip. Some of their studies investigated the relation between the MFM signal and the concentration of radicals in the macromolecule, determined independently by superconducting quantum interference device (SQUID) magnetometry and nuclear magnetic resonance (NMR) measurements, finding that it was proportional to the radical concentration and uncorrelated with the particle size. In the work by Tanaka et al. (2006), the MFM phase signal per unpaired electron is estimated by calculation and used to obtain values of radical concentration directly from MFM, finding a good agreement with the values obtained by SQUID. The group also synthesized dendrimer macromolecules with peripheric radical substituents and observed a hollow dot structure in the MFM images consistent with a magnetic signal arising from the surface of the particles.

The first account of MFM measurements on SMMs is due to the group of Prof J. Veciana in 2003 (Ruiz-Molina et al. 2003). The group prepared films of polycarbonate (PC) doped with the Mn_{12} SMM. The films were treated with vapors of different organic solvents to controllably induce aggregation of Mn_{12} molecules and surfacing of the paramagnetic particles on the film. The MFM measurements were performed in ambient conditions without external field, and the images show a positive phase shift (attractive force) in correspondence of the particles, even on the smaller aggregates with apparent lateral size of 40 nm. Following this result, in 2005, the group of Biscarini (Cavallini et al. 2005), in collaboration with the group of Prof Veciana, prepared PC/Mn_{12} films submicrometrically patterned by replica molding, using DVD discs as masters. The topography of the films showed nanometric structures organized along periodic lines, reproducing the pattern of digital information written in the DVD. After a short exposure of the films to organic vapors, the topography became smooth due to polymer diffusion in the swollen PC matrix, while the Mn_{12} molecules showed negligible lateral diffusion and concentrated on the surface of the film in the position of the initial protrusions, as highlighted by AFM tapping phase contrast and noncontact MFM images (Figure 2.12a). Interestingly, the MFM image shows both positive and negative phase shifts on opposite sides of each structure. Since the interaction with paramagnetic objects should always lead to positive phase shifts (attractive force), this finding suggests that the measured MFM signal contains also spurious contributions deriving

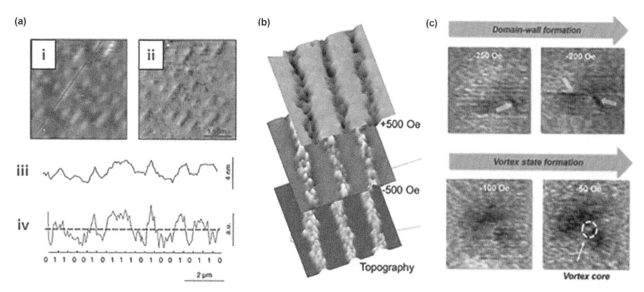

FIGURE 2.12 (a) AFM image (i) and profile (iii) compared with MFM image (ii) and profile (iv) of patterned Mn_{12} SMMs in PC matrix. (Reprinted with permission from Cavallini, M. et al., *Angewandte Chemie International Edition in English* 2005, 44, 888–892. Copyright (2005) John Wiley and Sons.) (b) MFM images of the magnetization reversal in surface-organized PBA nanoparticles with diameter ≈20 nm. (Reprinted with permission from Pinilla-Cienfuegos, E. et al., Particle and Particle Systems Characterization 2015, 693–700. Copyright (2015) John Wiley and Sons.) (c) MFM imaging of the magnetic vortex formation and core switching in a PBA nanoparticle with diameter ≈22 nm. (Reprinted with permission from Pinilla-Cienfuegos, E. et al., ACS Nano 2016, 10, 1764–1770 Copyright (2016) American Chemical Society.)

probably from a cross-talk with topographic information or to electrostatic interference.

Mn_{12} clusters were also studied by Sun et al. (2009), using a solution casting procedure to create films of Mn_{12} derivatives with a self-assembled honeycomb pattern of holes. The patterned films were measured by SQUID at 2 K, showing an open hysteresis loop analogous to the bulk compound, and by MFM at room temperature without external field, measuring a positive phase shift in the filled parts of the film with respect to the empty holes. Other examples of MFM applied to SMM systems in ambient conditions were published by Tangoulis et al. (2014), showing MFM phase contrast on Co_7 molecules deposited on HOPG from solution, and by Ruiz-Muelle et al. (2017), who imaged in a 400 Oe field Dy linear complexes covalently grafted to silicon surfaces with a polymeric linker.

Measurements conducted in ambient conditions offer limited sensitivity, due to both the paramagnetic state of the molecules and the instrumental noise, and do not give access to the rich magnetic phenomena occurring at low temperature. Although not a force microscopy, scanning Hall probe microscopy (SHPM), which is not treated here in detail, was the first SPM technique to be used to characterize a magnetic molecular nanostructure at low temperature (LT). Compared to MFM, SHPM shows a much lower resolution (≈500 nm (Ghirri et al. 2008)) due to the size of the Hall magnetic field sensor, which forbids imaging of nanometric objects. Nevertheless, this technique has the advantage of directly measuring the value of magnetic fields on the sample without perturbing its state with stray fields from the probe. Ghirri et al. (2008) performed SHPM on a pattern of lines

formed by PBA nanoparticles $(Cs_{0.7}Ni[Cr(CN)_6]_{0.9})$ grafted onto a silicon substrate; these superparamagnetic particles have a Curie temperature $T_c = 70$ K, below which the spins within each particle order ferromagnetically in a single domain. The authors imaged the inversion of the magnetization of the nanostructures at 20 K during a cycle where the magnetic field was brought from 1 T to -1 T. Furthermore, by analyzing the magnetic image with the help of micromagnetic simulations, they concluded that the nanoparticles were in a canted state at 20 K, in which the magnetization does not perfectly align with the external magnetic field but has a transverse component.

Recently, PBA $(K_{0.22}Ni(Cr(CN)_6)_{0.74})$ nanoparticles were also studied in great detail with low temperature (LT)-MFM by the group of Prof E. Coronado (Pinilla-Cienfuegos et al. 2015, 2016). Pinilla-Cienfuegos et al. (2015) reported high-resolution images of the magnetization reversal of nanoparticles both in isolated form and in organized lines formed by adhesion on aluminum oxide stripes. The nanoparticles had a Curie temperature of 40 K and a narrow size distribution around 20 nm, close to the limit of single domain behavior; therefore, each particle appeared as bright or dark spots in the image, depending on whether its magnetization was oriented parallel or opposite to the tip. The experimental temperature was set to 4.2 K for reasons of sensitivity, for reduction of drift and to increase the coercivity of the nanoparticles. A calibration procedure was used to guarantee that the fields applied in the experiment were smaller than the tip coercivity (750 Oe) and verify that the tip magnetization state remained constant throughout the measurements. MFM allowed the observation of individual switching events, which were affected by the size of the particles, the dipolar interactions between them and the stray field of the tip. To minimize the perturbation by the tip, the MFM images were taken at constant height of 100 nm from the surface plane: it was estimated that at this distance, the tip stray field on the sample was less than 300 Oe. Moreover, the authors showed that it was possible to determine the Curie temperature of the nanoparticles by performing MFM measurements of the remanent magnetization at different temperatures. The same group later published an LT-MFM study of Prussian blue nanoparticles with a size distribution from 13 to 25 nm at 4.2 K, around the critical size for single domain stability (≈22 nm) and exhibiting different magnetic reversal mechanisms depending on the particle size (Pinilla-Cienfuegos et al. 2016). Remarkably, the MFM measurement was able to image the vortex state of magnetization in larger particles, identify the vortex chirality and observe the switching of its core (Figure 2.12b).

Lorusso et al. (2013) published an LT-MFM study of gadolinium acetate ($[Gd_2(CH_3COO)\ 6(H_2O)_4]\cdot 4H_2O$) dots, with about 1.5 μm diameter and 12 nm height, deposited by dip-pen nanolithography. This work represents the first example of a semiquantitative magnetization measurement performed on a molecular system by MFM at low temperature. The authors measured the MFM frequency shift contrast at different values of field at 5 K and 9 K, showing that, by applying a scaling factor, the curves of MFM signal as a function of field were in agreement with the M(H) measurement performed by SQUID. The experimental scaling factor was close to the one estimated by calculation, assuming a point dipole tip and a paramagnetic response of the Gd compound with $g = 2$ and $S = 7/2$.

Finally, some of us (Serri et al. 2017) recently published an LT-MFM (10 K) and XMCD (7 K) study of microarrays of the SMM Terbium (III) double-decker (TbPc$_2$, Pc=phthalocyanine). The structures, fabricated by vacuum sublimation with a shadow mask, consisted of arrays of circular TbPc$_2$ dots with about 1.7 μm diameter and 24 nm height deposited on silicon and on thin films of PTCDA (Perylene-3,4,9,10-tetracarboxylic dianhydride) having a structural templating property. TbPc$_2$ has a strong magnetic anisotropy, with easy axis perpendicular to the phthalocyanine rings, and vacuum deposition on (oxidized) silicon results in an orientation of the easy axis in the plane of the substrate, while growth on a PTCDA layer leads to an out-of-plane easy axis. MFM images in magnetic field demonstrated the capability to discriminate between TbPc$_2$ dots with orthogonal anisotropy axes, since the magnetic probe is mostly sensitive to the out-of-plane magnetization of the sample and responds much more strongly above the dots grown on PTCDA. High-resolution images revealed details of the stray field distribution which were not previously observed in other studies on similar molecular systems. The MFM images of the dots on PTCDA at different fields were analysed with a statistical method (Figure 2.13). Each image was divided into three regions of different stray field intensity and sign, and the average of the MFM frequency

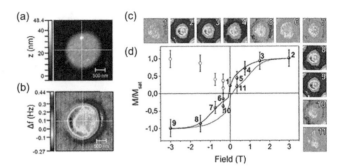

FIGURE 2.13 AFM (a) and MFM (b) image of a TbPc$_2$ dot on PTCDA. Evolution of the magnetic contrast as a function of field (c and d) at 5.9 K. (Reprinted with permission from Serri, M. et al., *Nano Letters* 2017, 17, 1899–1905. Copyright (2017) American Chemical Society.)

shift with its standard deviation was computed in each region. By calculating the difference of the signal between these regions, a curve representative of the MFM contrast of the dot was obtained. The agreement between the normalized M(H) curve measured on thin films by XMCD and the MFM contrast curve of the dots was good, although the open hysteresis was only observed in the XMCD measurement. We should highlight that the MFM experiment, due to the relatively slow speed of acquisition, measured the static response of TbPc$_2$ to magnetic field, which is paramagnetic; however, we can expect that future improvements in the scan rate may be able to observe the dynamic relaxation of magnetization that gives rise to the open hysteresis in SMMs.

Atomic force Microscopy Studies

It is worth to mention that MFM is not the only force microscopy tool that can be used to study magnetic molecular materials. It was recently reported by the group of A. Bousseksou and G. Molnár (Lopes et al. 2013; Hernández et al. 2014) that the temperature-induced spin transition in films of SCO material can be detected locally by AFM. Aggregates of SCO molecules can exhibit a large electron–lattice coupling, due to different bond lengths in HS and LS electronic configurations. In concomitance with the spin transition, the crystals show a microscopic contraction/expansion, which change the morphology (*e.g.* roughness and step heights) and the mechanical stiffness (Young modulus). Using AFM imaging and force distance measurements, the authors were able to monitor these changes locally and study the propagation of the spin transition through the material as a function of temperature. The SCO transition also results in a modification of the optical properties of the material, which was studied by the same authors (Lopes et al. 2013) using near-field scanning optical microscopy (NSOM) with sub-wavelength lateral resolution (<250 nm) (Dunn 1999), another SPM technique not treated in detail here. Other examples of the SCO complexes studied by SPM have been reported in Section 2.2.2.

2.4 Emerging Scanning Probe Techniques for the Investigation of Magnetic Molecules

New concepts in the field of scanning probe magnetic microscopy are being developed, delivering improvements in measurement sensitivity, resolution and speed allowing sensing down to the single electron spin. These efforts, often realized by the use of radio frequency (*rf*) signals coupled with an AFM or STM setup, aim to combine the spatial resolution of a scanning probe microscopy with the electronic spin sensitivity of the electron-spin resonance, for the detection and the manipulation of single spins of atoms and molecules on surface. Among these cutting-edge techniques, we currently believe that Magnetic Resonance Force Microscopy (MRMF), diamond nitrogen-vacancy (NV)-based techniques, magnetic exchange force microscopy (MExFM) and electron spin resonance STM techniques may bring an important contribution to the study of magnetic molecular systems in the near future.

2.4.1 Magnetic Resonance Force Microscopy

The MRMF is a technique able to perform magnetic resonance imaging (MRI) and spectroscopy at nanometric scales on electronic and nuclear spins. Since the first demonstration of mechanically detected molecular magnetic resonance in 1992, when Rugar et al. (1992) measured with a cantilever the electron spin resonance of a microscopic crystal of the organic spin label DPPH (2,2-Diphenyl-1-picrylhydrazyl), the capabilities of MRFM have improved significantly. It is now possible to achieve the detection of single electron spins (Rugar et al. 2004) and perform MRI tomography with sub 10 nm resolution using proton NMR (Degen et al. 2009), just to mention a few milestones. MRI uses spatial magnetic gradients and *rf* radiation to induce spin resonance in specific regions of space, where the field value matches the resonant condition imposed by the frequency of the *rf* radiation. The key to reduce the size of the resonant volume and improve the spatial resolution to the nanometric scale is to use strong magnetic gradients of the order of 10^6 T/m, which can be found in proximity (≈ 10 nm) of nanofabricated magnets. Two configurations have been tested for MRFM: in one, the sample is attached to a cantilever and scanned over a miniaturized magnet, and in the other, the magnet is attached on the cantilever and scanned over the sample. Several modulation schemes and *rf* pulse sequences have been proposed and demonstrated to produce an oscillating magnetic force between the resonant volume and the cantilever at the mechanical resonant frequency of the latter, in order to benefit from the amplification effect of the high Q cantilever. When aiming to resolve nanometric structures or individual spins, the magnetic forces can be as small as tens of aN. This requires employing ultrasoft cantilevers ($k \approx 10^{-4}$ N/m) and operating in UHV under cryogenic temperatures, in order to achieve very high Q values (of the order of 10^5) and to reduce thermal noise.

2.4.2 Nitrogen-Vacancy Centers in Diamond

An emerging scanning nanomagnetometry technique uses NV defects in diamonds as atomic probes of magnetic fields. Although this approach is based on optical detection, we believe it is worth mentioning it here with the other force microscopies for the stunning progress that NV technology has made in the past decade, which promises important applications in the study of molecular magnetic systems in the near future. The first demonstration of NV-based scanning magnetometry, published by Balasubramanian et al. (2008), achieved a spatial resolution of 20 nm and field resolution of 0.5 mT on a nickel structure. Tetienne et al. reported imaging of domain wall hopping in magnetic wires (Tetienne et al. 2014), while Grinolds et al. demonstrated room temperature imaging of a single spin using nitrogen vacancies (Grinolds et al. 2013). NV centers are photoluminescent and, crucially for the magnetic sensing, have a triplet spin ($S = 1$) ground state, where the $m_z = 0$ state lies lower in energy due to zero field splitting and the $m_z = \pm 1$ states are split by the Zeeman interaction with the magnetic field. By tuning the frequency of an *rf* field to the resonance of the $\Delta m_z = \pm 1$ transition, one is able to measure the field at the NV center. Differently from the MRFM, the resonant spin is in the NV probe and is detected optically by measuring the intensity of the photoluminescence, which decreases when resonance occurs. Very importantly, optical detection works with high sensitivity even at room temperature, broadening the scope of its applicability. NV probes can be attached to cantilevers to operate in noncontact AFM for topography measurement and control of the tip-sample distance, while measuring the local magnetic field, with minimal perturbation of the sample, since the use of magnetic probes is avoided.

2.4.3 Magnetic Exchange Force Microscopy

Magnetic exchange force microscopy was proposed to detect different spin orientations of the atoms on the surface of a sample. The technique is based on the quantum-mechanical exchange interaction, which results in a different repulsion potential between electrons with parallel or antiparallel spin configuration. In MExFM, a tip ending with an atom in a stable magnetic state is approached to the sample until the wavefunction of the tip atom starts to overlap with that of the atoms on the surface of the sample. This interaction is extremely short range and decays exponentially on a scale of about 0.1 nm; therefore, the experiments require atomically clean surfaces on the tip and the sample, which only an UHV environment can guarantee, and an ultra-low noise setup. MExFM was successfully used to image the spin structure of systems such as the antiferromagnetic insulator

NiO (Kaiser et al. 2007), Fe layers on W(001) (Schmidt et al. 2009) and Ir(111) (Grenz et al. 2017; Hauptmann et al. 2018).

2.4.4   Spin detection by STM

The first attempt to provide a magnetic sense to an STM was carried out by the electron spin noise (ESN)–STM where the noise of the tunneling current was analyzed to detect the precession frequency of spin centers in a constant magnetic field on silicon or on radical molecules (Manassen et al. 1989, 2000; Durkan & Welland 2002). On molecules, *rf* signals combined with STM technique were alternatively used to detect the vibrational modes and the spin of molecular radicals by ESN–STM (Messina et al 2009, Mugnaini et al. 2009). A slightly different approach was tempted for the first time on a magnetic molecule by Müllegger et al. (2014). They showed that the detection of nuclear and electronic magnetic transitions of a TbPc$_2$ SMM on the Au(111) surface was achieved by sweeping the frequency of the *rf* signal (sent in addition to the DC bias voltage) and monitoring the *dI/dV(V)*. In order to explain such experimental observation, the same group showed that the effect of the *rf* application and consequently of the electric field at the tunneling junction may cause the deformation of the TbPc$_2$ molecular structure. This suggests that the spin-phonon coupling plays a role in the detection mechanism of the molecular magnetic transitions (Müllegger et al. 2015).

Recent experiments on single atoms have pushed the capabilities of an STM equipped with an *rf*-circuit towards the realization of electronic paramagnetic resonance (EPR) experiments at the atomic scale (Natterer et al. 2017; Baumann et al. 2015). The measurements were performed on single Fe and Ho atoms by application of an *rf* signal in addition to the DC bias that was swept while monitoring the tunnel current by an SP tip at different magnetic fields. To date, the challenge to consolidate the use and the understanding of these techniques on magnetic molecules is still open and it remains an extremely interesting field to investigate.

References

Adler, H. et al., 2015. Interface properties of VOPc on Ni(111) and graphene/Ni(111): Orientation-dependent charge transfer. *The Journal of Physical Chemistry C*, 119(16), pp. 8755–8762.

Affronte, M., 2009. Molecular nanomagnets for information technologies. *Journal of Materials Chemistry*, 19(12), pp. 1731–1737.

Aguilà, D. et al., 2016. Switchable Fe/Co Prussian blue networks and molecular analogues. *Chemical Society Reviews*, 45(1), pp. 203–224.

Albrecht, T.R. et al., 1991. Frequency modulation detection using high-Q cantilevers for enhanced force microscope sensitivity. *Journal of Applied Physics*, 69(2), pp. 668–673.

Ametek, n.d. 7230 general purpose DSP Lock-in amplifier, Manual. Available at: www.ameteksi.com/-/media/ameteksi/download.

Amokrane, A. et al., 2017. Role of π-radicals in the spin connectivity of clusters and networks of tb double-decker single molecule magnets. *ACS Nano*, 11(11), pp. 10750–10760.

Angeloni, L. et al., 2016. Removal of electrostatic artifacts in magnetic force microscopy by controlled magnetization of the tip: Application to superparamagnetic nanoparticles. *Scientific Reports*, 6(1), art. n. 26293.

Ara, F. et al., 2016. A scanning tunneling microscopy study of the electronic and spin states of bis(phthalocyaninato)terbium(III) (TbPc $_2$) molecules on Ag(111). *Dalton Transactions*, 45(42), pp. 16644–16652.

Aromí, G. et al., 2012. Design of magnetic coordination complexes for quantum computing. *Chemical Society Reviews*, 41(2), pp. 537–546.

Bairagi, K. et al., 2016. Molecular-scale dynamics of light-induced spin cross-over in a two-dimensional layer. *Nature Communications*, 7, art. n. 12212.

Balasubramanian, G. et al., 2008. Nanoscale imaging magnetometry with diamond spins under ambient conditions. *Nature*, 455(7213), pp. 648–651.

Bardeen, J., 1961. Tunnelling from a many-particle point of view. *Physical Review Letters*, 6(2), pp. 57–59.

Barraud, C. et al., 2010. Unravelling the role of the interface for spin injection into organic semiconductors. *Nature Physics*, 6(8), pp. 615–620.

Baumann, S. et al., 2015. Electron paramagnetic resonance of individual atoms on a surface. *Science*, 350(6259), pp. 417–420.

Beni, A. et al., 2006. Optically induced valence tautomeric interconversion in cobalt dioxolene complexes. *Journal of the Brazilian Chemical Society*, 17(8), pp. 1522–1533.

Binnig, G. & Quate, C.F., 1986. Atomic force microscope. *Physical Review Letters*, 56(9), pp. 930–933.

Binnig, G. et al., 1982a. Surface studies by scanning tunneling microscopy. *Physical Review Letters*, 49(1), pp. 57–61.

Binnig, G. et al., 1982b. Tunneling through a controllable vacuum gap. *Applied Physics Letters*, 40(2), pp. 178–180.

Bleuzen, A. et al., 2000. Photoinduced ferrimagnetic systems in Prussian blue analogues C-X(I)Co-4[Fe(Cn)(6)](Y) (C-I = Alkali Cation). 1. Conditions to observe the phenomenon. *Journal of the American Chemical Society*, 122(28), pp. 6648–6652.

Bogani, L. & Wernsdorfer, W., 2008. Molecular spintronics using single-molecule magnets. *Nature Materials*, 7(3), pp. 179–186.

Bonizzoni, C. et al., 2017. Coherent coupling between Vanadyl Phthalocyanine spin ensemble and microwave photons: Towards integration of molecular spin qubits into quantum circuits. *Scientific Reports*, 7(1), art. n. 13096.

Brede, J. et al., 2010. Spin- and energy-dependent tunneling through a single molecule with intramolecular spatial resolution. *Physical Review Letters*, 105(4), art. n. 047204.

Burgess, J.A.J. et al., 2015. Magnetic fingerprint of individual Fe$_4$ molecular magnets under compression by a scanning tunnelling microscope. *Nature Communications*, 6, art. n. 8216.

Caneschi, A. et al., 2001. Pressure- and temperature-induced valence tautomericinterconversion in a O-Dioxolene adduct of a cobalt-tetraazamacrocycle complex. *Chemistry-A European Journal*, 7(18), pp. 3926–3930.

Caputo, M. et al., 2016. Manipulating the topological interface by molecular adsorbates: Adsorption of co-phthalocyanine on Bi$_2$Se$_3$. *Nano Letters*, 16(6), pp. 3409–3414.

Carpinelli, J.M. et al., 1996. Direct observation of a surface charge density wave. *Nature*, 381(6581), pp. 398–400.

Catala, L. & Mallah, T., 2017. Nanoparticles of Prussian blue analogs and related coordination polymers: From information storage to biomedical applications. *Coordination Chemistry Reviews*, 346, pp. 32–61.

Cavallini, M. et al., 2005. Magnetic information storage on polymers by using patterned single-molecule magnets. *Angewandte Chemie International Edition in English*, 44(6), pp. 888–892.

Chen, C.J., 1993. *Introduction to Scanning Tunneling Microscopy*, Oxford University Press.

Chen, X. et al., 2008. Probing superexchange interaction in molecular magnets by spin-flip spectroscopy and microscopy. *Physical Review Letters*, 101(19), art. n. 197208.

Coronado, E. & Epsetin, A.J., 2009. Molecular spintronics and quantum computing. *Journal of Materials Chemistry*, 19(12), pp. 1670–1671.

De Cian et al. 1985. Synthesis, structure, and spectroscopic and magnetic properties of lutetium(III) phthalocyanine derivatives: LuPc$_2$·CH$_2$Cl$_2$ and [LuPc(OAc)(H$_2$O)$_2$]·H$_2$O·2CH$_3$OH *Inorganic Chemistry*, 24(20), pp. 3162–3167.

Degen, C.L. et al., 2009. Nanoscale magnetic resonance imaging. *Proceedings of National Academy of Sciences of United States of America*, 106(5), pp. 1313–1317.

Deng, Z. et al., 2015. Self-assembly of bis(phthalocyaninato)terbium on metal surfaces. *Physica Scripta*, 90(9), art. n. 098003.

Dey, S.K. et al., 2007. Supramolecular self-assembled polynuclear complexes from tritopic, tetratopic, and pentatopic ligands: Structural, magnetic and surface studies. *Inorganic Chemistry*, 46(19), pp. 7767–7781.

Dreiser, J. et al., 2016. Out-of-plane alignment of Er(trensal) easy magnetization axes using graphene. *ACS Nano*, 10(2), pp. 2887–2892.

Dubout, Q. et al., 2015. Controlling the spin of Co atoms on Pt(111) by hydrogen adsorption. *Physical Review Letters*, 114(10), art. n. 106807.

Dunn, R.C., 1999. Near-field scanning optical microscopy. *Chemical Reviews*, 99(10), pp. 2891–2928.

Durkan, C. & Welland, M.E., 2002. Electronic spin detection in molecules using scanning-tunneling- microscopy-assisted electron-spin resonance. *Applied Physics Letters*, 80(3), pp. 458–460.

Erler, P. et al., 2015. Highly ordered surface self-assembly of Fe$_4$ single molecule magnets. *Nano Letters*, 15(7), pp. 4546–4552.

Félix, G. et al., 2014. Non-extensivity of thermodynamics at the nanoscale in molecular spin crossover materials: A balance between surface and volume. *Physical Chemistry Chemical Physics: PCCP*, 16(16), pp. 7358–7367.

Ferlay, S. et al., 1995. A room-temperature organometallic magnet based on Prussian blue. *Nature*, 378(6558), pp. 701–703.

Fowler, R.H. & Nordheim, L., 1928. Electron emission in intense electric fields. *Proceedings of the Royal Society A: Mathematical, Physical and Engineering Sciences*, 119(781), pp. 173–181.

Franke, K.J., Schulze, G. & Pascual, J.I., 2011. Competition of superconducting phenomena and kondo screening at the nanoscale. *Science*, 332(6032), pp. 940–944.

Fu, Y.-S. et al., 2012. Reversible chiral switching of bis(phthalocyaninato) terbium(III) on a metal surface. *Nano Letters*, 12(8), pp. 3931–3935.

Fukuzaki, E. & Nishide, H., 2006. Room-temperature high-spin organic single molecule: Nanometer-sized and hyperbranched poly[1,2,(4)-phenylenevinyleneanisylaminium]. *Journal of the American Chemical Society*, 128(3), pp. 996–1001.

Ganzhorn, M. & Wernsdorfer, W., 2014. Molecular quantum spintronics using single-molecule magnets. In J. Bartolomé, F. Luis, & J. F. Fernández, eds. *Molecular Magnets: Physics and Applications*. Berlin, Heidelberg, Germany: Springer Berlin Heidelberg, pp. 319–364.

Garnica, M. et al., 2013. Long-range magnetic order in a purely organic 2D layer adsorbed on epitaxial graphene. *Nature Physics*, 9(6), pp. 368–374.

Gatteschi, D., Sessoli, R. & Villain, J., 2006. *Molecular Nanomagnets*, Oxford University Press.

Ghirri, A. et al., 2008. Magnetic imaging of cyanide-bridged co-ordination nanoparticles grafted on FIB-patterned si substrates. *Small*, 4(12), pp. 2240–2246.

Giaever, I., 1960a. Electron tunneling between two superconductors. *Physical Review Letters*, 5(10), pp. 464–466.

Giaever, I., 1960b. Energy gap in superconductors measured by electron tunneling. *Physical Review Letters*, 5(4), pp. 147–148.

Giaever, I. & Megerle, K., 1961. Study of superconductors by electron tunneling. *Physical Review*, 122(4), pp. 1101–1111.

Goodwin, C.A.P. et al., 2017. Molecular magnetic hysteresis at 60 kelvin in dysprosocenium. *Nature*, 548, pp. 439–442.

Gopakumar, T.G. et al., 2012. Electron-induced spin crossover of single molecules in a bilayer on gold.

Angewandte Chemie International Edition, 51(25), pp. 6262–6266.

Gragnaniello, L. et al., 2017. Uniaxial 2D superlattice of Fe$_4$ molecular magnets on graphene. *Nano Letters*, 17(12), pp. 7177–7182.

Grenz, J. et al., 2017. Probing the nano-skyrmion lattice on Fe/Ir(111) with magnetic exchange force microscopy. *Physical Review Letters*, 119(4), art. n. 047205.

Grinolds, M.S. et al., 2013. Nanoscale magnetic imaging of a single electron spin under ambient conditions. *Nature Physics*, 9(4), pp. 215–219.

Gutlich, P., Spiering, H. & Hauser, A., 1999. *Spin Transition in Iron(II) Compounds*.

Hamers, R.J., Tromp, R.M. & Demuth, J.E., 1986. Surface electronic structure of Si (111)-(7 × 7) resolved in real space. *Physical Review Letters*, 56(18), pp. 1972–1975.

Hao, H.W., Baró, A.M. & Sáenz, J.J., 1991. Electrostatic and contact forces in force microscopy. *Journal of Vacuum Science & Technology B: Microelectronics and Nanometer Structures*, 9(2), art. n. 1323.

Hartmann, U., 1999. Magnetic Force Microscopy. *Annual Review of Materials Science*, 29, pp. 53–87.

Hatter, N. et al., 2015. Magnetic anisotropy in Shiba bound states across a quantum phase transition. *Nature Communications*, art. n. 8988.

Hauptmann, N. et al., 2018. Revealing the correlation between real-space structure and chiral magnetic order at the atomic scale. *Physical Review B*, 97(10), p. 100401.

Heinrich, A.J. et al., 2004. Single-atom spin-flip spectroscopy. *Science*, 306(5695), pp. 466–469.

Heinrich, B.W., Pascual, J.I. & Franke, K.J., 2018. Single magnetic adsorbates on s-wave superconductors. *Progress in Surface Science*, 93(1), pp. 1–19.

Heinrich, B.W. et al., 2013. Protection of excited spin states by a superconducting energy gap. *Nature Physics*, 9, pp. 765–768.

Hernández, E.M. et al., 2014. AFM imaging of molecular spin-state changes through quantitative thermomechanical measurements. *Advanced Materials*, 26(18), pp. 2889–2893.

Hewson, A.C., 1993. *The Kondo Problem to Heavy Fermions*, Cambridge, UK: Cambridge University Press.

Hirjibehedin, C.F., Lutz, C.P. & Heinrich, A.J., 2006. Spin coupling in engineered atomic structures. *Science*, 312(5776), pp. 1021–1024.

Ho, W., 2002. Single-molecule chemistry. *The Journal of Chemical Physics*, 117(24), art. n. 11033.

Hug, H.J. et al., 1998. Quantitative magnetic force microscopy on perpendicularly magnetized samples. *Journal of Applied Physics*, 83(11), pp. 5609–5620.

Iacovita, C. et al., 2008. Visualizing the spin of individual cobalt-phthalocyanine molecules. *Physical Review Letters*, 101(11), art. n. 116602.

Ishikawa, N., 2007. Single molecule magnet with single lanthanide ion. *Polyhedron*, 26(9–11), pp. 2147–2153.

Ishikawa, N. et al., 2003. Lanthanide double-decker complexes functioning as magnets at the single-molecular level. *Journal of the American Chemical Society*, 125(29), pp. 8694–8695.

Ishikawa, N. et al., 2004. Upward temperature shift of the intrinsic phase lag of the magnetization of Bis(phthalocyaninato)terbium by ligand oxidation creating an $S = 1/2$ spin. *Inorganic Chemistry*, 43(18), pp. 5498–5500.

Israelachvili, J.N., 2012. *Intermolecular and Surface Forces*, Amsterdam, The Netherlands: Academic Press.

Jaafar, M., Asenjo, A. & Vázquez, M., 2008. Calibration of coercive and stray fields of commercial magnetic force microscope probes. *IEEE Transactions on Nanotechnology*, 7(3), pp. 245–250.

Jaafar, M. et al., 2011. Distinguishing magnetic and electrostatic interactions by a Kelvin probe force microscopy-magnetic force microscopy combination. *Beilstein Journal of Nanotechnology*, 2, pp. 552–560.

Jaklevic, R.C. & Lambe, J., 1966. Molecular vibration spectra by electron tunneling. *Physical Review Letters*, 17(22), pp. 1139–1140.

Jasper-Toennies, T. et al., 2017. Robust and selective switching of an Fe^{III} spin-crossover compound on Cu$_2$N/Cu(100) with Memristance Behavior. *Nano Letters*, 17(11), pp. 6613–6619.

Kahle, S. et al., 2012. The quantum magnetism of individual manganese-12-acetate molecular magnets anchored at surfaces. *Nano Letters*, 12(1), pp. 518–521.

Kahn, O., 1993. *Molecular Magnetism*, Weinheim, Germany: VCH.

Kaiser, U., Schwarz, A. & Wiesendanger, R., 2007. Magnetic exchange force microscopy with atomic resolution. *Nature*, 446(7135), pp. 522–525.

Katoh, K. et al., 2009. Direct observation of lanthanide(III)-phthalocyanine molecules on Au(111) by using scanning tunneling microscopy and scanning tunneling spectroscopy and thin-film field-effect transistor properties of Tb(III)- and Dy(III)-phthalocyanine molecules. *Journal of the American Chemical Society*, 131(29), pp. 9967–9976.

Khajetoorians, A.A. et al., 2013. Spin excitations of individual fe atoms on Pt(111): Impact of the site-dependent giant substrate polarization. *Physical Review Letters*, 111(15), art. n. 157204.

Kim, H. et al., 2013. Switching and sensing spin states of co-porphyrin in bimolecular reactions on Au(111) using scanning tunneling microscopy. *ACS Nano*, 7(10), pp. 9312–9317.

Komeda, T. et al., 2011. Observation and electric current control of a local spin in a single-molecule magnet. *Nature Communications*, 2, art. n. 217.

Komeda, T. et al., 2014. Variation of kondo temperature induced by molecule–substrate decoupling in film formation of Bis(phthalocyaninato)terbium(III) molecules on Au(111). *ACS Nano*, 8(5), pp. 4866–4875.

Kondo, J., 1964. Resistance minimum in dilute magnetic alloys. *Progress of Theoretical Physics*, 32(1), pp. 37–49.

Lanzilotto, V. et al., 2016. The challenge of thermal deposition of coordination compounds: Insight into the case of an Fe_4 single molecule magnet. *Chemistry of Materials*, 28(21), pp. 7693–7702.

Launay, J.-P. & Verdaguer, M., 2013. *Electrons in Molecules*, Oxford University Press.

Létard, J.-F. et al., 1999. Critical temperature of the LIESST effect in iron(II) spin crossover compounds. *Chemical Physics Letters*, 313(1–2), pp. 115–120.

Lodi Rizzini, A. et al., 2011. Coupling single molecule magnets to ferromagnetic substrates. *Physical Review Letters*, 107(17), art. n. 177205.

Lohau, J. et al., 1999. Quantitative determination of effective dipole and monopole moments of magnetic force microscopy tips. *Journal of Applied Physics*, 86(6), pp. 3410–3417.

Lopes, M. et al., 2013. Atomic force microscopy and near-field optical imaging of a spin transition. *Nanoscale*, 5(17), art. n. 7762.

Lorusso, G. et al., 2013. Surface-confined molecular coolers for cryogenics. *Advanced Materials*, 25(21), pp. 2984–2988.

Loth, S., Lutz, C.P. & Heinrich, A.J., 2010. Spin-polarized spin excitation spectroscopy. *New Journal of Physics*, 12(12), p. 125021.

Loth, S. et al., 2010. Measurement of fast electron spin relaxation times with atomic resolution. *Science*, 329(5999), pp. 1628–1630.

Luo, Y. et al., 2015. STM study of the adsorption of single-molecule magnet Fe_4 on Bi (111) surface. *Surface Review and Letters*, 22(05), art. n. 1550060.

Malavolti, L. et al., 2013. Magnetism of $TbPc_2$ SMMs on ferromagnetic electrodes used in organic spintronics. *Chemical Communications*, 49(98), art. n. 11506.

Malavolti, L. et al., 2015. Magnetic bistability in a submonolayer of sublimated Fe_4 single-molecule magnets. *Nano Letters*, 15(1), pp. 535–541.

Manassen, Y., Mukhopadhyay, I. & Rao, N., 2000. Electron-spin-resonance STM on iron atoms in silicon. *Physical Review B*, 61(23), pp. 16223–16228.

Manassen, Y. et al., 1989. Direct observation of the precession of individual paramagnetic spins on oxidized silicon surfaces. *Physical Review Letters*, 62(21), pp. 2531–2534.

Mannini, M. et al., 2009. Magnetic memory of a single-molecule quantum magnet wired to a gold surface. *Nature Materials*, 8(3), pp. 194–197.

Mannini, M. et al., 2010. Quantum tunnelling of the magnetization in a monolayer of oriented single-molecule magnets. *Nature*, 468(7322), pp. 417–421.

Messina, P., et al., 2007. Spin Noise Fluctuations from Paramagnetic Molecular Adsorbates on Surfaces. *Journal of Applied Physics* 101 (5) pp. 053916.

Michinobu, T., Inui, J. & Nishide, H., 2003. Magnetic force microscopic images of nanometer-sized polyradical particles. *Polymer Journal*, 35(1), pp. 71–75.

Michinobu, T., Inui, J. & Nishide, H., 2010. Two-dimensionally extended organic high-spin poly(aminium cationic radical)s and their magnetic force microscopic images. *Polymer Journal*, 42(7), pp. 575–582.

Miyamachi, T. et al., 2012. Robust spin crossover and memristance across a single molecule. *Nature Communications*, 3, art. n. 938.

Miyasaka, M., Saito, Y. & Nishide, H., 2003. Magnetic force microscopy images of a nanometer-sized, purely organic high-spin polyradical. *Advanced Functional Materials*, 13(2), pp. 113–117.

Morita, S., Giessibl, F.J. & Wiesendanger, R., 2009. *Noncontact Atomic Force Microscopy*, Berlin, Germany: Springer.

Mugnaini, V. et al., 2009. Towards the detection of single polychlorotriphenylmethyl radical derivatives by means of Electron Spin Noise STM. *Solid State Sciences*, 11(5), pp. 956–960.

Müllegger, S. et al., 2014. Radio frequency scanning tunneling spectroscopy for single-molecule spin resonance. *Physical Review Letters*, 113(13), p. 133001.

Müllegger, S. et al., 2015. Mechanism for nuclear and electron spin excitation by radio frequency current. *Physical Review B*, 92(22), art. n. 220418.

Natterer, F.D. et al., 2017. Reading and writing single-atom magnets. *Nature*, 543(7644), pp. 226–228.

Nishide, H. et al., 2001. A nanometer-sized high-spin polyradical: Poly(4-phenoxyl-1,2-phenylenevinylene) planarly extended in a non-kekul fashion and its magnetic force microscopic images. *Journal of the American Chemical Society*, 123(25), pp. 5942–5946.

Niu, T., Zhang, J. & Chen, W., 2014. Molecular ordering and dipole alignment of vanadyl phthalocyanine monolayer on metals: The effects of interfacial interactions. *The Journal of Physical Chemistry C*, 118(8), pp. 4151–4159.

Niu, T. et al., 2012. Substrate reconstruction mediated unidirectionally aligned molecular dipole dot arrays. *The Journal of Physical Chemistry C*, 116(21), pp. 11565–11569.

Olsson, L. et al., 1998. A method for in situ characterization of tip shape in ac-mode atomic force microscopy using electrostatic interaction. *Journal of Applied Physics*, 84(8), pp. 4060–4064.

Palacio, F. et al., 1997. High-temperature magnetic ordering in a new organic magnet. *Physical Review Letters*, 79(12), pp. 2336–2339.

Pan, Y. et al., 2017. Scanning tunnelling spectroscopy and manipulation of double-decker phthalocyanine molecules on a semiconductor surface. *Journal of Physics: Condensed Matter*, 29(36), art. n. 364001.

Persson, B.N.J. & Baratoff, A., 1987. Inelastic electron tunneling from a metal tip: The contribution from resonant processes. *Physical Review Letters*, 59(3), pp. 339–342.

Pinilla-Cienfuegos, E. et al., 2015. Imaging the magnetic reversal of isolated and organized molecular-based nanoparticles using magnetic force microscopy. *Particle and Particle Systems Characterization*, 32(6) pp. 693–700.

Pinilla-Cienfuegos, E. et al., 2016. Switching the magnetic vortex core in a single nanoparticle. *ACS Nano*, 10(2), pp. 1764–1770.

Raşa, M., Kuipers, B.W.M. & Philipse, A.P., 2002. Atomic force microscopy and magnetic force microscopy study of model colloids. *Journal of Colloid and Interface Science*, 250(2), pp. 303–315.

Ratner, M., 2013. A brief history of molecular electronics. *Nature Nanotechnology*, 8(6), pp. 378–381.

Rau, I.G. et al., 2014. Reaching the magnetic anisotropy limit of a 3d metal atom. *Science*, 344(6187), pp. 988–992.

Repp, J. et al., 2005. Molecules on insulating films: Scanning-tunneling microscopy imaging of individual molecular orbitals. *Physical Review Letters*, 94(2), art. n. 026803.

Robles, R. et al., 2012. Spin doping of individual molecules by using single-atom manipulation. *Nano Letters*, 12(7), pp. 3609–3612.

Rolf-Pissarczyk, S. et al., 2017. Dynamical negative differential resistance in antiferromagnetically coupled few-atom spin chains. *Physical Review Letters*, 119(21), art. n. 217201.

Rugar, D., Mamin, H.J. & Guethner, P., 1989. Improved fiber-optic interferometer for atomic force microscopy. *Applied Physics Letters*, 55(25), pp. 2588–2590.

Rugar, D., Yannoni, C.S. & Sidles, J.A., 1992. Mechanical detection of magnetic resonance. *Nature*, 360(6404), pp. 563–566.

Rugar, D. et al., 2004. Single spin detection by magnetic resonance force microscopy. *Nature*, 430(6997), pp. 329–332.

Ruiz-Molina, D. et al., 2003. Single-molecule magnets on a polymeric thin film as magnetic quantum bits. In R. Vajtai et al., eds. *Nanotechnology*. International Society for Optics and Photonics, p. 594. Available at https://doi.org/10.1117/12.499143.

Ruiz-Muelle, A.B. et al., 2017. Covalent immobilization of dysprosium-based metal–organic chains on silicon-based polymer brush surfaces. *New Journal of Chemistry*, 41(15), pp. 7007–7011.

Sangregorio, C. et al., 1997. Quantum tunneling of the magnetization in an iron cluster nanomagnet. *Physical Review Letters*, 78(24), pp. 4645–4648.

van Schendel, P.J.A. et al., 2000. A method for the calibration of magnetic force microscopy tips. *Journal of Applied Physics*, 88(1), pp. 435–445.

Schmidt, R. et al., 2009. Probing the magnetic exchange forces of iron on the atomic scale. *Nano Letters*, 9(1), pp. 200–204.

Schreiber, S. et al., 2008. Magnetic force microscopy of superparamagnetic nanoparticles. *Small*, 4(2), pp. 270–278.

Schwöbel, J. et al., 2012. Real-space observation of spin-split molecular orbitals of adsorbed single-molecule magnets. *Nature Communications*, 3, art. n. 953.

Scott, G.D. & Natelson, D., 2010. Kondo resonances in molecular devices. *ACS Nano*, 4(7), pp. 3560–3579.

Selloni, A. et al., 1985. Voltage-dependent scanning-tunneling microscopy of a crystal surface: Graphite. *Physical Review B* 31, 2602(R).

Serrano, G. et al., 2016. Bilayer of terbium double-decker single-molecule magnets. *The Journal of Physical Chemistry C*, 120(25), pp. 13581–13586.

Serrano, G. et al., 2018. Magnetic bistability of TbPc$_2$ submonolayer on a graphene/SiC(0001) conductive electrode. *Nanoscale*, 10(6), pp. 2715–2720.

Serri, M. et al., 2017. Low-temperature magnetic force microscopy on single molecule magnet-based microarrays. *Nano Letters*, 17(3), pp. 1899–1905.

Sievers, S. et al., 2012. Quantitative measurement of the magnetic moment of individual magnetic nanoparticles by magnetic force microscopy. *Small*, 8(17), pp. 2675–2679.

Sk, R., Mulani, I. & Deshpande, A., 2018. Emergent properties of the organic molecule-topological insulator hybrid interface: Cu-phthalocyanine on Bi$_2$Se$_3$. *The Journal of Physical Chemistry C*, 122(40), pp. 22996–23001.

Smykalla, L., Shukrynau, P. & Hietschold, M., 2012. Investigation of ultrathin layers of bis(phthalocyaninato) Lutetium(III) on graphite. *Journal of Physical Chemistry C*, 116(14), pp. 8008–8013.

Stepanow, S. et al., 2010. Spin and orbital magnetic moment anisotropies of monodispersed bis(phthalocyaninato)terbium on a copper surface. *Journal of the American Chemical Society*, 132(34), pp. 11900–11901.

Stroscio, J.A. & Kaiser, W.J., 1993. *Scanning Tunneling Microscopy*, Academic Press.

Stroscio, J.A., Eigler, D.M. & Eigler, D.M., 1991. Atomic and molecular manipulation with the scanning tunneling microscope. *Science (New York, N.Y.)*, 254(5036), pp. 1319–1326.

Sun, H. et al., 2009. Self-organized honeycomb structures of Mn$_{12}$ single-molecule magnets. *The Journal of Physical Chemistry B*, 113(44), pp. 14674–14680.

Takada, M. & Tada, H., 2004. Low temperature scanning tunneling microscopy of phthalocyanine multilayers on Au(1 1 1) surfaces. *Chemical Physics Letters*, 392(1–3), pp. 265–269.

Takamatsu, S. et al., 2007. Significant increase of the barrier energy for magnetization reversal of a single-4f-ionic single-molecule magnet by a longitudinal contraction of the coordination space. *Inorganic Chemistry*, 46(18), pp. 7250–7252.

Tanaka, M., Saito, Y. & Nishide, H., 2006. Magnetic force microscopy as a new tool to evaluate local magnetization of organic radical polymers. *Chemistry Letters*, 35(12), pp. 1414–1415.

Tangoulis, V. et al., 2014. From molecular magnets to magnetic nanomaterials—Deposition of Co 7 single-molecule magnet; theoretical investigation of the exchange interactions. *European Journal of Inorganic Chemistry*, 2014(16), pp. 2678–2686.

Ternes, M., 2015. Spin excitations and correlations in scanning tunneling spectroscopy. *New Journal of Physics*, 17(6), art. n. 063016.

Tersoff, J. & Hamann, D.R., 1983. Theory and application for the scanning tunneling microscope. *Physical Review Letters*, 50(25), pp. 1998–2001.

Tersoff, J. & Hamann, D.R., 1985. Theory of the scanning tunneling microscope. *Physical Review B*, 31(2), pp. 805–813.

Tesi, L. et al., 2016. Quantum coherence in a processable vanadyl complex: New tools for the search of molecular spin qubits. *Chemical Science*, 7(3), pp. 2074–2083.

Tetienne, J.P. et al., 2014. Nanoscale imaging and control of domain-wall hopping with a nitrogen-vacancy center microscope. *Science*, 344(6190), pp. 1366–1369.

Thiele, S. et al., 2014. Electrically driven nuclear spin resonance in single-molecule magnets. *Science*, 344(6188), pp. 1135–1138.

Urdampilleta, M. et al., 2011. Supramolecular spin valves. *Nature Materials*, 10(7), pp. 502–506.

Vellekoop, B. et al., 1998. On the determination of the internal magnetic structure by magnetic force microscopy. *Journal of Magnetism and Magnetic Materials*, 190(1–2), pp. 148–151.

Vitali, L. et al., 2008. Electronic structure of surface-supported bis(phthalocyaninato) terbium(III) single molecular magnets. *Nano Letters*, 8(10), pp. 3364–3368.

Wäckerlin, C. et al., 2016. Giant hysteresis of single-molecule magnets adsorbed on a nonmagnetic insulator. *Advanced Materials*, 28(26), pp. 5195–5199.

Warner, B. et al., 2015. Tunable magnetoresistance in an asymmetrically coupled single-molecule junction. *Nature Nanotechnology* 10, pp. 259–263.

Wiberg, J. et al., 2009. Effect of anchoring group on electron injection and recombination dynamics in organic dye-sensitized solar cells. *Journal of Physical Chemistry C*, 113(9), pp. 3881–3886.

Wiesendanger, R., 2009. Spin mapping at the nanoscale and atomic scale. *Reviews of Modern Physics*, 81(4), pp. 1495–1550.

Young, R., Ward, J. & Scire, F., 1971. Observation of metal-vacuum-metal tunneling, field emission, and the transition region. *Physical Review Letters*, 27(14), pp. 922–924.

Young, R., Ward, J. & Scire, F., 1972. The topografiner: An instrument for measuring surface microtopography. *Review of Scientific Instruments*, 43(7), pp. 999–1011.

Zadrozny, J.M. et al., 2015. Millisecond coherence time in a tunable molecular electronic spin qubit. *ACS Central Science*, 1(9), pp. 488–492.

Zhang, J. et al., 2014. Single molecule tunneling spectroscopy investigation of reversibly switched dipolar vanadyl phthalocyanine on graphite. *Applied Physics Letters*, 104(11), art. n. 113506.

Zhang, Y. et al., 2010. Bis(phthalocyaninato)yttrium grown on Au(111): Electronic structure of a single molecule and the stability of two-dimensional films investigated by scanning tunneling microscopy/spectroscopy at 4.8 K. *Nano Research*, 3(8), pp. 604–611.

Zhang, Y. et al., 2015. Low-temperature scanning tunneling microscopy study on the electronic properties of a double-decker DyPc$_2$ molecule at the surface. *Physical Chemistry Chemical Physics*, 17(40), pp. 27019–27026.

Zhang, Y. et al., 2018. Detection and manipulation of charge states for double-decker DyPc$_2$ molecules on ultrathin CuO films. *ACS Nano*, 12(3), pp. 2991–2997.

Zhang, Y.F. et al., 2009. Low-temperature scanning tunneling microscopy investigation of bis(phthalocyaninato)yttrium growth on Au(111): From individual molecules to two-dimensional domains. *The Journal of Physical Chemistry C*, 113(22), pp. 9826–9830.

Zhao, A. et al., 2005. Controlling the Kondo effect of an adsorbed magnetic ion through its chemical bonding. *Science*, 309(5740), pp. 1542–1544.

Zhong, Q. et al., 1993. Fractured polymer/silica fiber surface studied by tapping mode atomic force microscopy. *Surface Science*, 290(1–2), pp. L688–L692.

Ultrafast Optical Pump-Probe Scanning Probe Microscopy/Spectroscopy

Hidemi Shigekawa and
Shoji Yoshida
University of Tsukuba

3.1   Introduction

The understanding and control of quantum dynamics, such as the transition and transport in nanoscale structures, are the key factors for continuing the advances in nanoscale science and technology. However, with size reduction, the differences in the electronic properties of materials and current devices, for example, those caused by the structural nonuniformity in each element, have an ever-increasing effect on macroscopic functions. For example, atomic-scale defects have markedly changed the entire situation: defects, which were once considered as a problem to be avoided, are now actively designed and controlled to realize desired functions. The fluctuation in the distribution of dopant materials governs the characteristic properties of macroscopic functions. Therefore, for further advances in science and technology, the development of a method for exploring the transient dynamics of local quantum functions in organized small structures is essential.

The spatial resolution of scanning tunneling microscopy (STM) is excellent. We can analyze local structures and electronic properties such as the local density of states with atomic resolution. Since the invention of STM, the direct imaging of atomic-scale structures has been lifting the veil from various long-standing problems and extending the frontiers of science and technology (Binning et al. 1982; Bhushan 2010; Wisendanger 1994). In STM, a sharp tip is placed above the target material, and information immediately below the probe tip is obtained through measurement of the tunnel current, spin, force, and so forth (Figure 3.1). Since the tunnel current logarithmically depends on the tip-sample distance, a 0.1 nm change in the distance produces a one-order change in the current. Therefore, if the STM tip is scanned over the sample surface while the tunnel current is kept constant using piezoelectric elements, three-dimensional imaging of the sample surface can be realized with atomic resolution.

In basic STM, the bias voltage applied between the STM tip and the sample is adjusted and the corresponding change in tunneling current is measured, giving information about the local electronic structures with atomic resolution. With the modulation of additional parameters, such as temperature, magnetic field, and tip-sample distance depending on the purpose, further information can be obtained. However, since the temporal resolution of STM

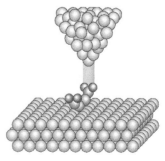

FIGURE 3.1   Schematic illustration of STM.

is limited, in general, to less than 100 kHz owing to the circuit bandwidth (Mamin et al. 1994; Wintterlin et al. 1997; Kemiktarak et al. 2007), the ultrafast dynamics in materials has been beyond its field of vision. In contrast, the advances in ultrashort-pulse laser technology have opened the door to the world of ultrafast phenomena. A prominent method is optical pump-probe (OPP) measurement, which has enabled ultrafast dynamics to be observed in the femtosecond range. However, the spatial resolution of such optical methods is generally limited by the wavelength, which may be averaged over the light spot area. Therefore, since the invention of STM in 1982, one of the most challenging goals has been to combine STM with ultrashort-pulse laser technology to simultaneously realize ultimate spatial and temporal resolutions. This issue has a long history, and many attempts have been made to achieve this goal using various approaches (Terada et al. 2010a; Shigekawa et al. 2010a).

In the following section, we focus on the development of laser-combined STM based on the OPP method.

3.2 Laser-Combined Scanning Tunneling Microscopy

3.2.1 Optical Pump-Probe Method and Optical Pump-Probe STM

In the OPP method, a sample is illuminated by a train of pulse pairs with a certain delay time. First, pulses are used as a pump to excite the sample, and second, pulses are used as a probe to observe the relaxation of the excited state (Figure 3.2a). When carriers excited by the pump pulse remain in the excited states, the absorption of the probe pulse is suppressed, that is, absorption bleaching occurs. Thus, if the reflectivity of the second pulse, for example, is measured as a function of delay time, we can obtain information on the relaxation of the excited state induced by the pump pulse through the change in the reflectivity of the probe pulse (Figure 3.2b). In this case, the time resolution is only limited by the pulse width, namely, to the femtosecond range.

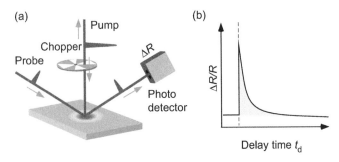

FIGURE 3.2 (a) Schematic illustration of OPP technique. (b) Normalized change in reflectivity $\Delta R/R$ as a function of delay time t_d.

In the new microscopy technique of OPP–STM, the sample surface below the STM tip is excited by a paired-pulse train with a certain delay time t_d, similar to that in the OPP method, but the signal is the tunnel current I, instead of change in the reflectivity of the probe pulse, as a function of delay time (Figure 3.3a). Namely, we probe the change in the number of carriers excited by the second pulse due to absorption bleaching through the change in the tunnel current as a function of the delay time. The optical pulses give rise to current pulses in the raw tunneling current I^*, which reflects the excitation and relaxation of the sample. If these current pulses decay rapidly compared with the timescale of the STM preamplifier bandwidth, they are temporally averaged in the preamplifier and cannot be detected directly in the signal I. Even in such a case, the relaxation dynamics can be probed through the t_d dependence of I (Shah 1999).

In OPP–STM measurement, we usually employ pump and probe pulses with equal intensity. When t_d is sufficiently long, the paired optical pulses with the same intensity independently induce two current pulses with the same height (I^*) as shown by A in Figure 3.3b. In contrast, when t_d is short and the second pulse illuminates the sample in the excited state caused by the first pulse, the second current pulse may have a different height dependence on t_d (B and C in Figure 3.3b). Since the change in the total current I^* induced by the first and second pulses causes a change in the average tunneling current I, as shown in Figure 3.3c, as a function of the delay time, we can elucidate the dynamics of the excited states through the change in tunneling current. In this case, we can obtain the temporal resolution of the OPP method together with the spatial resolution of STM.

3.2.2 How to Measure a Weak Signal

Since signals are weak, we need to use the lock-in detection method in OPP–STM. The excitation is oscillated at a certain frequency and the corresponding change in the signal is measured. In general, laser intensity is modulated in the OPP method, which, however, causes thermal expansion of the STM tip and sample (Grafström 2002). Since a 0.1 nm change in the tip-sample distance produces a one-order change in the tunneling current, it is difficult to detect a weak signal under this condition. A promising option is to modulate the delay time between the pump and probe pulses instead of the intensity. Using this modulation technique, the laser intensity is not changed and a modulation frequency independent of the noise frequency originating from thermal expansion can be chosen (Takeuchi et al 2004b, 2006; Terada et al. 2010a, b).

To realize a microscopy technique that enables us to visualize carrier dynamics in a nanometer-scale potential landscape, we developed a method for rectangular modulation of the delay time by using a pulse-picking technique. As shown in Figure 3.4, pulse trains are generated by two synchronized Ti:sapphire lasers at a 90 MHz repetition rate

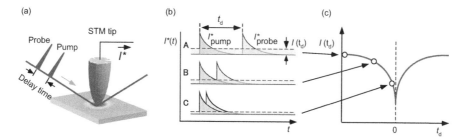

FIGURE 3.3 (a) Schematic illustration of OPP–STM. (b) Measurement mechanism. (c) OPP–STM spectrum corresponding to the mechanism in (b).

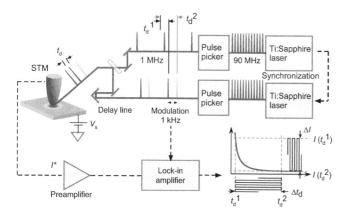

FIGURE 3.4 Schematic illustration of OPP–STM system. Rectangular modulation of the delay time is shown in the lower right of the figure.

(11 ns intervals) with a pulse width of 150 fs in this case. The relative timing of the two pulse trains is controlled by a synchronizing circuit, which provides a delay time that ranges from 0 to 11 ns with a time jitter. Each train is guided to a pulse picker that can selectively transmit 1 pulse per 90 pulses, resulting in the reduced repetition rate of 1 MHz. The pulse selection enables the production of an additional longer delay time that can be adjusted in multiples of 11 ns. Consequently, t_d can be adjusted continuously from zero to a large value as needed. This control of the delay allows nanometer-scale structures to be probed with a wide range of relaxation lifetimes.

In the pulse-picking method, t_d can be modified by large and discrete amounts by changing the timing of selecting pulses that are transmitted by pulse pickers, making the method suitable for modulating t_d in OPP–STM. For the modulation of t_d between t_d^1 and t_d^2 in a rectangular waveform, the lock-in detection of the tunneling current gives the value $\Delta I \left(t_d^1, t_d^2 \right) \equiv I \left(t_d^1 \right) - \left(t_d^2 \right)$ as shown in the lower right of Figure 3.4. As t_d^2 is set to a large value compared with the relaxation time of the probed dynamics, $\Delta I \left(t_d^1, t_d^2 \right)$ can be approximated as $\Delta I \left(t_d^1 \right) \equiv I \left(t_d^1 \right) - I \left(\infty \right)$, where $I(\infty)$ is the tunneling current for a delay time that is sufficiently long for the excited state to be relaxed. Therefore, $\Delta I \left(t_d^1 \right)$ is accurately obtained through the lock-in detection of I by sweeping t_d^1. In addition, since the modulation can be performed at a high frequency (1 kHz in our case), the

measurement is not significantly affected by low-frequency fluctuations in the laser intensity and tunneling current. Accordingly, this method reduces the measurement time and, hence, enables the spatial mapping of time-resolved (TR) signals (Terada et al. 2011; Yoshida et al. 2012; Yamashita et al. 2005).

3.3 Probing of Carrier Dynamics by OPP-STM

In this section, some examples of the OPP–STM measurement of semiconductors are shown. In STM for a semiconductor, a nanoscale metal-insulator-semiconductor (MIS) junction is formed by the STM tip, tunneling gap, and sample. In the case of a reverse-bias voltage between the STM tip and sample, bias voltage leakage causes tip-induced band bending (TIBB). Under optical illumination, the redistribution of photocarriers reduces the electric field, thereby reducing the band bending, i.e., surface photovoltage (SPV) (McEllistrem et al. 1993; Takeuchi et al. 2004a, b; Yoshida et al. 2007) is generated.

Figure 3.5 shows schematic illustrations to explain the mechanism. TIBB appears under a reverse-bias voltage (a). Under photoillumination, the redistribution of the photocarriers reduces the band bending, i.e., SPV is generated (b). Then, the excited state subsequently relaxes to the original state through two processes (c). One is the decay of photocarriers on the bulk side via recombination, drift, and diffusion, known as bulk-side decay. The other is the decay of minority carriers trapped at the surface via recombination and thermionic emission, called surface-side decay (Yokota et al. 2013).

When the second pulse arrives during the bulk-side decay, the density of carriers induced by the second pulse decreases owing to absorption bleaching. In such a case, the SPV, and thus the total tunneling current, changes as a function of the delay time. On the other hand, when the second pulse arrives during the surface-side decay, fewer photocarriers are trapped at the surface owing to the existence of SPV induced by the first pulse, resulting in a change in the dependence of the total tunneling current on the delay time. Therefore, by measuring the tunneling current as a function of delay time, we can obtain information about the carrier dynamics in both processes.

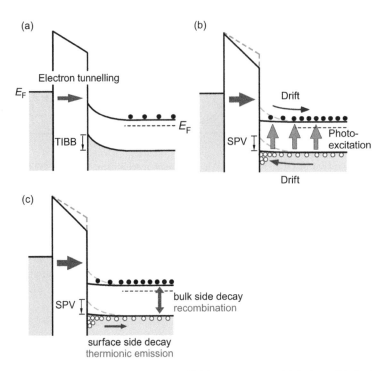

FIGURE 3.5   Band structures of a semiconductor under OPP–STM measurement (a) before, (b) during, and (c) after illumination.

In the next section, the decay processes of excited carriers probed by bulk-side decay are first discussed. Then, local carrier dynamics is discussed using the results obtained by the observation of surface-side decay.

3.3.1   Carrier Dynamics in a GaAs/AlGaAs/LT-GaAs Heterostructure

Figure 3.6 shows the result obtained for a heterostructure consisting of GaAs, AlGaAs, and low-temperature-grown (LT-GaAs). Figure 3.6a shows a schematic illustration of the

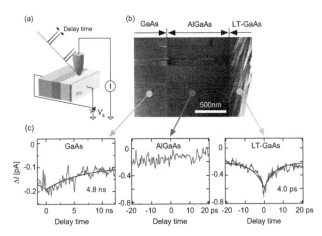

FIGURE 3.6   (a) Schematic illustration of OPP–STM measurement setup. (b) STM image of the sample of a GaAs/AlGaAs/LT-GaAs heterostructure sample (LT-GaAs: low-temperature-grown GaAs). (c) OPP–STM spectra obtained above GaAs, AlGaAs, and LT-GaAs.

experimental setup. Figure 3.6b and c show an STM image and OPP–STM spectra obtained at each region, respectively (Terada et al. 2011). The lifetimes of photocarriers in GaAs and LT-GaAs obtained by the OPP method are 6 ns and 1.5 ps, respectively. Since the excitation energy is lower than the gap energy of AlGaAs, no signal is acquired there.

In OPP–STM, measurement can be carried out wherever required because the probe method is STM. The decay constants obtained by OPP–STM are 4.8 ns for GaAs and 4.0 ps for LT-GaAs, which are in good agreement with those obtained by the OPP method. Of course, there is a slight difference because OPP–STM picks up local information.

Using this microscopy technique, the imaging of carrier dynamics information becomes possible. Figure 3.7a shows the photocarrier dynamics obtained along a line crossing an AlGaAs/GaAs interface. Here, instead of measuring a full spectrum by changing the delay time, the STM tip was scanned with the delay time fixed at 300, 600 fs, and so forth. Namely, the lines in Figure 3.7a show the carrier density at each delay time after photoexcitation. The carrier density decreases the delay time. If a two-dimensional (2D) scan is performed over the surface, as schematically shown in Figure 3.7b, 2D map of the time-dependent signal can be obtained, as shown in Figure 3.7c. A decay constant map can be obtained from the full series of 2D maps by fitting the change in density at each point.

3.3.2   Modulation of Carrier Dynamics in GaAs Pin Structure

Information on the carrier dynamics modulated by a local potential is important for understanding nanoscale physics and its application to the development of current devices

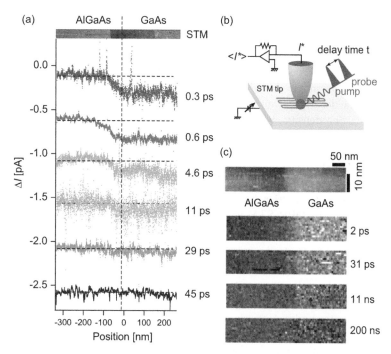

FIGURE 3.7 (a) One-dimensional OPP–STM signals obtained along a line crossing the /LT-GaAs interface of the sample shown in Figure 3.6b. (b) Schematic illustration of two-dimensional scan over a sample. (c) Two-dimensional OPP–STM images obtained at four delay times (0, 31 ps, 11 ns, 200 ns).

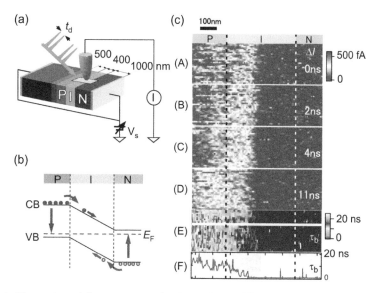

FIGURE 3.8 (a) Schematic illustration of the experimental setup of OPP–STM over a GaAs pin structure. (b) Schematic illustration of the inner potential of a GaAs pin structure. (c) Four OPP–STM images obtained above the GaAs pin sample at $t_d = 0$, 2, 4, and 11 ns. (E) Map of decay constant obtained from a series of two-dimensional OPP–STM images. (F) Cross section along the white line in (E).

with nanoscale structures. Here, an example of the carrier dynamics obtained for a GaAs pin structure is discussed (Yoshida et al. 2012). Figure 3.8a and b, respectively, show schematic illustrations of the experimental setup and the inner potential of the pin structure. In the p- and n-type regions, since the band is flat, recombination is the main process for the decay of photocarriers. In contrast, since there is a slope in the inner potential in the i-region, there

must be some effect of the inner potential on the carrier dynamics in the region.

Figure 3.8c (A–D) shows four images taken from a series of 2D maps of TR signals. Since a reverse-bias voltage between the STM tip and sample is necessary to measure the TR signal, only the left half of each figure is considered here. To observe the dynamics in the right half, we must change the sign of the bias voltage.

The carrier density decreases with the time after photoexcitation. The decay-constant map is obtained from the full series of 2D maps by fitting the change in carrier density at each point. Figures E and F in Figure 3.8c show the decay-constant map and the cross section along the white line in e, respectively. As expected, the decay constant decreases in the i-region. Namely, the carrier density decreases via drift and diffusion rather than via recombination in this region.

Another noteworthy point is that there is a fluctuation in the decay-constant map, suggesting the effect of local structures, such as atomic-scale defects, on the carrier dynamics. The examination of such phenomena by OPP–STM will be discussed in Section 3.3.3 and 3.3.4.

3.3.3 Atomic-Level Analysis

When some metals are deposited on a semiconductor surface, gap states are formed, which also modulate carrier dynamics. In STM on a semiconductor under a reverse-bias, TIBB occurs owing to the leakage of the applied bias voltage (McEllistrem et al. 1993; Takeuchi et al. 2004a, b; Yoshida et al. 2007), as explained in Section 3.2 (Figure 3.5). When a sample surface is illuminated using this condition, holes are trapped at the surface to reduce the band bending, thus generating SPV. If there is a gap state, holes trapped at the surface recombine with the electrons tunneling from the STM tip at the gap state as shown in Figure 3.9a. There are two limitations in this process, i.e., the injection of tunneling current from the STM tip and the capture rate of holes at the gap states. For a sufficient amount of tunneling current, the capture rate becomes the limiting process, which can be adjusted by changing the tip-sample distance. OPP–STM was applied to directly measure the hole capture rate at the atomic scale.

Figure 3.9c and d show STM images of manganese and iron atoms deposited on a GaAs(110) surface, respectively. Ga atoms are replaced by them as shown in Figure 3.9b. TR-STM measurements were carried out by placing the STM tip above these structures. Figure 3.9e and f show the spectra obtained. The decay constants were 1.6 and 14.3 ns for Mn/GaAs and Fe/GaAs, respectively. Although their structures are similar, the hole capture rate at the Mn site is one order faster, which is caused by the difference in their energy levels. Single atomic-level analysis is thus possible, which is expected to play an important role in analyzing the effects of dopants and atomic-level defects in semiconductor technologies.

Understanding and control of the quantum dynamics in nanoparticles and their interfaces with surrounding materials are important and play essential roles in various fields, such as semiconductor devices, catalysis, and energy transfer. In the OPP–STM measurement described above, the size dependence of carrier dynamics can be analyzed. The decay process should depend on the gap-state density, namely, the nanoparticle size. As expected, the decay constant was observed to increase with the decreasing nanoparticle size (Terada et al. 2010b; Yoshida et al. 2013b).

3.3.4 Effect of Atomic Step on Carrier Dynamics

Understanding and control of the effects of atomic steps and dislocations on carrier dynamics are important issues in material physics and the development of current devices, such as power devices using SiC and GaN (Bergman et al. 2001; Zhang et al. 2003). Here, OPP–STM is carried out on a GaAs surface step, and its effect on the carrier dynamics is observed from the surface-side decay, which is sensitive to the local carrier dynamics.

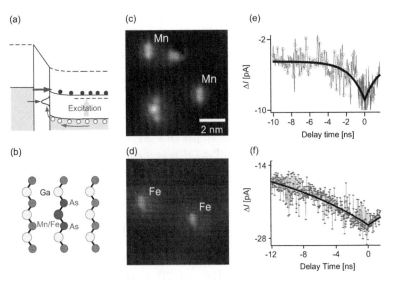

FIGURE 3.9 (a) Schematic illustration of the band structure during OPP–STM measurement above a semiconductor with a gap state. (b) Schematic illustration of (Mn/Fe)/GaAs structure. A Ga atom is replaced by a Mn/Fe atom. (c and d) STM images of GaAs surfaces with Mn and Fe atoms deposited on them, respectively. (e and f) OPP–STM spectra obtained above a Mn atom and an Fe atom in (c) and (d), respectively.

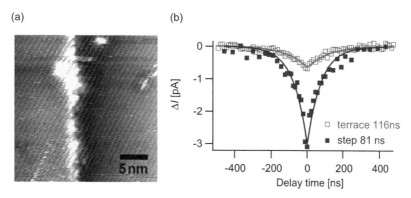

FIGURE 3.10 (a) STM image of a GaAs surface with a surface step. (b) OPP–STM spectra obtained above the step (■) and in a terrace far from the step (□).

Figure 3.10a shows an STM image of the GaAs surface with an atomic step, and Figure 3.10b shows the spectra obtained at an area without the step and above the step. To obtain accurate spectra, the laser intensity was adjusted to enhance the effect of the step (Terada et al. 2010a, b). The large amplitude of the signal at the atomic step is due to the large SPV originating from the Coulomb potential caused by the negative charges at the defect (Yoshida et al. 2008). The decay constants away from and at the atomic step were obtained to be 118 and 81 ns, respectively, clearly showing the effect of the atomic-step defect.

An atomic step forms gap states owing to the existence of dangling bonds, which act as traps to enhance the recombination of carriers. In OPP–STM, electrons tunneling from the STM tip combine with holes trapped at the gap state, as in the case of (Mn, Fe)/GaAs discussed in Section 3.3.3. Under a condition without TIBB [ ], holes recombine with electrons in the valence band via gap states, which is considered to occur at dislocations by a similar mechanism.

3.4　Probing Spin Dynamics by OPP-STM

Using circularly polarized light for pump and probe optical pulses, spin dynamics, which has been studied by, for example, spin-polarized STM (SP-STM) (Loth et al. 2010; Brede et al. 2012), can be observed, as has been carried out by the OPP method with optical orientation techniques (Takeuchi et al. 1990; Mirlin 1984). The mechanism

is similar to that of absorption bleaching. Spins are optically oriented by circularly polarized light and their dynamics are probed by STM. Here, as an example, right-handed circularly polarized light is used for the pump and probe pulses and down spins are excited (Figure 3.11a), which are randomized with time. When down spins excited by the first pulse remain in the excited states, the excitation of down spins by the second pulse as a function of the delay time is suppressed, by absorption bleaching (Figure 3.11b). In this case, the number of carriers excited by the second pulse increases with increasing delay time. The change in the number of excited carriers with the delay time produces a TR signal reflecting the spin dynamics (Yoshida et al. 2014).

First, we show the results of measurements above quantum wells (QWs) in Figure 3.12a, where the randomization of spins oriented in QWs was observed (Figure 3.12b). The sample was grown on a GaAs(100) surface from left to right, and two QWs with widths of 6 and 8 nm were formed. A 200 nm GaAs layer was placed as a spacer to isolate the two QWs. The sample was cleaved to prepare a clean (110) surface. Since OPP–STM is an STM, we can deduce where the location of the QWs are. Therefore, after observing the surface, the STM tip was placed above the QWs and measurements were carried out. The spectra shown in Figure 3.12c and d show the decay of spin orientation observed for the 6 and 8 nm QWs, respectively. The spin lifetime in the GaAs substrate was about 12 ps. In the QWs, the spin lifetime increased with increasing QW width and was as 68 ps for the 6 nm QW and 112 ps for the 8 nm QW.

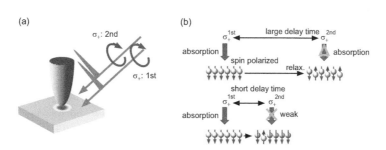

FIGURE 3.11 (a) Schematic illustration of OPP–STM setup with pump-probe pulses of circularly polarized light. (b) Mechanism of absorption bleaching for spin excitation.

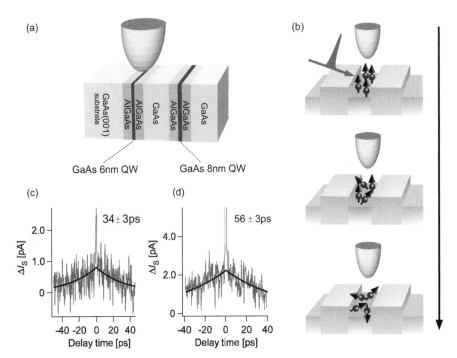

FIGURE 3.12 (a) Experimental setup of OPP–STM for GaAs QWs. (b) Schematic illustration of relaxation of spins oriented in a QW. (c and d) OPP–STM spectra obtained above QW of 6 and 8 nm width, respectively.

When a magnetic field is applied, spin precession occurs around the axis of the magnetic field. Thus, the number of carriers excited by the second pulse depends on the rotation angle owing to absorption bleaching. Namely, the signal intensity decreases with oscillation at the Larmor frequency as shown in Figure 3.13. From the relationship between the observed frequency and the applied magnetic field, $\omega L = g\mu_B B/h$, the local g factor can be evaluated. In addition, the decay constant of the spin orientation indicates the environmental conditions of electrons, for example, whether the electrons are free or trapped by defects.

The combination of this TR technique of examining spin dynamics with SP-STM (Loth et al. 2010; Brede et al. 2012) is an interesting target.

3.5 Phase-Controlled OPP-STM

3.5.1 Optical Pulses

In this section, we discuss OPP–STMs that can be realized using new laser technologies. In ordinary laser pulses, some cycles are included as shown in Figure 3.14a, whose phase is called the carrier envelope phase (CEP) (Jones et al. 2000). The CEP is random and fluctuates in pulses, which is the reason why the pulse width limits the time resolution. Recently, new laser technologies have become applicable, where the CEP is the same and locked in all pulses, as shown in Figure 3.14b. Furthermore, the CEP can be controlled as shown in Figure 3.14c. In Figure 3.14b and c, CEPs of zero and π are shown, respectively. On the basis of such control of the CEP, a new microscopy technique, THz-STM, has been developed, which is explained in 3.5.2.

3.5.2 Terahertz–STM (THz–STM)

One of the new microscopy techniques that has been extensively studied is THz–STM (Cocker et al. 2013, 2016; Yoshioka et al. 2016). Figure 3.15a shows a schematic illustration of THz–STM. THz pulses are generated by irradiating a crystal with a titanium–sapphire laser. As

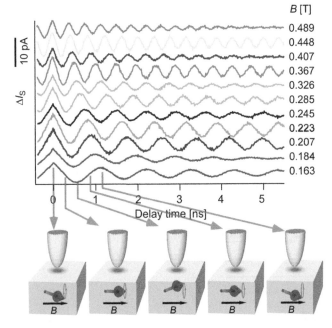

FIGURE 3.13 OPP–STM signal of spin precession obtained for a GaAs sample under several magnetic fields. Schematic of spin precession producing the oscillation of signal is shown in the lower part.

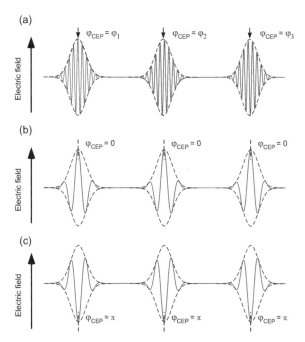

FIGURE 3.14 (a) Ordinary ultrashort laser pulses that include several waves in each pulse. The phase of such a wave in a pulse is called the carrier envelope phase (CEP). The CEP is random and fluctuates in each pulse. (b and c) Newly developed ultrashort pulses in which the CEP is fixed (zero and π for (b) and (c), respectively) and the same in all pulses.

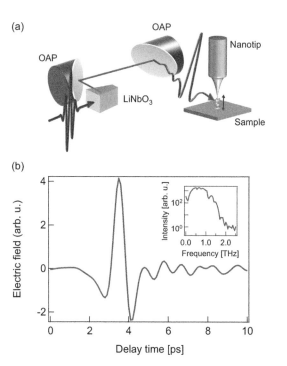

FIGURE 3.15 (a) Schematic illustration of THz–STM. (b) Example of a THz electric field. The inset shows its spectrum in the frequency space. OAP: off-axis parabolic mirror.

shown in Figure 3.15b, the THz pulse obtained has an electric field of almost a single cycle, and the wavelength depends on the pulse width of the excitation laser. Here, an

example of a THz pulse with a width of about 1 ps generated by a 130 fs pulse laser is shown.

In the case of a THz pulse, the CEP is locked automatically and can be controlled as shown in Figure 3.16a. The electric field has opposite directions for CEP values of zero and π, for example, which can be used to apply a bias voltage between the STM tip and the sample. In addition, since the bias voltage is applied as a very short pulse, a high bias voltage, which generally causes damage, can be applied in this case. Figure 3.16b shows an image of a graphite obtained by THz–STM. Atomic resolution was achieved.

In the OPP—STM described in the previous sections, TR signals were measured on the basis of the mechanism of absorption bleaching. In contrast, in THz–STM, a bias voltage can be applied at any delay time, thus behaves similarly to a stroboscope, allowing snapshot to be taken. There are several ways of carrying out TR measurements. One way is to combine infrared (IR) laser pulses with THz pulses, as shown in Figure 3.16c, in which an IR pulse is used as a pump pulse to excite the sample and a THz pulse is used as a probe to observe the dynamics induced by the IR pulse. Furthermore, when CEP-controlled pulses with a single electric field are used for pump and probe pulses, sub-cycle STM measurement can be achieved, in which the shorter pulse is used as a probe pulse. Namely, dynamics controlled by the electric field in a CEP-controlled pump pulse can be examined in detail by a CEP-controlled probe pulse with sub-cycle time resolution.

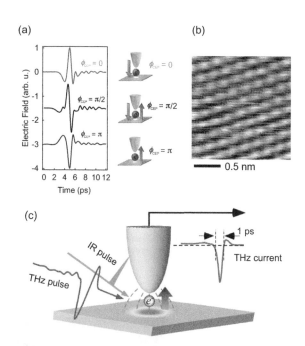

FIGURE 3.16 (a) THz electric field for CEP = zero, $\pi/2$ and π. The schematic illustrations on the right side show the directions of the electric fields. (b) OPP–STM image obtained for graphite, in which a THz electric field was used as a bias voltage. (c) Schematic of OPP–THz–STM.

3.5.3 Analysis of Laser Pulse Shape

The electric field below an STM tip is modulated from the incident field, as shown in Figure 3.17a, and also revealed by the antenna theory (Wang et al. 2004). Thus, to obtain accurate results using THz–STM, we need a method of evaluating the laser beam shape after modulation during STM measurement. In general, the original THz pulse shape is measured by an electro-optic (EO) sampling method, which is a technique for analyzing the shape of an electric field (Figure 3.17b). When a THz pulse and a shorter IR pulse enter a crystal, the polarization plane of the IR pulse is rotated by an amount depending on the intensity of the simultaneously existing THz electric field. Thus, a THz waveform can be probed if the rotation angle is measured while varying the delay time.

Next we discuss how to observe a THz pulse modulated at the tip apex. The combination of an IR pulse and a THz pulse is used for the measurement, as shown in Figure 3.17c. The tip apex is irradiated by the shorter pulse (517 nm, here) while the barrier height is reduced by the THz pulse, as shown in Figure 3.17d, which produces photoemission upon the shorter-pulse irradiation. Since the photocurrent depends on the intensity of the THz pulse, the THz pulse can be measured by varying the delay time (Wimmer et al. 2014; Herink et al. 2014).

Figure 3.18 shows an example obtained by the measurement (Yoshida et al. 2019). Figure 3.18a shows an image of the tungsten tip used for this measurement. Measurements were carried out by changing the IR spot position on the STM tip. Figure 3.18b shows a map of the photocurrent. The measurements were carried out at each pixel, and the

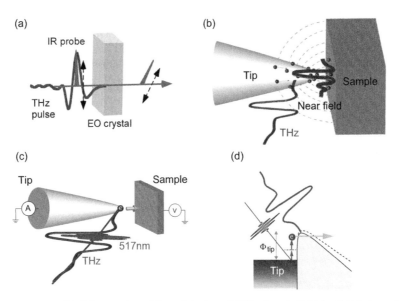

FIGURE 3.17 Schematics showing (a) EO sampling, (b) modulation of THz electric field by STM tip to Near field indicate incident and modulated THz pulses, respectively (c) measurement setup to probe THz electric field using photoemission, and (d) measurement mechanism in which hot electrons are produced by an IR pulse while the barrier height is modulated by a THz pulse.

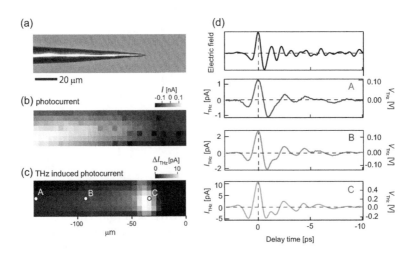

FIGURE 3.18 (a) Image of the STM tip used in an experiment to measure the laser pulse shape. (b) Map of photocurrent intensity. (c) Map of THz pulse intensity at $t_d = 0$. (d) Incident electric field (top) and THz electric fields obtained at positions A, B, and C in (c).

values were mapped in the color scale. THz waveforms were measured by varying the delay time at each pixel. Figure 3.18c shows a map of the signal intensity at zero delay, and Figure 3.18d shows the THz waveform obtained at the three positions (A, B, and C in Figure 3.18c), which have different shapes from that of the original THz pulse shown in the top of Figure 3.18d. At the tip apex, the signal intensity is strongest according to the enhancement by the tip, while the photocurrent is weak because the size of the tip is smaller than the spot size of the IR pulse. This illustrates the importance of checking pulse shape under the measurement conditions. This technique is applicable to the evaluation of near field in other experiments.

3.6 Other Techniques

The combination of the OPP technique with multiprobe STM (Hasegawa 2007) is an attractive application. Using this microscopy, we can probe local dynamics under well-controlled nanoscale operating conditions, enabling photo-induced ultrafast dynamics in organized small structures to be understood in more detail. For example, anisotropy in conductivity can be analyzed by changing the directions of the two probes. Furthermore, the depth profile of dynamics may be analyzed using multiple probes. When circularly polarized light is used, spin dynamics can be probed, as described in Section 3.4. The technique in combination with SP-STM is expected to be used to examine spin dynamics in more detail.

To increase the generality of this microscopy technique and its applicability to various types of materials and structures, additional techniques are desired. For example, as an application of TR-STM based on a method similar to that used for the analysis of a molecular electronic structure, two-photon absorption measurement was used together with TR-STM (Wu & Ho 2010). For further advances in OPP–STM, the development of direct techniques for detecting photocurrent in transient dynamics on the nanoscale is expected to play an important role in determining the optical characteristics of materials and devices. OPP–STM measurements of the transient photocurrent dynamics in the layered n-type semiconductor n-WSe$_2$ have been carried out (Yoshida et al. 2013a, b). Since WSe$_2$ has an indirect bulk band gap, the recombination lifetime of photoexcited carriers is significantly long (~10 μs) compared with the diffusion process. Therefore, under a forward bias voltage, i.e., a negative sample bias voltage, excited electrons are considered to directly tunnel to the STM tip as photocurrent, which has been clearly observed.

When the energies of pump and probe pulses are chosen appropriately, excited states may be included in the analysis similar to in the OPP method. The development of these techniques is considered to increase the generality of this microscopy technique, making it more practical and applicable to various phenomena. The essential mechanism of this microscopy technique involves the nonlinear interference between the excitations in the transient tunneling current generated by the two laser pulses; therefore, the introduction of new ideas is desirable to achieve further advances. By performing a stochastic analysis, the transition rate of a molecular conformation was clearly observed (Li et al. 2017).

Near-field scanning optical microscopy (NSOM) is used not only for spectroscopy but also for manufacturing nanoscale structures (Ohtsu 1998, 2003, 2008), and STM combined with synchrotron radiation has been opening the door to the characterization of atomic species by exciting core-level electrons (Okuda et al. 2009; Saito et al. 2006; Cummings et al. 2012).

3.7 Summary

Spectroscopy techniques for nanoscale analysis developed by the combination of laser technologies with STM have been overviewed. For the combination of STM with ultrashort-pulse laser technologies, the spatial resolution of STM and the temporal resolution of an ultrashort-pulse laser in the femtosecond range are simultaneously achieved by the delay-time modulation method developed using a pulse-picking technique. On the basis of the examples shown in this chapter, the development of other new techniques is expected to achieve further advances in this field.

References

Bergman, J. P., Lendenmann, H., Nilsson, P. A. et al. 2001. Crystal defects as source of anomalous forward voltage increase of 4H-SiC diodes. *Mater. Sci. Forum* 353–356: 299–302.

Bhushan, B. 2010. *Scanning Probe Microscopy in Nanoscience and Nanotechnology.* Berlin, Heidelberg, Germany: Springer.

Binning, G., Rohrer, H., Gerber, Ch. et al. 1982. Surface studies by scanning tunneling microscopy. *Phys. Rev. Lett.* 49: 57–61 (1982).

Brede, J., Chilian, B., Khajetoorians, A. A. et al. 2012. Atomic-scale spintronics. In *Handbook of Spintronics*, ed. D. Awschalom, J. Nitta, Y. Xu, 757–784. Berlin, Heidelberg, Germany: Canopus Academic Publishing and Springer.

Cocker, T. L., Jelic, V, Gupta, M. et al. 2013. An ultrafast terahertz scanning tunnelling microscope. *Nat. Photonics.* 7: 620–625.

Cocker, T. L., Peller, D., Yu, P. et al. 2016. Tracking the ultrafast motion of a single molecule by femtosecond orbital imaging. *Nature.* 539: 263–267.

Cummings, M. L., Chien, T. Y., Preissner, C. et al. 2012. Combining scanning tunneling microscopy and synchrotron radiation for high-resolution imaging and spectroscopy with chemical, electronic, and magnetic contrast. *Ultramicroscopy* 112: 22–31.

Grafström, S. 2002. Photoassisted scanning tunneling microscopy. *J. Appl. Phys.* 91: 1717–1753.

Hasegawa, S. 2007. Multi-probe scanning tunneling microscopy. In *Scanning Probe Microscopy*, ed. S. Kalinin, A. Gruverman, 480–505. New York: Springer.

Herink G., Wimmer, L., and Ropers. C. 2014. Field emission at terahertz frequencies: AC-tunneling and ultrafast carrier dynamics. *New. J. Phys.* 16: 123005.

Jones, D. J., Diddams, S. A., Ranka, J. K. et al. 2000. Carrier-envelope phase control of femtosecond mode-locked lasers and direct optical frequency synthesis. *Science* 288: 635–639.

Kemiktarak, U., Ndukum, T., Schwab, K. C. et al. 2007. Radio-frequency scanning tunnelling microscopy. *Nature* 450: 85–88.

Li, S., Chen, S., and Li, J. 2017. Joint space-time coherent vibration driven conformational transitions in a single molecule. *Phys. Rev. Lett.* 119: 176002.

Loth, S., Etzkorn, M., Lutz, C. P. et al. 2010. Measurement of fast electron spin relaxation times with atomic resolution. *Science* 329: 1628–1630.

Mamin, H. J., Birk, H., Wimmer, P. et al. 1994. High-speed scanning tunneling microscopy: Principles and applications. *J. Appl. Phys.* 75: 161–168.

McEllistrem, M., Haase, G., Chen, D. et al. 1993. Electrostatic sample-tip interactions in the scanning tunneling microscope. *Phys. Rev. Lett.* 70: 2471–2474.

Mirlin, D. N. 1984. Optical alignment of electron momenta in GaAs-type semiconductors. In *Optical Orientation*, ed. F. Meier, B. P. Zakharchenya, 133–172. Elsevier Science Publishers, North-Holland, Amsterdam.

Ohtsu, M. 1998. *Near-Field Nano/Atom Optics and Spectroscopy*. Berlin, Heidelberg, Germany: Springer-Verlag.

Ohtsu. M. 2003. *Progress in Nano-Electro-Optics II*. Berlin, Heidelberg, Germany: Springer-Verlag.

Ohtsu. M. 2008. *Progress in Nano-Electro-Optics VI*. Berlin, Heidelberg, Germany: Springer-Verlag.

Okuda, T., Eguchi, T., Akiyama, K. et al. 2009. Nanoscale chemical imaging by scanning tunneling microscopy assisted by synchrotron radiation. *Phys. Rev. Lett.* 102: 105503.

Saito, A., Maruyama, J., Manabe, K. et al. 2006. Development of a scanning tunneling microscope for in situ experiments with a synchrotron radiation hard-X-ray microbeam. *J. Shynchrotron Rad.* 13: 220.

Shah. J. 1999. *Ultrafast Spectroscopy of Semiconductors and Semiconductor Nanostructures*. Berlin, Heidelberg, Germany: Springer.

Shigekawa, H., Takeuchi, O., Terada, Y. et al. 2010. *Series: Handbook of Nanophysics*, ed. K. Sattler, Boca Raton, FL: Taylor & Francis. vol. 6, *Principles and Methods*.

Takeuchi, A., Muto, S., Inata, T. et al. 1990. Direct observation of picosecond spin relaxation of excitons in GaAs/AlGaAs quantum wells using spin-dependent optical nonlinearity. *Appl. Phys. Lett.* 56: 2213–2215.

Takeuchi, O., Yoshida, S., and Shigekawa, H. 2004a. Light-modulated scanning tunneling spectroscopy for nanoscale imaging of surface photovoltage. *Appl. Phys. Lett.* 84: 3645–3647.

Takeuchi, O., Aoyama, O., Oshima, R. et al. 2004b. Probing subpicosecond dynamics using pulsed laser combined scanning tunneling microscopy. *Appl. Phys. Lett.* 85: 3268–3270.

Takeuchi, O., Aoyama, M., Kondo, H. et al. 2006. Nonlinear dependences in pulse-pair-excited scanning tunneling microscopy. *Jpn. J. Appl. Phys.* 45: 1926–1930.

Terada, Y., Yoshida, S., Takeuchi, O. et al. 2010a. Laser-combined STM for probing ultrafast transient dynamics. *J. Phys. Cond. Mat.* 22: 264008–264015.

Terada, Y., Yoshida, S., Takeuchi, O. et al. 2010b. Real-space imaging of transient carrier dynamics by nanoscale pump–probe microscopy. *Nat. Photonics* 4: 869–874.

Terada, Y., Yoshida, S., Takeuchi, O. et al. 2011. Laser-combined scanning tunneling microscopy on the carrier dynamics in low-temperature-grown GaAs/AlGaAs/GaAs. *Adv. Opt. Tech.* 2011: 510186–510196.

Wang, K., Mittleman, D. M., van der Valk, N. C. J. et al. 2004. Antenna effects in terahertz apertureless near-field optical microscopy. *Appl. Phys. Lett.* 85: 2715–2717.

Wimmer, L., Herink, G., Solli, D. R. et al. 2014. Terahertz control of nanotip photoemission. *Nat. Photonics* 10: 432–436.

Wintterlin, J., Trost, J., Renisch, S. et al. 1997. Real-time STM observations of atomic equilibrium fluctuations in an adsorbate system: O/Ru(0001). *Surf. Sci.* 394: 159–169.

Wisendanger, R. 1994. *Scanning Probe Microscopy and Spectroscopy*. Cambridge: Cambridge University Press.

Wu, S. W. and Ho, W. 2010. Two-photon-induced hot-electron transfer to a single molecule in a scanning tunneling microscope. *Phys. Rev. B* 82: 085444–085451.

Yamashita, M., Shigekawa, H., and Morita, R. 2005. *Mono-Cycle Photonics and Optical Scanning Tunneling Microscopy-Route to Femtosecond Angstrom Technology*. Berlin, Heidelberg, Germany: Springer.

Yokota, M., Yoshida, S., and Mera, Y. 2013. Bases for time-resolved probing of transient carrier dynamics by optical pump-probe scanning tunneling microscopy. *Nanoscale* 5: 9170–9175.

Yoshida, S., Kanitani, Y., Oshima, R. et al. 2007. Microscopic basis for the mechanism of carrier dynamics in an operating p−n junction examined by using light-modulated scanning tunneling spectroscopy. *Phys. Rev. Lett.* 98: 026802–026805.

Yoshida, S., Kanitani, Y., Takeuchi, O. et al. 2008. Probing nanoscale potential modulation by defect-induced gap states on GaAs(110) with light-modulated scanning tunneling spectroscopy. *Appl. Phys. Lett.* 92: 102105–102107.

Yoshida, S., Terada, Y., Oshima, R. et al. 2012. Nanoscale probing of transient carrier dynamics modulated in a GaAs–PIN junction by laser-combined scanning tunneling microscopy. *Nanoscale* 4: 757–761.

Yoshida, S., Terada, Y., Yokota, M. et al. 2013a. Direct probing of transient photocurrent dynamics in p-WSe$_2$

by time-resolved scanning tunneling microscopy. *App. Phys. Exp.* 6: 016601–016604.

Yoshida, S., Yokota, M., Takeuchi, O. et al. 2013b. Single-atomic-level probe of transient carrier dynamics by laser-combined scanning tunneling microscopy. *App. Phys. Exp.* 6: 032401.

Yoshida, S., Aizawa, Y., Wang, Z. et al. 2014. Probing ultrafast spin dynamics with optical pump-probe scanning tunnelling microscopy. *Nat. Nanotechnol.* 9: 588–593.

Yoshida, S., Hirori, H., Tachizaki, T. et al. 2019. Sub-cycle transient scanning tunneling spectroscopy with visualization of enhanced terahertz near-field. *ACS Photonics.* 6: 1356–1364.

Yoshioka, K., Katayama, I., Minami, Y. et al. 2016. Real-space coherent manipulation of electrons in a single tunnel junction by single-cycle terahertz electric fields. *Nat. Photonics.* 10: 762–765.

Zhang, A. P., Rowland, L. B., Kaminsky, E. B. et al. 2003. Correlation of device performance and defects in AlGaN/GaN high-electron mobility transistors. *Eastman J. Electron. Mater.* 32: 388–394.

Triggering Chemical Reactions by Scanning Tunneling Microscopy

Johannes A.A.W. Elemans
Radboud University

4.1 Introduction

For more than three decades, the scanning tunneling microscope (STM) has proven itself as an indispensable tool to image solid surfaces at the highest detail possible, as well as atoms and molecules adsorbed on them. Since its invention in 1981 (Binnig & Rohrer 1982), the STM has had a huge impact on scientific research, because of its ability to image matter at sub-nanometer resolution. While in its early days, STM was mainly used by physicists to investigate the atomic structure of metals and alloys, often under extreme conditions of ultrahigh vacuum (UHV) and ultralow temperatures; nowadays, the technique is much more widely applied. It is suitable to image large molecules adsorbed to a surface, and it is no longer restricted to extreme environments: it has become fairly straightforward to obtain atomic scale images of matter at room temperature, in air, under high gas pressures, or in liquids (Elemans & De Feyter 2009).

The principle of operation of an STM is relatively simple (Figure 4.1). It consists of an atomically sharp, conductive tip, which is brought at nanometer distance from an atomically flat conductive surface. At sufficiently close distance of these electrodes, electrons are able to tunnel from the tip to the sample, or vice versa. When a bias voltage (typically between 0 and 1 V) is applied between tip and sample, a net tunneling current (typically between 1 pA and 1 nA) will start flowing, which can be measured. This current is highly dependent on the distance between the electrodes, making it also very sensitive to height differences between the tip and the sample. When the tip is mounted into a piezo tube, this height can be very accurately controlled, even at sub-Ångstrom range. When subsequently the surface is scanned while keeping the tip at a constant height, height differences on the surface as small as atoms will cause changes

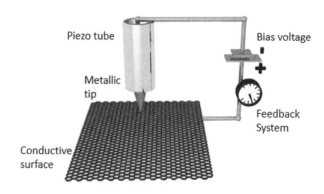

FIGURE 4.1 Schematic representation of an STM setup. (Reprinted with permission from den Boer et al., 2016. *Eur. Polymer J.* 83: 390–406. Copyright (2016) by Elsevier.)

in tunneling current, and thus, a topography map at atomic resolution can be obtained, at least in principle. As an alternative to constant height scanning, often a scanning mode is applied that keeps the current constant instead of the height. This means that during scanning, the distance between the tip and sample is continuously adjusted, a process that is controlled by a feedback mechanism. What results is a map of apparent heights, which is equally detailed as that obtained by scanning in the constant height mode, but generally safer, since tip crashes due to direct contact between tip and sample are prevented.

The interpretation of STM images is not always straightforward. Apparent height images are a convolution of 'real height' but also of the conductive properties of a surface and adsorbents on top of it. This means that when the tunneling current is also used as a feedback, a more conductive part of a surface may appear higher in an STM image than it is topographically. Despite this duality, it also enables the additional use of STM as a tool to analyze the local density of states (LDOS) of a sample. Or, in case molecules are adsorbed, STM can image different orbitals or detect variations in oxidation state of metal centers.

In the first decades of its existence, STM was mainly used to investigate surfaces, atoms, and molecules in a static state. Later on, it turned out that it can also serve as an excellent tool to study dynamic phenomena in real time. In this regard, the imaging of chemical reactivity with STM has become a very popular topic, in particular, in the chemistry community (Elemans et al. 2009; Münninghoff & Elemans 2017). The ability to image starting materials, products and intermediates of a reaction, all at the submolecular scale, is highly appealing. In addition, it may provide highly detailed information about a reaction mechanism at a single-molecule level, which may be hidden at the ensemble level. The importance of such research was highlighted in 2007 by the awarding of the Nobel Prize to Gerhard Ertl "for his studies of chemical processes on solid surfaces".

In addition to just observing and imaging a chemical reaction at a surface, an STM device can also be used to control or trigger a reaction. With an STM tip, atoms and molecules can be moved mechanically along a surface, bringing them in close proximity for a reaction to occur. It can also be used to inject electrons into chemical bonds, which then can break or form. Finally, applying simply a voltage difference between tip and sample may induce reactivity, since an STM can be considered as a (very nonstandard) electrochemical device.

This overview will focus on the triggering of chemical reactions with STM. It is divided in three sections, each describing a different environment in which such manipulations can be carried out: (i) in UHV, an ultraclean environment in which all possible contaminations can be eliminated, and the dynamics of atoms and molecules can be slowed down tremendously because of the possibility to work at ultralow temperatures (4 K or lower); (ii) under electrochemical control in polar liquids and electrolytes, in which additional electrodes in the STM setup allow full control over the electrochemical potential between tip and surface; and (iii) in apolar liquids or air, in which typically close-packed monolayers of molecules are subjected to chemical reactions under relatively mild conditions that are also encountered in a laboratory environment.

4.2 Triggering Reactions with STM in Ultrahigh Vacuum

A chemical reaction generally starts with the making or the breaking of a chemical bond. Pioneering work in which an STM tip was used to break chemical bonds of molecules adsorbed to a surface was carried out by Avouris and coworkers in 1992 (Dujardin et al. 1992). They scanned a silicon surface, onto which decaborane molecules were adsorbed, while applying a large voltage of ∼8 V between the STM tip and the surface. As a result, a large current density is delivered to the sample, corresponding to about

10^6 electrons per second passing to each atomic site of the silicon lattice. When these high-energy electrons passed through adsorbed decaborane molecules, many of these were observed to split into smaller fragments. The concept of breaking chemical bonds by the STM tip has subsequently been applied in a wide variety of dissociation reactions of molecules in UHV, such as acetylene (Gaudioso et al. 1999), pyridine and benzene (Laubon & Ho 2000b), and molecular oxygen (Martel et al. 1996; Stipe et al. 1997). Also dehydrogenation (the dissociation of hydrogen atoms) of molecules has been accomplished at the single-molecule level by STM, for example of ammonia (Maier et al. 2015) and *trans*-2-butene (Kim et al. 2002).

Dissociation and dehydrogenation reactions on a surface can be performed with an extreme level of control, as has been demonstrated by the groundbreaking work of the group of Ho. They were able to trigger chemical reactions at the same, single molecule, by applying highly local voltage pulses by the STM tip. In this way, an ethylene molecule on a Ni(110) surface was converted by a first voltage pulse to acetylene and then by a second pulse to single carbon atoms (Gaudioso et al. 1999). In follow-up work, single hydrogen or deuterium atoms could be removed stepwise from an acetylene molecule adsorbed to Cu(100), via the injection of low-energy electrons, to first yield ethynyl and then dicarbon (Figure 4.2) (Lauhon & Ho 2000a).

Compared to bond breaking, bond formation using an STM tip has been described much less frequently. Ho and coworkers deposited CO molecules that were connected to an STM tip on top of an iron atom on a Ag(110) surface and formed Fe(CO) and, subsequently, $Fe(CO)_2$ (Lee & Ho 1999). Similarly, they were able to oxidize CO to CO_2 on the same surface, by depositing a CO molecule from the tip onto an adsorbed oxygen atom (Hahn & Ho 2001).

Chemical bond dissociation and formation does not have to be limited to isolated single molecules but may have an effect on neighboring molecules. Yates and coworkers positioned dimethyldisulfide molecules into close-packed linear chains on an Au(111) surface and triggered the dissociation of an S–S bond of the first molecule in the chain by means of a voltage pulse by the STM tip (Maksymovych et al. 2008). As a consequence, a domino-type cascade of chemical reactions occurred, involving S–S bond recombinations and dissociations along the chain of molecules.

In addition to triggering chemical reactions very locally, the injection of 'hot electrons' by the STM tip into a

FIGURE 4.2 STM images and corresponding schematic representations of (a) bis-deuteroacetylene, (b) deuteroethynyl and (c) dicarbon. (Reprinted with permission from Lauhon et al., 2000. *Phys. Rev. Lett.* 84: 1527–30. Copyright (2000) by the American Physical Society.)

substrate can lead to a lateral travelling of these electrons through the substrate, which can subsequently give rise to chemical reactions in molecules that are adsorbed at remote sites (Macleod et al. 2009). Following this approach, the S–S bonds of thousands of the same dimethyldisulfide molecules as described above, but adsorbed at an Au(111) surface at a distance of up to 100 nm from the STM tip, could be dissociated by a single 2.5 V pulse into the bare metal substrate (Maksymovych et al. 2007).

Not unexpectedly, the nature of the substrate can play an important role in such nonlocal triggering of reactions. For example, the nonlocal dissociation of C–F bonds in copper hexadecafluorophthalocyanine molecules triggered by the injection of high-energy electrons into the substrate was two orders of magnitude more efficient on Ag(111) than on Au(111) (Chen et al. 2009). The difference was explained by a larger overlap of the orbitals of the benzene rings of the phthalocyanines with gold, when compared to silver. As a result, when a hot electron is injected into the benzene π^*-orbital to form an excited anion state, its lifetime is shorter on gold than on silver, which lowers the chance of a dissociation reaction in the former case.

Also chemical bond formation reactions can be triggered nonlocally. When very long voltage pulses of about 30 s were applied to a double layer of C_{60} molecules on a silicon surface, ring-like features composed of polymerized C_{60} were formed around the pulse position (Figure 4.3) (Nouchi et al. 2006). The ring shape was explained by assuming that charge carriers can induce both polymerization and depolymerization reactions between the C_{60} molecules, in which the depolymerizations require a higher energy. After the voltage pulse, the charge carriers start travelling through the substrate, away from the tip in a radial fashion. Initially, they have a high energy and the breaking of bonds dominates the making of bonds. But along their path they lose energy, and after a while, polymerization becomes more favorable than depolymerization. When they travel even further, their energy eventually becomes so low that they can no longer induce either of the reactions. A second voltage pulse, of higher energy, turned out to increase the

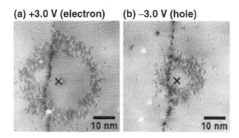

(a) +3.0 V (electron)   **(b) –3.0 V (hole)**

FIGURE 4.3 STM images of the formation of rings of polymerized C_{60} molecules upon application of (a) positive (injection of electrons) or (b) negative (injection of holes) voltage pulses to a double layer of C_{60} molecules adsorbed onto a Si(111) − 7 × 7 surface. (Reprinted with permission from Nouchi et al., 2006. *Phys. Rev. Lett.* 97: 196101. Copyright (2006) by the American Physical Society.)

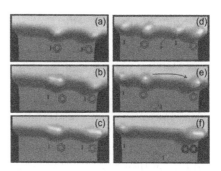

FIGURE 4.4 All steps of the Ullmann reaction induced and imaged by STM. (a) STM image of two iodobenzene molecules adsorbed at a step edge of Cu(111) (molecular structures are drawn in the image). (b–c) Dissociation of the iodobenzene molecules into phenyl radicals and iodine atoms. (d) Picking up and moving away an iodine atom with the tip. (e) Lateral movement of the phenyl radicals along the step edge. (f) Injection of tunneling electrons via a voltage pulse, making a covalent bond between the phenyl fragments. (Reprinted with permission from Hla et al., 2000. *Phys. Rev. Lett.* 85: 2777–80. Copyright (2000) by the American Physical Society.)

inner diameter of the ring, which supported this hypothesis. When holes were injected instead of electrons, smaller rings were formed, which was explained by the fact that holes dissipate their energy faster when travelling through the substrate.

The group of Rieder was the first who was able to trigger and image all individual steps of a multistep chemical reaction with the STM tip (Hla et al. 2000, 2001). They investigated an Ullmann reaction, in which two iodobenzene molecules are coupled to biphenyl at the step edge of a Cu(111) surface in UHV at 20 K (Figure 4.4). First, the two iodobenzene molecules were dissociated into iodine atoms and benzene radicals, by injecting tunneling electrons into the molecules via highly local voltage pulses with the tip. Next, the fragments were moved along the step edge in a controlled fashion by the same tip, and in the final step, two adjacently positioned benzene radicals were covalently connected, again by applying a voltage pulse via the tip. The success of the coupling reaction was proven by the ability to laterally move the biphenyl molecules along the surface with the tip.

4.3   Triggering Reactions with STM under Electrochemical Control in Polar Liquids

In the previous section, it was demonstrated that by means of voltage pulses that are applied via the STM tip, chemical reactions can be triggered in single molecules or groups of molecules very locally. However, when it is desired to induce reactions in larger groups, or even complete monolayers of molecules, the relatively local nature of a voltage pulse is inefficient. An alternative approach, which can address layers of adsorbed molecules on a much larger scale, is to use an electrochemical STM (EC-STM, also referred to as *in situ* STM). It typically operates in aqueous solutions or in organic solvents containing an electrolyte. The STM tip needs to be isolated in order to prevent leakage current effects, which is usually accomplished by coating it with an insulating polymer film. The STM bias voltage is not directly applied between tip and sample, but their potentials are set relative to a so-called reference electrode, which has a known redox potential. The electrochemical potentials of sample and tip can then be deduced by comparing them to the known potential of the reference electrode. Furthermore, a counter electrode is present, which supplies the electrons that are required to drive a chemical reaction that takes place at the sample (or sometimes at the tip). With the setup, ultimate control can be obtained over the potential of the surface. Thus, an EC-STM can be used as an electrochemical cell equipped with a microscope, which can trigger processes that occur at the electrode/liquid interface and image these processes at molecular or atomic scale. An EC-STM is therefore also an ideal tool to detect, identify, and manipulate molecules that can change their oxidation state.

With EC-STM, chemical reactions can be triggered in adsorbed molecules across an entire surface, by setting its electrochemical potential at a certain, desired value. In this way, for example, aniline molecules adsorbed at the interface of an Au(111) surface and an aqueous 0.1 M sulfuric acid solution could be oxidatively polymerized by increasing the electrochemical potential of the surface (Figure 4.5) (Yang et al. 2007). The polymerization process was evident from the STM images, in which regular patterns of aniline monomers were gradually converted to extended wire-like structures, corresponding to the polyaniline chains. Within these chains, the monomeric components, which were oriented in a zig-zag pattern, could still be resolved (Figure 4.5b).

In contrast to the local nature of bias voltage pulsing via the STM tip, in an EC-STM also nonlocal voltage pulses of the electrochemical potential can be applied, which can influence large parts of a surface, the molecules adsorbed thereon, and molecules dissolved in the supernatant solution. Following this approach, 3-butoxy-4-methylthiophene (**BuOMT**) could be electropolymerized at the interface of Au(111) and an electrolyte consisting of a dichloromethane solution containing molecular iodine and NBu_4PF_6 (Sakaguchi et al. 2004). When electrochemical voltage pulses of 1.4 V were applied to the system, STM images showed the formation of well-defined linear polythiophene wires, which were organized epitaxially along the three-symmetry axes of the underlying gold substrate (Figure 4.6a). Both the surface coverage and the lengths of the polymeric wires increased when more pulses were applied (Figure 4.6b). Polymers with lengths up to 75 nm, corresponding to a polymerization degree of $\sim$200, were observed. A mechanistic study of the consecutive reaction steps (Figure 4.6c) revealed that the presence of iodine in the solution was essential for both the formation and the adsorption of the polythiophenes. After the **BuOMT** monomers had been dissolved in the electrolyte solution, UV-vis spectroscopy revealed the formation of trimeric thiophene species, presumably via an oxidation reaction mediated by the co-dissolved I_2. A subsequent voltage pulse of 0.77 V oxidized these trimers to radical cations, which were proposed to act as polymerization nuclei. These adsorbed at the Au(111) surface, aided by an adsorbed adlayer of hexagonally organized iodine atoms. Subsequently, propagation occurred from the nuclei and thiophene chains grew along the solid/liquid interface. In follow-up work, thiophene-based block copolymers could be synthesized by following a similar approach (Sakaguchi et al. 2005). First, voltage pulses induced the polymerization of 3-octyl-4-methylthiophene (**C8MT**) at the I_2-covered Au(111) surface. Next, this surface was placed in another electrolyte solution, containing the structurally related monomer 3-octyloxy-4-methylthiophene (**C8OMT**) and again voltage pulses were applied. These activated both the **C8OMT** in solution, and the termini of the **C8MT** polymers at the surface. As a result, heteropolymeric

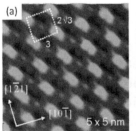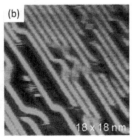

FIGURE 4.5 (a) STM image of aniline molecules adsorbed at the interface of Au(111) and aqueous 0.1 M H_2SO_4 at an electrochemical potential of 0.9 V (versus a reversible hydrogen electrode). (b) Appearance of polyaniline chains when the electrochemical potential is raised to 1.05 V. (Reprinted with permission from Yang et al., 2007. *J. Am. Chem. Soc.* 129: 8076–7. Copyright (2007) by the American Chemical Society.)

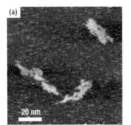

FIGURE 4.6 (a) EC-STM image of polymers formed from **BuOMT** at the interface of Au(111) and an electrolyte consisting of a dichloromethane solution containing molecular iodine and NBu_4PF_6 after five electrochemical voltage pulses of 1.4 V for a period of 150 ms. (b) Idem, after 13 pulses. (c) Schematic representation of the proposed mechanism. (Reprinted with permission from Sakaguchi et al., 2004. *Nat. Mater.* 3: 551–7. Copyright (2004) by Springer Nature.)

wires were formed, in the sense that, **C8OMT** wire fragments propagated from the **C8MT** wires already present at the surface. The two polymer types could be clearly identified because of their difference in brightness in the STM images. When both the **C8MT** and **C8OMT** monomers were present in solution and voltage pulses were applied, only multiblock heteropolymers were formed and no random copolymers. This observation indicates that preferentially homopolymers are formed, which are occasionally covalently linked via a heterojunction to form a block copolymer.

An EC-STM is also a suitable tool to control the electronic properties of adsorbates with well-defined redox states. In this regard, porphyrins and their related derivatives are interesting molecules. They can complex a large variety of transitional metals, which may exist in a range of oxidation states. In addition, also the free porphyrin ligand displays a rich redox chemistry. Borguet and coworkers investigated the redox behavior of monolayers of *meso*-tetrapyridyl porphyrins adsorbed at the interface of Au(111) and aqueous 0.1 M H_2SO_4 (He & Borguet 2007). When a slightly negative electrochemical potential (−0.1 V versus a saturated calomel electrode) was applied to the surface, these molecules adsorbed to the surface in a reduced state in which two electrons were added to the porphyrin ring (Ye et al. 2006). In the corresponding STM images, all the porphyrins had a similar, bright appearance. When subsequently a short electrochemical voltage pulse of 0.3 V was applied to the system, dark signatures appeared in the STM image. When the pulses had a longer duration or a larger magnitude, the population of these dark signatures increased, and for that reason, they were assigned to oxidized porphyrin species.

Just like porphyrins, structurally related phthalocyanine molecules often readily self-assemble into well-defined arrays at a surface. When adsorbed at the interface of Au(111) and an aqueous 0.1 M perchloric acid solution in an EC-STM setup, iron phthalocyanines (**FePc**s, Figure 4.7) appeared in two distinct contrasts, dark and bright, in the STM images, when the surface potential exceeded 0.3 V (Gu et al. 2016). The abundance of these contrasts was strongly dependent on the environment: under an inert nitrogen atmosphere, mostly dark signatures were present, whereas under an oxygen atmosphere, the bright signatures were

dominant. Based on these observations and on theoretical calculations, the dark signatures were assigned to native **FePc** molecules, while the bright ones were identified as **FePc**-O_2 complexes. Upon lowering the surface potential under oxygen atmosphere, the bright signatures started to convert to dark ones, which was attributed to the initiation of an oxygen reduction reaction ($O_2 + 4H^+ + 4e \rightarrow H_2O$) and the concomitant dissociation of O_2 from the **FePc**-O_2 complexes. The process was found to be fully reversible: upon increasing the surface potential, the bright signatures returned. As expected, under a nitrogen atmosphere, these signature changes were not observed. Further analysis of the **FePc**-covered electrode surface by cyclic voltammetry revealed a shift in the offset potential for the oxygen reduction reaction, which indicates that the monolayer of **FePc** molecules is catalytic.

4.4 Triggering Reactions with STM in Nonpolar Liquids or in Air

In addition to UHV or polar liquid environments, nonpolar organic liquids atop solid substrates have become increasingly popular environments to investigate self-assembly and functionality of molecules (Elemans et al. 2009). And also in organic liquids, or even in air, chemical reactions can be triggered locally by the STM tip or more globally by adjusting the bias voltage between tip and surface. Solid/liquid interfaces are generally excellent environments to construct closely packed (2D crystal-like) arrays of organic molecules. Via rational synthetic design, such molecules can express a certain function or reactivity. For example, molecules with polymerizable groups, such as alkenes or alkynes, can be organized within a monolayer such that these groups are in each other's proximity. They can then be subjected to a so-called 'topochemical polymerization', as a result of which covalently connected 2D networks are formed. Such polymerizations can be initiated by irradiating the layer by light of a certain wavelength (Grim et al. 1997; Miura et al. 2003; Okawa 2001a, b) or by applying a local voltage pulse. As an example of the latter method, diacetylene moieties were incorporated in a long alkane chain containing a terephthalic acid group, and these molecules were self-assembled via drop casting into lamellar arrays at a graphite surface (Figure 4.8) (Miura et al. 2003). Within the arrays, the acetylene functions were perfectly aligned, and when during scanning a voltage pulse was applied, their polymerization was initiated. The success of the reaction was apparent from the follow-up STM scan, which clearly showed the presence of a bright line along the lamellar axis, corresponding to a highly electron-conducting polydiacetylene chain. The length of these chains could be controlled elegantly by creating defects, which acted as chain termination points, with the STM tip into the self-assembled monolayer.

At a solid/liquid interface of graphite and 1-phenyloctane, related diacetylene-functionalized molecules did not only

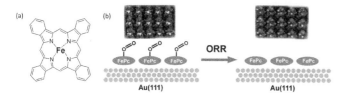

FIGURE 4.7 (a) Structure of **FePc**. (b) STM images and schematic representation of the oxygen reduction reaction mediated by **FePc** and imaged in the EC-STM setup. (Reprinted with permission from Gu et al., 2016. *ACS Nano* 10: 8746–50. Copyright (2016) by the American Chemical Society.)

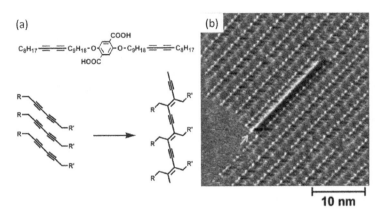

FIGURE 4.8 (a) Structure of the compound containing diacetylene functions (top) and a representation of the polymerization of the diacetylenes (bottom). (b) STM image of a self-assembled monolayer of these diacetylene compounds at the graphite/air interface. The arrow indicates the location where a voltage pulse was applied and from which a bright polydiacetylene chain has grown. (Reprinted with permission from Miura et al., 2003. *Langmuir* 19: 6474–82. Copyright (2003) by the American Chemical Society.)

form self-assembled monolayers but also bilayers and additional layers on top of that (Takajo et al. 2007). Compared to the domains of the first monolayer, the domains of the additional layers grew very large (up to micrometers) because they interacted much weaker with the first monolayer than the first monolayer with the underlying graphite substrate. When a voltage pulse was applied to the system, remarkably, only the diacetylene molecules in the top layer were topochemically polymerized, while the molecules in the bottom layer remained unaffected. This difference in reactivity was explained by assuming a model in which the formed polydiacetylene backbone is lifted from the alkyl substituents by 0.14 nm. Such a lifting is possible in the top layer but not in the bottom layer(s) due to steric confinement.

By means of topochemical polymerization, also other types of potentially functional molecules could be attached covalently to a self-assembled monolayer, in a process named 'chemical soldering'. When, for example, on top of a layer of diacetylene molecules, a small cluster of phthalocyanine molecules was deposited, polymerization of the diacetylenes by a local voltage pulse led to the growth of a polymer chain, which reactive chain terminus linked covalently to a phthalocyanine it encountered at its path of polymerization (Okawa et al. 2012). In a similar way, polydiacetylene chains have been connected to C_{60}-molecules (Nakaya et al. 2014).

As was observed at interfaces of solid and aqueous or organic electrolytes in an EC-STM, also at the interface of a solid and an organic liquid, a predefined setting of the bias voltage between sample and tip can induce reactivity in molecules adsorbed at that interface. An example of this approach is the triggering of multistep redox reactions of manganese(III)porphyrins adsorbed at a graphite/1-octanoic acid interface (Figure 4.9) (den Boer et al. 2013).

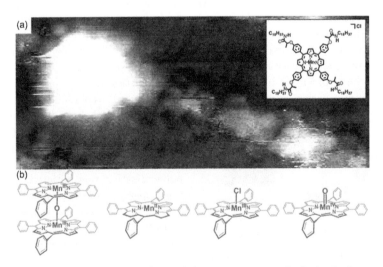

FIGURE 4.9 (a) STM image of a self-assembled monolayer of the manganese porphyrin in the inset at a graphite/1-octanoic acid interface. Four different signatures are observed, which correspond to the porphyrin species that are shown directly underneath as the structures in panel (b). (Reprinted with permission from den Boer et al., 2016. *Eur. Polymer J.* 83: 390–406. Copyright (2016) by Elsevier.)

Manganese porphyrins are capable to complex molecular oxygen and subsequently activate and dissociate the strong oxygen–oxygen bond. For that reaction to occur, however, first a reduction of the manganese center from Mn(III) to Mn(II) has to take place. It turned out that if the graphite surface in the STM setup was sufficiently negatively biased (below −700 mV), this reduction spontaneously occurred and the Mn(II)porphyrins appeared as darker signatures than the Mn(III)porphyrins in the STM images. After their reduction, these Mn(II)porphyrins appeared to be highly reactive for oxygen binding and dissociation, resulting in the formation of Mn(IV)=O species that appeared even brighter than the Mn(III)porphyrins. Interestingly, the unique spatial resolution of STM revealed that the Mn(IV)=O species were preferentially generated in pairs of two, at adjacent locations at the surface. From this observation, it was concluded that they were the product of a homolytic dissociation of O_2, followed by distribution of the two oxygen atoms over neighboring manganese porphyrins. In the absence of molecular oxygen in the system, or at bias voltages less negative than −700 mV, none of these signature changes were observed. Occasionally, also very large, bright signatures were observed, in particular at increased concentrations of manganese porphyrins in the supernatant solution. These signatures were identified as Mn porphyrin μ-oxo sandwich-type dimers, formed by coupling of an additional Mn porphyrin from solution to Mn(IV)=O complex at the surface. The formation of these μ-oxo-bridged dimers could also be directly triggered, in a highly local fashion, by applying a voltage pulse of several volts via the STM tip to the monolayer (Li et al. 2014). During this pulse, a strong inhomogeneous field is generated that can induce (local) charges in polarizable particles, which make them move to the region of highest field density, i.e., the tunneling gap between tip and surface. This leads to an enhanced local concentration, which is assumed to favor the close by formation of the μ-oxo-bridged dimers at the surface. In a radius of approximately 10 nm around the location of the pulse, none of these dimers were present as a result of desorption of part of the monolayer due to pulse and subsequent re-adsorption of non-reacted molecules from the supernatant.

In a similar approach, the group of Hipps reported the complexation of molecular oxygen to cobalt porphyrins adsorbed at the interface of HOPG and 1-phenyloctane and also showed that once adsorbed at a surface reactive, molecules can behave quite differently compared to when they are in solution (Friesen et al. 2012). Cobalt(II) octaethylporphyrin has been shown to bind O_2 in solution only at low temperatures or when an electron-donating axial ligand is coordinated to the porphyrin metal center. At the solid/liquid interface, the metal porphyrins appeared in two different signatures in the STM images, bright and dark. The bright signatures were identified as the native CoOEP complexes and the darker ones as CoOEP–O_2 adducts. This assignment was based on an STM experiment conducted in argon atmosphere (i.e., oxygen-deficient),

in which the bright structures predominantly covered the surface. Further evidence for the assignment of the dark topographies came from varying the partial pressure of oxygen in the system, which had direct impact on the relative population of the various signatures. In fact, the number of dark signatures at the surface exactly followed the Langmuir absorption isotherm. In a follow-up experiment, the partial oxygen pressure was fixed, the temperature of the system varied, and the relative abundance of CoOEP–O_2 species at each set temperature was analyzed. From these data, the enthalpy and entropy of the oxygen binding reaction could be derived as being $\Delta H = 68 \pm 10$ kJ/mol and $\Delta S = 297 \pm 30$ J/mol K. These values indicate that adsorption of the cobalt porphyrins to the (biased) surface triggers strong complexation of O_2. This unexpected behavior was attributed to the surface acting as an electron-donating reservoir that stabilizes the polarized cobalt-oxygen bond, much alike the stabilization by electron-donating axial ligands that coordinated to the cobalt porphyrins in solution. Similar axial-ligand-like behavior of the biased surface was observed for the coordination behavior of the nickel analogue of CoOEP NiOEP (Nandi et al. 2016). While this metal complex does not axially coordinate nitrogen-based ligands in solution, it readily binds them when it is adsorbed at the 1-phenyloctane/HOPG interface. The coordination of a ligand like imidazole is favored because it behaves as a π-acceptor of charge that is donated, via the nickel porphyrin, from the biased surface.

4.5 Conclusion and Outlook

This overview has shown that an STM is not only an excellent tool to image atoms and molecules but also to manipulate them in chemical reactions. By injecting high-energy electrons via a voltage pulse, chemical bonds can be broken or formed in a very local fashion. Following this concept, quite some examples of polymerization reactions between molecules have already been described, in a variety of environments. Alternatively, when high-energy electrons (or holes) are injected into the surface, they can travel over distances of hundreds of nanometers to trigger reactions at a remote location. And when one has the ability to control the electrochemical potential, by using a dedicated EC-STM setup, the reactivity of atoms or molecules of an entire monolayer on a surface can be triggered. In addition to providing insight into reaction mechanisms, at the highest detail possible, STM can also be utilized to manipulate reactivity. This ability may open the way to the bottom-up chemical design of molecular surfaces. Nowadays, nearly every desired molecule can be synthesized, and already to a large extent, the self-assembly of molecules on a surface over large distances can be predicted and controlled (Elemans 2016). By employing relatively weak but directional supramolecular interactions between molecular building blocks, specific and potentially functional patterned surfaces can be obtained. The disadvantage of such organized layers, however, is often their relatively weak adsorption to the underlying substrate. As a result, they can be easily disrupted by external factors such as increases in temperature or the presence of solvents. To solve these problems, the covalent connection of molecules within self-assembled monolayers, for example, by polymerization triggered by the STM tip may increase their robustness and applicability.

Despite its potential, the use of STM to manipulate reactions still comes with some challenges. In particular, when high-energy voltage pulses are applied as a trigger, undesired side effects may occur: molecules can, for example, decompose or desorb from the surface. This probably implies that for every type of reaction or set of molecules, specific reaction conditions have to be found to provide the desired effect. One must realize that currently the investigation of chemical reactions with STM, including their triggering, is still a research area that is very much in its infancy. However, given the importance of functional surfaces in everyday life, it can be expected that obtaining fundamentally new insights in their structure and reactivity, at the highest detail possible, may lead to the future rational design of new materials and catalysts.

References

Binnig, G. and Rohrer, H. 1982. Tunneling though a controllable vacuum gap. *Appl. Phys. Lett.* 40: 178–80.

Chen, L., Li, H. and Wee, A. T. S. 2009. Nonlocal chemical reactivity at organic-metal interfaces. *ACS Nano* 3: 3684–90.

den Boer, D., Li, M., Habets, T. et al. 2013. Detection of different oxidation states of individual manganese porphyrins during their reaction with oxygen at a solid/liquid interface. *Nat. Chem.* 5: 621–7.

Dujardin, G., Walkup, R. and Avouris, P. 1992. Dissociation of individual molecules with electrons from the tip of a scanning tunneling microscope. *Science* 255: 1232–5.

Elemans, J. A. A. W. 2016. Externally applied manipulation of molecular assemblies at solid-liquid interfaces revealed by STM. *Adv. Funct. Mater.* 26: 8932–51.

Elemans, J. A. A. W. and De Feyter, S. 2009. Structure and function revealed with submolecular resolution at the liquid-solid interface. *Soft Matter* 5: 721–35.

Elemans, J. A. A. W., Lei, S. and De Feyter, S. 2009. Molecular and supramolecular networks on surfaces: from 2D crystal engineering to reactivity. *Angew. Chem. Int. Ed.* 48: 7298–332.

Friesen, B. A., Bhattarai, A., Mazur U. and Hipps, K. W. 2012. Single molecule imaging of oxygenation of cobalt octaethylporphyrin at the solution/solid interface: thermodynamics from microscopy. *J. Am. Chem. Soc.* 134: 14897–904.

Gaudioso, J., Lee, H. J. and Ho, W. 1999. Vibrational analysis of single molecule chemistry: ethylene dehydrogenation on Ni(110). *J. Am. Chem. Soc.* 121, 8479–85.

Grim, P. C. M., De Feyter, S., Gesquière, A. et al. 1997. Submolecularly resolved polymerization of diacetylene molecules on the graphite surface observed with scanning tunneling microscopy. *Angew Chem. Int. Ed. Engl.* 36: 2601–3.

Gu, J.-Y., Cai, Z.-F., Wang D. and Wan, L.-J. 2016. Single-molecule imaging of iron-phthalocyanine-catalyzed oxygen reduction reaction by *in situ* scanning tunneling microscopy, *ACS Nano* 10: 8746–50.

Hahn, J. R. and Ho, W. 2001. Oxidation of a single carbon monoxide molecule manipulated and induced with a scanning tunneling microscope. *Phys. Rev. Lett.* 87: 166102.

He, Y. and Borguet, E. 2007. Dynamics of porphyrin electron-transfer reactions at the electrode-electrolyte interface at the molecular level. *Angew. Chem. Int. Ed.* 46: 6098–901.

Hla, S.-W., Bartels, L., Meyer, G. et al. 2000. Inducing all steps of a chemical reaction with the scanning tunneling microscope tip: towards single molecule engineering. *Phys. Rev. Lett.* 85: 2777–80.

Hla, S.-W., Meyer, G. and Rieder, K.-H. 2001. Inducing single-molecule chemical reactions with a UHV-STM: a new dimension for nano-science and technology. *Chem. Phys. Chem.* 2: 361–6.

Kim, Y., Komeda, T. and Kawai, M. 2002. Single-molecule surface reaction by tunneling electrons. *Surf. Sci.* 502–503: 7–11.

Lauhon, L. J. and Ho, W. 2000a. Control and characterization of a multistep unimolecular reaction. *Phys. Rev. Lett.* 84: 1527–30.

Lauhon, L. J. and Ho, W. 2000b. Single-molecule chemistry and vibrational spectroscopy: pyridine and benzene on Cu(001). *J. Phys. Chem. A* 104: 2463–7.

Lee, H. J. and Ho, W. 1999. Single-bond formation and characterization with a scanning tunneling microscope. *Science* 286: 1719–22.

Li, M., den Boer, D., Iavicoli, P. et al. 2014. Tip-induced chemical manipulation of metal porphyrins at the liquid-solid interface. *J. Am. Chem. Soc.* 136: 17418–21.

MacLeod, J. M., Lipton-Duffin, J., Fu, C. et al. 2009. Inducing nonlocal reactions with a local probe. *ACS Nano* 3: 3347–51.

Maier, S., Stass, I., Feng, Y. et al. 2015. Dehydrogenation of ammonia on Ru(0001) by electronic excitations. *J. Phys. Chem. C* 119: 10520–5.

Maksymovych, P., Dougherty, D. B., Zhu, X.-Y. et al. 2007. Nonlocal dissociative chemistry of adsorbed molecules induced by localized electron injection into metal surfaces. *Phys. Rev. Lett.* 99: 016101.

Maksymovych, P., Sorescu, D. C., Jordan, K. D. et al. 2008. Collective reactivity of molecular chains self-assembled on a surface. *Science* 322: 1664–7.

Martel, R., Avouris, P. and Lyo, I.-W. 1996. Molecularly adsorbed oxygen species on Si(111)-(7 × 7): STM-induced dissociative attachment studies. *Science* 272: 385–8.

Miura, A., De Feyter, S., Abdel-Mottaleb, M. S. et al. 2003. Light- and STM-tip-induced formation of one-dimensional and two-dimensional organic nanostructures. *Langmuir* 19: 6474–82.

Münninghoff, J. A. W. and Elemans, J. A. A. W. 2017. Chemistry at the square nanometer: reactivity at liquid/solid interfaces revealed with STM. *Chem. Commun.* 53: 1769–88.

Nakaya, M., Okawa, Y., Joachim, C. et al. 2014. Nanojunction between fullerene and one-dimensional conductive polymer on solid surfaces. *ACS Nano* 8: 12259–64.

Nandi, G., Chilukuri, B., Hipps, K. W. and Mazur, U. 2016. Surface directed reversible imidazole ligation to nickel(II) octaethylporphyrin at the solution/solid interface: a single molecule level study. *Phys. Chem. Chem. Phys.* 18: 20819–29.

Nouchi, R., Masunari, K., Ohta, T. et al. 2006. Ring of C_{60} polymers formed by electron or hole injection from a scanning tunneling microscope tip. *Phys. Rev. Lett.* 97: 196101.

Okawa, Y. and Aono, M. 2001a. Linear chain polymerization initiated by a scanning tunneling microscope tip at designated positions. *J. Chem. Phys.* 115: 2317–22.

Okawa, Y. and Aono, M. 2001b. Materials science. Nanoscale control of chain polymerization. *Nature* 409: 683–4.

Okawa, Y., Akai-Kasaya, M., Kuwahara, Y. et al. 2012. Controlled chain polymerisation and chemical soldering for single-molecule electronics. *Nanoscale* 4: 3013–28.

Sakaguchi, H., Matsumura, H. and Gong, H. 2004. Electrochemical epitaxial polymerization of single-molecular wires. *Nat. Mater.* 3: 551–7.

Sakaguchi, H., Matsumura, H., Gong, H. et al. 2005. Direct visualization of the formation of single-molecule conjugated copolymers. *Science* 310: 1002–6.

Stipe, B. C., Rezaei, M. A., Ho, W. et al. 1997. Single-molecule dissociation by tunneling electrons. *Phys. Rev. Lett.* 78: 4410–3.

Takajo, D., Okawa, Y., Hasegawa, T. et al. 2007. Chain polymerization of diacetylene compound multilayer films on the topmost surface initiated by a scanning tunneling microscope tip. *Langmuir* 23: 5247–50.

Yang, L. Y. O., Chang, C., Liu, S. et al. 2007. Direct visualization of an aniline admolecule and its electropolymerization on Au(111) with *in situ* scanning tunneling microscope. *J. Am. Chem. Soc.* 129: 8076–7.

Ye, T., He, Y and Borguet, E. 2006. Adsorption and electrochemical activity: an *in situ* electrochemical scanning tunneling microscopy study of electrode reactions and potential-induced adsorption of porphyrins. *J. Phys. Chem. B* 110: 6141–7.

The Circular Mode AFM: A New Experimental Approach for Investigating Nano-Tribology

Olivier Noel and
Nguyen Anh Dung
Le Mans University

Pierre-Emmanuel Mazeran
University of Technology of Compiegne

5.1 Introduction

Macroscopic mechanisms may find clarifications[1] or new concepts[2-3] from investigations at the nanoscale. It is particularly the case for tribology as friction and wear in a macroscopic contact result from the global response of a huge population of micro-contacts generated by the surface roughness in the contact interface. Consequently, the elementary mechanisms, which are responsible for friction and wear, are intricate to identify and to understand. A way to explore friction mechanisms is to consider contact models. For example, friction in a mono-asperity contact is obviously easier to apprehend as the contact area and pressure can easily be estimated.

On the experimental side, the Surface Force Apparatus (SFA) invented by Tabor and Winterton[4] in 1969 is the first device measuring forces in a mono-asperity contact in between two flat surfaces in air or vacuum. Its resolution is up to the nN and the control of the separation distance in between the surfaces is to an accuracy of 1 Å. In 1988, the SFA was developed and implemented by Israelachvili et al.[5] for measuring lateral forces to get dynamic properties of thin liquid films confined in between two surfaces whose thickness was controlled with an accuracy of 1 Å. The SFA is certainly one of the most accurate techniques for the ones concerned by force resolutions and separation distance control to confine a lubricant and to assess its tribological properties for instance. The cost of such accuracy is to acquire specific skills to avoid interferences from external factors during the measurements, as the sensitivity of this technique is extreme. In addition, the interaction radius is sized in between 10 and 40 μm. Eventually, the technique also meets a severe limitation associated to the nature of the samples, which are often limited to atomically smooth surfaces such as mica.[6]

In the 1980s, Binnig, Quate and Gerber[7] have developed the atomic force microscope (AFM) for imaging the topography of material surfaces with an atomic resolution. The principle of the AFM imaging is based on the measurement with a high resolution of interactions in between a nanometer-sized tip attached at the end of a cantilever and the surface atoms of the sample. In the commercial AFM contact-mode imaging, the normal interaction forces bend the cantilever and a feedback loop maintains at a constant value the vertical deflection of the cantilever, during a linear back and forth scanning. In recent years, other modifications of original AFM were made to measure lateral forces by means of friction (or lateral) force microscopy[8] (FFM; LFM). The principle of LFM is to record the torsion of the cantilever while scanning and to compute afterward the friction force. The LFM has been intensely used for friction experiments. Nevertheless, LFM is designed for imaging. It measures the torsion of the cantilever, which is a combined response of the normal and friction forces during the back and forth scanning. With appropriate data treatments, it is possible to generate a friction image, i.e. the friction force as a function of the location. Anyhow, this image is of moderate interest for nano-tribology researches. Actually, nano-tribologists need to understand the behavior of friction force with experimental parameters or conditions, such as the normal load or the sliding speed to have a better understanding of friction mechanisms. Obviously, the LFM mode may help to get some information, but it requires acquiring many time-consuming LFM images.

This chapter introduces a new experimental setup based on the AFM to investigate nano-tribological properties: the patented circular AFM mode.[9] It is demonstrated that this new AFM mode could be considered as a technological breakthrough in terms of practicability, accuracy and potentiality, for a new approach to face challenges in nano-tribology. It is complementary to the classical LFM modes, which is usually implemented for measuring tribological properties, with its advantages and its limitations. Some applications of the circular AFM mode are also reported.

5.2 Principle of the LFM and Limitations

5.2.1 Principle of the LFM

The LFM applied to nano-tribology consists in imposing a linear back and forth relative displacement of the contact in between the AFM tip and the sample, while applying a normal load on the contact.[10] During the single line scan, the lateral force signal related to the torsion of the cantilever is registered (Figure 5.1a). Then, it gives the trace and retrace lateral force signals from which the friction force can be computed (Figure 5.1b). It also gives access to the dissipated energy (area of the curve for one cycle) and the contact lateral stiffness.

By varying the normal load on the contact, it is possible to draw the friction law curve that is the evolution of the friction force with load (Figure 5.2). The friction laws are of interest to determine a friction coefficient if the friction mechanism follows an Amontons' law[11] or to have a better understanding of the friction behavior. However, such curve is fastidious and time consuming to obtain with a high accuracy.

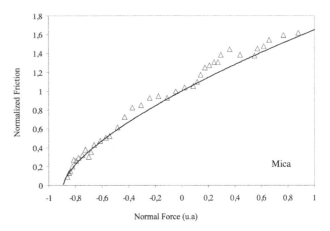

FIGURE 5.2 Friction law in between a nitride silicon AFM tip and a mica surface[12]: Normalized friction (i.e "friction force/friction force for a zero load") vs. the normal load. The black line is the fit of the experimental data by using the Derjaguin, Muller, and Toporov (DMT) theory.[13]

The AFM was originally dedicated to imaging. Therefore, the LFM mode used for nano-tribology is dependent on certain technical limitations.

5.2.2 Limitations

One of the main limitations of the LFM is the linear back and forth displacement, which induces stop periods when the friction scan changes its direction (Figure 5.3). Consequently, the sliding speed is not constant during the scan and the friction force measurements are done in a nonstationary regime. Besides, with LFM imaging mode, friction may be obtained by using a time-consuming data treatment.

Despite its undeniable qualities, the AFM is not used in all its potentiality. Indeed, the miniaturization of the systems makes it possible to increase very considerably their resonant frequency and to use these systems over much larger frequency ranges. This property is used in Tapping™

(a)

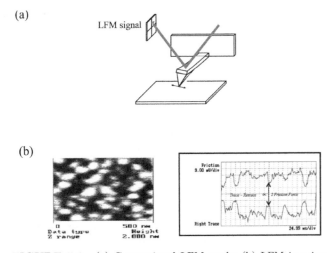

(b)

FIGURE 5.1 (a) Conventional LFM mode; (b) LFM imaging of hydrophilic silanized nano-domains embedded in a hydrophobic silanized matrix (left image) and trace and retrace LFM scanning profiles: the "Trace minus Retrace" (TMR) lateral force value is proportional to the friction force.

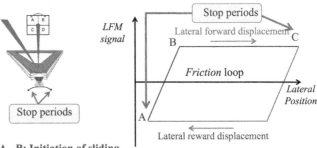

A - B: Initiation of sliding
A : Initiation of shear
B-C : Sliding of the contact
B : Initiation of sliding
C : End of sliding

FIGURE 5.3 Limitations of the commercial LFM mode for conducting friction force measurements.

or Peak Force™ modes (Bruker) where the lever is actuated at frequencies of the order of tens or hundreds of kHz.

In practice, the feedback system has a much lower dynamic range that is limited by the piezoelectric actuator resonance frequency. Typically, a few lines per second are produced. This considerably limits the dynamics of the device in terms of measurement (number of measurements per unit of time) compared to the real potential of the machine that is only limited by the resonance frequency of the piezoelectric actuator. For commercial AFM (for instance, AFM with a J-piezo-actuator from Bruker), this resonance frequency is typically 400 Hz. Consequently, in routine studies, tribological experiments are carried out with a scanning frequency of few Hz. Finally, when the probe rests and inverts its sliding direction, it may result strong changes in some probe-sample interactions, for example, shear stress in the case of imaging or nano-tribological measurements.

Limitations in Terms of Applications

The previous limitations are mainly detrimental to applications, which necessitate being in a stationary regime for the measurement. Particularly, in the case if the nature of the contact may evolve during the stop periods. Furthermore, the experimental sliding speeds are very low compared to sliding speeds used at the industrial level. Eventually, the screening of different conditions (such as in varying the sliding speed) or friction properties as friction laws is difficult to get with the LFM mode applied to tribology, in a reasonable time.

To overcome the previous limitations for tribology encountered with the conventional AFM, an original experimental setup based on a classical AFM was developed in merely modifying the electric voltages applied to the piezoelectric actuator and coupling a lateral excitation to a normal loading. This new experimental setup is called the circular mode AFM (CM-AFM).

5.3 The Circular Mode AFM[9]: A Fast and Accurate Nano-Tribometer

5.3.1 Principle

As its name suggests, the principle of the CM-AFM is to animate the AFM tip/sample contact with a high-frequency circular relative displacement (typically 100–500 Hz) in the plane of the sample. This motion is easily obtained by combining a sinusoidal voltage in the $[-X,X]$ direction and a sinusoidal voltage dephased of $\pi/2$ in the $[-Y,Y]$ direction of the piezoelectric actuator (Figure 5.4). The speed of the circular motion is controlled by adjusting the amplitude and the frequency of the voltages applied to the piezo-actuator in the X and Y directions. Sliding speeds up to 1 mm/s or more are easily reached (compared to tens of μm/s for conventional AFM experiments). The only limitation in the value of the sliding speed is the resonance frequency of the

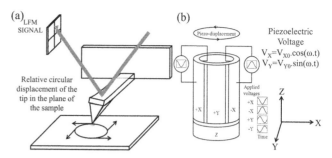

FIGURE 5.4 Schematic principle of the CM-AFM. (a) Representation of the circular motion of the relative contact. (b) Schematic of the applied voltages to the AFM scanner. (Reproduced from Nasrallah, H., Mazeran, P.-E., and Noel, O. 2001. *Review of Scientific Instruments* 82: 113703 with the permission of AIP publishing.)

piezo-actuator (typically 400 Hz for a J-piezo-actuator from Bruker).

The sliding speed, V, is given by the radius, R, (or amplitude of the voltage applied to the actuator) and by the frequency, f, of the circular motion: $V = 2\pi R f$. Since the frequency of the circular motion is close to the piezo-actuator resonance, it is necessary to experimentally quantify the value of the sliding speed. It is done by measuring from a topographic image, the diameter of different circles obtained by wear of a surface using different voltages applied to the actuator (Figure 5.5).

The LFM signal measured by the photodiodes is associated to the torsion of the AFM cantilever induced by the lateral force. When the tip or sample rotates, the friction force constantly changes its direction and the LFM signal, which measures the X component of the lateral force (see Figure 5.4) that has a sinusoidal shape. The amplitude of the LFM signal which, is proportional to the friction force (Figure 5.6), is directly and instantaneously measured without post data treatment by a lock-in amplifier. Furthermore, the lock-in amplifier naturally filters the other harmonics in the signal and offers unique signal-to-noise ratio.

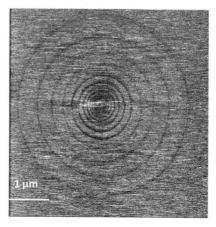

FIGURE 5.5 Imprints of the circles of different diameters generated by using the CM-AFM on a mica surface.

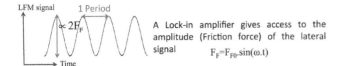

FIGURE 5.6   LFM signal generated by the CM-AFM. The friction force is directly measured with a lock-in amplifier.

5.3.2   Coupling the CM-AFM with the AFM Force Spectroscopy Mode

In the context of measurements without servo-control, as in the AFM force spectroscopy mode, it is perfectly possible to carry out friction measurements with the CM-AFM as a function of the normal load. It is implemented in simultaneously imposing a low-frequency vertical displacement (to vary the normal load) with a high-frequency lateral displacement (to generate friction). With a conventional AFM, this experimental procedure cannot be completed. Still, it is just possible to measure the lateral interaction forces for a given normal load. In these conditions, it takes a few hours to get a complete friction law curve with a low resolution (see Figure 5.2).

In coupling the force spectrosocopy mode with the CM-AFM, the displacement of the cantilever becomes helicoidal (Figure 5.7) and the behavior of the friction force with the normal load is immediate.

To summarize, when the CM-AFM is coupled with the force spectroscopy mode, it is possible to instantaneously and simultaneously obtain (i) the adhesion force and the

attractive force, and (ii) the friction law (friction/load dependence) with a high resolution, all this at a constant and continuous sliding speed. More generally, measurements made with a harmonic solicitation make it possible to obtain a much richer response, insofar as the physical signal measuring the interaction can be analyzed through the response (amplitude and phase) of the different harmonics of the signal.

The CM-AFM has several totally original advantages over the commercial modes (Table 5.1): (i) the continuous and instantaneous acquisition of the relationship between the lateral force and the normal force (friction law) with a high resolution and accuracy, (ii) a wide range of sliding speeds (up to 1 mm/s or more depending on the resonance frequency of the piezoelectric actuator) and thus a higher dynamic range, which makes it possible to detect events over much shorter times, (iii) a constant and continuous sliding speed (measurements are carried out in quasi-stationary mode) and (iv) the possibility of simultaneous and instantaneous acquisition of the interaction force (adhesion, indentation and perforation forces in the case of soft or fragile samples...) as a function of the sliding speed.

5.3.3   Limitations of the Circular Mode AFM?

During the circular displacement, the relative distance between the tip and the sample, and thus the vertical bending of the cantilever, is not constant due to the tilt and the roughness of the sample. The high frequency used to generate the circular mode does not allow maintaining a constant bending of the cantilever because the cut-off frequency of the servo-loop is too low. Typically, the CM-AFM is dedicated to low roughness samples. Its aim is to measure an average friction as a function of the load independently of the roughness of the sample. In experiments without servo-loop, as it is the case in experiments combining the force spectroscopy mode with the CM-AFM, the vertical bending is detected on the normal force curves only if the vertical scanning speed is too high.

For some experiments, the tilt of the sample might be problematic. It could be eliminated in positioning the sample on a micro-mechanical table (Small Dual-Axis Goniometer, Figure 5.8). Another solution is to inject a sinusoidal voltage on the X and Y electrode of the piezo-electric actuator to counteract the tilt of the sample.

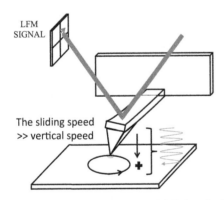

FIGURE 5.7   Combination of the CM-AFM and of the force spectroscopy mode. This experimental setup allows acquiring instantaneously and simultaneously the friction law and the adhesion force in a sliding nano-contact.

TABLE 5.1   Comparison of conventional AFM Modes, the CM-AFM and the CM-AFM Combined with the Spectroscopy Force Mode: Features, Advantages and Drawbacks

	Conventional Mode	CM-AFM	CM-AFM Combined with the Force Spectrum Mode
Sliding speeds	Low scanning (typically, sliding speeds ranging from 1 to 100 µm/s)	High range of sliding speeds (>1 mm/s)	High range of sliding speeds (>1 mm/s)
Advantages/Drawbacks (in italic)	Measurements done in a nonstationary regime; fastidious computation of the friction law	Measurements done in a stationary regime	Instantaneous and simultaneous (in one procedure) measurement of the adhesion force, friction coefficient for a given sliding speed; high-resolved friction law obtained in a short time

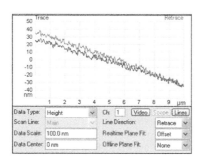
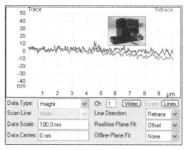

FIGURE 5.8 Principle of the experimental adjustment of the tilt of the sample by using a micro-mechanical table (photo inserted on the right graph): Topographic profile with no flatten correction before the adjustment (left image); Topographic profile with no flatten correction after the adjustment (right image).

The CM-AFM is firstly dedicated for investigating homogeneous samples. For heterogeneous materials, it is necessary to adjust the radius of the circular displacement to stay in a homogenous zone. To remain in sliding friction, the circular radius must be well higher than the contact radius. If not, the relative contact displacement does not generate sliding but only shearing.

5.4 Applications in Nano-Tribology

5.4.1 Dynamic Friction

The CM-AFM is compulsory to assess the dynamic friction in the case where the nature of the contact quickly evolves. It is particularly verified if capillary forces[14-15] or viscoelastic properties[15-17] are involved in the friction processes. In particular, it was shown that capillary forces might form in the contact in between hydrophilic surfaces.[18] These capillary forces are dependent on the contact time[19] and the nucleation time of a capillary bridge formed between nano-asperities in the contact which is estimated to be in the millisecond time scale.[20] If the stop periods last more than

this timescale, then the capillary bridge has time to form or to evolve. Thus, the friction force measurement is dependent on the formation of these capillary bridges. Accordingly, it is necessary to reach a stationary regime to perform relevant measurements.

Noel et al.[21] have used the CM-AFM combined to the spectroscopy force mode to measure the evolution of the adhesion force in a sliding nano-contact (Figure 5.9). It was shown that, the adhesion force evolves from capillary forces to van der Waals forces when the sliding speed increases. In the intermediate regime (Figure 5.9-regime II), the adhesion force decreases linearly with the logarithm of the sliding speed. It was also demonstrated that the sliding speed corresponding to the beginning of the intermediate regime (V_{start}) is influenced by the hydrophily of the surfaces in contact. In others words, the more the hydrophily of the surfaces in contact is, the highest is the radius of the capillary meniscus and the lower is V_{start}.

By using the combination of the CM-AFM with the force spectroscopy mode, it is easy to obtain high-resolved friction law in a very short time. Indeed, during the experiment, the friction force is measured while the normal load is varied due

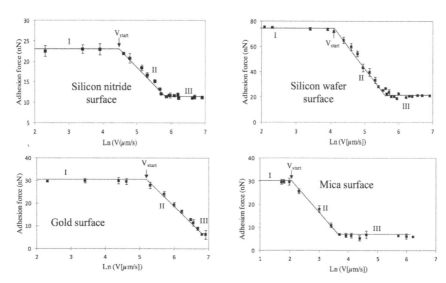

FIGURE 5.9 Adhesion force vs. the logarithm of the sliding speed in between a nitride silicon AFM tip and, respectively, (from left top to right bottom) a silicon nitride, silicon wafer, gold and mica surface. The adhesion force follows three regimes referred as I, II and III. V_{start} is the sliding speed corresponding to the beginning of the intermediate regime, II.

to the combination of the lateral (friction force measurement) and vertical (normal load variations) displacements of the AFM tip in contact with the surface (see difference of resolution between Figures 5.2 and 5.10). If the Amontons' friction law is verified, the friction coefficient is given by the slope of the curve (Figure 5.10).

In these conditions, the evolution of the friction coefficient with the sliding speed is drawn in a reasonable time, and it can be compared with the evolution of the adhesion force that is obtained simultaneously in one experimental procedure (Figure 5.11).

In the case of viscoelastic materials such as polymers, the adhesion forces are likely to deform the surfaces in the vicinity of the contact, and they generate an adhesive neck. In friction force measurements with conventional AFM modes, the adhesive neck formation time is comparable to that of the duration of the stop periods generated while the AFM tip changes its direction during scanning. In this condition, the CM-AFM appears as essential in order that the nature of the contact does not evolve during the lateral force measurements. Combined to the spectroscopy

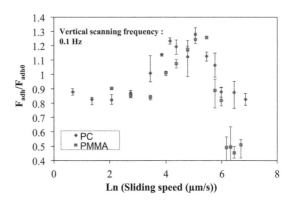

FIGURE 5.12 Adhesion force vs. the logarithm of the sliding speed for a polycarbonate (diamonds) and a polymethylmethacrylate (squares)/nitride silicon contact. Data are acquired by using the CM-AFM combined to the spectroscopy force mode.

force mode, the CM-AFM allows studying the behaviors of the adhesion force in a sliding nano-contact for a compliant material such as the polymethacrylate of methyl (PMMA) or the polycarbonate (PS) (Figure 5.12).

The behavior of the adhesion force with the sliding speed in Figure 5.12 is explained by a transition in the contact mechanics. At low sliding speed, i.e. at low strain rate, the material is compliant, and the contact obeys to a Johnson, Kendall and Roberts (JKR) contact mechanics.[22] At a moderate sliding speed, i.e. at a moderate strain rate, the material behaves like a rigid material and the contact mechanics obeys to a DMT contact mechanics.[22] Finally, at high sliding speed, the material is rigid enough to not strain: the polymer chains cannot orientate themselves and the van der Waals forces disappear.

5.4.2 Nano-Wear

Wear is a dynamic field considering its environmental and economical impact. Due to the emergence of the nanotechnologies and microsystems, nano-wear has attracted the attention of researchers. Also, new concepts of wear processes at the nanoscale were experimentally evidenced.[2,3,23] However, most of the studies reported in the literature and using the conventional AFM are based on bi-dimensional representations of the worn material. Indeed, a three-dimensional well-defined wear track is difficult to obtain in a short time with a conventional AFM. Likewise, due to the low scanning speed, nano-wear experiments may be subjected to the drift of the actuator. Otherwise, a dynamic molecular simulation on metallic materials has showed that in some conditions, wear laws at the macroscale could be retrieved at the nanoscale.[24] Therefore, the nano-wear mechanisms are still controversial and it needs new experimental evidenced.

Recently, the CM-AFM was implemented for studying nano-wear on a copper-based composite material.[25,26] During the experiments, the servo-control insures that the normal load is constant during the process. Consequently, the shear force applied in the contact is constant in all the

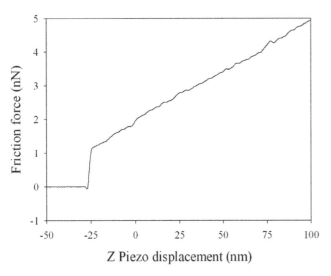

FIGURE 5.10 High-resolved friction law for a nitride silicon/gold contact acquired with the CM-AFM coupled to the spectroscopy force mode. The acquisition time is 10 s.

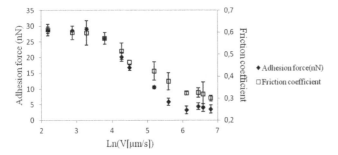

FIGURE 5.11 Evolution of the adhesion force and of the friction coefficient as a function of the logarithm of the sliding speed for a nitride silicon/gold contact. Data are acquired by using the CM-AFM combined to the spectroscopy force mode.

directions and it is not dependent on the lateral and longitudinal stiffness of the cantilever.

The CM-AFM presents several advantages for studying nano-wear: (i) the sliding velocity is much higher than conventional sliding speed, achieving significant wear volume in a short time. A reliable quantification of the wear volume based on three-dimensional imaging is then accessible; (ii) the drift of the piezoelectric actuator is limited because of the high scanning velocity but also because the average X and Y voltage applied on the scanner is null; (iii) the circular wear track can be easily detected and cannot be confused with surface scratches; (iv) during the process, the wear particles are expelled from the rubbed area; thus, it limits the presence of a third body in the contact interface; (v) probing the material in all directions of the surface material, thus making it possible to highlight the possible anisotropy of the material regarding wear (Figure 5.13); (vi) if worn during the experiment, the AFM tip is submitted to an isotropic wear.

Eventually, nano-wear behaviors with different experimental parameters can also be assessed[26] (Figures 5.14 and 5.15).

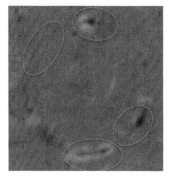

FIGURE 5.13 AFM topographic images, after image processing, of nano-wear anisotropies revealed by the CM-AFM used for nano-wear applications.

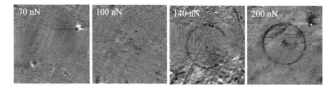

FIGURE 5.14 AFM topographic images, after image processing, of the wear of a copper-based composite material as a function of the normal load, by using the CM-AFM and a silicon nitride AFM tip.

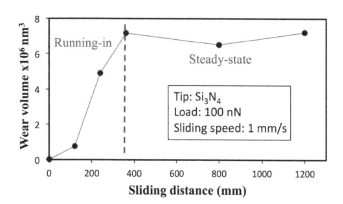

FIGURE 5.15 Wear volume of a copper-based composite material as a function of the sliding distance. The description by Barwell[27] of the running-in and steady-state regimes for wear of metals at the macroscopic scale is recovered at the nanoscopic scale.

Some of the main advantages for nano-wear applications are gathered in Table 5.2.

5.5 Others Applications and On-Going Works

The AFM indisputably imposed on the community of Life Sciences as a technique of choice for the study of biological objects.[28–30] For nano-mechanical measurements on biomembranes, AFM is conventionally used as a nanoindenter (normal sample loading) to estimate stiffness moduli of the biological system. The bearing force of the AFM tip is increased to determine their breaking strength, giving access to the thickness and mechanical strength of the substrate-supported lipid biomembranes. Such experiments are referred as punch-through experiments.[31] Currently, the CM-AFM is developed to carry out experiments in liquids. One of the targets is to apply the CM-AFM for concerns in the fields of biology and health. For example, in lipid and cell-membrane studies, the high-frequency lateral displacement of the AFM probe generates a lateral force which is the combination of a frictional force and a viscoelastic shear force related to the deformation of the membrane by the sample-tip contact. As a result, the lateral force contains a viscoelastic response and is therefore out of phase with the displacement. By playing on the normal force applied, these two components of the lateral force evolve. The simultaneous coupling with the force spectroscopy mode (vertical loading) makes it possible to measure simultaneously the breaking forces of the biomembranes and the lateral components of the friciton force. To summarize, the circular mode coupled

TABLE 5.2 Comparison between the conventional AFM Modes and the CM-AFM Combined to the Spectroscopy Force Mode for Wear Applications: Features, Drawbacks and Advantages

	Conventional Mode	CM-AFM
Solicitation velocity	Low scanning or sliding velocity (typically, ranging from 1 to 100 µm/s)	High sliding velocity (>6 mm/s)
Advantages/Drawbacks	High scanner drift; low wear; high shear force when the scan changes its direction	Limiting scanner drift; high wear in a limiting time; well-defined wear track; isotropic wear of the probe if any; anisotropic wear revealed if any; local probing

TABLE 5.3 Comparison between the Commercial AFM Modes and the CM-AFM Combined to the Spectroscopy Force Mode for Biological Applications: Features, Drawbacks and Advantages

	Classical Punch through Experiments	CM-AFM + Force Spectroscopy Mode
Accessible data	**Classical data:** • Membrane thickness • Penetration force • Adhesion force • No measurement of the lateral force	**Classical accessible data +** • Solid and viscous friction of the membrane • Friction laws
Solicitation speed	Governed by the low approach and retract speed (<1 µm/s)	Governed by the high sliding speed (>1 mm/s)

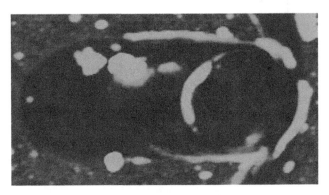

FIGURE 5.16 Nano-polishing of a Gallium Arsenide (AsGa) surface using the topographic CM-AFM imaging.

to the force spectroscopy mode thus offers (i) the coupling in real time of the vertical and horizontal stress measurements which give access to a measurement of the properties of the different layers of the biological sample; (ii) ease and speed of measurement (we obtain an almost instantaneous measurement of lateral forces and friction laws); (iii) dynamic measurement that separates the different contributions of the lateral force and (iv) measurement made at different sliding speeds (up to 1 mm/s), which allows seeing the evolution of the speed of loading on the mechanical response of the membrane.

Table 5.3 gathers some advantages of the CM-AFM for biological applications compared to the classical punch-through experiment.[31]

In conclusion, the CM-AFM is a new powerful AFM mode that allows fast and accurate measurements of mechanical and tribological properties. It offers new opportunities such as, for example, accessing to the adhesion force in a nano-sliding contact. It also affords numerous advantages of the CM-AFM for academic and industrials applications in many fields in nano-sciences. Some other applications in biology or polishing (Figure 5.16) taking part of the advantages of the CM-AFM are obvious and are still waiting for to be developed.

Bibliography

1. Mo, Y., Turner, K.T., Szlufarska, I. 2009. Friction laws at the nanoscale. *Nature* 457(7233): 1116–1119.
2. Gotsmann, B. and Lantz, M.A. 2008. Atomistic wear in a single asperity sliding contact. *Physical Review Letters* 101(12): 125501.
3. Jacobs, T.D.B. and Carpick, R.W. 2013. Nanoscale wear as a stress-assisted chemical reaction. *Nature Nanotechnology* 8(2): 108–112.
4. Tabor, D. and Winterton, R.H.S. 1968. Surface forces: direct measurement of normal and retarded van der Waals Forces. *Nature* 219(5159): 1120–1121.
5. Israelachvili, J.N., McGuiggan, P.M, and Homola, A.M. 1988. Dynamic properties of molecularly thin liquid films. *Science* 240 (4849): 189–191.
6. Bushan, B. *Fundamentals of Tribology and Bridging the Gap between the Macro- and Micro/Nanoscale.* Kluwer Academic Publishers, 2001, ISBN 978-0-7923-6837-3.
7. Binnig, G., Quate, C.F., and Gerber, C. 1986. Atomic force microscope. *Physical Review Letters* 56(9): 930.
8. Mate, C.M., McClelland, G.M., Erlandsson, R., and Chiang, S. 1987. Atomic-scale friction of a tungsten tip on a graphite surface. *Physical Review Letters* 59(17): 1942.
9. Noel, O., Mazeran, P.-E., and Nasrallah, H. Processes for surface measurement and modification by scanning probe microscopy functioning in continuous curvilinear mode, scanning probe microscopy and device permitting their implementation of said methods. US Patent 8997261 B2 (March 31, 2015).
10. Gnecco, E. and Meyer, E. *Fundamentals of Friction and Wear on the Nanoscale.* Springer: Berlin, Germany, 2015.
11. Amontons, in Table générale des matières contenues dans l' "Histoire" et dans les "Mémoires de l'Académie royale des sciences", par la Compagnie des libraires, tome 2. *Années* 1699–1710: 25–28. https://gallica.bnf.fr/ark:/12148/bpt6k3507z/f34. image.
12. PhD manuscript, 2000. Imagerie et Caractérisation Nanomécanique des Surfaces par Microscopie à Force Atomique. O. Piétrement p95. https://hal.archives-ouvertes.fr/tel-01802584.
13. Carpick, R.W., Ogletree, D.F., and Salmeron, M. 1997. Lateral stiffness: A new nanomechanical measurement for the determination of shear strengths with friction force microscopy. *Applied Physics Letter* 70: 1548–1550.
14. Bingelli, M. and Mate C.M. 1994. Influence of capillary condensation of water on nanotribology studied by force microscopy. *Applied Physics Letter* 65: 415.

15. Schumacher, A., Kruse, N., and Prins, R. 1996. Influence of humidity on friction measurements of supported MoS2 single layers. *Journal of Vacuum Science Technology B* 14: 1264.

16. Hammerschmidt, J.A., Gladfelter, W.L., and Haugstad, G. 1999. Probing polymer viscoelastic relaxations with temperature-controlled friction force microscopy. *Macromolecules* 32: 3360–3367.

17. Kagata, G., Gong, J.P., and Osada, Y. 2002. Friction of gels. 6. Effects of sliding velocity and viscoelastic responses of the network. *The Journal of Physical Chemistry B* 106(18): 4596–4601.

18. Bocquet, L., Charlaix, E., Ciliberto, S., and Crassous, J. 1998. Moisture-induced ageing in granular media and the kinetics of capillary condensation. *Nature* 396: 735–737.

19. Wei, Z. and Zhao, Y.-P. 2004. Experimental investigation of the velocity effect on adhesion forces with an atomic force microscope. *Chinese Physics Letters* 21: 616.

20. Szoszkiewicz, R. and Riedo, E. 2005. Nucleation time of nanoscale water bridges. *Physical Review Letters* 95: 135502.

21. Noel, O, Mazeran, P.E, and Nasrallah, H. 2012. Sliding velocity dependence of adhesion in a nanometer-sized contact. *Physical Review Letters* 108: 015503.

22. Johnson, K.L. *Contact Mechanics*. Cambridge University Press, Cambridge, UK.

23. Bennewitz, R. and Dickinson, J.T. 2008. Fundamental studies of nanometer-scale wear mechanisms. *MRS Bulletin* 33(12): 1174–1180.

24. Eder, S.J., Feldbauer, G., Bianchi, D., Cihak-Bayr, U., Betz, G., and Vernes, A. 2015. Applicability of macroscopic wear and friction laws on the atomic length scale. *Physical Review Letters* 115(2): 025502.

25. Noel, O., Vencl, A., and Mazeran, P.E. 2017. Exploring wear at the nanoscale with circular mode. *Beilstein Journal of Nanotechnology* 8: 2662–2668.

26. Vencl, A., Mazeran, P.-E., Bellafkih, S., and Noel, O. 2018. Assessment of wear behaviour of copper-based nanocomposite at the nanoscale. *Wear* 414–415: 212–218.

27. Barwell, F.T. 1958. Wear of metals. *Wear* 1(4): 317–332.

28. Dufrêne, Y.-F. 2002. Atomic force microscopy, a powerful tool in microbiology. *Journal of Bacteriology* 184(19): 5205–5213.

29. Goldsbury, C.S., Scheuring, S., and Kreplak, L. 2009. Introduction to atomic force microscopy (AFM) in biology. *Current Protocols in Protein Science* 58: 17.7.1–17.7.19.

30. Giocondi, M.C., Yamamoto, D., Lesniewska, E., Milhiet, P.E., Ando, T., and Le Grimellec, C. 2010. Surface topography of membrane domains. *Biochimica et Biophysica Acta* 1798(4): 703–718.

31. Alessandrini, A. and Facci, P. 2012. Nanoscale mechanical properties of lipid bilayers and their relevance in biomembrane organization and function. *Micron* 43: 1212.

Chemical Imaging with Fluorescent Nanosensors

Seren Hamsici, Robert Nißler,
Florian A. Mann, and
Daniel Meyer
Göttingen University

Sebastian Kruss
Göttingen University
Center for Nanoscale Microscopy and
Molecular Physiology of the Brain (CNMPB)

6.1 Basics of Chemical Imaging

Chemical imaging uses analytical methods at different locations of a sample to map chemical properties with spatial resolution such as Raman spectroscopy, mass spectroscopy and fluorescent sensors (Figure 6.1) In general, chemical imaging requires two things. First, a method that delivers chemical information, and second, an imaging concept that allows one to scan over or through a sample. Chemical imaging methods reveal a lot of information about the chemical composition and its dynamics, which is useful for different applications ranging from materials science, biology, agriculture to medicine. Classical chemical imaging approaches rely on spectroscopy methods, and they guarantee a very specific fingerprint of a sample. Nanomaterials have unique

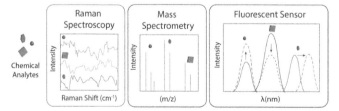

FIGURE 6.1 Analytical methods that have the potential for chemical imaging. Many different methods exist that provide chemical information about a sample (e.g. composition or concentration of an analyte). Prominent examples are Raman spectroscopy or mass spectrometry (MS). An alternative is fluorescent sensors/probe that change their fluorescence signature in the presence of a specific analyte. The fluorescence signal can change in intensity (decrease or increase) or shift/change its spectrum (e.g. wavelength maximum) in response to the analyte. If the fluorescent building block is a nanomaterial, such sensors are called nanosensors.

properties, and therefore, they are useful as building blocks for nanoscale sensors/probes (nanosensors). Research in this area has attracted a lot of interest during recent years (Kruss et al. 2013a and b). This chapter provides an introduction to chemical imaging with nanosensors, which is a growing field with promising applications.

In this first section, we will provide an overview about classical chemical imaging concepts to establish the perspective that is necessary to understand the benefits of nanosensors. Afterwards, we will conceptualize imaging with nanosensors and discuss design, kinetic requirements, functionalization strategies and possible applications with a main focus on life science applications.

6.1.1 Chemical Imaging Approaches

Chemical imaging allows to investigate the chemical composition of a sample (Figure 6.2). Vibrational spectroscopic techniques such as Raman spectroscopy-based imaging offers a fast and nondestructive approach to fingerprint the chemical composition of samples such as different types of biological cells (Rosch et al. 2005).

In Raman spectroscopy, monochromatic light from a laser source interacts with vibrations of the molecule. When light is scattered, most of the scattered photons from the molecule have the same energy/wavelength (Rayleigh scattering) as the excitation light. However, a very small fraction of the scattered light ($1/10^6$) has a different energy (Lorenz et al. 2017). This scattering is called Raman scattering (Raman 1928). The shift in wavelength of the scattered light provides information about molecular vibrations and consequently the structure of the molecule (Krafft et al. 2003). Examples of Raman spectroscopy include distinguishing gram-positive and gram-negative bacteria by detecting

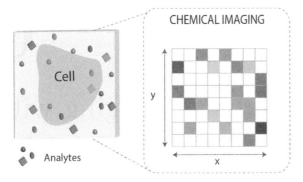

FIGURE 6.2 Concept of chemical imaging. In general, chemical imaging means that spatially resolved chemical information is collected from a sample. Most analytical methods measure in bulk, for example, a cuvette filled with the sample and provide no spatial information. In contrast, a chemical image allows one to resolve different parts of a sample similar to an image with pixels. In this scheme, a biological cell is shown and different analytes in and around it. On the right, the chemical image is shown and the color could represent spectra, composition or concentration of the sample depending on the exact method. Of course chemical imaging is not limited to non-changing samples. The chemical image could also change over time similar to a movie, which is very important if the dynamics contain a lot of information, for example, in chemical signaling between cells.

changes in molecular composition of the cell envelopes (Maquelin et al. 2002), which is possible because certain components are more or less specific to certain species of bacteria (Stöckel et al. 2015). Furthermore, plasmids were detected to monitor antibiotic resistance genes in *E. coli* (Walter et al. 2011). Besides biochemical information, by the aid of combination of optical microscopy, it is also possible to get morphological information about cells or tissues, which can be linked with pathological conditions. One experimental challenge is the weakness of the Raman Effect. Therefore, surface-enhanced Raman spectroscopy (SERS) helps to enhance Raman scattering in high-sensitivity molecular detection. SERS utilizes the fact that the Raman signal becomes much more prominent when the analyte molecules are located close to metallic (nano)particles (e.g. gold or silver), which dramatically enhances the local electrical field of the excitation light (Kneipp et al. 1997). The small distances required for SERS are usually achieved by using nanoparticles or nanopatterned surfaces. Nanomaterials, in this regard, are very versatile for analytical applications due to their unique size-dependent tunable electronic, optical, magnetic and mechanical properties (Biju 2014; Kruss et al. 2012, 2013a, b).

In clinics, rapid, sensitive and specific analysis of fresh blood samples for pathogens is crucial, because delayed antibiotic therapy dramatically increases mortality. SERS detects pathogens directly from blood and overcomes many challenges such as low concentrations of pathogens, the presence of complex blood matrix and high protein content (Cheng et al. 2013). Identification of *Staphylococcus aureus*

and *Escherichia coli* directly from whole uncultured blood was accomplished with high sensitivity and specificity by using SERS enabling physicians to get accurate and rapid treatment against bacterial pathogens (Boardman et al. 2016). Furthermore, SERS can be applied to monitor cellular events such as secretion, apoptosis and the change of the cellular microenvironment including pH, reactive oxygen species and redox potential (Cialla-May et al. 2017). It is also possible to design theranostic platforms that combine targeting, drug delivery (Hossain et al. 2015), imaging (Liu et al. 2015) and photothermal therapy (Zhang et al. 2014) with a combination of SERS and imaging techniques. These examples demonstrate that SERS-based platforms are a promising approach for basic science and clinical applications.

Another technique capable of chemical imaging is mass spectrometry (MS). MS has an exquisite chemical resolution, but it normally lacks spatial resolution. However, when MS spectra are collected from different locations of a sample, one can map the distribution of metabolites, surface lipids or proteins (Zimmerman et al. 2008). In this approach, the sample of interest is bombarded with an ion beam to induce ionization. The mass-to-charge ratio *(m/z)* of the charged objects distinguishes the different components (Shariatgorji et al. 2014). After the collection of individual mass spectra in different locations, they can be spatially arranged again as a pixel array (i.e. image). For any given observed spectrum, the molecular distributions across the surface are translated into an image (Watrous et al. 2011). The ionization source and ionization efficacy determine different aspects such as the achievable spatial resolution (Watrous & Dorrestein 2011). Commonly used techniques include matrix-assisted laser desorption/ionization (MALDI), nanoscale secondary ion MS (NanoSIMS) and desorption electrospray ionization (DESI) depending on their focused ionization source, which are light, primary ions or electrospray solvents, respectively (Dunham et al. 2017). Among them, SIMS imaging is a well-known technique to study the distribution of atoms and small molecules in tissues and even single cells at spatial resolutions down to the nanometer range (Spengler 2015). The characterization of the sample is firstly conducted by bombarding the sample surface with a primary ion beam. The secondary ions from the sample are separated, and their m/z ratios are measured. The primary ion beam has an impact on the deposition of chemically reactive species on the sample surface, which directly affects the generation of secondary ions. Consequently, it is important to choose the right primary ion species, which depends on the sample of interest (Musat et al. 2012). In NanoSIMS, very often specific isotopes (^{15}N, ^{13}C, ^{18}O) are used. They get into the sample, for example, by feeding nutrients containing these isotopes to cells. These approaches have been very useful to track metabolic activity (Saka et al. 2014). Isotope-modified nutrients are imaged by determining isotopic enrichment in comparison to a standard sample. Applications include estimation of single cell N_2 and carbon fixation rates of cyanobacteria (Foster et al. 2013), quantification of the uptake of nitrogen sources for individual cells within biofilm formation (Renslow et al. 2016) and determination of total ammonium and inorganic carbon assimilation quantified by metabolic activities of each single anaerobic phototrophs (Musat et al. 2008). Despite its powerfulness towards elemental distribution in nanometer scale, there are some limitations such as sample preparation (culturing with specific labeled substrates), destructiveness and the possibility of metabolism difference between single cells (Gao et al. 2016).

6.2 Chemical Imaging with Fluorescent Sensors/Probes

6.2.1 Introduction

Traditional analytical techniques as discussed in the previous chapter require expensive instrumentation and complicated sample preparation, which makes chemical imaging difficult for on-site and real-time detection. Because of these drawbacks, many efforts have been put forth to develop alternative methods for chemical imaging. Fluorescence microscopy is a powerful technique in the life sciences. In comparison with spectroscopic methods, it has the intrinsic advantage to collect images. However, it typically provides only the information that the fluorophore is in a certain location. The fluorophores are normally used only as labels and are not designed to report about their chemical environment. Thus, fluorescent sensors or probes are needed to use fluorescence microscopy methods for chemical imaging.

In general, sensors contain three main elements: a receptor, a transducer and a detection unit (Morrison et al. 2008). The receptor (recognition element) specifically interacts with the target analyte. This interaction is translated into a measurable signal (transducer) such as the production of chemicals, heat release or changes of the optical properties such as fluorescence (Fan et al. 2008). In the case of fluorescent sensors, detection of the optical signal is performed by the microscopy setup (camera). Therefore, the term 'probe' is more precise for the materials we discuss in this chapter. However, we will use the term 'nanosensor' because it is more often used in the literature.

Sensors are often categorized according to their main elements, which are either the method of signal transduction (i.e. optical, mechanical, or electrical) or by the type of receptor employed (i.e. catalytic [enzyme] or affinity based [antibody, aptamer, etc.]) (Ahmed et al. 2014). In order to generate sensors, it is important to consider the molecular interaction of the receptor and the analyte by mediating a particular reaction (Goode et al. 2015). Optical (nano)sensors are ideally suited and adaptable platforms to identify different types of analytes due to their compactness, flexibility, high sensitivity and selectivity. The readout of such sensors can take place *via* different approaches such as fluorescence, chemiluminescence, and light absorbance (Ansari et al. 2010). Molecular emission such as fluorescence

and phosphorescence is about 1,000 times superior in sensitivity than most spectrophotometric methods and at the same time enables lower limits of detection for the desired analytes (Yu et al. 2015). Even at low concentrations, it is possible to identify the spatial distribution of analytes within biological samples.

Fluorescent nanosensors are a rapidly developing area of research to visualize specific molecular processes occurring in cells continuously and create an accurate representation of biological events (Basabe-Desmonts et al. 2007; Kruss et al. 2013a, b). The large total surface area, as well as the small amounts of analytes and buffers required for nanoscale device-based assays, render them attractive for chemical imaging applications (Zhu et al. 2016). In addition, they present ideal surface properties for analyte or recognition element immobilization or offer increased detection capacity through phenomena such as high quantum yield and signal-to-noise ratio (Zhang et al. 2016). The quantum yield of a fluorescent material describes how efficient the energy from absorbed photons is translated into emitted photons and is important for the signal intensity of a sensor. Additionally, the signal-to-noise ratio of a sensor depends on the absorption cross section and photostability of the fluorescent material (Carter et al. 2014). By combining fluorescent labeling and the capability of sensing, it is possible to follow chemical properties and molecules with a temporal and spatial resolution (Figure 6.3). Target molecules, which are not intrinsically fluorescent can then be studied label-free according to their distribution in samples. Figure 6.4 shows different interesting applications such as imaging of pH values, temperature, oxygen or glucose levels. Furthermore, biologically important small molecules such as hydrogen peroxide (H_2O_2), nitric oxide (NO) or neurotransmitters can also be detected with fluorescent nanosensors despite of their low concentration and short lifetime.

6.2.2 Kinetic Requirements for Spatiotemporal Imaging

Which properties determine the performance of a nanosensor in chemical imaging and what are the spatial and temporal resolution limits? Obviously, its properties need to fit somehow to the spatiotemporal and chemical resolution that is necessary to gain useful information from a sample (Surana & Krishnan 2013). For example, if one wants to study fast changes of analyte concentrations, the nanosensors need to be fast enough to respond and follow these changes. In order to do that, binding of the analyte, photophysical changes and imaging have to be fast enough or match the sample dynamics. On the other side, if these processes are too fast, information can be lost because there are not enough photons or the detection is not possible on this timescale (camera speed). To predict the image of many nanosensors, it is necessary to account for all these things, which can be done in a simulation and various theoretical models (Meyer et al. 2017).

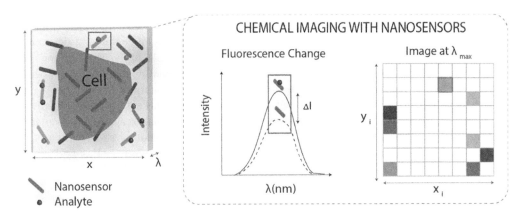

FIGURE 6.3 Basic concept of chemical imaging with nanosensors. Nanosensors are able to report about chemical changes in their environment. If they are designed to change their fluorescence in the presence of an analyte, they report about the local concentration of this analyte. In this schematic, a biological cell and multiple nanosensors are shown. Here, carbon nanotubes are shown but the concept holds for any nanomaterial. If a molecule of interest (analyte) is present, each individual nanosensor reports about the presence of this analyte (indicated by the square). The fluorescence could, for example, increase, when the analyte binds. Consequently, if many of these nanosensors are detected at a specific wavelength (e.g. the wavelength at the maximum intensity λ_{max}) as a collective, the corresponding image provides information about the presence/concentration of this analyte. Every pixel that corresponds to one or more sensors provides chemical information, and therefore, this method is a versatile chemical imaging method.

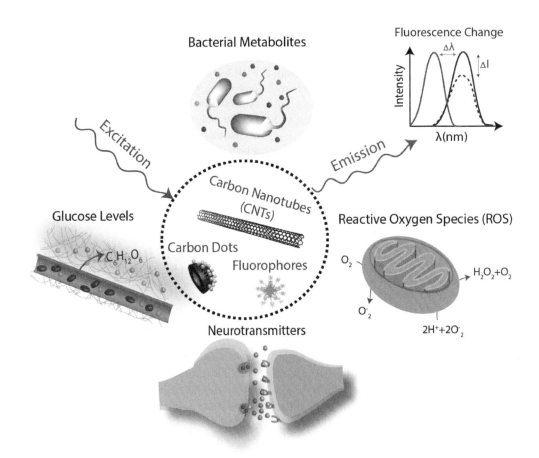

FIGURE 6.4 Fluorescent nanomaterials as building blocks for nanosensors and chemical imaging. Fluorescent (nano)materials can serve as central building blocks for nanosensors. A requirement is that the fluorescence spectrum changes in response to specific chemicals or conditions. Such nanosensors can be powerful tools to study complex processes in which spatial, temporal and chemical aspects are of importance. For example, detection and chemical imaging of mitochondrial activity via reactive oxygen species, glucose levels, bacterial metabolites and neurotransmitters.

One may simplify the interaction between analyte A and sensor S by the following reaction equation

$$A + S \underset{k_{\text{off}}}{\overset{k_{\text{on}}}{\rightleftharpoons}} AS \qquad (6.1)$$

where A is either freely diffusing or it can bind to the sensor and form a complex (AS). More precisely, an analyte binds to a specific binding site of the sensor as it might have multiple binding sites. Therefore, each binding site can be occupied (state 1) or in free state (state 0). Equation 6.1 describes a simple chemical reaction, and if one wants to know how the concentration of the different species changes, rate laws and kinetic considerations have to be used. However, in the single molecule world, it is not possible to predict the concentration in a deterministic way. An analogy is the decay of a radioactive isotope. When many atoms are observed it is possible to predict how much of the isotope is left after a certain amount of time. For one atom, it is not possible and it is purely stochastic. Consequently, the description of nanosensors has to be done by using stochastic kinetics. If the binding and unbinding rates k_{on} and k_{off} of a bimolecular reaction (e.g. between A and S) are known, the probability functions of binding and unbinding are described as follows:

Binding	Unbinding
$P_b(\tau) = \int_0^\tau c k_{\text{on}} e^{-c k_{\text{on}} t} dt$	$P_{db}(\tau) = \int_0^\tau k_{\text{off}} e^{-k_{\text{off}} t} dt$

These equations describe the probability that binding/unbinding happen in a time interval $[0;\tau]$. To implement such an approach in a simulation, a Monte-Carlo simulation is typically used (Gillespie 1978), which means that random numbers (reflecting the stochasticity) are used to decide if the reaction took place.

In short, if one wants to know whether a sensor is occupied after a given time τ, one can apply this algorithm.

For a non-occupied sensor (site), the probability $P_b(\tau)$ is compared to a random number R between 0 and 1. Only if $P_b(\tau) > R$, the reaction takes place. Vice versa for an occupied sensor (site), the condition $P_{db}(\tau) > R$ has to be true. For small time steps τ, this algorithm allows to predict the sensor occupation for a changing concentration profile $c(t)$ for the rate constants k_{off} and k_{on} (Figure 6.6a). If the sensor has only one binding site, one gets a binary response (Figure 6.6a). In contrast, for multiple binding sites, the sensor occupation trace will become more continuous (Figure 6.5b). A reasonable assumption is that sensor occupation correlates with the fluorescence response. In the easiest case, one can assume a linear relation.

This simulation approach can be extended to multiple sensors. Furthermore, experimental conditions need to be considered such as the Abbe limit of light microscopy or the frame rate of the camera.

A visualization of this approach can be seen in Figure 6.5 in which the abovementioned ideas were implemented in a stochastic simulation to analyze and predict the theoretical response image of a sensor array to a concentration gradient (Meyer et al. 2017). The motivation of this work are sensors that are capable to image neurotransmitter release from cells. It is extremely important to image these processes and inevitable to understand the human brain (see examples below).

This simulation has been used to investigate nanosensors in the context of a biologically meaningful event. Here, release of the neurotransmitter dopamine through exocytosis from a hypothetical neuron was used. The concentration gradient $c(x,y,t)$ of a single event release event was first calculated using a standard diffusion simulation. Then, sensors in different patterns were exposed/overlayed with this concentration profile (Figures 6.5 and 6.6c). Consecutively, a time-dependent simulation of each binding site of a sensor was performed according to the equations mentioned above and with the values of k_{on}, k_{off} and the currently

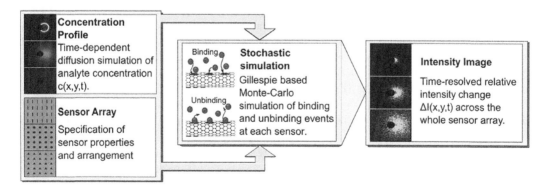

FIGURE 6.5 Simulation to calculate the response image of a sensor array. To understand how an image of nanosensors exposed to a complex spatiotemporal analyte pattern looks, simulations are needed. First, a spatiotemporal concentration profile of the analyte is simulated/used (e.g. the diffusion profile of a neurotransmitter after release from a neuronal cell). A sensor array is exposed to this spatiotemporal concentration profile. Then, binding and unbinding of the analyte to each sensor is calculated for every time point using a stochastic kinetic (Monte-Carlo) algorithm. In the last step, the results from each sensor are merged together to account for factors such as the optical resolution limit or the time resolution of an experimental setup. (Reprinted with permission from Meyer et al. 2017. Copyright (2018) American Chemical Society.)

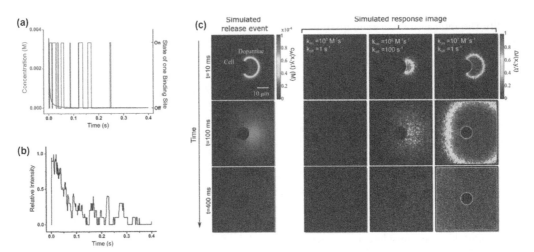

FIGURE 6.6 Transition from single-sensor responses to collective chemical images of nanosensors. (a) In the molecular world, one can only calculate probabilities for binding or unbinding (stochasticity). When a single nanosensor is exposed to a certain concentration profile (decaying curve), the state of each binding site of a sensor (on–off curve) can switch between a bound state (On = 1) and a free state (Off = 0) depending on the speed of this reaction described by the rate constants (k_{on}, k_{off}). (b) For several binding sites, the sensors response can then be translated into a response intensity. (c) The single-sensor response curves can be used to calculate the response image of an entire sensor array. By generating a theoretical (in this case very dense) sensor area and overlaying it with the concentration profile of a hypothetical event (here release of a signaling molecule (dopamine) on the right side of a cell), the response intensity of each sensor can be calculated (right) and combined with optical restrictions (diffraction pattern, exposure times) to predict the chemical image of experiments. Depending on the binding affinities/rate constants of the sensors, the response causes undersaturation or oversaturation of the image. Obviously, these images encode different amounts of information. Therefore, kinetics plays a central role and have to be considered in the design of nanosensors. (Reprinted with permission from Meyer et al. 2017. Copyright (2018) American Chemical Society.)

existing analyte concentration c as parameters (Figure 6.6a and b).

In a simulation, a nanosensor response trace has no resolution limit. However, in reality, the individual nanosensor responses (i.e. emitted photons) face optical restrictions (diffraction pattern, exposure times etc.). They were mimicked to predict the outcome of a real experiment. Figure 6.6c shows an example of this process and the importance of the chosen values of k_{on}, k_{off} and the properties of the setup once more. Too high values of k_{off} combined with low k_{on} values could lead to an undersaturation (Figure 6.6c, left) of the image. Vice versa, too strong binding affinities/too low dissociation constants $K_d = k_{on}/k_{off}$ cause oversaturation (Figure 6.6c, right) – only the right combination of rate constants leads to an image that corresponds to the simulated concentration pattern (left) (Figure 6.6c, middle).

Note also that these examples only refer to the temporal dynamics between sensor and analyte. In biological experiments, spatial distribution needs to be considered in many events, such as release of analytes by cells, because it occurs simultaneously at different positions (Figure 6.7). For example, an important question to ask is if a sensor array could resolve if there are one or two release events. Only sensors with the right appropriate properties (rate constants) will be able to achieve that.

This simulation approach highlights the most important parameters of nanosensors. It can be used to interpret images of given nanosensors. Additionally, it provides

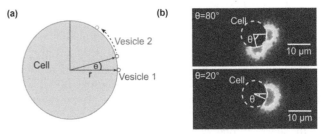

FIGURE 6.7 Spatial and temporal resolution aspects in chemical imaging. (a) A typical imaging problem/question in biology: Cells are releasing molecules via exocytosis (fusion of a vesicle with the cell membrane). A central question is if those two release events and are resolvable by a given method (here described by the cell radius r and the angle between two events θ). In other words, how close can two vesicles be to resolve them and "see" that there are two release events. (b) Example of a resolvable and a non-resolvable event. Depending on the kinetics of the sensors and the time resolution of the experimental setup, the resolution of the response image may vary. In any case, a critical angle can be defined on which two release positions are not distinguishable anymore. (Reprinted with permission from Meyer et al. 2017. Copyright (2018) American Chemical Society.)

a rational approach to design sensors (simulation-guided design of nanosensors). One could simulate the chemical imaging problem to predict the optimal properties of the sensors. Based on that, one can adjust the chemical functionalization strategy (see below).

6.2.3 Applications of Fluorescent Sensor Platforms

In this section, we discuss several examples of chemical imaging with nanosensors. One focus is chemical functionalization because it determines many of the sensor's properties (selectivity, rate constants etc.). We start with organic fluorophores because they have been already used in very important applications, which depicts what is possible. Then we report about nanosensors and focus on fluorescent carbon nanotube-based sensors because they make the unique advantages of nanomaterials very clear.

Organic Dyes for Chemical Imaging

A major field of interest are methods to measure the pH on surfaces or biological systems with high spatial and temporal resolution (Figure 6.8). Schreml and coworkers developed a 2D pH sensor and monitored the pH changes of skin during wound healing (Schreml et al. 2011). A ratiometric combination of an inert reference phosphorescence ruthenium(II)tris-(4,7-diphenyl-1,10-phenanthroline) [Ru(dpp)$_3$] particles and the pH-sensitive fluorophore fluorescein isothiocyanate (FITC) enabled the discrimination of pH values. By separating the photophysical readout of both dye particles in different light channels, it was possible to process the ratiometric luminescence intensities (RGB) as pseudo color images of the current pH value (Schreml et al. 2012).

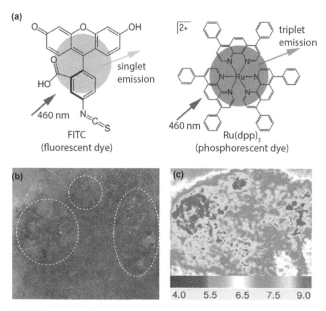

FIGURE 6.8 Sensor approach to visualize the pH during wound healing. (a) A pH-sensitive FITC (fluorescein isothiocyanate) and an inert phosphorescent reference such as ruthenium(II)tris-(4,7-diphenyl-1,10-phenanthroline) [Ru(dpp)$_3$] are used in a ratiometric approach to image the pH of a sample. (b) Image of a chronic wound. (c) Ratiometric chemical image of the pH in the selected region. The distribution of pH has important implications for the state of the wound healing. (Reproduced from Schreml et al. 2011).

Another prominent example of sensor-based chemical imaging is pressure-sensitive dyes. This technique relies on the quenching of the fluorophore by oxygen, while the oxygen concentration can be seen as proportional to the steady-state pressure (Bell et al. 2001). Typically, pyrene as well as metal-porphyrin derivatives are used for applications like coatings of wind tunnel models incorporated in oxygen-permeable polymers (Engler et al. 2002; Asai et al. 2002). Detailed reviews about this field are given in the study by Schäferling (2012).

Dopamine Imaging with Fluorescent Carbon Nanotubes

Single-walled carbon nanotubes (SWCNTs) have emerged as building blocks for nanosensor platforms, due to their unique mechanical, electrical, chemical and structural properties (Mu et al. 2014; Kruss et al. 2013a, b). They exhibit strong resonance Raman scattering and high optical absorption as well as photoluminescence in the near-infrared (nIR) range, properties that are of high interest for imaging in biological systems *in vitro* and *in vivo* (Zhou et al. 2009). In order to render them selective for a given target, functionalization is a crucial step for SWCNTs. Chemical functionalization of SWCNTs is necessary to unbundle them and make them water-soluble. A common approach is to functionalize them non-covalently with macromolecules such as DNA. Recently, specifically designed peptide barrels have been used to expand the chemical repertoire (Mann et al. 2018). For biomedical applications, it is furthermore important to achieve biocompatibility and control the interactions with biological samples (Bianco et al. 2005). Another, hybrid DNA/peptide approach (Figure 6.9) has recently been used to decorate DNA-functionalized SWCNTs with the peptide motif Arg-Gly-Asp (RGD) and tune the interactions with cell-surface receptors (Polo et al. 2018).

Neurotransmitters are an important class of biomolecules and govern the communication between neurons. Therefore, their detection is highly important for understanding underlying mechanisms of neurological brain disorders such as Alzheimer's disease, Parkinson's disease and epilepsy. Even though there are different methods available to detect them, these methods often lack either spatial, chemical or temporal resolution (Polo & Kruss 2016a).

Functionalization of SWCNTs with different polymers enables the creation of new sensor platforms having distinct selectivity and specificity towards different neurotransmitters. This process is known as Corona Phase Molecular Recognition (CoPhMoRe) (Kruss et al. 2014). It describes the process of structural confinement that arises when a (bio-)polymer such as DNA gets in contact with the hydrophobic SWCNT surface. These (bio-)polymers are not specific for a given target themselves. However, when they non-covalently adsorb onto the SWCNT surface, they can form structures, which would not be present in free solution. In order to understand the selectivity and sensitivity within neurotransmitter detection, the impact of DNA sequence was tested. One goal was to distinguish the

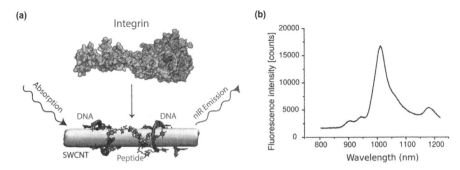

FIGURE 6.9 Functionalized nIR fluorescent SWCNTs for integrin targeting. (a) Molecular dynamics (MD) simulation of DNA-functionalized SWCNTs. In the middle of the DNA sequence is a small peptide sequence (RGD), which binds to specific parts (white circle) of cell surface receptors (integrins e.g. $\alpha_{IIb}\beta_3$pocket). (b) nIR fluorescence emission spectrum of such a DNA-peptide/SWCNT hybrid. (Reproduced from Polo et al. 2018.)

homologous neurotransmitters dopamine, epinephrine and norepinephrine (Mann et al. 2017). While SWCNT-(AT)$_{15}$ did not show any response to dopamine or norepinephrine, it showed a small fluorescence change towards epinephrine (Figure 6.10a). On the other hand, it was discovered that wrapping a specific oligonucleotide sequence (e.g. (GT)$_{15}$ or (GT)$_{10}$) around SWCNT makes the detection of the neurotransmitter dopamine in a nanomolar range possible. The oligonucleotide (GT)$_{15}$ adsorbs on the SWCNT in a way enabling the selective binding of dopamine, while other oligonucleotides do not exhibit this effect or in a less pronounced fashion. This approach enables also the detection of dopamine in the presence of an equimolar concentration of the similar neurotransmitter norepinephrine (Figure 6.10b and c).

Addition of dopamine to such sensors can be studied and observed on the single nanosensor level (Figure 6.11b). The mechanism was further studied with molecular dynamics (MD) simulations. They show that the dopamine's hydroxyl groups bind to the DNA's phosphate backbone (Figure 6.11c). This interaction ultimately causes a shift in SWCNT-fluorescence intensity through structural alterations of the corona phase. Other studies investigated

the impact of redox chemistry of the analyte on the sensor response. Redox active compounds more often caused fluorescence changes, but there was no general correlation to redox potential (Polo & Kruss 2016b). The mechanism of fluorescence modulation is therefore best in agreement with a conformational change of the corona that affects the photophysics of the SWCNT.

Immobilization of such dopamine nanosensors on a glass surface creates a platform for spatiotemporal chemical imaging of dopamine efflux from neuroprogenitor cells after an external K$^+$ stimulus (Figure 6.11a). SWCNT sensors were excited in the visible range and the nIR image collected. The cells were stimulated and the image of an array of sensors at different time points collected. It enabled for the first time to "see" dopamine release from cells and map the locations (Figure 6.11d and e). These results show that SWCNT-based fluorescent sensors provide a superior spatiotemporal resolution and enable chemical imaging of dopamine. Their nIR fluorescence is another major advantage as there is much less fluorescence background in this region of the spectrum. Together with their ultrahigh photostability, these sensors demonstrate the promises and advantages of nanosensor-based chemical imaging.

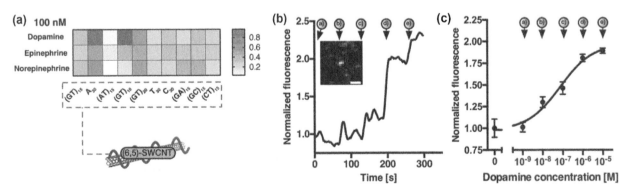

FIGURE 6.10 Selectivity and specificity of SWCNT sensors for different neurotransmitters. (a) SWCNT-based sensors were functionalized with different DNA sequences to modify and tune their selectivity towards different neurotransmitters. Here, different catecholamines (dopamine, epinephrine and norepinephrine) were tested at a concentration of 100 nM. The heat map shows clearly that there are strong differences in sensitivity and selectivity, which means that these parameters depend strongly on the exact chemical functionalization. (b and c) Detection of dopamine by a single nanosensor (addition of 1 nM (a) to 10 uM (e) in the presence of the dopamine homologue norepinephrine), demonstrates high selectivity of the sensor. (Reproduced from Mann et al. 2017.)

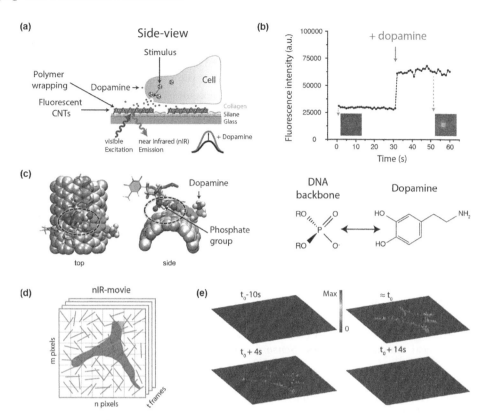

FIGURE 6.11 Real-time detection/imaging of dopamine efflux from cells. (a) Schematic of the chemical imaging concept. Dopamine-sensitive sensors based on SWCNTs are placed on a surface. On top, cells are cultivated that are known to release dopamine upon stimulation. (b) The addition of dopamine increases the fluorescence intensity of a single DNA functionalized SWCNT. (c) The mechanism of fluorescence modulation is attributed to interactions between the DNA backbone and the hydroxyl groups of the dopamine. The phosphate groups are pushed closer to the surface of the SWCNTs, which leads to an increase in fluorescence quantum yield of the SWCNTs. (d) Schematic of the chemical imaging idea. The cells are cultivated on top of the nanosensor-coated surface. When stimulated, they release dopamine and every sensor or in other words every pixel reports about its local concentration. The fluorescence image is therefore a chemical image of the dopamine concentration. (e) Response images that show dopamine release after stimulation at t_0 at the cell borders. After some time, dopamine has diffused away and the response corresponding to the dopamine concentration goes back to the base level. (Reproduced from Kruss et al. 2017.)

mRNA-Imaging Using Aptamer-Modified Gold Nanoparticles

Developments in RNA/DNA synthesis and functionalization have led to the idea of selecting new nucleic acid-based recognition units called aptamers (Tombelli et al. 2007). Aptamers are either single-stranded DNA or RNA molecules containing 20–80 nucleotides and can be selected for different targets starting from a huge library of molecules containing randomly created sequences for targeting purposes (Tombelli et al. 2005). These small-structured RNAs have the ability to bind to viral or host proteins with high affinity and specificity and also lots of features such as small size, flexibility, chemical simplicity, and reversible denaturation make aptamers superior when compared with other receptors like antibodies (Sullivan 2002). In order to make aptamers gain a high affinity for target molecules, a selection process called systematic evolution of ligands by exponential enrichment (SELEX) was discovered (Tuerk & Gold 1990). This strategy allows the isolated aptamers to recognize cells without prior knowledge of the target

molecules. Aptamers selected by these new SELEX techniques bind to certain molecules with high affinity and specificity and they can also be chemically modified with relative ease (Du et al. 2015; Wu et al. 2014).

By using DNA aptamer-modified nanomaterials, like gold NP, it was reported to image specific mRNA in living cells (Prigodich et al. 2012). Those multiplexed nanoflares are decorated with fluorophore-labeled complimentary nucleotide sequences, while an interaction with the target mRNA results in a fluorescence change. In this case, the fluorophores are initially quenched by the gold NP whereby the displacement of the complementary-RNA-fluorophores forces a fluorescence increase.

6.3 Conclusions and Future Perspectives

Nanosensor research combines the investigation of molecular recognition, nanomaterial synthesis and photophysics with microscopy, which makes the field quite multidisciplinary. In this chapter, a number of different approaches for chemical imaging has been discussed ranging from traditional techniques to novel platforms with relevant examples. Chemical imaging with fluorescent nanosensors is an emerging concept that takes advantage of nanomaterial properties (high photostability, unique spectral ranges such as the nIR, rich surface chemistry, sensitivity, ...). The imaging part adds the spatial dimension and enables one to study dynamics in highly interesting samples such as biological systems. In summary, chemical imaging with fluorescent nanosensors has the potential to enable major scientific breakthroughs and will become an important tool to study complex biological and chemical systems.

References

Ahmed A, Rushworth JV, Hirst NA et al. 2014. Biosensors for whole-cell bacterial detection. *Clin Microbiol Rev.* 27:631–646.

Ansari AA, Alhoshan M, Alsalhi MS et al. 2010. Prospects of nanotechnology in clinical immunodiagnostics. *Sensors.* 10:6535–6581.

Asai K, Amao Y, Iijima Y et al. 2002. Novel pressure-sensitive paint for cryogenic and unsteady wind-tunnel testing. *J Thermophys Heat Transf.* 16:109–115.

Basabe-Desmonts L, Reinhoudt DN, Crego-Calama M. 2007. Design of fluorescent materials for chemical sensing. *Chem Soc Rev.* 36:993–1017.

Bell JH, Schairer ET, Hand LA et al. 2001. Surface pressure measurments using luminiscent coatings. *Annu Rev Fluid Mech.* 33:155–206.

Bianco A, Kostarelos K, Partidos D et al. 2005. Biomedical applications of functionalised carbon nanotubes. *Chem Commun.* 571–577.

Biju V. 2014. Chemical modifications and bioconjugate reactions of nanomaterials for sensing, imaging, drug delivery and therapy. *Chem Soc Rev.* 43:737–962.

Boardman AK, Wong WS, Premasiri WR et al. 2016. Rapid detection of bacteria from blood with surface-enhanced Raman spectroscopy. *Anal Chem.* 88:8026–8035.

Carter KP, Young AM, Palmer AE. 2014. Fluorescent sensors for measuring metal ions in living systems. *Chem Rev.* 114:4564–4601.

Cheng IF, Chang HC, Chen TY et al. 2013. Rapid (<5 min) identification of pathogen in human blood by electrokinetic concentration and surface-enhanced raman spectroscopy. *Sci Rep.* 3:1–8.

Cialla-May D, Zheng X-S, Weber K et al. 2017. Recent progress in surface-enhanced Raman spectroscopy for biological and biomedical applications: From cells to clinics. *Chem Soc Rev.* 46:3945–3961.

Du F, Guo L, Qin Q et al. 2015. Recent advances in aptamer-functionalized materials in sample preparation. *Trends Anal Chem.* 67:134–146.

Dunham SJB, Ellis JF, Li B et al. 2017. Mass spectrometry imaging of complex microbial communities. *Acc Chem Res.* 50:96–104.

Engler RH, Mérienne MC, Klein C et al. 2002. Application of PSP in low speed flows. *Aerosp Sci Technol.* 6:313–322.

Fan X, White IM, Shopova SI et al. 2008. Sensitive optical biosensors for unlabeled targets: A review. *Anal Chim Acta.* 620:8–26.

Foster RA, Sztejrenszus S, Kuypers MMM. 2013. Measuring carbon and N_2 fixation in field populations of colonial and free-living unicellular cyanobacteria using nanometer-scale secondary ion mass spectrometry. *J Phycol.* 49:502–516.

Gao D, Huang X, Tao Y. 2016. A critical review of NanoSIMS in analysis of microbial metabolic activities at single-cell level. *Crit Rev Biotechnol.* 36:884–890.

Gillespie DT. 1978. Monte Carlo simulation of random walks with residence time dependent transition probability rates. *J Comput Phys.* 28:395–407.

Goode JA, Rushworth JVH, Millner PA. 2015. Biosensor regeneration: A review of common techniques and outcomes. *Langmuir.* 31:6267–6276.

Hossain MK, Cho HY, Kim KJ et al. 2015. In situ monitoring of doxorubicin release from biohybrid nanoparticles modified with antibody and cell-penetrating peptides in breast cancer cells using surface-enhanced Raman spectroscopy. *Biosens Bioelectron.* 71:300–305.

Kneipp K, Wang Y, Kneipp H et al. 1997. Single molecule detection using surface-enhanced Raman scattering (SERS). *Phys Rev Lett.* 78:1667–1670.

Krafft C, Knetschke T, Siegner A et al. 2003. Mapping of single cells by near infrared Raman microspectroscopy. *Vib Spectrosc.* 32:75–83.

Kruss S, Erpenbeck L, Amschler K et al. 2013a. Adhesion maturation of neutrophils on nanoscopically presented platelet glycoprotein ibα. *ACS Nano.* 7:9984–9996.

Kruss S, Hilmer AJ, Zhang J et al. 2013b. Carbon nanotubes as optical biomedical sensors. *Adv Drug Deliv Rev.* 65:1933–1950.

Kruss S, Landry MP, Vander Ende E et al. 2014. Neurotransmitter detection using corona phase molecular recognition on fluorescent single-walled carbon nanotube sensors. *J Am Chem Soc.* 136:713–724.

Kruss S, Srot V, Van Aken PA et al. 2012. Au-Ag hybrid nanoparticle patterns of tunable size and density on glass and polymeric supports. *Langmuir.* 28:1562–1568.

Liu Y, Ashton JR, Moding EJ et al. 2015. A plasmonic gold nanostar theranostic probe for in vivo tumor imaging and photothermal therapy. *Theranostics.* 5:946–960.

Lorenz B, Wichmann C, Stöckel S et al. 2017. Cultivation-free raman spectroscopic investigations of bacteria. *Trends Microbiol.* 25:413–424.

Mann FA, Herrmann N, Meyer D et al. 2017. Tuning selectivity of fluorescent carbon nanotube-based neurotransmitter sensors. *Sensors.* 17:1521–1534.

Mann FA, Horlebein J, Meyer NF et al. 2018. Carbon nanotubes encapsulated in coiled-coil peptide barrels. *Chem—A Eur J.* 24:1–6.

Maquelin K, Kirschner C, Choo-smith L et al. 2002. Identification of medically relevant microorganisms by vibrational spectroscopy. *J Microbiol Methods.* 51: 255–271.

Meyer D, Hagemann A, Kruss S. 2017. Kinetic requirements for spatiotemporal chemical imaging with fluorescent nanosensors. *ACS Nano.* 11:4017–4027.

Morrison DWG, Dokmeci MR, Demirci U. 2008. Clinical applications of micro- and nanoscale biosensors. In: *Biomedical Nanostructures*, eds. Gonsalves KE, Halberstadt CR, Laurencin CT et al., 433–454. Hoboken, NJ: John Wiley & Sons.

Mu B, Zhang J, Mcnicholas TP et al. 2014. Recent advances in molecular recognition based on nanoengineered platforms. *Acc Chem Res.* 47:979–988.

Musat N, Foster R, Vagner T et al. 2012. Detecting metabolic activities in single cells, with emphasis on nanoSIMS. *FEMS Microbiol Rev.* 36:486–511.

Musat N, Halm H, Winterholler B et al. 2008. A single-cell view on the ecophysiology of anaerobic phototrophic bacteria. *Proc Natl Acad Sci.* 105:17861–17866.

Polo E, Kruss S. 2016a. Nanosensors for neurotransmitters. *Anal Bioanal Chem.* 408:2727–2741.

Polo E, Kruss S. 2016b. Impact of redox-active molecules on the fluorescence of polymer-wrapped carbon nanotubes. *J Phys Chem C.* 120:3061–3070.

Polo E, Nitka TT, Neubert E et al. 2018. Control of integrin affinity by confining RGD peptides on fluorescent carbon nanotubes. *ACS Appl Mater Interfaces.* 10:17693–17703.

Prigodich AE, Randeria PS, Briley WE et al. 2012. Multiplexed nanoflares: MRNA detection in live cells. *Anal Chem.* 84:2062–2066.

Raman CV. 1928. A new radiation. *Indian J Phys.* 2:387–398.

Renslow RS, Lindemann SR, Cole JK et al. 2016. Quantifying element incorporation in multispecies biofilms using nanoscale secondary ion mass spectrometry image analysis. *Biointerphases.* 11:1–9.

Rosch P, Harz M, Schmitt M et al. 2005. Chemotaxonomic identification of single bacteria by micro-raman spectroscopy: Application to clean-room-relevant biological contaminations. *Appl Environ Microbiol.* 71:1626–1637.

Saka SK, Vogts A, Kröhnert K et al. 2014. Correlated optical and isotopic nanoscopy. *Nat Commun.* 5:1–8.

Schäferling M. 2012. The art of fluorescence imaging with chemical sensors. *Angew Chemie Int Ed.* 51:3532–3554.

Schreml S, Meier RJ, Weiß KT et al. 2012. A sprayable luminescent pH sensor and its use for wound imaging in vivo. *Exp Dermatol.* 21:951–953.

Schreml S, Meier RJ, Wolfbeis OS et al. 2011. 2D luminescence imaging of pH in vivo. *Proc Natl Acad Sci.* 108:2432–2437.

Shariatgorji M, Svenningsson P, Andrén PE. 2014. Mass spectrometry imaging, an emerging technology in neuropsychopharmacology. *Neuropsychopharmacology.* 39:34–39.

Spengler B. 2015. Mass spectrometry imaging of biomolecular information. *Anal Chem.* 87:64–82.

Stöckel S, Stanca AS, Helbig J et al. 2015. Raman spectroscopic monitoring of the growth of pigmented and non-pigmented mycobacteria. *Anal Bioanal Chem.* 407:8919–8923.

Sullivan CKO. 2002. Aptasensors: The future of biosensing. *Anal Bioanal Chem.* 372:44–48.

Surana S, Krishnan Y. 2013. Cellular and subcellular nanotechnology. *Cell Subcell Nanotechnol.* 991:9–23.

Tombelli S, Minunni M, Mascini M. 2005. Analytical applications of aptamers. *Biosens Bioelectron.* 20:2424–2434.

Tombelli S, Minunni M, Mascini M. 2007. Aptamers-based assays for diagnostics, environmental and food analysis. *Biomol Eng.* 24:191–200.

Tuerk C, Gold L. 1990. Systematic evolution of ligands by exponential enrichment: RNA ligands to bacteriophage T4 DNA polymerase. *Science.* 249:505–510.

Walter A, Reinicke M, Bocklitz T et al. 2011. Raman spectroscopic detection of physiology changes in plasmid-bearing Escherichia coli with and without antibiotic treatment. *Anal Bioanal Chem.* 400:2763–2773.

Watrous JD, Alexandrov T, Dorrestein PC. 2011. The evolving field of imaging mass spectrometry and its impact on future biological research. *J Mass Spectrom.* 46:209–222.

Watrous JD, Dorrestein PC. 2011. Imaging mass spectrometry in microbiology. *Nat Rev Microbiol.* 9:683–694.

Wu J, Zhu Y, Xue F et al. 2014. Recent trends in SELEX technique and its application to food safety monitoring. *Microchim Acta.* 181:479–491.

Yu Q, Wang Q, Li B et al. 2015. Technological development of antibody immobilization for optical immunoassays: Progress and prospects. *Crit Rev Anal Chem.* 45:62–75.

Zhang JJ, Cheng FF, Li JJ et al. 2016. Fluorescent nanoprobes for sensing and imaging of metal ions: Recent advances and future perspectives. *Nano Today.* 11: 309–329.

Zhang W, Wang Y, Sun X et al. 2014. Mesoporous titania based yolk–shell nanoparticles as multifunctional theranostic platforms for SERS imaging and chemo-photothermal treatment. *Nanoscale.* 6:14514–14522.

Zhou F, Resasco DE, Chen WR. 2009. Cancer photothermal therapy in the near-infrared region by using single-walled carbon nanotubes. *J Biomed Opt.* 14:1–7.

Zhu H, Fan J, Du J et al. 2016. Fluorescent probes for sensing and imaging within specific cellular organelles. *Acc Chem Res.* 49:2115–2126.

Zimmerman TA, Monroe EB, Tucker KR et al. 2008. Imaging of cells and tissues with mass spectrometry: Adding chemical information to imaging. *Methods Cell Biol.* 89:361–390.

Nanometer-Scale and Low-Density Imaging with Extreme Ultraviolet and Soft X-ray Radiation

Przemysław W. Wachulak
Military University of Technology

7.1 Introduction

The discovery and developments of visible light microscopy undoubtedly moved forward the advancements in various branches of science. Since its discovery, it significantly benefited biology, medicine, physics, material science and many other fields, allowing for major contributions in almost all aspects of life. Work initiated by Hans and Zacharias Janssen, Robert Hooke, Anton van Leeuwenhoek and later by Ernst Abbe, among others, allowed finally to construct diffraction-limited visible-light optical systems, capable of imaging features with a few hundreds of nanometers spatial resolution.

However, as the wavelength of visible radiation is between 400 and 700 nm, it limits the possibility to go beyond 200 nm spatial resolution in classical, diffraction-limited systems. This is due to the spatial resolution, expressed by the Rayleigh criterion [1] shown in Eq. 7.1:

$$\delta_{\text{Ray}} = \frac{a \cdot \lambda}{\text{NA}}, \qquad (7.1)$$

where λ is the illumination wavelength, NA is the numerical aperture of the objective or maximum half-angle subtended by the marginal rays during the recording (in case of lensless-type imaging methods, such as holography, coherent diffraction imaging (CDI) or ptychography) and a is a constant from 0.34 to 1, depending on the coherence parameter $m = \text{NA}_c/\text{NA}_o$ [2] (ratio of NA of the condenser and objective, or more generally, NA of the illumination of the sample to NA of the collecting of the radiation and recording).

Herein, we would like to show the application of the extreme ultraviolet (EUV) and soft X-ray (SXR) radiation for high-resolution imaging. Most popular sources of such radiations will be briefly mentioned. Synchrotrons and compact sources, such as high-order harmonic generation (HHG) sources, capillary discharge lasers (CDLs), laser-produced plasma (LPP) sources based on gas, liquid, and solid targets, are the sources most often employed for such purpose. Following that, two main types of imaging will be presented, namely: coherent and incoherent type, depending on the degree of spatial and temporal coherence of the EUV and SXR radiation, illuminating the sample under investigation. Coherent sources, such as HHG, CDL, free electron lasers (FELs), are employed to perform holography (Gabor type, Fourier type), Talbot imaging, diffraction imaging, ptychography and X-ray coherence tomography (XCT). The incoherent sources, such as laser-plasma sources and synchrotrons, are widely used for full-field imaging, scanning microscopy, tomography and contact microscopy.

For each technique, the principle of operation will be explained (description and scheme), their advantages and drawbacks will be mentioned and compared to the other techniques and the examples of their use will be shown.

Moreover, references related to the presented material are provided to direct the reader to seek further, extended information.

7.2 Extreme Ultraviolet and Soft X-ray Radiation

According to Eq. 7.1, to overcome the limitation of low spatial resolution in photon-based imaging systems is to decrease the illumination wavelength, thus, the wide interest of the scientific community in microscopy at the EUV and SXR wavelengths (Figure 7.1).

Among various wavelengths from the EUV ($\lambda = 10$–120 nm) and SXR ($\lambda = 0.1$–10 nm) spectral regions [3], the particularly suitable range of wavelengths for biological imaging is the so-called "water-window" spectral range. It is between 2.3 and 4.4 nm, where a high contrast is obtainable due to a difference in absorption of different constituents of the biological specimen, namely carbon and water present in the biological samples.

On the other hand, an intrinsic property of the EUV radiation is high absorption in any material. A few hundreds of nanometers of any solid material [4], and a few millimeters of any gas [5], is sufficient to absorb most of the EUV radiation incident upon such material. This radiation is dedicated to performing imaging of very thin samples, with transmissions in other wavelength ranges being too high to obtain optical contrast.

7.3 Coherent and Incoherent EUV and SXR Sources

Among various source parameters, including the wavelength of emission, source size, emission divergence, number of photons, and polarization, very important are also coherence parameters. Typical thermal sources, such as one depicted in Figure 7.2a, emit radiation with low coherence. The coherence should be divided, however, into two types: spatial and temporal coherence. Spatial coherence is related to the ability of two rays, emerging from different locations in space to interfere, for example, different spatial locations of the

emitting surface of the source. Thus, if the rays are emitted from a smaller area in space, the spatial coherence of the source increases, see Figure 7.2b. Temporal coherence, on the other hand, considers the ability of two rays separated in time but originating from the same region of space to interfere. Thus, spatial coherence depends on the source size and distance from the source, while the temporal coherence depends on the monochromaticity of the source emission [1], Figure 7.2c. Of course, the source can be both, spatially and temporary coherent. In the ideal case, such source emits a monochromatic radiation from a point-like spatial location, Figure 7.2d.

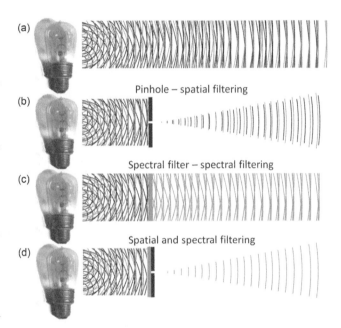

FIGURE 7.2 Schematically depicted emission from spatially and temporally incoherent thermal source (a). Spatial filtering using a pinhole increases the spatial coherence of the source (b), while spectral filtering increases the temporal coherence of the source (c). Spectral and spatial filtering produces a temporally and spatially coherent beam (d), which in this case is a spherical wave front of monochromatic radiation. Please notice greatly diminished power after the filtering is performed. Based on Ref. [6].

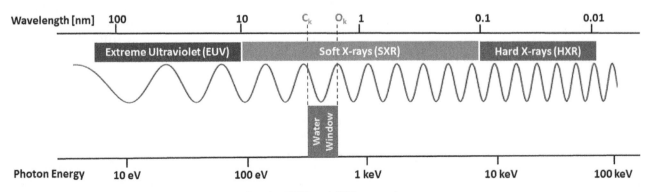

FIGURE 7.1 The electromagnetic spectrum, indicating EUV and SXR spectral range.

7.3.1 Incoherent Sources of EUV/SXR Radiation

In the incoherent sources, there is practically no spatial and temporal phase relationship between the two electromagnetic waves, originating from nonoverlapping spatiotemporal coordinates. Thus, upon their overlap, no interference phenomena will occur. This means that there is no modulation intensity in the region where the waves are superimposed. Of course, if the source is fully coherent, the modulation between those two waves is maximal; however, in case of partial coherence, the modulation will be reduced, accordingly to the degree of coherence. Herein, we will discuss the sources of incoherent EUV and SXR radiation with very low spatial and temporal coherence, which, for all practical reasons, might be considered incoherent.

Synchrotron Sources

Synchrotron radiation is generated by charged particles traveling at relativistic speeds in applied magnetic fields. This causes the particles to travel along curved paths [7]. The synchrotron radiation is produced in storage rings, Figure 7.3, where charged particles, such as electrons or positrons, are kept circulating at constant energy. The electrons are circulating inside the storage ring, composed by a series of magnets (bending magnets) separated by straight sections, where insertion devices are placed (undulators and wigglers). As the electrons are deflected through the magnetic field created by the magnets in bending magnets, short undulators or wigglers, they give off electromagnetic radiation, so that a beam of synchrotron light is produced.

Large-scale synchrotron facilities, which are state-of-the-art high brightness and tunable sources of EUV and SXR radiation, were already extensively utilized for nanoimaging and the implementation of full-field microscopes with spatial resolution of 12 nm [8] or 14 nm utilizing undulator radiation at a wavelength of $\lambda = 1.38$ nm, and third-order zone-plate diffraction [9], also including magnetic material imaging [10–13]. SXR microscopy with synchrotrons has been successfully employed mainly in transmission mode, either using diffractive optics, such as zone plates [14–16], raster scanning of the sample by focused SXR beam [17–19] or as a contact microscopy, where the sample is placed on top of a recording medium, such as a photoresist, and illuminated by SXR beam to make a "picture" of the specimen in the surface of the recording medium [20–22]. Synchrotron radiation at $\lambda = 2.4$ nm was used for imaging frozen-hydrated samples at atmospheric pressure, where details inside cells of algae as small as 35 nm were visible [23] or to examine rapidly frozen mouse 3T3 cells and obtain excellent cellular morphology at better than 50 nm lateral resolution, using transmission SXR microscope [24]. The synchrotron-based microscope in the "water-window" spectral range was developed to image frozen-hydrated specimens with a thickness of up to 10 μm at temperatures of around 100 K [25]. The various capabilities of full-field transmission X-ray microscopy, 3D X-ray tomography, Zernike phase contrast, quantification of absorption, and chemical identification via X-ray fluorescence and X-ray absorption near edge structure imaging are now also possible with synchrotron light [26] for characterization of biomaterials.

Even though in the synchrotrons the source size may be small and the use of wigglers and undulators improves the temporal coherence by spectral narrowing of the emitted radiation, the synchrotron sources are, generally, considered to be low-coherence or incoherent sources.

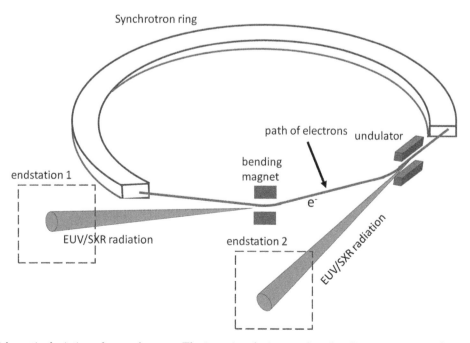

FIGURE 7.3 Schematic depiction of a synchrotron. The insertion devices, such as bending magnets, wigglers and undulators, are changing the straight path of electrons producing electromagnetic radiation which is directed to end stations for various experiments.

Laser-Plasma Sources (Gas Liquid, Solid Targets)

Laser-plasma sources operate by the interaction of laser pulses with a target to create a plasma [27], Figure 7.4. The simplest targets are solid-state bulk material [28], (Figure 7.4a). More elaborated solid-state targets are in the form of tapes [29], which provide a fresh surface of a metal foil such as Ta, Ti as well as Cu with the thickness of the order of microns [30], or microfoams [31]. The problem with a solid target, however, is the debris (ablation material) production, which limits significantly the use of that target together with fragile and expensive optics, such as zone plates or X-ray mirrors and for applications such as EUV lithography. Moreover, laser plasma from solid targets easily reach critical density:

$$n_{cr} = \frac{\varepsilon_0 m_e}{q_e^2}\omega^2, \qquad (7.2)$$

where q_e is electron charge, m_e is electron mass, ω is an angular frequency of the driving field and ε_0 is permittivity of a free space. Above the critical density, the incoming laser wave front is reflected back towards its source, prohibiting a further energy deposition into the plasma and reducing the effectiveness of energy conversion.

Liquid targets (Figure 7.4b) are also often employed in laser–matter interaction experiments, especially in the generation of short-wavelength radiation, for high spatial resolution microscopy [32] and tomography [33], for

example. Liquid targets, although provide a mass-limited target, require complex and expensive delivery system and a proper synchronization of droplet delivery to the laser focus.

Another possibility is of course to use gaseous targets. These targets are relatively simple in construction, providing regenerated (fresh) target for each laser–matter interaction; however, the target density is smaller than the solid or liquid targets, limiting the efficiency of the pump to EUV/SXR radiation conversion. Among gaseous targets, different solutions were reported so far. Over the years, gaseous targets have evolved from simple single-stream gas jets (Figure 7.4c), double-stream gas puff targets (Figure 7.4d), hollow-core fibers [34] elongated gas-jet targets [35], modulated diameter capillaries for phase matching required for HHG [36], gas cells [37], a combination of two gaseous targets for the purpose of coherent superposition and efficiency enhancement [38], array of gas jets [39], up to recently modulated, dual-gas multijet gas puff targets [40,41].

LPP sources, emitting short-wavelength radiation in the SXR and EUV spectral ranges, offer an important alternative to be the drivers dedicated to compact imaging systems [42]. They allow overpassing the limited accessibility of large facilities [43], maintaining a comparable spatial resolution [44]. Over the last few years, many efforts have been made to perform nanometer spatial-resolution imaging in the EUV and SXR spectral ranges employing

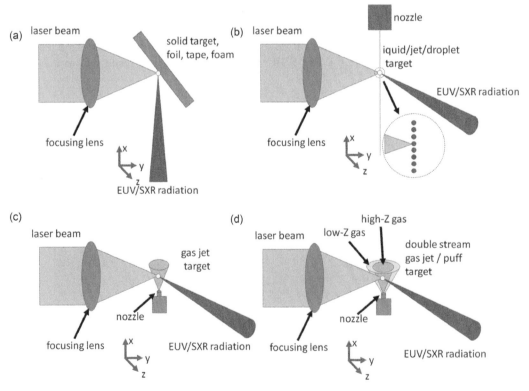

FIGURE 7.4 Schematic depiction of laser-plasma sources based on solid target in a form of bulk material, foil, tape or foam (a). Liquid jet or droplet source, where a focused laser beam interacts with a single droplet producing a plasma (b). Single-stream gas jet/puff laser-plasma source (c) and double-stream gas puff target laser-plasma source (d). The second gas is used for increasing the density of the first (inner, high Z-gas) to increase the target density.

compact sources. An SXR source, emitting at $\lambda = 2.88$ nm, based on a liquid jet nitrogen target was employed recently for microscopy in the "water-window" range with a sub-50 nm spatial resolution [45]. Very nice results were recently demonstrated [42], using the liquid nitrogen target-based system and 1.3 kHz repetition rate Nd:YAG laser, to record images of biological samples and nanostructures with a half-pitch spatial resolution of 40–50 nm.

Discharge Sources

Discharge sources operate on the principle of plasma generation in low-density targets, gasses, by means of high-voltage discharge, Figure 7.5. Typical discharge between electrodes may not achieve sufficiently high temperatures to emit radiation in the EUV and SXR region unless the discharge current is sufficiently high. If that condition is fulfilled, the electric current in highly conducting plasma generates a magnetic field that compresses the plasma and produces so-called pinching effect [46], or Z-pinch, referring to the direction of the current flow (along z- or optical axis).

The Z-pinch sources are based on the application of Lorentz force, in which a conductor carrying electrical current and placed in magnetic field experiences this force. The total Lorentz force is the volume integral over the charge distribution:

$$\vec{F} = \iiint \left(\rho \vec{E} + \vec{J} \times \vec{B} \right) dV, \qquad (7.3)$$

where ρ is the charge density (charge per unit volume). $\vec{E}$ is electric field vector, $\vec{J}$ is a current density vector and $\vec{B}$ is a magnetic field vector.

In Z-pinch discharge sources, the conductor is a plasma. When the electrical current flows through the plasma, the magnetic field compresses the plasma through the Lorentz force, increasing plasma pressure and its temperature. Those sources are now commercially available [47], i.e. xenon target based, operating with repetitions of a few kHz, and are used for nanoimaging and have possible application in the inspection of EUV masks [48].

7.3.2 Coherent Sources of EUV/SXR Radiation

In the coherent sources, there is strong phase relationship between the two electromagnetic waves, meaning that upon their overlap, the interference phenomena will occur and the intensity modulation in the interference pattern will be very high. Herein, we will show some examples of sources, emitting coherent EUV and SXR radiation.

Free Electron Lasers

FELs produce coherent, femtosecond X-ray pulses with peak spectral brightness a few orders of magnitude higher compared to third-generation synchrotron sources. In the FEL, a relativistic electron beam is traversing undulator or wiggler (with a modulated magnetic field) producing electromagnetic radiation, which builds up in the resonant cavity, similarly to laser scheme. This approach is very common in the visible and infrared (IR) region, where an oscillator configuration is employed, in which radiation is propagating in an optical cavity and interacting with electron bunches on successive, multiple passes. FEL's technique can be implemented on different types of accelerators, such as storage rings or linear accelerators. Storage rings, because of the electron beam recirculation, provide longer electron bunches (10–30 ps). Linear accelerators, on the other hand, are single pass systems, which provide much shorter bunches, in the range of 10 fs–10 ps duration. Those short-duration pulses are of interest for ultrashort radiation pulse generation or for applications of high-electron beam densities [49].

In order to reach the resonant wavelength, Eq. 7.4, coherence and high energy output, the employment of high beam energies is required. The fundamental wavelength of an undulator can be expressed as:

$$\lambda_1 = \frac{\left(1 + K^2/2 \right)}{2\gamma^2} \lambda_u, \qquad (7.4)$$

where $K \leq 1$ is electron deflection parameter (for undulator), $\gamma = E_e / \left(m_e c^2 \right)$, λ_u is an undulator magnetic period, E_e is the electron energy, m_e is the electron mass and c is the speed of light.

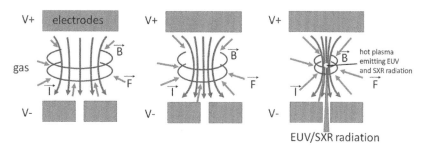

FIGURE 7.5 A basic explanation of the Z-pinch process. A high voltage is applied to electrodes. The region between electrodes is filled with a gas in which a Z-pinch plasma will be created. If the product of the voltage and the charge is higher than the ionization energy of the gas, the gas ionizes, which allows a high current flow across the gap between electrodes. The current flow produces a perpendicular magnetic field. The magnetic field compresses the target material to form a sufficiently hot plasma, which emits EUV and SXR radiation.

The mirrors for FEL optical cavities operating in the EUV and SXR ranges have very low reflectivity. This problem can be eliminated by the employment of very long undulators (i.e., 100 m in length range for $\lambda = 0.1$ nm) and high-electron beam density for ensuring a sufficient gain. The undulators are specially designed configurations of alternatively polarized magnetic fields, in which electron bunches are traveling through an undulator on an oscillating path in a way that the radiation they emit can interact with themselves. The radiation, spontaneously emitted by the wiggling electrons, is amplified later during the passage of the electron bunches through the undulator, generating short, monochromatic light pulses. Thus, the principle of operation of FEL in the EUV and SXR region is based on "self-amplified spontaneous emission" (SASE) technique, which, on one hand, eliminates low reflectivity mirrors but, on the other hand, requires a very high-quality electron beam and a long undulator [50] (Figure 7.6).

In long undulators, a microbunching takes place, which leads to a very high amplification of the radiation, many times more than in short undulators. Relativistic electron bunches passing an undulator have a much higher spatial extension than the wavelength of the radiation emitted by them. Thus, some electrons are losing energy to the light wave, falling back within the bunch and some of them are gaining energy from the wave. In the consequence, there is a concentration of electrons at particular spatial positions, called the microbunches. These microbunches are one wavelength distance apart from each other. It causes a massive amplification because at the microbunch forming point all electrons are moving, radiating the energy and amplifying coherently. The microbunching process stops when Coulomb forces restrict a further compression of the microbunches. At this point, called the regime of saturation, energy is transferred to the radiation back and forth by the microbunches [51].

The output of FEL lasers is being constantly improved; however, some typical parameters, i.e. the DESY's FLASH FEL, in Hamburg, Germany, dedicated to EUV and SXR studies, are photon energy range 24–295 eV, corresponding to a wavelength of 4.2–51 nm, photon beam size (FWHM) 65–95 μm, 3–27 μrad beam divergence, pulse duration 30–300 fs, average pulse energy, depending on the pulse duration and wavelength $E = 1$–500 μJ. FLASH also produces bright emission at the third harmonic of the fundamental mode (currently, 2018) down to 1.7 nm wavelength [52]. The pulse duration of the fundamental and harmonics is in the tens of femtosecond range, opening up the possibility of studying deep inner-shell atomic and molecular dynamics on a sub-femtosecond timescale or determination of crystallographic structure through diffraction lens-less imaging [53]. Similar experiments were conducted in the United States, at the SLAC Linac Coherent Light Source (LCLS) [54], where with an X-ray pulse with a photon energy of 8.52 keV ($\lambda = 1.45$ Å) it was possible to investigate protein crystal structure with 2.9 Å spatial resolution [55]. The FEL is tunable and has a very wide range of tunability, currently ranging from microwaves, through terahertz radiation, IR, visible spectrum, up to ultraviolet and X-rays [56].

High-Order Harmonics Generation

HHG is a nonlinear process, in which intense laser pulse, femtosecond range in duration, illuminates a target, Figure 7.7. During the interaction of laser pulses with the target material, the high harmonics (much higher than second, third or fourth, as in the visible wavelength range) are produced. Harmonic generation is a process, where laser light at frequency ω produces a radiation at new frequencies $n \cdot \omega$, where n is an integer number. The process was discovered by Franken in 1961 [57]. The first high harmonic generation was observed in 1977 [58] due to the interaction of CO_2 laser pulses with plasma from a solid target. The much more popular approach, used today, employs gaseous targets. It was first demonstrated by McPherson [59] and later by Ferray [60], where the plateau in the intensity of high harmonics was found, instead of continuous drop, as predicted by decreasing probability of absorbing n photons in a multiple-photon absorption process. This plateau is extremely useful because it spans hundreds of eV, extending into SXR region, ending abruptly afterward at energy called harmonic cut-off, currently in keV range [61]. The maximal harmonic photon energy E_c related to the minimum wavelength λ_c given by the cut-off law [62] is given by:

$$E_c = \frac{h \cdot c}{\lambda_c} = I_p + 3.17 U_p, \qquad (7.5)$$

where I_p is the ionization potential of the target atoms and $U_p = q_e^2 E_0^2 / (4 m_e \omega_0^2)$ is the ponderomotive energy, E_0 is a linearly polarized electric field amplitude, q_e is electron

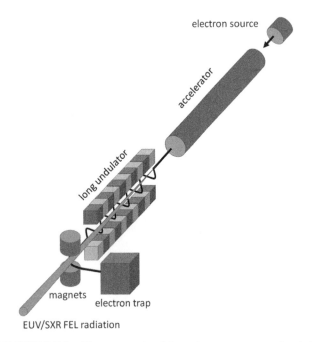

FIGURE 7.6 Electrons emitted from the source are accelerated to relativistic speeds. The relativistic electrons passing through a long undulator and emitting FEL radiation.

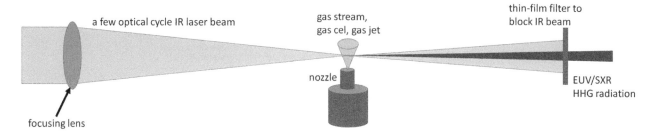

FIGURE 7.7 A scheme of a typical HHG setup using a gas target.

charge, m_e is electron mass and ω_0 is an angular frequency of the driving field.

The HHG process is most often explained using a semi-classical approach [63]. During the interaction of an atom with a high electric field of the incident laser pulse, an electron is released from an atom through the ionization process. It is assumed that the electron has zero initial velocity and is subsequently accelerated by the laser beam's electric field. After a half optical cycle, the electron reverses its direction because the electric field has changed and will accelerate back towards its ion. During a recombination process with the ionized atom, the atom returns to its ground state and a high-energy photon is released.

HHG sources are in a sense tunable since their emission wavelength can be selected by choosing an emission from a certain harmonic, the gas medium can also be changed to change the emission spectrum and saturation intensity [64]. The properties of HHG beam strongly depend on the incident driving laser field properties; thus, the harmonic generation emission is highly spatially and temporally coherent [65]. Moreover, due to the nonlinearity of the HHG process, HHG pulses are often shorter than the driving laser pulse, allowing to reach the duration range of attoseconds. The harmonics are generated co-linearly with driving laser beam and have low divergence and Gaussian beam profile. This means that HHG is a laboratory source of coherent, kilo-electronvolt X-rays, which operates on timescales relevant to chemical and biological processes.

The HHG radiation was also extensively used for nanoimaging; however, due to its coherence properties, the HHG sources were mostly used together with coherent imaging techniques, such as Gabor [66] and Fourier holography [67,68], diffraction imaging [69] or ptychography [70].

Capillary Discharge Lasers

In a CDL, which is presented in Figure 7.8, long capillaries with large length-to-diameter ratios $L/D > 100$ are used. Inside of those capillaries, an amplification medium is produced for EUV radiation generation by a direct current discharge excitation of a gas filling up the capillary. The capillary geometry provides a small volume and a resistance to Ohmic heating. Heat conduction to the capillary walls responsible for rapid plasma cooling during the decaying of the excitation current pulse allowed a large recombination rate and a population inversion.

An example of this type of source is Ne-like Ar CDL [71]. It can be configured to produce pulses with the energy of approximately 0.1 mJ and about 1.2 ns FWHM duration and operated at repetition rates of several Hz producing EUV average powers in excess of 1 mW with a high degree of spatial and temporal coherence. The laser operates at $\lambda = 46.9$ nm 3s $^1P_1$–3p $^1S_0$ transition of neon-like Ar. An alumina capillary 3.2 mm in diameter and 28 cm in length is filled with Ar. The gas is excited with a current pulse $I = 24$ kA, a 10%–90% rise time of approximately 25 ns and a first half-cycle duration of approximately 110 ns [72,73]. For the excitation current pulse, a multistage Marx generator, charged up to voltages of ~50 kV (single stage) is used. The current pulse compresses the plasma column achieving a plasma channel where a population inversion is created [74,75]. A continuous flow of Ar is injected in the front of the capillary.

The peak-to-peak beam divergence of the EUV beam is approximately 4 mrad and the laser emits unpolarized light. Spatial coherence of the source is often estimated using Young's double-slit or double-pinhole experiment. In an ideal double-slit experiment, in which both slits

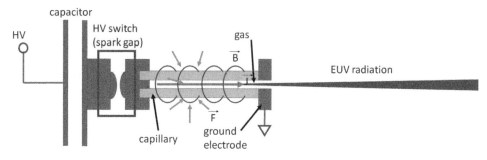

FIGURE 7.8 Scheme of the EUV CDL.

are uniformly illuminated, the modulus of the complex coherence factor is equal to the interference fringe visibility $|\mu_{12}| = V$ [76].

For 27 cm long capillary, the coherence radius R_C of the transverse coherence of the source may be defined following the convention of coherence area

$$A_C = \pi R_C^2 = \iint |\mu_{12}(\Delta x, \Delta y)|^2 \, d\Delta x \cdot d\Delta y, \qquad (7.6)$$

where $\mu_{12}(\Delta x, \Delta y)$ is the complex coherence factor and by fitting a Gaussian profile into the visibility data:

$$|\mu_{12}(\Delta x, \Delta y)| = V = \exp\left[\frac{-(\Delta x^2 + \Delta y^2)}{2R_C^2}\right] = \exp\left(\frac{-\Delta r^2}{2R_C^2}\right) \qquad (7.7)$$

where Δr is the distance between two point sources of electromagnetic waves producing the interference pattern, equivalent to slit separation. After such procedure, the $R_c \approx 0.6$ mm was estimated at 1.8 m from the capillary exit [77]. This value can be increased by increasing the length of the plasma [78].

Considering that the inverse relative bandwidth (IRB) of the source is:

$$\text{IRB} = \frac{\lambda}{\Delta\lambda}, \qquad (7.8)$$

where $\Delta\lambda$ is the wavelength emission range, IRB $\approx 2.5\text{--}3\cdot10^4$ [79], the longitudinal coherence length is

$$l_c = \frac{\lambda^2}{\Delta\lambda} = \lambda \cdot \text{IRB} \qquad (7.9)$$

For this particular source, it is equal ~1.3 mm.

It was demonstrated that the radiation from a CDL operating at a wavelength of $\lambda = 46.9$ nm EUV permits to obtain images with a spatial resolution better than 55 nm [80], and that the spatial resolution of holographic images, employing the same wavelength, can be improved up to sub-50 nm [81].

GRIP Lasers

Grazing incidence pumping (GRIP) configuration EUV and SXR sources are employing solid-state targets in a form of bulk materials. In such configuration, the pre-pulse irradiates the target at close to normal angle to form a plasma, and then, the main pulse irradiates the target at a small angle, relative to the surface of the target heating up the plasma (Figure 7.9). This approach reduces significantly the necessary pumping energy and enables operation at higher repetition rate [82]. Such geometry inherently provides traveling wave pumping, which allows matching the speed of the traveling wave

$$v = \frac{c}{\cos(\Theta)}, \qquad (7.10)$$

(where c is the speed of light and Θ is the grazing incidence angle of the laser pump—the main pulse) and amplified beam. Moreover, it increases the path length of the rays in the gain region of the plasma. For this reason, the fraction of the pump energy absorbed in the gain region is increased, improving, in turn, the source efficiency and yield.

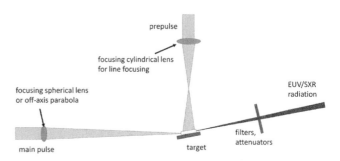

FIGURE 7.9 Schematic illustration of the GRIP pumping configuration for production of EUV and SXR radiation.

Such sources, currently producing saturated gain down to 6.85 nm wavelength and amplification down to 5.85 nm wavelength [83], exhibit high spatial and temporal coherence [84], which was exploited for incoherent, full-field EUV nanoimaging [85]. Moreover, a saturated Ni-like Ag X-ray laser [86] was employed for nanosecond temporal resolution Fourier holography [87] as well. A solid-state target-based tabletop EUV laser, emitting 13.2 nm wavelength radiation from Ni-like Cd ions, was also employed for imaging with a sub-38 nm spatial resolution [88].

Remarks

As can be seen, there is a variety of different approaches to the generation of EUV and SXR radiation. Some of them were mentioned herein, however, there are also other types of sources, i.e. optical field ionization lasers [89]. Moreover, from EUV and SXR nanoimaging standpoint, large-scale facilities such as synchrotron and FEL sources are mainly used due to their most striking advantages, such as large photon flux, tunability, and high brightness. Although these sources have obvious and unquestionable advantages and are used for state-of-the-art pioneering experiments, they are also very expensive, cannot be commercialized and are hardly accessible to the potential user. Thus, nowadays, a possibility of employment of compact tabletop and desk-top EUV and SXR sources, to perform various, novel imaging techniques, is being explored and is of high interest. For this reason, in the following paragraphs, various EUV and SXR nanoimaging techniques will be presented to give the reader a glimpse of current possibilities. This is by no means a complete overview of possible imaging techniques but a starting point to explore this interesting topic.

7.4 Incoherent EUV/SXR Imaging

The future development of nanotechnology requires tools for imaging of various objects with spatial resolution in the nanometer scale. Photon-based (bosonic-type) imaging at short wavelength vs. electron, or recently, neutron imaging has additional advantages due to a different interaction of photons with matter. Atomic resonance frequencies, leading to very high absorption coefficients at EUV and

SXR wavelengths, provide an enhanced optical contrast in those wavelength ranges, which is often exploited by various imaging methods and techniques.

The existing EUV and SXR imaging techniques can be divided very generally into two categories: coherent and incoherent. Incoherent imaging techniques are somewhat complimentary to coherent ones. Although the optical transfer function (OTF) for incoherent illumination is capable to transfer two times higher spatial frequencies than for the case of coherent illumination [76], the incoherent type of illumination often requires the use of additional optics, such as zone plate lenses. In that case, the spatial resolution of, for example, zone plate microscopy, might be inferior to coherent imaging techniques, which often does not require any additional optics, but, on the other hand, the coherence of the source can be problematic due to coherence effects present in the image, such as "twin-image" problem in holography.

The particularly suitable range of wavelengths for biological imaging is the so-called "water-window" spectral range. X-ray sources, emitting in the "water-window" region between 2.3 and 4.4 nm wavelength [90], are thus important for biological applications. High contrast in this spectral range is obtained due to a difference in absorption of different constituents of the biological specimen. While water, present in the biological sample, has a relatively small absorption coefficient in this spectral range, carbon, due to much higher absorption, gives very good contrast in the image. Thus, this spectral range is perfectly suitable for imaging of biological specimen.

The EUV range, however, offers another advantage, which is high absorption in very thin layers of any solid material [91] or just a few millimeters of gas at a fraction of atmospheric pressure is also sufficient to produce an absorption contrast [92] during the imaging process. Thus, this region is also often employed for material science application [93] such as, for example, studying micro- and nano-cracks in thin film materials [94]. A wavelength of 13.5 nm is also of particular interest related to EUV lithography [95,96]. Thus, many nanoimaging applications are related to a mask inspection [97], i.e. Zernike phase contrast microscope for EUV lithographic mask inspection [98].

7.4.1 Radiography (Shadowgraphy)

In EUV/SXR radiography, an object investigated is illuminated by an EUV or SXR radiation that partially penetrates the object and is recorded by a charge-coupled device (CCD) camera, which stores an information about local intensity changes due to a partial absorption of this radiation in the sample, as schematically depicted in Figure 7.10.

That forms a 2-D image of the object on the CCD camera. A typical experimental setup for EUV radiography, to measure a density of multijet gas targets for laser–matter interaction [99], is schematically shown in Figure 7.11.

In such setup, an IR laser pulse from Nd:YAG laser is focused onto a double-stream gas puff target to form a plasma, which is a source of radiation for radiography. The target is produced by pulsed injection of a working gas into a hollow stream of low Z-gas (helium) using an electromagnetic double-nozzle valve system. Depending on the spectral emission that is needed to be achieved, different gasses are supplied to the nozzle, to achieve efficient emission in the EUV (argon, xenon) or in the SXR region (nitrogen, krypton, argon). The emitted radiation is typically polychromatic, spanning a large wavelength range. Thus, it is filtered using thin film filters and/or multilayer mirrors to

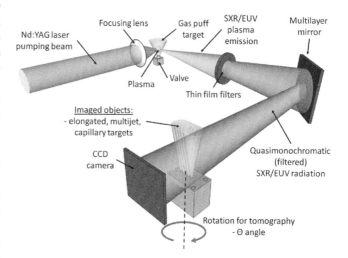

FIGURE 7.11 Experimental arrangement for the EUV/SXR radiography of gaseous object and its density estimation. (With permission from Ref. [99].)

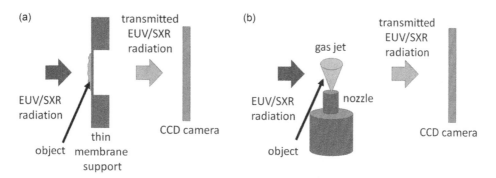

FIGURE 7.10 Schematic depiction of radiography imaging of thin solid objects (a) and gas jets (b).

provide monochromatic or quasi-monochromatic radiation. The radiation is partially and locally absorbed in an object being investigated, placed in the vicinity of the EUV/SXR-sensitive CCD camera. The monochromaticity allows for easier density estimation.

An example of such monochromatization of the EUV radiation is shown in Figure 7.12a. Typical radiogram of the gaseous object (in this case multijet gas puff target) [99] is depicted in Figure 7.12b and the density profiles of the multijet gas target are depicted in Figure 7.12c. The density can be evaluated from the radiograms by converting them to 2-D transmission maps and using the Eq. 7.11:

$$\rho(x,y) = \frac{-\ln\left[Tr(x,y)\right]}{\mu_a \cdot d(y)} \cdot m_{at}, \qquad (7.11)$$

where $Tr(x,y)$ is a 2-D transmission map of the object, $\mu_a = 2r_0 \cdot \lambda \cdot f_2$ is an atomic photoabsorption cross section, $r_0 = 2.82 \cdot 10^{-15}m$ is the classical electron radius, λ is the illumination wavelength, f_2 is the imaginary part of the atomic scattering factor, $d(y)$ is the path length on which the EUV beam is absorbed in the object, measured in direction of the EUV beam absorption, m_{at} is the atomic mass of the object (gas puff) material and (x,y) are spatial coordinates at the detector plane, horizontal and vertical, respectively.

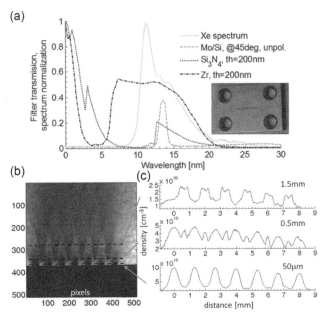

FIGURE 7.12 The spectrum obtained using transmission grating spectrometer (TGS). The emission in the EUV spectral range from Xe/He source gas puff target and transmission curves for all filters (a). The EUV radiation after filtering with a wavelength centered at 13.5 nm is depicted with a dashed line and was used for subsequent object (gas puff target) illumination. The EUV shadowgram (raw transmission data) of a single gas multijet gas puff targets produced with seven orifice nozzles at ten bar argon backing pressure (b). Multijet gas puff target density profiles for various distances from the nozzle: 50 μm, 0.5 mm and 1.5 mm (c) showing density modulations across primary and secondary jets.

7.4.2 Tomography

Tomography is an imaging technique, which is capable of three-dimensional imaging of objects, to extract information about them and render their three-dimensional representations. The schematic description of EUV/SXR tomographic imaging is depicted in Figure 7.13. From the experimental point of view, it is essentially identical to shadowgraphy (see Figure 7.11 for comparison) with one difference which is the requirement to precisely rotate the investigated object to obtain necessary projections for three-dimensional object reconstruction.

Next, the set of 2-D projections is combined and slices (images with additional orthogonal dimension to dimensions of the projections) are computed. In the next step, all slices are combined into a 3-D rendering of the object. The easiest way to obtain slices is by application of an Inverse Radon Transform (IRT); however, much better results are obtained using dedicated commercial or a freeware software, such as ImageJ with TomoJ toolbox [100] that uses weighted back-projection or iterative reconstruction methods together with post-processing segmentation.

To ensure a proper pixel segmentation number of equiangular projections, N_p has to fulfill the condition:

$$N_p \geq \frac{\pi}{2} \cdot N_d, \qquad (7.12)$$

where N_d is a number of detector pixels in the plane of object rotation. The angular step Θ can be calculated as $\Theta \leq 2\pi/N_p$. From a practical point of view, however, if obtained projections have high signal-to-noise (SNR) ratio, satisfactory results are obtained even with a much smaller number of projections [33]. In that work, a 3-D representation of diatom was reconstructed from only 53 projections covering 180° using a filtered back-projection algorithm and $\lambda = 3.37$ nm illumination wavelength from ethanol droplet target laser-plasma source.

As an example of EUV tomography approach, to visualize in 3-D low-density object is reported in Ref. [101],

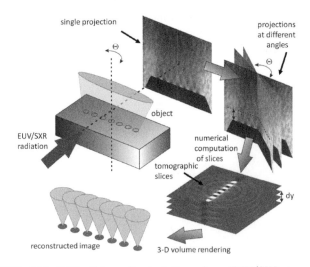

FIGURE 7.13 Schematic depiction of an EUV/SXR tomographic imaging.

where the radiograms (projections) were obtained by the EUV light illuminating the object (GPT) at $\lambda = 13.5$ nm [102,103], which is locally and partially absorbed by the gaseous object, forming an intensity image at the detector plane, further downstream the EUV beam. It is similar to shadowgraphy; however, in tomography, for each rotation angle Θ of the object, individual shadowgram of the target was recorded using the CCD camera. Next, an IRT or other dedicated software is used to obtain the so-called CT slices from two-dimensional shadowgrams. In the next step, the CT slices are stacked to form a volumetric rendering and

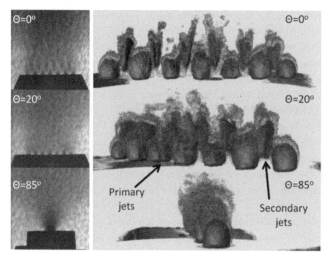

FIGURE 7.14 Shadowgrams of multijet GPT obtained at different angles (left) and the three-dimensional visualization (right) of the reconstructed object showing primary and secondary gas jets. (With permission from Ref. [99].) For comparison, they were depicted at the same angles.

visualization, such as the one in Figure 7.14, for a multijet gas object, with a density 3–4 orders of magnitude lower than the density of water. For high-quality reconstruction, SNR of the projections should be high. Moreover, the quality is better if more projections with a smaller angular step are used for volumetric reconstruction.

7.4.3 Full-Field Imaging

A full-field microscope obtains an entire image of the object at once, as a 2-D distribution of intensity recorded in the image plane of the optical system. An example of full-field EUV/SXR microscopes with on-axis a) and off-axis b) condenser configurations are schematically depicted in Figure 7.15, respectively.

The sources for those microscopes are based on a compact Nd:YAG laser, $\lambda = 1,064$ nm, 0.5–0.7 J pulse energy, 3–4 ns pulse duration and 1–10 Hz repetition rate. The laser beam is focused onto a double-stream gas puff target, which is schematically depicted in Figure 7.4d. The target is produced by an electromagnetic valve with double nozzles. Laser pulses irradiate the gaseous target producing laser plasma, which emits radiation in the EUV and SXR regions of the electromagnetic spectrum. Depending on the spectral range and the type of the microscope, the gas puff target source can be optimized for efficient emission either in the "water window" or for the EUV spectral range. The gas pressures, nozzle position with respect to the laser focus, valve timing with respect to the laser pulse were optimized in order to maximize the photon flux at the sample plane. The inner nozzle injects a small amount of working gas (high Z-gas—N_2 for SXR full-field microscope, and Ar for the EUV full-field microscope), while the outer nozzle injects an outer

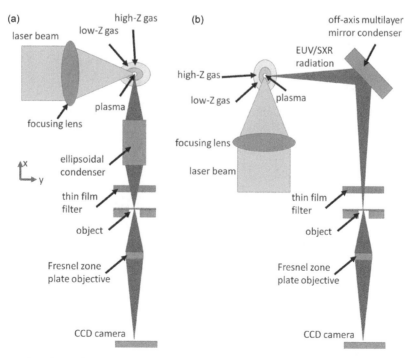

FIGURE 7.15 Two examples of EUV/SXR microscope configurations: on-axis (a) and off-axis (b).

gas (low Z-gas, He). The outer gas is used to narrow down the flow of the working gas, reducing its density gradient along the nozzle axis, to obtain higher target density by confining the target gas flow and, in turn, allow for higher photon yield in the EUV/SXR region.

In typical (exemplary, [104]) SXR microscope operating in the "water window" range, to focus the SXR radiation from the nitrogen plasma an ellipsoidal, axisymmetrical SXR on-axis condenser is used (Figure 7.15a). A titanium filter (typically 200–500 nm thick) selects the He-like nitrogen line at $\lambda = 2.88$ nm from nitrogen plasma emission. Filtered SXR radiation illuminates the sample/object. A Fresnel Zone Plate (FZP) objective is used to form a magnified image onto a back-illuminated SXR-sensitive CCD camera fulfilling the imaging condition

$$x^{-1} + y^{-1} = f^{-1}, \tag{7.13}$$

where x is the object to FZP distance and y is the FZP to CCD distance.

A silicon nitride FZP (400 nm thick, 250 μm in diameter, outer zone width of 30 nm), with a focal length $f = 2.6$ mm at 2.88 nm wavelength, is used as an objective. The focal length of the FZP lens can be calculated from the basic equation for the progression of the radii of the FZP zones:

$$r_n = \sqrt{n \cdot \lambda \cdot f + \frac{n^2 \cdot \lambda^2}{4}} \approx \sqrt{n \cdot \lambda \cdot f}, \tag{7.14}$$

where r_n is the radius of the consecutive zone, n is the zone number and λ is the wavelength for which the FZP is designed and is equal to

$$f = \frac{D \cdot \Delta r}{\lambda}, \tag{7.15}$$

where D is FZP diameter and Δr is the outer zone width of FZP equal to $\Delta r = r_n - r_{n-1}$. The NA of the FZP is then

$$\text{NA} = n_0 \cdot \sin\Theta \sim \frac{D}{2f} = \frac{\lambda}{2 \cdot \Delta r}, \tag{7.16}$$

where n_0 is the index of refraction of a vacuum, which defines the theoretical spatial resolution for FZP equal to

$$\delta_{\text{Ray}} = \frac{0.61\lambda}{\text{NA}} = 1.22\Delta r. \tag{7.17}$$

In the example experiment, the image plane was located 57 cm from the zone plate resulting in the geometrical magnification of the system $M = y/x$ of ~220×. Such a system is capable of achieving a half-pitch spatial resolution of ~60 nm. More details about the microscope can be found in Ref. [105].

Another example of a full-field microscope operates in the EUV range [106]. In such microscope, argon plasma radiation was collected and spectrally filtered using an ellipsoidal off-axis mirror coated with Mo/Si multilayers (Figure 7.15b). The condenser can be a mirror, however, Schwarzschild optics [107] or the zone plates are also being

used [108]. The condenser reflects 13.5 ± 0.5 nm radiation at 45° incidence angle. To eliminate longer wavelengths from the Ar plasma ($\lambda > 16$ nm), a 250 nm thick Zr filter is used. The sample was imaged using a poly-methyl methacrylate (PMMA) FZP objective (diameter 200 μm and the outer zone width $\Delta r = 50$ nm) onto the CCD camera. The geometrical magnification of the objective was 410×. Such a system, in this configuration, achieves a half-pitch spatial resolution of 48 nm [109]. More details about this system can be found in Ref. [110]. Synchrotron-based systems can achieve even better spatial resolution. Using the Advanced Light Source (ALS) synchrotron facility, a 10-nm spatial resolution imaging was demonstrated at 1.75 nm wavelength both in full field and in scanning mode [111].

A few examples of "water window" images of organic samples, acquired with the full-field SXR microscope are depicted in Figure 7.16a and b show a sample of CT 26 fibroblast from Mus musculus colon carcinoma (strain BALB/c), prepared on top of a 30 nm thick Si_3N_4 membrane. A direct comparison between the image acquired with a traditional optical microscope (Figure 7.16a) and the SXR microscope image (Figure 7.16b), acquired with 200 SXR pulses, at a source repetition rate of 10 Hz, exposure time of 22 s and detector (CCD) temperature of −20°C, are shown. The SXR image shows improved spatial resolution due to the employment of a shorter wavelength, beyond the diffraction limit of the optical-visible

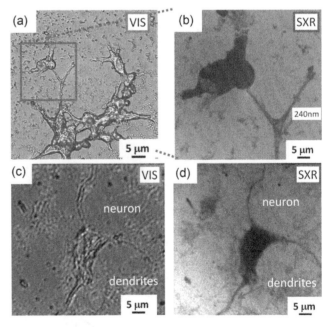

FIGURE 7.16 CT 26 fibroblast cells. Comparison of an optical image (a) and the SXR images (b) obtained with the "water window" microscope (square in (a)) that shows small features of the order of 240 nm. In (c) and (d) there is a comparison of the mouse hippocampal neuron, acquired with an optical microscope (40× objective, mag. 400×) and SXR microscope, respectively, showing a single neuron with dendrites branching out.

light microscopes. Some internal and external structures can be distinguished due to phase contrast in the visible light microscopy images and due to the modulation in the absorption of the SXR light through the sample in the SXR images. A second example consists on hippocampal neurons from E17 mouse embryos, cultured for 10 days on poly-D-lysine-coated, 50 nm thick Si_3N_4 substrates before fixation in 4% paraformaldehyde (PFA) in 20% phosphate-buffered sucrose (PBS), followed by dehydration from 100% to 70% ethanol and air drying. The neurons were also imaged using both microscopes: visible light microscope (40× objective, 400× magnification)—Figure 7.16c) and the SXR microscope (sample to CCD magnification of 410×, exposure of 200 SXR pulses, 20 s)—Figure 7.16d. The comparison with the optical image (Figure 7.16c) shows an improvement of the spatial resolution and absorption contrast, employing the "water-window" radiation. In this case, it is possible to observe that the high absorption coefficient at SXR wavelengths enhances the optical contrast to the point that barely visible neuron in Figure 7.16c is much better visualized in Figure 7.16d. In the SXR image, it is possible to distinguish the neuron and dendrites that are blurred due to inferior resolution and phase-type imaging using the optical microscope.

7.4.4 Scanning Microscopy

Scanning EUV and SXR microscopes are simpler in construction. In the scanning microscopes, instead of the condenser-objective assembly that was used to form an image in full-field microscopes, a focusing optics (Fresnel zone plate [112], ellipsoidal mirror) is used to focus an EUV/SXR beam to a small, preferably nanometer in diameter spot, see Figure 7.17.

Such spot, often called a probe, is used to illuminate a very small spatial region of the specimen to get a signal, either in transmission or in reflection mode, measured by a detector (photodiode, multichannel plate, etc.). A 2-D image of the sample is then obtained by a raster scanning of the specimen, usually placed on top of two-axis translation stages with nanometer positioning accuracy and repeatability, preferably piezoelectrically actuated.

At each probe position, various signals can be detected to map the elemental, chemical or crystallographic properties of the specimen. The signals detected from the sample are most often the transmitted EUV and SXR radiation, photoelectrons or fluorescence photons. This technique requires high photon flux and small source size (collimated EUV and SXR beams) to be able to focus down to a single nanometer spot. Thus, the scanning EUV and SXR microscopes are typically developed with the use of large-scale facilities [113].

This technique provides a 2-D map of the specimen's response to the EUV and SXR illumination. The spatial resolution in this approach is limited by the beam spot size on the specimen, related to the quality of the focusing optics and EUV/SXR beam, and is of the order of several tens of nanometers [114] or better [115]. This approach, even though is simpler in construction and requires only one piece of expensive EUV/SXR optics, is more time consuming to obtain a single image. This is especially important if the image focusing is required and the dose absorbed by the sample has to be limited.

7.4.5 Contact Microscopy

Another interesting and already well-established technique for obtaining high-resolution images of samples is projection imaging called contact microscopy. This method uses EUV/SXR radiation transmitted through the sample to expose a high-resolution photoresist underneath (typically PMMA for high resolution), being in contact with the sample. Afterward, the photoresist is chemically developed (usually a mixture of methyl isobutyl ketone [MIBK] and isopropyl alcohol [IPA] 1:1 to 1:3 solution), see Figure 7.18, and scanned with an atomic force microscope AFM or SEM to obtain an image of the relief structure—imprint. Using this method, first imprints of human blood platelets, fibroblasts and hydrated biological cells [116] were obtained.

In an example of the SXR contact microscope [117], the Ar plasma emission from laser-plasma source was tailored to the "water-window" spectral range by employing 200 nm thick Si_3N_4 window. Using that filter, most energy will reside in the wavelength range from 2.8 to 4 nm, well within the "water-window" range. The broadband SXR radiation

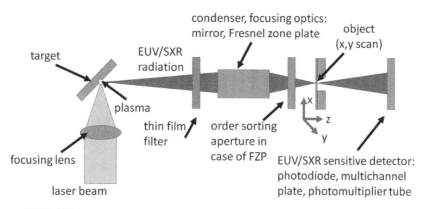

FIGURE 7.17 Scheme of EUV/SXR scanning microscope.

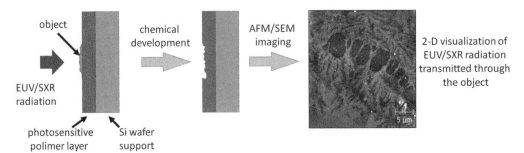

FIGURE 7.18   A flowchart for EUV/SXR contact microscopy.

from Ar/He gas puff target has a sufficient number of photons to expose the high-resolution photoresist (500 nm thick PMMA on top of a silicon wafer) acting as a detector. The object investigated is then placed in contact with the PMMA. The light that is locally transmitted by the object's structure illuminates the photoresist and changes its physical and chemical structure. After the irradiation, the photoresist is chemically developed and modulation of the light intensity absorbed by the object is converted to a modulation of the thickness of the resist, which is later converted to an image using the AFM or SEM microscopes. More details about this system can be found in Ref. [118].

It is not straightforward to assign the limit on spatial resolution in contact microscopy; however, there are several factors that influence the spatial resolution δ_{CM} as can be seen in the following equation (assuming all contributions have Gaussian profiles)

$$\delta_{CM} = \sqrt{\delta_s^2 + \delta_{phr}^2 + \delta_{tip}^2}, \qquad (7.18)$$

where δ_s is a factor related to shadowing (both diffraction and geometrical), δ_{phr} is a factor related to the spatial resolution of the photoresist and δ_{tip} is a factor related to the diameter of the AFM probe used to perform photoresist scanning after the exposure.

An example of using such a system for nanoimaging of biological samples is presented in Figure 7.19. The contact microscope images of the T24 cell lines (transitional cancer

cell of the urine bladder, ATCC) (a) and HCV29 cells (bladder carcinoma) (b), were acquired using 200 pulses of SXRs with an exposure time of 20 s.

Following exposure to the SXRs, the photoresist was cleaned using a sodium hypochlorite (5.25% w/w) solution to remove all residues of the sample. The photoresist was then developed in MIBK:IPA (1:2 v/v) for 120 s. The development process results in a high-resolution relief map of the cellular structures within the photoresist, which was then viewed at high magnification by means of an AFM microscope. The AFM scan was made in a semi-contact mode over an area (field of view—FOV) of 80×80 μm^2 with 512 pixels per line (a) and 100×100 μm^2 with 1,024 pixels per line (b), respectively. The micrographs of the contact microscope images of the T24 cancer cells are shown in Figure 7.19a and HCV29 cells in Figure 7.19b. In these images, the overall structures of the cells including the central dense part (nucleus) are clearly visible with a spatial resolution of approximately 80 nm half-pitch.

7.5   Coherent EUV/SXR Imaging

Coherent imaging techniques in the EUV and SXR spectral range employ spatially and temporary coherent radiation to perform nanoscale imaging using various techniques. The main advantage of coherent techniques is typically a lack of any EUV and SXR optics, which is usually a limiting factor in nanoimaging systems. This is a significant advantage, however, occupied by, typically, numerical reconstruction processes, either single iteration Fourier based, or iterative phase-retrieval algorithms.

7.5.1   Holography

Holography is one of the coherent imaging techniques. In holography, an interference of two beams, one scattered, reflected or transmitted through the investigated object, and the other, undisturbed, called reference beam, encodes amplitude and phase information of the investigated object in a form of an interference pattern with intensity modulation.

Shortly after the introduction of holography by Gabor [119], the feasibility of holographic imaging in the X-ray region with a resolution superior to that obtained by optical microscopy was discussed by Baez [120]. The first

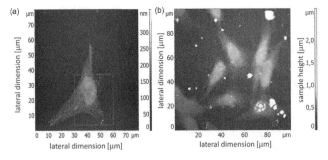

FIGURE 7.19   SXR contact microscopy images of a fixed transitional cancer cell of the urine bladder (T24), the field of view (FOV) = 80×80 μm^2 (a) and HCV29 (bladder carcinoma) the FOV of 100×100 μm^2 (b). Images were obtained with SXR exposure of 200 pulses, 20 s and the photoresist was scanned using AFM microscope in semi-contact mode.

demonstration of holographic imaging at SXR wavelengths achieved a spatial resolution of 5 μm, making use of an early laboratory-size SXR laser pumped by a fusion-class laser [121]. Subsequently, synchrotron radiation has been used in holographic imaging of biological samples and nanostructures [122]. Later, significantly more compact tabletop SXR lasers and high harmonic sources have been used to realize demonstrations of holographic imaging [123,124], with demonstrated resolution approaching the wavelength of illumination [82].

Gabor Holography

In Gabor configuration, often employed for EUV and SXR holography, the two beams are in-line, which is schematically depicted in Figure 7.20. For that, the two beams must have a phase relationship and should be able to interfere; thus, the spatial and temporal coherence of the EUV/SXR source has to be sufficient. While in the diffraction imaging only one beam is required (object beam), during the detection, the phase is lost and needs to be retrieved using phase-retrieval algorithms, in holography the reference beam encodes the phase of the object beam, forming the interference pattern.

If the object $o(x,y)$ and reference $r(x,y)$ wave fronts illuminating the detector are both expressed as two complex fields having an amplitude and the phase:

$$o(x, y) = |o(x, y)| \, e^{j\phi(x,y)} \tag{7.19}$$

$$r(x, y) = |r(x, y)| \, e^{j\psi(x,y)} \tag{7.20}$$

then the interference between these two complex fields occurring at the location of the detector can be expressed as the intensity of the sum of the two fields:

$$
\begin{aligned}
I(x, y) &= |r(x, y) + o(x, y)|^2 \\
&= [r(x, y) + o(x, y)] \, [r(x, y) + o(x, y)]^* \\
&= |r(x, y)|^2 + |o(x, y)|^2 + r^*(x, y) \cdot o(x, y) \\
&\quad + o^*(x, y) \cdot r(x, y),
\end{aligned} \tag{7.21}
$$

where * denotes conjugation. The first two terms are the intensities of both interfering beams, while the last term depends also on their phases. That is why the recording

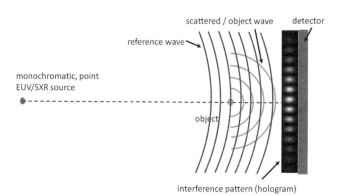

FIGURE 7.20 Gabor holography scheme often employed for EUV and SXR imaging.

medium sensitive only to the intensity is capable of storing the information of the intensity and the phase in the form of an interference pattern.

This pattern can be later optically or numerically processed (reconstructed) to retrieve the representation of the object. The hologram's lateral resolution is ultimately determined by the wavelength of the illumination λ and the NA that for a given detector is set by the object-recording medium distance. For highest spatial resolution, CCD cameras, due to relatively large pixel size, are often replaced by much higher spatial resolution photoresists, similar to detection system employed in contact microscopy, and the resist, after exposure and chemical development, is scanned using AFM. However, other factors are also very important, such as the temporal and spatial coherence of the illuminating source, the resolution of the medium in which the hologram is recorded, and the digitization process used are also factors that influence the ultimate resolution.

To take spatial coherence into account, the interference fringes produced by the reference beam and the beam diffracted from the point object will be stored in the recording medium only within the coherence area, with coherence radius R_C. Beyond that region, one can assume that the interference will not occur. This leads to spatial resolution limited by the spatial coherence to be

$$\delta_{sc} = \frac{a\lambda}{\sin\left[\arctan\left(R_c z_p^{-1}\right)\right]}, \tag{7.22}$$

where z_p is an object to detector distance and a—see Eq. 7.1.

Similarly, the temporal coherence limits the recording NA as well. The optical path difference between the reference and the diffracted beam has to be smaller than the coherence length l_c of the illumination source in order to observe the interference, thus spatial resolution limited by the temporal coherence is:

$$\delta_{tc} = \frac{a\lambda(z_p + l_c)}{\sqrt{l_c(2z_p + l_c)}}, \tag{7.23}$$

An important limitation to the spatial resolution is also the process of the digitization of the hologram. If the detector is a CCD camera, the limitation is its pixel size (in micron range); however, if the recording medium is a photoresist surface, then the resist resolution is much better than the diffraction limit in the EUV range and comparable to the SXR wavelengths, so in this case, more important is the digitization NA. This yields spatial resolution limited by the digitization process:

$$\delta_d = \frac{a\lambda}{\sin\left[\arctan\left(0.5 \cdot s \cdot z_p^{-1}\right)\right]}, \tag{7.22a}$$

where s is the AFM scan size, providing that the AFM sampling $pix_{\text{AFM}} = s/N \ll \delta_d$ (N—the number of points per line in the AFM scan).

Finally, there are also additional contributions, such as photoresist resolution and AFM tip curvature, so the final spatial resolution can also be estimated (assuming again Gaussian-type contributions) using the simplified equation:

$$\delta_{GH} = \sqrt{\delta_{sc}^2 + \delta_{tc}^2 + \delta_d^2 + \delta_{phr}^2 + \delta_{tip}^2}. \qquad (7.23a)$$

An example of this technique is a Gabor holography of carbon nanotubes (CNTs) [82]. A sample composed of CNTs with diameters between 50 and 80 nm and a length between 10 and 20 μm placed on a 100 nm thick silicon membrane, Figure 7.21a. The membrane acts as a support to locate the object at approximately 2.6 μm away from a 120 nm thick layer of PMMA photoresist providing a recording NA practically equal to 1. The limited area of the AFM scan that digitizes the hologram after photoresist development (9.9 × 9.9 μm²) reduces the effective value to NA = 0.88 (reconstruction NA). The CNTs were illuminated by a compact λ = 46.9 nm capillary discharge Ne-like Ar EUV laser [72]. To assure a proper reconstruction, the hologram recorded as a relief modulation in the photoresist surface was exposed with a dose within the linear response regime of the photoresist. The developed photoresist (MIBK:IPA 1:3) surface was mapped with an AFM to generate digitized holograms, Figure 7.21b, that were reconstructed by numerically simulating the illumination with an EUV readout wave. The amplitude and the phase distribution of the field in the image plane were obtained calculating the field emerging from the hologram illuminated by a plane reference wave and numerically back-propagating the fields with a Fresnel propagator [125]. The reconstructed image, Figure 7.21c, was found by taking the two-dimensional inverse fast Fourier transform (2D-IFFT) of the product of the spatial frequency of the Fresnel propagator and the 2D-FFT of the hologram.

A spatial resolution of 45.8 ± 1.9 nm was achieved, comparable to λ = 46.9 nm. This method allows for diffraction-limited EUV imaging; however, the use of photoresists and chemical development is somewhat impractical. Gabor configuration is the simplest one, does not require any optics and allows for very high NA hologram recordings; moreover, in the surface of the photoresist, much larger area than later digitized by AFM might be stored, which allows gigapixel images to be recorded in this scheme at EUV/SXR wavelengths. For example, a hologram 2 × 2 mm² in size, having a resolution of 50 nm is equivalent to a 1.6 gigapixel image of the sample. Of course, it has to be digitized later, but the information about the investigated object is recorded. This argument is also true for contact microscopy approach, presented earlier. The method is ultimately limited by the resolution of the photoresist, which for PMMA is ~10 nm.

Fourier Holography

Conventional mask-based Fourier transform holography (FTH) uses a small pinhole (or many pinholes [126]), which typically is of the order of 1 micron in diameter, fabricated into the object mask to create the reference wave, unlike the open area in Gabor configuration (Figure 7.22). Mask-based FTH has already been applied to image magnetic structures [127], tomography [128] and for sequential femtosecond imaging [129]. For highest spatial resolution reconstructions, a pinhole as small as possible (diameter $d \sim \lambda$) is necessary. On the other hand, for strong intensity modulation in the hologram and a high-quality reconstruction, the reference and object beams ought to have similar intensities [130]. Larger pinhole, on the other hand, limits the NA of the reference wave, so single-pinhole mask-based FTH ultimately limits the FOV. Another possibility is to use multiple reference holes [88], or a zone plate, which is used to create the reference wave, while a central opening in the zone plate allows for the incident beam to directly illuminate the object [131], as depicted in Figure 7.23a. The rapid and deterministic reconstruction and high-resolution capability make FTH an interesting imaging method.

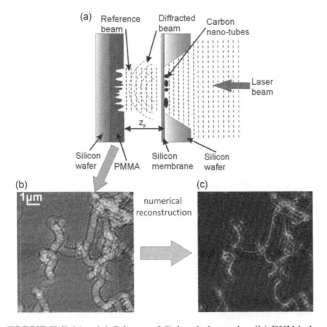

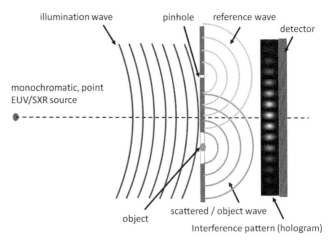

FIGURE 7.21 (a) Scheme of Gabor holography, (b) EUV hologram of CNTs and (c) numerical reconstruction. White lines represent the reconstructed CNTs.

FIGURE 7.22 Fourier holography scheme for EUV and SXR imaging with one reference wave encoding the amplitude and phase information from the object wave.

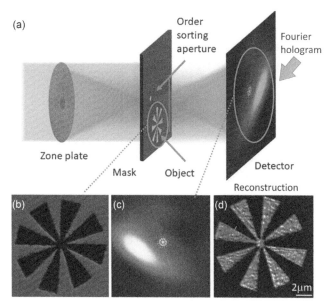

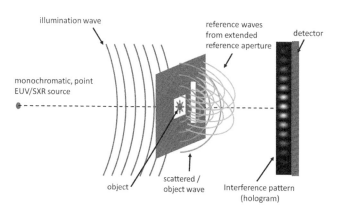

FIGURE 7.24 HERALDO scheme for EUV and SXR imaging with extended (arbitrary) reference wave encoding the amplitude and phase information from the object wave.

FIGURE 7.23 Schematic of the FTH setup with high NA reference obtained in the first order of FZP (a). An SEM of the object (b). Hologram—interference pattern collected on a CCD (c). The numerical intensity reconstruction (d) obtained from the hologram in (c) [132].

A similar approach to estimate limits of the spatial resolution, such as the one for Gabor holography, can be followed to find the resolution limit in FTH. Due to the fact that reference holes are typically a few microns away from the object and the object is orders of magnitude further from the detector (typically a few mm to a few cm), the equations for δ_{sc} and δ_{tc} hold. In FTH, there is no AFM digitization, but the CCD itself limits the readout NA, so the equation for δ_d is similar. There is no AFM tip curvature and photoresist resolution contribution, but a serious limitation may be a CCD pixel size, so it has to be taken into account, due to a fact that readout sampling influences the reconstruction, obtained by an inverse Fourier transform.

An example of nanoscale imaging using FTH [132] is depicted in Figure 7.23a. A Fresnel zone plate is used to create the reference wave from the first-order diffracted EUV ($\lambda = 46.9$ nm wavelength) beam and transmit the incident light to illuminate the object (Figure 7.23b). The reference and object waves interfere at the detector plane, producing the interference pattern—hologram (Figure 7.23c). The representation of the object is numerically reconstructed by IFFT of the interference pattern (Figure 7.23d).

HERALDO Holography

The problem of low FOV in single-pinhole FTH can be mitigated through the use of an extended reference aperture in technique, depicted in Figure 7.24, called holography extended reference by autocorrelation linear differential operation (HERALDO) [133], which has extended the FOV. HERALDO is an extension of FTH, which allows for larger reference apertures to be used while maintaining the same nanoscale resolution of conventional FTH.

HERALDO has been experimentally shown to provide high spatial resolution [134] and the capability for single-shot exposure with nanoscale temporal resolution [135]. Due to the reconstruction process, the high spatial frequencies in a hologram are amplified. Typically, the lower SNR at the higher spatial frequencies and resulting amplification from the application of a polynomial product results in a degradation of the reconstruction. Moreover, this makes HERALDO technique especially sensitive to noise.

7.5.2 Talbot Imaging

Generalized Talbot Imaging (GTI) allows for imaging any periodic, repeatable unit cells using self-imaging principle (Figure 7.25). This method allows for imaging of arbitrary, although periodical, features on the detector with a spatial resolution better than 100 nm on areas around a few mm^2.

It is based on Talbot effect, where a periodic structure (originally it was a grating) is illuminated by a spatially and temporally coherent beam allows obtaining images of this structure at certain distances being a multiple of some basic distance called the Talbot length. At these distances, all the phase shifts originating from a beam diffraction on periodic cells, or structures, will cancel out; thus, due to a local interference, they reproduce the image of the original object, without any additional EUV/SXR optics involved.

The GTI originated from merging two ideas. The combination of nanopatterning techniques, employing photosensitive polymers (photoresists) and coherent EUV sources for Talbot mask illumination allows the creation of complex, periodic imaging patterns along the optical axis at a distance

$$z = n \cdot z_T = n\frac{2d^2}{\lambda}, \qquad (7.24)$$

every Talbot distance z_T, without complex optical systems. d is a period of 1-D repeating structure (grating) and n is multiple of the Talbot distance. This equation also holds if the periodicity of the GTI mask is the same in two orthogonal directions [136].

In the simplest case, for Talbot imaging, spatial and temporal coherence of the EUV/SXR beam is required.

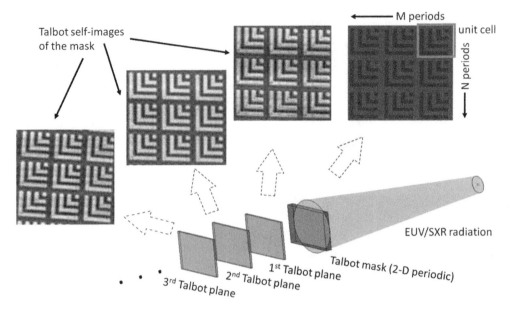

FIGURE 7.25 Scheme for GTI imaging using EUV/SXR radiation.

The recent development of compact high-flux EUV lasers provides an opportunity to efficiently use coherent imaging, such as GTI, for patterning nanometer scale features over large areas [137]. Talbot self-imaging can also be performed with broadband X-ray beams [138], where temporal coherence is reduced.

In Talbot imaging, proof-of-principle experiment [136] masks, fabricated in silicon nitride membrane by an e-beam, were illuminated by a coherent EUV beam emitted from the EUV laser. The size of each mask was 0.6×0.6 mm^2, and the first Talbot distance was 1 mm. A single elementary cell was $4,845 \times 4,845$ nm^2 in size and the smallest features were 140 nm in width.

This unit cell was repeated 124 times in each orthogonal directions, thus increasing the effective NA to 0.28. At $\lambda = 46.9$ nm illumination wavelength, it yields theoretical spatial Rayleigh resolution, expressed by

$$\delta_{\text{Ray}} = \frac{0.5 \cdot \lambda}{\text{NA}}, \qquad (7.25)$$

of 84 nm. A value of $a = 0.5$ was used due to the assumption of square mask aperture, instead of a typical circular aperture, yielding $a = 0.61$ for incoherent illumination.

The self-images of the mask were stored by a high-resolution detector, PMMA in this case, 50 nm in thickness, which has a much better spatial resolution than a CCD chip with relatively large pixel size. To obtain spatial resolution near the theoretical limit, the entire mask aperture needed to be illuminated by a fully spatial coherent EUV beam. Three different Talbot masks with different unit cells were fabricated and imaged onto a photoresist surface tuning precisely the distance between the mask and the detector around the Talbot length up to six Talbot lengths, as can be seen in Figure 7.26. The smallest structures, approximately 140 nm in size, were faithfully recorded on the detector, suggesting spatial resolution not far from a theoretical limit.

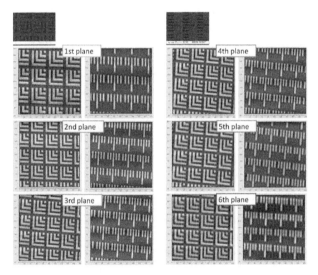

FIGURE 7.26 Example of self-imaging (up to a sixth Talbot plane) of two GTI masks with EUV radiation.

Almost identical self-images of the masks were visible up to six Talbot distances.

Talbot self-imaging technique in the EUV spectral range has potential applications in lithography processes. The main advantage of this method, which is important from a lithography point of view is the fact that small defects in the mask, which might undermine the entire lithographic process in a classical stepper-based lithographic scheme, in a Talbot mask will be averaged out in the entire FOV. Because the "signal" from the entire mask is much larger than the "noise" coming from the defect, the image that is formed at the Talbot length is defect-free. Talbot self-imaging effect opens new application possibilities for nanometer-scale imaging of periodical structures such as FLASH memories, magnetic materials, where repeatable, well-defined structures are repeated over larger areas.

7.5.3 Coherent Diffraction (Lensless) Imaging

In a typical CDI scheme, a spatially and temporary coherent beam of EUV or SXR radiation is focused onto a sample using usually a mirror (spherical, ellipsoidal, etc.). Thin film filters are used to block unwanted wavelengths of EUV/SXR radiation to provide monochromaticity and improve the temporal coherence. Light scattered from the object forms a diffraction pattern a few centimeters from the sample, which is detected using a CCD camera. In this process, the phase information of the diffracted beam is lost; only the intensity of the diffracted beam is recorded by the detector (Figure 7.27).

Three basic requirements must be satisfied performing lensless diffraction imaging. First, the object must be illuminated by a coherent wave front; second, the object must be isolated, which means that it must be surrounded by a substantial region with no scatterers. The region may be either opaque or transparent. Third, the scattered light must be collected with sufficient spatial resolution to oversample the diffraction pattern and to record it at a spatial frequency higher than the Nyquist frequency [139]. This is satisfied by a proper geometry of the imaging system so that the smallest diffraction speckle can be sampled linearly by two or more CCD pixels. The linear oversampling ratio is given by

$$Os = \frac{\lambda z}{pD},\qquad(7.26)$$

where z is the sample to CCD distance, λ is the wavelength, D is the sample diameter and p is the pixel size of the CCD camera. Good results and fast convergence of the phase-retrieval code are obtained at Os value close to 10 [140].

Afterward, a low-pass filter is applied to the autocorrelation function obtained from a Fourier transform of the diffracted intensity. Then the diffraction pattern is interpolated from a plane onto a spherical surface to remove the radial stretching and skew of the speckles near the edge of the diffraction pattern (corresponding to high diffracted

angles). Finally, a phase retrieval of the complex electron density is carried out by using the phase-retrieval algorithm, such as guided hybrid-input-output (GHIO) algorithm [141]. Such an approach allows achieving diffraction-limited resolution using a tabletop EUV diffraction microscope [142].

An example of coherent diffraction lensless imaging is shown in Figure 7.28. A binary object was made in the silicon nitride membrane using e-beam lithography (Figure 7.28b). Such an object was illuminated using a coherent output of capillary discharge EUV laser, in a scheme in Figure 7.28a, at the wavelength of 46.9 nm.

Three separate diffraction patterns were acquired with different exposures and stitched together by matching boundary conditions to improve the dynamic range of the detection. Additionally, due to large NA detection system, the resulting diffraction pattern was curvature corrected, as it would be projected on a spherical detector and not on a flat CCD chip (Figure 7.28c). Later the diffraction pattern was numerically reconstructed using GHIO algorithm (Figure 7.28d).

Initially, CDI method was employed for nanoimaging at synchrotrons and FELs, for example, for femtosecond snapshot like single pulse imaging [143] or for 3-D reconstruction based on 2-D CDI measurements of a giant mimivirus with FEL source [144] with a full period of 125 nm spatial resolution [145]. Currently, such an approach, using compact tabletop sources, allows for ∼20 nm spatial resolution using, for example, HHG sources [146].

7.5.4 Ptychography

Ptychography is a technique [147] that was developed to solve a lost-phase problem in diffraction imaging. The idea was to determine the relative phases between adjacent diffraction patterns, as can be seen in Figure 7.29. A focused EUV/SXR beam illuminates the object placed, i.e. on top of a thin silicon nitride membrane. The focus spot size might be of the order of a few hundreds of nanometers. The object diffracts the EUV/SXR radiation, and downstream the beam, a single diffraction pattern is collected. Then the object is raster scanned point by point to record a number of diffraction patterns. The scanning step is usually smaller than the diameter of the focus. For the reconstruction, numerical algorithms are used. Those are, typically, "ePIE" algorithm [148], difference map approach [149] or conjugate gradient [150].

Ptychographic imaging method bridges the gap between CDI and scanning transmission EUV/SXR microscope by measuring a set of diffraction patterns at each point of a scan. The advantages of this method are very high spatial resolution (for hard X-rays ∼10–20 nm) for test samples, the spatial resolution limited by geometry, not by optics, lack of optics allows for such high resolution and the method provides contrast for phase samples.

With the development of spatially coherent sources, sensitive detectors, such as CCD cameras, and most importantly, taking advantage of modern computing capabilities, the

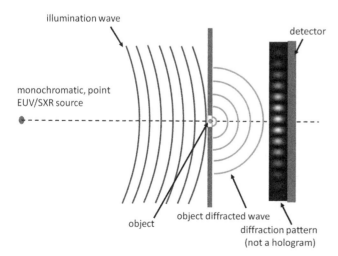

FIGURE 7.27 A general scheme for CDI.

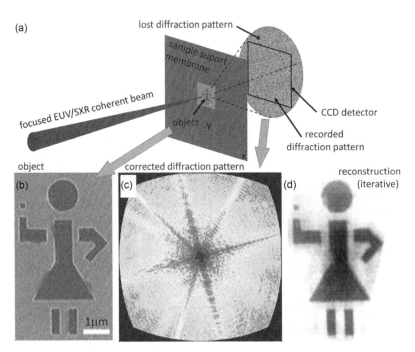

FIGURE 7.28 Lensless imaging using coherent SXR laser beams at 46.9 nm. (a) Scheme of the experimental setup, (b) SEM image of the object (scale bar = 1 μm), (c) coherent SXR diffraction pattern after curvature correction, (d) reconstructed image using phase-retrieval code, from Ref. [142].

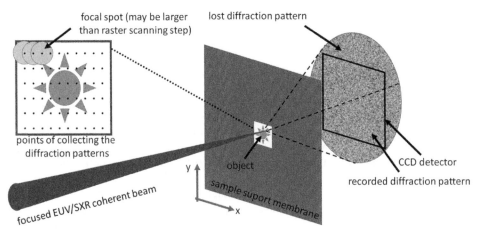

FIGURE 7.29 A scheme for ptychographic EUV and SXR imaging.

ptychographic imaging has resulted in the development of EUV and SXR microscopes with increased spatial resolution [151], electron microscopes [152], operating beyond the limitations of CDI or Fourier holography, without the need for imaging optics but with the requirement of acquisition of multiple diffraction patterns and quite significant numerical processing resources.

The ptychography was already used for imaging of biological samples [153], pores in the Ta_2O_5 structure [154] or zone-plate outer zones buried under a thin layer of other material [149], among others. It is foreseen that this technique will soon be capable of imaging samples in hard X-ray wavelength range with a spatial resolution approaching 1 nm [155].

7.5.5 X-ray Coherence Tomography

Optical coherence tomography (OCT) allows one to image an internal structure of the investigated sample in a noninvasive and nondestructive way [156]. Typically, to image a sample, such as multilayer optics, a cross section using a focused ion beam (FIB) is made and the cut is then inspected using an SEM microscope. This, however, damages the sample, which for some specific applications, such as testing and inspection of multilayer optics for the EUV lithography, poses a significant problem.

To overcome the limitation of such destructive measurements, noninvasive probing techniques, such as the OCT, were developed and are used nowadays [157]. The OCT is performed by splitting an illumination beam into two, namely, a reference beam and an object beam. The object beam is scattered from the sample being investigated, while the reference beam, reflected from the first layer, is superimposed with the object beam and interferes with it. Due to a broad bandwidth of the illumination source and, in turn, small coherence length, the interference occurs only if the delay between the two beams is within a coherence time, related to the coherence length through a speed of light constant. Thus, it allows one to probe the internal structure of the sample in the direction of the optical axis by the detection of the interference between two beams. It is done using an interferometer, by changing the delay between the beams (in a time-domain OCT) or by recording the interference in the spectral domain (frequency-domain OCT). The interferometer can also operate in on- and off-axis configuration. Such type of imaging turns out to be a very successful approach to image objects, which cannot be imaged with comparable precision, resolution and in a noninvasive way otherwise. One such object is an eye, and due to its advantages, the OCT nowadays became a standard technique to in vivo imaging of the internal structure of an eye, such as a retina [158], to early detect any pathological changes [159].

The main challenge of visible and near-IR OCT is the difficulty of observing opaque materials and the axial resolution of typically sub-micron to a few microns. In the OCT, the axial resolution depends on the coherence length expressed by Eq. 7.9, where λ is the illumination wavelength and $\Delta\lambda$ is the source bandwidth (i.e. FWHM spectral emission) and IRB is an inverse relative bandwidth. For typical broadband near-IR source of $\lambda = 1$ µm and $\Delta\lambda = 400$ nm ($IRB = 2.5$), the $l_c = 2.5$ µm, which for many applications is not sufficient. Moreover, many materials exhibit very low transmission in the VIS and near-IR ranges, so it is difficult to scatter the electromagnetic field out of refractive index discontinuities in the structure of the investigated samples.

As can be seen from Eq. 7.9, even though the IRB has a relatively small value, the central illumination wavelength λ is quite large. Thus, a straightforward way to improve the axial resolution is a further reduction of the central wavelength down to the X-ray region, to perform OCT, so-called XCT [160]. To facilitate that, broadband EUV and SXR sources can be used, including synchrotrons, laser-driven sources based on HHG and LPPs [161], in a scheme depicted in Figure 7.30. A broadband EUV/SXR radiation is focused using focusing optics onto a sample. The radiation reflected from the sample is collected using collection optics onto an entrance slit of the grazing incidence spectrometer (GIS) or transmission grating spectrometer (TGS), which records the spectrum of reflected radiation for the particular spatial location. Then the sample is shifted to a new location and another spectrum is collected for each location. After processing of the reflectivity spectrum, the depth profile can be obtained for each spatial location, which allows for very easy 3-D visualization of the investigated sample.

In those wavelength ranges, two possible transmission windows for XCT have been identified so far, namely a "silicon window" ($\lambda = 12$–40 nm), dedicated for material

science applications, and a well-known "water-window" range, dedicated for biology ($\lambda = 2.3$–4.4 nm). So far, the XCT allowed one to obtain an axial resolution of 18 nm in the EUV range, 8 nm in the "water-window" range employing a synchrotron radiation source [162], 24 nm axial resolution using a laser-driven HHG source [163] and 2 nm axial resolution with SXR radiation extending beyond "water window", at $\lambda = 2$–5 nm from Kr plasma [164].

The XCT in the "water window" spectral range, due to a central wavelength of $\lambda = 3.3$ nm and $\Delta\lambda = 2.1$ nm ($IRB = 1.6$), theoretically allows for the $l_c = 5.3$ nm. This makes possible to resolve in-depth structures of the order of a few nanometers. Moreover, the "water-window" radiation can be used for nanoimaging of biological samples, which exhibit a high natural contrast in this wavelength range due to differences in absorption between carbon and oxygen, present in the biological samples [165].

An example of the experimental approach to the XCT is imaging of Mo/Si multilayer structure in the wavelength range from 2 to 5 nm [164]. In this experiment, depicted in Figure 7.31, due to an interaction of the Nd:YAG laser pulses with the gaseous target, a krypton LPP is created, which emits SXR radiation. The radiation from krypton plasma illuminates the sample under investigation (Mo/Si multilayer structure), which reflects the SXR radiation at 45° with respect to the sample surface, towards a GIS SXR spectrometer.

A spatial coherence of the SXR radiation from the LPP source was limited by the size of the Kr plasma. According to Eq. 7.27

$$R_C = \frac{\lambda z}{2D} \tag{7.27}$$

the radius of coherence R_c was estimated to be in the range of 0.6–1.5 µm, at the location of the sample z, for the diameter of the plasma D of 0.5 mm and the wavelength range of the SXR emission from $\lambda = 2.2$ nm to 5.5 nm. Moreover, from the emission bandwidth of $\Delta\lambda \sim 2$ nm, the coherence

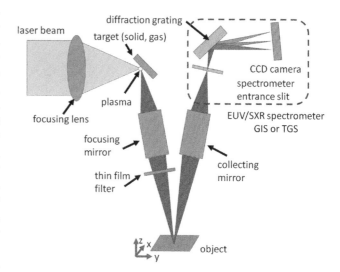

FIGURE 7.30 A general optical setup for laser-plasma OCT in the EUV/SXR spectral region.

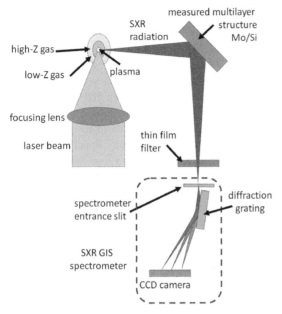

FIGURE 7.31 An experimental approach to XCT imaging of Mo/Si multilayer structure in the wavelength range from 2 to 5 nm wavelength.

length l_c from Eq. 7.9 is $l_c = 2\text{–}12.5$ nm. Such temporal coherence allows having a constructive interference between closest layers, and at the same time, it is not too large to limit the cross-interference terms between layers spaced further apart.

The Mo/Si sample reflectivity is presented in Figure 7.32a in the SXR wavelength range from 1.5 to 5.5 nm. The period of the Mo/Si structure was 10 nm with 6 nm of Si and 4 nm of Mo. There were 40 periods in total on top of the SiO_2 substrate.

Afterward, the experimental reflectivity curve is converted into k-space using

$$k = \frac{4\pi}{\lambda}\sqrt{n_D^2 - \sin^2\alpha}, \qquad (7.28)$$

where n_D is a wavelength-dependent index of refraction of the dominant material and α is incidence angle measured to

the normal of the surface of the sample. A reconstruction of a depth information from the Mo/Si structure (Figure 7.32b), based on experimental data (bottom curve) and compared to simulation (top curve). For visualization, only the first five periods are depicted. The numbers indicate horizontal position of each peak for comparison between theoretical and experimental data. Finally, a visualization of the depth structure of the sample in Figure 7.32c is showing a comparison between theoretical and experimental data.

Such an approach allows for an axial (depth) spatial resolution of the order of 2 nm, regarding the width of the discontinuity of the index of refraction. However, the position of the discontinuity can be found with a much better accuracy of less than 1 nm. Moreover, no optics are required; however, the samples typically exhibit a reflectivity of the order of 10^{-3} in the SXR range at α angles of $10°\text{–}45°$; thus, a reflectivity spectrum acquisition might be challenging.

7.6 Summary and Conclusions

Optical microscopy methods allow the observation of objects with sizes of several hundreds of nanometers due to the diffraction limit related to the wavelength of radiation used in the imaging process. The use of radiation with 10–100 times shorter wavelength allows direct improvement in spatial resolution, at least several times, obtained in the process of imaging. Most of the EUV/SXR imaging methods, presented here, allow obtaining spatial resolution better than 100 nm, fast exposure times and optical contrast in the short-wavelength range, which opens the possibility to acquire an additional information about the object. The main advantage of the presented methods, compared to the widely used optical microscopy, is thus improved spatial resolution and different interaction with matter due to more energetic photons.

An electron beam microscopy (SEM, TEM) has the best spatial resolution [166], which results from a very small wavelength of electrons used for imaging, of the order of 7 pm for electrons accelerated by a voltage of 30 kV. Such a wavelength allows for imaging, in modern

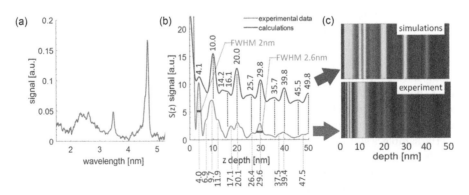

FIGURE 7.32 A measured reflectivity of the Mo/Si multilayer structure in the SXR wavelength range from 1.5 to 5.5 nm with the wavelengths of each peak indicated (a). A reconstruction of a depth information from the Mo/Si structure (b), based on experimental data (bottom curve) and compared to simulation (top curve). Visualization of depth structure of the sample (c) showing a comparison between theoretical and experimental data.

electron microscopes, with a spatial resolution better than 1 nm. The use of electrons as information carriers has its obvious advantages but also has drawbacks, resulting from a different kind of interaction of photons and electrons with the matter. Electron microscopy can, therefore, allow obtaining complementary information about the same sample but also often leads to too strong interaction of electrons with the sample. The methods that employ EUV radiation allow the observation of very delicate samples, often destroyed or disrupted by a high-energy electron beam. Photon-based microscopy in the EUV/SXR wavelength range also allows the observation of the samples coated with photoresist, sensitive to electron beam exposure, which cannot be observed by electron microscopy without influence on the photoresist. The advantage is also in the fact, that EUV/SXR microscopy does not need any special preparation of samples. Such sample preparation is often necessary for other imaging techniques such as electron microscopy to dissipate a surface charge.

Atomic force microscopy is another very useful method of using sharp probes with very small spatial dimensions to scan the sample surface with very high accuracy. Resolution of this method is also very good; for the best probes using

CNTs, the resolution is a few nm. The disadvantage of this method is the high sensitivity to contamination of the probe, reflected in a significant deterioration of spatial resolution and image quality. This method does not allow for imaging in transmission mode and only allows for examination of the sample surface.

The presented EUV/SXR imaging methods, depicted in terms of achievable spatial resolution in Figure 7.33, are complementary to existing and most commonly used imaging methods. Please notice that all the methods employing high-energy photons are shifted to the right on the spatial resolution scale, by roughly one to two orders of magnitude, comparing to the visible light methods. In addition, each of the presented EUV/SXR imaging methods also has its own characteristics. None of them is an ideal one but can be used for a particular kind of circumstances and for a particular purpose and application. Diffraction imaging, because its scalability to shorter wavelengths and its usability in the hard X-ray region of electromagnetic spectrum, may be used for imaging of nanometer scale noncrystalline structures. Holography and Talbot imaging have potential applications in the semiconductor industry. Both methods are noncontact and do not require additional

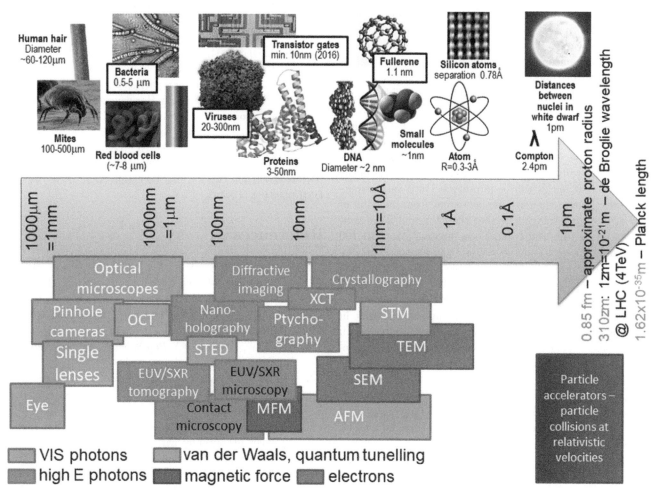

FIGURE 7.33 A chart depicting spatial resolution vs. well known and most common imaging methods.

steps, such as lithography mask inspection and cleaning in the classical lithographic, stepper-based process. Additionally, both methods are immune to small damage to the hologram of defects in the Talbot mask. To reconstruct the hologram, only its portion is required, while in case of Talbot mask, small defects are averaged out in the imaging process over the entire FOV. The full-field EUV microscopy using Fresnel zone-plate objectives may be used in high-resolution imaging of objects for applications in biology, material science and nanotechnology or semiconductor industry for actinic lithographic mask inspection. The EUV radiography can be used for imaging of gaseous targets (with densities a few orders of magnitude smaller than the density of water) and allows to directly visualize various types and configurations of gas targets/objects and can be scalable to nanometer spatial resolution, providing the source size is small.

One important thing, however, is to notice that even though the EUV and SXR imaging have certain advantages, to be fair, one must present also a few significant drawbacks. First of all, the EUV and SXR radiation must propagate in a vacuum, thus the necessity to use vacuum installations to provide a suitable vacuum environment for the EUV/SXR beam propagation. Such an environment is not suitable to house samples, especially biological ones, due to pressure reduction, lack of oxygen and water vaporization. Another important factor is the fact that the resolution of optics-based EUV and SXR imaging systems is severely limited by low-performance optics. Even though the optics itself is fabricated with the best possible accuracy (zone positioning in zone plates, zone profiles, etc.) and lowest possible surface roughness and figure error (in the case of grazing incidence mirrors), etc., yet still, the absorption of the material, low diffraction efficiency and low reflectivity limits significantly the performance of the EUV and SXR optics. A significant improvement will be possible if the overall throughput of the EUV and SXR optics will be improved. Third very important factor, maybe the most important one from the point of view of biological imaging, is that the EUV and SXR radiation is an ionizing radiation. Those photons carry 10–100 times more energy than the visible light photons and can easily ionize the material of the sample introducing a physical damage. This is the reason why EUV and SXR radiation that exposes the photoresist in holographic imaging or contact microscopy, presented herein, leaves the imprint in the photoresist surface after the developing procedure. Such damage also occurs in the biological material as well, especially in the DNA [167] and in many applications cannot be ignored.

As was presented, for EUV/SXR imaging, various types of sources were employed. At this point, still there is no ideal solution for the source and it is necessary to find a good compromise in the desktop SXR/EUV imaging systems, between the performance, namely high photon flux, possibility to obtain high spatial resolution images with low exposure time and the complexity, size and cost of these setups, which is still the main limitation of short-wavelength photon-based microscopes disallowing their widespread. Thus, further extensive research related to source development and novel image techniques is of high importance.

The idea of imaging in the EUV and SXR region, where the thinnest layer of any material, approximately 100–500 nm, absorbs the majority of radiation, is still very interesting and novel to the scientific community but also a difficult one. In my opinion, we still may expect much progress in this novel and interesting field in the years to come.

Acknowledgments

The author acknowledges Prof. Henryk Fiedorowicz and Dr. Andrzej Bartnik from the Institute of Optoelectronics, Military University of Technology, Prof. Mario C. Marconi, Prof. Carmen Menoni, Prof. Randy Bartels and Prof. Jorge Rocca, from Colorado State University, for their contributions to the work examples, presented herein. The author also thanks all collaborators, with whom he had a chance to work over the last years on some of the experiments presented in this text.

This work is partially supported by the National Science Centre, Opus programme, grant agreement numbers UMO-2015/17/B/ST7/03718 and UMO-2015/19/B/ST3/00435 and Beethoven program, number 2016/23/G/ST2/04319, the National Centre for Research and Development, award number LIDER/004/410/L-4/12/NCBR/2013, European Union FP7 Erasmus Mundus Joint Doctorate Program EXTATIC, under framework partnership agreement FPA-2012-0033, European Union's Horizon 2020 research and innovation program LaserlabEurope IV (654148), III (284464) and II (228334), Foundation for Polish Science HOMING Programme 2009, HOM/2009/14B, Defense Threat Reduction Agency Program, HDTRA-1-08-10-BRCWMD-BAA (USA) and Program NER, NSF Award DMI-0508484 (USA).

References

1. M. Born and E. Wolf, *Principles of Optics*, 7th ed. (Cambridge University Press, Cambridge, 1999).
2. J. M. Heck and D. T. Attwood, "Resolution determination in X-ray microscopy: an analysis of the effects of partial coherence and illumination spectrum", *Journal of X-Ray Science and Technology* 8, 95–104 (1998).
3. The International Organization for Standardization., "ISO 21348 Definitions of Solar Irradiance Spectral Categories," 6–7 (2007).
4. M. C. Marconi and P. W. Wachulak, "Extreme ultraviolet lithography with table top lasers", *Progress in Quantum Electronics* 34, 4, 173–190 (2010).
5. P. W. Wachulak, Ł. Wegrzynski, A. Bartnik, T. Fok, R. Jarocki, J. Kostecki, M. Szczurek, and H., Fiedorowicz, "Characterization of a dual-gas

multi-jet gas puff target for high-order harmonic generation using extreme ultraviolet shadowgraphy", *Laser and Particle Beams*, 31, 2, 219–227 (2013).

6. A. L. Schawlow, "Laser light", *Scientific American*, 219, 120 (1968).

7. J. Als-Nielsen and D. McMorrow, *Elements of Modern X-ray Physics* (John Wiley & Sons, USA, 2011). doi:10.1002/9781119998365

8. W. Chao, J. Kim, S. Rekawa, P. Fischer, and E. H. Anderson, "Demonstration of 12 nm resolution Fresnel zone plate lens based soft X-ray microscopy", *Optics Express* 17, 20, 17699 (2009).

9. S. Rehbein, S. Heim, P. Guttmann, S. Werner, and G. Schneider, "Ultrahigh-resolution soft-X-ray microscopy with zone plates in high orders of diffraction", *Physical Review Letters* 103, 110801 (2009).

10. B. L. Mesler, P. Fischer, W. Chao, E. H. Anderson, and D. H. Kim, "Soft x-ray imaging of spin dynamics at high spatial and temporal resolution", *Journal of Vacuum Science and Technology*, B 25, 2598 (2007).

11. P. Fischer, D. H. Kim, B. L. Mesler, W. Chao, and E. H. Anderson, "Magnetic soft X-ray microscopy: Imaging spin dynamics at the nanoscale", *Journal of Magnetism and Magnetic Materials*, 310, 2, 2689 (2007).

12. P. Fischer, D. H. Kim, W. Chao, J. A. Liddle, E. H. Anderson, and D. T. Attwood, "Soft X-ray microscopy of nanomagnetism", *Materials Today*, 9, 1–2, 26 (2006).

13. D. H. Kim, P. Fischer, W. Chao, E. Anderson, M. Y. Im, S. Ch. Shin, and S. B. Choe, "Magnetic soft x-ray microscopy at 15 nm resolution probing nanoscale local magnetic hysteresis", *Journal of Applied Physics*, 99, 08H303 (2006).

14. B. Nieman, D. Rudolph, and G. Schmahl, "Soft X-ray imaging zone plates with large zone numbers for microscopic and spectroscopic applications", *Optics Communication*, 12, 160 (1974).

15. B. Nieman, D. Rudolph, and G. Schmahl, "X-ray microscopy with synchrotron radiation", *Applied Optics*, 15, 1883 (1976).

16. G. Schneider, B. Riemann, P. Guttmann, D. Rudolph, and G. Schmahl, "Cryo X-ray microscopy", *Synchroton Radiation News*, 8, 19 (1995).

17. C. Jacobsen, S. Williams, E. Anderson, M. T. Browne, C. J. Buckley, D. Kern, J. Kirz, M. Rivers, and X. Zhang, "Diffraction-limited imaging in a scanning transmission X-ray microscope", *Optics Communications*, 86, 351 (1991).

18. C. Jacobsen, J. Kirz, and S. Williams, "Resolution in soft X-ray microscopes", *Ultramicroscopy*, 47, 55 (1992).

19. J. Kirz, C. Jacobsen, S. Lindaas, S. Williams, X. Zhang, E. Anderson, and M. Howells, *Soft X-ray Microscopy at the National Synchrotron Light Source*, published in Synchrotron Radiation in the Biosciences (Oxford University Press, UK, 1994), pp. 563.

20. G. Poletti, F. Orsini, D. Batani, "Study of multicellular living organisms by SXCM (Soft X-Ray Contact Microscopy)", *Solid State Phenomena*, 107, 7–10 (2005).

21. A. C. Cefalas, P. Argitis, Z. Kollia, E. Sarantopoulou, T. W. Ford, A. D. Stead, A. Marranca, C. N. Danson, J. Knott, and D. Neely, Technical Report RAL-TR-98-007, TMR Large-Scale Facilities Access Programme, National Hellenic Research Foundation, Greece (1998).

22. M. Kado, H. Daido, Y. Yamamoto, K. Shinohara, and M. C. Richardson, *Proceedings of 8th International Conference on X-ray Microscopy, IPAP Conference Series*, 7, 41–43 (2005).

23. G. Schneider, "Cryo X-ray microscopy with high spatial resolution in amplitude and phase contrast", *Ultramicroscopy*, 75, 85–104 (1998).

24. W. Meyer-Ilse, D. Hamamoto, A. Nair, S. A. Lelièvre, G. Denbeaux, L. Johnson, A. L. Pearson, D. Yager, M. A. Legros, and C. A. Larabell, "High resolution protein localization using soft X-ray microscopy", *Journal of Microscopy*, 201, 3, 395–403 (2001).

25. J. Maser, A. Osanna, Y. Wang, C. Jacobsen, J. Kirz, S. Spector, B. Winn, and D. Tennant, "Soft X-ray microscopy with a cryo scanning transmission X-ray microscope: I. Instrumentation, imaging and spectroscopy", *Journal of Microscopy*, 197, 1, 68–79 (2000).

26. J. C. Andrews, F. Meirer, Y. Liu, Z. Mester, and P. Pianetta, "Transmission X-Ray microscopy for full-field nano imaging of biomaterials", *Microscopy Research & Technique*, 4, 7, 671–681 (2011).

27. C. R. Stumpfel and J. L. Robitaille, "Investigation of the early phases of plasma production by laser irradiation of plane solid targets", *Journal of Applied Physics*, 43, 902 (1972). doi:10.1063/1.1661303.

28. A. Sjögren, M. Harbst, C. G. Wahlström, S. Svanberg, and C. Olsson, "High-repetition-rate, hard x-ray radiation from a laser-produced plasma: Photon yield and application considerations", *Review of Scientific Instruments*, 74, 4, 2300–2311 (2003).

29. E. Fill, J. Bayerl, and R. Tommasini, "A novel tape target for use with repetitively pulsed lasers", *Review of Scientific Instruments*, 73, 2190 (2002). doi:10.1063/1.1468685.

30. T. Nayuki, Y. Oishi, T. Fujii, K. Nemoto, T. Kayoiji, Y. Okano, Y. Hironaka, K. G. Nakamura, K. Kondo, and K. Ueda, "Thin tape target driver for laser ion accelerator", *Review of Scientific Instruments*, 74, 3293 (2003). doi:10.1063/1.1578156.

31. H. Shin, J. Dong, and M. Liu, "Nanoporous structures prepared by an electrochemical deposition process", *Advanced Materials*, 15, 19, 1610 (2003).

32. H. Legall, G. Blobel, H. Stiel, W. Sandner, C. Seim, P. Takman, D. H. Martz, M. Selin, U. Vogt, H. M. Hertz, D. Esser, H. Sipma, J. Luttmann, M. Höfer, H. D. Hoffmann, S. Yulin, T. Feigl, S. Rehbein, P. Guttmann, G. Schneider, U. Wiesemann, M. Wirtz, and W. Diete, "Compact x-ray microscope for the water window based on a high brightness laser plasma source", *Optics Express*, 20, 16, 18362–18369 (2012). doi:10.1364/OE.20.018362.

33. M. Bertilson, O. von Hofsten, U. Vogt, A. Holmberg, and H. M. Hertz, "High-resolution computed tomography with a compact soft x-ray microscope", *Optics Express*, 17, 13, 11057–11065 (2009). doi:10.1364/OE.17.011057.

34. E. Constant, D. Garzella, P. Breger, E. Mével, Ch. Dorrer, C. Le Blanc, F. Salin, and P. Agostini, "Optimizing high harmonic generation in absorbing gases: model and experiment", *Physical Review Letters*, 82, 1668 (1999).

35. J.-F. Hergott, M. Kovacev, H. Merdji, C. Hubert, Y. Mairesse, E. Jean, P. Breger, P. Agostini, B. Carre, and P. Salieres, "Extreme-ultraviolet high-order harmonic pulses in the microjoule range", *Physical Review A*, 66, 021801 (2002).

36. A. Paul, R. A. Bartels, R. Tobey, H. Green, S. Weiman, I. P. Christov, M. M. Murnane, H. C. Kapteyn, and S. Backus, "Quasi-phase-matched generation of coherent extreme-ultraviolet light", *Nature*, 421, 51 (2003).

37. I. J. Kim, C. M. Kim, H. T. Kim, G. H. Lee, Y. S. Lee, J. Y. Park, D. J. Cho, and C. H. Nam, "Highly efficient high-harmonic generation in an orthogonally polarized two-color laser field", *Physical Review Letters*, 94, 243901 (2005).

38. J. Seres, V. S. Yakovlev, E. Seres, Ch. Streli, P. Wobrauschek, Ch. Spielmann, and F. Krausz, "Coherent superposition of laser-driven soft-X-ray harmonics from successive sources", *Nature Physics*, 3, 878 (2007).

39. A. Pirri, C. Corsi, and M. Bellini, "Enhancing the yield of high-order harmonics with an array of gas jets", *Physical Review A*, 78, 011801(R) (2008).

40. M. G. Pullen, N. S. Gaffney, C. R. Hall, J. A. Davis, A. Dubrouil, H. V. Le, R. Buividas, D. Day, H. M. Quiney, and L. V. Dao, "High-order harmonic generation from a dual-gas, multi-jet array with individual gas jet control", *Optics Letters*, 38, 20, 4204–4207 (2013).

41. A. Willner, F. Tavella, M. Yeung, T. Dzelzainis, C. Kamperidis, M. Bakarezos, D. Adams, M. Schulz, R. Riedel, M. C. Hoffmann, W. Hu, J. Rossbach, M. Drescher, N. A. Papadogiannis, M. Tatarakis, B. Dromey, and M. Zepf, "Coherent control of high harmonic generation via dual-gas multijet arrays", *Physical Review Letters*, 107, 175002 (2011).

42. B. Li, T. Higashiguchi, T. Otsuka, W. Jiang, A. Endo, P. Dunne, and G. O'Sullivan, "'Water window'

sources: Selection based on the interplay of spectral properties and multilayer reflection bandwidth", *Applied Physics Letters*, 102, 041117 (2013).

43. M. A. Le Gros, G. Mcdermott, B. P. Cinquin, E. A. Smith, M. Do, W. L. Chao, P. P. Naulleau, and C. A. Larabell, "Biological soft X-ray tomography on beamline 2.1 at the Advanced Light Source", *Journal of Synchrotron Radiation*, 21, 6, 1–8 (2014).

44. S. Marino, S. Palanco, M. Gabás, R. Romero, and J. R. Ramos-Barrado, "Laser nano- and micro-structuring of silicon using a laser-induced plasma for beam conditioning", *Nanotechnology*, 26, 5, 55303 (2015).

45. K. W. Kim, Y. Kwon, K. Y. Nam, J. H. Lim, K. G. Kim, K. S. Chon, B. H. Kim, D. E. Kim, J. G. Kim, B. N. Ahn, H. J. Shin, S. Rah, K. H. Kim, J. S. Chae, D. G. Gweon, D. W. Kang, S. H. Kang, J. Y. Min, K. S. Choi, S. E. Yoon, E. A. Kim, Y. Namba, and K. H. Yoon, "Compact soft x−ray transmission microscopy with sub−50 nm spatial resolution", *Physics in Medicine and Biology*, 51, 6, N99–N107 (2006).

46. O. Buneman, The bennett pinch. Drummond J. E. (eds). *Plasma Physics*. LOC 60-12766 (McGraw-Hill, Inc., New York, 1961), p. 202.

47. EUV Light Source, Electrodeless Z-Pinch™ 10 Watt EUV Source, Energetiq Technology, Inc., 7 Constitution Way, Woburn, MA 01801, online link: http://www.insight-scientific.com/energetiq/EQ10_series_DataSheet.pdf.

48. L. Juschkin, R. Freiberger, and K. Bergmann, "EUV microscopy for defect inspection by dark-field mapping and zone plate zooming", *Journal of Physics*: Conference Series, 186, 012030 (2009). doi:10.1088/1742-6596/186/1/012030.

49. M. E. Couprie, "New generation of light sources: Present and future", *Journal of Electron Spectroscopy and Related Phenomena*, 196, 3–13 (2014). doi:10.1016/j.elspec.2013.12.007.

50. S. Mobilio, C. Meneghini, and F. Boscherini, eds., *Synchrotron Radiation—Basics, Methods and Applications* (Springer, Switzerland, 2015).

51. WWU Munster, The free electron laser FLASH. 2019. www.uni-muenster.de/Physik.PI/Zacharias/en/research/fel/free_electron_lasers.html

52. Deutsches Elektronen-Synchrotron DESY, FEL XUV beamline parameters available online: http://photon-science.desy.de/facilities/flash/flash_parameters/index_eng.html

53. H. N. Chapman, Structure determination using X-ray free-electron laser pulses. Wlodawer A., Dauter Z., and Jaskolski M. (eds) Protein crystallography, pp.295–324, 2017. *Methods in Molecular Biology*, vol. 1607. (Humana Press, New York, 2017).

54. SLAC Linac Coherent Light Source (LCLS). 2019. https://lcls.slac.stanford.edu/

55. M. R. Sawaya, D. Cascio, M. Gingery, J. Rodriguez, L. Goldschmidt, J.-P. Colletier, M. M. Messerschmidt, S. Boutet, J. E. Koglin, G. J. Williams, A. S. Brewster, K. Nass, J. Hattne, S. Botha, R. B. Doak, R. L. Shoeman, D. P. DePonte, H.-W. Park, B. A, Federici, N. K. Sauter, I. Schlichting, and D. S. Eisenberg, Protein crystal structure obtained at 2.9 Å resolution from injecting bacterial cells into an X-ray free-electron laser beam, *Proceedings of the National Academy of Sciences of the United States of America*, 111, 12769–12774 (2014). doi:10.1073/pnas.1413456111.

56. "New Era of Research Begins as World's First Hard X-ray Laser Achieves "First Light"". SLAC National Accelerator Laboratory. April 21, 2009. http://home.slac.stanford.edu/pressreleases/2009/20090421.htm

57. P. A. Franken, A. E. Hill, C. W. Peters, and G. Weinreich, "Generation of optical harmonics", *Physical Review Letters*, 7, 118 (1961).

58. N. H. Burnett, H. A. Baldis, M. C. Richardson, and G. D. Enright, "Harmonic generation in CO2 laser target interaction", *Applied Physics Letters*, 31, 172–174 (1977). doi:10.1063/1.89628.

59. A. McPherson, G. Gibson, H. Jara, U. Johann, T. S. Luk, I. A. McIntyre, K. Boyer, and C. K. Rhodes, "Studies of multiphoton production of vacuum-ultraviolet radiation in the rare gases", *Journal of the Optical Society of America B.* 4, 595 (1987).

60. M. Ferray, A. L'Huillier, X. F. Li, L. A. Lompre, G. Mainfray, and C. Manus, "Multiple-harmonic conversion of 1064 nm radiation in rare gases", *Journal of Physics B: Atomic, Molecular and Optical Physics*, 21, 3 (1988).

61. J. Seres, E. Seres, A. J. Verhoef, G. Tempea, C. Streli, P. Wobrauschek, V. Yakovlev, A. Scrinzi, C. Spielmann, and F. Krausz, "Laser technology: Source of coherent kiloelectronvolt X-rays", *Nature*, 433, 596 (2005). doi:10.1038/433596a

62. J. L. Krause, K. J. Schafer, and K. C. Kulander, "High-order harmonic generation from atoms and ions in the high intensity regime", *Physical Review Letters*, 68, 24, 3535–3538 (1992).

63. P. Corkum, "Plasma perspective on strong-field multiphoton ionization", *Physical Review Letters*, 71, 13, 1994–1997 (1993).

64. T. Brabec and F. Krausz, "Intense few-cycle laser fields: Frontiers of nonlinear optics", *Reviews of Modern Physics*, 72, 545 (2000).

65. A. L'Huillier, K. J. Schafer, and K. C. Kulander, "Theoretical aspects of intense field harmonic generation", *Journal of Physics B: Atomic, Molecular and Optical Physics*, 24, 15 (1991).

66. R. A. Bartels, A. Paul, H. Green, H. C. Kapteyn, M. M. Murnane, S. Backus, I. P. Christov, Y. Liu, D. Attwood, and C. Jacobsen, "Generation of spatially

67. coherent light at extreme ultraviolet wavelengths", *Science*, 297, 376–378 (2002).

67. G. O. Williams, A. I. Gonzalez, S. Künzel, L. Li, M. Lozano, E. Oliva, B. Iwan, S. Daboussi, W. Boutu, H. Merdji, M. Fajardo, and Ph. Zeitoun, "Fourier transform holography with high harmonic spectra for attosecond imaging applications", *Optics Letters*, 40, 13, 3205–3208 (2015).

68. O. Kfir, S. Zayko, C. Nolte, M. Sivis, M. Möller, B. Hebler, S Sai Phani Kanth Arekapudi, D. Steil, S. Schäfer, M. Albrecht, O. Cohen, S. Mathias, and C. Ropers, "Nanoscale magnetic imaging using circularly polarized high-harmonic radiation", *Science Advances*, 3, eaao4641 (2017).

69. A. Ravasio, D. Gauthier, F. R. N. C. Maia, M. Billon, J.-P. Caumes, D. Garzella, M. Geleoc, O. Gobert, J.-F. Hergott, A.-M. Pena, H. Perez, B. Carre, E. Bourhis, J. Gierak, A. Madouri, D. Mailly, B. Schiedt, M. Fajardo, J. Gautier, P. Zeitoun, P. H. Bucksbaum, J. Hajdu, and H. Merdji, "Single-shot diffractive imaging with a table-top femtosecond soft X-ray laser-harmonics source", *Physical Review Letters*, 103, 028104 (2009).

70. P. D. Baksh, M. Odstrčil, H. S. Kim, S. A. Boden, J. G. Frey, and W. S. Brocklesby, "Wide-field broadband extreme ultraviolet transmission ptychography using a high-harmonic source", *Optics Letters*, 41, 7, 1317–1320 (2016). doi:10.1364/OL.41.001317

71. J. J. Rocca, D. C. Beetle, and M. C. Marconi, "Proposal for soft-x-ray and XUV lasers in capillary discharges", *Optics Letters*, 13, 565 (1988).

72. D. Macchietto, B. R. Benware, and J. J. Rocca, "Generation of millijoule-level soft-x-ray laser pulses at a 4-Hz repetition rate in a highly saturated tabletop capillary discharge amplifier", *Optics Letters*, 24, 1115, (1999).

73. B. R. Benware, C. D. Macchietto, C. H. Moreno, and J. J. Rocca, "Demonstration of a high average power tabletop soft X-ray laser", *Physical Review Letters*, 81, 5804 (1998).

74. J. J. Rocca, "Table-top soft x-ray lasers", *Review of Scientific Instruments*, 70, 3799 (1999).

75. J. J. Rocca, D. P. Clark, J. L. A. Chilla, and V. N. Shlyaptsev, "Energy extraction and achievement of the saturation limit in a discharge-pumped tabletop soft X-ray amplifier", *Physical Review Letters*, 77, 1476 (1996).

76. J. W. Goodman, *Statistical Optics* (Wiley, New York, 1985), pp. 171–187.

77. Y. Liu, M. Seminario, F. G. Tomasel, C. Chang, J. J. Rocca, and D. T. Attwood, "Achievement of essentially full spatial coherence in a high-average-power soft-x-ray laser," *Physical Review A*, 63, 033802, (2001).

78. M. C. Marconi, J. L. A. Chilla, C. H. Moreno, B. R. Benware, and J. J. Rocca, "Measurement of the spatial coherence buildup in a discharge pumped

table-top soft x-ray laser", *Physical Review Letters*, 15, 2799 (1997).

79. L. Urbanski, L. M. Meng, M. C. Marconi, M. Berril, O. Guilbaud, A. Klisnick, and J. J. Rocca, "Line width measurement of a capillary discharge soft X-ray laser", *Proceedings of SPIE—The International Society for Optical Engineering*, 8140 (2011). doi:10.1117/12.892929.

80. P. W. Wachulak, C. A. Brewer, F. Brizuela, C. S. Menoni, W. Chao, E. H. Anderson, R. A. Bartels, J. J. Rocca, and M. C. Marconi, "Simultaneous determination of feature size and resolution in soft x-ray microscopy images", *Journal of the Optical Society of America B*, 25, 7, B20 (2008).

81. P. W. Wachulak, M. C. Marconi, R. Bartels, C. S. Menoni, and J. J. Rocca, "Soft x-ray laser holography with wavelength resolution", *Journal of the Optical Society of America B*, 25, 11, 1811–1814 (2008).

82. Y. Wang, M. A. Larotonda, B. M. Luther, D. Alessi, M. Berrill, V. N. Shlyaptsev, and J. J. Rocca, "Demonstration of high-repetition-rate tabletop soft-x-ray lasers with saturated output at wavelengths down to 13.9 nm and gain down to 10.9 nm", *Physical Review A*, 72, 053807 (2005).

83. A. Rockwood, Y. Wang, S. Wang, M. Berrill, V. N. Shlyaptsev, and J. J. Rocca, "Compact gain-saturated x-ray lasers down to 6.85 nm and amplification down to 5.85 nm", *Optica*, 5, 3, 257–262 (2018). doi:10.1364/OPTICA.5.000257

84. F. Pedaci, Y. Wang, M. Berrill, B. Luther, E. Granados, and J. J. Rocca, "Highly coherent injection-seeded 13.2 nm table-top soft x-ray laser," *Optics Letters*, 33, 491, (2008).

85. G. Vaschenko, C. Brewer, F. Brizuela, Y. Wang, M. A. Larotonda, B. M. Luther, M. C. Marconi, J. J. Rocca, and C. S. Menoni, "Sub- 38nm resolution tabletop microscopy with 13nm wavelength laser light", *Optics Letters*, 31, 9 (2006).

86. H. T. Kim, I. W. Choi, N. Hafz, J. H. Sung, T. J. Yu, K.-H. Hong, T. M. Jeong, Y.-C. Noh, D.-K. Ko, K. A. Janulewicz, J. Tümmler, P. V. Nickles, W. Sandner, and J. Lee, "Demonstration of a saturated Ni-like Ag x-ray laser pumped by a single profiled laser pulse from a 10-Hz Ti:sapphire laser system", *Physical Review A*, 77, 023807 (2008).

87. H. T. Kim, I. J. Kim, C. M. Kim, T. M. Jeong, T. J. Yu, S. K. Lee, J. H. Sung, J. W. Yoon, H. Yun, S. C. Jeon, I. W. Choi, and J. Lee, "Single-shot nanometer-scale holographic imaging with laser-driven x-ray laser", *Applied Physics Letters*, 98, 121105 (2011).

88. G. Vaschenko, F. Brizuela, C. Brewer, M. Grisham, H. Mancini, C. S. Menoni, M. C. Marconi, J. J. Rocca, W. Chao, J. A. Liddle, E. H. Anderson, D. T. Attwood, A. V. Vinogradov, I. A. Artioukov, Y. P. Pershyn, and V. V. Kondratenko, "Nanoimaging

with a compact extreme-ultraviolet laser", *Optics Letters*, 30, 16, 2095–2097 (2005).

89. S. Sebban, T. Mocek, D. Ros, L. Upcraft, Ph. Balcou, R. Haroutunian, G. Grillon, B. Rus, A. Klisnick, A. Carillon, G. Jamelot, C. Valentin, A. Rousse, J. P. Rousseau, L. Notebaert, M. Pittman, and D. Hulin, "Demonstration of a Ni-like Kr optical-field-ionization at 32.8 nm", *Physical Review Letters*, 89, 253901 (2002).

90. L. B. Da Silva, J. E. Trebes, R. Balhorn, S. Mrowka, E. Anderson, D. T. Attwood, T. W. Barbee Jr., J. Brase, M. Corzett, J. Gray, J. A. Koch, C. Lee, D. Kern, R. A. London, B. J. MacGowan, and D. L. Mathews, "X-ray laser microscopy with an rat sperm nuclei", *Science*, 258, 269 (1992).

91. D. Attwood, A. Sakdinawat, *X-Rays and Extreme Ultraviolet Radiation* (Cambridge University Press, UK, 2017), online ISBN: 9781107477629. doi:10.1017/CBO9781107477629).

92. P. W. Wachulak, A. Bartnik, Ł. Węgrzyński, T. Fok, J. Kostecki, M. Szczurek, R. Jarocki, and H. Fiedorowicz, "Characterization of pulsed capillary channel gas puff target using EUV shadowgraphy", *Nuclear Instruments and Methods in Physics Research B*, 345, 15–21 (2015). doi:10.1016/j.nimb.2014.12.060

93. G. Tomassetti, A. Ritucci, A. Reale, and P. Zuppella, "Capillary discharge-pumped EUV laser and its applications", *AIP Conference Proceedings*, 899, 349 (2007). doi:10.1063/1.2733180

94. P. W. Wachulak, A. Bartnik, A. Baranowska-Korczyc, D. Pánek, P. Brůža, J. Kostecki, Ł. Węgrzyński, R. Jarocki, M. Szczurek, K. Fronc, D. Elbaum, and H. Fiedorowicz, "Study of crystalline thin films and nanofibers by means of the laser-plasma EUV-source based microscopy", *Radiation Physics and Chemistry*, 93, 54–58 (2013). doi:10.1016/j.radphyschem.2013.02.019

95. C. Mbanaso and G. Denbeaux, EUV lithography. Bhushan B. (eds). *Encyclopedia of Nanotechnology, pp. 1-77* (Springer, Dordrecht, The Netherlands, 2016, DOI: https://doi.org/10.1007/978-94-017-9780-1˙391).

96. N. I. Chkhalo and N. N. Salashchenko, "Projection XEUV-nanolithography", *Nuclear Instruments and Methods in Physics Research Section A*, 603, 147–149 (2009).

97. M. Goldstein and P. Naulleau, "Actinic microscope for extreme ultraviolet lithography photomask inspection and review", *Optics Express*, 20, 14, 15752–15768 (2012). doi:10.1364/OE.20.015752

98. Y. G. Wang, R. Miyakawa, A. Neureuther, and P. Naulleau, "Zernike phase contrast microscope for EUV mask inspection", *Proceedings of SPIE*, 9048, 904810 (2014). doi:10.1117/12.2048180

99. P. W. Wachulak, "The novel gas puff targets for laser-matter interaction experiments", *Review*

of Scientific Instruments, 87, 091501 (2016). doi:10.1063/1.4962012.

100. C. Messaoudi, T. Boudier, C. O. Sanchez Sorzano, and S. Marco, "TomoJ: tomography software for three-dimensional reconstruction in transmission electron microscopy", *BMC Bioinformatics*, 8, 288 (2007).

101. P. W. Wachulak, Ł. Węgrzyński, Z. Zápraný, A. Bartnik, T. Fok, R. Jarocki, J. Kostecki, M. Szczurek, D. Korytár, and H. Fiedorowicz, "Extreme ultraviolet tomography using a compact laser–plasma source for 3D reconstruction of low density objects", *Optics Letters*, 39, 3, 532–535 (2014), also published in *The Virtual Journal for Biomedical Optics*, 9, 4 (2014).

102. H. Fiedorowicz, A. Bartnik, R. Jarocki, J. Kostecki, J. Krzywinski, J. Mikołajczyk, R. Rakowski, A. Szczurek, and M. Szczurek, "Compact laser plasma EUV source based on a gas puff target for metrology applications", *Journal of Alloys and Compounds*, 401, 99 (2005).

103. R. Rakowski, A. Bartnik, H. Fiedorowicz, F. de Gaufridy de Dortan, R. Jarocki, J. Kostecki, J. Mikołajczyk, L. Ryc, M. Szczurek, and P. Wachulak, "Characterization and optimization of the laser-produced plasma EUV source at 13.5nm based on a double-stream Xe/He gas puff target", *Applied Physics B*, 101, 773 (2010).

104. P. W. Wachulak, A. Torrisi, A. Bartnik, D. Adjei, J. Kostecki, Ł. Wegrzynski, R. Jarocki, M. Szczurek, and H. Fiedorowicz, "Desktop water window microscope using a double-stream gas puff target source", *Applied Physics B*, 118, 4, 573–578 (2015).

105. P. Wachulak, A. Torrisi, M. F. Nawaz, A. Bartnik, D. Adjei, Š. Vondrová, J. Turňová, A. Jančarek, and J. Limpouch, "A compact 'water window' microscope with 60 nm spatial resolution for applications in biology and nanotechnology", *Microscopy and Microanalysis*, 21, 5, 1214–1223 (2015).

106. P. W. Wachulak, A. Bartnik, H. Fiedorowicz, and J. Kostecki, "A 50nm spatial resolution EUV imaging–resolution dependence on object thickness and illumination bandwidth", *Optics Express*, 19, 10, 9541–9550 (2011).

107. C. A. Brewer, F. Brizuela, P. Wachulak, D. H. Martz, W. Chao, E. H. Anderson, D. T. Attwood, A. V. Vinogradov, I. A. Artyukov, A. G. Ponomareko, V. V. Kondratenko, M. C. Marconi, J. J. Rocca, and C. S. Menoni, "Single-shot extreme ultraviolet laser imaging of nanostructures with wavelength resolution", *Optics Letters*, 33, 518 (2008).

108. W. Chao, E. Anderson, G. P. Denbeaux, B. Harteneck, J. A. Liddle, D. L. Olynick, A. L. Pearson, F. Salmassi, Ch. Y. Song, and D. T. Attwood, "20-nm-resolution soft x-ray microscopy demonstrated by use of multilayer test structures", *Optics Letters*, 28, 21, 2019–2021 (2003). doi:10.1364/OL.28.002019

109. P. W. Wachulak, A. Torrisi, A. Bartnik, Ł. Węgrzyński, T. Fok, and H. Fiedorowicz, "A desktop extreme ultraviolet microscope based on a compact laser-plasma light source", *Applied Physics B*, 123, 25, 1–5 (2016).

110. A. Torrisi, P. Wachulak, Ł. Węgrzyński, T. Fok, A. Bartnik, T. Parkman, Š. Vondrová, J. Turňová, and B. J. Jankiewicz, "A stand-alone compact EUV microscope based on gas-puff target source", *Journal of Microscopy*, 265, 2, 251–260 (2017).

111. W. Chao, P. Fischer, T. Tyliszczak, S. Rekawa, E. Anderson, and P. Naulleau, "Real space soft x-ray imaging at 10 nm spatial resolution", *Optics Express*, 20, 9, 9777–9783 (2012).

112. C. Jacobsen, S. Williams, E. Anderson, M. Browne, C. Buckley, D. Kern, J. Kirz, M. Rivers, and X. Zhang, "Diffraction-limited imaging in a scanning transmission x-ray microscope," *Optics Communications*, 86, 3–4, 351–364 (1991).

113. J. B. Kortright, S.-K. Kim, G. Ohldag, G. Meigs, and A. Warwick, "Magnetization imaging using scanning transmission x-ray microscopy," *AIP Conference Proceedings*, 507, 1, 49–54 (2000).

114. B. Hornberger, M. Feser, and C. Jacobsen, "Quantitative amplitude and phase contrast imaging in a scanning transmission X-ray microscope", *Ultramicroscopy*, 107, 8, 644–655 (2007).

115. R. Früke, J. Kutzner, T. Witting, H. Zacharias, and Th. Wilhein, "EUV scanning transmission microscope operating with high-harmonic and laser plasma radiation", *Europhysics Letters*, 72, 6, 915–921 (2005). doi:10.1209/epl/i2005-10340-7

116. M. Kado, M. Kishimoto, S. Tamotsu, K. Yasuda, M. Aoyama, and K. Shinohara, X-Ray Lasers and Coherent X-Ray Sources: Development and Applications X 8849, 1–7 (2013), Proceedings of a meeting held 27–29 August 2013, San Diego, CA, USA. At SPIE Optical Engineering + Applications.

117. M. G. Ayele, P. W. Wachulak, J. Czwartos, D. Adjei, A. Bartnik, Ł. Wegrzynski, M. Szczurek, L. Pina, and H. Fiedorowicz, "Development and Characterization of laser- plasma soft X-ray source for contact microscopy", *Nuclear Instruments and Methods in Physics Research Section B*, 411, 35–43 (2017). doi:10.1016/j.nimb.2017.03.082

118. M. G. Ayele, J. Czwartos, D. Adjei, P. Wachulak, I. U. Ahad, A. Bartnik, Ł. Wegrzynski, M. Szczurek, R. Jarocki, H. Fiedorowicz, M. Lekka, K. Pogoda, and J. Gostek, "Contact microscopy using a compact laser produced plasma soft X-ray source", *Acta Physica Polonica A*, 129, 2, 237–240 (2016).

119. D. Gabor, "A new microscope principle", *Nature*, 161, 777–778 (1948).

120. A. V. Baez, "A study in diffraction microscopy with special reference to X-Rays", *Journal of the Optical Society of America*, 42, 756–762 (1952).

121. J. E. Trebes, S. B. Brown, E. M. Campbell, D. L. Matthews, D. G. Nilson, G. F. Stone, and D. A. Whelan, "Demonstration of X-ray holography with an X-ray laser", *Science*, 238, 517–519 (1987).

122. A. Rosenhahn, R. Barth, X. Cao, M. Schurmann, M. Grunze, and S. Eisebitt, "Vacuum-ultraviolet Gabor holography with synchrotron radiation", *Ultramicroscopy*, 107, 1171–1177 (2007).

123. R. I. Tobey, M. E. Siemens, O. Cohen, M. M. Murnane, H. C. Kapteyn, and K. A. Nelson, "Ultrafast extreme ultraviolet holography: dynamic monitoring of surface deformation", *Optics Letters*, 32, 286–288 (2007).

124. P. Wachulak, M. C. Marconi, R. Bartels, C. S Menoni, and J. J. Rocca, "Volume extreme ultraviolet holographic imaging with numerical optical sectioning", *Optics Express*, 15, 10622–10628 (2007).

125. U. Schnars and W. P. O. Juptner, "Digital recording and numerical reconstruction of holograms". *Measurement Science & Technology*, 13, R85–R101 (2002).

126. W. F. Schlotter, R. Rick, K. Chen, A. Scherz, J. Stöhr, J. Lüning, S. Eisebitt, Ch. Günther, W. Eberhardt, O. Hellwig, and I. McNulty, "Multiple reference Fourier transform holography with soft x rays", *Applied Physics Letters*, 89, 163112 (2006). doi:10.1063/1.2364259

127. S. Eisebitt, J. Lüning, W. Schlotter, and M. Lörgen, "Lensless imaging of magnetic nanostructures by X-ray spectro-holography," *Nature*, 432, 885–888 (2004).

128. E. Guehrs, A. M. Stadler, S. Flewett, S. Frömmel, J. Geilhufe, B. Pfau, T. Rander, S. Schaffert, G. Büldt, and S. Eisebitt, "Soft x-ray tomoholography," *New Journal of Physics*, 14, 013022 (2012).

129. C. Günther, B. Pfau, R. Mitzner, B. Siemer, S. Roling, H. Zacharias, O. Kutz, I. Rudolph, D. Schondelmaier, R. Treusch, and S. Eisebitt, "Sequential femtosecond X-ray imaging," *Nature Photonics*, 5, 99–102 (2011).

130. J. W. Goodman, *Introduction to Fourier Optics*, 3rd ed. (Roberts and Company Publishers, USA, 2005).

131. I. McNulty, J. Kirz, C. Jacobsen, E. H. Anderson, M. R. Howells, and D. P. Kern, "High-resolution imaging by fourier transform X-ray holography", *Science (New York, N.Y.)*, 256, 1009–1012 (1992).

132. E. B. Malm, N. C. Monserud, Ch. G. Brown, P. W. Wachulak, H. Xu, G. Balakrishnan, W. Chao, E. Anderson, and M. C. Marconi, "Tabletop singleshot extreme ultraviolet Fourier transform holography of an extended object", *Optics Express* 21, 8, 9959–9966 (2013) also published in *The Virtual Journal of Biomedical Optics*, 8, 5, (2013).

133. M. Guizar-Sicairos and J. R. Fienup, "Holography with extended reference by autocorrelation linear differential operation", *Optics Express*, 15, 17592–17612 (2007).

134. D. Zhu, M. Guizar-Sicairos, B. Wu, A. Scherz, Y. Acremann, T. Tyliszczak, P. Fischer, N. Friedenberger, K. Ollefs, M. Farle, J. Fienup, and J. Stöhr, "High-resolution X-Ray lensless imaging by differential holographic encoding", *Physical Review Letters*, 105, 043901 (2010).

135. D. Gauthier, M. Guizar-Sicairos, X. Ge, W. Boutu, B. Carré, J. Fienup, and H. Merdji, "Singleshot femtosecond X-Ray holography using extended references", *Physical Review Letters*, 105, 1–4 (2010).

136. A. Isoyan, F. Jiang, Y. C. Cheng, F. Cerrina, P. Wachulak, L. Urbanski, J. Rocca, C. Menoni, and M. Marconi, "Talbot lithography: Self-imaging of complex structures", *Journal of Vacuum Science Technology B*, 27, 6, 2931–2937 (2009).

137. P. W. Wachulak, L. Urbanski, M. G. Capeluto, D. Hill, W. S. Rockward, C. C. Iemmi, E. H. Anderson, C. S. Menoni, J. J. Rocca, and M. C. Marconi, "New opportunities in interferometric lithography using extreme ultraviolet tabletop lasers", *Journal of Micro/Nanolithography MEMS MOEMS*, 8, 021206 (2009).

138. J. M. Kim, I. H. Cho, S. Y. Lee, H. C. Kang, R. Conley, C. Liu, A. T. Macrander, and D. Y. Noh, "Observation of the Talbot effect using broadband hard x-ray beam", *Optics Express*, 18, 24, 24975 (2010).

139. J. Miao, D. Sayre, and H. N. Chapman, "Phase retrieval from the magnitude of the Fourier transforms of nonperiodic objects", *Journal of the Optical Society of America a-Optics Image Science and Vision*, 15, 1662–1669 (1998).

140. J. Miao, T. Ishikawa, E. H. Anderson, and K. O. Hodgson, "Phase retrieval of diffraction patterns from noncrystalline samples using the oversampling method", *Physical Review B*, 67, 174104 (2003).

141. J. Miao, C. C. Chen, C. Y. Song, Y. Nishino, Y. Kohmura, T. Ishikawa, D. Ramunno-Johnson, T. K. Lee, and S. H. Risbud, "Three-dimensional GaN-Ga2O3 core shell structure revealed by X-ray diffraction microscopy," *Physical Review Letters*, 97, 215503 (2006).

142. R. L. Sandberg, C. Song, P. W. Wachulak, D. A. Raymondson, A. Paul, B. Amirbekian, E. Lee, A. E. Sakdinawat, C. La-O-Vorakiat, M. C. Marconi, C. S. Menoni, M. M. Murnane, J. J. Rocca, H. C. Kapteyn, and J. Miao, "High numerical aperture tabletop soft x-ray diffraction microscopy with 70-nm resolution", *Proceedings of the National Academy of Sciences of the United States of America*, 105, 1, 24–27 (2008).

143. H. N. Chapman, A. Barty, M. J. Bogan, S. Boutet, M. Frank, S. P. Hau-Riege, S. Marchesini,

B. W. Woods, S. Bajt, W. H. Benner, R. A. London, E. Plönjes, M. Kuhlmann, R. Treusch, S. Düsterer, T. Tschentscher, J. R. Schneider, E. Spiller, T. Möller, C. Bostedt, M. Hoener, D. A. Shapiro, K. O. Hodgson, D. van der Spoel, F. Burmeister, M. Bergh, C. Caleman, G. Huldt, M. M. Seibert, F. R. N. C. Maia, R. W. Lee, A. Szöke, N. Timneanu, and J. Hajdu, "Femtosecond diffractive imaging with a soft-X-ray free-electron laser", *Nature Physics*, 2, 839–843 (2006).

144. T. Ekeberg, M. Svenda, C. Abergel, F. R. N. C. Maia, V. Seltzer, J. Claverie, M. Hantke, O. Jönsson, C. Nettelblad, G. van der Schot, M. Liang, D. P. DePonte, A. Barty, M. M. Seibert, B. Iwan, I. Andersson, N. D. Loh, A. V. Martin, H. Chapman, C. Bostedt, J. D. Bozek, K. R. Ferguson, J. Krzywinski, S. W. Epp, D. Rolles, A. Rudenko, R. Hartmann, N. Kimmel, and J. Hajdu, "Three-dimensional reconstruction of the giant mimivirus particle with an X-ray free-electron laser", *Physical Review Letters*, 114, 098102 (2015).

145. M. M. Seibert, T. Ekeberg, F. R. N. C. Maia, M. Svenda, J. Andreasson, O. Jonsson, D. Odic, B. Iwan, A. Rocker, D. Westphal, et al., "Single mimivirus particles intercepted and imaged with an X-ray laser", *Nature*, 470, 78–81 (2011).

146. M. D. Seaberg, D. E. Adams, E. L. Townsend, D. A. Raymondson, W. F. Schlotter, Y. Liu, C. S. Menoni, L. Rong, C. Chen, J. Miao, H. C. Kapteyn, and M. M. Murnane, "Ultrahigh 22 nm resolution coherent diffractive imaging using a desktop 13 nm high harmonic source", *Optics Express*, 19, 23, 22470–22479 (2011). doi:10.1364/OE.19.022470

147. W. Hoppe, "Beugung im inhomogenen Primärstrahlwellenfeld. I. Prinzip einer Phasenmessung von Elektronenbeugungsinterferenzen", *Acta Crystallographica Section A*, 25, 4, 495 (1969). doi:10.1107/S0567739469001045

148. A. M. Maiden and J. M. Rodenburg, "An improved ptychographical phase retrieval algorithm for diffractive imaging", *Ultramicroscopy*, 109, 10, 1256–1262 (2009).

149. P. Thibault, M. Dierolf, A. Menzel, O. Bunk, C. David, and F. Pfeiffer, "High-resolution scanning x-ray diffraction microscopy", *Science*, 321, 5887, 379–382 (2008). doi:10.1126/science.1158573.

150. M. Guizar-Sicairos and J. R. Fienup, "Phase retrieval with transverse translation diversity: a nonlinear optimization approach", *Optics Express*, 16, 10, 7264–7278 (2008). doi:10.1364/OE.16.007264

151. M. Dierolf, A. Menzel, P. Thibault, P. Schneider, C. M. Kewish, R. Wepf, O. Bunk, and F. Pfeiffer, "Ptychographic X-ray computed tomography at the nanoscale", *Nature*, 467, 436–439 (2010). doi:10.1038/nature09419

152. M. J. Humphry, B. Kraus, A. C. Hurst, A. M. Maiden, and J. M. Rodenburg, "Ptychographic electron microscopy using high-angle dark-field scattering for sub-nanometre resolution imaging", *Nature Communications*, 3, 730 (2012). doi:10.1038/ncomms1733

153. R. N. Wilke, M. Priebe, M. Bartels, K. Giewekemeyer, A. Diaz, P. Karvinen, and T. Salditt. " Hard X-ray imaging of bacterial cells: nano-diffraction and ptychographic reconstruction", *Optics Express*, 20, 17, 19232–19254 (2012). doi:10.1364/OE.20.019232

154. M. Holler, A. Diaz, M. Guizar-Sicairos, P. Karvinen, E. Färm, E. Härkönen, M. Ritala, A. Menzel, J. Raabe, and O. Bunk, "X-ray ptychographic computed tomography at 16 nm isotropic 3D resolution", *Scientific Reports*, 4, 3857 (2014). doi:10.1038/srep03857

155. H. Chapman, "A new phase for X-ray imaging", *Nature*, 467, 409 (2010).

156. D. Huang, E. A. Swanson, C. P. Lin, J. S. Schuman, W. G. Stinson, W. Chang, M. R. Hee, T. Flotte, K. Gregory, C. A. Puliafito, and J. G. Fujimoto, "Optical coherence tomography", *Science*, 254, 5030, 1178–1181 (1991).

157. W. Drexler and J. G. Fujimoto, *Optical Coherence Tomography*, (Springer International Publishing, Basel, Switzerland , 2015).

158. P. Keane, P. J. Patel, S. Liakopoulos, F. M. Heussen, S. R. Sadda, and A. Tufail, "Evaluation of age-related macular degeneration with optical coherence tomography", *Survey of Ophthalmology*, 57, 389–414 (2012).

159. M. Wojtkowski, B. L. Sikorski, I. Gorczynska, M. Gora, M. Szkulmowski, D. Bukowska, J. Kałuzny, J. G. Fujimoto, and A. Kowalczyk, "Comparison of reflectivity maps and outer retinal topography in retinal disease by 3-D Fourier domain optical coherence tomography", *Optics Express*, 17, 5, 4189 (2009).

160. S. Fuchs, A. Blinne, C. Rödel, U. Zastrau, V. Hilbert, M. Wünsche, J. Bierbach, E. Frumker, E. Förster, and G. G. Paulus, "Optical coherence tomography using broad-bandwidth XUV and sof X-ray radiation", *Applied Physics B*, 106, 789–795 (2012).

161. P. Jaeglé, *Coherent Sources of XUV Radiation, Soft X-Ray Lasers and High-Order Harmonic Generation* (Springer, New York, 2006), ISBN: 978-0-387-23007-8. doi:10.1007/978-0-387-29990-7

162. S. Fuchs, Ch. Rödel, A. Blinne, U. Zastrau, M. Wünsche, V. Hilbert, L. Glaser, J. Viefaus, E. Frumker, P. Corkum, E. Förster, and G. G. Paulus, "Nanometer resolution optical coherence tomography using broad bandwidth XUV and soft x-ray radiation", *Scientific Reports*, 6, 20658 (2016). doi:10.1038/srep20658

163. S. Fuchs, M. Wünsche, J. Nathanael, J. J. Abel, Ch. Rödel, J. Biedermann, J. Reinhard, U. Hübner, and G. G. Paulus, "Optical coherence tomography with

nanoscale axial resolution using a laser-driven high-harmonic source", *Optica*, 4, 8, 903 (2017).

164. P. Wachulak, A. Bartnik, and H. Fiedorowicz, "Optical coherence tomography (OCT) with 2 nm axial resolution using a compact laser plasma soft X-ray source", *Scientific Reports*, 8, 8494 (2018). doi:10.1038/s41598-018-26909-0, https://rdcu.be/QZQP

165. R. Feder, E. Spiller, J. Topalian, A. N. Broers, W. Gudat, B. J. Panessa, Z. A. Zadunaisky, and J. Sedat, "High-resolution soft x-ray microscopy", *Science*, 197, 4300, 259–260 (1977). doi:10.1126/science.406670

166. C. O. Girit, J. C. Meyer, R. Erni, M. D. Rossell, C. Kisielowski, L. Yang, C.-H. Park, M. F. Crommie, M. L. Cohen, S. G. Louie, and A. Zettl, "Graphene at the edge: Stability and dynamics," *Science*, 323, 5922, 1705–1708, (2009).

167. D. Adjei, A. Wiecheć, P. Wachulak, M. Ayele, J. Lekki, W. Kwiatek, A. Bartnik, M. Davidkova, L. Vyšín, L. Juha, L. Pina, and H. Fiedorowicz, "DNA strand breaks induced by soft X-ray pulses from a compact laser plasma source", *Radiation Physics and Chemistry*, 120, 17–25 (2015). doi:10.1016/j.radphyschem.2015.11.021.

8

X-ray Imaging of Single Nanoparticles and Nanostructures

Daniel Rolles
Kansas State University

8.1 Introduction

X-ray based spectroscopy and imaging methods are versatile tools for characterizing nanostructures and nanomaterials. Because of the element specificity and the chemical sensitivity of the X-ray absorption process and because of the X-rays' large penetration depths, X-ray absorption spectroscopy methods such as (near-edge) X-ray absorption fine structure (NEXAFS/XAFS) and related techniques such as X-ray emission (XES) or X-ray photoelectron spectroscopy (XPS) are commonly used to determine the chemical composition of nanostructures, even if they are buried inside larger structures. Furthermore, due to the short wavelength of X-ray radiation, various X-ray imaging techniques can be used to image nanostructures with atomic resolution. Although the breakthrough of many of these methods came with the widespread availability of high-brightness, tunable X-rays from large-scale synchrotron radiation facilities, a variety of other, more compact X-ray sources makes most of them also suitable for smaller laboratory environments.

Despite their success in material sciences, nanosciences, and life sciences, there are some drawbacks associated to these techniques, which limit their applicability to nonidentical and non-reproducible nanoparticles. These drawbacks mainly result from the fact that in order to obtain sufficient signal levels, X-ray measurements on nanoparticles typically have to be performed on an ensemble of particles, which are, furthermore, often deposited on a substrate for easier sample handling. However, many nanoparticles, especially during their growth process, are not homogenous in size, composition, and morphology. Ensemble measurements only yield average quantities, from which little information can be obtained about a specific particle or about the size,

composition, and morphology variations that occur within the ensemble. Moreover, many of the nanoparticle properties, such as their morphology and their reactivity, strongly depend on the environment. Measurements performed on deposited particles therefore do not yield the same information as measurements of nanoparticles in their native aerosol environment. For example, aerosolized soot particles formed in a combustion process can change significantly when deposited on a surface; yet, it is, predominantly, the aerosol that is important, e.g., for atmospheric chemistry and climate impact.

In order to be able to characterize and image single, unsupported nanoparticles, X-rays with a sufficiently short wavelength are required to provide sub-nanometer resolution and with sufficient photon fluence (number of photons per unit area) to yield enough scattering signal *before* radiation damage occurs. These requirements are fulfilled by new, accelerator-based, intense and short-pulse X-ray sources, called X-ray free-electron lasers (XFELs), which started operation in the last decade and provide unprecedented opportunities for nanoscience. In the following, a brief introduction into nanoparticle imaging with XFELs will be given, followed by a description of representative examples that highlight the versatility of these techniques, and an outlook into possible future applications.

8.2 Nanoparticle Imaging with X-ray Free-Electron Lasers

8.2.1 X-ray Free-Electron Lasers

In conventional synchrotron radiation facilities, X-ray beams with high-brightness and tunable photon energy are produced by sending relativistic electron bunches, which are circulating inside a storage ring, through periodic magnetic structures called *undulators*. The accelerated motion of the relativistic electrons in the magnetic field generates X-rays, which are emitted along the propagation direction of the electron beam. The superposition of X-rays produced from each magnetic period of the undulator results in a strong enhancement of the emission around a specific photon energy, for which the undulator is tuned by varying the gap between the poles and, thus, the magnetic field of the undulator (Winick & Bienenstock 1978). XFELs, such as the Linac Coherent Light Source (LCLS) in the United States, SACLA in Japan, and the European XFEL in Germany, are based on a similar working principle as synchrotron radiation facilities, except that the electron bunches, which are typically produced by a photocathode and which are accelerated by a linear accelerator, do not circulate inside a storage ring. Instead, they are sent, in a single pass, through a long chain of undulators with a total length of several tens up to over one hundred meters. When the electrons enter this long undulator, they initially emit incoherent synchrotron radiation at the resonant undulator photon energy. As the radiation and the electron bunch co-propagate along the undulator, the electrons interact collectively with the radiation they emit, which imprints a microscopic substructure on the electron bunch with the periodicity on the resonant wavelength. The coherent enhancement of the X-rays emitted from each of these microbunches leads to an exponential gain along the XFEL undulator, which can produce X-ray pulses with several millijoules pulse energy, corresponding to $\sim 10^{13}$ X-ray photons per pulse. In most XFELs, statistical, coherent fluctuations in the radiation field at the beginning of the undulator are coherently enhanced in a process termed "self-amplified spontaneous emission (SASE)", and these statistical fluctuations are therefore imprinted in the temporal and spectral profile of the output pulses. These fluctuations can be reduced when using a coherent seed pulse to start the lasing process, and several of such seeding

schemes have been demonstrated and are now in use at some of the XFEL facilities.

Since the X-ray pulses produced by XFELs not only contain an enormous number of photons but also have pulse durations of typically only a few up to a few hundred femtoseconds, the peak brilliance (flux per unit source size and unit solid angle within a certain photon bandwidth) of XFELs is 10^7–10^{10} times greater than that of modern synchrotron storage rings. When focused to spot sizes of 1 μm and below, XFEL beams can reach intensities 10^8–10^{10} times greater than other laboratory X-ray sources (McNeil and Thompson 2010; Waldrop 2014) (Figure 8.1).

8.2.2 Coherent X-ray Diffractive Imaging with XFELs

Because of their short wavelength – an X-ray photon with a photon energy of 12 kiloelectronvolts (keV; 1 eV = 1.602×10^{-19} J) has a wavelength of approximately 0.1 nm – X-rays are commonly used for high-resolution structure determination. One of the most prominent examples is X-ray crystallography, where the structure of protein crystals or of other crystalline samples is routinely determined with sub-nanometer resolution from the diffraction patterns formed by elastically scattered X-ray photons. However, many samples, such as many nanostructures and nanoparticles, and even most biological systems, do not form (sufficiently large) crystals, and most functional studies are not compatible with a crystalline environment that may alter the reaction of interest or change the morphology of the nanoparticle. In these cases, other diffractive imaging techniques have to be used for atomic-scale structure determination.

Since the manufacturing of high-resolution X-ray lenses is extremely challenging and the best achievable resolution using X-ray lenses has, to date, been limited to about 15 nm (Chao et al. 2005), *lensless imaging* methods (see Figure 8.2) have been developed, where the far-field

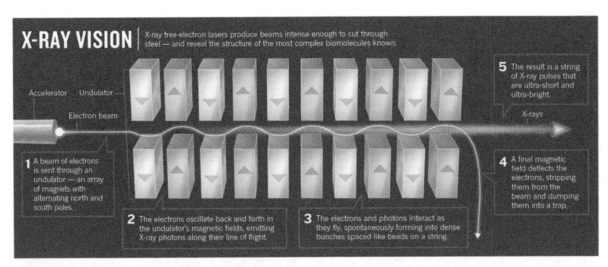

FIGURE 8.1 Working principle of an XFEL. (Taken from Waldrop 2014.)

(a) **Lens-based imaging**

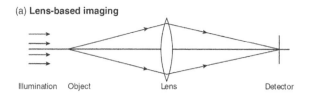

(b) **Lensless imaging**

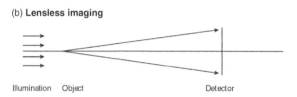

FIGURE 8.2 In lensless imaging methods, the role of the lens in a conventional microscope (a), which focuses the light scattered by the object onto the detector plane through addition of an appropriate spherical phase to the scattered wave field, is taken over by a computational process. This process converts the intensity distribution measured in the far field (b) into a real-space image using iterative phasing algorithms. (Taken from Barty 2010.)

diffraction intensity is recorded on a large-area X-ray detector, and computational algorithms are used to form an image (Miao et al. 1999; Nugent 2010). Instead of the Bragg spots that are produced by a periodic, crystalline sample, a single (non-periodic) nano-object produces a continuous diffraction pattern, which has a relatively simple and direct mathematical relation to the object. However, the inversion of the diffraction pattern in order to retrieve the complete real-space information about the object requires knowledge of both the intensity and the phase of the diffraction signal, and the latter is typically lost by the measurement, which only records intensity information. This phase problem does not occur in lens-based imaging methods, where phasing is performed by the lens. In lensless imaging methods, this role has to be performed by computational phase-retrieval methods (Marchesini 2007), or by holographic approaches, in which the scattered X-rays interfere with a known reference wave, and the resulting hologram contains both phase and intensity information such that the object can directly be reconstructed mathematically.

Although most of the lensless imaging techniques were initially developed at synchrotron radiation facilities, where coherent X-ray beams are typically produced by employing small pinholes, their breakthrough for single-particle imaging (SPI) required the extraordinary (spatially) coherent, nearly monochromatic photon flux and short pulse durations provided by XFELs and their smaller predecessors in the vacuum ultraviolet (VUV) and soft X-ray domain, such as the free-electron laser in Hamburg (FLASH) (Barty 2010; Barty et al. 2013; Barty 2016). While the large coherent and monochromatic photon flux is necessary to provide sufficient elastic scattering signal from a single nano-object to apply the coherent diffractive imaging algorithms, the short pulse durations are required in order to overcome the radiation damage that occurs as a consequence of

the intense photon pulse. Since ionization cross sections are typically significantly larger than the elastic scattering cross sections (as an example, the ratio of ionization to elastic scattering cross section for atomic carbon at 12 keV is 10:1), a focused XFEL pulse strongly ionizes a nano-object within a few femtoseconds, resulting in a subsequent breakup into many atomic and molecular neutral and ionic fragments that starts on a timescale of a few tens of femtoseconds (see Figure 8.3).

The underlying idea of single-particle and single-molecule coherent diffractive imaging experiments with XFELs is that an intense, few-femtosecond X-ray pulses can be used to "outrun" radiation damage by recording the diffraction image before the object is destroyed by the intense photon pulse (Neutze et al. 2000). This concept, which permits the use of a radiation dose much higher than conventional damage limits and which became known as "diffraction before destruction" (Chapman, Carleman, & Timneanu 2014), was first demonstrated experimentally at the FLASH facility in Hamburg, Germany, in 2006. The sample was a lithographically prepared silicon nitride microstructure, a few microns in diameter, which was completely vaporized by the focused, 25-fs-duration free-electron laser pulse, but not before a diffraction pattern of the object was captured on the X-ray camera, allowing a reconstruction of the diffraction image with 15 nm spatial resolution (Chapman et al. 2006a).

Following this initial demonstration, further proof-of-principle experiments were performed, e.g., on nanometer to micron-sized aerosols (Bogan 2010) and on various biological objects such as cells and viruses (Seibert et al. 2010). However, the achievable spatial resolution of these experiments was limited by the wavelengths available at FLASH, which can produce soft X-rays up to 300 electronvolts (eV), corresponding to a wavelength of approximately 4 nm. The next breakthrough for XFEL-based nanoparticle imaging was the start of operation of the LCLS at the SLAC National Accelerator Laboratory in the United States, where intense X-ray pulses with photon energies initially up to 2 keV and later up to 10 keV were available. This allowed coherent imaging of single nanocrystals, viruses, and other nanoparticles with sub-nanometer resolution, including the first biomolecular structure determination from an XFEL, performed using nanocrystals of the membrane protein complex photosystem I (PSI) as small as 200 nm in size (Chapman et al. 2011); and the first report of high-quality, high-resolution diffraction data obtained from a noncrystalline biological sample, a single mimivirus particle, with a single X-ray pulse (Seibert et al. 2011).

8.3 Recent Examples of Nanoparticle Imaging with XFELs

While one of the main motivations for XFEL-based imaging experiments was (and still is) the dream of imaging single

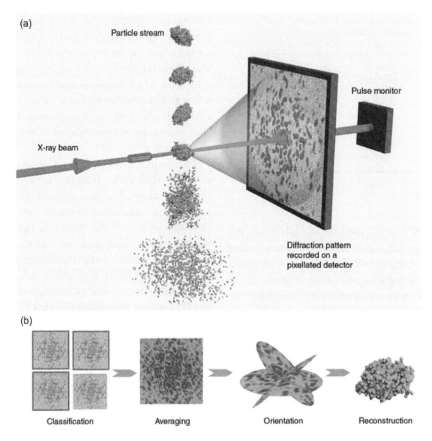

FIGURE 8.3 The concept of "diffraction before destruction" for three-dimensional SPI with an XFEL: (a) A stream of aerosolized molecules is injected into the XFEL beam, which has sufficient pulse energy for a diffraction pattern to be collected in a single pulse before radiation damage destroys the particle. (b) The individual diffraction patterns from many random particle orientations are then sorted and averaged, and the three-dimensional real-space image of the particle's electron density is reconstructed using phase-retrieval techniques. (Taken from Gaffney and Chapman 2007.)

(bio-)molecules (Neutze et al. 2000), coherent X-ray imaging concepts can be, and have been, applied to a wide range of other nanoparticles and nanostructures, such as rare-gas, metal, and core-shell clusters (Bogan 2010; Gorkhover et al. 2012; Rupp et al. 2012; Pedersoli et al. 2013; Barke et al. 2015; Gorkhover et al. 2016); soot nanoparticles (Loh et al. 2012); and self-assembled nanostructures in helium droplets (Gomez et al. 2014). The broad spectrum of possibilities for imaging and time-resolved studies of single nanoparticles, which XFELs can provide, thus open up new opportunities for nanosciences that have not yet been fully embraced by the nanoscience community at large. The following sections give some examples for the versatility of the XFEL-based imaging approaches. As more XFEL facilities around the globe are now starting their user operation, while several others are currently still under construction, other variations of these themes as well as new applications will certainly appear in the near future.

8.3.1 Imaging Nanoparticle Aerosols

In order to image single, unsupported aerosol particles, the aerosols are injected into a vacuum chamber using a particle injector based on an aerodynamic lens (Bogan et al. 2008), which strips the particles of possible solvents and is able to deliver a collimated beam of single nanoparticles for a wide range of particle sizes from 1 nm to 10 μm (see Figure 8.4). This substrate-free particle delivery allows an investigation of the morphology and size distribution of unsupported sub-micrometer-sized particulate matter, which plays an important role, e.g. in combustion studies, toxicology, climate science, and atmospheric chemistry, in their native form. By combining single-shot nanoparticle X-ray imaging with mass spectrometry of the ionic fragments produced after the interaction of the intense X-ray pulse with the nanoparticle, the particles' chemical composition can be determined along with their morphology (Loh et al. 2012). In addition to soot particles, other examples of aerosol particles that have been studied with single-particle X-ray imaging methods are individual gold nanoclusters and carbon fibers (see Figure 8.5). Furthermore, it was demonstrated that the three-dimensional shape of silver nanoparticles with less than 100 nm in diameter can be determined when using X-rays will a relatively long wavelength of 13.5 nm (corresponding to a photon energy of 92 eV) (see Figure 8.6), and that diffraction signal with (near-) atomic resolution can be measured from viruses of less than 100 nm in size with 7-keV hard X-rays (Munke et al. 2016).

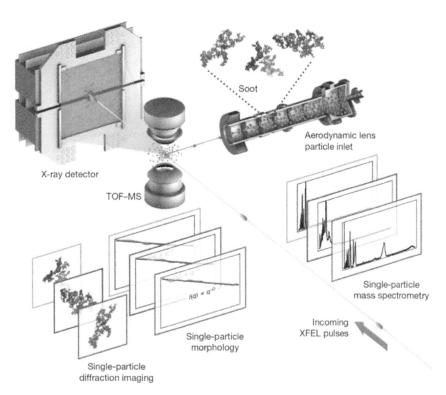

FIGURE 8.4 Schematic of concurrent imaging, morphology, and spectroscopy of single soot particles in flight. The aerosol particles are injected into a vacuum chamber through an aerodynamic lens. Coherent XFEL pulses intersect the stream of nanoparticles, producing a single-shot diffraction pattern and a large number of ion fragments whenever an XFEL pulse hits a particle. Diffraction patterns are recorded on a large-area X-ray detector, and ionic fragments are captured by a time-of-flight mass spectrometer (TOF-MS). (Taken from Loh et al. 2012.)

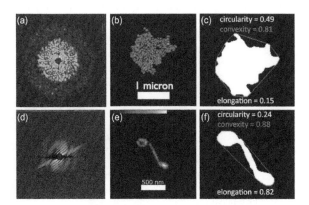

FIGURE 8.5 Single-shot imaging of individual aerosolized nanoparticles in flight. By applying a phase-retrieval algorithm to the coherent diffraction patterns from a gold nanocluster (a) and a carbon fiber (d), real-space images of the particles' morphology (b and e) are obtained, from which information about the shape of the aerosol particles in free flight can be extracted (c and f). (Taken from Bogan et al. 2010.)

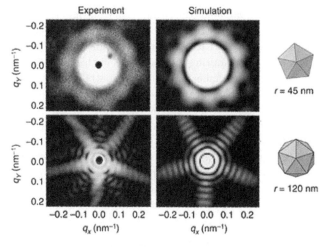

FIGURE 8.6 Single-shot diffraction patterns of individual silver nanoparticles recorded at a wavelength of 13.5 nm. The experimental patterns are shown along with the simulated patterns for the particle sizes and shapes shown on the right. (Adapted from Barke et al. 2015.)

8.3.2 Three-Dimensional Imaging of Nanostructures

Mathematically, coherent diffractive imaging can be readily extended in order to fully reconstruct three-dimensional objects. However, in this case, it is necessary to record diffraction patterns of the same object illuminated from many different angles, which either requires many different exposures of the same object or a large number of diffraction images of (nearly) identical objects. According to the Fourier slice theorem, the Fourier transform of a projection

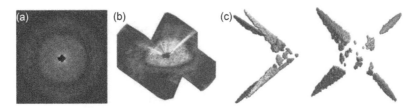

FIGURE 8.7 Experimental diffraction data collected from a three-dimensional test object at a synchrotron radiation source. (a) Diffraction pattern recorded at a single orientation. (b) Diffraction intensities collected at different angles merged into a 3D volume. (c) Reconstructed 3D object. (Taken from Chapman et al. 2006b.)

in real space is equivalent to the Fourier transform of a slice through reciprocal space. The three-dimensional structure of an object can, therefore, be assembled from the real-space projections that are reconstructed from the Fourier-space diffraction images.

For the case of coherent diffractive imaging, this was demonstrated experimentally, e.g., on engineered nanostructure using a synchrotron radiation source (Chapman et al. 2006b), as shown in Figure 8.7 and, with XFELs, for the case of single iron oxide nanoparticles (Loh et al. 2010), individual gold nanocrystals (Xu et al. 2014), as well as for biological nano-objects (Ekeberg et al. 2015). In the latter case, a three-dimensional image of a giant mimivirus, which is approximately 450 nm in diameter, was assembled from 198 two-dimensional diffraction patterns of nearly identical viruses that were injected into vacuum as an aerosol stream. The two-dimensional diffraction patterns recorded at random orientations were computationally sorted according to their orientation, reconstructed using a phase-retrieval algorithm, and the resulting real-space projections were put together to form a three-dimensional image of the virus with a resolution of slightly over 100 nm. The main factor limiting the resolution in this case was the relatively small number of diffraction patterns that could be collected during the experiment. Improved sample injection techniques and higher repetition rates of the XFELs have now made it possible to collect orders of magnitude more diffraction images in the same amount of beamtime, such that it should be possible to obtain three-dimensional reconstructions of particles ranging from a few ten to a few hundred nanometers with sufficient resolution to resolve their shape as well as any relevant internal structure.

8.3.3 Nanoparticle X-ray Holography

While all of the examples discussed above have used computational phase-retrieval methods to reconstruct the real-space image from the single-shot diffraction pattern, a powerful method to circumvent the phase problem in lensless imaging is X-ray holography. If the scattered X-rays produced from the object are made to interfere with a known reference wave, the resulting hologram contains both phase and intensity information such that the object can be reconstructed by direct Fourier inversion. A technical challenge in applying this elegant technique, which is also known as Fourier-transform holography, for single-shot

X-ray diffraction experiments lies in the necessity to place a known reference scatterer close to the object and to record the interference of both object and reference waves with sufficient signal-to-noise ratio and resolution.

In the first demonstration of single-shot X-ray holography, the object, a magnetic nanostructure, was placed behind a lithographically manufactured mask with a micrometer-sized sample aperture and a nanometer-sized hole that defined a reference beam (Eisebitt et al. 2004). A later variation of this setup placed the object and several reference scatterers in close proximity on the same supporting membrane and used two time-delayed X-ray pulses to record two holographic snapshots of the same structure with a 50 fs delay of one with respect to the other (see Figure 8.8), demonstrating the possibility of sequential holographic X-ray imaging for the study of femtosecond dynamics on a nanometer length scale (Günther et al. 2011).

Other holographic X-ray imaging experiments also used two time-delayed diffraction images of the same object but relying on the known structure of the object as a reference

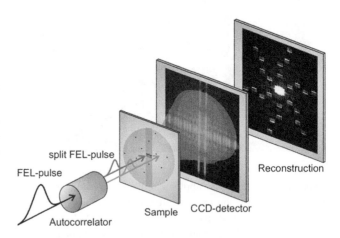

FIGURE 8.8 Schematic setup for femtosecond time-resolved X-ray holography. The XFEL pulse is split into two time-delayed pulses using a mirror-based autocorrelator. Both pulses are brought to overlap, at a small angle, on the sample, consisting of an edged microstructure of an object with several nearby reference scatters. From the two spatially separated X-ray holograms, which are recorded on a regular X-ray CCD camera, the real-space images of the object at two different time delays with respect to each other can be reconstructed via Fourier transformation. (Adapted from Günther et al. 2011.)

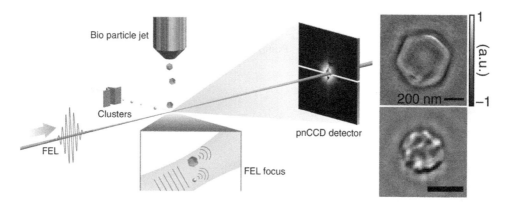

FIGURE 8.9 Schematic of an in-flight X-ray holography setup. Xenon nanoclusters with diameters of 30–120 nm are injected as reference scatterers together with aerosolized mimiviruses. When one of the 1.2-keV (1-nm wavelength) XFEL pulses illuminates both, a cluster and a mimivirus, in close proximity to each other, an X-ray hologram is formed that is recorded with an X-ray CCD camera. The two panels on the right show the reconstructed images of a mimivirus (top) and a xenon cluster, obtained by Fourier transformation of one of the holographic diffraction patterns. (Adapted from Gorkhover et al. 2018.)

for the holographic reconstruction of the structural change induced in the object by the first pulse (Chapman et al. 2007). A subsequent experiment demonstrated that the efficiency of Fourier-transform holography with X-rays can be increased by more than three orders of magnitude if a uniformly redundant array is placed as a reference scatterer next to the sample (Marchesini et al. 2008).

In order to apply the Fourier-transform holography technique to single, unsupported nanoparticles and to allow for the rapid collection of several hundreds or thousands of X-ray holograms, which is necessary for imaging non-reproducible nanoparticles (see Section 3.1) or for assembling three-dimensional reconstructions of reproducible objects (see Section 3.2), the method of in-flight holography was developed. In that case, heavy-element nanoparticles are used as reference X-ray scatterers that are injected together with the aerosol containing the objects (see Figure 8.9). The first demonstration of this technique used (nearly) spherical rare-gas nanoclusters as reference X-ray scatterers to encode the relative phase information into the diffraction patterns of a virus. From these X-ray holograms, an unambiguous three-dimensional map of the 450-nm virus and the 30–120-nm nanoclusters was obtained with a lateral resolution of less than 20 nm (Gorkhover et al. 2018). Possible future applications of this in-flight X-ray holography method include direct imaging of aerosol nucleation and droplet formation in their native environments without ensemble averaging effects, which opens novel avenues to study air pollution, combustion, cloud formation, and catalytic processes on the nanoscale at the single-particle level.

8.3.4 Time-Resolved Studies of Femtosecond and Picosecond Dynamics on the Nanoscale

Apart from imaging the atomic structure and morphology of nanoparticles and nanostructures, a key feature enabled

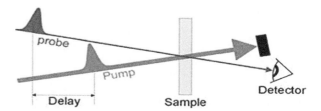

FIGURE 8.10 Schematic of a pump-probe experiment to observe ultrafast dynamics in a sample. The "pump" pulse triggers the dynamics, while the "probe" pulse takes snapshot "pictures" at different delay times Δt between the pump and probe pulses, thus creating stroboscopic images that make up a "movie" of the dynamics in the sample.

by the short and intense XFEL pulses is the ability to resolve dynamic processes on the femtosecond timescale. Such time-resolved measurements are often performed in a pump-probe scheme (see Figure 8.10). A first light pulse (either from the XFEL or from a synchronized optical laser) initiates ("pumps") the dynamics, and a second light pulse "probes" its evolution at various time delays after the pump pulse, thus recording a series of stroboscopic snapshots of the process that can be assembled into a "molecular movie". Since XFELs pulses are on the order of a few to a few hundred femtoseconds, which is approximately 1,000 times shorter than the X-ray pulses produced by a synchrotron, the achievable temporal resolution is on the same timescale as molecular vibrations and atomic and molecular motion during chemical reactions, making it possible to record snapshots that can "freeze" this motion in time.

Examples of dynamic processes in nanoparticles and nanostructures that have been studied using a combination of pump-probe techniques with XFEL-based X-ray imaging include X-ray damage (Chapman et al. 2007; Günther et al. 2011) and X-ray or laser-induced nanoplasma formation (Barty et al. 2008; Gorkhover et al. 2016) as well as light-induced structural changes in protein nanocrystals

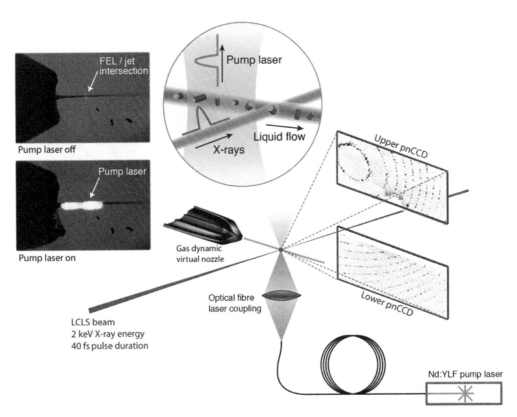

FIGURE 8.11   Pump-probe experiment using an optical Nd:YLF laser synchronized with the pulse from an XFEL for time-resolved structure determination on PSI–ferredoxin cocrystals by femtosecond serial nanocrystallography. The light-sensitive protein nanocrystals are injected as a liquid jet of few micron diameter, which is intersected by the optical laser and the XFEL beam. The insets on the left show the thermal glow of the 2-keV X-rays interacting with the liquid jet (top) and the scatter of the optical pump laser from the liquid jet (bottom). The X-ray scattering pattern of each nanocrystal is detected by two large-area CCD detectors, and the individual diffraction images are further processed to reveal the changes in the crystal structure as a function of delay between the optical pump pulse and the XFEL probe pulse. (Taken from Aquila et al. 2012.)

(Aquila et al. 2012; Kupitz et al. 2014; Nango et al. 2016; Tenboer et al. 2016). One of the first pump-probe experiments on biological nanocrystals performed at an XFEL was the observation of light-induced changes in protein complexes that are involved in the photosynthetic light harvesting in plants (see Figure 8.11). In this experiment, large and irreversible structural changes of the protein crystal were observed by X-ray diffraction at time delays of 5–10 μs after excitation with a laser pulse in the visible (Aquila et al. 2012).

Another time-resolved X-ray imaging experiment studied the formation of a laser-induced nanoplasma in superheated nanoparticles on a timescale of 100 fs up to a few picoseconds. In this case, single xenon clusters with diameters ranging from 30 to 60 nm were superheated by an intense, near-infrared femtosecond laser pulse, and the structural evolution of the sample was imaged with a subsequent single X-ray pulse of 0.8 nm wavelength (Gorkhover et al. 2016). Ultrafast surface softening on the nanometer scale at the plasma/vacuum interface was observed within 100 fs of the heating pulse (see Figure 8.12). Subsequently, similar experiments with increased temporal and spatial resolution were also performed with silicon dioxide and metal nanoparticles, and improved synchronization schemes between the optical and X-ray pulses combined with shorter pulse XFEL and laser pulse durations have now pushed the possible temporal resolution of such pump-probe experiments to well below 100 fs.

8.4   Summary and Future Perspectives

The increasing availability of intense, femtosecond soft and hard X-ray pulses produced by XFELs has opened up new opportunities for imaging the structure and dynamics of nanoparticles and nanostructures with few-femtosecond temporal and sub-nanometer spatial resolution. While the power of XFEL-based coherent imaging has, by now, been firmly established for (protein) nanocrystalography, the potential for single nanoparticle imaging in other scientific fields is still less explored. Yet, a number of showcase experiments have demonstrated that nanoparticle imaging with XFELs offers capabilities that no other technique can currently provide, such as the possibility to study the morphology, atomic composition, and temporal evolution of single, unsupported nanoparticles in their native aerosol environment with unprecedented resolution.

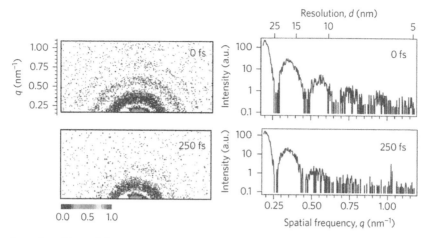

FIGURE 8.12 Time-dependent X-ray diffraction patterns of superheated xenon clusters for the near-infrared (NIR) heating pulse and the X-ray probe pulse arriving simultaneously (top) and with a delay of 250 fs between the NIR laser and the XFEL pulse (bottom). The left column shows the single-shot diffraction patterns, the right column the radial projection of each of the diffraction patterns. The higher-order diffraction rings disappear into the noise level with increasing time delay between the NIR and XFEL laser pulses, indicating that the hard-sphere shape of the pristine sample is transformed into solid core and an expanding outer shell with an exponentially decaying density profile. (Adapted from Gorkhover et al. 2016.)

In order to systematically explore these possibilities and to further develop the technical capabilities in a community-wide effort, the SPI initiative was recently established at LCLS (Aquila et al. 2015). Within this project, it was shown, e.g., that meaningful diffraction signal can be measured from single viruses of less than 100 nm in size, although too few patterns could be recorded to allow for a three-dimensional reconstruction (Munke et al. 2016). Many of the experimental challenges resulting from low hit rates and long data acquisition times will soon be alleviated, however, by the next generation of high-repetition-rate XFELs such as the European XFEL and LCLS-II. They are designed to deliver X-ray pulses with up to 27 kHz and 1 MHz, respectively, thus outperforming current XFELs, which typically operate at a repetition rate of 120 Hz or less, by several orders of magnitude (Wiedorn et al. 2018). Apart from these improvements of the light sources, further instrumental developments are still needed in order to fully realize the scientific opportunities given by the combination of single-particle X-ray imaging with spectroscopic techniques such as mass spectrometry or XES. This combination allows element-specific information to be collected along with morphological data, thus providing a unique and powerful tool for nanoparticle characterization.

With continuing advances of the time-resolved techniques at XFELs, constant progress is made towards the dream of recording X-ray "movies" that make it possible to observe how nanostructures form, evolve, and react in real time. This will enable studying, e.g., the growth of nanoparticles and their chemical reactions, such as the real-time photochemistry of dust and soot nanoparticles that play a role in combustion and climate science or the operation of nanoscopic devices such as single-molecules photoswitches. In time-resolved biological studies, XFEL sources will enable the atomic-scale imaging of biochemical reactions as they progress, and nanophotonics and nanoplasmonics research can benefit from studying the plasmonic response of isolated nanoparticles without a substrate that complicates the theoretical description and makes the comparison between theory and experiment more difficult.

Acknowledgments

The author is supported by the Chemical Sciences, Geosciences, and Biosciences Division, Office of Basic Energy Sciences, Office of Science, US Department of Energy, Grant No. DE-FG02-86ER13491. Helpful discussions with many colleagues and collaborators on the subject of X-ray imaging of nanoparticles and nanostructures with XFELs are gratefully acknowledged. Special thanks go to the dedicated scientific and technical staff of the free-electron laser facilities worldwide, without whom none of this work would have been possible.

References

Aquila, A., Hunter, M.S., Doak, R.B. et al. 2012. Time-resolved protein nanocrystallography using an X-ray free-electron laser. *Opt. Express* 20: 2706–16.

Aquila, A., Barty, A., Bostedt, C. et al. 2015. The Linac Coherent Light Source single particle imaging roadmap. *Struct. Dyn.* 2: 041701.

Barke, I., Hartmann, H., Rupp, D. et al. 2015. The 3D-architecture of individual free silver nanoparticles captured by X-ray scattering, *Nat. Commun.* 6: 6187.

Barty, A., Boutet, S., Bogan, M.J. et al. 2008. Ultrafast single-shot diffraction imaging of nanoscale dynamics. *Nat. Photonics* 2: 415–9.

Barty, A. 2010. Time-resolved imaging using x-ray free electron lasers. *J. Phys. B* 43: 194914.

Barty, A., Küpper, J., and Chapman, H.N. 2013. Molecular imaging using X-Ray free-electron lasers. *Annu. Rev. Phys. Chem.* 64: 415–35.

Barty, A. 2016. Single molecule imaging using X-ray free electron lasers, *Curr. Opin. Struct. Biol.* 40: 186–94.

Bogan, M.J., Benner, W.H., Boutet, S. et al. 2008. Single particle X-ray diffractive imaging. *Nano Lett.* 8: 310–6.

Bogan, M. J. 2010. Imaging aerosol particles. *J. Phys. B* 43: 194013.

Bogan, M.J., Boutet, S., Chapman, H.N. et al. 2010. Aerosol imaging with a soft X-ray free electron laser. *Aerosol Sci. Technol.* 44: 1–6.

Chao, W., Harteneck, B.D., Liddle, J.A., Anderson, E.H., and Attwood, D.T. 2005. Soft X-ray microscopy at a spatial resolution better than 15 nm. *Nature* 435: 1210–3.

Chapman, H.N., Barty, A., Bogan, M.J. et al. 2006a. Femtosecond diffractive imaging with a soft-X-ray free-electron laser. *Nat. Phys.* 2: 839–43.

Chapman, H.N., Barty, A., Marchesini, S. et al. 2006b. High-resolution ab initio three-dimensional X-ray diffraction microscopy. *J. Opt. Soc. Am. A* 23: 1179–200.

Chapman, H.N., Hau-Riege, S.P., Bogan, M.J. et al. 2007. Femtosecond time-delay X-ray holography. *Nature* 448: 676–9.

Chapman, H. N., Fromme, P., Barty, A. et al. 2011. Femtosecond X-ray protein nanocrystallography. *Nature* 470: 73–7.

Chapman, H.N., Carleman, C., and Timneanu, N. 2014. Diffraction before destruction. *Phil. Trans. R. Soc. B* 369: 20130313.

Eisebitt, S., Luning, J., Schlotter, W.F. et al. 2004. Lensless imaging of magnetic nanostructures by X-ray spectroholography. *Nature* 432: 885–8.

Ekeberg, T. Svenda, M. Abergel, C. et al. 2015. Three-dimensional reconstruction of the giant mimivirus particle with an X-ray free-electron laser. *Phys. Rev. Lett.* 114: 098102.

Gaffney, K.J. and Chapman, H.N. 2007. Imaging atomic structure and dynamics with ultrafast x-ray scattering. *Science* 316: 1444–8.

Gomez, L.F. Ferguson, K. Cryan, J. et al. 2014. Shapes and vorticities of superfluid helium nanodroplets. *Science* 345: 906–9.

Gorkhover, T., Adolph, M., Rupp, D. et al. 2012. Identification of twinned gas phase clusters by single-shot scattering with intense soft x-ray pulses. *Phys. Rev. Lett.* 108: 245005.

Gorkhover, T., Schorb, S., Coffee, R. et al. 2016. Femtosecond and nanometer visualization of structural dynamics in superheated nanoparticles. *Nat. Photonics* 10: 93–7.

Gorkhover, T., Ulmer, A., Ferguson, K. et al. 2018. Femtosecond X-ray Fourier holography imaging of free-flying nanoparticles. *Nat. Photonics* 12: 150–3.

Günther, C.M., Pfau, B., Mitzner, R. et al., 2011. Sequential femtosecond X-ray imaging. *Nature Photonics* 5: 99–102.

Kupitz, C. Basu, S. Grotjohann, I. et al. 2014. Serial time-resolved crystallography of photosystem II using a femtosecond X-ray laser. *Nature* 513: 261–5.

Loh, N.D., Bogan, M.J., Elser, V. et al. 2010. Cryptotomography: reconstructing 3D Fourier intensities from randomly oriented single-shot diffraction patterns. *Phys. Rev. Lett.* 104: 225501.

Loh, N.D., Hampton, C.Y., Martin, A.V. et al. 2012. Fractal morphology, imaging and mass spectrometry of single aerosol particles in flight. *Nature* 486: 513–7.

Marchesini, S. 2007. A unified evaluation of iterative projection algorithms for phase retrieval. *Rev. Sci. Instrum.* 78: 011301.

Marchesini, S., Boutet, S., Sakdinawat, A.E. et al. 2008. Massively parallel X-ray holography. *Nat. Photonics* 2: 560–3.

McNeil, B.W.J. and Thompson, N.R. 2010. X-ray free-electron lasers. *Nat. Photonics* 4: 814–21.

Miao, J., Charalambous, P., Kirz, J., and Sayre, D. 1999. Extending the methodology of X-ray crystallography to allow imaging of micrometre-sized non-crystalline specimens. *Nature* 400: 342–4.

Munke, A., Andreasson, J., Aquila, A. et al. 2016. Coherent diffraction of single Rice Dward virus particles using hard X-rays at the Linac Coherent Light Source. *Sci. Data* 3: 160064.

Nango, E. Royant, A. Kubo, M. et al. 2016. A three-dimensional movie of structural changes in bacteriorhodopsin. *Science* 354: 1552–7.

Neutze, R., Wouts, R., van der Spoel, D., Weckert, E., and Hajdu, J. 2000. Potential for biomolecular imaging with femtosecond X-ray pulses. *Nature* 406: 752–7.

Nugent, K.A. 2010. Coherent methods in the X-ray sciences. *Adv. Phys.* 59: 1–99.

Pedersoli, E., Loh, N.D., Capotondi, F. et al. 2013. Mesoscale morphology of airborne core-shell nanoparticle clusters: X-ray laser coherent diffraction imaging. *J. Phys. B* 46: 164033.

Rupp, D., Adolph, M., Gorkhover, T. et al. 2012. Identification of twinned gas phase clusters by single-shot scattering with intense soft x-ray pulses. *New. J. Phys.* 14: 055016.

Seibert M. M., Boutet, S., Svenda, M. et al. 2010. Femtosecond diffractive imaging of biological cells. *J. Phys. B* 43: 194015.

Seibert, M. M., Ekeberg, T., Maia, F.R. et al. 2011. Single mimivirus particles intercepted and imaged with an X-ray laser. *Nature* 470: 78–81.

Tenboer, J., Basu, S., Zatsepin, N. et al. 2016. Time-resolved serial crystallography captures high-resolution intermediates of photoactive yellow protein. *Science* 346: 1242–6.

Waldrop, M. 2014. The big guns. *Nature* 505: 604–6.

Wiedorn, M.O., Oberthür, D. Bean, R. et al. 2018. Megahertz serial crystallography. *Nat. Commun.* 9: 4025.

Winick, H. and Bienenstock, A. 1978. Synchrotron radiation research. *Ann. Rev. Nucl. Part. Sci.* 28: 33–113.

Xu, R., Jiang, H., Song, C. et al. 2014. Single-shot three-dimensional structure determination of nanocrystals with femtosecond X-ray free-electron laser pulses. *Nat. Commun.* 5: 4061.

Helium Ion Microscopy

Alex Belianinov,
Olga S. Ovchinnikova, and
Artem A. Trofimov
Oak Ridge National Laboratory

Kyle T. Mahady
University of Tennessee

Philip D. Rack
Oak Ridge National Laboratory
University of Tennessee

9.1 Introduction

The interest in controlling matter at the molecular and the atomic scales led to the development of instrumentation that allowed nanoscale objects to be imaged and modified. Common examples of high-resolution imaging tools include the transmission electron microscope (TEM) (Champness, 2001), scanning tunneling microscope (STM) (Binnig and Rohrer, 1983), the atomic force microscope (AFM) (Binnig et al., 1986), the scanning electron microscope (SEM) (Goldstein et al., 2003), and the focused ion beam (FIB) (Giannuzzi, 2006). The ideas behind these date back to the 1930s with independent efforts by Ernst Ruska (Knoll and Ruska, 1932) and Manfred von Ardenne (1938). The SEM has been a mainstay in many laboratories for over 50 years, but it is reaching a fundamental limit in the smallest possible focused diameter of the electron beam. As a result, alternative and modified techniques, such as scanning transmission electron microscopy (STEM) or FIB, have been emerging. While ion beams can potentially produce more damage to the material than electron microscopy, it also avoids many of the shortcomings of electron microscopy. The diameter of the focused spot is directly proportional to the wavelength of incident radiation, and, since the ions are much heavier, they operate at substantially smaller wavelengths than electrons with the same energy (Joy, 2013).

As a special class of FIB tools, the helium ion microscope (HIM) offers higher resolution than SEMs (Hill et al., 2008) in combination with a multitude of additional capabilities. The HIM outperforms gallium FIB (Ga^+ FIB) with respect to a smallest resolvable feature size (Cybart et al., 2015; Jijin et al., 2011), and it is capable of imaging uncoated and insulating samples with outstanding image quality and depth of field (Boden et al., 2012). The tool also offers a direct path to defect engineering and tuning material properties (Fox et al., 2015; Stanford et al., 2016c; Yoon et al., 2016), and finally, a chemical imaging option via secondary ion mass spectrometry (SIMS) (Hlawacek and Gölzhäuser, 2016; Pillatsch et al., 2013).

The distinguishing feature of the HIM is the gas field ion source (GFIS). Figure 9.1 is a simple schematic of the GFIS column used in the Zeiss ORION NanoFab. In a GFIS, a sharp metal tip in vacuum is cryogenically cooled, and a positive voltage is applied to ionize the imaging gas (helium or neon in this iteration of the instrument). As a result, the ions are accelerated away from the tip towards a negatively biased extractor, and the beam is subsequently collimated, shaped, and rastered over the samples (Hlawacek and Gölzhäuser, 2016). This approach is distinctly different from liquid metal ion source (LMIS) used for gallium ions or the thermal field emitter (Schottky source) and the cold field emitter (CFE).

At the apex of the GFIS source, most of the helium is ionized by only a few top atoms increasing the current density (ion current per emitted atom) and source brightness. Typically, a tip is sharpened until there are just three atoms remaining, arranged in a triangular pattern (trimer). This configuration maximizes the He^+ ion current because the threefold symmetric arrangement is stable, and offers strong emission in the vertical direction, *i.e.* parallel to the axis of the ion column (Notte et al., 2007). The emitted current also increases with decreasing temperature (about 5% per Kelvin); thus, the GFIS ion source must be cooled between 60 and 90 K, which via a solid nitrogen cooling

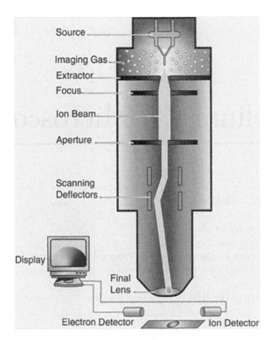

FIGURE 9.1 Simplified cross-sectional diagram of a Zeiss ORION HIM identifying key components. (Reprinted by permission from *Springer Nature* and Copyright Clearance Center Joy, 2013.)

system, operated at a vacuum level sustained at about 3 Torr. Furthermore, the cooling improves the collection and surface accommodation of the imaging gas (Müller and Bahadur, 1956).

Specific functional conditions of the SFIS, *i.e.* the atomically sharp tungsten needle, are 5–50 kV acceleration, at temperatures of 60–90 K, maintained under ultrahigh vacuum (UHV), but operated at 10^{-7}–10^{-6} Torr of helium or neon gas. Optimized source brightness is 5×10^9 A/cm²·sr, with reduced source brightness (at 30 kV) of 1×10^9 A/cm²·sr·V with ca. 1 eV energy spread. The virtual source size (for helium) is less than 0.25 nm, with the total emitted current of ca. 150 pA and the de Broglie wavelength of ca. 0.080 pm (Belianinov et al., 2017b). Practically, helium spot size is ~0.5 and 1.9 nm for neon.

Once the imaging gas is ionized by the GFIS tip, ions are accelerated towards the sample. When an ion strikes the surface, it initiates a multistage process that includes many channels of energy dissipation, usually producing secondary electrons (SE1). Likewise, in SEM, SEs are produced as a result of an ionization event initiated by an incident electron. SEs escape from the surface and are subsequently captured by an SE detector, similar to the one developed by Everhart and Thornley (1960). SE generation is crucial, since electron detection is the operating imaging mode in both the SEM and the HIM.

As electrons (SEM) or ions (HIM) interact with the samples and generate SEs (SE1), they may backscatter generating additional SE2. The presence of parasitic SE2 is one of the main limiting factors for spatial resolution and quality of an image in SEM. For electrons, the yield of the SE2 increases as the backscattered energy decreases.

Consequently, in most materials and at most incident beam energies, the electron-produced SE2 (eSE2) yield is significantly higher than the eSE1 yield. On the contrary, ions, due to their size, do not backscatter as much, and ion-produced SE2 (iSE2) generation is insignificant (Ramachandra et al., 2009; Ziegler et al., 2010). This allows HIM images to offer enhanced surface detail and improved resolution (Hlawacek et al., 2014).

Subsequently, when an ion or an electron is injected into the bulk, its energy loss is determined by *stopping power*, measured in electron volts per angstrom. The magnitude of the stopping power depends on the energy of the incoming particle and on the physical properties of the sample (Bethe, 1930; Bethe and Ashkin, 1953). Since ions are much heavier than electrons, for the same material and at the same incident beam energy, they travel at ca. 1% or less of electron velocity resulting in the stopping power that is an order of magnitude higher (Ziegler, 1988; Joy and Luo, 1989; Ziegler et al., 2010). This would imply generation of more SEs, and thus, more current at the detector; however, other considerations also play a role, such as an average energy loss per SE excited and the depth at which electrons are excited (see Section 9.1 for details).

Another difference between electrons and ions is the consequence of changing the incident beam energy. Increasing electron energy increases beam brightness but reduces the number of SEs proportionally. This behavior is due to a decrease in the stopping power and larger electron penetration into the specimen. Therefore, eSE signal is not greatly affected by change in energy. For ions, on the other hand, increase in energy results in both greater brightness and greater stopping power (Ramachandra, 2009). The magnitude of the iSE yield is almost always higher than the corresponding eSE yield (Dapor, 2011; Ramachandra et al., 2009). As a result, at the same beam energy, ions will be more surface sensitive and yield more signal, producing better images.

Furthermore, HIM imaging offers greater depth of field. SEM images typically have a shorter depth of field due to a larger angle of beam convergence. Therefore, a smaller fraction of the surface is in focus (Ramachandra, 2009; Wight et al., 2008). On the contrary, the depth of field in the HIM is larger due to a reduced convergence angle, which can be enhanced further by decreasing the aperture size.

Despite their differences, the SEM and the HIM face similar imaging challenges; namely, surface charging and sample damage. Surface charging is caused by imbalance in the charge supplied by the incident beam and the charge dissipation by the sample. As a result, imaging artifacts may appear obscuring true structure. Typically, for SEM, the samples are coated with a thin conductive layer, which may obscure finer details (Goldstein et al., 2003). In the HIM, the incident beam is positively charged, and a diffuse electron beam can effectively neutralize charge accumulation. This is particularly useful for imaging biological samples, which may be incompatible with metal coatings. (Hlawacek and Gölzhäuser, 2016). The authors note that in SEM, a similar type of an approach can be utilized; where a gas flooding with N_2 or a small radioactive source can supply the positive charge negating the charging effects of electrons; however, this is difficult and cumbersome to implement practically in many laboratory settings.

Besides surface charging, imaging is complicated by sample damage. In SEM, even relatively low electron dose can be expected to chemically alter or destroy organics such as polymers and biological samples. Ions, which are thousands of times heavier than electrons, can induce even more damage. Reports of gold specimens irradiated with 25–150 keV He^+ ions with a dose of ca. 6×10^{17} ions/cm^2 at 100°C show voids (Brown et al., 1973). Blister formation in molybdenum targets irradiated with 7–80 keV He^+ ions at ambient and high temperatures have also been observed (Erents and McCracken, 1973). The process started after passing a critical dose of ca. 5×10^{17} ions/cm^2 with blister size increasing with energy but not ion dose. It was also shown (Lehtinen et al., 2011) that an ion impact on the material can lead to a formation of defects such as vacancies (single, double, or complex) and even result in amorphization of a crystalline material structure (some atoms in the crystal are shifted from their lattice positions but no atoms are sputtered). On the other hand, ions striking the surface generates SEs with high efficiency, and therefore, the HIM can be operated at a lower current than the SEM, which can reduce sample damage.

Moreover, for readily sputtered materials, use of He^+ ions is an ideal fabrication approach. Since He^+ ions suffer little lateral scatter, it is capable of precisely cutting holes or fabricating slots no more than a few nanometers in diameter and spacing. Helium, when compared to Ga^+, can remove material from a surface at a significantly slower rate, providing a method to shape, mark, and pattern materials at the nanoscale. Specifically, a He^+ beam at 1 pA emission removes two atomic layers per second over an area of a few nm^2, offering very high patterning precision (Buchheim et al., 2016; Joy, 2013).

An additional fabrication approach in the HIM is via direct deposition of materials onto samples. FIB-induced deposition (FIBID) is similar to electron beam-induced deposition (EBID) where an active precursor can be introduced into the chamber via a precise gas injector. An ion beam, directed into a region within the precursor stream, will deposit the material at the point of contact with the surface. Outside the primary ion irradiated area, volatile compounds vaporize rapidly and will be removed by the chamber pump (Utke et al., 2008). This technology makes it possible to rapidly fabricate nanoscale, three-dimensional structures in a competitively short time. Possibilities for aforementioned nanofabrication were recently explored on the example of 2D van der Waals ferrielectric system such as copper indium thiophosphate (CIPS) (Belianinov et al., 2015, 2016).

The HIM has also been demonstrated to be a powerful tool for defect engineering (Stanford et al., 2017d). The technique has been used to controllably tune the electrical properties from semiconducting, to insulating, to metallic by introducing defects into multilayer MoS_2 and WSe_2 (Fox et al., 2015; Stanford et al., 2016c, 2017a; Chhowalla et al., 2013). As active investigations to characterize additional materials continue, manipulation of more properties and more possibilities can be expected to emerge.

Although it has taken decades of active research to recognize, develop, and commercialize a HIM, nowadays, HIM offers high resolution, high SE yield, long depth of focus, and surface sensitivity (Ramachandra et al., 2009). Currently, it relies on several key technologies that had to come together to make this tool a reality. These include cryogenics, UHV, gas purity, vibration control, ion optics, and detectors. It is evident that ion beams and the SFISs are no longer a restricted novelty but rather an emerging player in areas of high-resolution imaging, nanofabrication, and direct write. HIM has also been employed for non-imaging applications where it has been used to directly pattern a substrate, make nanopores, induce patterned chemical processes, implant stress (Arora et al., 2007), or perform material analysis. The technique has already shown an unprecedented level of nanoscale control of material properties, opening new avenues to materials research (Jesse et al., 2016). Following sections will provide the reader with necessary theoretical background on ion interaction with material and highlight the latest progress in the areas of imaging, material fabrication, and composition analysis using helium ion microscopy (HIM).

9.2 Ion Beam Interaction with Matter

As mentioned earlier, the key component of the HIM is GFIS. As imaging gas atoms (helium or neon) become ionized near the tip apex of the source, the resulting ions are projected away from the tip by the strong electric field. Correspondingly, when the gas in the GFIS column is directed towards the sample, ions impinging on to a target penetrate some finite depth into the bulk, exhausting their energy, through a variety of interactions. The depth to which they penetrate is generally on the order of tens to hundreds of nanometers, for ions with energies in the range of thousands to tens of thousands of electronvolts. Note that the depth can vary greatly depending on the ion species and the material. As the ions travel through the target, they are scattered randomly, giving rise to what is called an "interaction volume", which is the influenced region of the material. Furthermore, the interacting ions excite electrons within the bulk, which may escape the surface, forming the basis of the detected signal in the HIM. However, as ions possess far greater mass than electrons, they interact with the target lattice, generating damage in the form of vacancy-interstitial pairs, amorphization, surface sputtering, as well as impurities, as ions accumulate/implant in the target.

The stopping of ions in solids is principally divided into two mechanisms, which occur simultaneously (Ziegler et al.,

2010). The first mechanism is called nuclear, or elastic stopping, and it is a result of collisions between the primary ion and the atomic cores. The second mechanism is called electronic, or inelastic stopping, and is due to the interaction between the primary ion and the electrons within the bulk. The strength of these mechanisms is expressed as stopping power, which is defined as the spatial derivative of the energy, shown in Eq. 9.1.

$$S(E) = \frac{dE}{dx} \qquad (9.1)$$

As Eq. 9.1 implies, the stopping power is a function of the ion energy averaged over the material. The energy loss over any path length of an ion trajectory may vary depending on how energetic the individual interactions are, as discussed in more detail below. The stopping power is conventionally given in units of eV/Å; or eV/(g/cm²), as this may easily be used to calculate the stopping power for materials with varying densities. Over the ion energy range of 1 to ~100 keV, the electronic and nuclear stopping powers may be similar in magnitude. The relative strength varies greatly with ion species – for lighter ions (*e.g.*, helium) the stopping process tends to be dominated by electronic stopping,

while for heavier ions (*e.g.*, neon), electronic stopping may be a fraction in the magnitude of nuclear stopping. Nuclear stopping also tends to dominate at very low energies, where even helium loses much of its energy to elastic collisions (see Figure 9.2). Importantly for the HIM, the total stopping power (nuclear and electronic) of ions tends to be far higher than for electrons, leading to a smaller interaction volume than is observed for energetic electrons (see Figure 9.3). Because of the smaller interaction volume, electrons of the target are excited within a smaller radius from the initial point of impact of the ion, which in general produces more SEs.

Nuclear stopping process arises from elastic collisions between primary ions and atomic nuclei, and these collisions are the primary means by which ion bombardment leads to target damage. A useful approach to study elastic collisions is the binary collision approximation (BCA). In the BCA model, the nuclear stopping process is reduced to a series of binary collisions, taking place between the primary ion and the nearest atom in the target. During each binary collision, the energy transferred from the primary ion to the stationary target atom may be taken to be elastic, as illustrated by Eq. 9.2.

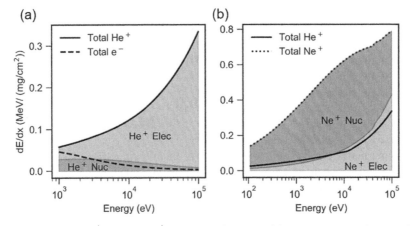

FIGURE 9.2 Stopping powers for (a) He⁺ and (b) Ne⁺ in copper. (Data for (a) is from the ESTAR and ASTAR databases (Berger et al., 2017). Data for Figure (b) was obtained from stopping and range of ions in matter (SRIM) (Ziegler, 2013).)

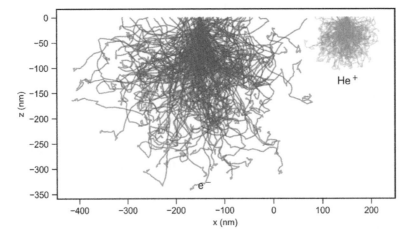

FIGURE 9.3 Trajectories of 200 10-keV electrons and helium ions in a copper substrate. The electron trajectories were simulated using Casino (Joly et al., 2015). The He⁺ trajectories were simulated using EnvizION (Mahady et al., 2017a).

$$\Delta E = \frac{4EM_1M_2}{(M_1 + M_2)^2} \sin^2 \frac{\Theta}{2} \tag{9.2}$$

Here, M_1 and M_2 are the masses of the primary ion and target atom, respectively, E is the primary ion energy before the energy transfer, and Θ is the scattering angle from the collision in center-of-mass coordinates. The scattering angle associated with the binary collision of a He$^+$ ion and a copper atom is shown in Figure 9.4, where it is expressed as a function of the ion energy E, and the impact parameter p, which is the directness of the impact. The scattering angle and consequently the transferred energy, are maximized for direct, head-on impacts between the primary ion and stationary target atom (*i.e.* at small impact parameters). It can also be seen that as the energy goes down, the scattering angle increases. Therefore, as a primary ion moves, its path in the material is generally fairly straight until it exhausts much of its energy becoming progressively more tortuous from more frequent, large-angle scattering events. Note that Figure 9.2 shows the nuclear stopping power as a function of only the primary ion energy, species, and target material; the actual energy lost by the ion in any given collision depends on both the specific atomic nucleus participating in the collision and the impact parameter, so that the instantaneous energy loss may vary considerably from the average stopping power.

If a collision transfers sufficient energy from the primary ion to a target atom, it will produce a vacancy-interstitial pair. Inside the bulk, this displacement threshold may be tens of electronvolts; if an atom in the target is imparted less than the displacement threshold, it will rapidly relax back to its lattice site. In the case where a target atom has imparted more energy than the displacement threshold, it may begin a recoil cascade: the atom will begin moving through the bulk, undergoing a similar process as the incident ion, colliding with stationary target atoms, which may also be displaced from their lattice site. Recoils generated near the surface may be ejected but only if they have sufficient energy to overcome the surface-binding energy. The surface-binding

energy varies by material and recoil species, and it is generally on the order of a few electronvolts. Energy from collisions that are below the displacement threshold convert to phonon production. The trajectories of several 10-keV Ne$^+$ ions are illustrated in Figure 9.5a, with the associated recoil cascades in Figure 9.5b; Ne$^+$ ions were chosen for illustrative purposes, as more recoil atoms are generated relative to He$^+$ ions due to much higher nuclear stopping power (Figure 9.2). Sputtered atoms are visible as the straight recoil trajectories exiting the surface. Surface features of the material may vary from flats, whether due to material removal, deposition, or initial geometry. The surface topography interacts with the ion/recoil cascades in a complex way, as illustrated in Figure 9.5d and e. Here, the Ne$^+$ ions impinge on a nanoscale valley and sputter dramatically more material near the top of the valley, than when striking near the valley bottom, or even striking a flat surface.

Electronic stopping is the dominant energy loss mechanism for light ions, and, importantly, the energy lost in this manner leads to the production of SEs. While nuclear stopping is the dominant mechanism for the deflection of moving atoms, electronic stopping acts to dissipate energy. According to the Bethe theory (Bethe, 1941; Sternglass, 1957), the number of electrons excited per unit of distance traveled can be expressed as demonstrated in Eq. 9.3.

$$n_e = \frac{1}{\delta} S_{EE}(E) \tag{9.3}$$

where δ is the average energy loss of a moving atom per SE excited, S_{EE} is the electronic stopping power, and E is the instantaneous ion energy. The parameter δ depends on the moving atom species and material (Ramachandra et al., 2009). Once excited, an SE will undergo motion approximated by a diffusion process (Sternglass, 1957). If the electron reaches the surface with sufficient energy to

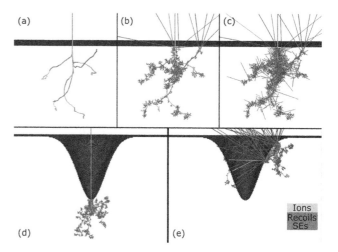

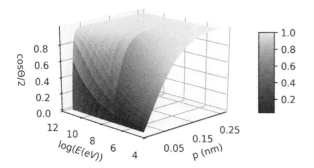

FIGURE 9.4 Scattering angle for a He$^+$ ion undergoing an elastic binary collision with a Cu atom, as a function of ion energy E and impact parameter p. The scattering angle is given in center-of-mass coordinates. The scattering angle is computed using the "Magic Formula" method described and employed by SRIM (Ziegler et al., 2010).

FIGURE 9.5 (a) Illustration of trajectories of five 10-keV Ne$^+$ ions with (b) the associated recoil cascades and (c) the trajectories of excited SEs. The ion trajectories and recoil cascades associated with striking various points in a high aspect-ratio valley are shown in (d) and (e). The trajectories were simulated using EnvizION (Mahady et al., 2017a).

overcome the material's surface potential, it can escape, forming the SE signal. Equation 9.3 implies that greater electronic stopping power tends to lead to a greater signal; however, other considerations also play a role, such as the parameter δ and the depth at which electrons are excited. Simulated SE trajectories generated from an incident energetic ion are shown in Figure 9.5c.

The discussion of models presented here is based on SRIM (Ziegler et al., 2010), as this has been the *de facto* standard Monte Carlo code for simulating ion-solid interactions for decades. SRIM (Biersack and Eckstein, 1984; Biersack and Haggmark, 1980; Ziegler et al., 2010) simulates a range of targets, including varying phases of matter, compounds, and multilayered targets, and it has been widely used to study the HIM (Bell et al., 2009a; Livengood et al., 2007, 2009; Sijbrandij et al., 2008). In this code, Monte Carlo simulations replace the governing partial differential equations for ion-solid interactions with a series of discrete, randomized steps. At each step of the simulation, a moving atom travels the distance Δx, often taken to be the local mean of the material interatomic distance. The moving atom is then assumed to undergo a binary collision with a target atom, with the impact parameter randomized. Electronic energy loss may be included by modeling the stopping powers as a continuous force. After each collision, the energy of the moving atom and direction of travel are updated, and the simulation continues until the atom exhausts its energy or leaves the computational domain (*e.g.*, by being sputtered). As energetic collisions generate displacements, the method recursively simulates each recoil atom in the cascade. It is important to note, however, that SRIM lacks an SE simulation or the ability to simulate cumulative damage to the target from sputtering, which makes it limited for HIM applications.

A variety of more generalized Monte Carlo simulation codes have been developed for FIB to study ion-solid interactions; we briefly mention some of the most prominent, which have been applied to study the HIM. The IoniSE (Ramachandra et al., 2009) code was developed to simulate ion-induced SE emission. The Monte Carlo methods of Inai et al. (2007) and Ohya et al. (Ohya, 2003; Ohya and Ishitani, 2004) have been used to study SE emission in ion microscopy for various targets and ion species. EnvizION was developed to study the HIM, and it uses voxels to represent a target, allowing it to simulate ion beam-induced deposition (IBID) (Smith et al., 2010; Timilsina et al., 2013), sputtering (Mahady et al., 2017a; Timilsina and Rack, 2013; Timilsina et al., 2014), SE emission (Mahady et al., 2017b), and gas-assisted etching, in a dynamically evolving target.

9.3　Ion imaging

Analogous to SEM, the most common application for the HIM is imaging. SE signal is typically used, although, backscattered-helium ions (BSHe) can be employed as well.

FIGURE 9.6 Standard SEM SE image (a) vs HIM SE image (b). Note the drastically different contrast mechanisms that make different materials easy to distinguish. In both cases, the image is constructed from low-energy (<10 eV) SEs. The SEM image was acquired with a 5 keV electron beam energy, while the HIM image used a 20 keV helium beam. (Reprinted with permission from Ward et al., 2006. © 2006 American Vacuum Society.)

The low-energy spread of the generated SEs, the small interaction volume, the sub-nm probe size, and a limited number of high angle scattering ions all contribute to high-resolution imaging. An example of HIM advantageous can be seen in Figure 9.6. Figure 9.6a is an SEM image of an alignment cross, which has a good resolution; however, it lacks material contrast. The image in Figure 9.6b is an HIM image, where it is clear that the material in the center of the alignment cross is different from the material on the outside of the cross. One can also see small dark areas of contamination on the sidewalls outside of the cross that are not readily apparent in the SEM image (Ward et al., 2006). For more examples comparing SEM and HIM imaging also see Boden et al. (2012).

9.3.1　Charge Neutralization

A serious challenge in SEM is surface charge accumulation on insulating materials. This masks native surface structure and complicates image interpretation due to artifacts. Common approaches to reduce charging in SEM conductive sample coating, which helps to overcome charging but can obstruct surface features (Schürmann et al., 2015). Alternatively, environmental or variable pressure SEMs are capable of introducing a gas, such as nitrogen, into the chamber, which leads to collisions of SEs with the gas molecules and results in ionization events that produce positive ions neutralizing the negatively charged regions. The technique, however, suffers from a decreased signal-to-noise ratio (SNR) because of the scattering of the incident beam by the gas molecules.

HIM imaging also charges insulating samples, but the surface charge is positive, due to the positively charged incident (He$^+$) beam. In the HIM, this produces low signal darker images, due to a larger potential barrier for SEs. This can be overcome with a diffuse electron flood gun. The electrons balance the charge of the helium ions, and high

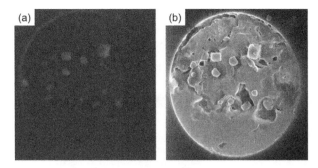

FIGURE 9.7 Images illustrating that charge neutralization during HIM imaging of insulating materials can be achieved using the electron flood gun: (a) an image of the end of an optical fiber comprising a silicon core clad in silica; (b) the same area with the flood gun operating. (Reprinted by permission from Springer Nature and Copyright Clearance Center Hlawacek and Gölzhäuser, 2016.)

resolution and large depth of field are maintained, while imaging insulating samples. An example of charge neutralization with the electron flood gun is shown in Figure 9.7, where the silicon core of an optical fiber is imaged with and without the flood gun (Hlawacek and Gölzhäuser, 2016).

It is worth mentioning that the charge compensation mechanism in HIM is efficient enough, so that charging artifacts are negligible. A good example of charge neutralization is in imaging a chromium photomask. With flood gun applied to the sample surface between He⁺ line scans, the image SNR improved dramatically. In the charge neutralized image (Figure 9.8b), the transition from the chromium edge to the quartz can be clearly seen and the particle defect located in the center of the chromium line can be much better delineated compared to when the flood gun was off (Figure 9.8a) (Hlawacek and Gölzhäuser, 2016). For more examples showing flood gun capabilities, also see (Bazou et al., 2011).

9.3.2 Soft Materials

Due to its effective charge neutralization, the HIM has a lot of potential in imaging soft insulating materials such as biologicals or polymers. The technique has been shown to be quite effective in studying butterfly wings microstructure (Boden et al., 2012), cancer cells (Bazou et al., 2011), articular cartilage network (Vanden Berg-Foels et al., 2012), and kidney cell parasites (de Souza and Attias, 2015). Recent fascinating examples of bioimaging have been illustrated by Leppänen et al., who imaged a bacteriophage attached to the host cell at different stages of the infection (Leppänen et al., 2017). In Figure 9.9a and b, examples of infected bacterial cells with phages are shown. HIM was able to visualize individual phages attached to the cells. Moreover, the authors milled the materials *in situ*, something that is quite challenging to do with a gallium beam. In addition, neon and helium milling enabled subsurface imaging of bacterial colonies and, in some cases, also imaged structures inside individual bacteria after cross sectioning (Figure 9.9c).

In addition to biological samples, the HIM was recently used to study polymers to visualize molecular organization at nanoscale. Bai et al. (2015) investigated perpendicular block copolymer lamellae in high aspect-ratio gratings produced by solvent vapor annealing at room temperature and by thermal annealing at 150°C. Cross-section images of samples with various thickness were obtained by imaging with the HIM, as shown in Figure 9.10a and b, confirming author's attempt to produce perpendicular lamellae in high aspect-ratio templates by thermal annealing. Another example of HIM usage for polymer research can be found in (Borodinov et al., 2018), where Borodinov et al. utilized plasma etching, HIM imaging, image analysis, and molecular dynamics simulation to highlight the internal structure of bottlebrush polymers in thin films at high resolution, without conductive coatings and extensive sample preparation. Figure 9.10c shows surface topography for the brush

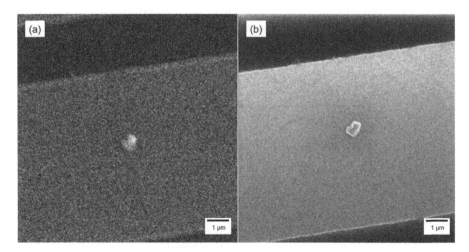

FIGURE 9.8 Patterned chromium on quartz photomask imaged with an HIM. (a) Effects of beam-induced surface charge degrade image SNR and resolution. (b) Charge neutralization significantly improved the SNR and resolution. (Reprinted by permission from Springer Nature and Copyright Clearance Center Hlawacek and Gölzhäuser, 2016, © 2016.)

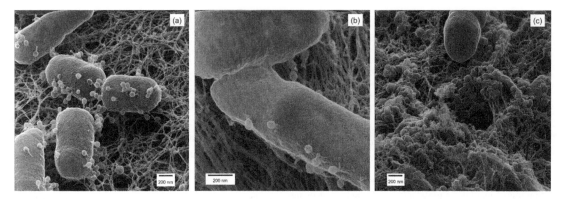

FIGURE 9.9 (a) HIM image showing bacterial cells with ongoing phage infections. (b) Higher magnification visualization of individual T4-phages attached onto the bacterial cell surface, some with contracted tails indicating genome injection. (c) He-milled bacteria showing cut-off surface and a half-away cut phage particle on top of it. (Images are courtesy of Miika Leppänen and Ilari Maasilta, University of Jyväskylä, Finland.)

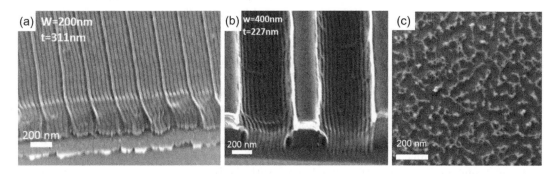

FIGURE 9.10 (a and b) Perpendicular lamellae produced by thermal annealing. A 70° tilted HIM images of self-assembled lamella of polystyrene/polydimethylsiloxane. Samples were annealed at 150°C for 24 h. (a) Sidewall is antireflective coating (ARC), the bottom surface is Pt; ARC grating depth 350 nm, trench width 200 nm, lamellae thickness 311 nm. (b) Both sidewall and bottom surface are ARC; depth 450 nm, trench width 400 nm, lamellae thickness 227 nm. (Reproduced with permission of Royal Society of Chemistry (RSC) Pub in the format Book via Copyright Clearance Center from Bai et al., 2015.) (c) HIM micrograph of the as-cast bottlebrush polymer film highlighting the backbone segments. (Reprinted with permission from Borodinov et al. 2018. © 2018 American Chemical Society.)

polymer films that clearly highlights the backbone segments due to the difference in height of the topographical features.

Another interesting examples of imaging capabilities of HIM are shown in Figure 9.11, where Figure 9.11a and b displays organic porosity of a shale formation visualized as a part of the exploratory study of the influence of nanometer-sized pores on a methane gas storage capacity (Eberle et al., 2016), and Figure 9.11c and d shows collection of graphene-like layers all stacked on top of each other from the imaging study of thermal desulfurization (TDS) of petroleum coke.

9.4 Material Fabrication Using the Ion Beam

FIB nanomachining has been used for decades to perform direct write micro- and nanoscale processing. Conveniently, the He$^+$ and Ne$^+$ GFIS has extended the capability/resolution of FIB direct-write additive and subtractive nanomachining (Alkemade et al., 2012; Stanford et al., 2017a) with the goal of atomic-scale processing (Jesse et al., 2016). As was mentioned in the previous section,

ion beams can readily remove material and modify it at the nanoscale. This is critical for fabrication of functional nanoscale features for nanodevices. The He$^+$ and Ne$^+$ ions can remove material at a slower rate than gallium, providing a method to design, mark, and sculpt at the nanoscale (Joy, 2013). Moreover, FIBs allow direct defect engineering *via* ion implantation, direct write using gaseous and liquid precursors, and direct material deposition. Generally, helium allows higher resolution work, whereas neon offers milling opportunities. Below we discuss how the He$^+$/Ne$^+$ has been utilized to sculpt materials for advanced devices.

9.4.1 Semiconductor Milling

Semiconductor processing technology has been a driving force for the improvement in computing power and efficiency. Some of the traditional techniques (*e.g.*, Ga$^+$ FIB) are currently reaching fundamental limits in terms of SE yield, sputter yield, and spot size (Orloff et al., 2003). The He-FIB limits are yet to be reached. As shown in Figure 9.12, material can be readily removed in a specific region, and

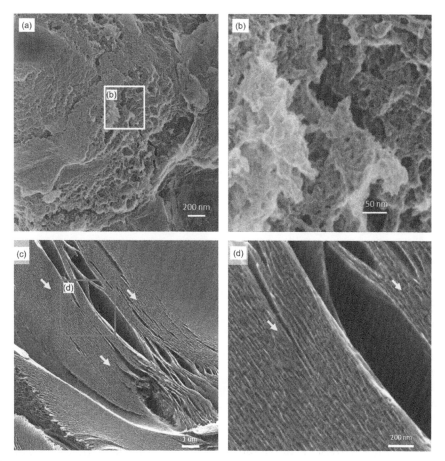

FIGURE 9.11 (a) HIM showing organic porosity in the shale of Marcellus Formation, northern Appalachian Basin; (b) enlarged image of the area indicated in (a). (Reprinted with permission from Eberle et al., 2016. © 2016, American Chemical Society.) (c) HIM images of the petroleum coke samples imaged as is after shaft calciner processing, with TDS; (d) enlarged image of the area indicated in (c). Arrows highlight the damage to the petroleum coke structure with micro-cracks caused by the TDS. Note how much layering and/or ordering is in the sample structure.

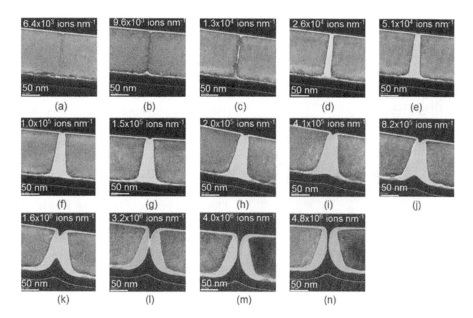

FIGURE 9.12 High-resolution TEM (HRTEM) micrographs of the He^+ milled areas in the membrane Si samples at various doses ranging from (a) 6.4×10^3 to (n) 4.8×10^6 ions/nm. (Reprinted with permission from Tan et al., 2014. © 2014, American Vacuum Society.)

depth and width can be successfully altered using variation in the exposure dose and number of scans. In the case of Si membrane, samples on silicon oxide, initial changes are induced at the dose of 6.4×10^3 ions/nm (Figure 9.12a). With increasing dose, the crystalline lattice starts to show disruption at the center region of the ion beam (where the current density is highest), followed by the substrate amorphization, when the damage density exceeds a certain threshold (appears as bright regions in Figure 9.12c). At higher doses, the amorphous regions close to the surface of the sample grow through the depth of the sample as well as laterally, since the outer part of the probe current also creates enough subsurface damage to reach the damage threshold (Tan et al., 2014).

Another notable example of direct writing with a focused helium ion beam was to use it for synthesis of platinum in an aqueous solution, where Ievlev et al. used two irradiation steps' procedure and achieved a spatial resolution down to 15 nm (Ievlev et al., 2017).

Additionally, helium-based measurements are used for writing structures in resists and have demonstrated sub-10-nm resolution imaging on samples prone to charging, such as photomasks and photoresist (Winston et al., 2009). Moreover, using HIM, Li et al. (2012) fabricated nanoimprint lithography templates in thin hydrogen silsesquioxane (HSQ) layers, as shown in Figure 9.13a. With these templates, the authors successfully demonstrated 4 nm half-pitch line pattern into a layer of UV resist (Figure 9.13b), achieving a better resolution than previously reported for nanoimprint lithography with conventionally designed templates (Wu et al., 2008).

Neon, with three times the SE yield and two-to-three times lower sputter yield relative to Ga^+, has demonstrated very good nanomachining performance in circuit

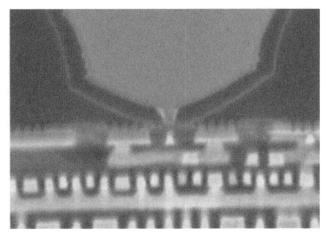

FIGURE 9.14 Ne^+ beam-induced via nanomachining on 22 nm Fin-FET devices. Ne^+ beam machined via through the silicon substrate end pointing on poly (metal-gate) line. (Reprinted by permission from Springer Nature and Copyright Clearance Center Hlawacek and Gölzhäuser, 2016, © 2016.)

edit; providing a high level of SNR and level of control unobtainable with gallium (Tan et al., 2011). An example of Ne^+ circuit edit results is shown in Figure 9.14. A 22 nm fin field-effect transistor (Fin-FET) device and a 10 keV neon beam were used to perform the edits. In the shown application, Ne^+ beam was used to edit between the p-device and n-device transistor fins. After successful end pointing on the very narrow contact layer and Ga^+-IBID, dielectric step was completed, the contact layer was re-exposed by a Ne^+ beam with a 40 nm (FWHM) via. This is a particularly interesting example, as it demonstrates neon ability to access the poly-gate, source, and drain levels without damaging the device (Hlawacek and Gölzhäuser, 2016).

Another example of neon milling includes TEM lamella preparation, which is sensitive to damage and Ga^+ contaminants. The neon ion beam enables finer control during milling (reduced neon sputtering rate of 30%–50% that of gallium together with the neon spot size of ca. 2 nm) and can be useful in samples where gallium contamination is not acceptable (Pekin et al., 2016; Wei et al., 2015). The high-resolution capability of the Ne beam is shown in Figure 9.15, where Figure 9.15a and c are high-resolution TEM (HRTEM) images of gallium-finished and HIM-modified regions of silicon, respectively, and Figure 9.15b and d are corresponding selected area diffraction patterns from the region. Figure 9.15a displays a noisy image relative to Figure 9.15c from the HIM-modified region of the sample. The inset fast Fourier transform (FFT) of the images also show the increase in high-frequency information attained from the modified region. The uniform background contrast of the modified area is indication of more uniform thickness. These images show that the amorphous layer of material on the sample, which contributes to background noise only, is reduced by HIM modification. Similarly, the diffraction pattern in Figure 9.15b shows less information than that

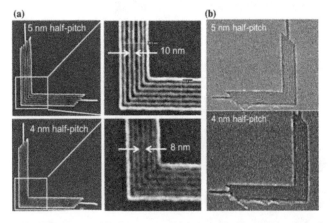

FIGURE 9.13 SEM images of (a) 5 nm half-pitch (top) and 4 nm half-pitch (bottom) nested L's on a 12-nm thick HSQ layer fabricated with HIM, and (b) nanoimprint lithography on a UV resist using the HSQ templates of (a). The "grainy" appearance of the patterns is attributed to a thin Pt coating, applied to minimize sample charging during SEM imaging. (Reprinted by permission from Springer Nature and Copyright Clearance Center Li et al., 2012, © 2016.)

FIGURE 9.15 HRTEM (a) and diffraction (b) information from gallium-finished region. HRTEM (c) and diffraction information (d) from helium-modified region. (Published under a Creative Commons Attribution International License from Fox et al., 2012.)

in Figure 9.15d. The extended high-frequency information in the diffraction pattern recorded from the HIM-modified region in Figure 9.15d indicates that this area of the sample is thinner, while still retaining its high-quality crystalline structure. The sample was measured along the [110] direction.

Neon milling does have a few undesired side effects. Neon implantation can occur at higher dose, forming bubbles in the material. Also, neon use can be limited due to larger interaction volume. However, despite these drawbacks, thorough investigations have shown the advantage of neon GFIS sources over gallium in a broad range of applications such as circuit timing and fault isolation (Rahman et al., 2013).

9.4.2 Defect Engineering in Two-Dimensional Materials

Defects and impurities are known to govern material properties (Tilley, 2008). As defect-free specimens are exceedingly rare, defect engineering attempts to maximize the positive aspects of defecting the material. Defect control with atomic-scale precision is crucial for the functional nanoscale materials, demanded by modern applications. Motivated by the needs of future International Technology Research for Semiconductor roadmap notes, researchers at Intel have pioneered much of the exploratory work for He$^+$ and Ne$^+$ nanomachining. Some examples include the extent and the evolution of beam-induced damage during milling (Livengooda et al., 2007, 2009), He$^+$ sputtering artifacts in Si membrane pores (Tan et al., 2014), and development of a laser-assisted sputtering (Stanford et al., 2016a)

and reactive etching processes (Stanford et al., 2016b) to mitigate damage and accelerate etching. Many other groups have also explored nanoscale He$^+$ sputtering in various materials, such as Si (Fox et al., 2012; Rudneva et al., 2013), Au (Veligura et al., 2013; Melli et al., 2013; Kollmann et al., 2014; Scipioni et al., 2010; Wang et al., 2013), Ni (Yudong et al., 2012), and SiN$_x$ membranes (Jijin et al., 2011; Marshall et al., 2012; Sawafta et al., 2014).

Not surprisingly, introduction of graphene as a unique, high-mobility, zero-gap semiconductor ignited a lot of interest in 2D materials. Various studies have investigated high-resolution milling of supported and suspended graphene for various applications (Hang et al., 2014; Lemme et al., 2009; Bell et al., 2009b; Kalhor et al., 2014; Abbas et al., 2014; Buchheim et al., 2016; Yoon et al., 2016). Detailed investigation of the relationship between the helium ion dose and the defect density, coupled with Raman spectroscopy, demonstrated HIM's precise control over the defect density, transforming semi-metallic graphene to insulating carbon by means of ion dose and energy (Hang et al., 2014; Nakaharai et al., 2013). The generation of defects in graphene has also been utilized to create graphene-based electronic devices in a single material system, as well as in a stacked van der Waals heterogeneous structure system with other 2D nanomaterials, such as hexagonal boron nitride (h-BN) (Nanda et al., 2015). Recently, the effects of Ne-focused ion beam on the fidelity of graphene nanostructures have also been explored (Vighter et al., 2016). As shown in Figure 9.16, graphene nanostructures, supported on silicon and a layer of insulating silicon dioxide (Si/SiO$_2$), were fabricated with Ne$^+$. The iSE images of the milled pads were used as a qualitative indication of electrical conductivity. In Figure 9.16a and c, both pads have the same strip width of ca. 400 nm, while the width in Figure 9.16b is ca. 300 nm.

Now 2D materials with other functionalities, such as semiconductors, insulators, and correlated electron materials, are becoming ubiquitous. Defects in transition-metal dichalcogenides (TMDs) can change band structure and electronic behavior. As a result, TMDs can have a variety of properties suitable for optical devices (Ross et al., 2014),

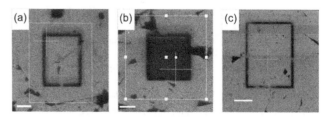

FIGURE 9.16 Graphene nanostructures supported on SiO$_2$ that have been fabricated using direct-write Ne$^+$ lithography. (a–c) iSE images of Ne$^+$ beam-milled pads demonstrating electrical conductivity (scale bars are 2 μm). (Published under Creative Commons Attribution 4.0 International License from Vighter et al., 2016.)

high-performance transistors (Yi et al., 2015), *etc.* A variety of methods have been tried to control point defects in TMDs including helium ion-beam irradiation. Different doses of He^+ have been used to introduce defects into multilayer MoS_2 and WSe_2 tuning material resistivity and electrical properties from semiconducting, to insulating, to metallic (Figure 9.17a) (Fox et al., 2015; Stanford et al., 2016c; Iberi et al., 2016). These phase transitions were due to preferential sulfur sputtering over a dose range of 10^{13}–10^{18} ions/cm^2. WSe_2 nanoribbons (ca. 9 nm) have also been used with He^+ milling and XeF_2 gas-assisted etching (Stanford et al., 2017b). Furthermore, single-silicon nanowires were shown to change electrical and thermal properties with helium irradiation (Figure 9.17b) (Zhao et al., 2017).

One of the most recent works demonstrated a strategy for generating atomically thin circuitry using metallic-like WSe_2 and WS_2 to direct-write edge contacted transistor and direct-write logic gates on single TMD flakes (Stanford et al., 2017c). For more details on defect engineering in TMDs, reader is referred to an exhaustive review by Lin et al. (2016).

In addition to graphene and TMDs, helium ion interaction with other low-dimensional materials were investigated as well, such as novel layered 2D van der Waals ferrielectric system (Belianinov et al., 2015) and CIPS. The fabrication of surface microstructures and tuning of ferroelectric behavior were explored by the helium-ion beam. The local helium-ion modifications improved conductivity and induced ferrielectric switching, expanding local control and offering flexibility for designing heterogeneous 2D material devices (Belianinov et al., 2016). More recently, a variation of chemical properties induced by helium ion beam

was demonstrated using several 2D transition-metal thiophosphates (TPS) with a general formula $Cu_{1-x}In_{1+x/3}P_2S_6$. Specifically, ion irradiation changes copper and oxygen concentrations, and overall system undergoes chemical phase separation, while preserving layered structure of the initial compound (Figure 9.18) (Belianinov et al., 2017a). As research to characterize additional members of this large material family continues, even more properties and possibilities can be expected to emerge.

Atomic scale precision of the HIM has recently been used in magnetic materials. Emerging need for smaller magnets, well below the 100 nm scale, with a variety of shapes, for sensors and data storage has been driving this area. Ion-induced modification using Ne^+ has been used to change the magnetic properties of $Fe_{60}Al_{40}$ thin films at the nanoscale via generated chemical disordering. Bali et al. (2014) showed that the irradiation with Ne^+ can create antisite defects (Fe_{Al} and Al_{Fe}) and transform $Fe_{60}Al_{40}$ from B2-structure to A2. This type of chemical disorder drastically changes local environment of iron resulting in the magnetic phase transition and an increased saturation magnetization, enabling positive magnetic patterning (Figure 9.19).

Focused Ne^+ beam was also recently used for direct magnetic writing. An attempt to generate a magnetic stripe pattern was performed on the B2-$Fe_{60}Al_{40}$ thin film. Induction of a magnetization has been established in the irradiated regions, where 500 nm wide stripes were formed. Moreover, expected non-ferromagnetic gaps between the magnetized stripes were also confirmed (Röder et al., 2015).

Research in HIM-assisted modification of magnetism is still new but can be expected to grow in modification of other relevant materials.

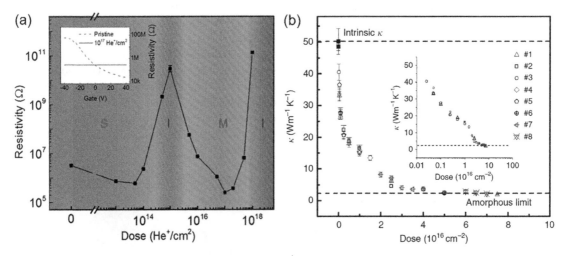

FIGURE 9.17 (a) Double log plot of electrical resistivity vs He^+ dose for a substrate supported, mechanically exfoliated bilayer MoS_2 flake. The letters S, I, and M correspond to regions with semiconducting, insulating, and metallic-like behavior, respectively. Inset: resistivity measurement as a function of gate bias. The resistivity of the pristine sample (dashed line) was widely tunable. The sample irradiated with 10^{17} He^+/cm^2 showed little gate response (solid line). (Reprinted with permission from Fox et al., 2015. © 2018 American Chemical Society.) (b) Measured thermal conductivity (κ) of eight individual samples (#1–#8) vs dose. Inset: the same data plotted on a logarithmic scale. The solid black square denotes the thermal conductivity of intrinsic nanowires (namely, with zero dose) that is derived from averaging calculated ones by fitting several intrinsic portions for the eight samples. (Published under Creative Commons Attribution 4.0 International License from Zhao et al., 2017.)

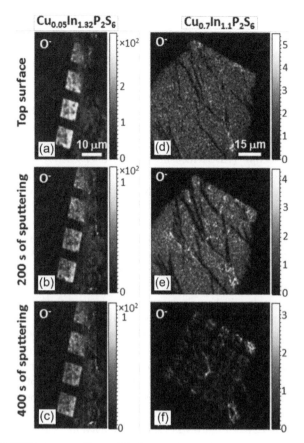

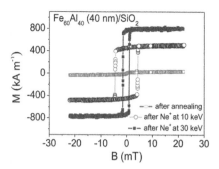

FIGURE 9.19 Magnetic hysteresis loops for $Fe_{60}Al_{40}$ films of 40-nm thickness measured after vacuum annealing at 773 K (empty squares) and after irradiation of the annealed films with Ne^+ at 10 keV (circles) and 30 keV (solid squares). A fluence of 6×10^{14} ions/cm² was used for the irradiation exposures. (Reprinted with permission from Bali et al., 2014. © 2018 American Chemical Society.)

FIGURE 9.18 Time-of-flight secondary ion mass spectrometry (ToF-SIMS) results of O^- signal on helium ion irradiated regions for $Cu_{0.05}In_{1.32}P_2S_6$ (panels a–c) and $Cu_{0.7}In_{1.1}P_2S_6$ (panels d–f) at the surface of the sample. (a) O^- signal for $Cu_{0.05}In_{1.32}P_2S_6$ with no Cs sputtering; total ion count 1.97×10^6 ions/cm², (b) O^- signal for $Cu_{0.05}In_{1.32}P_2S_6$ after 200 s of Cs sputtering; total ion count 6.72×10^5 ions/cm², (c) O^- signal for $Cu_{0.05}In_{1.32}P_2S_6$ after 400 s of Cs sputtering; total ion count 6.45×10^6 ions/cm², (d) O^- signal for $Cu_{0.7}In_{1.1}P_2S_6$ with no Cs sputtering; total ion count 5.92×10^4 ions/cm², (e) O^- signal for $Cu_{0.7}In_{1.1}P_2S_6$ after 200 s of Cs sputtering; total ion count 5.23×10^4 ions/cm², (f) O^- signal for $Cu_{0.7}In_{1.1}P_2S_6$ after 400 s of Cs sputtering; total ion count 1.15×10^4 ions/cm². (Published under Creative Commons Attribution 4.0 International License from Belianinov et al., 2017a.)

9.4.3 Three-Dimensional Focused Ion Beam-Induced Deposition

In addition to imaging and patterning, FIB can be used for direct deposition of selected precursors. These are typically volatile compounds such as organometallic or halide gases. Using a gas injection system FIBID opens doors for creation of 3D structures at the nanoscale or milling enhancement by locally introducing etchants.

The deposition process is based on the interaction between the ion beam, the injected material, and the surface. At first, the precursor is deposited on the sample, but the sticking coefficient is enhanced in the beam-irradiated region. At the point of contact of the ion beam

and a stream of volatile molecules with the surface deposition occurs. It is important to note that for this to happen, the sputter yield must be smaller than the deposition yield. Given the combination of gas, volatile compounds, and sample surface requires an understanding of many fundamental aspects including precursor chemistry, surface diffusion, and adsorption–desorption mechanics of the precursor on a given surface.

Alkemade et al. showcase early work on studying the growth mechanisms of He^+ beam-induced deposition of PtC_x (x is typically = 5) nanopillars grown from the $MeCpPtMe_3$ ($C_9H_{16}Pt$) precursor in constant current (Alkemade et al., 2010) and pulsed current (Ping et al., 2010) modes. Subsurface damage during nanopillar growth has also been investigated (Drezner et al., 2012). Ogowa et al. have shown $W(CO)_6$ nanopillars grown with the He^+ beam (Kohama et al., 2013; Hayashida et al., 2012). Later, the electrical/chemical/structural properties of PtC_x were compared for both Ne^+ and He^+, where enhanced conductivity was realized for Ne^+ due to Pt grain coarsening induced by enhanced nuclear energy loss during the Ne^+ growth (Wu et al., 2013); however, neon bubbling occurred due to the neon dose required for growth. Very high conductivity (ca. 60 μOhm·cm) and high-resolution (10 nm) lines were achieved using He^+ and the $Co_2(CO)_8$ precursor, as very efficient carbon desorption was realized (Wu et al., 2014).

A survey of research on electron-induced deposition (EBID) and focused ions (FIBID) showed similar metal content for several precursors, such as $Me_3PtCpMe$, WF_6, and $Co_2(CO)_8$ (Figure 9.20) (Fowlkes et al., 2016; Belianinov et al., 2017b). Comparing He FIBID and EBID shows deposit composition of 75–93 at.% vs 85 at.% for tungsten and 16–20 at.% vs 13–21 at.% for platinum (Fowlkes et al., 2016; Gamo and Namba, 1990; Matsui et al., 1989; Rotkina et al., 2003; Wu et al., 2013). This can be attributed to the crucial role that the SEs play in the chemistry of precursor dissociation. For more details, an overview of compositions for various precursors with

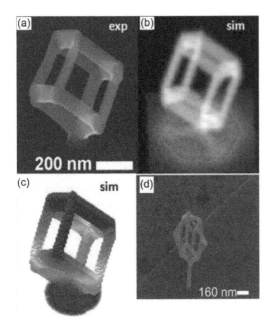

FIGURE 9.20 EBID and FIBID. (a) SEM image of EBID growth of a cubic frame. The segment length was 200 nm, and the cubic frame was rotated vertically by 54.7°. The rotation made all segment angles equal at 35.3°, requiring a dwell time (τ_d) of 11.0 ms at a pixel point pitch (Λ) of 1 nm. There were 2,008 individual pixel dwells required to write the cubic mesh. (b) Virtual SEM image of the simulation results that mimic the experiment in (a). (c) Simulation result rendered in the form of deposit surface voxels (each volume element is the amount of material at a given dwell time), where the color map is normalized to the vertical z-coordinate. (d) An example of helium FIBID-grown deltahedron, a ten-face solid, with a Me$_3$PtCpMe precursor. (Belianinov et al., 2017b, reproduced with permission.)

various beams consult Silvis-Cividjian and Hagen et al. (2006). Additionally, helium ion and electron deposition from Me$_3$PtCpMe resulted in roughly the same size of the Pt grains embedded in the carbonaceous matrix (Alkemade and Miro, 2014).

Most recently, HIM together with a Protochips liquid cell were combined to crystallize, grow, and image thiamethoxam (TMX) *in situ*. TMX is a promising member of a new set of insecticide compounds, called neonicotinoids. Authors illustrated growth and morphology of the TMX crystals at different He$^+$ exposure doses and demonstrated controlled crystallization of the soft material in liquid environment (Belianinov et al., 2018).

9.5   HIM as Analytical Tool

With a continuous progress in materials science and engineering, analysis of materials structure and properties becomes progressively challenging. As a result, multimodal physical and chemical nano-analysis (*e.g.*, simultaneous correlated physical and chemical information) are gaining interest.

9.5.1   Backscattering Spectrometry in the HIM

SEs are not the only product of incident ion beam interaction with a material – some fraction of the ions backscatter, enabling Rutherford backscattering spectrometry (Hlawacek et al., 2014; Hlawacek and Gölzhäuser, 2016). While SE images are rich in morphological contrast, they are poor in chemical contrast (aside from energy dispersive X-ray spectroscopy (EDS) measurements), backscattered-helium images are poor in topography but rich in elemental differentiation. Early attempts to perform backscatter spectrometry were possible in the HIM energy regime with a solid-state detector (SSD) (Sijbrandij et al., 2008, 2010). However, limited energy resolution and reduced mass separation capability of the SSD indicated that a more sophisticated detection scheme is required. Experimentally demonstrated ToF backscattering (ToF-BS) spectrometry showed the feasibility of chemical analysis with a lateral resolution down to 55 nm, close to physical limits (Klingner et al., 2016). An experimental evaluation of the overall lateral resolution in ToF-BS mode is shown in Figure 9.21. The test sample is a piece of glassy carbon coated with rectangular patch of Ni (25 μm × 25 μm × 110 nm). The edge of a Ni patch is imaged in SE mode without pulsing the primary beam (Figure 9.21a) and in pulsed beam ToF-BS mode (Figure 9.21b). Figure 9.21c shows averaged line scans across the Ni edge from images (a) and (b). The areas that were used for averaging are indicated as dashed rectangles in (a) and (b) as well as the directions of line profiles (arrows).

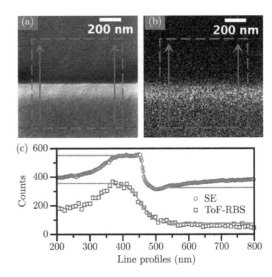

FIGURE 9.21 (a) SE image without pulsing the ion beam and (b) ToF-BS image of a Ni patch on the test sample described in the text; (c) derived line profiles of the Ni edge. Line profiles are measured and averaged across the Ni edge according to the rectangles plotted in (a) and (b) and error functions fitted at the edges. The edge resolution (80%–20%) was determined to be 10.9 nm in SE mode and 53.7 nm in ToF-BS mode using 55 ns beam pulses. (Reprinted from Klingner et al., 2016, © 2016, with permission from Elsevier.)

The blanking direction in these measurements was 52° with respect to the Ni edge. The edge resolution (80%−20%) evaluates to 10.9 nm in SE mode (without pulsing the beam) and 53.7 nm in ToF-BS mode using beam pulses of 55 ns length.

9.5.2 Secondary Ion Mass Spectrometry in HIM

One of the most recent multimodal chemical imaging techniques is SIMS with the HIM. Developed only a few years ago, it is based on a commercially available Zeiss ORION NanoFab (Wirtz et al., 2015). The combination of the high-resolution imaging and the chemical sensitivity of SIMS allows the detection of the ion distribution with a spatial resolution of ca. 15 nm, which makes the Zeiss ORION NanoFab an attractive prospect for chemical imaging with SIMS (Dowsett et al., 2012; Pillatsch et al., 2013; Wirtz et al., 2012).

HIM–SIMS relies on a compact double-focusing magnetic sector, which can be used to analyze a wide mass range with mass resolution of ∼500. Dowsett and Wirtz et al. (Dowsett and Wirtz, 2017; Dowsett et al., 2012) have developed a low-field high-efficiency ion extraction system, with the intent of reducing the probe degradation and maintaining the sub-10 nm Ne probe size. The ion extraction system is fully retractable, and the HIM can be used in SE mode without being affected by the SIMS. Schematic of the system is shown in Figure 9.22a. The post-acceleration accelerates secondary ions to the required voltage for the spectrometer, which is based on a modified Mattauch–Herzog configuration with a 450-mm-long focal plane. The length of the focal plane was optimized to collect ions with m/q extending to 250 amu/e with a mass resolution greater than 400. Also, collection of ions with m/q extending to ca. 500 can be achieved with lower accelerating voltages at the expanse of mass resolution. Lastly, a system of transfer lenses transports the beam and focuses it onto the entrance plane of the spectrometer (Figure 9.22b) (Dowsett and Wirtz, 2017).

Four different masses can be detected in parallel by a multi-collection system with four detectors (one fixed and three movable).

Despite the name, the typical gas that was used for HIM–SIMS is neon. Helium has a low sputter yield, while neon loses most of its energy close to the surface through the collision cascade, and the surface area contributing to the sputtering of secondary particles is noticeably larger (Pillatsch et al., 2013; Rahman et al., 2012). Consequently, neon sputter yield is considerably higher and comparable to cesium and oxygen beams commonly used in other SIMS instruments.

While technology is still young, the HIM–SIMS promises great analytical possibilities. Three modes of chemical imaging currently exist: (i) mass-resolved spectra for identification of surface chemical species, (ii) depth chemical profiling, and (iii) mass-selected high-resolution chemical mapping (Kim et al., 2018). For mass-resolved spectra, the standard mass range is between 1 and 250 amu but can be extended to 500 amu at the expense of mass resolution. Depth chemical profiling can be performed on areas smaller than few hundred nanometers and is applicable for process control during nano-machining, when layered structures are created or inspected. Finally, mass-selected mapping is one of the most interesting features of HIM–SIMS, where four independent chemical images can be created based on user-specified masses of interest. Probe size smaller than the dimensions of the collision cascade offers SIMS maps with highest achievable resolution (Wirtz et al., 2016).

The tool was recently used to explore the chemical composition of hybrid organic–inorganic perovskites (HOIPs). Gratia et al. (2016, 2017) mapped the elemental distribution of mixed perovskite films demonstrating that a part of the film intrinsically segregates into iodide-rich perovskite nanodomains on a length scale extending to a few hundred nanometers (Figure 9.23). Thus, the homogeneity of the film is disrupted, leading to a variation in the optical properties at the micrometer scale. Figure 9.23a shows the SEM image of the surface of the mixed cation/mixed

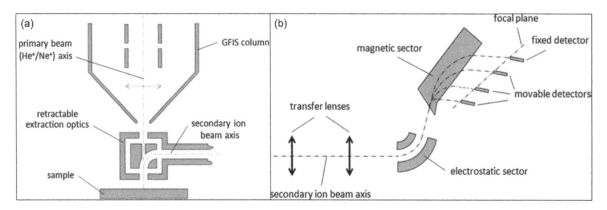

FIGURE 9.22 (a) Schematic of the extraction region of the HIM. Some electrodes are simplified or removed for clarity. The entire extraction system is removable, so SE imaging and milling operations are unaffected. (b) Schematic of the double-focusing magnetic sector spectrometer. The spectrometer has a long straight focal plane, allowing parallel detection of a wide range of masses. (Reprinted with permission from Dowsett and Wirtz, 2017. © 2018 American Chemical Society.)

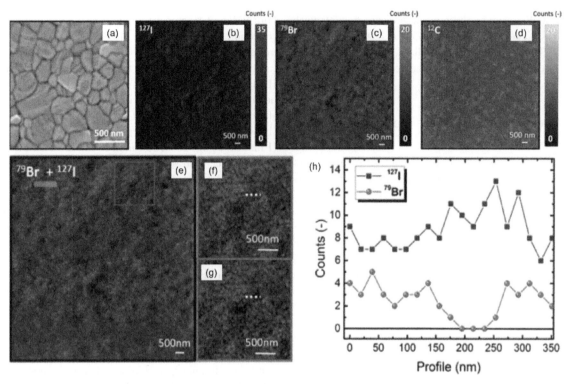

FIGURE 9.23 Elemental nanoscale HIM–SIMS mapping. (a) SEM surface image of $(FAPbI_3)_{0.85}(MAPbBr_3)_{0.15}$ perovskite deposited on a mesoporous TiO_2 scaffold. (b–d) HIM–SIMS elemental mapping of ^{127}I, ^{79}Br, and ^{12}C across a $10 \times 10\ \mu m^2$ area. (e) Overlap of the ^{79}Br and ^{127}I signals. The spots have low ^{79}Br and high ^{127}I intensities. (f and g) Zoomed views of (f) combined ^{79}Br and ^{127}I and (g) ^{79}Br alone. (h) Profiles of the ^{79}Br and ^{127}I signals across the white dashed lines in (f) and (g). (Reprinted with permission from Gratia et al., 2016. © 2018 American Chemical Society.)

halide $(FAPbI_3)_{0.85}(MAPbBr_3)_{0.15}$ film (MA = methylammonium; FA = formamidinium). Figure 9.23b–d shows the elemental distributions of ^{127}I, ^{79}Br, and ^{12}C across a $10 \times 10\ \mu m^2$ area, indicating compositional inhomogeneity at the nanoscale. Overlap of the elemental maps of ^{127}I and ^{79}Br in Figure 9.23e demonstrates that most of the regions with low amount of ^{79}Br correspond to regions with a high ^{127}I content. A zoomed view of such region, as well as the profiles of the ^{79}Br and ^{127}I signals along the dashed white lines are shown in Figure 9.23f–h.

Lui et al. (2018) combined mass spectrometry with scanning probe microscopy and optical spectroscopy in a unique multimodal chemical imaging to show that $CH_3NH_3PbI_3$ perovskite material is ferroelastic (and not ferroelectric). By combining experimental results with first principles simulations, it was revealed that the ferroelasticity drives chemical segregation. Authors observed twin domain contrast in SEM and employed Ne^+ HIM–SIMS to show that this is due to dissimilar chemical composition, which results in different local electronic band structures and conductivity. Figure 9.24a shows the HIM–SIMS chemical maps for $CH_3NH_3^+$, where the grains and grain boundaries are clearly visible. The $CH_3NH_3^+$ concentration varies between grains producing twin domains. As illustrated in the enlarged images, Figure 9.24b and c, grains with both high (Figure 9.24b) and low (Figure 9.24c) $CH_3NH_3^+$ concentration are observed with twin domains of

comparable widths to those observed by authors in piezoresponse force microscopy (PFM). This suggests a difference in chemical composition between the twin domains in the sample.

9.6 Conclusions

Despite being a relatively new technique, HIM with its unique GFIS quickly gained popularity in imaging, defect engineering, and milling. Particularly attractive features of this technology are the small interaction volume of He and Ne, small beam spot size, and relatively slow sputtering rates, when compared to other FIBs. The tool also offers outstanding imaging quality on insulating samples, a key feature for the investigation of soft, polymeric, and biological materials. In addition, the combination with the SIMS offers chemical imaging with nanometer spatial resolution. It can be expected that HIM will continue to have a strong influence on broad range of scientific disciplines, including chemistry, materials science, and biology.

Acknowledgments

This work was partially funded by the Center for Nanophase Materials Sciences (CNMS), which is a DOE Office of Science User Facility.

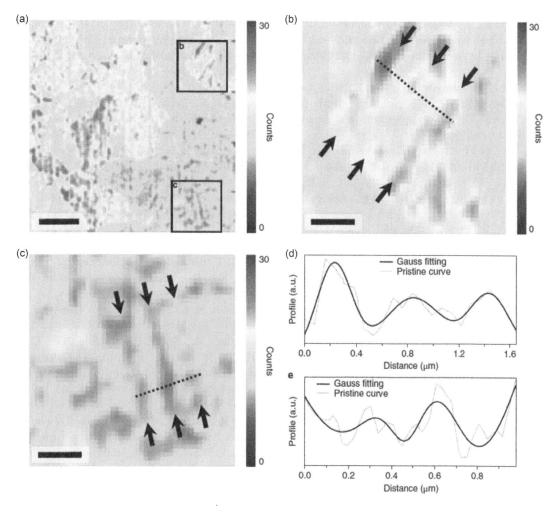

FIGURE 9.24 HIM–SIMS mapping. (a) $CH_3NH_3^+$ distribution. (b and c) Enlarged images of the areas indicated in (a). (d and e) Profile analyses of the dashed lines in (b) and (c), respectively. The average domain widths calculated from the lines are (b) 270 nm and (c) 163 nm. Scale bars: (a) 2 μm and (b, c) 0.5 μm. (Reprinted by permission from Springer Nature and Copyright Clearance Center: Springer, Nature Materials, Chemical nature of ferroelastic twin domains in $CH_3NH_3PbI_3$ perovskite, Liu et al., 2018, © 2018.)

References

Abbas, A. N., Liu, G., Liu, B., Zhang, L., Liu, H., Ohlberg, D., Wu, W. & Zhou, C. 2014. Patterning, characterization, and chemical sensing applications of graphene nanoribbon arrays down to 5 nm using helium ion beam lithography. *ACS Nano*, 8, 1538–1546.

Alkemade, P. F. A., Chen, P., Veldhoven, E. V. & Maas, D. 2010. Model for nanopillar growth by focused helium ion-beam-induced deposition. *Journal of Vacuum Science & Technology B*, 28, C6F22–C6F25.

Alkemade, P. F. A., Koster, E. M., van Veldhoven, E. & Maas, D. J. 2012. Imaging and nanofabrication with the helium ion microscope of the Van Leeuwenhoek Laboratory in Delft. *Scanning*, 34, 90–100.

Alkemade, P. F. A. & Miro, H. 2014. Focused helium-ion-beam-induced deposition. *Applied Physics A*, 117, 1727–1747.

Arora, W. J., Sijbrandij, S., Stern, L., Notte, J., Smith, H. I. & Barbastathis, G. 2007. Membrane folding by helium ion implantation for three-dimensional device fabrication. *Journal of Vacuum Science & Technology B: Microelectronics and Nanometer Structures Processing, Measurement, and Phenomena*, 25, 2184–2187.

Bai, W., Gadelrab, K., Alexander-Katz, A. & Ross, C. A. 2015. Perpendicular block copolymer microdomains in high aspect ratio templates. *Nano Letters*, 15, 6901–6908.

Bali, R., Wintz, S., Meutzner, F., Hübner, R., Boucher, R., Ünal, A. A., Valencia, S., Neudert, A., Potzger, K., Bauch, J., Kronast, F., Facsko, S., Lindner, J. & Fassbender, J. 2014. Printing nearly-discrete magnetic patterns using chemical disorder induced ferromagnetism. *Nano Letters*, 14, 435–441.

Bazou, D., Behan, G., Reid, C., Boland, J. J. & Zhang, H. Z. 2011. Imaging of human colon cancer cells using He-Ion scanning microscopy. *Journal of Microscopy*, 242, 290–294.

Belianinov, A., Burch, M. J., Hysmith, H. E., Ievlev, A. V., Iberi, V., Susner, M. A., Mcguire, M. A., Maksymovych,

P., Chyasnavichyus, M., Jesse, S. & Ovchinnikova, O. S. 2017a. Chemical changes in layered ferroelectric semiconductors induced by helium ion beam. *Scientific Reports*, 7, 16619.

Belianinov, A., Burch, M. J., Kim, S., Tan, S., Hlawacek, G. & Ovchinnikova, O. S. 2017b. Noble gas ion beams in materials science for future applications and devices. *MRS Bulletin*, 42, 660–666.

Belianinov, A., He, Q., Dziaugys, A., Maksymovych, P., Eliseev, E., Borisevich, A., Morozovska, A., Banys, J., Vysochanskii, Y. & Kalinin, S. V. 2015. CuInP$_2$S$_6$ room temperature layered ferroelectric. *Nano Letters*, 15, 3808–3814.

Belianinov, A., Iberi, V., Tselev, A., Susner, M. A., Mcguire, M. A., Joy, D., Jesse, S., Rondinone, A. J., Kalinin, S. V. & Ovchinnikova, O. S. 2016. Polarization control via He-ion beam induced nanofabrication in layered ferroelectric semiconductors. *ACS Applied Materials & Interfaces*, 8, 7349–7355.

Belianinov, A., Pawlicki, A., Burch, M., Kim, S., Ievlev, A., Fowler, J. & Ovchinnikova, O. 2018. In situ liquid cell crystallization and imaging of thiamethoxam by helium ion microscopy. *Journal of Vacuum Science & Technology B*, 36, 051803.

Bell, D. C., Lemme, M. C., Stern, L. A. & Marcus, C. M. 2009a. Precision material modification and patterning with He ions. *Journal of Vacuum Science & Technology B: Microelectronics and Nanometer Structures Processing, Measurement, and Phenomena*, 27, 2755–2758.

Bell, D. C., Lemme, M. C., Stern, L. A., Williams, J. R. & Marcus, C. M. 2009b. Precision cutting and patterning of graphene with helium ions. *Nanotechnology*, 20, 455301.

Berger, M. J., Coursey, J. S., Zucker, M. A. & Change, J. (eds.) July 2017. *Stopping-Power & Range Tables for Electrons, Protons, and Helium Ions; NIST Standard Reference Database 124*, National Institute of Standards and Technology: Gaithersburg, MD.

Bethe, H. A. 1930. Zur Theorie des Durchgangs schneller Korpuskularstrahlen durch Materie. *Annalen der Physik (in German)*, 397, 325–400.

Bethe, H. A. 1941. On the theory of secondary emission. *Physical Review*, 59, 940.

Bethe, H. A. & Ashkin, J. 1953. Passage of radiations through matter. In: Segré, E. (ed.) *Experimental Nuclear Physics*. New York: John Wiley & Sons, 17–53.

Biersack, J. P. & Eckstein, W. 1984. Sputtering studies with the Monte Carlo program TRIM. SP. *Applied Physics A*, 34, 73–94.

Biersack, J. P. & Haggmark, L. G. 1980. A Monte Carlo computer program for the transport of energetic ions in amorphous targets. *Nuclear Instruments and Methods*, 174, 257–269.

Binnig, G., Quate, C. F. & Gerber, C. 1986. Atomic force microscope. *Physical Review Letters*, 56, 930–933.

Binnig, G. & Rohrer, H. 1983. Scanning tunneling microscopy. *Surface Science*, 126, 236–244.

Boden, S. A., Asadollahbaik, A., Rutt, H. N. & Bagnall, D. M. 2012. Helium ion microscopy of Lepidoptera scales. *Scanning*, 34, 107–120.

Borodinov, N., Belianinov, A., Chang, D., Carrillo, J.-M., Burch, M. J., Xu, Y., Hong, K., Ievlev, A. V., Sumpter, B. G. & Ovchinnikova, O. S. 2018. Molecular reorganization in bulk bottlebrush polymers: direct observation via nanoscale imaging. *Nanoscale*, 10, 18001–18009.

Brown, R. D., Rao, P. & Ho, P. S. 1973. Observation of damage in helium ion irradiated gold films using transmission electron microscopy. *Radiation Effects*, 18, 149–155.

Buchheim, J., Wyss, R. M., Shorubalko, I. & Park, H. G. 2016. Understanding the interaction between energetic ions and freestanding graphene towards practical 2D perforation. *Nanoscale*, 8, 8345–8354.

Champness, P. E. 2001. *Electron Diffraction in the Transmission Electron Microscope*. Routledge, London, UK.

Chhowalla, M., Shin, H. S., Eda, G., Li, L.-J., Loh, K. P. & Zhang, H. 2013. The chemistry of two-dimensional layered transition metal dichalcogenide nanosheets. *Nature Chemistry*, 5, 263.

Cybart, S. A., Cho, E. Y., Wong, T. J., Wehlin, B. H., Ma, M. K., Huynh, C. & Dynes, R. C. 2015. Nano Josephson superconducting tunnel junctions in YBa2Cu3O7-δ directly patterned with a focused helium ion beam. *Nature Nanotechnology*, 10, 598.

Dapor, M. 2011. Secondary electron emission yield calculation performed using two different Monte Carlo strategies. *Nuclear Instruments and Methods in Physics Research Section B: Beam Interactions with Materials and Atoms*, 269, 1668–1671.

De Souza, W. & Attias, M. 2015. New views of the toxoplasma gondii parasitophorous vacuole as revealed by Helium Ion Microscopy (HIM). *Journal of Structural Biology*, 191, 76–85.

Dowsett, D. & Wirtz, T. 2017. Co-registered in situ secondary electron and mass spectral imaging on the helium ion microscope demonstrated using lithium titanate and magnesium oxide nanoparticles. *Analytical Chemistry*, 89, 8957–8965.

Dowsett, D., Wirtz, T., Vanhove, N. & Pillatsch, L. 2012. Secondary ion mass spectrometry on the helium ion microscope: A feasibility study of ion extraction. *Journal of Vacuum Science & Technology B*, 30, 06F602.

Drezner, Y., Greenzweig, Y., Fishman, D., Veldhoven, E. V., Maas, D. J., Raveh, A. & Livengood, R. H. 2012. Structural characterization of He ion microscope platinum deposition and sub-surface silicon damage. *Journal of Vacuum Science & Technology B*, 30, 041210.

Eberle, A. P. R., King, H. E., Ravikovitch, P. I., Walters, C. C., Rother, G. & Wesolowski, D. J. 2016. Direct measure of the dense methane phase in gas shale organic porosity by neutron scattering. *Energy & Fuels*, 30, 9022–9027.

Erents, S. K. & Mccracken, G. M. 1973. Blistering of molybdenum under helium ion bombardment. *Radiation Effects*, 18, 191–198.

Everhart, T. E. & Thornley, R. F. M. 1960. Wideband detector for micro-microampere low-energy electron currents. *Journal of Scientific Instruments*, 37, 246.

Fowlkes, J. D., Winkler, R., Lewis, B. B., Stanford, M. G., Plank, H. & Rack, P. D. 2016. Simulation-guided 3D nanomanufacturing via focused electron beam induced deposition. *ACS Nano*, 10, 6163–6172.

Fox, D. S., Chen, Y., Faulkner, C. C. & Zhang, H. 2012. Nano-structuring, surface and bulk modification with a focused helium ion beam. *Beilstein Journal of Nanotechnology*, 3, 579–585.

Fox, D. S., Zhou, Y., Maguire, P., O'Neill, A., Ó'Coileáin, C., Gatensby, R., Glushenkov, A. M., Tao, T., Duesberg, G. S. & Shvets, I. V. 2015. Nanopatterning and electrical tuning of MoS2 layers with a subnanometer helium ion beam. *Nano Letters*, 15, 5307–5313.

Gamo, K. & Namba, S. 1990. Microfabrication using focused ion beams. *Euro III-Vs Review*, 3, 41–42.

Giannuzzi, L. A. 2006. *Introduction to Focused Ion Beams: Instrumentation, Theory, Techniques and Practice*. Springer, New York.

Goldstein, J., Newbury, D. E. & Joy, D. C. 2003. *Scanning Electron Microscopy and X-Ray Microanalysis*. New York: Kluwer Academic/Plenum Publishers.

Gratia, P., Grancini, G., Audinot, J.-N., Jeanbourquin, X., Mosconi, E., Zimmermann, I., Dowsett, D., Lee, Y., Grätzel, M., de Angelis, F., Sivula, K., Wirtz, T. & Nazeeruddin, M. K. 2016. Intrinsic halide segregation at nanometer scale determines the high efficiency of mixed cation/mixed halide perovskite solar cells. *Journal of the American Chemical Society*, 138, 15821–15824.

Gratia, P., Zimmermann, I., Schouwink, P., Yum, J.-H., Audinot, J.-N., Sivula, K., Wirtz, T. & Nazeeruddin, M. K. 2017. The many faces of mixed ion perovskites: Unraveling and understanding the crystallization process. *ACS Energy Letters*, 2, 2686–2693.

Hang, S., Moktadir, Z. & Mizuta, H. 2014. Raman study of damage extent in graphene nanostructures carved by high energy helium ion beam. *Carbon*, 72, 233–241.

Hayashida, M., Iijima, T., Fujimoto, T. & Ogawa, S. 2012. Position-controlled marker formation by helium ion microscope for aligning a TEM tomographic tilt series. *Micron*, 43, 992–995.

Hill, R., Notte, J. & Ward, B. 2008. The ALIS He ion source and its application to high resolution microscopy. *Physics Procedia*, 1, 135–141.

Hlawacek, G. & Gölzhäuser, A. 2016. *Helium Ion Microscopy*. Springer International Publishing, Switzerland.

Hlawacek, G., Veligura, V., van Gastel, R. & Poelsema, B. 2014. Helium ion microscopy. *Journal of Vacuum Science & Technology B, Nanotechnology and Microelectronics: Materials, Processing, Measurement, and Phenomena*, 32, 020801.

Iberi, V., Liang, L., Ievlev, A. V., Stanford, M. G., Lin, M.-W., Li, X., Mahjouri-Samani, M., Jesse, S., Sumpter, B. G., Kalinin, S. V., Joy, D. C., Xiao, K., Belianinov, A. & Ovchinnikova, O. S. 2016. Nanoforging single layer MoSe2 through defect engineering with focused helium ion beams. *Scientific Reports*, 6, 30481.

Ievlev, A. V., Jakowski, J., Burch, M. J., Iberi, V., Hysmith, H., Joy, D. C., Sumpter, B. G., Belianinov, A., Unocic, R. R. & Ovchinnikova, O. S. 2017. Building with ions: towards direct write of platinum nanostructures using in situ liquid cell helium ion microscopy. *Nanoscale*, 9, 12949–12956.

Inai, K., Ohya, K. & Ishitani, T. 2007. Simulation study on image contrast and spatial resolution in helium ion microscope. *Journal of Electron Microscopy*, 56, 163–169.

Jesse, S., Borisevich, A. Y., Fowlkes, J. D., Lupini, A. R., Rack, P. D., Unocic, R. R., Sumpter, B. G., Kalinin, S. V., Belianinov, A. & Ovchinnikova, O. S. 2016. Directing matter: toward atomic-scale 3D nanofabrication. *ACS Nano*, 10, 5600–5618.

Jijin, Y., David, C. F., Lewis, A. S., Colin, A. S., Jason, H., Zheng, R., Lu-Chang, Q. & Adam, R. H. 2011. Rapid and precise scanning helium ion microscope milling of solid-state nanopores for biomolecule detection. *Nanotechnology*, 22, 285310.

Joy, D. C. 2013. *Helium Ion Microscopy: Principles and Applications*. New York: Springer.

Joy, D. C. & Luo, S. 1989. An empirical stopping power relationship for low-energy electrons. *Scanning*, 11, 176–180.

Joly, D. C., Poirier-Demers, N. & Demers, H. 2015. Casino 3.3; [Online]. Available: www.gel.usherbrooke.ca/casino/index.html.

Kalhor, N., Boden, S. A. & Mizuta, H. 2014. Sub-10nm patterning by focused He-ion beam milling for fabrication of downscaled graphene nano devices. *Microelectronic Engineering*, 114, 70–77.

Kim, S., Dyck, O., Ievlev, A. V., Vlassiouk, I. V., Kalinin, S. V., Belianinov, A., Jesse, S. & Ovchinnikova, O. S. 2018. Graphene milling dynamics during helium ion beam irradiation. *Carbon*, 138, 277–282.

Klingner, N., Heller, R., Hlawacek, G., Borany, J. V., Notte, J., Huang, J. & Facsko, S. 2016. Nanometer scale elemental analysis in the helium ion microscope using time of flight spectrometry. *Ultramicroscopy*, 162, 91–97.

Knoll, M. & Ruska, E. 1932. Das Elektronenmikroskop. *Zeitschrift für Physik (in German)*, 78, 318–339.

Kohama, K., Iijima, T., Hayashida, M. & Ogawa, S. 2013. Tungsten-based pillar deposition by helium ion microscope and beam-induced substrate damage. *Journal of Vacuum Science & Technology B*, 31, 031802.

Kollmann, H., Piao, X., Esmann, M., Becker, S. F., Hou, D., Huynh, C., Kautschor, L.-O., Bösker, G., Vieker, H., Beyer, A., Gölzhäuser, A., Park, N., Vogelgesang, R., Silies, M. & Lienau, C. 2014. Toward plasmonics with nanometer precision: Nonlinear optics of helium-ion milled gold nanoantennas. *Nano Letters*, 14, 4778–4784.

Lehtinen, O., Kotakoski, J., Krasheninnikov, A. V. & Keinonen, J. 2011. Cutting and controlled modification of graphene with ion beams. *Nanotechnology*, 22, 175306.

Lemme, M. C., Bell, D. C., Williams, J. R., Stern, L. A., Baugher, B. W. H., Jarillo-Herrero, P. & Marcus, C. M. 2009. Etching of graphene devices with a helium ion beam. *ACS Nano*, 3, 2674–2676.

Leppänen, M., Sundberg, L.-R., Laanto, E., de Freitas Almeida, G. M., Papponen, P. & Maasilta, I. J. 2017. Imaging bacterial colonies and phage–bacterium interaction at sub-nanometer resolution using helium-ion microscopy. *Advanced Biosystems*, 1, 1700070.

Li, W.-D., Wu, W. & Williams, R. S. 2012. Combined helium ion beam and nanoimprint lithography attains 4 nm half-pitch dense patterns. *Journal of Vacuum Science & Technology B*, 30, 06F304.

Lin, Z., Carvalho, B. R., Kahn, E., Lv, R., Rao, R., Terrones, H., Pimenta, M. A. & Terrones, M. 2016. Defect engineering of two-dimensional transition metal dichalcogenides. *2D Materials*, 3, 022002.

Liu, Y., Collins, L., Proksch, R., Kim, S., Watson, B. R., Doughty, B., Calhoun, T. R., Ahmadi, M., Ievlev, A. V., Jesse, S., Retterer, S. T., Belianinov, A., Xiao, K., Huang, J., Sumpter, B. G., Kalinin, S. V., Hu, B. & Ovchinnikova, O. S. 2018. Chemical nature of ferroelastic twin domains in $CH_3NH_3PbI_3$ perovskite. *Nature Materials*, 17(11):1013–1019.

Livengood, R. H., Greenzweig, Y., Liang, T. & Grumski, M. 2007. Helium ion microscope invasiveness and imaging study for semiconductor applications. *Journal of Vacuum Science & Technology B: Microelectronics and Nanometer Structures Processing, Measurement, and Phenomena*, 25, 2547–2552.

Livengood, R. H., Tana, S., Greenzweig, Y., Notte, J. & Mcvey, S. 2009. Subsurface damage from helium ions as a function of dose, beam energy, and dose rate. *Journal of Vacuum Science & Technology B: Microelectronics and Nanometer Structures Processing, Measurement, and Phenomena*, 27, 3244–3249.

Mahady, K. T., Shida, T., Greenzweig, Y., Livengood, R., Raveh, A., Fowlkes, J. & Rack, P. D. 2017b. Monte Carlo simulations of secondary electron emission due to ion beam milling. *Journal of Vacuum Science & Technology B*, 35, 041805.

Mahady, K. T., Tan, S., Greenzweig, Y., Livengood, R., Raveh, A. & Rack, P. 2017a. Monte Carlo simulations of nanoscale Ne + ion beam sputtering: investigating the influence of surface effects, interstitial formation, and the nanostructural evolution. *Nanotechnology*, 28, 045305.

Marshall, M. M., Yang, J. & Hall, A. R. 2012. Direct and transmission milling of suspended silicon nitride membranes with a focused helium ion beam. *Scanning*, 34, 101–106.

Matsui, S., Ichihashi, T. & Mito, M. 1989. Electron beam induced selective etching and deposition technology. *Journal of Vacuum Science & Technology B: Microelectronics Processing and Phenomena*, 7, 1182–1190.

Melli, M., Polyakov, A., Gargas, D., Huynh, C., Scipioni, L., Bao, W., Ogletree, D. F., Schuck, P. J., Cabrini, S. & Weber-Bargioni, A. 2013. Reaching the theoretical resonance quality factor limit in coaxial plasmonic nanoresonators fabricated by helium ion lithography. *Nano Letters*, 13, 2687–2691.

Müller, E. W. & Bahadur, K. 1956. Field ionization of gases at a metal surface and the resolution of the field ion microscope. *Physical Review*, 102, 624–631.

Nakaharai, S., Iijima, T., Ogawa, S., Suzuki, S., Li, S.-L., Tsukagoshi, K., Sato, S. & Yokoyama, N. 2013. Conduction tuning of graphene based on defect-induced localization. *ACS Nano*, 7, 5694–5700.

Nanda, G., Goswami, S., Watanabe, K., Taniguchi, T. & Alkemade, P. F. 2015. Defect control and n-doping of encapsulated graphene by helium-ion-beam irradiation. *Nano Letters*, 15, 4006–4012.

Notte, J., Ward, B., Economou, N., Hill, R., Percival, R., Farkas, L. & Mcvey, S. 2007. An introduction to the helium ion microscope. *AIP Conference Proceedings*, 931, 489–496.

Ohya, K. 2003. Comparative study of target atomic number dependence of ion induced and electron induced secondary electron emission. *Nuclear Instruments and Methods in Physics Research Section B: Beam Interactions with Materials and Atoms*, 206, 52–56.

Ohya, K. & Ishitani, T. 2004. Monte Carlo simulation of topographic contrast in scanning ion microscope. *Journal of Electron Microscopy*, 53, 229–235.

Orloff, J., Swanson, L. & Utlaut, M. 2003. *High Resolution Focused Ion Beams: FIB and its Applications: Fib and Its Applications: The Physics of Liquid Metal Ion Sources and Ion Optics and Their Application to Focused Ion Beam Technology*. Springer Science & Business Media, Switzerland.

Pekin, T. C., Allen, F. I. & Minor, A. M. 2016. Evaluation of neon focused ion beam milling for TEM sample preparation. *Journal of Microscopy*, 264, 59–63.

Pillatsch, L., Vanhove, N., Dowsett, D., Sijbrandij, S., Notte, J. & Wirtz, T. 2013. Study and optimisation of SIMS performed with He^+ and Ne^+ bombardment. *Applied Surface Science*, 282, 908–913.

Ping, C., Emile Van, V., Colin, A. S., Huub, W. M. S., Diederik, J. M., Daryl, A. S., Philip, D. R. & Paul, F. A. A. 2010. Nanopillar growth by focused helium ion-beam-induced deposition. *Nanotechnology*, 21, 455302.

Rahman, F. F., Mcvey, S., Farkas, L., Notte, J. A., Tan, S. & Livengood, R. H. 2012. The prospects of a subnanometer focused neon ion beam. *Scanning*, 34, 129–134.

Rahman, F. F., Notte, J. A., Livengood, R. H. & Tan, S. 2013. Observation of synchronized atomic motions in the field ion microscope. *Ultramicroscopy*, 126, 10–18.

Ramachandra, R. 2009. *A Study of Helium Ion Induced Secondary Electron Production*. Doctor of Philosophy PhD Dissertation, University of Tennessee.

Ramachandra, R., Griffin, B. & Joy, D. 2009. A model of secondary electron imaging in the helium ion scanning microscope. *Ultramicroscopy*, 109, 748–757.

Röder, F., Hlawacek, G., Wintz, S., Hübner, R., Bischoff, L., Lichte, H., Potzger, K., Lindner, J., Fassbender, J. & Bali, R. 2015. Direct depth- and lateral- imaging of nanoscale magnets generated by ion impact. *Scientific Reports*, 5, 16786.

Ross, J. S., Klement, P., Jones, A. M., Ghimire, N. J., Yan, J., Mandrus, D., Taniguchi, T., Watanabe, K., Kitamura, K. & Yao, W. 2014. Electrically tunable excitonic light-emitting diodes based on monolayer WSe2 pn junctions. *Nature Nanotechnology*, 9, 268–272.

Rotkina, L., Lin, J.-F. & Bird, J. 2003. Nonlinear current-voltage characteristics of Pt nanowires and nanowire transistors fabricated by electron-beam deposition. *Applied Physics Letters*, 83, 4426–4428.

Rudneva, M., van Veldhoven, E., Malladi, S. K., Maas, D. & Zandbergen, H. W. 2013. Novel nanosample preparation with a helium ion microscope. *Journal of Materials Research*, 28, 1013–1020.

Sawafta, F., Carlsen, A. & Hall, A. 2014. Membrane thickness dependence of nanopore formation with a focused helium ion beam. *Sensors*, 14, 8150.

Schürmann, M., Frese, N., Beyer, A., Heimann, P., Widera, D., Mönkemöller, V., Huser, T., Kaltschmidt, B., Kaltschmidt, C. & Gölzhäuser, A. 2015. Helium ion microscopy visualizes lipid nanodomains in mammalian cells. *Small*, 11, 5781–5789.

Scipioni, L., Ferranti, D. C., Smentkowski, V. S. & Potyrailo, R. A. 2010. Fabrication and initial characterization of ultrahigh aspect ratio vias in gold using the helium ion microscope. *Journal of Vacuum Science & Technology B*, 28, C6P18–C6P23.

Sijbrandij, S., Notte, J., Scipioni, L., Huynh, C. & Sanford, C. 2010. Analysis and metrology with a focused helium ion beam. *Journal of Vacuum Science & Technology B*, 28, 73–77.

Sijbrandij, S., Thompson, B., Notte, J., Ward, B. W. & Economou, N. P. 2008. Elemental analysis with the helium ion microscope. *Journal of Vacuum Science & Technology B: Microelectronics and Nanometer Structures Processing, Measurement, and Phenomena*, 26, 2103–2106.

Silvis-Cividjian, N. & Hagen, C. W. 2006. Electron-beam–induced nanometer-scale deposition. *Advances in Imaging and Electron Physics*. Elsevier.

Smith, D. A., Joy, D. C. & Rack, P. D. 2010. Monte Carlo simulation of focused helium ion beam induced deposition. *Nanotechnology*, 21, 175302.

Stanford, M. G., Lewis, B. B., Iberi, V., Fowlkes, J. D., Tan, S., Livengood, R. & Rack, P. D. 2016a. In situ mitigation of subsurface and peripheral focused ion beam damage via simultaneous pulsed laser heating. *Small*, 12, 1779–1787.

Stanford, M. G., Lewis, B. B., Mahady, K., Fowlkes, J. D. & Rack, P. D. 2017a. Review Article: Advanced nanoscale patterning and material synthesis with gas field helium and neon ion beams. *Journal of Vacuum Science & Technology B*, 35, 030802.

Stanford, M. G., Mahady, K., Lewis, B. B., Fowlkes, J. D., Tan, S., Livengood, R., Magel, G. A., Moore, T. M. & Rack, P. D. 2016b. Laser-assisted focused He+ ion beam induced etching with and without XeF2 gas assist. *ACS Applied Materials & Interfaces*, 8, 29155–29162.

Stanford, M. G., Pudasaini, P. R., Belianinov, A., Cross, N., Noh, J. H., Koehler, M. R., Mandrus, D. G., Duscher, G., Rondinone, A. J. & Ivanov, I. N. 2016c. Focused helium-ion beam irradiation effects on electrical transport properties of few-layer WSe2: enabling nanoscale direct write homo-junctions. *Scientific Reports*, 6, 27276.

Stanford, M. G., Pudasaini, P. R., Cross, N., Mahady, K., Hoffman, A. N., Mandrus, D. G., Duscher, G., Chisholm, M. F. & Rack, P. D. 2017b. Tungsten diselenide patterning and nanoribbon formation by gas-assisted focused-helium-ion-beam-induced etching. *Small Methods*, 1, 1600060.

Stanford, M. G., Pudasaini, P. R., Gallmeier, E. T., Cross, N., Liang, L., Oyedele, A., Duscher, G., Mahjouri-Samani, M., Wang, K., Xiao, K., Geohegan, D. B., Belianinov, A., Sumpter, B. G. & Rack, P. D. 2017c. High conduction hopping behavior induced in transition metal dichalcogenides by percolating defect networks: Toward atomically thin circuits. *Advanced Functional Materials*, 27, 1702829.

Stanford, M. G., Pudasaini, P. R., Gallmeier, E. T., Cross, N., Liang, L., Oyedele, A., Duscher, G., Mahjouri-Samani, M., Wang, K., Xiao, K., Geohegan, D. B., Belianinov, A., Sumpter, B. G. & Rack, P. D. 2017d. High conduction hopping behavior induced in transition metal dichalcogenides by percolating defect networks: Toward atomically thin circuits. *Advanced Functional Materials*, 27, 1702829.

Sternglass, E. J. 1957. Theory of secondary electron emission by high-speed ions. *Physical Review*, 108, 1–12.

Tan, S., Klein, K., Shima, D., Livengood, R., Mutunga, E. & Vladár, A. 2014. Mechanism and applications of helium transmission milling in thin membranes. *Journal of Vacuum Science & Technology B*, 32, 06FA01.

Tan, S., Livengood, R. H., Hallstein, R., Shima, D., Greenzweig, Y., Notte, J. & Mcvey, S. 2011. *Neon Ion Microscope Nanomachining Consideration. ISTFA 2011*, San Jose, CA. ASM International, 40–45.

Tilley, R. J. D. 2008. *Defects in Solids*. Hoboken, NJ: John Wiley & Sons.

Timilsina, R. & Rack, P. D. 2013. Monte Carlo simulations of nanoscale focused neon ion beam sputtering. *Nanotechnology*, 24, 495303.

Timilsina, R., Smith, D. A. & Rack, P. D. 2013. A comparison of neon versus helium ion beam induced deposition via Monte Carlo simulations. *Nanotechnology*, 24, 115302.

Timilsina, R., Tan, S., Livengood, R. & Rack, P. 2014. Monte Carlo simulations of nanoscale focused neon ion beam sputtering of copper: elucidating resolution limits and sub-surface damage. *Nanotechnology*, 25, 485704.

Utke, I., Hoffmann, P. & Melngailis, J. 2008. Gas-assisted focused electron beam and ion beam processing and fabrication. *Journal of Vacuum Science & Technology B: Microelectronics and Nanometer Structures Processing, Measurement, and Phenomena*, 26, 1197–1276.

Vanden Berg-Foels, W. S., Scipioni, L., Huynh, C. & Wen, X. 2012. Helium ion microscopy for high-resolution visualization of the articular cartilage collagen network. *Journal of Microscopy*, 246, 168–176.

Veligura, V., Hlawacek, G., Berkelaar, R. P., van Gastel, R., Zandvliet, H. J. W. & Poelsema, B. 2013. Digging gold: keV He+ ion interaction with Au. *Beilstein Journal of Nanotechnology*, 4, 453–460.

Vighter, I., Anton, V. I., Ivan, V., Stephen, J., Sergei, V. K., David, C. J., Adam, J. R., Alex, B. & Olga, S. O. 2016. Graphene engineering by neon ion beams. *Nanotechnology*, 27, 125302.

von Ardenne, M. 1938. Das Elektronen-Rastermikroskop. *Zeitschrift für Physik (in German)*, 109, 553–572.

Wang, Y., Abb, M., Boden, S. A., Aizpurua, J., de Groot, C. H. & Muskens, O. L. 2013. Ultrafast nonlinear control of progressively loaded, single plasmonic nanoantennas fabricated using helium ion milling. *Nano Letters*, 13, 5647–5653.

Ward, B. W., Notte, J. A. & Economou, N. P. 2006. Helium ion microscope: A new tool for nanoscale microscopy and metrology. *Journal of Vacuum Science & Technology B: Microelectronics and Nanometer Structures Processing, Measurement, and Phenomena*, 24, 2871–2874.

Wei, D., Huynh, C. & Ribbe, A. 2015. Focused Ne+ ion beams for final polishing of TEM lamella prepared through Ga-FIB systems. *Microscopy and Microanalysis*, 21, 1409–1410.

Wight, S., Meier, D., Postek, M., Vladar, A., Small, J. & Newbury, D. 2008. He ion induced secondary electron and backscattered electron images compared side-by-side with electron beam induced secondary electron, backscattered, and transmission electron images. *Microscopy and Microanalysis*, 14, 1186–1187.

Winston, D., Cord, B. M., Ming, B., Bell, D., Dinatale, W., Stern, L., Vladar, A., Postek, M., Mondol, M. & Yang, J. 2009. Scanning-helium-ion-beam lithography with hydrogen silsesquioxane resist. *Journal of Vacuum Science & Technology B: Microelectronics and Nanometer Structures Processing, Measurement, and Phenomena*, 27, 2702–2706.

Wirtz, T., Dowsett, D. & Philipp, P. 2016. SIMS on the helium ion microscope: a powerful tool for high-resolution high-sensitivity nano-analytics. *Helium Ion Microscopy*. Springer International Publishing, Switzerland.

Wirtz, T., Philipp, P., Audinot, J. N., Dowsett, D. & Eswara, S. 2015. High-resolution high-sensitivity elemental imaging by secondary ion mass spectrometry: from traditional 2D and 3D imaging to correlative microscopy. *Nanotechnology*, 26, 434001.

Wirtz, T., Vanhove, N., Pillatsch, L., Dowsett, D., Sijbrandij, S. & Notte, J. 2012. Towards secondary ion mass spectrometry on the helium ion microscope: An experimental and simulation based feasibility study with He+ and Ne+ bombardment. *Applied Physics Letters*, 101, 041601.

Wu, H., Stern, L., Chen, J., Huth, M., Schwalb, C., Winhold, M., Porrati, F., Gonzalez, C., Timilsina, R. & Rack, P. 2013. Synthesis of nanowires via helium and neon focused ion beam induced deposition with the gas field ion microscope. *Nanotechnology*, 24, 175302.

Wu, H., Stern, L. A., Xia, D., Ferranti, D., Thompson, B., Klein, K. L., Gonzalez, C. M. & Rack, P. D. 2014. Focused helium ion beam deposited low resistivity cobalt metal lines with 10 nm resolution: implications for advanced circuit editing. *Journal of Materials Science: Materials in Electronics*, 25, 587–595.

Wu, W., Tong, W. M., Bartman, J., Chen, Y., Walmsley, R., Yu, Z., Xia, Q., Park, I., Picciotto, C., Gao, J., Wang, S.-Y., Morecroft, D., Yang, J., Berggren, K. K. & Williams, R. S. 2008. Sub-10 nm nanoimprint lithography by wafer bowing. *Nano Letters*, 8, 3865–3869.

Yi, Y., Wu, C., Liu, H., Zeng, J., He, H. & Wang, J. 2015. A study of lateral Schottky contacts in WSe 2 and MoS 2 field effect transistors using scanning photocurrent microscopy. *Nanoscale*, 7, 15711–15718.

Yoon, K., Rahnamoun, A., Swett, J. L., Iberi, V., Cullen, D. A., Vlassiouk, I. V., Belianinov, A., Jesse, S., Sang, X. & Ovchinnikova, O. S. 2016. Atomistic-scale simulations of defect formation in graphene under noble gas ion irradiation. *ACS Nano*, 10, 8376–8384.

Yudong, W., Boden, S. A., Bagnall, D. M., Rutt, H. N. & Groot, C. H. D. 2012. Helium ion beam milling to create a nano-structured domain wall magnetoresistance spin valve. *Nanotechnology*, 23, 395302.

Zhao, Y., Liu, D., Chen, J., Zhu, L., Belianinov, A., Ovchinnikova, O. S., Unocic, R. R., Burch, M. J., Kim, S., Hao, H., Pickard, D. S., Li, B. & Thong, J. T. L. 2017. Engineering the thermal conductivity along an individual silicon nanowire by selective helium ion irradiation. *Nature Communications*, 8, 15919.

Ziegler, J. F. 1988. The stopping and range of ions in solids. *Ion Implantation Science and Technology (Second Edition)*. 3–61, Academic Press, San Diego, CA.

Ziegler, J. F. 2013. Srim [Online]. Available: www.srim.org.

Ziegler, J. F., Ziegler, M. D. & Biersack, J. P. 2010. SRIM—The stopping and range of ions in matter (2010). *Nuclear Instruments and Methods in Physics Research Section B: Beam Interactions with Materials and Atoms*, 268, 1818–1823.

Nanofiber Characterization by Raman Scanning Microscopy

Alexander Khmaladze
SUNY University at Albany

10.1 Introduction

10.1.1 Raman Spectroscopy

Raman effect was predicted by Adolf Smekal in 1923 (and it is called Smekal-Raman effect in Germany). Grigory Landsberg and Leonid Mandelstam discovered the effect of the inelastic combinatorial scattering of light in crystals on 21 February, 1928, which was called "combinatorial scattering" due to the combination of frequencies of photons and molecular vibrations (Landsberg and Mandelstam 1928). They presented this fundamental discovery for the first time at a colloquium on 27 April, 1928 (which is why it was called "combinatorial scattering" in the Soviet Union). In the same year, C.V. Raman and K.S. Krishnan were looking for "Compton component" of scattered light in liquids and vapors. They found the same combinatorial scattering of light. Raman stated that "The line spectrum of the new radiation was first seen on 28 February 1928." However, because he published his results first (Raman 1928), it was Raman who received the Nobel Prize for this discovery in 1930.

In 1998, Raman effect was designated a National Historic Chemical Landmark by the American Chemical Society, in recognition of its significance as a tool for analyzing the composition of liquids, gases, and solids. The amount of time that passed between these two events is not surprising. In order for Raman spectroscopy to be a truly useful tool for chemical probing of samples, two inventions were needed: (a) a laser, high-intensity, low-spectral bandwidth light source and (b) a very sensitive light detector for capturing the few photons generated by Raman effect.

At the core of Raman effect is inelastic scattering of light. Generally, when a quantum system (a molecule in a crystal, liquid, etc.) is "pumped" by a light source to a "virtual" state, it then almost immediately relaxes back, usually to the ground state, and emits the photon of the same energy as the pumping light. Therefore, if one were to shine a laser pointer on the wall, the color of the spot would appear the same as the color of the laser light. This effect is referred to as elastic light scattering, or Rayleigh scattering. However, a small fraction of these transitions ends up in one of the higher vibrational states in the system (see Figure 10.1). By measuring the frequency difference between the elastically and inelastically scattered photons, one can deduce the energy gap between the vibrational states within the system and, therefore, characterize the system to determine the chemical composition of the sample. The Raman spectrum provides a "fingerprint" representing the set of bonds that exist in the material: the vibrational frequencies are characteristic of chemical bonds or groups of bonds. Relative intensities correspond principally to the species concentration (but can also be related to the orientation of the material or molecule with respect to the incoming laser polarization).

One of the important advantages of this technique is in the fact that when this method is applied to the chemical imaging of an unknown sample, the chemical map created

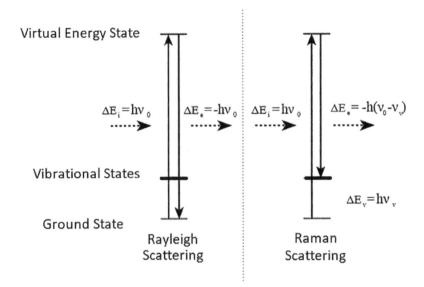

Virtual Energy State

$\Delta E_i = h\nu_0$ $\Delta E_e = -h\nu_0$ $\Delta E_i = h\nu_0$ $\Delta E_e = -h(\nu_0 - \nu_v)$

Vibrational States

$\Delta E_v = h\nu_v$

Ground State

Rayleigh Scattering   Raman Scattering

FIGURE 10.1   Rayleigh and Raman scattering.

by this method is diffraction limited by the pump wavelength, generally in the (near)-visible part of the spectrum (under 1 µm), which is much smaller than the wavelength of the infrared transitions between various vibrational states (typically tens of microns).

10.1.2   Scanning Microscopy

Laser scanning microscopy uses a laser to raster scan a sample and detect the intensity of the scattered light. A schematic of a laser scanning microscope is shown in Figure 10.2. The laser light is focused on a sample by the microscope objective, which controls the size of the laser spot. To obtain the scanned image of the sample, either the stage moves in X and Y direction or the beam is moved over the sample by using X and Y scanning mirrors. In both cases, a computer keeps track of where on the sample the beam is pointed, and the non-imaging detector (e.g. photodiode, photo-multiplier tube, or an avalanche photo diode) captures the scattered signal from each point of the sample. Therefore, the computer detects the scattered light intensity for each pixel, which creates a three-dimensional map of the sample. Invented by Marvin Minsky, confocal scanning

microscopes are now well known for clear image quality and 3D mapping capabilities.

While this method of image formation is different from a more traditional full-field imaging technique, where the system of lenses is used to form an image of the sample on the surface of a CCD chip, it is subject to principally the same limitations as the latter. Namely, it is diffraction limited by the wavelength of light used for scanning, since it is generally impossible to focus the beam to a smaller spot size, and the size of the laser spot is what determines the resolution of the scanning microscope. In some instances, however, it is possible to oversample the scanning and deduce some information about the object under study.

When the scanning scheme is coupled with Raman spectroscopic detection, the spectrometer is used as a non-imaging detector, and the Raman signal is spectrally dispersed and projected on the detector, which is either a matrix or a line detector. Each pixel of this detector then corresponds to a specific wavelength, determined by calibration with a known light source. The entire set of Raman wavelengths can, in principle, be detected at the same time. The only limitations are the size of the detection array and the sensitivity of the detector to a specific wavelength.

10.1.3   Core/Shell Nanofibers

Nanofibers are produced from either synthetic or natural materials providing environmental and physical signals to support the growth and development of tissues (Liu 2013). Synthetic polymeric fiber scaffolds deliver a more controllable environment than natural materials, as they provide mechanical and chemical properties needed for a specific application. One of the commonly used methods for producing micro- and nanofibrous scaffolds is electrospinning (Bhardwaj and Kundu 2010; Huang et al. 2003). Electrospinning involves dissolving a polymeric material in

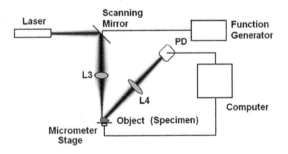

FIGURE 10.2   Basic components of a laser scanning microscope. L3 is a beam focusing lens and lens L4 is lens used to collect the light scattered from the sample.

a solvent and subjecting it to an electric field, which induces a charge repulsion opposing the liquid droplets' surface tension (Haider et al. 2015). Once the forces of surface tension of the liquid droplets are overcome, the polymer-solvent solution is ejected from the cone to the collector plate, where fibers are shaped in a nonwoven mesh (Doshi and Reneker 1993).

There are many forms of electrospinning, which include single-fluid and dual-fluid electrospinning. Dual-fluid electrospinning uses a co-axial spinneret and two dissimilar polymer solutions, drawn independently through a capillary, which then may produce a core/shell configuration nanofiber, containing one material (core) encapsulated within the other (shell) (Elahi et al. 2013).

Core/shell nanofibers offer advantages for several biological applications, such as drug delivery (Chourpa et al. 2005), tissue repair (Zhu et al. 2016), and tissue engineering (Ramakrishna et al. 2006), where it can produce biocompatible and biodegradable fibers with specific hydrophilicity and mechanical properties (Chen et al. 2010; Zhang et al. 2004). However, dual-fluid electrospinning needed for core/shell fiber synthesis is more complex than single-fluid electrospinning, as it is particularly sensitive to multiple environmental factors, such as humidity, temperature, solvent interactions, and intermixing, which can lead to blending of core/shell fiber materials (Wang et al. 2013). As a result, a reliable method for scaffold material characterization is needed.

Previously, characterization of the core/shell nanofiber structures has been accomplished by means of transmission electron microscopy (TEM) (Chen et al. 2010; Zhang et al. 2004, 2005b, 2006; Wang et al. 2013; Pakravan et al. 2012; Nguyen et al. 2011), scanning electron microscopy (SEM) (Xu et al. 2013; Wang et al. 2010), and atomic force microscopy (AFM) (Chen et al. 2010). TEM characterization method is based on utilizing different densities of core and shell materials, which results in each material transmitting different amounts of electrons, in turn leading to darker or lighter appearance of the fibers (Nguyen et al. 2011). SEM characterization of poly(glycerol sebacate) (PGS)/poly(L-lactic acid) (PLLA) core/shell nanofibers was performed by micro-sectioning in liquid nitrogen using a cryogenic microtome (Xu et al. 2013). Finally, AFM was employed to detect the difference in surface roughness of the core/shell fibers, compared to a nanofiber blend of the two materials (Chen et al. 2010).

Although these methods of fiber characterization are well established, they do not reveal the actual spatial distribution of the chemical content within the nanofiber scaffolds. Knowing the ratios of each material located in the core and shell can provide an understanding of morphological changes the fibers undergo with varying electrospinning parameters.

One characterization technique that can deliver such information is confocal Raman microscopy discussed earlier. Among other things, it eliminates the need for sample processing prior to imaging (Minsky 1988; Lutz et al. 2015).

It has been used for a number of different imaging research applications, from carbon nanotubes (CNTs) (Hennrich et al. 2005) to label-free live cell imaging (Caspers et al. 2003; Klein et al. 2012). Here, the discussion is focused on the characterization of PGS/polylactic-co-glycolic acid (PLGA) core/shell nanofibers using Raman spectroscopic mapping to analyze their core/shell chemical structure and morphology (Sfakis et al. 2017).

10.1.4 Nanofibers as a Growth Factor Delivery

In the field of tissue regeneration, implantable fiber scaffolds are used to direct cell organization and growth *in vivo* and to supplement tissues with growth factors, proteins, vitamins, peptides, etc. (Ozdemir et al. 2016). Such scaffolds have an advantage of delivering the molecules of interest specifically to the target tissue (Lee et al. 2010). This greatly reduces the amount of molecules metabolized by other organs, decreases the doses required to induce the desired action, and leads to fewer side effects. This is especially important for epithelial tissues, because they are avascular and, therefore, mainly sustained through the molecules that diffuse out of the extracellular matrix (ECM) (Ozdemir et al. 2016). Electrospun fibers, designed to mimic the topography of the ECM, are some of the best structures for delivery of drugs and growth factors, when compared to other carrier systems, such as nanoparticles, liposomes, and dendrimers. They have high loading efficiency due to their large surface area. They are also very straightforward to synthesize, and their properties can be fine-tuned in many ways to tailor their release profile (Valizadeh and Farkhani 2014; Zhang et al. 2005a). Previously, optimized elastin/PLGA hybrid nanofiber scaffolds were used for slow delivery of growth factors, using the epidermal growth factor (EGF) as a model. EGF is known to stimulate branching morphogenesis of embryonic mouse submandibular salivary glands (Häärä et al. 2009; Dhulekar et al. 2016).

The use of double-fluid electrospinning, aimed at enclosing the growth factor, or other molecules of interest, within emulsion droplets (inner phase), which are distributed within a continuous phase, and which in turn is further emulsified within another immiscible layer containing the bulk polymers of the nanofibers, has been shown (Valizadeh and Farkhani 2014; Zhang et al. 2005a). With this method, EGF is encased within the inner phase of a water-in-oil-in-water emulsion (w/o/w), while elastin and PLGA are present in the outer phase (Foraida et al. 2017).

10.2 Sample Preparation and Characterization Methods

10.2.1 PGS/PLGA Nanofiber Preparation

PGS/PLGA nanofibers were made using a core/shell coaxial spinneret. PGS pre-polymer was dissolved in hexafluoroisopropanol (HFIP) (16% w/w). The PLGA solution consisted

of 1% NaCl, 10 μL SRB dye, and 85:15 PLGA dissolved
in HFIP, making an 8% w/w solution (Soscia et al. 2013;
Cantara et al. 2012; Sequeira et al. 2012; Sfakis et al. 2016a).
Two independent syringe pumps were used with the two
polymeric solutions, connected to the co-axial spinneret.
The core and shell solution flow rates were varied to obtain
either a homogenous fiber mat or a non-homogenous fiber
mat containing PGS-rich beads. The non-homogenous fiber
mat was obtained with flow rates of 9 and 1.5 μL/min for
the PGS and PLGA solution, respectively (Figure 10.3a). A
homogenous PGS/PLGA fiber mat was obtained with flow
rates of 1.5 and 1.5 μL/min for the PGS and PLGA solution, respectively (Figure 10.3b). The two solutions did not
come in contact until they met at the end of the needle
tip, where a 12 kV voltage was applied. For Raman analysis, samples were spun for 5 s on a glass microscope slide
wrapped in aluminum foil to create low-density monolayer
of fibers.

Figure 10.4 shows the diagram of the electrospinning
apparatus and the expected internal structure of the
PGS/PLGA fiber. Fiber diameters were obtained by averaging 100 different fiber diameters from several SEM
images.

10.2.2 PLGA Nanofiber Parameters

Using a single-fluid electrospinning setup, 8% PLGA, 10 μL
SRB dye, and 1% NaCl (w/w) in HFIP were placed in a
syringe pump at a flow rate of 3 μL/min (Figure 10.3c).
The voltage and the distance of the needle from the collector
plate were 10 kV and 15 m, respectively. All samples were
electrospun on glass microscope slides wrapped in aluminum
foil and spun on for 5 s to create a low-density monolayer
of fibers. Fiber diameters were obtained by averaging 100
different fiber diameters from several SEM images.

10.2.3 Epidermal Growth Factor

EGF from murine submaxillary gland (Cat. No. 62229-50-9)
was purchased from Sigma Aldrich (St. Louis, MO).
Soluble (hydrolyzed) elastin (Cat. No. ES12) was purchased
from Elastin Products (Owensville, MO), poly(D-lactide-co-
glycolide) (PLGA), (Cat. No. B6006-1) with a molecular
weight of 95,000 Da, and a lactic to glycolic acid ratio of
85:15 was purchased from Birmingham Polymers (Pelham,
AL). HFIP (Cat. No. 105228) and ethyl cellulose (EC) (Cat.
No. 45-247499) were purchased from Sigma Aldrich (St.
Louis, MO).

FIGURE 10.3 SEM images of fibers used for this study. PGS/PLGA fiber mat with varying flow rates of (a) 9 μL/min/1.5 μL/min
and (b) 1.5 μL/min/1.5 μL/min, respectively. (c) SEM image of PLGA nanofiber mat. Scale, 2 μm. Average fiber diameters for Figure
10.1a, b and c were 366 ± 150 nm, 245 ± 60 nm and 166 ± 37 nm, respectively (Sfakis et al. 2017).

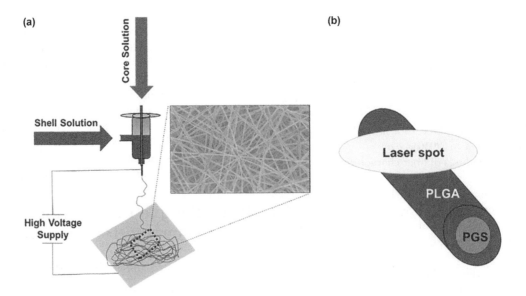

FIGURE 10.4 A diagram of the electrospinning apparatus with SEM image of the fiber mat (a), and the expected internal structure
of the fibers, with laser spot from Raman scanning microscope shown to scale (b) (Sfakis et al. 2017).

10.2.4 Preparation of Elastin–PLGA Nanofibers

Equal weights of 12% w/w soluble elastin in HFIP (hydrolyzed elastin from bovine neck ligament) and of 12% w/w PLGA in HFIP were mixed together to yield a solution containing 4% elastin and 4% PLGA (w/w). Sodium chloride (1% w/w) was added to the solution before electrospinning to minimize beading. The solution was pumped through a 3-mL syringe, which was connected to an automated microsyringe pump through PTTE tubing to a (25 G) needle. The needle tip and the aluminum foil-coated collector ground were both connected to a high voltage source (14 kV). The voltage was applied to the needle tip and the collector ground while pumping the blend polymer solution at a rate of 6 μL/min and a distance between the needle tip and the collector ground of 15 cm. The nanofibers were formed and collected on the collector ground. All samples were electrospun on glass microscope slides wrapped in aluminum foil and spun on for 5 s to create a low-density monolayer of fibers. Fiber diameters were obtained by averaging 100 different fiber diameters from several SEM images.

10.2.5 Growth Factor Loading by Blend Electrospinning

Blend electrospinning of growth factors and drugs is a conventional method where the loaded molecule is homogeneously distributed within the electrospinning solution and electrospun as one entity. Briefly, 100 μg of EGF from murine submaxillary gland (EGF) was reconstituted in 0.2 mL of 1% BSA in PBS to obtain a final concentration of 0.5 μg/μL of EGF. A volume of 30 μL of EGF/BSA solution therefore contains 15 μg of EGF, which was added to a mixture solution of 0.75 g 12% elastin in HFIP and 0.75 g PLGA in HFIP to yield a final concentration of 4% elastin/4% PLGA and 10 ng/mg of EGF. Sodium chloride (1% w/w) was added to the solution before electrospinning to minimize beading. All samples were electrospun on glass microscope slides wrapped in aluminum foil and spun on for 5 s to create a low-density monolayer of fibers. Fiber diameters were obtained by averaging 100 different fiber diameters from several SEM images. The obtained nanofibers were labeled (Blend).

10.2.6 Growth Factor Loading by Double Emulsion Electrospinning

A 30 μL of (0.5 μg/μL) solution of EGF in 1% BSA solution was added dropwise to 0.5 g of chloroform solution containing 2% (w/w) EC and mixed by vortexing for 5 min to yield a creamy emulsion. This emulsion was introduced dropwise to a mixture solution of 0.5 g of 12% elastin/HFIP solution and 0.5 g of 12% PLGA solution. The emulsion was further mixed by vortexing for 5 min. The final concentrations are 10 ng EGF/1 mg emulsion, 4% elastin (w/w), and 4% PLGA (w/w). All samples were electrospun on glass microscope slides wrapped in aluminum foil and spun on for

5 s to create a low-density monolayer of fibers. Fiber diameters were obtained by averaging 100 different fiber diameters from several SEM images. The nanofibers obtained were labeled (Emulsion).

10.2.7 SEM Characterization

SEM imaging of the scaffolds was carried out using a Zeiss 1550 field emission SEM (Leo Electron Microscopy Ltd., Cambridge, UK; Carl Zeiss, Jena, Germany), as described previously (Soscia et al. 2013). Briefly, the scaffolds were mounted on 1 cm^2 stubs and coated with approximately 5 nm of gold–palladium to minimize sample charging. Images were captured using an in-lens detector, 1–5 kV acceleration voltage, and a working distance of 2–6 mm. The microscope annotation software was used to apply scale bars, and ImageJ software was used to measure the fiber diameters from calibrated images. At least 4 scaffolds of each type were imaged and 100 nanofibers analyzed for average fiber diameter measurements.

Figure 10.5 shows representative SEM images of each EGF-loaded nanofiber type as compared to control nanofibers. As shown, the EGF-loaded nanofibers prepared by blend electrospinning (Figure 10.5b) double

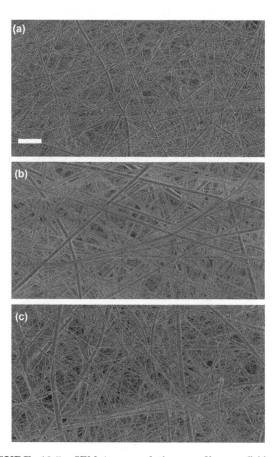

FIGURE 10.5 SEM images of the nanofiber scaffolds. (a) Unloaded nanofibers (control), (b) blend electrospun EGF (Blend), (c) emulsion electrospun EGF (Emulsion). Scale bar, 2 μm (Foraida et al. 2017).

emulsion electrospinning (Figure 10.5c) are predominantly homogeneous with an average nanofiber diameter around 370 nm. This diameter is larger than the control nanofibers with an average diameter of 308 nm (Figure 10.5a). The increase in the nanofiber diameter may be attributed to the changes in the solution characteristics upon adding EGF.

10.2.8   Confocal Raman Spectroscopy Characterization of Nanofibers

Raman spectra of PGS and PLGA polymers were measured using LabRAM HR Evolution confocal scanning microscope (HORIBA) with Synapse detector. The excitation wavelength was 473 nm. The diffraction grating of 300 gr/mm, together with 50× microscope objective and a confocal pinhole of 100 μm was used. Acquisition time was 5 s, and each spectral measurement consisted of three accumulations, which were averaged, and cosmic ray spikes were removed.

PLGA nanofiber was used to obtain a pure PLGA spectrum. To obtain a pure PGS spectrum, a PGS polymer film was used, since PGS alone does not form a fiber structure.

Further, Raman spectra of EGF, EC, and PLGA were measured using XploRA Plus confocal scanning microscope (HORIBA). The excitation wavelength was 532 nm. The diffraction grating of 600 gr/mm, together with 200 μm spectrometer slit, 40× microscope objective, and a confocal aperture of 100 μm were used. Acquisition time was 120 s times three accumulations for all measurements. Spike filter based on multiple accumulations was engaged. To obtain the spectra of individual fiber components, pure droplets of each component on aluminum foil were used.

The spatial distribution measurement (confocal Raman mapping) was performed using the same Raman microscope systems that were used for the collection of point spectra. The system settings were the same, except that the laser power was reduced by 50% to avoid burning the fibers. The confocal scanning was done either across or along the fiber. All spectra were subjected to the polynomial baseline

correction routine in LabSpec software (HORIBA). After the baseline correction, the spectral maps were saved as three-dimensional tables, where the X and Y coordinates correspond to the location of a particular pixel on the sample, and Z coordinate contains the spectral information (Raman intensity as a function of Raman wavelength shift). This process yielded the so-called "hyperspectral" Raman image.

Singular Value Decomposition (SVD) analysis was also applied to identify different chemical regions in the image and, thus, observe the organization of the fiber components. As previously described (Khmaladze et al. 2014), SVD-based fiber mapping was done using Map Analyzer, an in-home SVD code written in LabVIEW.

10.3   Nanofiber Characterization by Raman Scanning Microscopy

10.3.1   Raman Spectra of PLGA Nanofibers and PGS Films

Prior to Raman imaging of the core/shell nanofibers, PLGA single fibers and PGS films were initially used to establish differences in their spectral features. The 473 nm excitation resulted in a strong Raman signal from both types of polymer. Of particular advantage was the availability of the CH part of the spectrum (around 3,000 cm^{-1}), where both PGS and PLGA components had distinct and prominent spectral features, as can be seen in Figure 10.6. The PGS main peak was at 2,911 cm^{-1}, while PLGA peak was at 2,947 cm^{-1}. Both polymers had additional distinctive features in the 700–1,700 cm^{-1} range but at a much lower intensity (Figure 10.6).

10.3.2   Mapping of PLGA and PGS/PLGA Nanofibers

After the differences between the two polymeric components within a fiber were identified, Raman spectra of PGS/PLGA

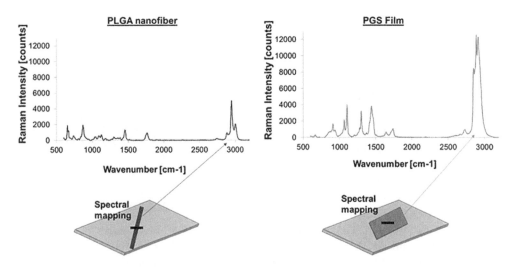

FIGURE 10.6   XY Raman mapping of PLGA nanofiber and PGS film (Sfakis et al. 2017).

nanofibers were acquired. Fiber inhomogeneities (likely due to unstable electrospinning parameters, producing PGS-rich droplets throughout the fiber strands) were observed when electrospinning with higher core flow rates. While these fibers were not the kind of fibers used for the tissue engineering applications, spectral analysis of these artifacts was very informative, because it demonstrated the ability of this technique to spectrally and spatially resolve the two different compounds at the submicron scale (Figure 10.7).

Figure 10.7a shows spectra of PLGA nanofiber and PGS film. Further, Figure 10.7b shows a zoomed in view of the distinct polymer peaks. A 50x optical image of PGS/PLGA nanofibers with a PGS-rich droplet in the middle is shown in Figure 10.7c. Finally, Figure 10.7d shows the spectrum from the center of the droplet, demonstrating that both polymers are present. This is to be expected, since the Raman signal comes from both the inner (PGS) and the outer (PLGA) parts of the fiber. It must be noted that the strength of Raman signal is a function of both the amount of material and its Raman cross section. Thus, the intensity of Raman peak cannot be taken as a direct measure of the amount of a particular component, i.e. a stronger PLGA signal in

Figure 10.7b does not necessarily indicate that there is more PLGA material than PGS material in the excitation volume.

PLGA-only nanofibers were produced to support characterization of PGS/PLGA nanofibers. They, too, were Raman mapped, both along and across a fiber (see Figure 10.8). The classical least squares (CLS) linear regression routine, which decomposed the spectra into pure PGS and PLGA base components, was applied to Raman mapping. As expected, for the spectral mapping taken along a single PLGA nanofiber, CLS plot shows no trace of the PGS spectra (Figure 10.8c). The small increase in the component contribution of PLGA is due to the scan slowly getting closer to the center of the fiber (which is slightly curved and not exactly aligned along the scan axis, as can be seen in Figure 10.8b). Figures 8d–e show the spectral mapping across a single PLGA nanofiber. Again, as expected, no significant traces of the PGS component were observed while scanning across the nanofiber (Figure 10.8f).

To confirm the core/shell structure of the PGS/PLGA nanofibers, spectroscopic Raman mapping across a typical fiber was performed on multiple samples. As in the case of a droplet, it was expected that the signal from the fiber

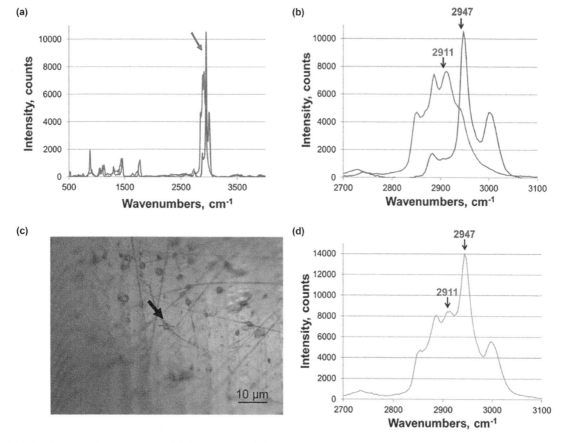

FIGURE 10.7 Raman reference spectra: (a) Spectra of PLGA nanofiber and PGS film. Arrow indicates the greatest difference between the polymer spectra. (b) Zoomed in view of distinct polymer peaks. Chosen peaks for PLGA and PGS were 2,947 cm^{-1} and 2,911 cm^{-1}, respectively; (c) 50× optical image of PGS/PLGA nanofibers with bubble in the middle. The line shows where Raman imaging was performed (indicated with an arrow). Scale, 10 μm. (d). Spectra obtained from core/shell PLGA/PGS nanofiber bubble structure, demonstrating that both polymers are present in the "bubble" spectrum, with peaks characteristic of PLGA and PGS indicated with arrows (Sfakis et al. 2017).

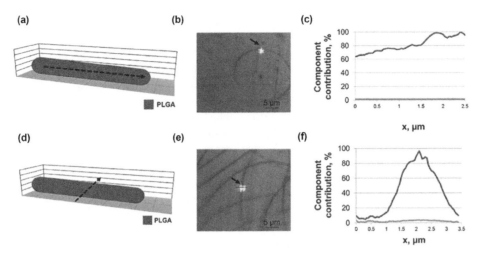

FIGURE 10.8   Raman spectra from PLGA nanofiber. (a) Schematic of PLGA fiber cross section. Raman mapping was performed along the dashed arrow. (b) Optical image showing sample surface mapping recorded parallel to the fiber (indicated with an arrow). Image taken with 50x objective. The dark gray line indicates the location of Raman scan. Scale, 5 μm. (c) Lateral profile confocal Raman scan showing polymer component contribution through the nanofiber. PLGA (dark gray) and PGS (light gray). (d) Schematic of PLGA fiber cross section. Dashed arrow indicates where Raman mapping was performed. (e) Optical image showing sample surface mapping recorded perpendicular to the fiber (indicated with an arrow). Image taken with 50x objective. The dark gray line indicates the location of Raman scan. Scale, 5 μm. (f) Confocal Raman scan across the fiber showing polymer component contribution through the nanofiber. PLGA (dark gray) and PGS (light gray) (Sfakis et al. 2017).

center would be a mixture of PGS and PLGA spectral signatures, while at the fiber edges, only the shell material PLGA would be present (see Figure 10.9). Figure 10.9a shows a schematic of the Raman mapping occurring across the fiber, and Figure 10.9b is the conventional microscope image of the PGS/PLGA fiber scan area, where the mapped line is shown in gray. Figure 10.9c is the result of CLS fitting of the scan spectra, indicating the distribution of the base components, PGS and PLGA, along the fiber cross section. The peak of PGS component in the CLS plot shows that there is more PGS in the center of the fiber than at the edges; therefore, its relative contribution is higher. The non-zero PGS contributions at the edges of the scan are due to the large size of the laser beam spot, compared to the fiber thickness.

Even when the beam is positioned at the fiber edge, some part of the excitation volume still contains core material.

Further, a scan along the PGS/PLGA nanofiber was performed (Figure 10.10). The CLS analysis of the scan (Figure 10.10c) shows no significant variation in the distribution of the base components, PLGA and PGS, indicating that this fiber is uniform in the axial direction.

10.3.3   SVD Analysis

SVD analysis was also applied to the line scans across the fiber, in order to confirm its chemical composition. SVD is a well-known method of partitioning the spectra from a hyperspectral data set into distinct groups, based on the major

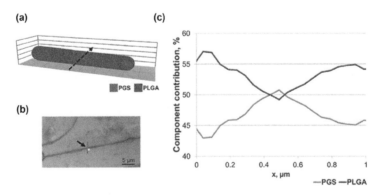

FIGURE 10.9   Raman spectra from core/shell nanofiber cross section. (a) Schematic of fiber cross section. Raman mapping was done perpendicular to nanofiber (indicated with a dashed arrow). (b) Optical image showing sample surface mapping recorded perpendicular to fiber (indicated with an arrow). Image taken with 50x objective. Scale, 5 μm. (c) Confocal Raman scan across the fiber showing polymer component contribution through a nanofiber (only the part of the scan crossing the fiber is shown; the sum of PLGA and PGS components is normalized to 100%). PLGA (dark gray) and PGS (light gray) (Sfakis et al. 2017).

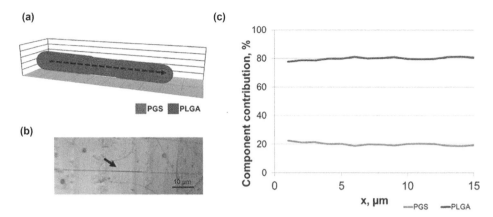

FIGURE 10.10 Raman spectra from core/shell nanofiber lateral section. (a) Schematic of fiber cross section. Raman mapping was parallel to nanofiber (indicated with a dashed arrow). (b) Optical image showing sample surface mapping parallel to a fiber (indicated with an arrow). Image taken with 50x objective. Scale, 10 μm. (c) Lateral profile confocal Raman scan showing polymer component contribution through a nanofiber (the sum of PLGA and PGS components is normalized to 100%). PLGA (dark gray) and PGS (light gray) (Sfakis et al. 2017).

contributing spectral line shapes, known as SVD components (Khmaladze et al. 2014; Sfakis et al. 2016b). The method works as follows: each spectrum collected from a line scan across the fiber is plotted as a single point on an SVD scatter plot, showing the magnitude of one SVD component versus another. This two-dimensional plot is a projection of multidimensional space on the basis of the two leading SVD components. Since the fiber contains two polymers, the two leading SVD components usually resemble the actual spectra of PLGA or PGS. The position of each point on the SVD scatter plot depends on whether that spectrum contains contributions from PLGA, PGS, or both. Therefore, the points group together depending on their chemical composition, making it possible to classify image regions based on their chemical composition. Once the pixels have been assigned to groups, the processing algorithm pseudocolors the original image accordingly, to create a chemical map of the sample (Figure 10.11).

Figure 10.11a shows the SVD scatter plot obtained from a core/shell fiber, where three distinct groups of spectra, which correspond to the core, shell, and image background, are visible. By examining the shapes of individual components, it was found that SVD components 1 and 2 closely resembled PLGA and PGS spectra. As expected, the signal from the center of the fiber included contributions from both the core and the shell (which lies above and below the core). Consequently, Figure 10.11a shows the contribution from both SVD components for the core area. The signal from the edges of the fiber shows contributions from PLGA only, and the signal from the points off the fiber has very small contributions of either SVD component. It is worth noting that since this method automatically extracts the line shapes of SVD components, it does not require the knowledge of PLGA and PGS spectra. Also, as it is frequently the case with SVD analysis, components 1 and 2 do not mimic the PLGA and PGS spectra exactly, which explains the slight angle between the axis of the plot and PLGA- and PGS-based spectral points in Figure 10.11a.

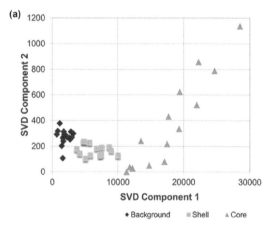

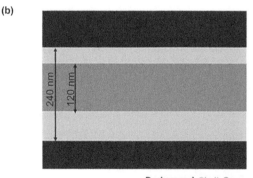

FIGURE 10.11 Raman spectroscopic imaging of a nanofiber cross section by SVD: (a) SVD scatter plot, showing clear separation of core and shell spectra; (b) 1-D hyperspectral Raman image of the structure of the fiber based on the SVD analysis. Dark gray indicates core material, light gray indicates shell material, and black indicates no chemical signature of either polymer (Sfakis et al. 2017).

Once the spectra have been classified, the assigned groups were mapped back onto the one-dimensional scan line (Figure 10.11b) to identify regions of different chemical composition within the sample.

Using this method, the thickness of PLGA shell/PGS core can be estimated. As mentioned before, it is impossible to resolve the features of the fiber directly due to the diffraction limit of the scanning laser spot size. As the laser spot is scanned across the fiber, the signal from the shell region will be detectable when the laser spot is only partially overlapping with the fiber, i.e. the thicknesses of the regions will appear wider by the size of the laser spot (~500 nm) on each side of the fiber. It is possible, however, to estimate the thickness of the core by knowing the distribution of core and shell regions from SVD and overall thickness of the fiber from SEM measurements. Figure 10.11b shows that the diameter of the core is 50% of the thickness of the fiber (about 240 nm), so it is approximately 120 nm.

10.3.4 Raman Spectroscopy Mapping of EGF Nanofibers

To identify the localization of EGF within the fibers prepared by blend and emulsion electrospinning, we mapped the nanofibers using Raman spectroscopy. Figure 10.12a shows spectra of PLGA, EC, and EGF films, which were subsequently used for fiber characterization, similarly to the above. Pure PLGA, EC, and EGF components were used as basis for CLS linear regression routine. Figure 10.12b shows

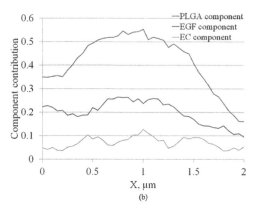

FIGURE 10.13 (a) Optical image showing sample surface mapping recorded perpendicular to the blend fiber, (b) the distribution of CLS components of PLGA, EC, and EGF, indicating a nearly uniform blend. Three-point moving average was applied to CLS components to reduce the appearance of random noise (Foraida et al. 2017).

the spectrum taken from the center of the blend nanofiber, which contains all three materials. The algorithm shows the presence of PLGA/EC/EGF, which together closely match the actual measured spectrum.

One-dimensional scan area, taken across the blend fiber, is shown in Figure 10.13. The distribution of CLS components is shown in Figure 10.13b, indicating no significant variations of each material within the blend. Since the size of the diffraction-limited laser spot is large in comparison to the size of the fiber, even when the center of the beam is positioned just off the edge of the fiber, some part of the excitation volume still contains the fiber and thus produce Raman signals, making fiber's Raman signature to appear thicker than the actual fiber.

Further, Figure 10.14a shows the scan taken across the emulsion fiber, which is expected to contain the "islands"

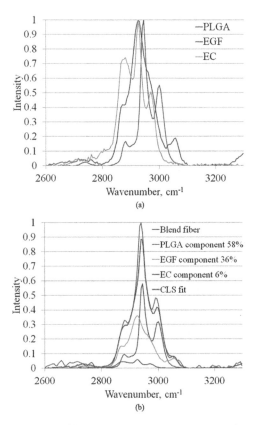

FIGURE 10.12 (a) Raman reference spectra of PLGA, EC, and EGF films and (b) spectra obtained from the center of the blend fiber, containing all three materials. The result of CLS fit of basic components of PLGA, EC, and EGF is also shown (Foraida et al. 2017).

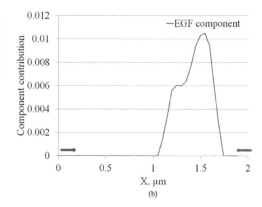

FIGURE 10.14 (a) Optical image showing sample surface mapping recorded perpendicular to the emulsion fiber and (b) the distribution of CLS component EGF, indicating the presence of an EGF "island" in the emulsion. The arrows indicate the location of the boundaries of the fiber (based on the EC signature, which is not shown). Three-point moving average was applied to CLS component to reduce the appearance of random noise (Foraida et al. 2017).

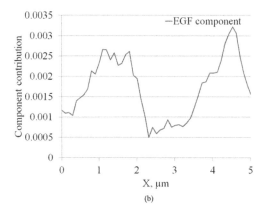

(b)

FIGURE 10.15 (a) Optical image showing sample surface mapping recorded along the emulsion fiber and (b) the distribution of CLS component EGF, indicating the presence of multiple EGF "islands" in the emulsion (indicated by arrows). Three-point moving average was applied to CLS component to reduce the appearance of random noise (Foraida et al. 2017).

of EGF embedded in the fiber. Figure 10.14b shows, while the part of the cross-section scan of the fiber (left side) contains almost no traces of EGF, the other side of the fiber contains clear traces of EGF. Finally, Figure 10.15a shows the scan taken along the emulsion fiber, where the variation of the EGF component (Figure 10.15b) provides a clear indication of the presence of EGF islands along the fiber.

These findings suggest that emulsion electrospinning was successful in incorporating EGF in the form of embedded islands enclosed within EC inside the nanofibers. Blend electrospinning, on the other hand, resulted in uniform distribution of EGF to both the surface and the bulk of the nanofibers.

10.4 Conclusion

This chapter shows the use of Raman spectroscopic mapping to study the structure of polymer nanofibers used for synthetic scaffolds in tissue engineering applications. The techniques were capable of confirming the core/shell structure of the PGS/PLGA nanofibers via direct observation of Raman signatures, associating each individual polymer and their distribution in the mixture with different spectra. Further, detailed analysis of nanofiber structure by applying the SVD algorithm was performed, enabling automated detection of core and shell polymer distribution within the fiber.

Raman spectroscopic mapping also showed that emulsion electrospinning was successful in localizing the growth factor within a double emulsion system (w/o/w), to produce EGF-loaded elastin-PLGA nanofiber substrates for epithelial tissues scaffolds. These scaffolds can provide sufficient nutrients and growth factors to the target tissue, especially in the initial stages of regeneration. Sustained release of these factors from an implantable device is

necessary to eliminate the need to systemically supplement multiple times a day. Here, EGF-loaded elastin-PLGA nanofibers were produced by blend and emulsion electrospinning.

While primarily developed for tissue engineering applications, this Raman scattering-based technique to analyze the morphology of nanofibers can be employed in many other applications where the exact distribution of chemicals within one-dimensional or two-dimensional nanostructures is required.

References

Bhardwaj, N. and Kundu, S. K. 2010. Electrospinning: A fascinating fiber fabrication technique. *Biotechnol. Adv.* 28: 325–47.

Cantara, S. I., Soscia, D. A., Sequeira, S. J., Jean-Gilles, R. P., Castracane, J. and Larsen, M. 2012. Selective functionalization of nanofiber scaffolds to regulate salivary gland epithelial cell proliferation and polarity. *Biomaterials* 33: 8372–82.

Caspers, P. J., Lucassen, G. W. and Puppels, G. J. 2003. Combined in vivo confocal Raman spectroscopy and confocal microscopy of human skin. *Biophys. J.* 85: 572–80.

Chen, R., Huang, C., Ke, Q., He, C., Wang, H. and Mo, X. 2010. Preparation and characterization of coaxial electrospun thermoplastic polyurethane/collagen compound nanofibers for tissue engineering applications. *Colloids Surf. B Biointerfaces* 79: 315–25.

Chourpa, I., Douziech-Eyrolles, L., Ngaboni-Okassa, L. et al. 2005. Molecular composition of iron oxide nanoparticles, precursors for magnetic drug targeting, as characterized by confocal Raman microspectroscopy. *Analyst* 130: 1395–403.

Dhulekar, N., Ray, S., Yuan, D. et al. 2016. Prediction of growth factor-dependent cleft formation during branching morphogenesis using a dynamic graph-based growth model. *IEEE/ACM Trans. Comput. Biol. Bioinform.* 13: 350–64.

Doshi, J. and Reneker, D. H. 1993. Electrospinning process and applications of electrospun fibers. *J. Electrostat.* 35: 151–60.

Elahi, F., Lu, W., Guoping, G. and Khan, F. 2013. Core-shell fibers for biomedical applications—A review. *Bioeng. Biomed. Sci. J.* 3: 1–14.

Foraida, Z. I., Sharikova, A., Peerzada, L., Khmaladze, A., Larsen, M. and Castracane, J. 2017. Double emulsion electrospun nanofibers as a growth factor delivery vehicle for salivary gland regeneration. *Proc. SPIE* 10352: 103520E.

Häärä, O., Koivisto, T. and Miettinen, P. J. 2009. EGF-receptor regulates salivary gland branching morphogenesis by supporting proliferation and maturation of epithelial cells and survival of mesenchymal cells. *Differentiation* 77: 298–306.

Haider, A., Haider, S. and Kang, I. K. 2015. A comprehensive review summarizing the effect of electrospinning parameters and potential applications of nanofibers in biomedical and biotechnology. *Arab. J. Chem.* 11: 1165–88.

Hennrich, F., Krupke, R., Lebedkin, S. et al. 2005. Raman spectroscopy of individual single-walled carbon nanotubes from various sources. *J. Phys. Chem. B* 109: 10567–73.

Huang, Z. M., Zhang, Y. Z., Kotaki, M. and Ramakrishna, S. 2003. A review on polymer nanofibers by electrospinning and their applications in nanocomposites. *Compos. Sci. Technol.* 63: 2223–53.

Khmaladze, A., Jasensky, J., Price, E. et al. 2014. Hyperspectral imaging and characterization of live cells by broadband coherent anti-stokes Raman scattering (CARS) microscopy with singular value decomposition (SVD) analysis. *Appl. Spectrosc.* 68: 1116–22.

Klein, K., Gigler, A. M., Aschenbrenner, T. et al. 2012. Label-free live-cell imaging with confocal Raman microscopy. *Biophys. J.* 102: 360–68.

Landsberg, G.S. and Mandelstam, L.I. 1928. New phenomenon in scattering of light (preliminary report). *J. Russian Physico-Chem. Soc. Phys.* 60: 335.

Lee, K., Silva, E. A. and Mooney, D. J. 2010. Growth factor delivery-based tissue engineering: general approaches and a review of recent developments. *J. Royal Soc. Interface* 8: 153–70.

Liu, H. 2013. Electrospining of nanofibers for tissue engineering applications. *J. Nanomater.* 2013: 1–31.

Lutz, A., De Graeve, I. and Terryn, H. 2015. Non-destructive 3-dimensional mapping of microcapsules in polymeric coatings by confocal Raman spectroscopy. *Prog. Org. Coat.* 88: 32–8.

Minsky, M. 1988. Memoir on inventing the confocal scanning microscope. *Scanning* 10: 128–38.

Nguyen, T. T. T., Chung, O. H. and Park, J. S. 2011. Coaxial electrospun poly(lactic acid)/chitosan (core/shell) composite nanofibers and their antibacterial activity. *Carbohydr. Polym.* 86: 1799–806.

Ozdemir, T., Fowler, E. W., Hao, Y. et al. 2016. Biomaterials-based strategies for salivary gland tissue regeneration. *Biomater. Sci.* 4: 592–604.

Pakravan, M., Heuzey, M. C. and Ajji, A. 2012. Core-shell structured PEO-chitosan nanofibers by coaxial electrospinning. *Biomacromolecules* 13: 412–21.

Ramakrishna, S., Fujihara, K., Teo, W., Yong, T., Ma, Z. and Ramaseshan, R. 2006. Electrospun nanofibers: solving global issues. *Mater. Today* 9: 40–50.

Raman, C. V. 1928. A new radiation. *Indian J. Phys.* 2: 387–398.

Sequeira, S. J., Soscia, D. A., Oztan, B. et al. 2012. The regulation of focal adhesion complex formation and salivary gland epithelial cell organization by nanofibrous PLGA scaffolds. *Biomaterials* 33: 3175–86.

Sfakis, L., Kamaldinov, T., Larsen, M., Castracane, J. and Khmaladze, A. 2016a. Quantification of Confocal images using LabVIEW for tissue engineering applications. *Tissue Eng. C* 22: 1028–37.

Sfakis, L., Costa, F., Tuschel, D. et al. 2016b. Core/shell nanofiber characterization by raman scanning microscopy. *Latin America Optics & Photonics Conference (LAOP)* LTu5A.5, Medellin, Colombia.

Sfakis, L., Sharikova, A., Tuschel, D. et al. 2017. Core/shell nanofiber characterization by raman scanning microscopy. *Biomed. Opt. Express* 8: 1025–35.

Soscia, D. A., Sequeira, S. J., Schramm, R. A. et al. 2013. Salivary gland cell differentiation and organization on micropatterned PLGA nanofiber craters. *Biomaterials* 34: 6773–84.

Valizadeh, A. and Farkhani, S. M. 2014. Electrospinning and electrospun nanofibers. *IET Nanobiotechnology* 8: 83–92.

Wang, C., Yan, K. W., Lin, Y. D. and Hsieh, P. C. H. 2010. Biodegradable core/shell fibers by coaxial electrospinning: Processing, fiber characterization, and its application in sustained drug release. *Macromolecules* 43: 6389–97.

Wang, N., Burugapalli, K., Wijesuriya, S. et al. 2013. Electrospun polyurethane-core and gelatin-shell coaxial fibre coatings for miniature implantable biosensors. *Biofabrication* 6: 015002.

Xu, B., Li, Y., Fang, X. et al. 2013. Mechanically tissue-like elastomeric polymers and their potential as a vehicle to deliver functional cardiomyocytes. *J. Mech. Behav. Biomed. Mater.* 28: 354–65.

Zhang, Y., Huang, Z., Xu, X., Lim, C. T. and Ramakrishna, S. 2004. Preparation of core—shell structured PCL-r-Gelatin Bi-component nanofibers by coaxial electrospinning. *Chem. Mater.* 16: 3406–09.

Zhang, Y., Lim, C. T., Ramakrishna, S. and Huang, Z.-M. 2005. Recent development of polymer nanofibers for biomedical and biotechnological applications. *J. Mater. Sci. Mater. Med.* 16: 933–46.

Zhang, Y. Z., Venugopal, J., Huang, Z. M., Lim, C. T. and Ramakrishna, S. 2005. Characterization of the surface biocompatibility of the electrospun PCL-Collagen nanofibers using fibroblasts. *Biomacromolecules* 6: 2583–89.

Zhang, Y. Z., Wang, X., Feng, Y., Li, J., Lim, C. T. and Ramakrishna, S. 2006. Coaxial electrospinning of (fluorescein isothiocyanate-conjugated bovine serum albumin)-encapsulated poly(ε-caprolactone) nanofibers for sustained release. *Biomacromolecules* 7: 1049–57.

Zhu, L., Liu, X., Du, L. and Jin, Y. 2016. Preparation of asiaticoside-loaded coaxially electrospinning nanofibers and their effect on deep partial-thickness burn injury. *Biomed. Pharmacother.* 83: 33–40.

Quantification of Nanostructure Orientation via Image Processing

Nasser Mohieddin Abukhdeir
University of Waterloo

11.1 Introduction

The alignment of one-dimensional (1D) nanostructures [1] has impacted nanoscience and nanotechnology through the development of novel methods for self-assembly driven alignment [2,3] and a broad range of applications from sensors [4,5] to field-emission devices [6]. These nanostructures are typically composed of nanorods (Figure 11.1a), nanowires (Figure 11.1b), and similar structures deposited on surfaces [1]. While the majority of applications involve deposition on surfaces, where the orientation of the nanostructures is constrained within 2D, there are many examples of 3D vertically aligned nanostructures [7,8], although methods for imaging 3D structures at the nanoscale are not widely accessible.

Orientational order corresponds to spatial correlation of the orientation of structures, which themselves might have variation of orientation along their arclength (nanowires). This can be quantified through the use of an orientational distribution function (ODF) [9,10],

$$\int_\theta \int_\phi f(\theta, \phi) w(\theta, \phi) \mathrm{d}\theta \mathrm{d}\phi = 1 \qquad (11.1)$$

for a domain (2D or 3D) which quantifies the probability of finding a nanostructure oriented along $\boldsymbol{m}(\theta, \phi)$ within a domain where $w(\theta, \phi)$ is a weighting function. Orientational order parameters (OOPs) are frequently used as convenient approximations for ODFs and correspond to scalar weights of expansions of orthogonal bases,

$$f(\theta, \phi) = \sum_{m=0}^{\infty} \sum_{n=0}^{\infty} S_{mn} \phi(\theta, \phi) \qquad (11.2)$$

where $\{S_{mn}\}$ corresponds to OOPs and varies between zero and one for an orthonormal basis. Figure 11.1c shows a sample ODF for an image of nanowires within a 2D domain and approximations of the ODF using increasing numbers of $\{S_i\}$.

In general, the degree of orientational order of 1D nanostructures has a significant effect on their properties [1] and, consequently, devices and applications which use them. For example, the degree of alignment has been shown to increase the charge mobility of nanorod-based field-effect transisters [12]. Thus, given the relation between orientational order and property/performance, the determination of structure/property relations is a vital task for nanoscience and nanotechnology researchers. In order to achieve this, significant advances in quantification of both orientational- and translational-ordered nanostructures have been made over the past decade [13].

In this chapter, the theory of orientational order-order, relevant to 1D nanostructures, and image processing methods needed apply this theory to the characterization of experimental images is summarized. Methods for reconstructing the ODF of a domain, through computation of a finite set of OOPs, are described. Image processing methods relevant to determination of 1D nanostructure alignment through the use of object-fitting and mathematical morphology are presented, along with computationally efficient methods for identification of overlapping nanostructures.

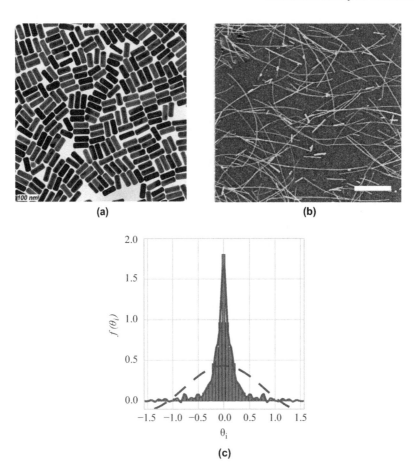

FIGURE 11.1 Sample experimental images of surfaces containing orientationally ordered (a) nanorods (scanning electronic microscopy (SEM) imaging) and (b) nanowires (transmission electron microscopy (TEM) imaging). (c) Sample high (solid line) and low (dotted line) order approximations of an ODF computed for a surface of orientationally-ordered 1D nanostructures superimposed on a histogram of individual nanostructure orientations. Reproduced with permission from refs. [10,11].

11.2   Background

11.2.1   Nanostructure Morphology

There are a broad range of nonspherical nanostructures that have been experimentally demonstrated to be aligned [1]. These can be grouped, in general, into nanostructures whose orientation can be described with a single unit vector $\boldsymbol{m}$ or a unit vector function $\boldsymbol{m}(s)$ which varies along the arclength of the pseudo-1D structure. These types of 1D nanostructures will be referred to as *nanorods* and *nanowires*, respectively.

In order to generalize the orientation vector to arbitrary nanostructure shape, it may be determined for 2D (noncircular) and 3D (nonspherical) structures given the shape density $\rho(\boldsymbol{r})$ [14],

$$\boldsymbol{m} = m^{-1} \int_V (\boldsymbol{r} - \boldsymbol{R}) \; \rho \; \mathrm{d}V \qquad (11.3)$$

$$\boldsymbol{m} = m^{-1} \int_A (\boldsymbol{r} - \boldsymbol{R}) \; \rho \; \mathrm{d}A \qquad (11.4)$$

where the center of mass is,

$$\boldsymbol{R} = m^{-1} \int_V \boldsymbol{r} \; \rho \; \mathrm{d}V \qquad (11.5)$$

$$\boldsymbol{R} = m^{-1} \int_A \boldsymbol{r} \; \rho \; \mathrm{d}A \qquad (11.6)$$

where m is the mass of the object. Figure 11.2 shows several examples of 2D nanostructures, the corresponding center of mass and orientation vectors. For all but the last case, a deformed nanowire, this approach is adequate.

Deformed nanowire orientation is not well approximated using a single-orientation vector $\boldsymbol{m}$. Instead, it requires an orientation vector function $\boldsymbol{m}(s)$, or an approximation of this function, varying along the *topological skeleton* (skeleton) of the nanowire, $\boldsymbol{r}(s)$. This skeleton may be determined in several different ways [15], with an intuitive approach for it to be composed of all points in an object which have more than one closest point to the object boundary. Figure 11.3 shows computational approximations of nanorod and nanowire skeletons from sample images. Computational artifacts resulting from noise in the sample images and bifurcation points due to the nonlinear shape are present in skeleton approximations. These complications may be dealt with using different types of post-processing methods. The orientation vector function is then found from the unit tangent vector of the skeleton,

FIGURE 11.2 Schematics of a sample set of nanorod and nanowire shapes with centroid and orientation vectors shown: (a) rectangle, (b) ellipse, (c) nanorod (Figure 11.1a), and (d) deformed nanowire (Figure 11.1b). Orientation vectors are centered using the object centroid.

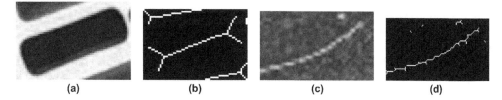

FIGURE 11.3 Images of an individual (a/c) nanorod/nanowire and its (b/d) topological skeleton.

$$m(s) = \left| \frac{\partial r}{s} \right|^{-1} \frac{\partial r}{s}. \qquad (11.7)$$

given linear mass density or *line density* of the object $\bar{\rho}(s)$. A scalar measure of the accuracy of the constant orientation vector approximation $m \approx m(s)$ may be found,

$$\alpha = l^{-1} \int_s m \cdot m(s) \mathrm{d}s \qquad (11.8)$$

where l is the length of the object, as $\alpha \to 1$ the constant orientation vector approximation is exact and as $\alpha \to 0$ is a poor approximation.

11.2.2 Orientational Distribution Functions and Order Parameters

The overall approach to characterizing and quantifying orientational order of nanostructured domains is based on a theoretical foundation developed through the study of the orientationally ordered *nematic* liquid crystalline (NLC) phase [9,16]. Figure 11.4 shows coarse-grained molecular simulation visualizations of both 2D (Figure 11.4a) and 3D (Figure 11.4b) NLC domains. Most relevant to 1D nanostructures, *nonpolar uniaxial* NLC order is traditionally quantified through the combination of a scalar S and unit vector n order parameters. The vector order parameter, or nematic director, quantifies the average orientation of objects (molecules for an NLC), and the scalar order parameter quantifies the degree to which the molecules are aligned in the direction of n. As will be elucidated in this section, this characterization of NLC orientation is a low-order approximation of 3D ODF of the domain.

For 3D orientationally ordered nanostructures, NLC order parameter theory may be directly applied [9]. For nonpolar uniaxial 1D nanostructures, the ODF of a domain has the form (spherical coordinates),

$$\int_0^\pi f(\theta) \sin\theta d\theta = 1 \qquad (11.9)$$

where θ is the angle between the average orientation vector n and the nanostructure alignment vector m. The ODF can be shown to have the form [9],

$$f(\theta) = \frac{1}{2} + \sum_{n=1}^\infty \frac{4n+1}{2} S_{2n} P_{2n}(\cos\theta) \qquad (11.10)$$

where $\{S_{2n}\}$ define an infinite set of scalar parameters and $P_{2n}(x)$ are Legendre polynomials. The NLC scalar-order parameter $S = S_2$ is a first-order approximation to the ODF of a domain, with Figure 11.1c showing an example of an ODF (2D) with a first-order approximation superimposed. A convenient approach to compute the average orientation vector n of a domain is through determining the major eigenvector of the *alignment tensor* [19],

$$\mathbf{Q} = <m_i m_i - \boldsymbol{\delta}> \qquad (11.11)$$

where $\{m_i\}$ is the set of orientation vectors of objects.

Similarly, the ODF for 2D orientationally ordered nonpolar 1D nanostructured domains is [10,20],

$$\int_0^{2\pi} f(\theta) d\theta = 1 \qquad (11.12)$$

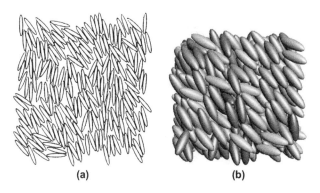

FIGURE 11.4 Coarse-grained molecular simulation visualizations of NLC order for (a) elliptic molecules in a 2D domain ($S_2 = 0.6$ from Eq. 11.10) and (b) ellipsoidal molecules in a 3D domain. Reproduced with permission from refs. [17,18].

which can be shown to have the form,

$$f(\theta) = \frac{1}{2\pi} + \frac{1}{\pi}\sum_{n=1}^{\infty} S_{2n}\cos 2n\theta. \qquad (11.13)$$

Similar to the 3D case, the 2D alignment tensor is,

$$\mathbf{Q} = <2\boldsymbol{m}_i\boldsymbol{m}_i - \boldsymbol{\delta}> \qquad (11.14)$$

whose major eigenvector corresponds to the average orientation vector $\boldsymbol{n}$.

Reconstruction of 3D and 2D ODFs for a domain of orientationally ordered nanostructures may be achieved through calculation of $\{S_{2n}\}$ up to an appropriate order, as shown in Figure 11.1c. Domain averaging of the scalar order parameters for 3D domains,

$$S_{2n} = <P_{2n}(\cos\theta_i)> = N^{-1}\sum_{i=1}^{N} P_{2n}(\cos\theta_i) \qquad (11.15)$$

and 2D domains,

$$S_{2n} = <\cos 2n\theta_i> = N^{-1}\sum_{i=1}^{N}\cos 2n\theta_i \qquad (11.16)$$

may be used to construct an approximation for $f(\theta)$ up to finite order n.

11.2.3 Image Processing and Mathematical Morphology

In order to quantify orientational-ordered nanostructured surfaces, more specifically, approximate the ODF through determination of OOP values for sample images of these surfaces, image processing methods must be applied to determine the set of orientation vectors $\{\boldsymbol{m}_i\}$ or orientation vector functions $\{\boldsymbol{m}_i(s)\}$ for nanorods and nanowires, respectively. In general, image processing involves four sequential tasks [21]: filtering, segmentation, object detection, and object analysis. The filtering task is typically required due to the presence of measurement uncertainty or "noise" in analytical techniques used to generate the image from the nanostructured surface. The segmentation task involves the mapping of the local image intensity to a lower dimensional space, typically a binary space which differentiates between the "foreground" (nanostructures) and "background" (substrate surface). The object detection task groups sets of foreground pixels into contiguous groups or objects. Finally, in the context of orientational-order quantification, the object analysis task involves the determination of the orientation of individual objects for computation of the ODF (Sections 11.3–11.4).

Modern image processing methods[1] typically combine filtering and segmentation tasks. The most simple approach

to image segmentation is thresholding, where each pixel of an image is compared to a global or spatially local threshold in order to compute a segmented binary image where the *foreground* corresponds to objects and the *background* corresponds to the surface/substrate. In addition to measurement noise, the intensity of the image may vary locally. Figure 11.5 shows sample nanorod images (based on Figure 11.1a) modified to include additional (Gaussian) noise, variation in the local image intensity, or a combination of both. These image modifications demonstrate typical features of nanostructured surface imaging which pose challenges to image segmentation.

Two standard approaches to image segmentation which combine filtering and thresholding are Otsu binarization [23] and adaptive thresholding [22]. Otsu binarization determines a global threshold through approximating the image intensity distribution as a bimodal and computing a threshold that lies between the two modes. This approach is robust for images in which the intensity does not vary significantly and noise is present. Figure 11.6 shows the result of applying both a mean global threshold and using Otsu binarization to a noise-enhanced nanorod image (Figure 11.5a). Otsu binarization results in a binary image in which the measurement noise is minimally present, compared to that using the mean threshold in which the background is poorly segmented. However, applying Otsu binarization method to an intensity variation-enhanced image (Figure 11.5b) demonstrates the limitation of using a global threshold, where large areas of the image are segmented inaccurately due to the use of a single global threshold (Figure 11.6c).

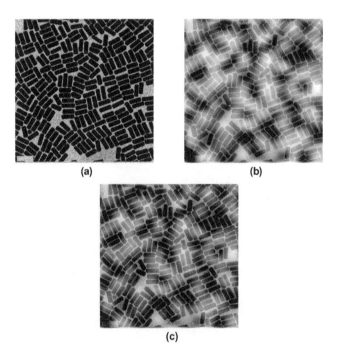

(a)   (b)

(c)

FIGURE 11.5 Modified versions of the sample nanorod image (Figure 11.1a) to artificially enhance the presence of (a) noise, (b) image intensity variation, and (c) combined noise and image intensity variation.

---

[1]The Open Computer Vision Library (OpenCV) [22] is used for all examples in this section.

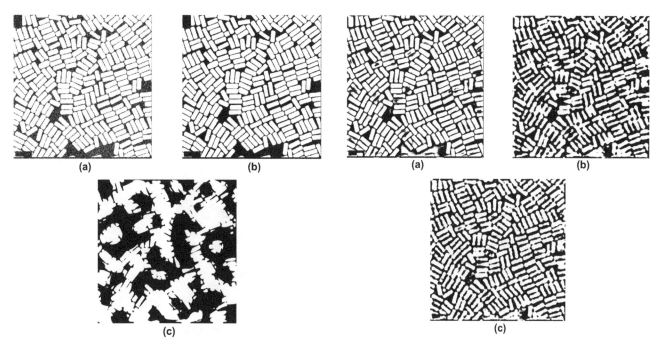

FIGURE 11.6 Binary images resulting from the application of different types of segmentation/thresholding to the modified nanorod images from Figure 11.5: (a) standard thresholding applied to the noise-enhanced nanorod image (Figure 11.5a), (b) Otsu binarization applied to the noise-enhanced image (Figure 11.5a), and (c) Otsu binarization applied to the noise and intensity variation-enhanced image (Figure 11.5c).

FIGURE 11.7 Binary images resulting from the application of different types of segmentation/thresholding to the modified nanorod images from Figure 11.5: (a) adaptive thresholding (mean) applied to the intensity-enhanced nanorod image (Figure 11.5b), (b) adaptive thresholding (mean) applied to the noise and intensity-enhanced image (Figure 11.5c), and (c) adaptive thresholding (Gaussian) applied to the noise and intensity-enhanced image (Figure 11.5c).

For images with varying intensity, adaptive thresholding is required for accurate image segmentation [24]. This method involves computing a local threshold for each pixel through a weighted average of the neighboring pixel values. The choice of weighting function, local window size (in pixels), and (potentially) other parameters introduces additional complexity to the use of this class of thresholding methods. Figure 11.7 shows example binary images resulting from applying adaptive thresholding to the noise and intensity variation-enhanced nanorod image (Figure 11.5c) using constant and Gaussian weighting functions. The former corresponds to choosing the local mean intensity value for thresholding. The local window size was chosen to be an approximation of the shortest length scale of the nanorod size (in pixels).

Focusing on Figure 11.7a, which shows the binary image resulting from applying mean adaptive thresholding to the intensity variation-enhanced nanorod image (Figure 11.5b), the local/adaptive thresholding approach is able to accurately segment the image into foreground and background, unlike global thresholding methods. Figures 11.7b–7c show the results of applying both mean and Gaussian adaptive thresholding to the noise and intensity-enhanced nanorod image (Figure 11.5c), where the Gaussian weighting function performs slightly better compared to the constant weighting function (local mean threshold) method.

Standard object detection methods involve *labelling* contiguous foreground regions within the domain, which correspond to individual objects depending on the accuracy of the segmentation task. This approach is appropriate for nonoverlapping nanostructures, such as nanorods, but not for overlapping objects which is frequently the case with nanowires. Figures 11.8a–c show the result of labelling the nanorod binary image. Individual nanorods are, for the most part, correctly identified using this standard technique. However, for the nanowire image, the presence of overlapping objects results in the grouping of multiple physical objects into single detected objects.

Focusing on nonoverlapping nanostructures, following object labeling, a discrete-space formulation of Eq. (11.5) can be used to find the centroid of each object [22]. In order to find the orientation vector for each object, shape-fitting methods are typically used instead of computation of the discrete-space formulation of Eq. (11.3). Shapes such as rectangles and ellipses may be used if appropriate for the deposited nanostructures. The shape-fitting approach that would yield an equivalent result to that of Eq. (11.3) involves fitting of a line to each object. Figure 11.8 illustrates the use of different shapes for fitting of the objects identified in the labeling task of the nanorod binary image. The accuracy of the centroid and orientation of the fitted shapes compared to that of the actual 1D nanostructures is highly dependent on the accuracy of the image segmentation and labeling tasks. As can be seen for object fitting of all three types (rectangle, ellipse, and line), the incorrect

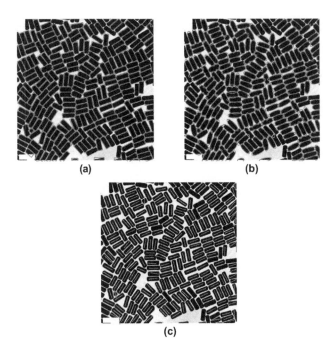

FIGURE 11.8 Examples of shape fitting (gray) of the segmented and labeled binary image (nanorods) using different shapes: (a) rectangles, (b) ellipses, and (c) lines.

grouping of multiple nanorods into a single object introduces a small but significant number of inaccurate nanostructure orientations. This motivates the use of more accurate image segmentation and labeling methods and/or an alternative object detection method.

Object identification for overlapping nanorod and nanowire objects is significantly more complex. The use of simple labeling methods results in predominantly grouping multiple 1D nanostructures into single detected objects. This motivates the use of more advanced object identification methods, such as mathematical morphology methods [15,25]. Mathematical morphology methods enable the identification of structures and geometry within an image without requiring the explicit identification of objects. Mathematical morphology methods are based on a basic set of operations (erosion, dilation, opening, and closing) and transforms (hit-or-miss, pruning, *etc.*), see ref. [15] for further details.

Mathematical morphology methods are able to identify *branchpoints* and *endpoints* in binary images, which are directly relevant to identifying overlapping nanorod and nanowire objects. The first step in *binary morphology*, mathematical morphology of binary images, is the computation of the *topological skeleton* (or skeleton) from the segmented binary image. The skeleton of a binary image is composed of an approximation of the medial axis of the foreground object(s). Figure 11.9 shows the process of computing the skeleton beginning with the binary segmented nanowire image (Figure 11.9a).

The preferred approach to compute the skeleton of 1D nanostructures involves the use of methods which preserve the topology of the original binary image, such as the use

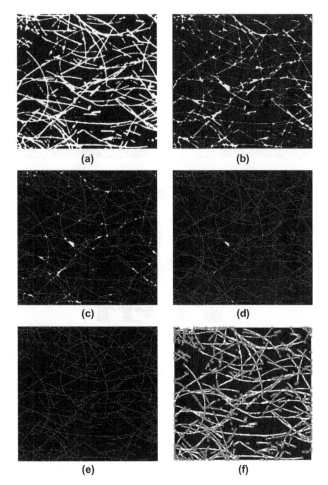

FIGURE 11.9 Example of the intermediate stages of skeletonization of the (a) nanowire sample binary image through the use of the hit-or-miss transform with appropriate structuring elements for skeletonization [15]: (b) 5 iterations, (c) 10 iterations, (d) 15 iterations, and (e) the resulting topological skeleton. (f) Branchpoints and endpoints (gray) identified using the hit-or-miss transform with appropriate structuring elements.

of the *hit-or-miss* transform [26]. The hit-or-miss transform involves the application of *structuring elements*, one associated with a local structure being present (foreground) and another being absent (background). Given the skeleton, additional applications of the hit-or-miss transform include using structuring elements which identify branchpoints and endpoints [15].

11.3 Orientational Order Quantification of Nonoverlapping Nanorod-Covered Surfaces

In this section, an algorithm is described and demonstrated for quantifying orientational order of surfaces with nonoverlapping nanorods (constant orientation $m(s) \approx m$). In order to compute global ODF/OOP values, an algorithm presented in ref. [10] can be summarized as:

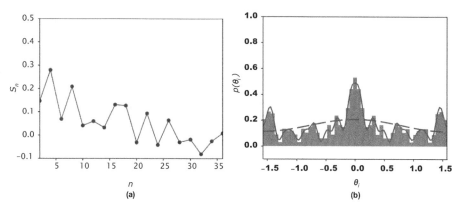

FIGURE 11.10 (a) Plot of OOP magnitude S_n versus order n and (b) the reconstruction of the ODF using a $n = 1$ order (dashed line) and $n = 35$ (solid line) superimposed on a histogram of individual nanorod orientations.

1. The set of nanorod orientation vectors is determined through filtering, thresholding, and object identification (Section 11.2.3).

2. The 2D alignment tensor of the set of objects is computed using Eq. (11.14), along with the average orientation vector $\boldsymbol{n}$, its major eigenvector.

3. A finite set of n OOP values $\{S_{2n}\}$ are computed using Eq. (11.16), such that the S_{2n} value falls below a user-specified tolerance.

4. The ODF of the set of nanorods is reconstructed by using a truncated form of Eq. (11.13).

Figure 11.10 shows the results of applying the algorithm to the sample nanorod image (Figure 11.1a). A suitable value of n to approximate the ODF of the sample nanorod image was found to be < 35. The result of approximating the ODF using a single OOP S_2 versus a set $\{S_{2n}\}$ shows that the use of low-order ODF approximations is not appropriate for aligned nanostructures [10].

Analysis of the OOP plot in Figure 11.10a indicates that the sample nanorod image (Figure 11.1a) has a relatively low degree of order with $S_2 \approx 0.15$. However, after analyzing the high-order approximation of the ODF in Figure 11.10b, secondary peaks near $\theta = \frac{\pi}{2}$ indicate that there is orientational order of nanorods in the domain orthogonal to the computed orientation vector $\boldsymbol{n}$. Visual inspection of Figure 11.1a qualitatively confirms this where there are impinging clusters of nanorods oriented orthogonally with respect to their neighbors.

11.4 Orientational Order Quantification of Overlapping Nanowire-Covered Surfaces

In this section, an algorithm is described and demonstrated for quantifying orientational order of surfaces with overlapping nanorods and nanowires where a constant orientation approximation ($\boldsymbol{m}(s) \approx \boldsymbol{m}$) is valid. First, an image processing method is described from ref. [10] which is able to decompose overlapping nanorods and nanowires into nonoverlapping ones. Then a modified method for determining global ODF and OOP values is presented which accounts for the difference in length of these nanostructures.

Overlapping nanorods and nanowires present challenges for both the application of object identification and shape-fitting methods which are adequate for nonoverlapping cases (Section 11.2.3). The use of mathematical morphology operations is one approach to overcome these challenges, where morphological features may be identified for further analysis. Two key morphological features, endpoints and branchpoints, are needed to achieve object identification and determine orientation vectors. Endpoints of 1D nanostructures may be used to either approximate a constant orientation vector $\boldsymbol{m}$ or as a starting point to compute an orientation vector function $\boldsymbol{m}(s)$. Branchpoints may be used to identify areas where nanostructures overlap, which is required for the identification of unique 1D nanostructures when overlap is present. Figure 11.9f shows the results of mathematical morphology analysis of overlapping nanowire images, with endpoints and branchpoints identified.

Even with endpoints and branchpoints identified, it is a complex task to uniquely identify overlapping nanostructures. Ref. [10] presents a simple method to circumvent this task through identifying and then *removing* branchpoints from the topological skeleton of an image. Thus, a new topological skeleton is created with branchpoints now having the background pixel value, and thus, the image contains only endpoints. This introduces an approximation, where if two nanostructures overlap in the original topological skeleton, they now correspond to four separate nonoverlapping nanostructures. This enables standard object identification methods to be used on the modified skeleton.

Another complexity that is introduced for both overlapping nanostructures and nanowires, even when not overlapping, is the variation of orientation along their length. One approach to quantifying orientation in this case is to use shape-fitting methods to compute the mean orientation vector as described in Section 11.2.3. However, for images with large numbers of nanostructures, this can be a computationally intensive task, which is typically the case for densely covered surfaces where overlapping is likely to occur. Given that the topological skeleton would be needed to remove branchpoints as previously described, a simple approach to compute the orientation vector for the branchpoint-less topological skeleton would be to use the orientation vector parallel to the end-to-end vector of each nanostructure segment. However, given that the resulting nanostructure segments will have substantially varying lengths, the method for determining OOP values in order to reconstruct the global ODF needs to be modified. This can be achieved through modifying the OOP Eq. (11.16) to account for each nanostructure having different lengths l_i and computing a weighted average:

$$S_{2n}^w = L^{-1} \sum_{i=1}^{N} l_i \cos 2n\theta_i \qquad (11.17)$$

where $L = \sum_i l_i$ is the total length of nanostructures present in the image. Figure 11.11 shows the OOP and ODF results of applying Eq. (11.17) to the nanowire skeleton with branchpoints removed. As with the nanorod sample, the nanowire sample has relatively low orientational order with $S_2 \approx 0.3$ but does not exhibit secondary peaks in the ODF at $\theta = \frac{\pi}{2}$, which is in agreement with visual inspection of the sample image.

Applying the mathematical morphology-enhanced method which can capture overlapping 1D nanostructures enables robust quantification of orientational order for *both* nonoverlapping and overlapping cases. Results of applying the method to the sample nanorod image (Figure 11.1a) are shown in Figure 11.12a. The resulting skeleton includes branchpoints at the ends of each of the skeletons associated with individual nanorods. A mathematical morphology operation that removes endpoints, or *pruning*, may be used to remove these artifacts from the skeletonization. Figure 11.12b shows the skeleton follow four pruning operations, which may now be used to determine the individual nanorod orientations for further analysis.

Figure 11.13 shows the OOP and ODF results from the mathematical morphology and object-fitting method in

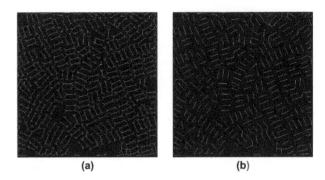

FIGURE 11.12 Topological skeletons of the sample nanorod image (Figure 11.1a) (a) before and (b) after four pruning operations.

Section 11.4. There is qualitative agreement for both the OOP (Figure 11.13a) and ODF (Figures 11.13b–11.13c) from both methods. However, there are non-negligible differences in the OOP values and the resulting ODFs. These differences could be addressed by more sophisticated object identification methods which correct the multiple identification errors that both the object fitting (Figure 11.8a) and mathematical morphology methods introduce (Figure 11.12b).

11.5   Summary

In this chapter, the theory of orientational order of 1D nanostructures and image processing methods for the application of this theory to analysis of experimental images was presented. These methods were then demonstrated on sample images of nanorod and nanowire-covered surfaces. While the focus of this chapter was on the analysis of surfaces (as opposed to volumes), this was not due to a limitation of the theory or image processing methods, but instead, the lack of experimental nanotomography data for 3D domains of 1D nanostructures.

The quantification of nanostructure orientation is a vital task to enable nanoscience and nanotechnology researchers

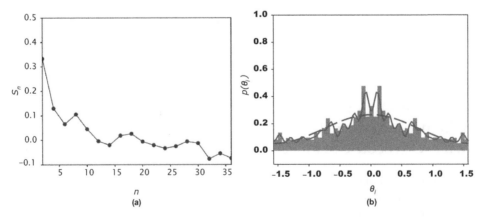

FIGURE 11.11   (a) Plot of OOP magnitude S_n versus order n and (b) the reconstruction of the ODF using a $n = 1$ order (dashed line) and $n = 35$ (solid line) superimposed on a histogram of individual nanowire segment orientations.

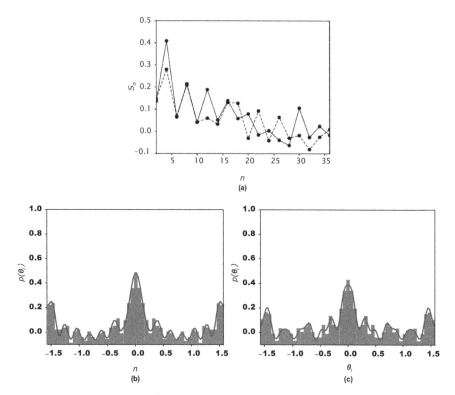

FIGURE 11.13 (a) Plot of OOP magnitude S_n versus order n using the standard shape-fitting approach and mathematical morphology approach (dotted line), (b–c) reconstruction of the ODF using $n = 35$ (solid line) superimposed on a histogram of individual nanorod orientations for the object-fitting and mathematical morphology approaches, respectively.

to identify structure/property relations. While appropriate measures for the quantification of orientational order of 2D and 3D domains have been established, the image processing of nanostructure images is still a challenging task. As has been demonstrated in Sections 11.3–11.4, mathematical morphology methods provide a significant improvement over traditional object-fitting methods. Even so, the significant amount of measurement noise and low resolution of nanoscale imaging techniques motivates further advances and refinement for robust characterization of nanostructured domains.

References

1. Bin Su, Yuchen Wu, and Lei Jiang. The art of aligning one-dimensional (1d) nanostructures. *Chemical Society Reviews*, 41(23):7832–7856, 2012.
2. Claudia Pacholski, Andreas Kornowski, and Horst Weller. Self-assembly of zno: from nanodots to nanorods. *Angewandte Chemie International Edition*, 41(7):1188–1191, 2002.
3. Leonid Vigderman, Bishnu P. Khanal, and Eugene R. Zubarev. Functional gold nanorods: synthesis, self-assembly, and sensing applications. *Advanced Materials*, 24(36):4811–4841, 2012.
4. Jin Huang and Qing Wan. Gas sensors based on semiconducting metal oxide one-dimensional nanostructures. *Sensors*, 9(12):9903–9924, 2009.
5. Kannan Balasubramanian. Challenges in the use of 1d nanostructures for on-chip biosensing and diagnostics: A review. *Biosensors and Bioelectronics*, 26(4):1195–1204, 2010.
6. Younan Xia, Peidong Yang, Yugang Sun, Yiying Wu, Brian Mayers, Byron Gates, Yadong Yin, Franklin Kim, and Haoquan Yan. One-dimensional nanostructures: synthesis, characterization, and applications. *Advanced Materials*, 15(5):353–389, 2003.
7. Zhong Lin Wang. Nanostructures of zinc oxide. *Materials Today*, 7(6):26–33, 2004.
8. Irene Gonzalez-Valls and Monica Lira-Cantu. Vertically-aligned nanostructures of zno for excitonic solar cells: a review. *Energy & Environmental Science*, 2(1):19–34, 2009.
9. Claudio Zannoni. Order parameters and orientational distributions in liquid crystals. In *Polarized Spectroscopy of Ordered Systems*, pp. 57–83. Springer, Dordrecht, 1988.
10. Jianjin Dong, Irene A Goldthorpe, and Nasser Mohieddin Abukhdeir. Automated quantification of one-dimensional nanostructure alignment on surfaces. *Nanotechnology*, 27(23):235701, 2016.
11. Veronika Kozlovskaya, Eugenia Kharlampieva, Bishnu P. Khanal, Pramit Manna, Eugene R. Zubarev, and Vladimir V. Tsukruk. Ultrathin layer-by-layer hydrogels with incorporated

gold nanorods as ph-sensitive optical materials. *Chemistry of Materials*, 20(24):7474–7485, 2008.

12. Baoquan Sun and Henning Sirringhaus. Solution-processed zinc oxide field-effect transistors based on self-assembly of colloidal nanorods. *Nano Letters*, 5(12):2408–2413, 2005.

13. Nasser Mohieddin Abukhdeir. Computational characterization of ordered nanostructured surfaces. *Materials Research Express*, 3(8):082001, 2016.

14. Jan Flusser, Tomas Suk, and Barbara Zitová. *2D and 3D Image Analysis by Moments*. John Wiley & Sons, Hoboken, NJ, 2016.

15. Edward R Dougherty, Roberto A Lotufo, and The International Society for Optical Engineering SPIE. *Hands-on morphological image processing*, volume 71. SPIE Optical Engineering Press, Washington, DC, 2003.

16. P.G. de Gennes and J Prost. *The Physics of Liquid Crystals*. Oxford University Press, New York, second edition, 1995.

17. J. A. Cuesta and D. Frenkel. Monte Carlo simulation of two-dimensional hard ellipses. *Physical Review A*, 42:2126–2136, 1990.

18. C. M. Care and D. J. Cleaver. Computer simulation of liquid crystals. *Reports on Progress in Physics*, 68(11):2665–2700, 2005.

19. A. Sonnet, A. Kilian, and S. Hess. Alignment tensor versus director: Description of defects in nematic liquid crystals. *Physical Review E*, 52(1):718–722, 1995.

20. J. P. Straley. Liquid crystals in two dimensions. *Physical Review A*, 4:675–681, 1971.

21. Lou Ross and John C Russ. *The Image Processing Handbook*, volume 17. Cambridge University Press, Cambridge, 2011.

22. Gary Bradski and Adrian Kaehler. *Learning OpenCV: Computer Vision with the OpenCV Library*. "O'Reilly Media, Inc.", Sebastopol, CA, 2008.

23. Nobuyuki Otsu. A threshold selection method from gray-level histograms. *IEEE Transactions on Systems, Man, and Cybernetics*, 9(1):62–66, 1979.

24. Wilhelm Burger and Mark J Burge. *Principles of Digital Image Processing: Advanced Methods*. Springer Science & Business Media, Dordrecht, 2013.

25. Robert M Haralick, Stanley R Sternberg, and Xinhua Zhuang. Image analysis using mathematical morphology. *Pattern Analysis and Machine Intelligence, IEEE Transactions on*, (4):532–550, 1987.

26. Frank Y Shih. *Image Processing and Mathematical Morphology: Fundamentals and Applications*. CRC press, Boca Raton, FL, 2009.

Chemometrics and Super-Resolution at the Service of Nanoscience: Aerosol Characterization in Hyperspectral Raman Imaging

Marc Offroy
Université de Lorraine, CNRS, LIEC

Ludovic Duponchel
Université de Lille, CNRS, LASIR

12.1 Introduction

Chemometrics is the science that uses mathematics and statistical methods not only to design or select optimal measurement procedures and experiments but also to provide maximum chemical information by analyzing data in chemistry (Brereton 2003, 2007). This chemical discipline is interdisciplinary that combines techniques from the signal processing (Otto et al. 2016), the data analysis (Massart et al. 1997, Vandeginste et al. 1998) and, of course, chemical models (Ruckebusch 2016). Its development is strongly related to the use of computers in chemistry. In the 1970s, some analytical groups were already working with mathematical tools on mainframe computers which are ascribed nowadays to chemometric methods (Sharaf et al. 1986). In 1972, the Swede, Svante Wold, and the American, Bruce R. Kowalski introduced the term "chemometrics", and since then, this science continues to grow (Geladi and Esbensen 2005). Indeed, in 1980s, the emergence of personal computers gives the opportunity to a new age for the acquisition, processing and interpretation of chemical data. In the 1990s, advances in computing gave industries and research centers the opportunity to generate data at a relatively low economic cost. Nowadays, chemometricians are faced with

big data problems. Modern instruments generate an always larger amounts of data fairly automatically in a very short period of time. The complexity of the data is always higher. Unfortunately, the education of chemists in mathematics and statistics is not enough. Today, every scientist uses software in one form or another, but there is a necessity to understand deeper the mathematical methods used for interpreting chemical data. One objective of the chemometrician is to develop those aspects in order to make complicated mathematical methods more practicable. In this way, this chapter is to demonstrate the ramp up of chemometrics in physical chemistry in the field of nanoscience.

In physics, chemistry or biology, many researchers are interesting by nanoscience and need instrumental improvements to address the sub-micrometric analysis challenges. The originality of our approach is to use both the super-resolution concept and multivariate curve resolution–alternative least square algorithm (MCR-ALS) in confocal Raman imaging on atmospheric aerosols. The aim goals are to surmount the instrumental limits in Raman scattering and to characterize the chemical components at the level of the individual particles. A first part in this chapter will present the hyperspectral imaging and will explain the MCR-ALS algorithm in order to extract the relevant chemical information from data. A second one will explain the concept of the super-resolution whose discipline comes from the signal processing community. And the last one demonstrates the possibility of our approach to go beyond the diffraction limit on a Raman instrument in order to characterize some industrial aerosols.

12.2 Hyperspectral Imaging and Multivariate Curve Resolution

Spectroscopic imaging is a powerful analytical measurement technique that provides spatial distribution and chemical information. This instrumental technique provides not only significant molecular analysis (i.e. macroscopic characterization) but also gives a spatial representation of the "chemical" reality with map generation of the chemical components from samples. A confocal microscope coupled to a spectrometer with a data collection system is thus used in hyperspectral imaging. The microscope focuses and collects the scattered or non-absorbed photons from a specific area of the analyzed sample. A motorized stage is below the microscopes to allow movement in x, y or z directions. An autofocus system could be found on some microscopes in order to maximize its position control near the surface. Two different hyperspectral acquisition systems exist: Global imaging and mapping imaging (Krishnan et al. 1995, Salzer and Siesler 2009). The global imaging systems can only record a single monochromatic image due to a global illumination for a specific wavelength of the area sample, while the mapping systems record a data "cube" (i.e. x pixels $\times$ y pixels $\times$ λ wavelengths) due to a scan of area sample for multiple wavelengths. Thus, the most attractive method is the mapping even if the time of acquisition is

more important for a point-by-point or line-by-line scan. Nevertheless, the benefits are significant for the spatial resolution and the spectral intensity. The Raman instrument presented later uses a mapping system point-by-point illumination. Beyond the macroscopic characterizations, the coupling spectrometers with microscopes make possible map generation that represents the spatial distribution of chemical components from samples. Indeed, the purpose of hyperspectral imaging is to isolate a specific wavelength (i.e. selectivity) for a particular chemical component in order to deduct its spatial distribution (see Figure 12.1). However, this classical approach has several drawbacks. First, it is necessary to know *a priori* all the pure components in the sample. If this hypothesis is not true, it could be possible to select a nonselective spectral area and, thus, overestimate the concentrations of a component (i.e. the number of pixels generated for an area of interest could be false). In other words, the cartography is biased, and therefore, it is not representative of the analytic reality. Moreover, when there is a strong spectral overlap (e.g. wide bandwidth and/or sample complexity), it is impossible to identify a truly selective wavelength. In addition, it would be difficult to detect, identify and produce chemical maps for unexpected components. Despite these difficulties, this classical approach in hyperspectral imaging remains widespread, since it works in the majority of cases.

On complex sample, the use of chemometric tools could overcome all these drawbacks and a better representation of the analytic reality is given (de Juan et al. 2009). Indeed, several mathematical methods have demonstrated their effectiveness in hyperspectral imaging for the data analysis (e.g. PARAFAC, OPA, SIMPLISMA and MCR-ALS). It is possible to gather all of them under the term "multivariate curve resolution" (Tauler et al. 1995, Brown et al. 2009).

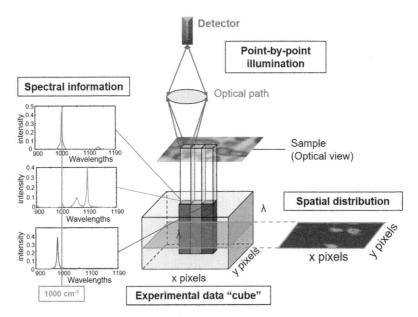

FIGURE 12.1 The experimental data "cube" is acquired by a mapping system. As a consequence, each pixel has a spectrum. If we know *a priori* that a compound has a selective wavelength at 1,000 cm^{-1}, it is then possible to integrate the signal for each pixel and its corresponding chemical map is generated.

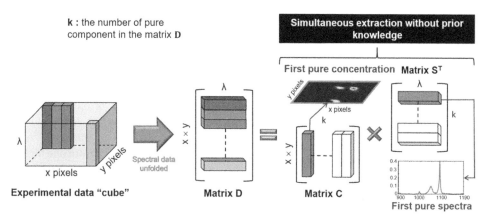

FIGURE 12.2 The multivariate curve resolution in hyperspectral imaging.

These approaches in hyperspectral imaging make possible to extract simultaneously and without any prior knowledge on the sample, the pure spectra of each components and their respective chemicals maps (de Juan et al. 2004, Duponchel et al. 2003). There is no need to select a specific wavelength to use these approaches (see Figure 12.2). The pure component's concentration distribution and spectral matrices, respectively, noted $\mathbf{C}$ and $\mathbf{S^T}$ are obtained by a bilinear decomposition of a matrix noted $\mathbf{D}$ obtained by unfolding the experimental data "cube". In this chapter, we will present the MCR-ALS because it is one of the most adaptable and powerful methodology.

12.2.1 The Multivariate Curve Resolution and Alternative Least Square (MCR-ALS) in Hyperspectral Imaging

MCR-ALS algorithm has been successfully applied in many research fields as chromatography (Tauler and Barceló 1999), electrophoresis (Godoy-Caballero et al. 2013), infrared spectroscopy (Laxalde et al. 2014), Raman (Liu et al. 2017), UV-Visible (Fernández et al. 2013), fluorescence (Bosco et al. 2007), nuclear magnetic resonance (Winning et al. 2008), mass spectroscopy (Hugelier et al. 2018) or Surface-Enhanced Raman Spectroscopy (Mamián-López and Poppi 2015). Hyperspectral imaging data acquired correspond to a data "cube" (i.e. three dimensions) x pixels $\times$ y pixels $\times$ λ wavelengths. The main idea of the MCR-ALS method is that each pixel can be described as a linear combination of the signal of a set of pure component spectra. The term "pure component" can be the pure signal of a chemical compound or a part of a sample (i.e. mixture) with a proper spectral signature. This basic assumption of MCR-ALS in spectroscopy is the validity of the multi-component Beer–Lambert law for a studied system. As we have seen on Figure 12.2, the data cube is unfolded into a matrix $\mathbf{D}$ (two dimensions) with rows (x pixels $\times$ y pixels) and columns (λ). Then, the MCR-ALS algorithm decomposes the matrix $\mathbf{D}(n \times m)$ into the product of the distribution maps (matrix $\mathbf{C}(n \times k)$) and the spectra of the pure component (matrix $\mathbf{S^T}(k \times m)$). In other words, a bilinear

model of k pure components can reproduce the matrix $\mathbf{D}$ containing n mixed spectra collected at m spectral channels with $n \times k$ concentration profiles and their $k \times m$ related spectral profiles (Eq. 12.1). The error contribution of the model is represented by the matrix $\mathbf{E}(n \times m)$.

$$\mathbf{D} = \mathbf{CS^T} + \mathbf{E} \tag{12.1}$$

The MCR-ALS methodology is an iterative algorithm that recovers the underlying spectroscopic bilinear model under the action of constraints on a single data set or multiset structures formed by several concatenated data matrices. The MCR-ALS algorithm is based on the following steps:

1. Estimation of the k pure components in the matrix $\mathbf{D}$ by rank analysis. To do this, a Singular Value Decomposition (SVD) can be applied (Golub and Reinsch 1970). In this determination, it is assumed that singular values associated with experimental noise are significantly lower than those associated with systematic chemical data variance. Otherwise, it is rank-deficient (Ruckebusch et al. 2006). It is a key step that leaves no stone unturned because it conditions the relevant MCR-ALS resolution.

2. The generation of initial estimates ($\mathbf{S^T}$) e.g. SIMPLe-to-use Interactive Self-modelling Mixture Analysis (SIMPLISMA) approach or Orthogonal Projection Approach (OPA) can be used for selecting the purest pixel spectra (Gourvénec et al. 2002). It is possible to estimate $\mathbf{C}$ for initialization, but it is not exposed in this chapter.

3. ALS calculation of $\mathbf{C}$ and $\mathbf{S^T}$ under constraints until proper reproduction of $\mathbf{D}$ i.e. until the criteria of convergence in the optimization process is achieved (de Juan et al. 2009).

The ALS step involves the operations $\mathbf{C} = \mathbf{DS}(\mathbf{S^T S})^{-1}$ and $\mathbf{S^T} = (\mathbf{C^T C})^{-1}\mathbf{CD}$ respectively. The end of the iterative process takes place when the product of the resolved concentration profiles and spectra reproduce the original data $\mathbf{D}$ without significant variation between the results of consecutive cycles (minimization process). The parameter used to measure the fit quality of the MCR-ALS model is the explained variance (r^2) and the lack of fit, lof (%), defined as follows:

$$r^2\,(\%) = 100 \times \left(1 - \frac{\sum_{i,j} e_{ij}^2}{\sum_{i,j} d_{ij}^2}\right) \qquad (12.2)$$

$$\text{lof}\,(\%) = 100 \times \sqrt{\frac{\sum_{i,j} e_{ij}^2}{\sum_{i,j} d_{ij}^2}} \qquad (12.3)$$

where e_{ij} are the elements of the **E** matrix and d_{ij} are the elements of the raw data set **D**. The subindexes i and j refer to the pixel and wave number, respectively. During the MCR-ALS steps, the algorithm allows introducing previous knowledge about the chemical data by the use of constraints on **C** and/or $\mathbf{S^T}$ profiles (i.e. in the concentration and/or spectral direction). Constraints can be very diverse (e.g. non-negativity, unimodality and closure) and can be used to all or to some of the components in the system analyzed (Bro et al. 1998). The use of constrains in the resolution process helps to decrease rotation ambiguity effects in the final results obtained. After the multivariate curve resolution analysis of matrix **D**, the pure distribution maps of the components can be obtained by folding back the stretched concentration profiles in **C** to recover the original 2D spatial structure of the image. At this step, the hyperspectral imaging with the MCR-ALS data analysis gives a better interpretation and knowledge of the chemical reality. The distribution of pure components contained in the sample is an asset for microscopic studies; nevertheless, the spatial resolution limit can be a real drawback when micronic or submicronic samples are analyzed.

Indeed, even if hyperspectral images are powerful analytical measurements, far-field imaging spectroscopy instruments didn't have the expected development despite of the interest in nanoscience e.g. on multi-channel detection, the first way to increase the resolution is the pixel-size reduction on the well-known charge-coupled device (CCD) sensors. However, the pixel size reduction increases unfortunately the shot noise. On single-channel detection, the spatial resolution limit is first and foremost dictated by the diffraction limit. The near-field spectroscopy was then invented to push back the spatial resolution limit. Nevertheless, this approach is complex to setup in both the instrumental and environmental condition (Paesler and Moyer 1996). In the 1980s, the signal processing community showed the possibility to overcome the limitations of optical systems. The image processing algorithms called super-resolution were born (Tsai and Huang 1984, Nguyen and Milanfar 2001, Farsiu et al. 2004b). The main advantage of such signal processing approach is that it costs less then replacing totally or a part of the instrument. The existing imaging systems can be still used even if it is not optimal. It is sometimes the only way to increase the spatial resolution since no alternative instrumental setup may be available.

12.3 The Super-Resolution Concept in Hyperspectral Imaging

The super-resolution is defined by the use of image processing methods from the signal processing research field to overcome limitations of optical systems. The main idea of super-resolution concept is the fusion of several low-resolution (LR) images of the same "object" from slightly shifted point of view to generate one higher-resolution image i.e. with higher spatial details (Farsiu et al. 2004a). The N LR images (i.e. low pixel density) are defined by $M_1 \times M_2$ pixels and are denoted $\{\mathbf{C}_k\}_{k=1}^{N}$. These LR images can be considered worse and different representations of a single image with much-higher resolution noted $\mathbf{C}_{\text{SR}}$ defined by $L_1 \times L_2$ pixels, where $L_1 > M_1$ and $L_2 > M_2$ for $1 \leq k \leq N$. The parameters L_1 and L_2 are user-defined that always satisfy the following rule of thumb:

$$L_1 L_2 \leq N M_1 M_2 \qquad (12.4)$$

Let $L_1 = n M_1$ and $L_2 = n M_2$ with n the resolution enhancement factor which must be selected has the highest value to satisfy Eq. 12.4. The super-resolution concept considers that each LR image's $\mathbf{C}_k$ is the result of a particular geometric warping, linear space-invariant blurring and uniform decimation (subsampling) performed on the ideal high-resolution image $\mathbf{C}_{\text{SR}}$ (Farsiu et al. 2004a). In general, we also consider that the k-th LR images are corrupted with an additive Gaussian noise. Thus, it is possible to express the steps previously described with an analytical model (Eq. 12.5):

$$\mathbf{C}_k = \mathbf{X}_k \mathbf{H}_k \mathbf{T}_k \mathbf{C}_{\text{SR}} + \mathbf{E}_k \qquad (12.5)$$

The matrix $\mathbf{T}_k$ sized $[L_1 L_2 \times L_1 L_2]$ corresponds to the geometric translation between the $\mathbf{C}_{\text{SR}}$ image and $\mathbf{C}_k$ images. The matrix $\mathbf{H}_k$ sized $[L_1 L_2 \times L_1 L_2]$ is the blur matrix defined by the optical system point spread function (PSF). The PSF describes the response of an imaging system onto a point source or point object (Williams 1998). In other words, it corresponds to the 2D distribution optical system of light in the focal plane. The matrix $\mathbf{X}_k$ sized $[M_1 M_2 \times L_1 L_2]$ is the decimation resulting in $\mathbf{C}_k$. It corresponds to the reduction of the number of observed pixels in the measured images and it shows some aliasing effects (Shannon 1948). The additive Gaussian noise observed in the k-th LR image is noted $\mathbf{E}_k$. Figure 12.3 displays the effect for each operation considered previously.

A generalization of this model can be applied to all N LR images (Eq. 12.6).

$$\begin{bmatrix} \mathbf{C}_1 \\ \vdots \\ \mathbf{C}_N \end{bmatrix} = \begin{bmatrix} \mathbf{X}_1 \mathbf{H}_1 \mathbf{T}_1 \\ \vdots \\ \mathbf{X}_N \mathbf{H}_N \mathbf{T}_N \end{bmatrix} \mathbf{C}_{\text{SR}} + \begin{bmatrix} \mathbf{E}_1 \\ \vdots \\ \mathbf{E}_N \end{bmatrix} \qquad (12.6)$$

Here, the matrix $\mathbf{C}_{\text{SR}}$ must be found from the set of LR images $\{\mathbf{C}_k\}_{k=1}^{N}$ and the estimation of $\mathbf{X}_k$, $\mathbf{H}_k$ and $\mathbf{T}_k$ matrices for all $k = 1, \ldots, N$. Super-resolution is, therefore, an inverse problem, and this mathematical problem is hard

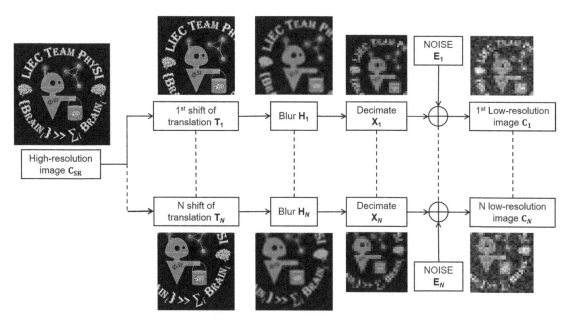

FIGURE 12.3 Representation of the analytical model used in the super-resolution concept. A LR image is a noisy, decimated, blurred and warped version of the original high-resolution image.

to resolve (Vogel 2002). Indeed, the number of LR images are often much smaller than the dimension of $\mathbf{C}_{\mathrm{SR}}$ i.e. an ill-posed problem. Furthermore, if $\mathbf{A} = \mathbf{X}_k \mathbf{H}_k \mathbf{T}_k$, then the matrix $\mathbf{A}$ is difficult to invert because the determinant is close to 0 i.e. ill-conditioned problem. It is inherent numerical instability of such inverse problem. As a result, small variations in $\{\mathbf{C}_k\}_{k=1}^N$ induce large errors on $\mathbf{C}_{\mathrm{SR}}$ estimations. Despite of all these difficulties, it is possible to simplify the analytical model of super-resolution concept in hyperspectral imaging.

Whatsoever the detection mode for our spectroscopic experiment, the instrumental setup is a spectrometer coupled to a microscope with a motorized table As a consequence, $\mathbf{T}_k$ matrices are just defined by the shift between LR images $\{\mathbf{C}_k\}_{k=1}^N$, with respect to one of them taken as a reference. These shifts are translation movements from the XY piezo-motorized stage between two acquisitions and are defined by the user. The main idea of the super-resolution concept is that these shifts must be lower than the pixel size of the LR image. The pixel size is also defined close to the intrinsic spatial resolution from the instrument setup. Furthermore, it is possible to consider that $\forall k, \mathbf{H}_k = \mathbf{H}$ because all LR images are obtained with the same optical system and the PSF remain constant all over the time experiment. Generally, this matrix is estimated from a spatial resolution evaluation, but even a rough guess of it can be used for the super-resolution algorithm e.g. Gaussian filter. The matrix $\mathbf{X}_k$ depends only on the decimation ratio between the high-resolution image and the resolved LR images i.e. the ratio between the numbers of pixels of the two. Ultimately, taking these facts into consideration, it is now possible to retrieve the high-resolution image $\mathbf{C}_{\mathrm{SR}}$ with a super-resolution algorithm.

In the last three decades, many different iterative and non-iterative methods have been proposed to resolve

the model. The practitioners of super-resolution usually explicitly or implicitly define a cost function to estimate $\mathbf{C}_{\mathrm{SR}}$ and the most common is the least-squares (LS) cost function which minimizes the L_2 or the L_1 norm of the residual vector (Farsiu et al. 2004a, Vandewalle et al. 2007). In our case, we choose L_1 norm (Eq. 12.7) because it has been shown to be very robust to data outliers compared to L_2 norm (Farsiu et al. 2004a).

$$\widehat{\mathbf{C}_{\mathrm{SR}}} = \underset{\mathbf{C}_{\mathrm{SR}}}{\mathrm{ArgMin}} \left[\sum_{k=1}^N \|\mathbf{XHT}_k \mathbf{C}_{\mathrm{SR}} - \mathbf{C}_k\|_1 \right] \quad (12.7)$$

However, for a singular model matrix like $\mathbf{A} = \mathbf{XHT}_k$, there is an infinite space of solutions minimizing Eq. 12.7. As a consequence, some form of regularization must be included in the cost function in order to stabilize the inverse problem (i.e. to find a stable solution of $\mathbf{C}_{\mathrm{SR}}$). In other words, a regularization term is necessary to constrain the space of solutions. A regularization term called bilateral total variation (BTV) was applied due to its robust performance while preserving the edge content common to real image sequences (Farsiu et al. 2004a). The Eq. 12.8 formulates the minimization criteria.

$$\widehat{\mathbf{C}_{\mathrm{SR}}} = \underset{\mathbf{C}_{\mathrm{SR}}}{\mathrm{ArgMin}} \left[\sum_{k=1}^N \|\mathbf{XHT}_k \mathbf{C}_{\mathrm{SR}} - \mathbf{C}_k\|_1 \right.$$
$$\left. + \lambda \sum_{l=0}^P \sum_{m=0}^P \alpha^{m+l} \left\| \mathbf{C}_{\mathrm{SR}} - S_x^l S_y^m \mathbf{C}_k \right\|_1 \right] \quad (12.8)$$

The scalar λ is for properly weighting the first term (L_1 norm) against the second term (regularization cost). The operators S_x^l and S_y^m correspond respectively to shifting the $\mathbf{C}_{\mathrm{SR}}$ by l pixels in horizontal direction and by m pixels in vertical direction. The scalar weight α is applied to give a spatially decaying effect to the summation of the regularization term. In order to validate the super-resolution concept coupled with MCR-ALS in Raman hyperspectral imaging, it is necessary to define and to measure the intrinsic spatial resolution of the instrumental setup.

12.4 The Criteria to Measure the Spatial Resolution in Hyperspectral Imaging

The spatial resolution is defined as the smallest observable distance between two separate objects (Lasch and Naumann 2006, Stelzer 1998). It is also necessary to make several numerous experimental methods to estimate it. The purpose of this part is not to find the best criterion but to propose two techniques to objectively and quantitatively estimate the improvement brought by the new concept of super-resolution: The *Rayleigh criterion* and the "step-edge" (Levenson et al. 2006, Offroy et al. 2010). An ideal sample called "target" sample is required for these two methods on spectroscopic measurements. It is considered as an ideal sample because it is spatially and chemically well defined. Thus, the "target" sample has a substrate where patterns are uniformly distributed and well defined with different sizes (see Figure 12.4). Of course, the composition of the substrate and patterns depend of the instrument and spectroscopy. Indeed, the materials used to build these ideal samples should be adjustable with the spectroscopic experiment applied. The main idea is to have the best contrast when the signal of the instrument goes on the substrate to the pattern.

For the Rayleigh criterion, spectral acquisitions are performed though well-defined consecutive "bars" (three or more) in the x or y direction. The size of a "bar" is equal to the distance between two consecutives "bars". Considering the signal profile along perpendicular axis to the bars, if the signal between three consecutive peaks (S) is less than $8/\pi^2$ (i.e. 81%) of the total intensity of the peak (S_{max}) then the three "bars" are resolved. The distance between the two signal maxima is a spatial resolution measurement.

For the "step-edge" criterion, a spectral acquisition is performed through a step on the "target" sample. Considering the "step-edge" as an input to on optical system, then the signal along a perpendicular direction (x or y) to the edge is the line spread function (LSF). The derivative of the LSF is the PSF in a considered direction. The PSF is defined as the response of the system to a point source observation. The Full-Width at Half-Maxima (FWHM) of the PSF give the spatial resolution for a direction. This approach is very attractive since it allows in a single experiment to determine directly a quantitative measure of the spatial resolution in a considered direction (x or y).

The next parts will demonstrate the super-resolution concept applied in Raman imaging spectroscopy to push back the spatial and chemical resolution of far-field instrument.

12.5 Aerosol Characterization in Hyperspectral Raman Imaging

Metallurgical industries contribute to the air-quality deterioration with dust emissions. These aerosol particles

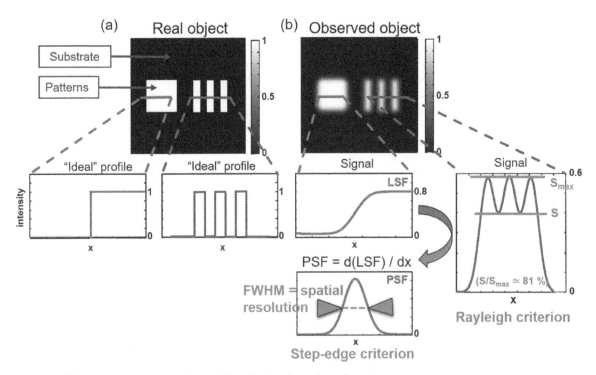

FIGURE 12.4 The criteria to measure the spatial resolution along the x direction.

aggregate and react with tropospheric reagents leading to complex heterogeneous and multiphasic phenomena which occur at aerosol size scale (Sobanska et al. 2006, 2014). The chemical composition and size of these particles have an impact on the aerosol's visibility in the atmosphere and on the climate (Forster et al. 2007). Detailed knowledge of the aerosol composition is thus needed for source identification and apportionment. Moreover, their bioavailability and toxicity properties depend on the size, morphology and chemical composition of individual aerosol particles (Batonneau et al. 2004). As a consequence, the Raman hyperspectral imaging is the perfect tool to answer on the chemical composition. Nevertheless, the spatial resolution considered as optimal for Raman imaging experiments is around 1 μm due to the photon wavelength. The diffraction limit prevents to detect micron or submicron-sized particles.

A potential solution for obtaining molecular information on micron and submicron-sized aerosols particles with in situ conditions is proposed with super-resolution and multivariate curve resolution in Raman hyperspectral imaging. Our proposed approach will be divided into two parts. Firstly, we will use a "target" sample to measure the intrinsic spatial resolution of the Raman instrument and to quantitatively evaluate the contribution of the super-resolution concept. Secondly, the super-resolution will be coupled with MCR-ALS analysis in post processing for the exploration of atmospheric particles.

12.5.1 Raman Instrumentation

A LabRAM HR confocal scanning spectrometer manufactured by Horiba Jobin Yvon Scientific Company has been used to acquire all Raman hyperspectral images. A confocal microscope Olympus BX 40 with high stability and a ×100 objective (NA = 0.9) is coupled with a spectrometer. This instrument is equipped with a 1,200 grooves/mm holographic grating that enables a spectral resolution around 0.5 cm^{-1}. A liquid nitrogen-cooled CCD detector is used in

the spectrometer. Raman backscattering is obtained with a 532 nm excitation wavelength (25 mW) supplied by a solid-state laser. The samples are placed under the microscope on a XYZ piezoelectric stage from Physik Instrument company. The displacements of this stage are up to 100 μm in the three directions with a step that can reach the nanometer (if necessary). The z direction of the table is not changed during the acquisition. It is only fixed at the beginning of the experiment in order to focus correctly on the surface of the sample. All the instrumental setup is on an optical table with a vibration control system ST-UT2 from Newport Corporation, USA. It uses pneumatic cells inside the table's legs to detect and to damp the low-frequency vibrations due to the environment (e.g. people walking, building equipment, passing automobiles, and subway). This device can be on or off when required for the experiment.

12.5.2 Increased Spatial Resolution in Hyperspectral Raman Imaging

A "target" sample has been specially created for Raman hyperspectral imaging. The selected material was a Silicon (Si) wafer, and it is considered as an "ideal" sample. Indeed, it is a perfect candidate due to its good Raman scattering properties. Silicon gives a localized and intense spectral contribution at 520.7 cm^{-1}, while Au patterns have no Raman backscattering. Electron lithography has been used to draw submicron Au patterns on a Si wafer surface (see Figure 12.5). First a poly methyl methacrylate (PMMA) electron-sensitive resist solution was spin-coated onto Si sample and baked to leave a hardened thin film on the surface (around 500 nm). An electron beam has scanned the sample following the pattern's mask and caused PMMA chain scissions. This fragmented polymer chains were then dissolved by a developer solution. A 50 nm thickness of gold was then deposited by evaporation over the entire sample surface. The final stage called "lift-off" was the immersion of the sample in acetone to remove any remaining traces of

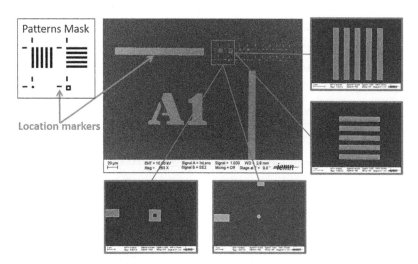

FIGURE 12.5 The target sample mask and the Scanning Electron Microscope (SEM) general view with homothetic Au submicron patterns on a Si wafer.

PMMA and finally reveal the gold patterns. The structure of the sample was inspired by the Microscopy Resolution Test Chart (NBS 101A). Namely, five bars regularly spaced horizontally and vertically. The distance between two bars corresponds to the width of a bar. A circle with a diameter equal to the bar width and a square with a contour equal to the bar width has been also added. This mask is replicated in a homothetic way in order to observe widths from 1 μm to 62.5 nm. Furthermore, markers are added to locate the different patterns when they become too small and almost invisible with the microscope.

The "target" sample has for purposes two objectives. The first one is to measure the intrinsic spatial resolution along the x and y directions of the sample plane with a quantitative criterion for different instrumental setups. And the second one is to estimate the "new" spatial resolution for the same instrumental setups after super-resolution concept. The purpose is also to find the best experimental setup for which the spatial resolution is the best before and after a super-resolution algorithm.

Measure of the Intrinsic Spatial Resolution

In confocal systems, pinholes have a direct effect on the spatial resolution. Hence, four experiments with different pinhole sizes were performed to measure the intrinsic spatial resolution of the Raman. Furthermore, a fifth experiment

has been made in order to observe the influence of the anti-vibration control system on the optical table. The "step-edge" criterion was applied on a location marker from the Si wafer (see Figure 12.6). The mapping size is 10 × 60 pixels in x direction and a 60 × 10 in y direction, with a step size of 100 nm. The spatial-resolution calculations have been done on eight LSF derivative profiles for each mapping.

Table 12.1 presents the results of the intrinsic spatial-resolution for five Raman experiments. First, the spatial resolution is significantly different in the x and y directions when the pinhole size is between 300 and 1,000 μm (experimental #1 to #3). This is explained by the fact that the Raman instrument is not in a real confocal mode with these experimental setups. Only a pinhole size equal to 200 μm gives the same spatial resolution in each directions. Second, the experiments #4 and #5 show fairly equal spatial resolution which demonstrates that an active anti-vibration system does not change the intrinsic spatial resolution for conventional Raman measurements. As a result, the intrinsic spatial resolution in each direction can be considered to be around 0.59 μm. Moreover, the standard deviation on each values confirms the excellent repeatability of the "step-edge" criterion. Indeed, the standard deviation does not exceed 20 nm. It is also interesting to note that the intrinsic spatial resolution experimentally is much worse

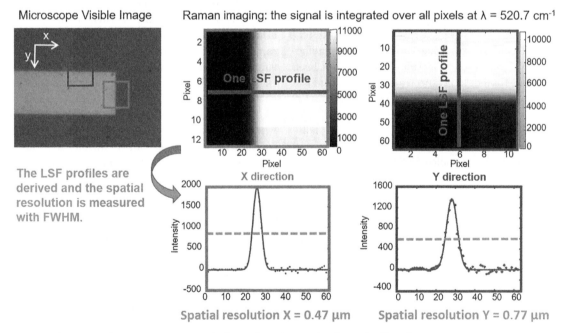

FIGURE 12.6   The "step-edge" criterion was applied on a location marker from the Si wafer.

TABLE 12.1   The Intrinsic Spatial Resolution Measurements of the Raman Instrument for Different Experimental Setups

		Intrinsic Spatial Resolution		
Experiments	Pinhole Size (μm)	Anti-vibration System State	Along x-Direction (μm)	Along y-Direction (μm)
#1	1,000	On	0.47 ± 0.01	0.77 ± 0.01
#2	500	On	0.46 ± 0.02	0.78 ± 0.02
#3	300	On	0.52 ± 0.01	0.66 ± 0.01
#4	200	On	0.59 ± 0.02	0.56 ± 0.01
#5	200	Off	0.57 ± 0.01	0.62 ± 0.01

than the one calculated theoretically with the diffraction limit formula for a confocal system (see Eq. 12.9).

$$\text{Diffraction limit} \approx 0.46 \times \frac{\lambda}{\text{NA}} \quad (12.9)$$

This equation predicts a spatial resolution around 272 nm for the 532 nm wavelength laser (excitation) and a $\times 100$ objective (NA = 0.9). These first results provide strong evidence that the spatial resolution must always be estimated for having an unbiased "vision" of the instrumental performance.

Characterization of the Improved Spatial Resolution

The spatial resolution of the super-resolved image was measured following the previous methodology i.e. "step-edge" criterion on location marker. A set of LR images was obtained by inducing a sub-pixel shift (lower than the pixel size) in the x and y directions between them. As it happens, 100 Raman data cubes over a location marker with a pixel size of 1 μm, shifted from 100 nm from each other in x and y directions, were acquired. Then, all data cubes were integrated at 520.7 cm^{-1} to obtain 100 LR Raman images. The super-resolution algorithm (i.e. "Norm L1 + BTV") was then used to generate a high-resolution image of a location marker. Table 12.2 presents the spatial resolution calculated from the super-resolved image with the "step-edge" criterion on the same five experiments' conditions. Even if there is still a significant difference between the x and y directions for a pinhole's size from 300 to 1,000 μm, the new spatial resolution is always better than the intrinsic spatial resolution. As observed before, only a 200 μm pinhole size provides an equivalent spatial resolution in each direction. In this best condition (i.e. confocal mode), the new spatial resolution is around 200 nm in each direction and it corresponds to improvement of at least 65% compared with the intrinsic spatial resolution. We can state that these results are by all means the most spectacular because we also push back the diffraction limit of this Raman instrument. This gain is far from negligible when submicron samples are considered. We also notice that the spatial resolution values of experiment #5 are not indicated in Table 12.2. Indeed, the super-resolution failed to converge to a stable solution. An anti-vibration system seems necessary.

In order to prove again the assertion for which the anti-vibration system is necessary for applying the super-resolution concept, we have redone another set of experiments. The intrinsic spatial resolution is around 0.59 μm,

it seems interesting to use the super-resolution concept on the pattern with the corresponding widths i.e. 0.5 μm. We then focused on the pattern with five vertical bars. The same experimental conditions of the experiments #1 and #5 were redone on this particular pattern. To demonstrate the power of the super-resolution concept, we have redone the experiment #1 because it is the worst confocal condition for the Raman instrument. Figure 12.7 presents the super-resolved images for these two experiments on a 5-bar pattern. As one can see, experiment #5 does not allow us to retrieve the pattern while experiment #1 does. This fact indicates that even using a piezo stage is not enough when applying a super-resolution algorithm. Indeed, the highest reproducibility in position is guaranteed by the anti-vibration system.

When for the first time our super-resolution results had been presented for other spectroscopic techniques, people thought the concept was just "over-sampling". In order to dispel these doubts, Figure 12.8 compares a super-resolution image from the experiment #1 with a new experimental-integrated image with the same pixel density. The new mapping step over the 5-bar pattern is 100 nm with a 1,000 μm pinhole size and activated anti-vibration system. The spatial resolution and the contrast of the super-resolution image are significantly improved than the experimental over-sampled image. The LSF profiles confirm this observation.

This first part focused on the "target" sample. Without this step, it is possible to use the super-resolution concept, but it is difficult to know if the results are the best. Indeed, the study with this "ideal" sample allows us to determine the optimal experimental conditions in order to use our approach (i.e. pinhole size = 200 μm and anti-vibration system table stat "on"). Furthermore, the "target" sample gives us the opportunity to find the correct regularization term used in our model. The next part will demonstrate the advantage to use both super-resolution and MCR-ALS on concrete example.

12.5.3 Spectral and Spatial Aerosol Characterization in Raman Hyperspectral Imaging

The aerosol sample was collected in the courtyard of a lead-acid battery-recycling factory located in Toulouse, France. The dust particles were collected by the impaction during the middle of afternoon on silver sheets using a cascade impactor (Dekati Ltd, Finland) operating at a flow rate of 10 L/min. The advantage of a silver sheet is that it

TABLE 12.2 The Spatial Resolution of the Instrument after Applying a Super-Resolution Algorithm for Different Experimental Setups

		Spatial Resolution after Using Super-Resolution		
Experiments	Pinhole Size (μm)	Anti-vibration System State	New Spatial Resolution Along x-Direction (μm)	New Spatial Resolution Along y-Direction (μm)
#1	1,000	On	0.18 ± 0.01	0.26 ± 0.02
#2	500	On	0.19 ± 0.01	0.33 ± 0.01
#3	300	On	0.20 ± 0.02	0.25 ± 0.01
#4	200	On	0.19 ± 0.01	0.22 ± 0.01
#5	200	Off	–	–

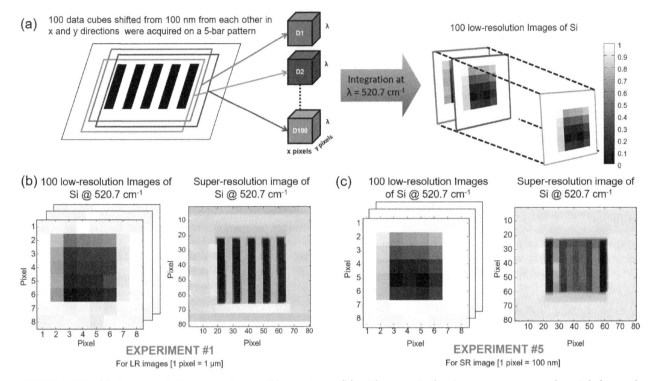

FIGURE 12.7   (a) Super-resolution concept on a 5-bar pattern, (b) with an anti-vibration system state and a pinhole equal to 1,000 μm (experiment #1), and (c) without an anti-vibration system and a pinhole equal to 200 μm (experiment #5).

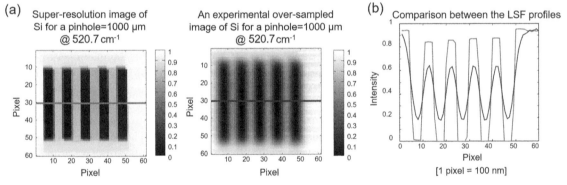

FIGURE 12.8   (a) Comparison between a super-resolution image from the experiment #1 with a new experimental-integrated image with the same pixel density and (b) comparison between their LSF profiles.

is transparent in Raman. The impactor setup is three successive stages with aerodynamic cut-off diameters (AD) which correspond to the size segregation of 2.5 μm < AD < 1 μm (stage 1), 1 μm < AD < 0.1 μm (stage 2) and AD < 0.1 μm (stage 3). The collection time was 1 h to ensure the particle dispersion on the collecting substrate. In this study, only the stage 2 was studied to demonstrate the feasibility of the analytical methodology considering submicron-sized aerosols. Unlike the target sample, we still work with 532 cm^{-1} laser excitation but we add a 0.6 density filter. The problem due to the lead oxidation is then avoided. The sample was then first explored to select a region of interest between 900 and 1,200 cm^{-1} (see Figure 12.9). We then optimized the experimental conditions on this area to obtain the best signal-to-noise ratio. The acquisition time for one data cube (i.e. dimension equal to 6 × 12 pixels

with a pixels equal to 1 μm) is around 40 min. Thus, we produced 25 data cubes shifted 0.2 μm from each other in the x and y directions. Figure 12.9 presents the 1,800 Raman spectra. Pre-processing by weighted least squares (WLS) was used on the raw spectra to remove the fluorescent effect (Piqueras et al. 2011). The WLS baseline algorithm uses an automatic approach to find which points are most likely due to baseline alone. This method is based on a recursive local fitting of the entire spectrum with a baseline obtained by applying a Whittaker smoother. In each iterative cycle, the wavelength channels that provide positive residuals in the previous iteration are down weighted in the baseline fitting operation. The fitted baseline obtained after a small number of fitting cycles is used to correct the spectrum. This method tunes two parameters: one is the smoothness of the fit and the other one is associated with

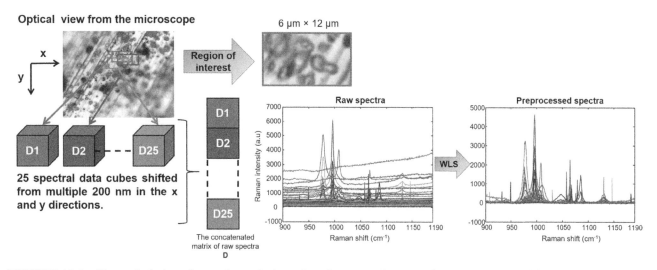

FIGURE 12.9 The optical view of aerosol sample from the microscope. A region of interest is chosen. Then 25 data cubes were acquired in x and y directions. The fluorescent effect on spectra is removed by WLS on the concatenated matrix.

the penalty imposed to the points giving positive residuals in the fit. This pre-process is also called asymmetric weighted least-square (AsLS) and was proposed originally by Eilers et al. (2003) to substrate baseline shifts in chromatography and was already particularly useful in Raman spectra.

MCR-ALS was performed on the 25 sub-matrices **D** structured as a column-wise augmented matrix, since spectra direction was common to all data cubes i.e. from 900 to 1,190 cm^{-1}. Therefore, the decomposition is carried out on 1,800 WLS-corrected spectra ($6 \times 12 \times 25$). The SVD analysis detected a rank of six for the entire **D** matrix (i.e. six pure contributions). SIMPLISMA was applied to the full multiset **D** for the initial estimates of pure spectra ($\mathbf{S}_{ini}^{T}$). The non-negativity constraint was performed on concentration profiles and spectra. The normalization constraint was also applied on $\mathbf{S}^T$ in order to avoid scaling fluctuations in the profiles during optimization (two-norm is normalizing to unit length in our case). The results of the six extracted chemical components are shown in Figure 12.10.

One advantage of the MCR-ALS on a concatenated matrix is that 25 images are extracted directly for each contribution with the associated spectra and without a prior knowledge. Indeed, the refolded matrix **C** contains the 25 images for the six components (i.e. 25 chemical maps) and $\mathbf{S}^T$ have the six associated spectra. The pixel size of the 25 images is 1 μm. The lack of fit is 9.99% and the explained variance is 99.00% which is satisfactorily the right choice in the number of contributions needed to describe the aerosol sample. The second check to confirm the validity of the MCR-ALS resolution results is the interpretability of the images and pure spectra related to these contributions. First, the contribution number #6 is due to uncorrected fluorescence effect on the boundary of the spectral domain. Then contributions #1 to #5 are pure chemical compounds or mixtures. The identification of these five contributions has been made based on the literature. The particles collected are mainly composed of sea salt-derived particles i.e. Na_2SO_4 and $CaSO_4.H_2O$ or reacted

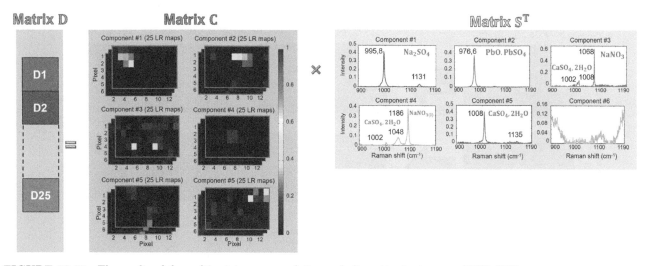

FIGURE 12.10 The results of the multivariate curve resolution and alternative least-square (MCR-ALS).

particles i.e. NaNO$_3$ in aggregation with submicronic metal-rich particles i.e. PbO.PbSO$_4$. However, internal mixing of sea salt-derived particles is observed (i.e. Na$_2$SO$_4$ and CaSO$_4$.2H$_2$O). The co-crystallization during the sea salt droplet's evaporation in the atmosphere explains the formation of mixed solid particles. The distribution maps show a rough localization of the pure chemical compounds. They are blurred and present low spatial resolution as expected from the large 1 μm pixel size in the original maps.

At this stage, the 25 distribution maps are shifted slightly from each other by 200 nm in x and y directions for each chemical compounds. The super-resolution process is then applied separately to each set of 25 images considered as low-resolved for obtaining a higher spatial resolution image. A super-resolved image is calculated from the fusion of 25 LR images (6 × 12 pixels) with a pixel size equal to 1 μm and shifted with a motion step of 200 nm between them in x and y directions. These new super-resolution images (30 × 60 pixels) then have a pixel size equal to 200 nm. As for the "target" sample analysis, we present in Figure 12.11 over-sampled images from each five chemical components which have been extracted with the same MCR-ALS and WLS procedures but on a single data cube with an oversampling

experimental step of 200 nm from the same aerosol sample area. In this way, the over-sampled and super-resolution images have the same pixel density for each pure contribution. Considering the results, one can notice directly the improvement in spatial resolution and contrast on the super-resolution results compared with over-sampling procedure. Indeed, the component #1 is much better described on the super-resolution image. The edges and the shape of the aerosol aggregates are better defined. It is also observed for area (A) and (B) on the component #2. This is even more evident on the component #4. A single particle is detected on the over-sampled map while two stuck particles seem to be observed on the super-resolution image. Some new aggregates can even be observed on the component #3 (C) or on the component #5 (D) and (E) for the super-resolution images that are impossible to observe on the over-sampled images.

These results clearly evidence that couples MCR-ALS methodology and the super-resolution post-processing lead to richer information on physicochemical description of single particles collected in industrial area than that obtained with classical Raman imaging approach. In particular, the significant improvement of the species distribution

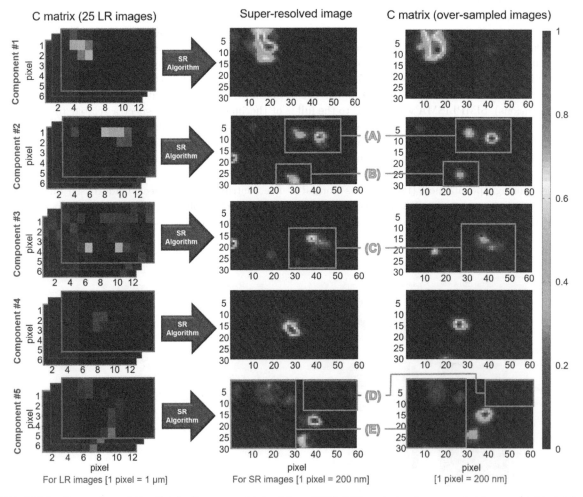

FIGURE 12.11 Comparison of the distribution maps obtained from MCR-ALS analysis of oversampled images (right plots), MCR-ALS multiset structure analysis of the LR images (left plots) and the super-resolution images (middle plots).

maps reveals aggregation of submicronic particles with a single composition whereas classical Raman imaging show a unique particle with a micronic size composed of several species in internal mixing.

12.6 Conclusion

In this chapter, we demonstrate the importance of chemometrics and super-resolution in nanoscience. Indeed, the aerosols can be more described in Raman hyperspectral imaging even if the spatial resolution is at the beginning around 0.59 μm for the considered far-field instrument. A first part was to demonstrate that the spatial resolution can be pushing back with the super-resolution concept. The study of a "target" sample shows an improvement of 65% to achieve 200 nm for the considered far-field spectrophotometer. A multivariate curve resolution called MCR-ALS was, in a second part, coupled with super-resolution in order to explore the heterogeneous structure of submicron particles for describing physical and chemical processes that may occur in the atmosphere.

Furthermore, we also demonstrated in these Raman experiments that the stability of the Raman system was a necessary condition for applying super-resolution post-processing. A high-performance anti-vibration system and a nano-displacement stage is essential. A high reproducibility in x and y directions during shifts is necessary to help the super-resolution algorithm to converge to a stable solution.

The advantage of chemometrics and super-resolution is that it could be applied in many other topics in nanoscience. Another purpose of this chapter was to present an application in a didactic way because those methodologies could be used more widely in the field of nanoscience.

Acknowledgments

We are grateful to Pr. Peyman Milanfar for assistance with super-resolution concept and Dr. Sophie Sobranska who shared the sample of atmospheric aerosols. We would like to thank Myriam Moreau in charge of the micro-spectrometry/micro-imaging Raman platform at LASIR, FRANCE and IEMN (Institut of Electronics, Microelectronics and Nanotechnology) at Université de Lille1, FRANCE, for technical support in order to create the "target" sample.

References

Batonneau, Y., Bremard, C., Gengembre, L. et al. 2004. Speciation of PM10 sources of airborne nonferrous metals within the 3-km zone of lead/zinc smelters. *Environ. Sci. Technol.* 38: 5281–5289.

Bosco, M., Callao, M.P., Larrechi, M.S. 2007. Resolution of phenol, and its di-hydroxyderivative mixtures by excitation–emission fluorescence using MCR-ALS:

Application to the quantitative monitoring of phenol photodegradation. *Talanta* 72: 800–807.

Brereton, R.G. 2003. *Chemometrics: Data Analysis for the Laboratory and Chemical Plant.* John Wiley & Sons, Ltd, Chichester.

Brereton, R.G. 2007. *Applied Chemometrics for Scientists.* John Wiley & Sons, Ltd., Chichester.

Bro, R., Sidiropoulos, N.D. 1998. Least squares algorithms under unimodality and non-negativity constraints. *J. Chemom.* 12: 223–247.

Brown, S.D., Tauler, R., Walczak, B. 2009. *Comprehensive Chemometrics: Chemical and Biochemical Data Analysis*, vol. 1–4. Elsevier Science, Amsterdam.

de Juan, A., Tauler, R., Dyson, R., Marcolli, C., Rault, M., Maeder, M. 2004. Spectroscopic imaging and chemometrics: A powerful combination for global and local sample analysis. *Trends Anal. Chem.* 23: 70–79.

de Juan, A., Maeder, M., Hancewicz, T., Duponchel, L., Tauler, R. 2009. Chemometric tools for image analysis. In: *Infrared and Raman Spectroscopic Imaging*, eds. R. Salzer, H.W. Siesler, 65–106. Wiley-VCH, Weinheim.

Duponchel, L., Elmi-Rayaleh, W., Ruckebusch, C., Huvenne, J.-P. 2003. Multivariate curve resolution methods in imaging spectroscopy: Influence of extraction methods and instrumental perturbations. *J. Chem. Inf. Compu. Sci.* 43: 2057–2067.

Eilers, P. 2003. A perfect smoother. *Anal. Chem.* 75: 3631–3636.

Farsiu, S., Robinson, M.D., Elad, M., Milanfar, P. 2004a. Fast and robust multiframe super-resolution. *IEEE Trans. Image Process.* 13: 1327–1344.

Farsiu, S., Robinson, D., Elad, M., Milanfar, P. 2004b. Advances and challenges in super-resolution. *Int. J. Imaging Syst. Technol.* 14: 47–57.

Fernández, C., Callao, M.P., Larrechi, M.S. 2013. UV–visible-DAD and ^{1}H-NMR spectroscopy data fusion for studying the photodegradation process of azo-dyes using MCR-ALS. *Talanta* 117: 75–80.

Forster, P., Ramaswamy, V., Artaxo, P. et al. 2007: Changes in atmospheric constituents and in radiative forcing. In: *Climate Change 2007: The Physical Science Basis*, Contribution of Working Group I to the Fourth Assessment Report of the Intergovernmental Panel on Climate Change (IPCC), eds. S.D. Solomon, D. Qin, M. Manning, Z. Chen, M. Marquis, K.B. Averyt, M. Tignor, H.L. Miller, 129–234. Cambridge University Press, Cambridge.

Geladi, P., Esbensen, K. 2005. The start and early history of chemometrics: Selected interviews: Part 1. *J. Chemom.* 4(5): 337–354.

Godoy-Caballero, M.D.P., Culzoni, M.J., Galeano-Díaz, T., Acedo-Valenzuela, M.I. 2013. Novel combination of non-aqueous capillary electrophoresis and multivariate curve resolution-alternating least squares to determine phenolic acids in virgin olive oil. *Anal. Chim. Acta* 763: 11–19.

Golub, G.H., Reinsch, C. 1970. Singular value decomposition and least squares solutions. *Numer. Math.* 14: 403–420.

Gourvénec, S., Massart, D.L., Rutledge, D.N. 2002. Determination of the number of components during mixture analysis using the Durbin-Watson criterion in the orthogonal Projection Approach and in the SIMPLe-to-use Interactive Self-modelling Mixture Analysis approach. *Chemom. Intell. Lab. Syst.* 61: 51–61.

Hugelier, S., Piqueras, S., Bedia, C., de Juan, A., Ruckebusch, C. 2018. Application of a sparseness constraint in multivariate curve resolution: Alternating least squares. *Anal. Chim. Acta* 1000: 100–108.

Krishnan, K., Powell, J.R., Hill, S.L. 1995. Infrared microimaging. In: *Pratical Guide to Infrared Microspectroscopy*, Pratical Spectroscopy Series 19, eds. H.J. Humeci, M. Dekker, 85–110. CRC Press, New York.

Lasch, P., Naumann, D. 2006. Spatial resolution in infrared microspectroscopic imaging of tissues. *BBA Biomembr.* 1758: 814–829.

Laxalde, J., Caillol, N., Wahl, F., Ruckebusch, C., Duponchel, L. 2014. Combining near and mid infrared spectroscopy for heavy oil characterization. *Fuel* 133: 310–316.

Levenson, E., Lerch, P., Martin, M. 2006. Infrared imaging: Synchrotrons vs. arrays, resolution vs. speed. *Infrared Phys. Technol.* 49: 45–52.

Liu, Y.-J., Andrè, S., Saint Cristau, L. et al. 2017. Multivariate statistical process control (MSPC) using Raman spectroscopy for in-line culture cell monitoring considering time-varying batches synchronized with correlation optimized warping (COW). *Anal. Chim. Acta* 952: 9–17.

Mamián-López, M.B., Poppi, R.J. 2015. SERS hyperspectral imaging assisted by MCR-ALS for studying polymeric. *Microchem. J.* 123: 243–251.

Massart, D.L., Vandeginste, B., Buydens, L., De Jong, S., Lewi, P., Smeyers-Verbeke, J. 1997. *Handbook of Chemometrics and Qualimetrics*, vol. 20 A, Elsevier, Amsterdam.

Nguyen, N., Milanfar, P. 2001. A computationally efficient super resolution image reconstruction algorithm. *IEEE Trans. Image Process.* 10: 573–583.

Offroy, M., Roggo, Y., Milanfar, P., Duponchel, L. 2010. Infrared chemical imaging: Spatial resolution evaluation and super-resolution concept. *Anal. Chim. Acta* 674: 220–226.

Otto, M. 2016. Signal processing and time series analysis. *Chemometrics: Statistics and Computer Application in Analytical Chemistry*. Wiley-VCH Verlag GmbH & Co. KGaA, Weinheim.

Paesler, M., Moyer, P.J. 1996. *Near-Field Optics: Theory, Instrumentation, and Applications*. John Wiley, New York.

Piqueras, S., Duponchel, L., Tauler, R., de Juan, A. 2011. Resolution and segmentation of hyperspectral biomedical images by multivariate curve resolution-alternating least squares. *Anal. Chim. Acta.* 705: 182–192.

Ruckebusch, C. 2016. *Resolving Spectral Mixtures with Applications from Ultrafast Time-Resolved Spectroscopy to Super-Resolution Imaging*, Data Handling in Science and Technology. Elsevier, Amsterdam.

Ruckebusch, C., de Juan A., Duponchel L., Huvenne J.P. 2006. Matrix augmentation for breaking rank-deficiency: A case study. *Chemom. Intell. Lab. Syst.* 80: 209–214.

Salzer, R., Siesler, H.W. 2009. *Infrared and Raman Spectroscopic Imaging*. Wiley-VCH, Weinheim.

Shannon, C.E. 1948. A mathematical theory of communication. *Bell Syst. Tech. J.* 27: 379–623.

Sharaf, M.A., Illman, D.L., Kowalski, B.R. 1986. *Chemometrics, Chemical Analysis Series*. Wiley, New York.

Sobanska, S., Falgayrac, G., Laureyns, J., Brmard, C. 2006. Chemistry at level of individual aerosol particle using multivariate curve resolution of confocal Raman image. *Spectrochim. Acta Part A* 64: 1102–1109.

Sobanska, S., Falgayrac, G., Rimetz–Planchon, J. et al. 2014. Resolving the internal structure of individual atmospheric aerosol particule by the combination of atomic microscopy, ESEM-EDX, Raman and ToF-SIMS imaging. *Microchem. J.* 114: 89–98.

Stelzer, E.H.K. 1998. Contrast, resolution, pixilation, dynamic range and signal-to-noise ratio: Fundamental limits to resolution in fluorescence light microscopy. *J. Microsc.* 189: 15–24.

Tauler, R., Barceló, D. 1999. Multivariate curve resolution applied to liquid chromatography: Diode array detection. *Trends Anal. Chem.* 12: 319–327.

Tauler, R., Smilde, A., Kowalski, B. 1995. Selectivity, local rank, three-way data analysis and ambiguity in multivariate curve resolution. *J. Chemom.* 9: 31–58.

Tsai, R.Y., Huang, T.S. 1984. Multi-frame image restoration and registration. *Adv. Comput. Vis. Image Process.* 1: 317–339.

Vandeginste, B., Massart, D.L., Buydens, L., De Jong, S., Lewi, P., Smeyers-Verbeke, J. 1998. *Handbook of Chemometrics and Qualimetrics*, vol. 20 B. Elsevier, Amsterdam.

Vandewalle, P., Sbaiz, L., Vandewalle, J., Vetterli, M. 2007. Super-resolution from unregistred and totally aliased signals using subspace methods. *IEEE Trans. Signal Process.* 55: 3687–3703.

Vogel, C.R. 2002. *Computational Methods for Inverse Problems*. Society for Industrial and Applied Mathematics, Philadelphia

Williams, T. 1998. *The Optical Transfer Function of Imaging Systems*. CRC Press, New York.

Winning, H., Larsen, F.H., Bro, R., Engelsen S.B. 2008. Quantitative analysis of NMR spectra with chemometrics. *J. Magn. Reson.* 190: 26–32.

Atomic Force Microscope Nanoscale Mechanical Mapping

Bernard Haochih Liu and
Nguyen Thi Phuong Linh
National Cheng Kung University

Alice Chinghsuan Chang
National Cheng Kung University
National Institute for Materials Science

13.1 Introduction

In 1986, the Nobel Prize in Physics was awarded to Prof. Ernst Ruska, Dr. Gerd Binnig, and Dr. Heinrich Rohrer, where Ruska received one half of the prize for his work that invented the first electron microscope (EM); Binnig and Rohrer shared the other half for their invention of the scanning tunneling microscope (STM). Both EM and STM are indispensable instruments that shaped the realm of nanotechnology. As Prof. Richard Feynman proposed in his famous speech, "There's Plenty of Room at the Bottom", in the 1959 American Physics Society annual meeting, researchers need tools for "manipulating and controlling things on a small scale", and the 1986 Nobel Prize credited both EM and STM techniques for providing a way to "write small" and "read small".

As the inventor of EM, Ruska had waited 55 years for the Nobel Prize; on the other hand, the two IBM scientists, Binnig and Rohrer, had designed the very first instrument in the family of the scanning probe microscope (SPM) merely 5 years ago. The STM operates on the principle of tunneling current between two nearby conductive surfaces (tip and sample) under electric bias; therefore, the STM requires a high-vacuum environment, atomically smooth specimen surface, and was limited to conductive materials. To extend the atomic-resolution imaging capability of STM to insulators, Binnig teamed up with Prof. Calvin Quate and Dr. Christoph Gerber while visiting-researching at Stanford University, to develop a new instrument, the "atomic force microscope" (AFM) in 1985. Ever since then, a variety of scanning probe techniques have been developed, and the properties investigated have been extended from surface morphology to those of electrical, magnetic, mechanical, chemical, optical, and biological natures.

At the core of the AFM/SPM technology, the cantilever probe acts as an ultra-sensitive force sensor reflecting the tip-sample interactions that connect to material properties. In addition to sensing, the AFM probe can be actuated to apply infinitesimal force on the specimen surface; such combined features of sensor and actuator make AFM an ideal tool to investigate the stress–strain relation of a material, that is, the mechanical properties of a material.

Based on a literature survey on the Web of Science using AFM and a variety of mechanical properties (e.g., modulus, elastic, hardness, and strength) as the keywords, we studied the technology trends of more than 200 highly cited papers for the past 20 years (Sirghi and Rossi, 2006, Elinski et al., 2017, Eaton et al., 2002, Sumarokova et al., 2018, Wang and Russell, 2017, Yang et al., 2017, Li et al., 2008, Zhou et al., 2013, Zeng and Tan, 2017, Calabri et al., 2008, Fuhrmann et al., 2011, Dokukin et al., 2016, Cross et al., 2008, Neugirg et al., 2016, Notbohm et al., 2011, Liu et al., 2010b, Mathur et al., 2000, Dai et al., 2007, Müller and Dufrene, 2010, Lahiji et al., 2010, Iyer et al., 2009, Huang et al., 2015, Rigato et al., 2015, Kuznetsova et al., 2007, Li et al., 2000, James et al., 2017, Eaton et al., 2008, Powell et al., 2017, Bahri et al., 2017, Takechi-Haraya et al., 2018, Rother et al., 2014, Carvalho et al., 2010, Ladjal et al., 2009, Giessibl, 2000, Arnoldi et al., 2000, Poloni et al., 2016, Guz et al., 2016, Corbin et al., 2015, Baumgartner et al., 2000, Best et al., 2001, Lekka et al., 2012a,b, Paxton et al., 2004, Nawaz et al., 2012, Cheng et al., 2015, Walczyk and Schonherr, 2014a, Chaudhuri et al., 2009, Jee and Lee, 2010, Seifert et al., 2015, Rheinlaender et al., 2011, Rebelo et al., 2013, Xu et al., 2009, Liu et al., 2010a, Pfahl et al., 2017, Cole et al., 2010, Staunton et al., 2016, Curry et al., 2017, Gannepalli et al., 2013, Iqbal et al., 2017, Hinterdorfer and Dufrene, 2006, Dimitriadis et al., 2002, Zhu et al.,

2011, Baker et al., 2016, Last et al., 2009, Rotsch et al., 1999, Walczyk and Schonherr, 2014b, Guo et al., 2017, Kumar et al., 2017, Deng et al., 2011, Zhang et al., 2012a, Lekka, 2016, Canetta et al., 2014, Rotsch and Radmacher, 2000, Labernadie et al., 2010, Calzado-Martin et al., 2016, Aguiar-Moya et al., 2017, Asay and Kim, 2006, Lu et al., 2018, Soofi et al., 2009, Iwamoto et al., 2009, Castellanos-Gomez et al., 2012, Guo et al., 2014, Hugel et al., 2001, Smolyakov et al., 2016, Arkhipov et al., 2009, Mathur et al., 2001, Ortiz and Hadziioannou, 1999, Namazu et al., 2000, Jahangir et al., 2014, Harris and Charras, 2011, Liu et al., 2011, Richter and Brisson, 2005, Reviakine and Brisson, 2000, Lin et al., 2011, Wiggins et al., 2006, Casdorff et al., 2017a, Alsteens et al., 2012, Lehaf et al., 2012, Xia et al., 2014, Garcia et al., 2006, Pfreundschuh et al., 2015, Chopinet et al., 2013, Hecht et al., 2015, Louise Meyer et al., 2010, McAllister et al., 2014, Liu et al., 2014a, Liu et al., 2013, Milani et al., 2011, Bettini et al., 2011, Zhang et al., 2014, Ferrari et al., 2010, Qin et al., 2014, Minary-Jolandan et al., 2012, Wang and Liu, 2017, Soenen et al., 2013, Cartagena and Raman, 2014, Pfreundschuh et al., 2014a, Pan et al., 2015, Chyasnavichyus et al., 2014, Raman et al., 2011, Faria et al., 2008, Lee et al., 2008, Jiang and Zhu, 2015, Yuan and Verma, 2006, Efremov et al., 2017, Domke and Radmacher, 1998, Yablon et al., 2014, Tuson et al., 2012, Carrion-Vazquez et al., 2000, Oberhauser et al., 2002, Wenger et al., 2007, Wu et al., 1998, Sun et al., 2017, Frank et al., 2007, Marshall Jr et al., 2001, Wu et al., 2005, Nowatzki et al., 2008, Takechi-Haraya et al., 2017, Sokolov et al., 2013, Loparic et al., 2010, Liparoti et al., 2017, Guyomarc'h et al., 2014, Oberhauser et al., 1998, Al-Rekabi and Contera, 2018, Pfreundschuh et al., 2014b, Dufrene et al., 2013, Chen, 2014, Ebenstein and Pruitt, 2006, Cross et al., 2007, Andriotis et al., 2014, Casdorff et al., 2017b, Sweers et al., 2011, Cerf et al., 2009, Plodinec et al., 2012, Nalam et al., 2015, Fakhrullina et al., 2017, Touhami et al., 2003, Dong et al., 2017, Yim et al.,

2010, Man et al., 2017, Ouasti et al., 2011, Schlesinger and Sivan, 2017, Wang et al., 2014, Tetard et al., 2010, Glaubitz et al., 2014, Dokukin and Sokolov, 2012a, Trtik et al., 2012, Chao and Zhang, 2011, Richter et al., 2003, Farahi et al., 2017, Camesano and Logan, 2000, Zhang et al., 2012b, Jorba et al., 2017, Mahaffy et al., 2004, Rabe et al., 2000, Schön et al., 2011, Dokukin and Sokolov, 2012b, Kim et al., 2017, Chyasnavichyus et al., 2015, Emad et al., 1998, Zhou et al., 2012, Zhong et al., 2014, Tang et al., 2009, Lorenzoni et al., 2017, Greving et al., 2012, Maherally et al., 2015, Lehenkari and Horton, 1999, Liu et al., 2014b, Adamcik et al., 2011, Darling et al., 2010, Zhang et al., 2016, Chen et al., 2017, Oberhauser et al., 2001, Butt et al., 1999, Hayashi and Iwata, 2015, Dulinska et al., 2006, Nguyen et al., 2016, Roduit et al., 2009, Bertolazzi et al., 2011, Sharma et al., 2010, Chlanda et al., 2018, Jing et al., 2006, Chen et al., 2009, Cuenot et al., 2004, Payamyar et al., 2014, Connizzo and Grodzinsky, 2017, Yao et al., 1999, Uhlig and Magerle, 2017, Young et al., 2011, Jones et al., 2002, Crichton et al., 2011, Fukuma, 2009, Ni and Li, 2006).

Figure 13.1 presents the categories of materials studied by AFM mechanical characterization. Generally, AFM has been adopted for the study of soft materials, such as biomaterials and polymers, as well as nanomaterials (thin films, nanoparticles, quantum dots, etc.) that require ultra-high resolutions. The nondestructive nature of AFM methods has been well recognized by the researchers to apply on soft materials; AFM is also capable of measuring harder materials such as metals and ceramics with success, using suitable control parameters and cantilever probes with comparable spring constants. Nevertheless, inconsistent testing protocols and the lack of quantitative measurements that can be linked and compared to other mechanical tests have stymied the use of AFM mechanical characterization. Additional details and possible solutions will be discussed in the second part of this chapter.

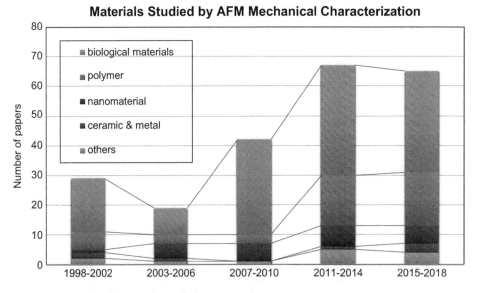

FIGURE 13.1   Materials studied by AFM mechanical characterization.

Figure 13.2 shows the AFM methods applied for mechanical property measurements by years. Since AFM manufacturers and developers tend to use different names to differentiate themselves from other competitors, we have categorized the methods based on their fundamental techniques. Among these techniques, the majority are based on the compression method, which finds the relation between the applied force by probe tip and the deformation of specimen surface. As the elastic modulus, or its reduced form, is the ratio of applied stress (force per area) and the strain (change of dimensions), one can calculate the modulus by the tip-sample interaction data. Conventional approaches include force curve, force volume, and nanoindentation; these techniques either require single-point testing or a time-consuming multiple scan of a selected area. In recent years, the use of a higher harmonic mode and a better feedback control scheme has extended the capability of single-point testing into a pixel-by-pixel force curve data acquisition during a single AFM morphology scan. The "Nanomechanical" method (e.g., Bruker PeakForce QNM, JPK QI mode, Park Systems Force Volume Imaging) provides quantitative force–distance data at each point of the AFM scan image, and thus it is possible to perform a mechanical mapping of the sample surface.

Other methods (Force Modulation, AM-FM, and Contact Resonance) are derivatives from the amplitude/frequency modulation techniques, whereas the bi-model signal is processed to infer various mechanical properties. There are also methods that use typical topographic modes (Contact, Tapping, etc.) whose phase image reveals the contrast of mechanical properties. Such methods are not quantitative, and the influences of several mechanical properties on the change of image phase angle cannot be decoupled.

The use of characterization methods for specific mechanical properties is summarized in Figure 13.3. It is clear that the "modulus" is the major focus in AFM mechanical measurement, followed by "adhesion force" and "strength". Note that usually each paper reported on multiple materials/scan modes/properties and the number of papers may not match among these three figures.

One significant difference of AFM techniques as compared to the compression/indentation test is that AFM probes have no well-defined contact area. In addition, the cantilever probe has elastic deformation during indentation, which makes the tip-sample force curve complicated. Consequently, the contact model chosen for the calculation of elastic modulus needs to consider the tip and sample materials, the status of the sample surface, the shape of the probe tips, etc., to ensure an accurate characterization result.

In the second part, we will discuss additional details on the measurement protocol for AFM mechanical mapping and the consideration of contact model selection.

13.2 AFM Mechanical Mapping and Practical Application

Able to image the nanoscale topography of specimens, AFM has been widely used to study localized mechanical properties by conducting nanoindentation tests at the interested position from the captured image. Along with developments in both AFM hardware and software, a matrix of

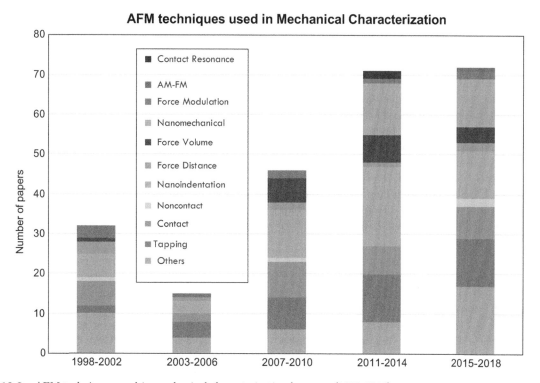

FIGURE 13.2 AFM techniques used in mechanical characterization by years (1998–2018).

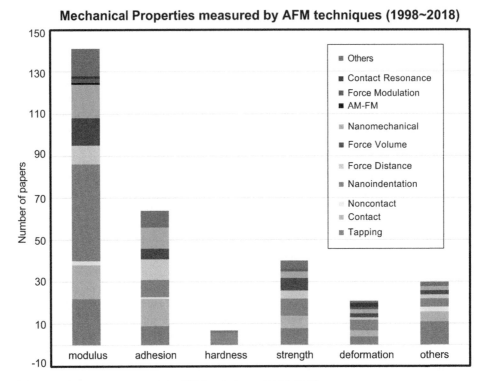

FIGURE 13.3   Mechanical properties measured by AFM techniques (1998–2018).

nanoindentation tests can be setup to reveal the mechanical distribution among sample surfaces. Such mode is known as "force volume" and is probably the earliest approach of mechanical mapping. Although the force images taken by force volume mode provide knowledge about the *in situ* mechanical performances on sample surfaces, the main drawbacks involving time-consuming scan and poor image resolution need to be overcome. Recently, AFM developers proposed a method, which applies a customized force as the signal for the feedback system during each engagement of AFM probe toward the sample surface, so that the speed of the mechanical measurements can be significantly improved and the data acquired with precise quantification (Dokukin and Sokolov, 2012b, Sahin and Erina, 2008).

During every contact of AFM probe and specimen, the force–distance curve is collected for the calculation of different mechanical properties of the specimen material. The commonly considered mechanical properties derived from AFM include elastic modulus, adhesion, deformation, and energy dissipation, and we will introduce the working principles of these mechanical properties and examples of the practical applications below.

13.2.1   Elastic Modulus

Elastic modulus E, also known as Young's modulus, indicates the resistance of elastic deformation of material under an external load. Such property can be evaluated from the gradient of a force–distance curve, where material possessing higher E is more resistant to the applied force so that a steeper force–distance curve can be found. To quantify E of

the tested specimen, the elastic deformation of a sample under a force becomes a challenge in the case of AFM nanoindentation, and a variety of contact mechanism models have been proposed for the definition of contact condition between tip and sample surface during nanoindentation. With an assumption of elastic behavior of material, those contact mechanism theories define the contact area between tip and sample and sample deformation during nanoindentation in accordance with tip geometries. With the longest history among all of the contact mechanism theories, the Hertz model (Eq. 13.1) is doubtless the most common one selected for the calculation of specimen E in AFM nanoindentation (Hertz, 1881):

$$E = 3F\left(1 - \nu^2\right)/4R^{1/2}\delta^{3/2} \qquad (13.1)$$

Equation 13.1 describes the contact between a spherical tip with tip radius R and derives material E from the relation of applied force F, indent depth δ, and Poisson's ratio ν of the tested material. In addition to the Hertz model, there are several models defining different contact mechanisms of tip and sample surface and will be discussed later in this chapter.

The elastic modulus characterization by AFM mechanical mapping has proven a useful tool in the study of living microorganisms (Chang et al., 2017, Chang, 2018). In a recent study, for example, *Staphylococcus aureus* (*S. aureus*) was treated in citric acid with different pH values using a 3D printed AFM fluid cell (Nguyen, 2018). The *S. aureus* is a Gram-positive bacterium and ubiquitous human pathogen that enters the body via broken skin or mucous membranes, causing medical complications such as a nosocomial infection of surgical wounds in hospitals (Kochan et al., 2018, Lowy, 1998, Cotter and Hill, 2003). Figure 13.4 shows the representative *in situ* time-lapsed 3D elastic modulus mapping of individual bacterium treated in citric acid with pH = 2.30. It is clear that at $t = 15$ min (Figure 13.4b) the modulus tends to increase the mapped area of values in the range [−30.8–11.6 nN] and [11.6–53.58 nN], while the range [−72.36 to −30.38 nN] decreased.

Additionally, at $t = 30$ min (Figure 13.4c), the range of the modulus of bacteria [−72.36 to −30.38 nN] increased while modulus values in the ranges [−30.8 to 11.6 nN] decreased; furthermore, the modulus in [11.6–53.58 nN] disappeared afterward. However, the modulus mapped area of values in the ranges [−30.8 to 53.58 nN] increased at around $t = 45$ min and $t = 60$ min in (Figure 13.4d,e); and at $t = 75$ min (Figure 13.4f), the elastic modulus mapping was dominated by the ranges [11.6–95.56 nN]. In summary, the modulus of the bacteria started to increase at $t = 30$ min while the modulus of the substrate decreased when the bacterial ones increased. The mechanism of the citric acid effect to the bacteria was proposed to be that of citric acid, being a weak acid, has potent antimicrobial activity as the undissociated form of weak acids can pass freely through the cell membrane. Additionally, the bacterial cytoplasmic pH is generally higher than that of the growth medium; the weak acid dissociates, releasing a proton which leads to acidification of the cytoplasm. Thus, the citric acid can dissociate into citrate and hydrogen ions, and citrate can start the citric acid (Krebs) cycle to produce NADH, H^+, and $FADH_2$ in order to synthesize adenosine triphosphate (ATP), and the by-product H_2O. More interestingly, Krebs cycle starts from sugar synthesis of every living specimen, particularly the transfer of a two-carbon acetyl group from acetyl-CoA to the four-carbon acceptor compound (oxaloacetate) to form a six-carbon compound (citrate); the citrate was produced by citric acid dissociation, which can also start the Krebs cycle.

Nonetheless, there is no NADH and $FADH_2$ inhibition for *S. aureus*, which means the amount of citric acid was not regulated by bacterial cells. Thus, citric acid can damage the *S. aureus* cell more in the treatment, and the bacteria-citric acid reactions brought about the substrate modulus mapping changes (Nguyen, 2018).

13.2.2 Adhesion

After conducting nanoindentation, the AFM probe withdraws from the sample surface and experiences attraction from sample that is attributed to the surface adhesion characteristic of material. The adhesion property can be directly obtained from the minimum value of force at the retract force–distance curve, where an evident peak of negative force is generally captured. The attraction can be explained as the effect of water meniscus on the sample surface, the hydrophobic characteristics of the sample surface (the commonly used silicon tip is hydrophobic), or the existence of adhesive substances on the sample surface.

In our previous study, the change of local adhesion properties of the *Streptococcus mutans* biofilm as a function of observation times was monitored, which was displayed in Figure 13.5 (Liu and Yu, 2017). Corresponding to the color scale, it is evident to see the adhesive evolution at the region between bacterial cells, where more than a 66 nN increase in adhesion can be observed. Such result was connected to the secretion of an adhesive substance, such as extracellular polymeric substance (EPS), for the formation of biofilm by

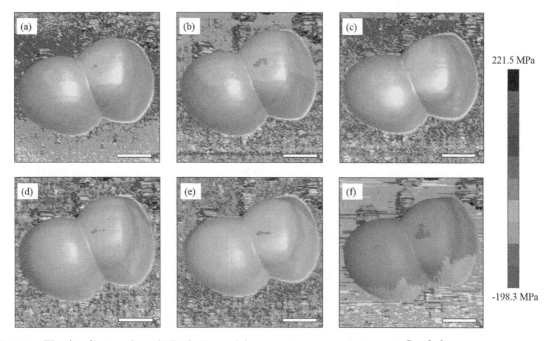

221.5 MPa

−198.3 MPa

FIGURE 13.4 The *in situ* time-lapsed 3D elastic modulus mapping on morphologies of *Staphylococcus aureus* treated in citric acid (pH = 2.30). The time stamps of (a)–(f) are 0, 15, 30, 45, 60, and 75 min, respectively (Nguyen, 2018).

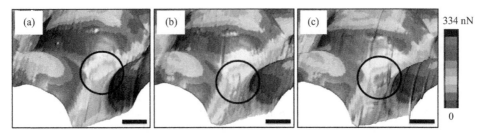

FIGURE 13.5 *In situ*, time-lapsed adhesion force maps of *S. mutans* cells. The mapping on bacterial cells was monitored at time sequences, which were captured at (a) 17 min, (b) 39 min, and (c) 62 min after sample preparation. The scale bars were 200 nm (Liu and Yu, 2017).

S. mutans cells. As a result, continuous adhesion mapping can be a suitable method for the investigation of cell activity and the secretion or metabolic mechanism of cells in biological fields.

13.2.3 Deformation

In nanoindentation, material is deformed under an external load, and its deformation can be determined from the extended force–distance curve as the distance into the sample surface, that is, the indent depth, at a certain applied force. Although the calculation methods are different, deformation somehow has a negative correlation with sample's elastic modulus E – at constant force, in which a sample with a higher E exhibited better mechanical resistance of elastic deformation, and consequently, a steeper force–distance curve with less indent depth will be found. Deformation mapping reveals the easiness or difficulty for an AFM probe to indent on a sample surface and can be used as an indicator for the structural stability of material.

In Figure 13.6, the *S. mutans* biofilm specimens were immersed into 75 wt% alcohol with various treatment time periods (Yang, 2015). With a longer treatment time, the biofilm cells show decreasing viability (data not shown) and the deformation performances of *S. mutans* biofilm exhibit an increasing trend, where Figure 13.6a, referring to a 60-s-treatment time, presents an average deformation of about 4 nm and the 150-s-treatment specimen (Figure 13.6c) has a deformation of approximately 20 nm under a constant applied force by AFM probe. The observation suggests that the deformation mapping can be used for the examination of the change in material structure or strength under various testing factors.

13.2.4 Dissipation Energy

When an AFM probe conducts nanoindentation on a sample surface, the energy of the probe can be absorbed by the nonelastic sample and results in a loss of energy that is defined as the area encircled by the extended and retracted force–distance curves. The ability of material to acquire energy from the probe during nanoindentation is called energy dissipation and is related to other mechanical properties. For instance, ideal elastic materials possess energy dissipation close to 0, where the collected extended and retracted force–distance curves are highly overlapped, and the adhesive specimen shows high energy dissipation, as the probe requires extra energy to break away from the sample surface. It has been noted that water has damping characteristics, so that the given energy will be trapped in the water-containing material.

Bacterial biofilm includes bacterial cells and a matrix that is composed of EPS, water, etc., and energy dissipation is expected to be higher in the region of biofilm matrix than the bacterial cells (Liu and Yu, 2017). Figure 13.7a shows the topography of the *S. mutans* biofilm and Figure 13.7b presents the corresponding energy-dissipation map. Apparently, the *S. mutans* cells have lower energy dissipation while the biofilm matrix, which contains approximately 70% water, exhibits higher energy dissipation behavior. Such differences between bacterial cells and the surrounding biofilm indicate that energy dissipation mapping could be adopted for the examination of water composition in samples and for the quantification of water absorption by materials.

Summarizing the mechanical properties of materials that can be characterized by AFM and the practical applications

FIGURE 13.6 *In situ*, time-lapsed deformation maps of *S. mutans* cells. Deformation mapping on bacterial cells treated with 75 wt% alcohol for (a) 60 s, (b) 120 s, and (c) 150 s. The scale bars were 500 nm (Yang, 2015).

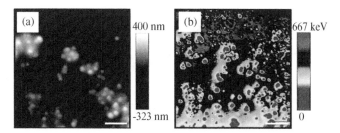

FIGURE 13.7 Topography and energy dissipation map of *S. mutans* biofilm. (a) The topography image of the cells and biofilm and (b) the corresponding energy dissipation map. The scale bars were 3 μm (Liu and Yu, 2017).

of AFM mechanical mapping, the precision and accuracy of the calculated mechanical properties are then considered. Noticeably, a slight change in experimental parameters can be greatly enlarged for nanoscale measurements and lead to a significant difference in the derived mechanical properties. In the following section, the common factors that may influence the assessment of practical mechanical properties of materials will be identified and the selection of proper testing parameters will be suggested.

13.3 Influential Factors for Mechanical Measurements

Before conducting AFM nanonidentation on materials for the quantification of mechanical properties, experimental uncertainties that contribute to AFM probes consisting of tip and cantilever should be excluded. Here, the focus is on the geometries of tip and spring constants of cantilever.

13.3.1 Tip Geometry

Different from conventional compression tests, which contact the opposite planes of specimens for applying loads and calculate the mechanical properties depending on the surface area of samples, the AFM probe is contacting the sample with its front part, and thus, the tip shape cannot be neglected. The contact area between the probe and sample varies as a function of indent depth, and several contact mechanism models have been proposed to fit different tip shapes for the computation of mechanical properties. To define the contact area, the geometrical parameters, including radius for a spherical tip, half-open angle for a conical tip, and cylindrical radius for a flat punch, are taken into account and expressed with various orders in the equations. Currently, the most commonly adopted models are probably the Hertz model for a spherical tip, the Sneddon model for a conical tip, and the flat-punch model for a cylindrical tip, as exhibited in Figure 13.8 (Hertz, 1881, Sneddon, 1965, Chang, 2018).

Even with the specific contact mechanism models established for various tip geometries, divergences of the derived mechanical properties measured by different tips have been

reported. Conducting nanoindentation on polymer and biological materials, the conical tip displayed higher elastic modulus of the sample than the spherical tip, and such a result contributes to the hardening effect of material, which is induced by stress concentration using the sharp tip (Chang et al., 2016, Rodriguez-Florez et al., 2013, Schwiedrzik and Zysset, 2015).

13.3.2 Spring Constant of Probe Cantilever

The cantilever of the AFM probe plays a role in quantifying the force during nanoindentation so that the mechanical properties of materials can be calculated. In accordance with Hooke's law (Eq. 13.2), force is the product of the spring constant k and the deflection of the cantilever, and k is proportional to the increment of the force unit. The increment of the force unit relates to resolution in both force and indent distance, which is critical for shallow indentation. Therefore, a probe with a compliant cantilever is considered to have a higher sensitivity than a hard one:

$$F = k \times d \tag{13.2}$$

Nevertheless, a soft cantilever is unable to deform stiff material, and the detected force–distance curve contains the bending degree of the cantilever rather than the indent distance into the sample surface (Figure 13.9a). In order to deform the tested material and conduct a valid nanoindentation, a stiffer cantilever (with higher k) that possesses higher mechanical resistance under external force is a solution, while the sensitivity of the probe in both force and distance control will decrease at the same time (Figure 13.9b,c). In conclusion, a competing relation exists between probe and sample, in which either the applied force dominantly bends the cantilever or makes the deformation of the material via a slightly bent or tough cantilever. It should also be noted that the proper selection of the AFM probe is to choose a cantilever with a small k in the condition that the stiffness of the probe beam is larger than that of the specimen.

After illustrating the effects of the AFM probe on the mechanical characterization of materials, the experimental factors for the nanoindentation processes are then discussed. At the approaching stage of the probe toward the sample surface, the condition of the sample surface and the distance of the probe into the sample surface can cause the results to diverge. Consequently, this should be carefully considered to eliminate unexpected influences by these parameters.

13.3.3 Contact Point in Nanoindentation

AFM nanoindentation is generally conducted for detecting mechanical properties in the surface/sub-surface range of materials, and therefore, the distance of the probe indenting into the sample surface plays an important role in E calculation (Dimitriadis et al., 2002), where the applied force and the corresponding indent distance (sample deformation) are adopted. To achieve an accurate indent depth,

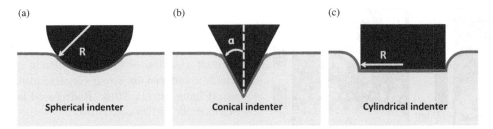

FIGURE 13.8 Schematics of common contact mechanisms. Nanoindentation on a homogeneous material using (a) spherical tip with tip radius R, (b) conical tip with half-open angle α, and (c) cylindrical tip with radius R (Chang, 2018).

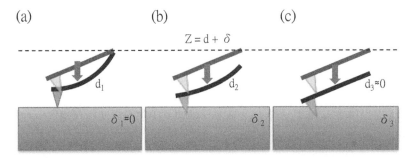

FIGURE 13.9 Various nanoindentation conditions by probes with different cantilevers. The probes in (a–c) show cantilevers with increasing stiffness. During the approaching process, the vertical movement of probe Z was the sum of cantilever deflection d and sample deformation δ (Chang, 2018).

determination of the contact point from the extended force–distance curve is necessary and will be displayed below.

It has been noted that a layer of water, which is referred to as water meniscus, exists on a material surface due to the environmental humidity and can pull the probe toward the specimen. During the approaching of the probe toward the specimen, the attraction of water pulls-in the probe, which causes bending of the cantilever, and thus, a sudden drop of force can be observed. Although the change in force is detected, the sample is not yet indented by the probe, and the correct selection of the contact point as the beginning of valid nanoindentation is essential for the removal of the influence of water. In Figure 13.10, we illustrated the relation between the force–distance curve and the probe cantilever, where the beam is unnaturally bent by water meniscus (point B) and the following approach makes the cantilever return to the original straight state (Chang, 2018). At point C, the force becomes 0 from the negative value and the compensation of the bent cantilever is considered. Because force steadily increases after point C, we can define point C as the contact point for the valid nanoindentation test, and the role of water in the mechanical measurement is eliminated.

13.3.4 Surface Effect of Materials

In addition to water meniscus, another issue concerning inconsistency of mechanical properties in the surface and bulk measurements is the surface effect, which contributes to the different molecular structures or compositions between material surface and interior. To investigate how the surface

effect obscures the practical mechanical performance of materials by surface examination, we designed an experiment in which the mechanical responses of three polymer specimens were subjected to AFM nanoindentation and compared.

In accordance with the mechanical performances of different polymer materials, such as polydimethylsiloxane (PDMS), polyvinyl alcohol (PVA), and polymethyl methacrylate (PMMA), the force–distance curve within an indent depth of 40 nm exhibited differences between these materials, as displayed in Figure 13.11. Moreover, we compared the force data of PDMS obtained from AFM nanoindentation (with an indent depth of 40 nm) and NI (with an indent depth of 400 nm) to investigate the working distance of the surface effect. Figure 13.12 shows similar shapes of force–distance curves, that is, the power law of the force–distance curve and eliminates the concern of surface effects. These results demonstrate the recognition of mechanical characteristics of materials by AFM nanoindentation and indicate that, even with inconstant surface and bulk mechanical properties, the surface examination still reveals the mechanical intrinsic properties of samples and constitutes a suitable method in material research.

13.3.5 Indent Depth

As mentioned previously, the AFM probe can be easily affected by the surface condition of materials, especially for shallow indentation (<100 nm), and thus, the nonideal contact of tip and sample will result in an inaccurate evaluation of mechanical properties. Generally, the geometry of the tip is not the ideal round- or conical-shaped, and the sample

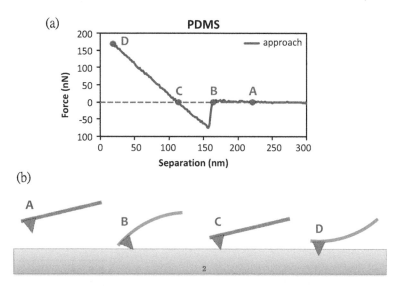

FIGURE 13.10 Connection of force–distance curve and cantilever bending. (a) The approaching force–distance curve and (b) the bending conditions of cantilever (Chang, 2018).

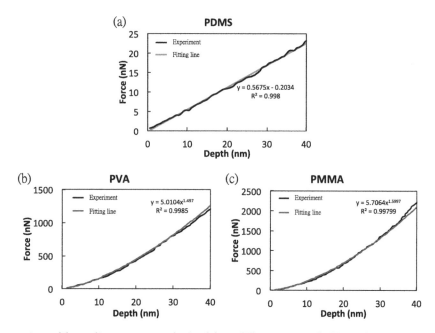

FIGURE 13.11 Comparison of force–distance curves obtained from different materials. Force–distance curves and fitting lines of (a) PDMS, (b) PVA, and (c) PMMA collected by AFM nanoindentation.

surface is not flat but undulate, which is known as surface roughness. The practical tip shape might be asymmetric, and the tip radius and half-open angle can differ from the nominal value provided by the manufacturer. The surface roughness of a sample leads to the existence of a void between the tip and sample surface, especially for shallow indentation, and can overestimate derived mechanical properties, such as elastic modulus and adhesion force.

To obtain the real mechanical behaviors of materials, information about the real shape of the tip and surface roughness of the sample is required. Common methods for the assessment of tip shape involve direct observation by electron microscopy and the topography imaging of a tip-shape characterization kit such as the TipCheck™ specimen, whose unique sharp needle-like surface structure can help with the blind construction of the tip geometry in three dimensions. Solutions to overcome the influence of surface roughness include: (i) adjusting indent depth and (ii) selection of a sharp probe. With knowledge of the surface roughness of a material, as a rule of thumb, the indent depth needs to be larger than ten times the roughness, so that the tip can be in complete contact with the sample surface (Baker, 1997). In the case of thin film in which the thickness cannot satisfy the target indent depth, the adoption of a sharp probe with a tip radius smaller than the surface roughness can be used to solve this issue. When conducting AFM

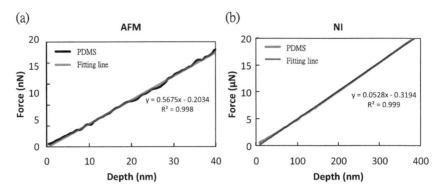

FIGURE 13.12 Force–distance curves of PDMS observed by nanoindentation. PDMS specimen was examined by (a) AFM with fitting and by (b) NI with fitting.

nanoindentation using a sharp probe, the sample surface can be considered to be locally flat and the effect of roughness can be neglected (Chang et al., 2016).

Now, most of the parameters concerning the characteristics of the AFM probe and the experimental factors during the nanoindentation process have been determined above. It was found that the conventional contact mechanism models were improper for nonideal elastic materials, and the selection of those equations could cause high uncertainty in the derived mechanical properties (Chang and Liu, 2018). In the following part, the influences of unsuitable fitting of equations to the force data of samples will be illustrated.

13.3.6 Intrinsic Nonelastic Characteristics of Materials

The contact area between the tip and sample is generally estimated with tip geometry, from which the models assume complete contact of the two objects at a known depth. Conventional contact mechanism models are based on the deformation of elastic materials, while the reality is that different mechanical behaviors exist between materials. In our prior study, it was noted that both bacterial cells and polymer sample PDMS perform distinct mechanical responses toward the external force using a spherical tip, whose force–distance curves exhibited a divergent power law from that of the Hertz model prediction (Figure 13.13). Consequently, the adoption of the Hertz model led to higher uncertainty in the evaluation of mechanical properties.

With the surface heterogeneity taken into account, such as adhesion of materials, new contact mechanism models are proposed, and the commonly used Derjaguin–Muller–Toporov (DMT), Johnson–Kendall–Roberts (JKR), and Maugis models are considered here (Derjaguin et al., 1975, Johnson et al., 1971, Maugis and Gauthier-Manuel, 1994). The role of adhesion in the interaction between tip and sample is shown in Figure 13.14 for each model, and the simplified equation of these models is compared to the Hertz model, summarized in Table 13.1. However, the high uncertainty of mechanical evaluation for the bacterial and PDMS samples could not be solved using these models due to the inconsistent power law of force–distance curves.

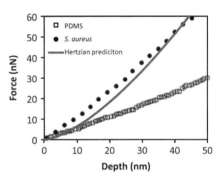

FIGURE 13.13 Divergence of force–distance curves from Hertz prediction. Both the force–distance curves of bacterial cells and PDMS obtained using spherical tips exhibit an inconsistent power law with the Hertz prediction (Chang and Liu, 2018).

Even though DMT, JKR, and Maugis models take material adhesion into account, their predicted contact area between tip and sample are all based on the spherical tip, and therefore, the simplified relation of force and distance is the same as that of Hertz theory. Because of the various geometrical parameters, the power law of the force–distance curve will be 2 when using a conical (pyramidal is also included) tip, 1.5 using a spherical tip, and 1 using a cylindrical tip.

The spherical tips were selected for the mechanical examination of the bacterial and PDMS samples, while the power law from the ideal prediction, 1.5, was found. We observed the mechanical intrinsic characteristic of the material, where this type of material exhibits incompressible and high surface-tension characteristics, and the sample deformation during nanoindentation should not be completely consistent with tip shape. A method to reduce uncertainty and assess the precise mechanics of the material is to conduct nanoindentation on the reference sample with known mechanical properties and revise the parameters of the existing equation. Details about the steps for model modification can be found in the literature (Chang et al., 2016, Chang and Liu 2018).

Considering all of the influential factors discussed in this chapter, a protocol for the examination and selection of experimental parameters is established and demonstrated in the next section.

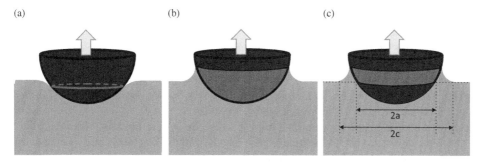

FIGURE 13.14 Schema of the roles of surface adhesion in three contact mechanism models. The attraction (a) along the perimeter is taken into account for the DMT model, (b) among the contact surface is considered by the JKR model, and (c) among the tip-specimen interface is calculated in the Maugis model (Chang, 2018).

TABLE 13.1 Power Laws of the Predicted Force Curves in Different Contact Mechanism Models

Models	Adhesion Contribution	Normalized Equations	Power Law
Hertz	No	$F_n = A_n^3$ $\delta_n = A_n^2$	$F_n = \delta_n^{1.5}$
DMT	Along the perimeter	$F_n = A_n^3 - 2$ $\delta_n = A_n^2$	$F_n \sim \delta_n^{1.5}$
JKR	Among the contact surface	$F_n = A_n^3 - A_n\sqrt{6A_n}$ $\delta_n = A_n^2 - \left(2\sqrt{6A_n}\right)/3$	$F_n \sim \delta_n^{1.5}$
Maugis	Tip-specimen interface	$F_n = A_n^3 - \lambda A_n^2\sqrt{m^2-1} + m^2\tan^{-1}\sqrt{m^2-1}$ $\delta_n = A_n^2 - \left(4\lambda A_n\sqrt{m^2-1}\right)/3$	$F_n \sim \delta_n^{1.5}$

Normalized force F_n, contact area A_n and indent depth δ_n were used for the presentation of the relation between F_n and A_n, and thus, the power law of the predicted force curves was obtained.
Source: Chang (2018).

13.4 Protocol for Accurate Mechanical Measurements

In Figure 13.15, a protocol for the proper selection of experimental parameters in AFM mechanical mapping is proposed and demonstrated (Chang and Liu, 2018, Chang,

2018). First, the capture of a valid force–distance curve should be achieved. We have shown the importance of the spring constant of the probe cantilever in nanoindentation. Specifically, a softer cantilever is more sensitive in both force and distance control, while it might not be able to indent into a specimen, that is, cause deformation and

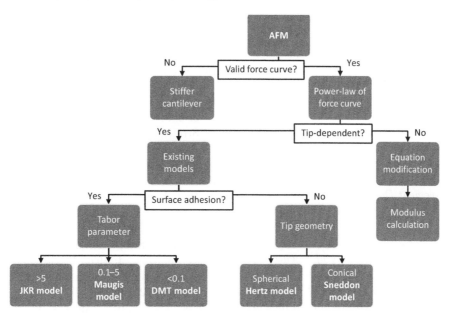

FIGURE 13.15 Dendrogram for the selection of experimental parameters for E evaluation. Positive correlation of applied force and indent depth is considered as the valid force curve. The detected force curve is checked to determine if its power law is consistent with the equation of the tip. Tabor parameter μ indicates the relation between sample stiffness and adhesion (Chang, 2018).

measure an accurate force curve. The suitability of the AFM probe relied on the stiffness of the tested material, and therefore, the examination of probe performance on each material should be confirmed prior to the nanoindentation tests. Then, the power law of the collected force–distance curve should be checked before fitting to the tip-dependent contact mechanism model. The inconsistent power law of the detected force–distance curve and adapted equation leads to the high deviation of the estimated E, and the actual mechanical properties of materials would be obscured (Chang, 2018). If the power law of the force–distance curve is the same as the ideal prediction, which is based on tip geometry, the surface adhesion from the specimen would then be considered. For the adhesive specimens, the Tabor parameter μ was used for the appropriate choice between DMT, JKR, and Maugis models. For materials without surface forces, the geometries of indenters dominated the choice, and the commonly adopted theories were the sphere-based Hertz model and the cone-based Sneddon model.

Different from the conventional testing methods which determine the contact model (thus its power law behavior) before the selection of experimental parameters, we demonstrated the importance of considering the actual mechanical behaviors between the AFM probe tip and the materials under test. With knowledge of the practical performance of samples during nanoindentation, measurement uncertainty can be significantly decreased, and the derived modulus E will be more consistent. For nanoindentation tests, especially by AFM, on soft, biological, and nano-materials, this proposed protocol proved that the mechanical characterization of materials and the mechanical properties comparison across samples and testing methods can be more accurate, effective, and meaningful.

References

Adamcik, J., Berquand, A. & Mezzenga, R. 2011. Single-step direct measurement of amyloid fibrils stiffness by peak force quantitative nanomechanical atomic force microscopy. *Applied Physics Letters*, 98, 193701.

Aguiar-Moya, J. P., Salazar-Delgado, J., García, A., Baldi-Sevilla, A., Bonilla-Mora, V. & Loría-Salazar, L. G. 2017. Effect of ageing on micromechanical properties of bitumen by means of atomic force microscopy. *Road Materials and Pavement Design*, 18, 203–215.

Al-Rekabi, Z. & Contera, S. 2018. Multifrequency AFM reveals lipid membrane mechanical properties and the effect of cholesterol in modulating viscoelasticity. *Proceedings of the National Academy of Sciences of the United States of America*, 115, 2658–2663.

Alsteens, D., Dupres, V., Yunus, S., Latge, J. P., Heinisch, J. J. & Dufrene, Y. F. 2012. High-resolution imaging of chemical and biological sites on living cells using peak force tapping atomic force microscopy. *Langmuir*, 28, 16738–16744.

Andriotis, O. G., Manuyakorn, W., Zekonyte, J., Katsamenis, O. L., Fabri, S., Howarth, P. H., Davies, D. E. & Thurner, P. J. 2014. Nanomechanical assessment of human and murine collagen fibrils via atomic force microscopy cantilever-based nanoindentation. *Journal of the Mechanical Behavior of Biomedical Materials*, 39, 9–26.

Arkhipov, A., Roos, W. H., Wuite, G. J. & Schulten, K. 2009. Elucidating the mechanism behind irreversible deformation of viral capsids. *Biophysical Journal*, 97, 2061–2069.

Arnoldi, M., Fritz, M., Bäuerlein, E., Radmacher, M., Sackmann, E. & Boulbitch, A. 2000. Bacterial turgor pressure can be measured by atomic force microscopy. *Physical Review E*, 62, 1034.

Asay, D. B. & Kim, S. H. 2006. Effects of adsorbed water layer structure on adhesion force of silicon oxide nanoasperity contact in humid ambient. *Journal of Chemical Physics*, 124, 174712.

Bahri, A., Martin, M., Gergely, C., Pugniere, M., Chevalier-Lucia, D. & Marchesseau, S. 2017. Atomic force microscopy study of the topography and nanomechanics of Casein Micelles captured by an antibody. *Langmuir*, 33, 4720–4728.

Baker, S. P. 1997. Between nanoindentation and scanning force microscopy: measuring mechanical properties in the nanometer regime. *Thin Solid Films*, 308, 289–296.

Baker, S. R., Banerjee, S., Bonin, K. & Guthold, M. 2016. Determining the mechanical properties of electrospun poly-epsilon-caprolactone (PCL) nanofibers using AFM and a novel fiber anchoring technique. *Materials Science and Engineering C: Materials for Biological Applications*, 59, 203–212.

Baumgartner, W., Hinterdorfer, P., Ness, W., Raab, A., Vestweber, D., Schindler, H. & Drenckhahn, D. 2000. Cadherin interaction probed by atomic force microscopy. *Proceedings of the National Academy of Sciences*, 97, 4005–4010.

Bertolazzi, S., Brivio, J. & Kis, A. 2011. Stretching and breaking of ultrathin MoS2. *ACS Nano*, 5, 9703–9709.

Best, R. B., Li, B., Steward, A., Daggett, V. & Clarke, J. 2001. Can non-mechanical proteins withstand force? Stretching barnase by atomic force microscopy and molecular dynamics simulation. *Biophysical Journal*, 81, 2344–2356.

Bettini, E., Eriksson, T., Boström, M., Leygraf, C. & Pan, J. 2011. Influence of metal carbides on dissolution behavior of biomedical CoCrMo alloy: Sem, Tem and AFM studies. *Electrochimica Acta*, 56, 9413–9419.

Butt, H.-J., Kappl, M., Mueller, H., Raiteri, R., Meyer, W. & Rühe, J. 1999. Steric forces measured with the atomic force microscope at various temperatures. *Langmuir*, 15, 2559–2565.

Calabri, L., Pugno, N., Menozzi, C. & Valeri, S. 2008. AFM nanoindentation: tip shape and tip radius of curvature effect on the hardness measurement. *Journal of Physics: Condensed Matter*, 20, 474208.

Calzado-Martin, A., Encinar, M., Tamayo, J., Calleja, M. & San Paulo, A. 2016. Effect of actin organization on the stiffness of living breast cancer cells revealed by peak-force modulation atomic force microscopy. *ACS Nano,* 10, 3365–3374.

Camesano, T. A. & Logan, B. E. 2000. Probing bacterial electrosteric interactions using atomic force microscopy. *Environmental Science & Technology,* 34, 3354–3362.

Canetta, E., Riches, A., Borger, E., Herrington, S., Dholakia, K. & Adya, A. K. 2014. Discrimination of bladder cancer cells from normal urothelial cells with high specificity and sensitivity: combined application of atomic force microscopy and modulated Raman spectroscopy. *Acta Biomaterialia,* 10, 2043–2055.

Carrion-Vazquez, M., Oberhauser, A. F., Fisher, T. E., Marszalek, P. E., Li, H. & Fernandez, J. M. 2000. Mechanical design of proteins studied by single-molecule force spectroscopy and protein engineering. *Progress in Biophysics and Molecular Biology,* 74, 63–91.

Cartagena, A. & Raman, A. 2014. Local viscoelastic properties of live cells investigated using dynamic and quasi-static atomic force microscopy methods. *Biophysical Journal,* 106, 1033–1043.

Carvalho, F. A., Connell, S., Miltenberger-Miltenyi, G., Pereira, S. N. V., Tavares, A., Ariëns, R. A. & Santos, N. C. 2010. Atomic force microscopy-based molecular recognition of a fibrinogen receptor on human erythrocytes. *ACS Nano,* 4, 4609–4620.

Casdorff, K., Keplinger, T., Bellanger, H., Michen, B., Schon, S. & Burgert, I. 2017a. High-resolution adhesion mapping of the odd–even effect on a layer-by-layer coated biomaterial by atomic-force-microscopy. *ACS Applied Materials & Interfaces,* 9, 13793–13800.

Casdorff, K., Keplinger, T. & Burgert, I. 2017b. Nano-mechanical characterization of the wood cell wall by AFM studies: comparison between AC- and QI mode. *Plant Methods,* 13, 60.

Castellanos-Gomez, A., Poot, M., Steele, G. A., Van Der Zant, H. S., Agrait, N. & Rubio-Bollinger, G. 2012. Elastic properties of freely suspended MoS2 nanosheets. *Advanced Materials,* 24, 772–775.

Cerf, A., Cau, J. C., Vieu, C. & Dague, E. 2009. Nanomechanical properties of dead or alive single-patterned bacteria. *Langmuir,* 25, 5731–5736.

Chang, A. C. 2018. *Development of Rapid, In-situ Multiproperty Characterization of Biological Materials Using Scanning Probe Microscopy.* Ph.D., National Cheng Kung University.

Chang, A. C., Liao, J.-D. & Liu, B. H. 2016. Practical assessment of nanoscale indentation techniques for the biomechanical properties of biological materials. *Mechanics of Materials,* 98, 11–21.

Chang, A. C. & Liu, B. H. 2018. Modified flat-punch model for hyperelastic polymeric and biological materials in nanoindentation. *Mechanics of Materials,* 118, 17–21.

Chang, A. C., Liu, B. H., Shao, P.-L. & Liao, J. D. 2017. Structure-dependent behaviours of skin layers studied by atomic force microscopy. *Journal of Microscopy,* 267, 265–271.

Chao, Y. & Zhang, T. 2011. Optimization of fixation methods for observation of bacterial cell morphology and surface ultrastructures by atomic force microscopy. *Applied Microbiology and Biotechnology,* 92, 381–392.

Chaudhuri, O., Parekh, S. H., Lam, W. A. & Fletcher, D. A. 2009. Combined atomic force microscopy and side-view optical imaging for mechanical studies of cells. *Nature Methods,* 6, 383–387.

Chen, J. 2014. Nanobiomechanics of living cells: a review. *Interface Focus,* 4, 20130055.

Chen, Q., Liu, J., Thundat, T., Gray, M. R. & Liu, Q. 2017. Spatially resolved organic coating on clay minerals in bitumen froth revealed by atomic force microscopy adhesion mapping. *Fuel,* 191, 283–289.

Chen, Y. Y., Wu, C. C., Hsu, J. L., Peng, H. L., Chang, H. Y. & Yew, T. R. 2009. Surface rigidity change of *Escherichia coli* after filamentous bacteriophage infection. *Langmuir,* 25, 4607–4614.

Cheng, X., Putz, K. W., Wood, C. D. & Brinson, L. C. 2015. Characterization of local elastic modulus in confined polymer films via AFM indentation. *Macromolecular Rapid Communications,* 36, 391–397.

Chlanda, A., Kijenska, E., Rinoldi, C., Tarnowski, M., Wierzchon, T. & Swieszkowski, W. 2018. Structure and physico-mechanical properties of low temperature plasma treated electrospun nanofibrous scaffolds examined with atomic force microscopy. *Micron,* 107, 79–84.

Chopinet, L., Formosa, C., Rols, M. P., Duval, R. E. & Dague, E. 2013. Imaging living cells surface and quantifying its properties at high resolution using AFM in QI mode. *Micron,* 48, 26–33.

Chyasnavichyus, M., Young, S. L. & Tsukruk, V. V. 2014. Mapping micromechanical properties of soft polymer contact lenses. *Polymer,* 55, 6091–6101.

Chyasnavichyus, M., Young, S. L. & Tsukruk, V. V. 2015. Recent advances in micromechanical characterization of polymer, biomaterial, and cell surfaces with atomic force microscopy. *Japanese Journal of Applied Physics,* 54, 08LA02.

Cole, J. T., Mitala, C. M., Kundu, S., Verma, A., Elkind, J. A., Nissim, I. & Cohen, A. S. 2010. Correction for Cole et al., Dietary branched chain amino acids ameliorate injury-induced cognitive impairment. *Proceedings of the National Academy of Sciences,* 107, 2373–2373.

Connizzo, B. K. & Grodzinsky, A. J. 2017. Tendon exhibits complex poroelastic behavior at the nanoscale as revealed by high-frequency AFM-based rheology. *Journal of Biomechanics,* 54, 11–18.

Corbin, E. A., Kong, F., Lim, C. T., King, W. P. & Bashir, R. 2015. Biophysical properties of human breast cancer cells measured using silicon MEMS resonators and atomic force microscopy. *Lab on a Chip,* 15, 839–847.

Cotter, P. D. & Hill, C. 2003. Surviving the acid test: responses of gram-positive bacteria to low pH. *Microbiology and Molecular Biology Reviews,* 67, 429–453.

Crichton, M. L., Donose, B. C., Chen, X., Raphael, A. P., Huang, H. & Kendall, M. A. 2011. The viscoelastic, hyperelastic and scale dependent behaviour of freshly excised individual skin layers. *Biomaterials,* 32, 4670–4681.

Cross, S. E., Jin, Y. S., Rao, J. & Gimzewski, J. K. 2007. Nanomechanical analysis of cells from cancer patients. *Nature Nanotechnology,* 2, 780–783.

Cross, S. E., Jin, Y. S., Tondre, J., Wong, R., Rao, J. & Gimzewski, J. K. 2008. AFM-based analysis of human metastatic cancer cells. *Nanotechnology,* 19, 384003.

Cuenot, S., Frétigny, C., Demoustier-Champagne, S. & Nysten, B. 2004. Surface tension effect on the mechanical properties of nanomaterials measured by atomic force microscopy. *Physical Review B,* 69, 165410.

Curry, N., Ghezali, G., Kaminski Schierle, G. S., Rouach, N. & Kaminski, C. F. 2017. Correlative STED and atomic force microscopy on live astrocytes reveals plasticity of cytoskeletal structure and membrane physical properties during polarized migration. *Frontiers in Cellular Neuroscience,* 11, 104.

Dai, G., Wolff, H. & Danzebrink, H.-U. 2007. Atomic force microscope cantilever based microcoordinate measuring probe for true three-dimensional measurements of microstructures. *Applied Physics Letters,* 91, 121912.

Darling, E. M., Wilusz, R. E., Bolognesi, M. P., Zauscher, S. & Guilak, F. 2010. Spatial mapping of the biomechanical properties of the pericellular matrix of articular cartilage measured in situ via atomic force microscopy. *Biophysical Journal,* 98, 2848–2856.

Deng, Y., Sun, M. & Shaevitz, J. W. 2011. Direct measurement of cell wall stress stiffening and turgor pressure in live bacterial cells. *Physical Review Letters,* 107, 158101.

Derjaguin, B. V., Muller, V. M. & Toporov, Y. P. 1975. Effect of contact deformations on the adhesion of particles. *Journal of Colloid and Interface Science,* 53, 314–326.

Dimitriadis, E. K., Horkay, F., Maresca, J., Kachar, B. & Chadwick, R. S. 2002. Determination of elastic moduli of thin layers of soft material using the atomic force microscope. *Biophysical Journal,* 82, 2798–2810.

Dokukin, M. E., Kuroki, H., Minko, S. & Sokolov, I. 2016. AFM study of polymer brush grafted to deformable surfaces: quantitative properties of the brush and substrate mechanics. *Macromolecules,* 50, 275–282.

Dokukin, M. E. & Sokolov, I. 2012a. On the measurements of rigidity modulus of soft materials in nanoindentation experiments at small depth. *Macromolecules,* 45, 4277–4288.

Dokukin, M. E. & Sokolov, I. 2012b. Quantitative mapping of the elastic modulus of soft materials with HarmoniX and PeakForce QNM AFM modes. *Langmuir,* 28, 16060–16071.

Domke, J. & Radmacher, M. 1998. Measuring the elastic properties of thin polymer films with the atomic force microscope. *Langmuir,* 14, 3320–3325.

Dong, Z., Liu, Z., Wang, P. & Gong, X. 2017. Nanostructure characterization of asphalt-aggregate interface through molecular dynamics simulation and atomic force microscopy. *Fuel,* 189, 155–163.

Dufrene, Y. F., Martinez-Martin, D., Medalsy, I., Alsteens, D. & Muller, D. J. 2013. Multiparametric imaging of biological systems by force–distance curve-based AFM. *Nature Methods,* 10, 847–854.

Dulinska, I., Targosz, M., Strojny, W., Lekka, M., Czuba, P., Balwierz, W. & Szymonski, M. 2006. Stiffness of normal and pathological erythrocytes studied by means of atomic force microscopy. *Journal of Biochemical and Biophysical Methods,* 66, 1–11.

Eaton, P., Fernandes, J. C., Pereira, E., Pintado, M. E. & Malcata, F. X. 2008. Atomic force microscopy study of the antibacterial effects of chitosans on *Escherichia coli* and *Staphylococcus aureus*. *Ultramicroscopy,* 108, 1128–1134.

Eaton, P., Smith, J. R., Graham, P., Smart, J. D., Nevell, T. G. & Tsibouklis, J. 2002. Adhesion force mapping of polymer surfaces: factors influencing force of adhesion. *Langmuir,* 18, 3387–3389.

Ebenstein, D. M. & Pruitt, L. A. 2006. Nanoindentation of biological materials. *Nano Today,* 1, 26–33.

Efremov, Y. M., Wang, W. H., Hardy, S. D., Geahlen, R. L. & Raman, A. 2017. Measuring nanoscale viscoelastic parameters of cells directly from AFM force–displacement curves. *Scientific Reports,* 7, 1541.

Elinski, M. B., Menard, B. D., Liu, Z. & Batteas, J. D. 2017. Adhesion and friction at graphene/self-assembled monolayer interfaces investigated by atomic force microscopy. *The Journal of Physical Chemistry C,* 121, 5635–5641.

Emad, A., Heinz, W. F., Antonik, M. D., D'costa, N. P., Nageswaran, S., Schoenenberger, C.-A. & Hoh, J. H. 1998. Relative microelastic mapping of living cells by atomic force microscopy. *Biophysical Journal,* 74, 1564–1578.

Fakhrullina, G., Akhatova, F., Kibardina, M., Fokin, D. & Fakhrullin, R. 2017. Nanoscale imaging and characterization of *Caenorhabditis elegans* epicuticle using atomic force microscopy. *Nanomedicine,* 13, 483–491.

Farahi, R. H., Charrier, A. M., Tolbert, A., Lereu, A. L., Ragauskas, A., Davison, B. H. & Passian, A. 2017. Plasticity, elasticity, and adhesion energy of plant cell walls: nanometrology of lignin loss using atomic force microscopy. *Scientific Reports,* 7, 152.

Faria, E. C., Ma, N., Gazi, E., Gardner, P., Brown, M., Clarke, N. W. & Snook, R. D. 2008. Measurement of elastic properties of prostate cancer cells using AFM. *Analyst,* 133, 1498–1500.

Ferrari, L., Kaufmann, J., Winnefeld, F. & Plank, J. 2010. Interaction of cement model systems with superplasticizers investigated by atomic force microscopy, zeta potential, and adsorption measurements. *Journal of Colloid and Interface Science*, 347, 15–24.

Frank, I. W., Tanenbaum, D. M., Van Der Zande, A. M. & McEuen, P. L. 2007. Mechanical properties of suspended graphene sheets. *Journal of Vacuum Science & Technology B: Microelectronics and Nanometer Structures*, 25, 2558–2561.

Fuhrmann, A., Staunton, J. R., Nandakumar, V., Banyai, N., Davies, P. C. & Ros, R. 2011. AFM stiffness nanotomography of normal, metaplastic and dysplastic human esophageal cells. *Physical Biology*, 8, 015007.

Fukuma, T. 2009. Wideband low-noise optical beam deflection sensor with photothermal excitation for liquid-environment atomic force microscopy. *Review of Scientific Instruments*, 80, 023707.

Gannepalli, A., Yablon, D. G., Tsou, A. H. & Proksch, R. 2013. Corrigendum: mapping nanoscale elasticity and dissipation using dual frequency contact resonance AFM. *Nanotechnology*, 24, 159501.

Garcia, R., Gomez, C. J., Martinez, N. F., Patil, S., Dietz, C. & Magerle, R. 2006. Identification of nanoscale dissipation processes by dynamic atomic force microscopy. *Physical Review Letters*, 97, 016103.

Giessibl, F. J. 2000. Atomic resolution on Si(111)-(7×7) by noncontact atomic force microscopy with a force sensor based on a quartz tuning fork. *Applied Physics Letters*, 76, 1470–1472.

Glaubitz, M., Medvedev, N., Pussak, D., Hartmann, L., Schmidt, S., Helm, C. A. & Delcea, M. 2014. A novel contact model for AFM indentation experiments on soft spherical cell-like particles. *Soft Matter*, 10, 6732–6741.

Greving, I., Cai, M., Vollrath, F. & Schniepp, H. C. 2012. Shear-induced self-assembly of native silk proteins into fibrils studied by atomic force microscopy. *Biomacromolecules*, 13, 676–682.

Guo, D., Li, J., Xie, G., Wang, Y. & Luo, J. 2014. Elastic properties of polystyrene nanospheres evaluated with atomic force microscopy: size effect and error analysis. *Langmuir*, 30, 7206–7212.

Guo, M., Tan, Y., Yu, J., Hou, Y. & Wang, L. 2017. A direct characterization of interfacial interaction between asphalt binder and mineral fillers by atomic force microscopy. *Materials and Structures*, 50, 141.

Guyomarc'h, F., Zou, S., Chen, M., Milhiet, P. E., Godefroy, C., Vie, V. & Lopez, C. 2014. Milk sphingomyelin domains in biomimetic membranes and the role of cholesterol: morphology and nanomechanical properties investigated using AFM and force spectroscopy. *Langmuir*, 30, 6516–6524.

Guz, N. V., Patel, S. J., Dokukin, M. E., Clarkson, B. & Sokolov, I. 2016. Biophysical differences between chronic myelogenous leukemic quiescent and proliferating stem/progenitor cells. *Nanomedicine*, 12, 2429–2437.

Harris, A. R. & Charras, G. T. 2011. Experimental validation of atomic force microscopy-based cell elasticity measurements. *Nanotechnology*, 22, 345102.

Hayashi, K. & Iwata, M. 2015. Stiffness of cancer cells measured with an AFM indentation method. *Journal of the Mechanical Behavior of Biomedical Materials*, 49, 105–111.

Hecht, F. M., Rheinlaender, J., Schierbaum, N., Goldmann, W. H., Fabry, B. & Schäffer, T. E. 2015. Imaging viscoelastic properties of live cells by AFM: power-law rheology on the nanoscale. *Soft Matter*, 11, 4584–4591.

Hertz, H. 1881. On the contact of elastic solids. *Z. Reine Angew. Mathematik*, 92, 156–171.

Hinterdorfer, P. & Dufrene, Y. F. 2006. Detection and localization of single molecular recognition events using atomic force microscopy. *Nature Methods*, 3, 347–355.

Huang, Q., Wu, H., Cai, P., Fein, J. B. & Chen, W. 2015. Atomic force microscopy measurements of bacterial adhesion and biofilm formation onto clay-sized particles. *Scientific Reports*, 5, 16857.

Hugel, T., Grosholz, M., Clausen-Schaumann, H., Pfau, A., Gaub, H. & Seitz, M. 2001. Elasticity of single polyelectrolyte chains and their desorption from solid supports studied by AFM based single molecule force spectroscopy. *Macromolecules*, 34, 1039–1047.

Iqbal, F., Pyczak, F., Neumeier, S. & Göken, M. 2017. Crack nucleation and elastic/plastic deformation of TiAl alloys investigated by in-situ loaded atomic force microscopy. *Materials Science and Engineering: A*, 689, 11–16.

Iwamoto, S., Kai, W., Isogai, A. & Iwata, T. 2009. Elastic modulus of single cellulose microfibrils from tunicate measured by atomic force microscopy. *Biomacromolecules*, 10, 2571–2576.

Iyer, S., Gaikwad, R. M., Subba-Rao, V., Woodworth, C. D. & Sokolov, I. 2009. Atomic force microscopy detects differences in the surface brush of normal and cancerous cells. *Nature Nanotechnology*, 4, 389–393.

Jahangir, R., Little, D. & Bhasin, A. 2014. Evolution of asphalt binder microstructure due to tensile loading determined using AFM and image analysis techniques. *International Journal of Pavement Engineering*, 16, 337–349.

James, S. A., Hilal, N. & Wright, C. J. 2017. Atomic force microscopy studies of bioprocess engineering surfaces - imaging, interactions and mechanical properties mediating bacterial adhesion. *Biotechnology Journal*, 12, 1600698.

Jee, A.-Y. & Lee, M. 2010. Comparative analysis on the nanoindentation of polymers using atomic force microscopy. *Polymer Testing*, 29, 95–99.

Jiang, T. & Zhu, Y. 2015. Measuring graphene adhesion using atomic force microscopy with a microsphere tip. *Nanoscale*, 7, 10760–10766.

Jing, G. Y., Duan, H. L., Sun, X. M., Zhang, Z. S., Xu, J., Li, Y. D., Wang, J. X. & Yu, D. P. 2006. Surface effects on elastic properties of silver nanowires: contact atomic-force microscopy. *Physical Review B*, 73, 235409.

Johnson, K., Kendall, K. & Roberts, A. Surface energy and the contact of elastic solids. *Proceedings of the Royal*

Society of London A: Mathematical, Physical and Engineering Sciences, 1971. The Royal Society, 301–313.

Jones, D. M., Smith, J. R., Huck, W. T. & Alexander, C. 2002. Variable adhesion of micropatterned thermoresponsive polymer brushes: AFM investigations of poly (N-isopropylacrylamide) brushes prepared by surface-initiated polymerizations. *Advanced Materials*, 14, 1130–1134.

Jorba, I., Uriarte, J. J., Campillo, N., Farre, R. & Navajas, D. 2017. Probing micromechanical properties of the extracellular matrix of soft tissues by atomic force microscopy. *Journal of Cellular Physiology*, 232, 19–26.

Kim, H., Yamagishi, A., Imaizumi, M., Onomura, Y., Nagasaki, A., Miyagi, Y., Okada, T. & Nakamura, C. 2017. Quantitative measurements of intercellular adhesion between a macrophage and cancer cells using a cup-attached AFM chip. *Colloids and Surfaces B: Biointerfaces*, 155, 366–372.

Kochan, K., Perez-Guaita, D., Pissang, J., Jiang, J.-H., Peleg, A. Y., Mcnaughton, D., Heraud, P. & Wood, B. R. 2018. In vivo atomic force microscopy–infrared spectroscopy of bacteria. *Journal of The Royal Society Interface*, 15, 20180115.

Kumar, S., Cartron, M. L., Mullin, N., Qian, P., Leggett, G. J., Hunter, C. N. & Hobbs, J. K. 2017. Direct imaging of protein organization in an intact bacterial organelle using high-resolution atomic force microscopy. *ACS Nano*, 11, 126–133.

Kuznetsova, T. G., Starodubtseva, M. N., Yegorenkov, N. I., Chizhik, S. A. & Zhdanov, R. I. 2007. Atomic force microscopy probing of cell elasticity. *Micron*, 38, 824–833.

Labernadie, A., Thibault, C., Vieu, C., Maridonneau-Parini, I. & Charriere, G. M. 2010. Dynamics of podosome stiffness revealed by atomic force microscopy. *Proceedings of the National Academy of Sciences of the United States of America*, 107, 21016–21021.

Ladjal, H., Hanus, J.-L., Pillarisetti, A., Keefer, C., Ferreira, A. & Desai, J. Atomic force microscopy-based single-cell indentation: experimentation and finite element simulation. *IEEE/RSJ International Conference on Intelligent Robots and Systems*, 2009. 1326–1332.

Lahiji, R. R., Xu, X., Reifenberger, R., Raman, A., Rudie, A. & Moon, R. J. 2010. Atomic force microscopy characterization of cellulose nanocrystals. *Langmuir*, 26, 4480–4488.

Last, J. A., Liliensiek, S. J., Nealey, P. F. & Murphy, C. J. 2009. Determining the mechanical properties of human corneal basement membranes with atomic force microscopy. *Journal of Structural Biology*, 167, 19–24.

Lee, C., Wei, X., Kysar, J. W. & Hone, J. 2008. Measurement of the elastic properties and intrinsic strength of monolayer graphene. *Science*, 321, 385–388.

Lehaf, A. M., Hariri, H. H. & Schlenoff, J. B. 2012. Homogeneity, modulus, and viscoelasticity of polyelectrolyte multilayers by nanoindentation: refining the buildup mechanism. *Langmuir*, 28, 6348–6355.

Lehenkari, P. & Horton, M. 1999. Single integrin molecule adhesion forces in intact cells measured by atomic force microscopy. *Biochemical and Biophysical Research Communications*, 259, 645–650.

Lekka, M. 2016. Discrimination between normal and cancerous cells using AFM. *Bionanoscience*, 6, 65–80.

Lekka, M., Gil, D., Pogoda, K., Dulinska-Litewka, J., Jach, R., Gostek, J., Klymenko, O., Prauzner-Bechcicki, S., Stachura, Z., Wiltowska-Zuber, J., Okon, K. & Laidler, P. 2012a. Cancer cell detection in tissue sections using AFM. *Archives of Biochemistry and Biophysics*, 518, 151–156.

Lekka, M., Pogoda, K., Gostek, J., Klymenko, O., Prauzner-Bechcicki, S., Wiltowska-Zuber, J., Jaczewska, J., Lekki, J. & Stachura, Z. 2012b. Cancer cell recognition–mechanical phenotype. *Micron*, 43, 1259–1266.

Li, H., Oberhauser, A. F., Fowler, S. B., Clarke, J. & Fernandez, J. M. 2000. Atomic force microscopy reveals the mechanical design of a modular protein. *Proceedings of the National Academy of Sciences*, 97, 6527–6531.

Li, Q. S., Lee, G. Y., Ong, C. N. & Lim, C. T. 2008. AFM indentation study of breast cancer cells. *Biochemical and Biophysical Research Communications*, 374, 609–613.

Lin, L.-Y., Kim, D.-E., Kim, W.-K. & Jun, S.-C. 2011. Friction and wear characteristics of multi-layer graphene films investigated by atomic force microscopy. *Surface and Coatings Technology*, 205, 4864–4869.

Liparoti, S., Sorrentino, A. & Speranza, V. 2017. Micromechanical characterization of complex polypropylene morphologies by HarmoniX AFM. *International Journal of Polymer Science*, 2017, 1–12.

Liu, B. H., Li, K.-L., Kang, K.-L., Huang, W.-K. & Liao, J.-D. 2013. In situ biosensing of the nanomechanical property and electrochemical spectroscopy of *Streptococcus mutans*-containing biofilms. *Journal of Physics D: Applied Physics*, 46, 275401.

Liu, B. H. & Yu, L.-C. 2017. In-situ, time-lapse study of extracellular polymeric substance discharge in *Streptococcus mutans* biofilm. *Colloids and Surfaces B: Biointerfaces*, 150, 98–105.

Liu, H., Wen, J., Xiao, Y., Liu, J., Hopyan, S., Radisic, M., Simmons, C. A. & Sun, Y. 2014a. In situ mechanical characterization of the cell nucleus by atomic force microscopy. *ACS Nano*, 8, 3821–3828.

Liu, J., Weller, G. E., Zern, B., Ayyaswamy, P. S., Eckmann, D. M., Muzykantov, V. R. & Radhakrishnan, R. 2010a. Computational model for nanocarrier binding to endothelium validated using in vivo, in vitro, and atomic force microscopy experiments. *Proceedings of the National Academy of Sciences of the United States of America*, 107, 16530–16535.

Liu, K., Song, Y., Feng, W., Liu, N., Zhang, W. & Zhang, X. 2011. Extracting a single polyethylene oxide chain from a single crystal by a combination of atomic force microscopy imaging and single-molecule force spectroscopy: toward the investigation of molecular interactions in their condensed states. *Journal of the American Chemical Society*, 133, 3226–3229.

Liu, S., Ng, A. K., Xu, R., Wei, J., Tan, C. M., Yang, Y. & Chen, Y. 2010b. Antibacterial action of dispersed single-walled carbon nanotubes on Escherichia coli and Bacillus subtilis investigated by atomic force microscopy. *Nanoscale*, 2, 2744–2750.

Liu, X. R., Deng, X., Liu, R. R., Yan, H. J., Guo, Y. G., Wang, D. & Wan, L. J. 2014b. Single nanowire electrode electrochemistry of silicon anode by in situ atomic force microscopy: solid electrolyte interphase growth and mechanical properties. *ACS Applied Materials & Interfaces*, 6, 20317–20323.

Loparic, M., Wirz, D., Daniels, A. U., Raiteri, R., Vanlandingham, M. R., Guex, G., Martin, I., Aebi, U. & Stolz, M. 2010. Micro- and nanomechanical analysis of articular cartilage by indentation-type atomic force microscopy: validation with a gel-microfiber composite. *Biophysical Journal*, 98, 2731–2740.

Lorenzoni, M., Evangelio, L., Fernández-Regúlez, M., Nicolet, C., Navarro, C. & Pérez-Murano, F. 2017. Sequential infiltration of self-assembled block copolymers: a study by atomic force microscopy. *The Journal of Physical Chemistry C*, 121, 3078–3086.

Louise Meyer, R., Zhou, X., Tang, L., Arpanaei, A., Kingshott, P. & Besenbacher, F. 2010. Immobilisation of living bacteria for AFM imaging under physiological conditions. *Ultramicroscopy*, 110, 1349–1357.

Lowy, F. D. 1998. Staphylococcus aureus infections. *New England Journal of Medicine*, 339, 520–532.

Lu, W. L., Li, J. M., Yang, J., Xu, C. G., Zhang, S. S., Yan, J., Zhang, T. T. & Zhao, H. H. 2018. Effects of astragalus polysaccharide on mechanical characterization of liver sinusoidal endothelial cells by atomic force microscopy at nanoscale. *Chinese Journal of Integrative Medicine*, 24, 455–459.

Mahaffy, R., Park, S., Gerde, E., Käs, J. & Shih, C. 2004. Quantitative analysis of the viscoelastic properties of thin regions of fibroblasts using atomic force microscopy. *Biophysical Journal*, 86, 1777–1793.

Maherally, Z., Smith, J. R., Ghoneim, M. K., Dickson, L., An, Q., Fillmore, H. L. & Pilkington, G. J. 2015. Silencing of Cd44 in glioma leads to changes in cytoskeletal protein expression and cellular biomechanical deformation properties as measured by AFM nanoindentation. *BioNanoScience*, 6, 54–64.

Man, J., Yang, H., Wang, Y., Yan, C. & Zhang, S. 2017. Nanotribological properties of nanotextured Ni-Co coating surface measured with AFM colloidal probe technique. *Journal of Laser Micro/Nanoengineering*, 12, 16–21.

Marshall Jr, G., Balooch, M., Gallagher, R., Gansky, S. & Marshall, S. 2001. Mechanical properties of the dentinoenamel junction: AFM studies of nanohardness, elastic modulus, and fracture. *Journal of Biomedical Materials Research: An Official Journal of The Society for Biomaterials and The Japanese Society for Biomaterials*, 54, 87–95.

Mathur, A. B., Collinsworth, A. M., Reichert, W. M., Kraus, W. E. & Truskey, G. A. 2001. Endothelial, cardiac muscle and skeletal muscle exhibit different viscous and elastic properties as determined by atomic force microscopy. *Journal of Biomechanics*, 34, 1545–1553.

Mathur, A. B., Truskey, G. A. & Reichert, W. M. 2000. Atomic force and total internal reflection fluorescence microscopy for the study of force transmission in endothelial cells. *Biophysical Journal*, 78, 1725–1735.

Maugis, D. & Gauthier-Manuel, B. 1994. JKR-DMT transition in the presence of a liquid meniscus. *Journal of Adhesion Science and Technology*, 8, 1311–1322.

McAllister, Q. P., Strawhecker, K. E., Becker, C. R. & Lundgren, C. A. 2014. In situ atomic force microscopy nanoindentation of lithiated silicon nanopillars for lithium ion batteries. *Journal of Power Sources*, 257, 380–387.

Milani, P., Gholamirad, M., Traas, J., Arneodo, A., Boudaoud, A., Argoul, F. & Hamant, O. 2011. In vivo analysis of local wall stiffness at the shoot apical meristem in Arabidopsis using atomic force microscopy. *Plant Journal*, 67, 1116–1123.

Minary-Jolandan, M., Tajik, A., Wang, N. & Yu, M. F. 2012. Intrinsically high-Q dynamic AFM imaging in liquid with a significantly extended needle tip. *Nanotechnology*, 23, 235704.

Müller, D. J. & Dufrene, Y. F. 2010. Atomic force microscopy as a multifunctional molecular toolbox in nanobiotechnology. In: Rodgers, P. (ed.) *Nanoscience and Technology: A Collection of Reviews from Nature Journals*. Singapore: World Scientific, 346 pp.

Nalam, P. C., Gosvami, N. N., Caporizzo, M. A., Composto, R. J. & Carpick, R. W. 2015. Nano-rheology of hydrogels using direct drive force modulation atomic force microscopy. *Soft Matter*, 11, 8165–8178.

Namazu, T., Isono, Y. & Tanaka, T. 2000. Evaluation of size effect on mechanical properties of single crystal silicon by nanoscale bending test using AFM. *Journal of Microelectromechanical Systems*, 9, 450–459.

Nawaz, S., Sanchez, P., Bodensiek, K., Li, S., Simons, M. & Schaap, I. A. 2012. Cell visco-elasticity measured with AFM and optical trapping at sub-micrometer deformations. *PLoS One*, 7, e45297.

Neugirg, B. R., Koebley, S. R., Schniepp, H. C. & Fery, A. 2016. AFM-based mechanical characterization of single nanofibres. *Nanoscale*, 8, 8414–8426.

Nguyen, A. V., Nyberg, K. D., Scott, M. B., Welsh, A. M., Nguyen, A. H., Wu, N., Hohlbauch, S. V., Geisse, N. A., Gibb, E. A., Robertson, A. G., Donahue, T. R. & Rowat, A. C. 2016. Stiffness of pancreatic cancer cells is associated with increased invasive potential. *Integrative Biology (Camb)*, 8, 1232–1245.

Nguyen, T. P. L. 2018. *Applications of 3D Printed AFM Fluid Cells on Biomaterials Using Atomic Force Microscopy*. Master thesis, National Cheng Kung University.

Ni, H. & Li, X. 2006. Young's modulus of ZnO nanobelts measured using atomic force microscopy and nanoindentation techniques. *Nanotechnology,* 17, 3591–3597.

Notbohm, J., Poon, B. & Ravichandran, G. 2011. Analysis of nanoindentation of soft materials with an atomic force microscope. *Journal of Materials Research,* 27, 229–237.

Nowatzki, P. J., Franck, C., Maskarinec, S. A., Ravichandran, G. & Tirrell, D. A. 2008. Mechanically tunable thin films of photosensitive artificial proteins: preparation and characterization by nanoindentation. *Macromolecules,* 41, 1839–1845.

Oberhauser, A. F., Badilla-Fernandez, C., Carrion-Vazquez, M. & Fernandez, J. M. 2002. The mechanical hierarchies of fibronectin observed with single-molecule AFM. *Journal of Molecular Biology,* 319, 433–447.

Oberhauser, A. F., Hansma, P. K., Carrion-Vazquez, M. & Fernandez, J. M. 2001. Stepwise unfolding of titin under force-clamp atomic force microscopy. *Proceedings of the National Academy of Sciences,* 98, 468–472.

Oberhauser, A. F., Marszalek, P. E., Erickson, H. P. & Fernandez, J. M. 1998. The molecular elasticity of the extracellular matrix protein tenascin. *Nature,* 393, 181.

Ortiz, C. & Hadziioannou, G. 1999. Entropic elasticity of single polymer chains of poly (methacrylic acid) measured by atomic force microscopy. *Macromolecules,* 32, 780–787.

Ouasti, S., Donno, R., Cellesi, F., Sherratt, M. J., Terenghi, G. & Tirelli, N. 2011. Network connectivity, mechanical properties and cell adhesion for hyaluronic acid/PEG hydrogels. *Biomaterials,* 32, 6456–6470.

Pan, J., Zhu, H., Hou, Q., Wang, H. & Wang, S. 2015. Macromolecular and pore structures of Chinese tectonically deformed coal studied by atomic force microscopy. *Fuel,* 139, 94–101.

Paxton, W. F., Kistler, K. C., Olmeda, C. C., Sen, A., St. Angelo, S. K., Cao, Y., Mallouk, T. E., Lammert, P. E. & Crespi, V. H. 2004. Catalytic nanomotors: autonomous movement of striped nanorods. *Journal of the American Chemical Society,* 126, 13424–13431.

Payamyar, P., Kaja, K., Ruiz-Vargas, C., Stemmer, A., Murray, D. J., Johnson, C. J., King, B. T., Schiffmann, F., Vandevondele, J., Renn, A., Gotzinger, S., Ceroni, P., Schutz, A., Lee, L. T., Zheng, Z., Sakamoto, J. & Schluter, A. D. 2014. Synthesis of a covalent monolayer sheet by photochemical anthracene dimerization at the air/water interface and its mechanical characterization by AFM indentation. *Advanced Materials,* 26, 2052–2058.

Pfahl, V., Phani, M. K., Büchsenschütz-Göbeler, M., Kumar, A., Moshnyaga, V., Arnold, W. & Samwer, K. 2017. Conduction electrons as dissipation channel in friction experiments at the metal-metal transition of LSMO measured by contact-resonance atomic force microscopy. *Applied Physics Letters,* 110, 053102.

Pfreundschuh, M., Alsteens, D., Hilbert, M., Steinmetz, M. O. & Muller, D. J. 2014a. Localizing chemical groups while imaging single native proteins by high-resolution atomic force microscopy. *Nano Letters,* 14, 2957–2964.

Pfreundschuh, M., Alsteens, D., Wieneke, R., Zhang, C., Coughlin, S. R., Tampe, R., Kobilka, B. K. & Muller, D. J. 2015. Identifying and quantifying two ligand-binding sites while imaging native human membrane receptors by AFM. *Nature Communications,* 6, 8857.

Pfreundschuh, M., Martinez-Martin, D., Mulvihill, E., Wegmann, S. & Muller, D. J. 2014b. Multiparametric high-resolution imaging of native proteins by force–distance curve-based AFM. *Nature Protocols,* 9, 1113–1130.

Plodinec, M., Loparic, M., Monnier, C. A., Obermann, E. C., Zanetti-Dallenbach, R., Oertle, P., Hyotyla, J. T., Aebi, U., Bentires-Alj, M., Lim, R. Y. & Schoenenberger, C. A. 2012. The nanomechanical signature of breast cancer. *Nature Nanotechnology,* 7, 757–765.

Poloni, L. N., Zhong, X., Ward, M. D. & Mandal, T. 2016. Best practices for real-time in situ atomic force and chemical force microscopy of crystals. *Chemistry of Materials,* 29, 331–345.

Powell, L. C., Hilal, N. & Wright, C. J. 2017. Atomic force microscopy study of the biofouling and mechanical properties of virgin and industrially fouled reverse osmosis membranes. *Desalination,* 404, 313–321.

Qin, C., Clarke, K. & Li, K. 2014. Interactive forces between lignin and cellulase as determined by atomic force microscopy. *Biotechnology for Biofuels,* 7, 65.

Rabe, U., Amelio, S., Kester, E., Scherer, V., Hirsekorn, S. & Arnold, W. 2000. Quantitative determination of contact stiffness using atomic force acoustic microscopy. *Ultrasonics,* 38, 430–437.

Raman, A., Trigueros, S., Cartagena, A., Stevenson, A. P., Susilo, M., Nauman, E. & Contera, S. A. 2011. Mapping nanomechanical properties of live cells using multi-harmonic atomic force microscopy. *Nature Nanotechnology,* 6, 809–814.

Rebelo, L. M., De Sousa, J. S., Mendes Filho, J. & Radmacher, M. 2013. Comparison of the viscoelastic properties of cells from different kidney cancer phenotypes measured with atomic force microscopy. *Nanotechnology,* 24, 055102.

Reviakine, I. & Brisson, A. 2000. Formation of supported phospholipid bilayers from unilamellar vesicles investigated by atomic force microscopy. *Langmuir,* 16, 1806–1815.

Rheinlaender, J., Geisse, N. A., Proksch, R. & Schaffer, T. E. 2011. Comparison of scanning ion conductance microscopy with atomic force microscopy for cell imaging. *Langmuir,* 27, 697–704.

Richter, R., Mukhopadhyay, A. & Brisson, A. 2003. Pathways of lipid vesicle deposition on solid surfaces: a combined QCM-D and AFM study. *Biophysical Journal,* 85, 3035–3047.

Richter, R. P. & Brisson, A. R. 2005. Following the formation of supported lipid bilayers on mica: a study combining AFM, QCM-D, and ellipsometry. *Biophysical Journal,* 88, 3422–3433.

Rigato, A., Rico, F., Eghiaian, F., Piel, M. & Scheuring, S. 2015. Atomic force microscopy mechanical mapping of micropatterned cells shows adhesion geometry-dependent mechanical response on local and global scales. *ACS Nano,* 9, 5846–5856.

Rodriguez-Florez, N., Oyen, M. L. & Shefelbine, S. J. 2013. Insight into differences in nanoindentation properties of bone. *Journal of the Mechanical Behavior of Biomedical Materials,* 18, 90–99.

Roduit, C., Sekatski, S., Dietler, G., Catsicas, S., Lafont, F. & Kasas, S. 2009. Stiffness tomography by atomic force microscopy. *Biophysical Journal,* 97, 674–677.

Rother, J., Noding, H., Mey, I. & Janshoff, A. 2014. Atomic force microscopy-based microrheology reveals significant differences in the viscoelastic response between malign and benign cell lines. *Open Biology,* 4, 140046.

Rotsch, C., Jacobson, K. & Radmacher, M. 1999. Dimensional and mechanical dynamics of active and stable edges in motile fibroblasts investigated by using atomic force microscopy. *Proceedings of the National Academy of Sciences,* 96, 921–926.

Rotsch, C. & Radmacher, M. 2000. Drug-induced changes of cytoskeletal structure and mechanics in fibroblasts: an atomic force microscopy study. *Biophysical Journal,* 78, 520–535.

Sahin, O. & Erina, N. 2008. High-resolution and large dynamic range nanomechanical mapping in tapping-mode atomic force microscopy. *Nanotechnology,* 19, 445717.

Schlesinger, I. & Sivan, U. 2017. New information on the hydrophobic interaction revealed by frequency modulation AFM. *Langmuir,* 33, 2485–2496.

Schön, P., Bagdi, K., Molnár, K., Markus, P., Pukánszky, B. & Julius Vancso, G. 2011. Quantitative mapping of elastic moduli at the nanoscale in phase separated polyurethanes by AFM. *European Polymer Journal,* 47, 692–698.

Schwiedrzik, J. J. & Zysset, P. K. 2015. The influence of yield surface shape and damage in the depth-dependent response of bone tissue to nanoindentation using spherical and Berkovich indenters. *Computer Methods in Biomechanics and Biomedical Engineering,* 18, 492–505.

Seifert, J., Rheinlaender, J., Novak, P., Korchev, Y. E. & Schaffer, T. E. 2015. Comparison of atomic force microscopy and scanning ion conductance microscopy for live cell imaging. *Langmuir,* 31, 6807–6813.

Sharma, S., Rasool, H. I., Palanisamy, V., Mathisen, C., Schmidt, M., Wong, D. T. & Gimzewski, J. K. 2010. Structural-mechanical characterization of nanoparticle exosomes in human saliva, using correlative AFM, FESEM, and force spectroscopy. *ACS Nano,* 4, 1921–1926.

Sirghi, L. & Rossi, F. 2006. Adhesion and elasticity in nanoscale indentation. *Applied Physics Letters,* 89, 243118.

Smolyakov, G., Thiebot, B., Campillo, C. C., Labdi, S., Severac, C., Pelta, J. & Dague, E. 2016. Elasticity, adhesion and tether extrusion on breast cancer cells provide a signature of their invasive potential. *ACS Applied Materials & Interfaces,* 8, 27426–27431.

Sneddon, I. N. 1965. The relation between load and penetration in the axisymmetric Boussinesq problem for a punch of arbitrary profile. *International Journal of Engineering Science,* 3, 47–57.

Soenen, H., Besamusca, J., Fischer, H. R., Poulikakos, L. D., Planche, J.-P., Das, P. K., Kringos, N., Grenfell, J. R. A., Lu, X. & Chailleux, E. 2013. Laboratory investigation of bitumen based on round robin DSC and AFM tests. *Materials and Structures,* 47, 1205–1220.

Sokolov, I., Dokukin, M. E. & Guz, N. V. 2013. Method for quantitative measurements of the elastic modulus of biological cells in AFM indentation experiments. *Methods,* 60, 202–213.

Soofi, S. S., Last, J. A., Liliensiek, S. J., Nealey, P. F. & Murphy, C. J. 2009. The elastic modulus of Matrigel as determined by atomic force microscopy. *Journal of Structural Biology,* 167, 216–219.

Staunton, J. R., Doss, B. L., Lindsay, S. & Ros, R. 2016. Correlating confocal microscopy and atomic force indentation reveals metastatic cancer cells stiffen during invasion into collagen I matrices. *Scientific Reports,* 6, 19686.

Sumarokova, M., Iturri, J. & Toca-Herrera, J. L. 2018. Adhesion, unfolding forces, and molecular elasticity of fibronectin coatings: an atomic force microscopy study. *Microscopy Research and Technique,* 81, 38–45.

Sun, Y., Hu, Z., Zhao, D. & Zeng, K. 2017. Mechanical properties of microcrystalline metal-organic frameworks (MOFs) measured by bimodal amplitude modulated-frequency modulated atomic force microscopy. *ACS Applied Materials & Interfaces,* 9, 32202–32210.

Sweers, K., Van Der Werf, K., Bennink, M. & Subramaniam, V. 2011. Nanomechanical properties of alpha-synuclein amyloid fibrils: a comparative study by nanoindentation, harmonic force microscopy, and Peakforce QNM. *Nanoscale Research Letters,* 6, 270.

Takechi-Haraya, Y., Goda, Y. & Sakai-Kato, K. 2018. Atomic force microscopy study on the stiffness of nanosized liposomes containing charged lipids. *Langmuir,* 34, 7805–7812.

Takechi-Haraya, Y., Sakai-Kato, K. & Goda, Y. 2017. Membrane rigidity determined by atomic force microscopy is a parameter of the permeability of liposomal membranes to the hydrophilic compound calcein. *AAPS PharmSciTech,* 18, 1887–1893.

Tang, C. Y., Kwon, Y.-N. & Leckie, J. O. 2009. The role of foulant–foulant electrostatic interaction on limiting flux for RO and NF membranes during humic acid fouling—Theoretical basis, experimental evidence, and AFM interaction force measurement. *Journal of Membrane Science,* 326, 526–532.

Tetard, L., Passian, A. & Thundat, T. 2010. New modes for subsurface atomic force microscopy through nanomechanical coupling. *Nature Nanotechnology*, 5, 105–109.

Touhami, A., Nysten, B. & Dufrêne, Y. F. 2003. Nanoscale mapping of the elasticity of microbial cells by atomic force microscopy. *Langmuir*, 19, 4539–4543.

Trtik, P., Kaufmann, J. & Volz, U. 2012. On the use of peak-force tapping atomic force microscopy for quantification of the local elastic modulus in hardened cement paste. *Cement and Concrete Research*, 42, 215–221.

Tuson, H. H., Auer, G. K., Renner, L. D., Hasebe, M., Tropini, C., Salick, M., Crone, W. C., Gopinathan, A., Huang, K. C. & Weibel, D. B. 2012. Measuring the stiffness of bacterial cells from growth rates in hydrogels of tunable elasticity. *Molecular Microbiology*, 84, 874–891.

Uhlig, M. R. & Magerle, R. 2017. Unraveling capillary interaction and viscoelastic response in atomic force microscopy of hydrated collagen fibrils. *Nanoscale*, 9, 1244–1256.

Walczyk, W. & Schonherr, H. 2014a. Characterization of the interaction between AFM tips and surface nanobubbles. *Langmuir*, 30, 7112–7126.

Walczyk, W. & Schonherr, H. 2014b. Dimensions and the profile of surface nanobubbles: tip-nanobubble interactions and nanobubble deformation in atomic force microscopy. *Langmuir*, 30, 11955–11965.

Wang, D., Liu, F., Yagihashi, N., Nakaya, M., Ferdous, S., Liang, X., Muramatsu, A., Nakajima, K. & Russell, T. P. 2014. New insights into morphology of high performance BHJ photovoltaics revealed by high resolution AFM. *Nano Letters*, 14, 5727–5732.

Wang, D. & Russell, T. P. 2017. Advances in atomic force microscopy for probing polymer structure and properties. *Macromolecules*, 51, 3–24.

Wang, M. & Liu, L. 2017. Investigation of microscale aging behavior of asphalt binders using atomic force microscopy. *Construction and Building Materials*, 135, 411–419.

Wenger, M. P., Bozec, L., Horton, M. A. & Mesquida, P. 2007. Mechanical properties of collagen fibrils. *Biophysical Journal*, 93, 1255–1263.

Wiggins, P. A., Van Der Heijden, T., Moreno-Herrero, F., Spakowitz, A., Phillips, R., Widom, J., Dekker, C. & Nelson, P. C. 2006. High flexibility of DNA on short length scales probed by atomic force microscopy. *Nature Nanotechnology*, 1, 137–141.

Wu, B., Heidelberg, A. & Boland, J. J. 2005. Mechanical properties of ultrahigh-strength gold nanowires. *Nature Materials*, 4, 525–529.

Wu, H., Kuhn, T. & Moy, V. 1998. Mechanical properties of L929 cells measured by atomic force microscopy: effects of anticytoskeletal drugs and membrane crosslinking. *Scanning: The Journal of Scanning Microscopies*, 20, 389–397.

Xia, D., Zhang, S., Hjortdal, J. Ø., Li, Q., Thomsen, K., Chevallier, J., Besenbacher, F. & Dong, M. 2014. Hydrated human corneal stroma revealed by

quantitative dynamic atomic force microscopy at nanoscale. *ACS Nano*, 8, 6873–6882.

Xu, X., Melcher, J., Basak, S., Reifenberger, R. & Raman, A. 2009. Compositional contrast of biological materials in liquids using the momentary excitation of higher eigenmodes in dynamic atomic force microscopy. *Physical Review Letters*, 102, 060801.

Yablon, D. G., Grabowski, J. & Chakraborty, I. 2014. Measuring the loss tangent of polymer materials with atomic force microscopy based methods. *Measurement Science and Technology*, 25, 055402.

Yang, R., Zaheri, A., Gao, W., Hayashi, C. & Espinosa, H. D. 2017. AFM identification of beetle exocuticle: bouligand structure and nanofiber anisotropic elastic properties. *Advanced Functional Materials*, 27, 1603993.

Yang, Y.-C. 2015. *Study of Biomechanical and Electrochemical Responses of* Streptococcus mutans *under Stressed Environment.* Master thesis, National Cheng Kung University.

Yao, X., Jericho, M., Pink, D. & Beveridge, T. 1999. Thickness and elasticity of gram-negative murein sacculi measured by atomic force microscopy. *Journal of Bacteriology*, 181, 6865–6875.

Yim, E. K., Darling, E. M., Kulangara, K., Guilak, F. & Leong, K. W. 2010. Nanotopography-induced changes in focal adhesions, cytoskeletal organization, and mechanical properties of human mesenchymal stem cells. *Biomaterials*, 31, 1299–1306.

Young, T. J., Monclus, M. A., Burnett, T. L., Broughton, W. R., Ogin, S. L. & Smith, P. A. 2011. The use of the PeakForcetmquantitative nanomechanical mapping AFM-based method for high-resolution Young's modulus measurement of polymers. *Measurement Science and Technology*, 22, 125703.

Yuan, Y. & Verma, R. 2006. Measuring microelastic properties of stratum corneum. *Colloids and Surfaces B: Biointerfaces*, 48, 6–12.

Zeng, Z. & Tan, J. C. 2017. AFM nanoindentation to quantify mechanical properties of nano- and micron-sized crystals of a metal-organic framework material. *ACS Applied Materials & Interfaces*, 9, 39839–39854.

Zhang, J., Wang, R., Yang, X., Lu, W., Wu, X., Wang, X., Li, H. & Chen, L. 2012a. Direct observation of inhomogeneous solid electrolyte interphase on MnO anode with atomic force microscopy and spectroscopy. *Nano Letters*, 12, 2153–2157.

Zhang, L., Yang, F., Cai, J. Y., Yang, P. H. & Liang, Z. H. 2014. In-situ detection of resveratrol inhibition effect on epidermal growth factor receptor of living MCF-7 cells by atomic force microscopy. *Biosensors and Bioelectronics*, 56, 271–277.

Zhang, T., Zheng, Y. & Cosgrove, D. J. 2016. Spatial organization of cellulose microfibrils and matrix polysaccharides in primary plant cell walls as imaged by multichannel atomic force microscopy. *Plant Journal*, 85, 179–192.

Zhang, X., Zhong, Y.-X., Yan, J.-W., Su, Y.-Z., Zhang, M. & Mao, B.-W. 2012b. Probing double layer structures of Au (111)–BMIPF 6 ionic liquid interfaces from potential-dependent AFM force curves. *Chemical Communications,* 48, 582–584.

Zhong, Y. X., Yan, J. W., Li, M. G., Zhang, X., He, D. W. & Mao, B. W. 2014. Resolving fine structures of the electric double layer of electrochemical interfaces in ionic liquids with an AFM tip modification strategy. *Journal of the American Chemical Society,* 136, 14682–14685.

Zhou, Z., Zheng, C., Li, S., Zhou, X., Liu, Z., He, Q., Zhang, N., Ngan, A., Tang, B. & Wang, A. 2013. AFM nanoindentation detection of the elastic modulus of tongue squamous carcinoma cells with different metastatic potentials. *Nanomedicine,* 9, 864–874.

Zhou, Z. L., Ngan, A. H., Tang, B. & Wang, A. X. 2012. Reliable measurement of elastic modulus of cells by nanoindentation in an atomic force microscope. *Journal of the Mechanical Behavior of Biomedical Materials,* 8, 134–142.

Zhu, Y., Dong, Z., Wejinya, U. C., Jin, S. & Ye, K. 2011. Determination of mechanical properties of soft tissue scaffolds by atomic force microscopy nanoindentation. *Journal of Biomechanics,* 44, 2356–2361.

Electron Holography for Mapping Electric Fields and Charge

Martha R. McCartney and
David J. Smith
Arizona State University

14.1 Introduction to Electron Holography

The macroscopic behavior of many materials is closely interconnected with the presence of nanoscale electrostatic fields and the accumulation of electric charge. Measurement and control of these intrinsic materials properties, especially as a function of growth and treatment conditions, is often an essential step along the path towards realizing practical applications. The transmission electron microscope (TEM) is an indispensable tool for materials characterization at the nanoscale, but it cannot directly provide access to details about fields and charges within and surrounding the object under examination. The crux of the problem is that the electrostatic (and magnetic) fields cause phase changes to the incident electron beam which are not observable using conventional TEM imaging modes that only provide intensity information. Electron holography overcomes this fundamental limitation. By using a coherent source, electron holography interference techniques enable the phase change of the electron beam traversing the sample relative to a vacuum (reference) wave to be determined. Reconstruction of the complex electron wave function from the recorded electron hologram allows quantitative measurement of electric charge and electric fields to be made with nanoscale resolution, subject to certain requirements and limitations that are elaborated below.

Electron holography was originally proposed as a means to circumvent the insurmountable limitation on electron microscope resolution imposed by spherical aberration (Gabor 1949). However, almost half a century passed before the first convincing aberration-correction experiments to successfully overcome the resolution limit using electron holography were reported (Orchowski et al. 1995). Subsequently, the technique has become widespread, in large part because the essential source of coherent electrons needed for electron interference is readily available in modern analytical and high-resolution TEMs equipped with field-emission electron guns (FEGs). Many possible variants of electron holography have been enumerated (Cowley 1992): the approach most commonly used and described in this chapter is a type of side-band holography, usually termed off-axis electron holography. The theoretical and experimental bases for off-axis electron holography are briefly outlined in the following section. Practical guidelines for using this technique, as well as requirements for preparing samples suitable for holographic observation, are then summarized. An overview is given of some examples where electron holography has been used for mapping charge and measuring electrostatic fields. Finally, ongoing problems and prospects for future developments of the technique are discussed. Further applications of electron holography to a wider variety of materials problems than could be mentioned here can be found elsewhere (see, for example, Dunin-Borkowski et al. 2004; McCartney and Smith 2007; Lichte et al. 2007; Midgley and Dunin-Borkowski 2009; Pozzi et al. 2014; Dunin-Borkowski et al. 2018).

14.2 Off-Axis Electron Holography

The theoretical basis for off-axis electron holography can be expressed relatively simply if the effects of dynamical electron diffraction are neglected. The phase change, $\delta\phi(x)$, of the electron wave that is transmitted by the object relative to that of the vacuum or reference wave is given by the expression:

$$\delta\phi(x,y) = C_E \int_0^t V(x,y)dz - \frac{e}{\hbar} \iint B_\perp(x,y)dxdz \quad (14.1)$$

where z is in the incident beam direction, x and y are directions in the plane of the sample, V is the total electrostatic potential of the sample, and $B_\perp$ is the component of the magnetic induction perpendicular to both x and z (Reimer 1991). The interaction constant C_E depends on the energy of the incident electron beam and is given by the expression

$$C_E = \frac{2\pi}{\lambda E} \frac{E + E_0}{2E + E_0} \quad (14.2)$$

where λ is the wavelength of the incident electron, and E and E_0 are the kinetic and rest mass energies, respectively, of the electron. C_E has the value of 0.00653 rad/(V·nm) for a TEM operated at 300 keV. In principle, these integrals should involve the entire path of the electron wave from the FEG source to the final hologram detector. Thus, fields external to the object, including specimen fringing fields and stray fields due to other microscope components, may influence the phase change that is finally measured. Hence, these must be taken into account when they cannot be ignored entirely, for example, by subtraction of a reference hologram with the sample removed from the holography field of view. In some specimens such as charged nanowires (NWs), because of the nonuniformity of the fields along the electron-beam direction, it may even be necessary to resort to three-dimensional tomographic holography, as described briefly below, although this process is usually quite time-consuming (Wolf et al. 2013).

For many practical situations, neither V nor B vary along the beam path within the sample thickness t, and it can often also be assumed that electrostatic and/or magnetic fringing fields outside the sample are small enough to be neglected. The expression above for the *phase change* can then be simplified to

$$\delta\phi(x,y) = C_E V(x)t(x) - \frac{e}{\hbar} \int B_\perp(x)t(x)dx \quad (14.3)$$

For nonmagnetic materials that are of direct relevance in this chapter, the second term will be zero. The reduced expression then provides the basis for measurement and quantification of electrostatic fields, which could include any variations in potential due to internal fields in addition to the mean inner potential (MIP) (i.e., composition and density) of the specimen.

Further differentiation of Eq. (14.3) leads to the *phase gradient*, as given by the expression

$$\frac{d\phi(x)}{dx} = C_E \frac{d}{dx}[V(x)t(x)] - \frac{e}{\hbar}B_\perp(x)t(x) \quad (14.4)$$

For a sample of uniform thickness and composition, the first term is zero, so that measurement of the phase gradient

would then provide a direct and quantitative measure of the in-plane magnetic induction. When the specimen thickness is locally varying, further phase imaging and processing is, however, necessary in order to differentiate between contributions from the MIP and from the magnetic field. Strategies that can be used for this specific purpose when necessary have been given elsewhere (Dunin-Borkowski et al. 1998).

The specimen geometries that are used with off-axis electron holography to determine the electrostatic and magnetic fields using phase shifts and phase gradients, respectively, are illustrated schematically in Figure 14.1. The sinusoidal interference fringes produced by overlapping the object and vacuum reference waves provide amplitude and potentially thickness information about the sample, while the desired phase information, which is related to the specimen fields and any isolated electric charge, is encoded in the relative fringe positions. Since the vacuum reference wave needs to be overlapped with the specimen area of interest, it is helpful if the edge of the specimen is within the field of view. An exception to this requirement is when the sample is situated on a thin and weakly scattering support film.

After digital recording of the electron hologram, further computer processing is required in order to reconstruct the (complex) wave function, from which the phase shifts caused by the specimen can be extracted and quantified. The procedure used for holographic reconstruction of the electrostatic fields associated with a doped Si NW is illustrated in Figure 14.2. The original hologram is shown in Figure 14.2a, with an enlargement showing local bending of the interference fringes due to the NW in Figure 14.2b. A fast Fourier transform (FFT) of the hologram is calculated, giving a central autocorrelation peak, corresponding to the image intensity, as well as two sidebands that contain the

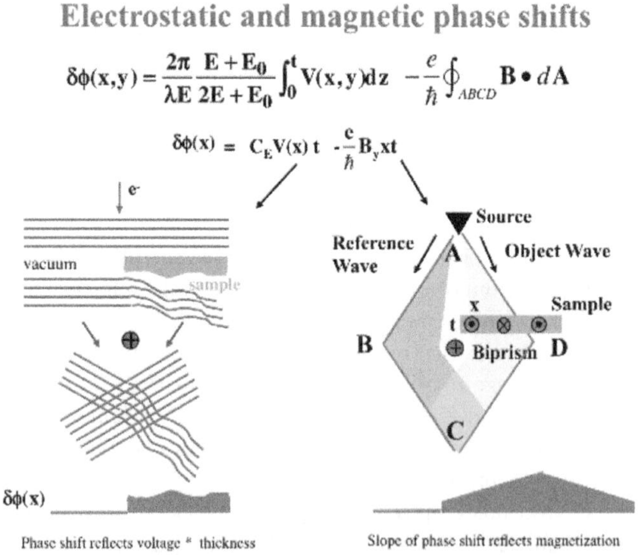

Electrostatic and magnetic phase shifts

$$\delta\phi(x,y) = \frac{2\pi}{\lambda E}\frac{E+E_0}{2E+E_0}\int_0^t V(x,y)dz - \frac{e}{\hbar}\oint_{ABCD} \mathbf{B}\bullet d\mathbf{A}$$

$$\delta\phi(x) = C_E V(x)t - \frac{e}{\hbar}B_y xt$$

FIGURE 14.1 Schematic diagram illustrating the origins of phase shifts used for the measurement of electrostatic and magnetic fields using the technique of off-axis electron holography.

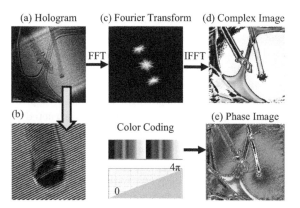

FIGURE 14.2 Schematic illustrating the steps involved in quantitative phase imaging using the technique of off-axis electron holography. (a) Electron hologram showing doped Si NW supported on holey carbon film; (b) enlargement showing bending of interference fringes at edge of Si NW; (c) FFT of hologram; (d) reconstructed complex image calculated by IFT from FFT sideband; (e) color-coded phase image showing electrostatic profile along and surrounding Si NW.

required phase information – Figure 14.2c. With suitable masking to select one of the sidebands, and by applying an inverse Fourier transform (IFT), a complex image is reconstructed – Figure 14.2d. The phase image, which is the arctangent of the ratio of the imaginary and real parts of the complex image, can then be calculated – Figure 14.2e. Further processing based on Eqs. (14.3) and (14.4), taking into account such factors as the specimen composition and thickness, will then enable the electrostatic potential and/or

magnetic fields within and surrounding the sample to be determined quantitatively with nanoscale or even atomic-scale spatial resolution. Computer programs suitable for these tasks have been developed by several investigators and are also commercially available. It is even possible to do hologram processing at rates of up to several frames per second, while seated in front of the microscope console rather than waiting until after the microscope session has been completed.

14.3 Practical Aspects

14.3.1 Instrumental

Several electron microscope components are essential for applying the off-axis electron holography technique. These are indicated in Figure 14.3. These include: (i) a FEG electron source that provides a highly coherent electron beam incident at the sample; (ii) a post-specimen electrostatic biprism, which consists of a thin (∼0.5 μm), positively charged wire normally located in the selected-area aperture plane, that causes overlap of the wave passing through the object with a reference wave passing through vacuum (Möllenstedt and Düker 1955); and (iii) a quantitative electron recording device having large dynamic range and excellent input–output linearity, such as a charge-coupled device (CCD) camera, which records the resulting interference pattern (i.e., hologram) in digital format, thereby facilitating subsequent hologram processing and interpretation. The so-called 'Lorentz' mini-lens, which is located below the

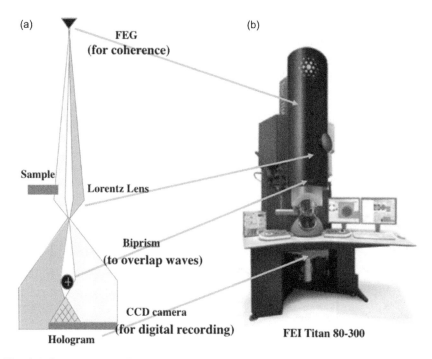

FIGURE 14.3 (a) Sketch indicating essential microscope components for off-axis electron holography: the FEG provides coherent incident illumination, the electrostatic biprism provides overlap of object and vacuum reference waves, and the CCD camera enables quantitative hologram recording; (b) photograph of Titan 80-300 aberration-corrected TEM equipped for off-axis electron holography research. The Lorentz lens provides additional flexibility by enabling observations to be made with a much larger field of view.

lower bore of the objective-lens pole piece, enables a larger field of view to be obtained, which proves to be highly useful for many important cases of electrostatic field imaging, such as dopant depletion regions in semiconductor devices. In addition, this lens also allows holographic imaging of magnetic specimens to be made at relatively high magnification (~50–100 kX times) with the normal objective lens switched off so that the specimen is located in a (almost) field-free environment (Dunin-Borkowski et al. 1998).

The electrostatic biprism represents the keystone of the off-axis electron holography technique since it enables the basic interference of the object and reference waves that is required to produce the electron hologram. Typically, the biprism consists of a thin conducting wire electrically insulated from the microscope column and able to sustain potentials that can be as high as 1,000 V for atomic-resolution studies (Lichte and Lehmann 2008). Quartz fibers coated with a thin metallic film are most commonly used but platinum wires are also supplied by one microscope manufacturer. For operational convenience, the biprism wire is usually mounted across an opening in a multi-aperture holder such as the regular TEM selected-area aperture strip, which provides for convenient insertion and removal from the path of the electron beam. The wire should preferably also be rotatable so that it can be aligned with salient specimen features of interest, and it should be mechanically stable since any drift or vibrations will reduce the interference-fringe contrast and hence degrade the final phase image quality.

14.3.2 Important Parameters

The biasing voltage applied to the biprism controls several important and inter-related parameters, namely the contrast and spacing of the interference fringes and the extent of fringe overlap, normally termed the field of view. The sequence of images shown in Figure 14.4 graphically illustrates the effect of increasing the biprism voltage on these parameters. Initially, at zero applied voltage, the shadow of the biprism wire is simply bounded by Fresnel diffraction fringes. At 40 V, the waves on each side of the wire start to overlap across a distance of ~6 nm, producing relatively coarse interference fringes that are still however affected by

Fresnel diffraction effects due to the wire. At 80 V, the fringe contrast and spacings have decreased, the overlap region has enlarged to ~18 nm, and the Fresnel diffraction effects are only visible at the perimeters of the overlap region. Higher applied voltages will continue to expand the field of view and produce finer fringe spacings that should in principle enable better spatial resolution in the reconstructed phase images, as explained further below. However, these gains will also be accompanied by decreased fringe contrast that will negatively impact image quality due to the poorer signal-to-noise content of the phase image. Higher beam coherence helps to ameliorate this latter effect but eventually some practical compromise may need to be made depending on the particular microscope and the specific application.

In practice, the interference fringe spacing varies inversely with the biprism voltage, and the proportionality depends on several experimental factors including the electron energy, the objective-lens focal length, and the relative location of the biprism (Lichte 1986). Calculation is possible when the relevant dimensions are known but measurements are easily made for any specific microscope geometry, as illustrated by the example shown in Figure 14.5a. Likewise, it is straightforward to measure and plot the width of the field of view as a function of the biprism voltage, as shown in Figure 14.5b. The fringe contrast (i.e., visibility) will eventually determine the attainable precision of phase measurements (de Ruijter and Weiss 1993). Thus, it is essential to avoid or at least minimize all possible contributing factors that have negative effects, which include mechanical and electrical stability and beam energy spread, as well as external stray magnetic fields (Cooper et al. 2007). As a check for future reference, it can be helpful to determine the fringe contrast for some specific configuration, which can be done using the expression

$$\Gamma = \frac{I_{\max} - I_{\min}}{I_{\max} + I_{\min}} \qquad (14.5)$$

where $I_{\max}$ and $I_{\min}$ are measured directly from an interference pattern recorded with the sample removed from the field of view.

Finally, it is necessary to appreciate the complicated interplay between spatial resolution and magnification,

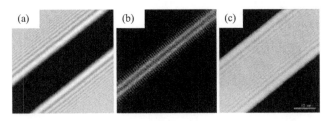

FIGURE 14.4 Images showing effect of biprism voltage on formation of electron hologram; (a) 0 V. Visible fringes are due to Fresnel diffraction effects around edge of biprism; (b) 40 V. Object and reference wave have overlap of ~6 nm; (c) 80 V. Some modulation at edge of overlap region due to Fresnel fringes. Overlap region increased to ~18 nm.

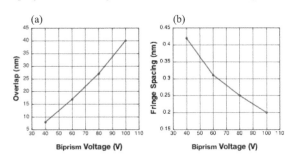

FIGURE 14.5 (a) Plot showing increase of overlap region (i.e., field of view) with increase in biprism voltage for Philips CM200FEG TEM operated at 200 kV. (b) Plot showing dependence of interference fringe spacing on biprism voltage for Philips CM200FEG TEM operated at 200 kV.

fringe spacing and detector pixel size, which has been discussed in detail elsewhere (Lichte 1986). Undersampling of the interference fringes can cause a dramatic loss of fringe visibility by factors of 3 or more (Smith and McCartney 1999) but can be avoided, or at least minimized, by using a sufficiently high image magnification. Our experience when studying nanoscale fields suggests that the period of the interference fringes should preferably be no more than one third of the targeted spatial resolution. Moreover, the image magnification should be adjusted such that each fringe period is sampled by at least 6 but preferably by as many as 10 detector pixels for studies requiring sensitive phase measurements.

14.3.3 Operating Guidelines

Electron holography studies are always preceded by a preliminary period wherein the electron microscope is operated in its normal mode. Standard alignments of the incident beam illumination are made, any required specimen tilting is carried out, and the image astigmatism is corrected. The specimen feature of interest needs to be located and brought into focus, and it should be confirmed that the edge of the specimen, which is required for the vacuum reference wave, will remain in the field of view once the electron holography begins.

Additional practical factors also need to be considered when doing off-axis electron holography. For example, extraneous effects that are not associated with genuine specimen features, such as charging, lens distortions, and nonlinearities in the CCD recording system, can easily perturb the phase information about the specimen that is contained in the displaced hologram fringes. Thus, it is standard practice to record a reference hologram with the specimen removed from the field of view and without making any other changes to the microscope imaging parameters. Phase images free of distortions and artifacts can then be obtained by carrying out a division of the object and reference waves (de Ruijter and Weiss 1993).

The coherence (monochromaticity) of the electron beam incident on the sample is a crucial parameter because it determines the all-important contrast (i.e., visibility) of the holographic interference fringes. The beam coherence should be maximized to the greatest extent possible by minimizing the beam convergence angle at the sample with a small condenser aperture and a small spot size. However, unlike conventional TEM imaging where rotationally symmetric illumination is normally required, the use of highly elliptical illumination is recommended for electron holography since this configuration allows significant gains in lateral beam coherence to be achieved. As indicated by the shape of the illumination spot shown in Figure 14.6, the condenser lens stigmators are adjusted so that the area illuminated in the specimen plane is very wide in the direction perpendicular to the biprism wire but narrow in the orthogonal direction parallel to the biprism wire with the condenser lens overfocused. Typically, the minor axis should be about 2–5

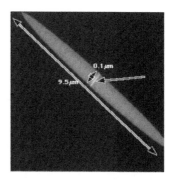

FIGURE 14.6 Photo showing highly elliptical incident illumination due to deliberate condenser lens astigmatism. Major and minor axes dimensions of 9.5 and 0.1 μm, respectively. Arrow indicates position of holographic interference fringes.

times larger than the region of overlap between the object and reference waves, and the major axis should be another 50–100 times larger (Smith and McCartney 1999).

14.4 Sample Preparation

Minimizing artifacts due to sample preparation is an important aspect of studying electric fields and charges using off-axis electron holography. The area intended for examination should be electron-transparent and close to vacuum in order to ensure an unperturbed reference wave. Specimens should preferably be prepared with uniform thickness, or with a well-defined thickness profile, in order to improve the precision of the data available in the reconstructed phase image. Wedge polishing is very useful in this regard (Han et al. 2006). Misleading phase shifts due to diffraction effects or bend contours can be minimized by tilting samples slightly away from a zone-axis orientation (Gan et al. 2015). However, some compromise in the amount of tilt often has to be made in order to avoid smearing-out of interface profiles in nanostructured devices that have two-dimensional design features (Formanek and Bugiel 2006a). Moreover, sample charging due to beam-induced emission of secondary electrons may degrade the potential profile measurement (McCartney 2005). Charge accumulation can be reduced by carbon coating of sample surfaces before commencing the holography studies (Li et al. 2003).

Thus, recognizing and avoiding sample-preparation artifacts represents an ongoing concern for electron-holography studies. Mechanical polishing followed by argon-ion milling is the standard method used for preparing cross-sectional specimens (Bravman and Sinclair 1984). However, early off-axis electron holography studies of an Si *p–n* junction showed that the ion milling gave rise to amorphous layers of up to 25 nm in thickness on both sample surfaces that were considered to be electrically inactive (Rau et al. 1999). Low-angle wedge polishing followed in turn by low-angle, low-energy ion milling and coating with a thin carbon film was shown to significantly mitigate these effects (Li et al. 2003), and high-quality potential maps of a *p*-MOS field-effect transistor prepared by this method were obtained (Han et al. 2006).

Different approaches to sample thinning are required when the device region of interest is located well below the sample surface (i.e., 0.5 to several microns). The focused-ion-beam (FIB) has developed into a site-specific, analytical tool that is widely used throughout the semiconductor industry for TEM sample preparation, being especially useful for failure analysis (Giannuzzi and Stevie 2005). However, FIB milling of Si-based specimens can result in inactive surface layers with thicknesses as large as ~100 nm (Dunin-Borkowski et al. 2005). These effects are attributable to energetic Ga ions penetrating deep into the sidewalls of the material during FIB milling (Twitchett et al. 2005). Procedures for minimizing these FIB-induced artifacts have been widely discussed and include *in situ* biasing (Twitchett et al. 2002), *in situ* annealing (Cooper et al. 2006), and low-energy backside ion milling (Ikarashi et al. 2008). In addition, uneven milling of wafers can create thickness corrugations, referred to as 'curtaining' (Dunin-Borkowski et al. 2005). These sample artifacts can be greatly alleviated, if not entirely avoided, by 'backside' milling, as well as by completing the thinning process with gentle, low-angle milling using conventional argon-ion milling (Formanek and Bugiel 2006b). A detailed discussion related to FIB milling of semiconductors and suggesting viable strategies for minimizing artifacts can be found elsewhere (Cooper et al. 2010).

14.5 Mapping Electric Charge and Electric Fields

Off-axis electron holography has evolved over many years into a highly useful technique for studying *and* quantifying electric (and magnetic) fields inside materials with high spatial resolution and precision. The theoretical basis is solid, and the requirements for instrument operation and sample preparation, as outlined in the preceding sections, are widely appreciated. This section provides a brief but representative overview of applications of the technique to some materials problems of contemporary interest in the nanoscience area. References are also given to examples published in the scientific literature to enable the reader to delve deeper into particular topics.

14.5.1 *p–n* Junctions

The electronic behavior of many types of semiconductor devices is strongly influenced by internal electrostatic fields that are associated with extended defects and the band offsets that are present at heterojunctions. For nanostructured devices, and in the presence of the high levels of dopants that are required for device operation, the shapes of these fields can be quite complicated, so that mapping multidimensional dopant distributions with nanoscale resolution becomes highly challenging. Secondary-ion mass spectrometry (SIMS) is unparalleled for measuring dopant concentrations down to the part per million level but its lateral resolution is inherently poor relative to device

dimensions (Diebold et al. 1996). Scanning spreading resistance microscopy requires extensive modeling before the resistance profiles measured for silicon devices can be interpreted in terms of dopant profiles (Eyben et al. 2003). Off-axis electron holography does not measure dopant distributions directly, but the technique has established a strong track record over the years for determining electrostatic potential profiles caused by the presence of dopants, which can then be interpreted in terms of dopant distributions (see, among many examples, McCartney et al. 1994; Rau et al. 1999; McCartney et al. 2002; Gribelyuk et al. 2002; Lenk et al. 2005; Cooper et al. 2007; Twitchett-Harrison et al. 2007; Ikarashi et al. 2008; Ikarashi et al. 2012; Cooper et al. 2013).

For doped semiconductor devices, the electrostatic potential consists of two terms: the MIP of the sample, usually designated as V_0 and the built-in potential difference across the junction between the *p*- and *n*-type regions, which is usually designated as V_{pn}.

$$\delta\phi(x) = C_E \left[V_0(x) + V_{pn}(x) \right] t(x) \tag{14.6}$$

Thus, mapping the electrostatic potential across the junction of a doped semiconductor device requires prior knowledge of the MIP *and* an independent measure of the local specimen thickness. The technique of convergent beam electron diffraction (probe size ~2–10 nm) is well suited for this latter task, at least for crystalline samples. Local thicknesses can typically be measured with an accuracy of ~5% over the specimen thickness range that is suitable for off-axis electron holography (Li et al. 2003). An alternative approach to determining thickness is based on manipulating the amplitude image, but this method also requires independent knowledge of the inelastic mean free path (McCartney and Gajdardziska 1993).

Microscope geometry and sample preparation severely restricted the earliest electron holography studies of doped semiconductors (Matteucci et al. 1979). Later device studies, which took advantage of digital recording, established conclusively that the internal electric field across an unbiased *p–n* junction due to a simple one-dimensional (1-D) dopant profile could be mapped *and* quantified with nanoscale resolution (McCartney et al. 1994). Mapping of the 2-D potential associated with metal-oxide-semiconductor (MOS) transistors later demonstrated spatial resolutions of better than 10 nm with a sensitivity of 0.1 V (Rau et al. 1999). Comparisons between experimental and simulated potential profiles showed a spatial resolution of ~6nm and a sensitivity of ~0.17 V for a 1-D Si *p–n* junction, as shown in Figure 14.7 (Gribelyuk et al. 2002). Comparable results for other devices can be found elsewhere (see, for example, Lenk et al. 2005; Han et al. 2008; Ikarashi et al. 2008).

More recent device studies have continued to push the envelopes of spatial resolution and/or sensitivity. A scaled MOSFET with 30-nm gate length was examined (Ikarashi et al. 2008), and observations of doped Si test samples, which took advantage of an aberration-corrected TEM with

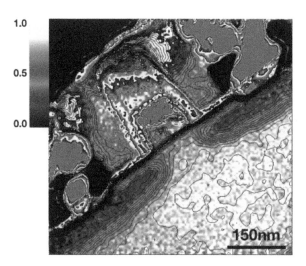

FIGURE 14.7 Reconstructed 2D phase map showing electrostatic potential variations associated with 0.13 μm p-FET device. (Adapted from Gribelyuk et al. 2002.)

exceptional mechanical stability, demonstrated that steps in electrostatic potential as small as 0.030 ± 0.003 V could be reliably measured (Cooper et al. 2007). Mention should also be made here of using a combined holography-tomography approach to visualize the three-dimensional (3D) electrostatic profile distribution across an electrically biased Si p–n junction (Twitchett-Harrison et al. 2007): this study revealed an extensive region of dopant deactivation attributable to the FIB milling that was used for sample preparation. Further challenges for studying deep-submicron electronic device structures with off-axis electron holography have also been discussed (Gribelyuk et al. 2008; Ikarashi et al. 2010).

14.5.2 III-Nitride Semiconductors

Electron holography has been used to good effect in studying sheet charge and electrostatic fields in III-nitride semiconductors. These materials feature large polarization fields as well as charge accumulation at interfaces that are attributable to their highly polar nature. Studies of $In_{0.13}Ga_{0.87}N$ quantum wells (QWs) as a function of layer thickness showed that the average electrostatic fields were

as high as 1.7 MV/cm across thinner QWs although the field strengths dropped sharply for greater well thicknesses (Stevens et al. 2004). Studies of GaN/InGaN/GaN QWs showed the accumulation of interfacial sheet charge (Cherns et al. 2001; Cai and Ponce 2002a), and a two-dimensional electron gas (2DEG) was measured in the vicinity of a GaN/AlGaN heterointerface (Cai et al. 2001). The presence of interfacial bound sheet charge was also confirmed for the more complicated case of a p-AlGaN/InGaN/n-AlGaN light-emitting diode structure (McCartney et al. 2000). Detailed studies were made of polarization fields and sheet charge in AlInN/AlN/GaN heterostructures, as shown in Figure 14.8 (Zhou et al. 2009). The potential profile calculated from the phase image corresponded to an electric field strength of ∼6.9 MV/cm, consistent with theoretical calculations. In addition, the positive curvature in the GaN layer near the AlN/GaN interface signaled the presence of the 2DEG, and further analysis revealed that the integrated 2DEG density was ∼2.1 × 10^{13}/cm². More recently, electrostatic potential profiles across one-monolayer-thick InN/GaN multiple quantum wells (MQWs) have been measured, and it was shown that the fields inside the GaN barriers decreased from ∼0.7 to ∼0.2 MV/cm as the layer thickness was increased from 5 to 20 nm, which agreed closely with model simulations (Zhou et al. 2013). These experimental studies of nitride-based materials reinforce the usefulness of off-axis electron holography for investigating important aspects of semiconductor heterojunctions and nanostructured devices that are inaccessible to other measurement techniques.

14.5.3 Nanowires and Nanotubes

Another fruitful area for electron holography studies has been the examination of nanostructures in the form of NWs and nanotubes. For example, an early study of Ge/Si core/shell NWs used the holography technique to quantify the amount of hole accumulation in the core region of undoped Ge-Si core-shell NWs that was caused by the valence-band offset between the two materials (Li et al. 2011). Observations of Si–Ge NWs and axial p–n junctions in Si NWs were also made, as illustrated by the example in Figure 14.9 (Gan et al. 2013a). Difficulties were

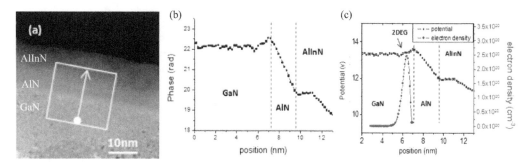

FIGURE 14.8 (a) Reconstructed phase image from an $Al_{0.85}In_{0.15}N/AlN/GaN$ high electron mobility transistor; (b) phase profile across the AlInN/AlN/GaN interface region; (c) potential profile (open squares) and electron distribution (filled circles) across the AlInN/AlN/GaN interface. (Adapted from Zhou et al. 2009.)

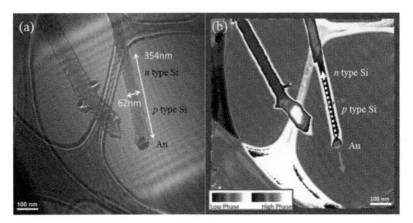

FIGURE 14.9 (a) Hologram of doped Si NW supported on holey carbon film; (b) reconstructed phase image visualized with pseudo-color. Scale bar shown as inset. (Adapted from Gan et al. 2013a.)

initially experienced with charging of the Au catalyst particles located at the ends of the Si NWs, but more consistent junction profiles were obtained by coating the samples with carbon before observation. Further study revealed that the surfaces of the NWs were equiphase or equipotential across the p–n junction, meaning that the surface potential was pinned at mid-gap, while comparisons with simulations enabled the Schottky barrier height as well as the active dopant concentrations to be estimated (Gan et al. 2013a). Observations of p–n heterojunctions in Si–Ge NWs proved to be more complicated due to severe NW kinking near the junction, and the average built-in potential was measured to be ∼0.6 V, which was slightly higher than the expected value of 0.4 V (Gan et al. 2013b). More recent electron holography studies have focused on the Schottky barriers formed by the Au catalyst particle at the NW tip (Gan et al. 2013a; He et al. 2013). Experiments and simulations aimed at improving the measurement accuracy and further investigating the role of experimental and materials parameters, such as the work function of the gold catalyst particles that cap the NWs, are still needed (Gan et al. 2016). Mapping of changes in dopant levels using electron holography was demonstrated for single Si NWs (den Hertog et al. 2009) and for core-shell NWs consisting of doped Ge shells surrounding intrinsic Ge cores (Chung and Rabenberg 2006). Attention has also been directed towards the occurrence of polarization-induced charge redistribution at homogeneous zincblende (ZB)/wurtzite (WZ) heterojunctions in InAs(Sb) nanopillars (Li et al. 2012) as well as ZnSe nanobelts (Li et al. 2014), and reasonable agreement with theoretical simulations was achieved in the latter case.

14.5.4 Quantum Dots and Charged Defects

The sensitivity of off-axis electron holography to changes in electrostatic potential suggests that the technique could be used to detect the presence of electric charge within suitably thinned specimens. Quantum dots (QDs) and structural defects such as dislocations are prime locations for

charge accumulation to occur, but accurate quantification of the amount of charge and the charge density can be difficult to achieve using electron holography because of overlap effects as well as stringent sample preparation requirements. Nevertheless, the electric fields across individual wurtzite GaN QDs embedded within an AlN matrix were successfully measured, as shown by the example in Figure 14.10, revealing local fields as high as 7.8 MV/cm, in close agreement with atomistic tight-binding calculations (Zhou et al. 2011). The sample geometry made this study especially challenging because the GaN QDs were entirely enclosed within the AlN matrix so that the separate MIP contributions from the two different materials, which were overlapping along the beam direction, had to be taken into account.

In another study, the accumulation of positive charge (i.e., holes) in individual Ge QDs was investigated (Li et al. 2009). These samples were prepared using wedge polishing because ion implantation and nonuniform sample thickness due to ion milling had to be avoided. These self-assembled Ge QDs were sandwiched between boron-doped Si layers, resulting in valence-band offsets that caused the holes to be confined in the QDs. The shapes of the dots and the amount of Si above and below each dot along the electron beam

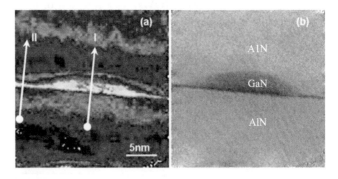

FIGURE 14.10 (a) Phase and (b) amplitude images calculated from the reconstructed hologram of wurtzite GaN QD embedded within an AlN matrix. The amplitude image helped to distinguish between the separate AlN and GaN MIP contributions. (Adapted from Zhou et al. 2011.)

direction had to be carefully taken into account before the amount of accumulated charge could be reliably extracted from the phase image. A one-dimensional Poisson equation was applied to the projected electrostatic potential so that the resulting number of holes per unit volume could be estimated. The calculated number of ~30 ± 3 holes/dot was remarkably close to the average number of 40 holes per each Ge QD measured using a macroscopic capacitance-voltage method.

Studying charged defects with electron holography seems even more challenging than quantifying charge in QDs, not just because of demanding sample preparation issues but also because of likely structural disorder and possible impurity segregation along the defect projections. Studies of threading screw dislocations in *n*-doped 4H-SiC showed the existence of negatively charged dislocation cores, and comparisons with between experiments and simulations showed that the density of trapped charge correlated with the doping level (Chung et al. 2011). Holography studies of undoped GaN indicated that edge, mixed, and screw dislocations were negatively charged (Cai and Ponce 2002b), whereas dislocations were positively charged in *p*-GaN (Cherns et al. 2002) and negatively charged in *n*-GaN (Cherns and Jiao 2001). Cross-sectional studies of charged dislocations in *n*-ZnO revealed negative charge near the defect core with charge densities ~2 electrons/nm (Müller et al. 2006). Finally, another remarkable study for very well-defined sample geometry and special imaging conditions has demonstrated the possibility of counting the charge on individual nanoparticles with a sensitivity of one elementary unit of charge (Gatel et al. 2013).

14.6 Problems and Future Prospects

Reliable sample preparation and artifact-free phase imaging will continue to be major concerns for future electron holography studies of nanoscale electric fields and charge, especially for investigating semiconductor devices with reduced feature sizes where surface charging and the presence of any native oxides are likely to have relatively greater impact. Other artifacts that will continue to complicate interpretation of electrostatic field measurements include those that arise from charging of the specimen due to emission of secondary electrons, the production of electron–hole pairs in the material, and their trapping at interfaces in the specimen. Methods to circumvent and/or account for these effects will need to be investigated in future experiments, in some cases, by taking advantage of *in situ* biasing holders, as outlined in the next paragraph.

Biasing holders are likely to play an increasingly valuable role in future research involving holographic measurements of electric fields and charge. Studies by ourselves (for example, Han et al. 2008) and others (see, for example, He et al. 2013; Yazdi et al. 2015) have demonstrated successful *in situ* biasing of doped Si-based junctions and NWs, in particular allowing direct measurement of changes in depletion width and junction voltage as a function of applied bias. Biasing is also critical in enabling the extent of dopant activation to be determined by using comparisons with device simulations (Gan et al. 2016). This ability to bias the sample during holography observation introduces an additional degree of freedom that can be decisive for identifying and eventually accounting for other artifacts, such as surface pinning and trapped charge (Cooper et al. 2010), which might otherwise complicate reliable hologram interpretation and quantification. Sample biasing can also be useful in enabling surface and interface charging effects due to sample preparation and/or electron-beam irradiation to be delineated from the built-in potentials and sheet charge that are being investigated. Moreover, the measurement of dopant profiles and trapped charge with nanometer-scale spatial resolution will be enhanced by the ability to modify the depletion regions by way of *in situ* forward/reverse biasing during observation. The knowledge provided by such holography studies could be pivotal in the process of developing next-generation electronic and optoelectronic devices.

Semiconductor NWs continue to attract much attention from the materials community, especially because of the novel photonic, electronic, and thermal properties that are caused by quantum confinement effects arising from their unique one-dimensional geometry (Hyun et al. 2013). Different methods for NW synthesis continue to be explored and techniques for characterization of structure-property relationships are of much interest and relevance. The realization of practical electronic devices based on semiconductor NWs is highly topical but also challenging since precise control over dopant incorporation *and* activation is required. Doping has been realized during NW growth, for example, by using gas-phase deposition (Agarwal et al. 2007) or ion implantation (Hoffman et al. 2009), but measuring the amount and distribution of *activated* dopant concentrations in the NW geometry is far from straightforward, given that considerable dopant diffusion is also likely to occur during thermal annealing (Koren et al. 2010). Dopant profiles cannot be determined directly using off-axis electron holography, but the technique can be used to measure the electrostatic potential profiles that are caused by the activated dopants, which can then be compared with actual device measurements and/or simulations (Gan et al. 2013a; Gan et al. 2016).

The desirability of carrying out full, three-dimensional tomographic mapping of electric fields and charge distributions should be self-evident given that many practical examples involve variable fields both within and outside the samples of interest. However, the practicalities of tomographic holography are arduous and time-consuming, given the need to acquire multiple holograms over a wide range of tilt (typically ± 70° in 2° steps), as well as reference holograms, albeit at less frequent intervals (Twitchett-Harrison et al. 2007). The effect of progressive changes in projected sample thickness on the sets of images also needs to be taken into account during the reconstruction process (Twitchett-Harrison et al. 2008), and holograms in the series that have been adventitiously recorded under diffracting conditions need to be manually discarded before carrying out the full tomographic reconstruction. Despite these formidable hurdles, successful three-dimensional mapping of several different semiconductor combinations have been reported, including an Si *p–n* junction (Twitchett-Harrison et al. 2008), as well as GaAs and GaAs–AlGaAs core-shell NWs (Wolf et al. 2011; Lubk et al. 2014), and also Pt nanoparticles (Tanagaki et al. 2012).

References

Agarwal P., Vijayaraghavan M.N., Neuilly F., Hijzen E. and Hurkx G.A.M. 2007. Breakdown enhancement in silicon nanowire *p-n* junctions. *Nano Lett.* 7: 896–899.

Bravman J. and Sinclair R. 1984. The preparation of cross-section specimens for transmission electron microscopy. *J. Electron Microsc. Tech.* 1: 53–61.

Cai J. and Ponce F.A. 2002a. Study of charge distribution across interfaces in GaN/InGaN/GaN single quantum wells using electron holography. *J. Appl. Phys.* 91: 9856–9862.

Cai J. and Ponce F.A. 2002b. Determination by electron holography of the electronic charge distribution at threading dislocations in epitaxial GaN. *Phys. Stat. Sol. (a)* 192: 407–411.

Cai J., Ponce F.A., Tanaka S., Omiya H. and Nakagawa Y. 2001. Mapping the internal potential across GaN/AlGaN heterostructures by electron holography. *Phys. Stat. Sol. (a)* 188: 833–837.

Cherns D. and Jiao C.G. 2001. Electron holography studies of the charge on dislocations in GaN. *Phys. Rev. Lett.* 87: 205504.

Cherns D., Mokhtari H., Jiao C.G., Averback R. and Riechert H. 2001. Profiling band structure in GaN devices by electron holography. *J. Cryst. Growth* 230: 410–414.

Cherns D., Jiao C.G., Mokhtari H., Cai J. and Ponce F.A. 2002. Electron holography studies of the charge on dislocations in GaN. *Phys. Stat. Sol. (b)* 234: 924–930.

Chung J. and Rabenberg L. 2006. Mapping of electrostatic potentials within core-shell nanowires by electron holography. *Appl. Phys. Lett.* 88: 013106.

Chung S., Berechman R.A., McCartney M.R. and Skowronski M. 2011. Electronic structure analysis of threading screw dislocations in 4H-SiC using electron holography. *J. Appl. Phys.* 109: 043906.

Cooper D., Twitchett A.C., Somodi P.K., Midgley P.A., Dunin-Borkowski R.E., Farrer, I. and Ritchie D.A. 2006. Improvement in electron holographic phase images of focused-ion-beam-milled GaAs and Si *p-n* junctions by *in situ* annealing. *Appl. Phys. Lett.* 88: 063510.

Cooper D., Truche R., Rivallin P., Hartmann J-M., Laugier F., Bertin F. and Chabli A. 2007. Medium resolution off-axis electron holography with millivolt sensitivity. *Appl. Phys. Lett.* 91: 143501.

Cooper D., Ailliot C., Barnes J.P., Hartmann J-M., Salles P., Benassayag G. and Dunin-Borkowski R.E. 2010. Dopant profiling of focused ion beam milled semiconductors using off-axis electron holography: reducing artifacts, extending detection limits and reducing the effects of gallium implantation. *Ultramicroscopy* 110: 383–389.

Cooper D., Rivallin P., Guegan G., Plantier C., Robin E., Guyot F. and Constant I. 2013. Field mapping of focused ion beam prepared semiconductor devices by off-axis and dark-field electron holography. *Semicond. Sci. Tech.* 28: 125013.

Cowley J.M. 1992. Twenty forms of electron holography. *Ultramicroscopy* 41: 335–348.

de Ruijter W.J. and Weiss J.K. 1993. Detection limits in quantitative off-axis electron holography. *Ultramicroscopy* 50: 269–283.

den Hertog M.I., Schmid H., Cooper D., Rouviere J.-L., Bjork M.T., Riel H., Rivallin P., Karg S. and W. Riess. 2009. Mapping active dopants in single silicon nanowires using off-axis electron holography. *Nano Lett.* 9: 3837–3843.

Diebold A., Kump M.R., Kopanski J.J. and Seiler D.G. 1996. Characterization of two-dimensional dopant profiles: status and review, *J. Vac. Sci. Technol. B* 14: 196–201.

Dunin-Borkowski R.E., McCartney M.R., Smith D.J. and Parkin S.S.P. 1998. Towards quantitative electron holography of magnetic thin films using *in situ* magnetization reversal. *Ultramicroscopy* 74: 61–73.

Dunin-Borkowski R.E., McCartney M.R. and Smith D.J. 2004. Electron holography of nanostructured materials. In *Encyclopedia of Nanoscience and Nanotechnology*, 3. H.S, Nalwa (ed.). American Scientific Publishers, Stevenson Ranch, CA. pp. 41–99.

Dunin-Borkowski R.E., Newcomb S.B., Kasama T., McCartney M.R., Weyland M. and Midgley P.A. 2005. Conventional and back-side focused ion beam milling for off-axis electron holography of electrostatic potentials in transistors. *Ultramicroscopy* 103: 67–81.

Dunin-Borkowski R.E., Kovács A., Kasama T., McCartney M.R. and Smith D.J. 2018. Electron holography. In *Science of Microscopy*. P.W. Hawkes and J.C.H. Spence (eds.) Springer, Switzerland.

Eyben P., Denis S., Clarysse T. and vanderVorst W. 2003. Progress towards a physical contact model for scanning spreading resistance microscopy. *Mater. Sci. Eng. B* 102: 132–137.

Formanek P. and Bugiel E. 2006a. On specimen tilt for electron holography of semiconductor devices. *Ultramicroscopy* 106: 292–300.

Formanek P. and Bugiel E. 2006b. Specimen preparation for electron holography of semiconductor devices. *Ultramicroscopy* 106: 365–375.

Gabor D. 1949. Microscopy by reconstructed wavefronts. *Proc. Roy. Soc. A* 197: 454–487.

Gan Z., Perea D.E., Yoo J., Picraux S.T., Smith D.J. and McCartney M.R. 2013a. Mapping electrostatic profiles across axial *p-n* junctions in Si nanowires using off-axis electron holography. *Appl. Phys. Lett.* 103: 153108.

Gan Z., Perea D.E., Picraux S.T., Smith D.J. and McCartney M.R. 2013b. Characterization of abrupt heterojunctions in SiGe NW using off-axis electron holography. *Microsc. Microanal.* 19(Supp. 2): 1386–1387.

Gan Z., DiNezza M., Zhang Y.-H., Smith D J. and McCartney M.R. 2015. Determination of mean inner potential and inelastic mean free path of ZnTe using off-axis electron holography and dynamical effects affecting phase determination. *Microsc. Microanal.* 21: 1406–1412.

Gan Z., Perea D., Yoo J., He Y., Colby R., Barker J. E., Gu M., Mao S.X., Wang C., Picraux S.T., Smith D.J. and McCartney M.R. 2016. Characterization of electrical properties in axial Si-Ge nanowire heterojunctions using off-axis electron holography and atom-probe tomography. *J. Appl. Phys.* 120: 104301.

Gatel C., Lubk A., Pozzi G., Snoeck E. and Hÿtch M. 2013. Counting elementary particles by electron holography. *Phys. Rev. Lett.* 111: 025501.

Giannuzzi L.A. and Stevie F.A. 2005. *Introduction to Focused Ion Beams.* Springer, New York.

Gribelyuk M.A., McCartney M.R., Li J., Murthy C.S., Ronsheim P., Doris B., McMurray J.S., Hegde S. and Smith D.J. 2002. Mapping of electrostatic potential in deep submicron CMOS devices by electron holography. *Phys. Rev. Lett.* 89: 025502.

Gribelyuk M.A., Domenicucci A.G., Ronsheim P.A., McMurray J.S. and Gluschenkov O. 2008. Application of electron holography to analysis of submicron structures. *J. Vac. Sci. Tech. B* 26: 408–414.

Han M.-G., Li J., Xie Q., Fejes P., Connor J., Taylor B. and McCartney M.R. 2006. Sample preparation for precise and quantitative electron holographic analysis of semiconductor devices. *Microsc. Microanal.* 12: 295–301.

Han M.-G., Smith D.J. and McCartney M.R. 2008. *In situ* electron holographic analysis of biased Si n^+-p junctions. *Appl. Phys. Lett.* 92: 143502.

He K., Cho J.-H., Jung Y., Picraux S.T. and Cumings J. 2013. Silicon nanowires: electron holography studies of doped p-n junctions and biased Schottky barriers. *Nanotechnology* 24: 115703.

Hoffman S., Bauer S., Ronning C., Stelzner Th., Michler J., Ballif C., Sivakov V. and Christiansen S.H. 2009. Axial *p-n* junctions realized in silicon nanowires by ion implantation. *Nano Lett.* 9: 1341–1346.

Hyun J.K., Zhang S. and Lauhon L.J. 2013. Nanowire heterostructures. *Annu. Rev. Mater. Res.* 43: 451–479.

Ikarashi N., Ikezawa T., Uejima K., Fukai T., Miyamura M., Toda A. and Hane M. 2008. Electron holography analysis of a shallow junction for planar-bulk metal-oxide-semiconductor field-effect transistors approaching the scaling limit. *J. Appl. Phys.* 103: 114514.

Ikarashi N., Toda A., Uejima K., Yako K., Yamamoto T., Hane M. and Sato H. 2010. Electron holography for analysis of deep submicron devices: present status and challenges. *J. Vac. Sci. Technol B* 28: C1D5–C1D10.

Ikarashi N., Takeda H., Yako K. and Hane M. 2012. *In situ* electron holography of surface potential response to gate voltage application in a sub-30-nm gate-length metal-oxide-semiconductor field-effect transistor. *Appl. Phys. Lett.* 100: 143508.

Koren E., Berkovitch N. and Rosenwaks Y. 2010. Measurement of active dopant distribution and diffusion in individual silicon nanowires. *Nano Lett.* 10: 1163–1167.

Lenk A., Lichte H., and Muehle U. 2005. 2D-mapping of dopant distribution in deep sub micron CMOS devices by electron holography using adapted FIB-preparation. *J. Electron Microsc.* 54: 351–359.

Li J., McCartney M.R. and Smith D.J. 2003. Semiconductor dopant profiling by off-axis electron holography. *Ultramicroscopy* 94: 149–161.

Li L., Ketharananthan S., Drucker J. and McCartney M.R. 2009. Study of hole accumulation in individual germanium quantum dots in *p*-type silicon by off-axis electron holography. *Appl. Phys. Lett.* 94: 232108.

Li L., Smith D.J., Dailey E., Madras P., Drucker J. and McCartney M.R. 2011. Observation of hole accumulation in Ge/Si core/shell nanowires using off-axis electron holography. *Nano Lett.* 11: 493–497.

Li L., Jin L., Wang J., Smith D.J., Yin W.-J., Yan Y., Song H., Choy W.C.H., and McCartney M.R. 2012. Polarization-induced charge distribution at homogeneous zincblende/ wurtzite heterostructural junctions in ZnSe nanobelts. *Adv. Maters.* 24: 1328–1332.

Li L., Gan Z., McCartney M.R., Liang H., Yu H., Yin W.-J., Yan Y., Gao Y., Wang J. and Smith D.J. 2014. Determination of polarization-fields across polytype interfaces in InAs nanopillars. *Adv. Mater.* 26: 1052–1057.

Lichte H. 1986. Electron holography approaching atomic resolution. *Ultramicroscopy* 20: 293–304.

Lichte H., Formanek P., Lenk A., Linck M., Matzeck C., Lehmann M. and Simon P. 2007. Electron holography: applications to materials questions. *Annu. Rev. Mater. Res.* 37: 539–588.

Lichte H. and Lehmann M. 2008. Electron holography - basics and applications. *Rep. Prog. Phys.* 71: 016102.

Lubk A., Wolf D., Prete P., Lovergine N., Niermann T., Sturm, S. and Lichte H. 2014. Nanometer-scale tomographic reconstruction of three-dimensional electrostatic potentials in GaAs/AlGaAs core-shell nanowires. *Phys. Rev. B* 90: 125404.

Matteucci G., Missiroli G.F., Pozzi G., Merli P.G. and Vecchi I. 1979. Interference electron-microscopy in thin-film investigations. *Thin Solid Films* 62: 5–17.

McCartney M.R. 2005. Characterization of charging in semiconductor device materials by electron holography. *J. Electron Microsc.* 54: 239–242.

McCartney M.R. and Gajdardziska-Josikovska M. 1993. Absolute measurement of normalized thickness, t/λ_I, from off-axis electron holography. *Ultramicroscopy* 53: 283–289.

McCartney M.R., Smith D.J., Hull R, Bean J.C., Völkl E. and Frost B.G. 1994. Direct observation of potential distribution across Si/Si *p-n* junctions using off-axis electron holography. *Appl. Phys. Lett.* 65: 2603–2605.

McCartney M.R., Ponce F.A., Cai J. and Bour D.P. 2000. Mapping electrostatic potential across an AlGaN/InGaN/GaN diode by electron holography. *Appl. Phys. Lett.* 76: 3055–3057.

McCartney M.R., Gribelyuk M.A., Li J, Ronsheim P., McMurray J.S. and Smith D.J. 2002. Quantitative analysis of one-dimensional dopant profile by electron holography. *Appl. Phys. Lett.* 80: 3213–3215.

McCartney M.R. and Smith D.J. 2007. Electron holography: phase imaging with nanometer resolution. *Annu. Rev. Mater. Res.* 37: 729–767.

Möllenstedt G. and Düker H. 1955. Fresnelscher interferenzversuch mit einem biprisma fur elektronenwellen. *Naturwiss.* 42: 41.

Midgley P.A. and Dunin-Borkowski R.E. 2009. Electron tomography and holography in materials science. *Nat. Mat.* 8: 271–280.

Müller E, Gerthsen D, Brückner P, Scholz F, Gruber T. and Waag A. 2006. Probing the electrostatic potential of charged dislocations in *n*-GaN and *n*-ZnO epilayers by transmission electron holography. *Phys. Rev. B* 73: 245316.

Orchowski A., Rau W.D. and Lichte H. 1995. Electron holography surmounts resolution limit of electron microscopy. *Phys. Rev. Lett.* 74: 399–402.

Pozzi G., Beleggia M., Kasama T. and Dunin-Borkowski R.E. 2014. Interferometric methods for mapping static electric and magnetic fields. *Comptes Rendus. Phys.* 15: 126–139.

Rau W.D., Schwander P., Baumann F.H., Hoppner W. and Ourmazd A. 1999. Two-dimensional mapping of the electrostatic potential in transistors by electron holography. *Phys. Rev. Lett.* 82: 2614–2617.

Reimer L. 1991. *Transmission Electron Microscopy.* Springer, Berlin.

Smith D.J. and McCartney M.R. 1999. Practical electron holography. In *Introduction to Electron Holography.* E. Völkl, L.F. Allard, and D.C. Joy (eds.) pp. 87–106. Kluwer, New York.

Stevens M., Bell A., McCartney M.R, Ponce F.A., Marui H. and Tanaka S. 2004. Effect of layer thickness on the electrostatic potential in InGaN quantum wells. *Appl. Phys. Lett.* 85: 4651–4653.

Tanagaki T., Aizawa S., Suzuki T. and Tonomura A. 2012. Three-dimensional reconstructions of electrostatic potential distributions with 1.5-nm resolution using off-axis electron holography. *J. Electron Microsc.* 61: 77–84.

Twitchett A.C. Dunin-Borkowski R.E. and Midgley P.A. 2002. Quantitative electron holography of biased semiconductor devices. *Phys. Rev. Lett.* 88: 238302.

Twitchett A.C., Dunin-Borkowski R.E., Hallifax R.J., Broom R.F., Midgley P.A. 2005. Off-axis electron holography of unbiased and reverse-biased focused ion beam milled Si *p-n* junctions. *Microsc. Microanal.* 11: 66–78.

Twitchett-Harrison A.C., Yates T.J.V., Newcomb S.B., Dunin-Borkowski R.E. and Midgley P.A. 2007. High-resolution three-dimensional mapping of semiconductor dopant potentials. *Nano Lett.* 7: 2020–2023.

Twitchett-Harrison A.C., Yates T.J.V. and Dunin-Borkowski R.E. 2008. Quantitative electron holographic tomography for the 3D characterisation of semiconductor device structures *Ultramicroscopy* 108: 1401–1407.

Wolf D., Lichte H., Pozzi G., Prete P. and Lovergine N. 2011. Electron holographic tomography for mapping the three-dimensional distribution of electrostatic potential in III-V semiconductor nanowires. *Appl. Phys. Lett.* 98: 264103.

Wolf D., Lubk A., Roder F. and Lichte H. 2013. Electron holographic tomography. *Curr. Opin. Solid State Mat. Sci.* 17: 126–134.

Yazdi S., Kasama T., Beleggia M., Yekta M.S., McComb D.W., Twitchett-Harrison A.C. and Dunin-Borkowski R.E. 2015. Towards quantitative electrostatic potential mapping of working semiconductor devices using off-axis electron holography. *Ultramicroscopy* 152: 10–20.

Zhou L., Cullen D.A., Smith D.J., McCartney M.R., Mouti A., Gonsherek M., Feltin E., Carlin J.F. and Grandjean N. 2009. Polarization field mapping of $Al_{0.85}In_{0.15}N/AlN/GaN$ heterostructure. *Appl. Phys. Lett.* 94:121909.

Zhou L., Smith D.J., McCartney M.R., Xu T. and Moustakas T.D. 2011. Measurement of electric field across individual wurtzite GaN quantum dots using electron holography. *Appl. Phys. Lett.* 99: 101905.

Zhou L., Dimakis E., Hathwar R., Aoki T., Smith D.J., Moustakas T.D., Goodnick S.M. and McCartney M.R. 2013. Measurement and effects of polarization fields on one-monolayer-thick InN/GaN multiple quantum wells. *Phys. Rev. B* 88: 125310.

Terahertz Spectroscopy for Nanomaterial Characterization

Xinlong Xu, Lipeng Zhu,
Yanping Jin, Yuanyuan Huang,
Zehan Yao, Chuan He, Longhui
Zhang, and Changji Liu
Northwest University

15.1 Introduction

THz region locates between microwave region and infrared region with the frequency between 0.1 and 10 THz. One THz (10^{12} Hz) corresponds to 33.3 cm^{-1} and 4.1 meV with the wavelength of 300 μm as shown in Figure 15.1. It is rich in science but less developed in technology as THz radiation is relatively low and the THz signals are often submerged by background blackbody radiation. On one hand, the research on the THz characterization of materials is particularly important as the working frequency of microelectronic devices developing from GHz to THz. On the other hand, many physical quantities such as carrier collision frequency, molecular vibration and rotation frequency, plasmon resonance frequency, excitonic transition frequency, Landau level transition frequency, quantum-well sublevel transition frequency, Cooper pair energy as well as intravalley dynamics are located in THz region with the time scale in picoseconds.

Due to the development of ultrafast laser technology, THz science and technology has flourished at the end of the 1980s based on coherent THz radiation and detection progresses by femtosecond lasers. This further promotes the development of new kind of THz spectroscopy, which is the so-called time-domain spectroscopy (TDS) as the signal is taken in time domain instead of in frequency domain. In THz spectroscopy, the THz radiation by a femtosecond laser is mainly focused on the photo-antenna effect pioneered by Auston (Auston et al. 1985, Auston et al. 1983) and Grischkowsky (Exter et al. 1989, Fattinger & Grischkowsky 1989) and optical rectification effect proposed by Zhang et al. (Hu et al. 1990, Xu et al. 1992) in the early time. The first method is

based on the acceleration of the photoexcited carriers in semiconductors such as GaAs by the electric fields (either external electric field or internal surface depletion field or Dember field and so on) (Wu et al. 2013a,b). The second method is based on the nonlinear optical effect in electro-optical crystals (such as ZnTe and LiNbO$_3$) to generate THz radiation through sub-picosecond optical rectification. A sub-picosecond optical rectification method can generate a wide spectrum as broad as 70 THz (Han and Zhang 2001). Recently, air-based filament plasma (Andreeva et al. 2016) by a femtosecond laser has also been used as THz radiation source with high power intensity and ultra-broad spectrum. High-intensity THz radiation can also be achieved by optimizing the phase-matching condition by tilted-pulse front excitation proposed by Hebling et al. (Fülöp et al. 2011).

Correspondingly, THz radiation by the femtosecond excitation can be detected by a photoconductive switch in semiconductors (Grischkowsky et al. 1988, Jepsen & Keiding 1995, Rudd et al. 2000) or by linear electro-optic effect in electro-optic crystals (Cai et al. 1998, Jepsen et al. 1996, Nahata et al. 1996a, Wu et al. 1996). The first method uses the THz electric field to drive the transient carriers generated by a probing beam in a photoconductive switch, and the output current is directly proportional to the THz electric field amplitude. The second method is based on the refractive index modulation by the THz electric field, which changes the refractive index ellipsoid of an electro-optic crystal for the probing beam. Recently, Zhang et al. (Ho et al. 2010) also proposed an air plasma-based THz detection with quite broadband response.

THz spectroscopy using coherent THz pulse has the following features. (i) THz–TDS is not sensitive to the

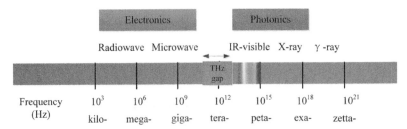

FIGURE 15.1   THz region (also called THz gap as lack of efficient THz components to manipulate THz waves) in electromagnetic spectrum.

blackbody radiation. The signal-to-noise ratio is as high as $10^4{:}1$ below 3 THz with good stability, which is much higher than that of Fourier-transform infrared spectroscopy (Han et al. 2001). (ii) THz–TDS can be used as an effective method for nondestructive and qualitative identification of materials in THz region (Brucherseifer et al. 2000, Fattinger et al. 1990, Han et al. 2000, Smye et al. 2001). The vibrational and rotational modes of many biological macromolecules, semiconductor materials, dielectric materials, ferroelectric materials, and multiferroic materials fall in the range of THz frequencies. THz spectroscopy can get the information of carriers in a noncontact method, which is more convenient and effective than that of Hall measurement (Mittleman et al. 1997). (iii) THz–TDS can be used to extract quantitatively the optical constants of materials in THz region with high reliability and precision without using the Kramers–Kronig relation as both the amplitude and phase information can be obtained simultaneously from the coherent THz pulse. Therefore, THz spectroscopy has been used for dielectric materials and semiconductors (Fattinger et al. 1990, Jeon & Grischkowsky 1997, Katzenellenbogen & Grischkowsky 1992, Schall et al. 1999), gas molecules (Cheville & Grischkowsky 1995, Mittleman et al. 1998), biological macromolecules (Brucherseifer et al. 2000, Fischer et al. 2002, Han et al. 2000, Smye et al. 2001), and superconductors (Ludwig et al. 1996). (iv) Additionally, THz electromagnetic radiation can transmit through nonpolar materials, which can be used for security check or integrated circuit inspection (Hu & Nuss 1995, Mittleman et al. 1996, Wu et al. 1996). (v) Combined with time-resolved spectroscopic technique, optical-pump THz-probe (OPTP) spectroscopy can be used to analyze dynamics of photoexcited carriers (Huber et al. 2001, Uhd Jepsen et al. 2001) in particular for the evolution of carriers, phonons, excitons, plasmons after photoexcitation. (vi) Materials can also generate THz radiation after interaction with femtosecond laser pulses. This can also be used for the nonlinear optical analysis of materials as well as the surface and interface information of materials (Huang et al. 2018, Huang et al. 2017, Zhang et al. 2017, Zhu et al. 2017).

The optical properties of nanomaterials in THz region are relatively less studied, while it is fertile for THz functional devices such as THz emitters, THz detectors, and THz polarizers. Nanomaterials show peculiar physical properties, such as small size effect, surface effect, and quantum tunneling effect when the physical dimensions are comparable to the electromagnetic wavelength, De Broglie wavelength, carrier coherence length, or Bohr radius of excitons. Many physical properties of nanomaterials such as phonon frequency and excitonic resonance frequency are located in THz region. THz wave interaction with nanomaterials will inevitably affect the amplitude, phase, polarization, and propagation of THz waves, which can be further used for manipulating THz waves.

The combination of THz technology and nanotechnology can be traced back to the year of 2002, in which Jeon et al. studied the complex conductivity of single-walled carbon nanotube films by THz–TDS (Jeon et al. 2002). They found that the carbon nanotube films could not be described by Drude model but by Maxwell–Garnett model, where the nanotubes are embedded in a dielectric host (Jeon et al. 2002). In 2004, Jeon et al. (Jeon et al. 2004) also studied the absorption and dispersion of aligned carbon nanotube films. In 2005, we also studied the dielectric constant change under the pump light and we observed the light-induced dielectric transparency in carbon nanotube films (Xu et al. 2005). Since then, THz spectroscopy has been used for ZnO nanostructures (Han et al. 2008), GaAs nanowires (Patrick et al. 2007), InAs and InP nanowires (Joyce et al. 2013), CdS_xSe_{1-x} nanobelts (Liu et al. 2014), and so on. In 2012, the first "TeraNano" international conference (Horiuchi 2012) was held, which indicated that the combination of THz science and nanotechnology became a new research field.

In this chapter, we present a tutorial review on THz photonics for nanomaterials. This includes the basic principles of THz–TDS for static properties of nanomaterials, OPTP spectroscopy for carrier dynamics of nanomaterials, and THz emission spectroscopy for nonlinear optical properties of nanomaterials. Then we move to the recent progresses on THz spectroscopy in zero-dimensional (0D) nanoparticles, one-dimensional (1D) nanowires and nanotubes, and two-dimensional (2D) graphene and other 2D materials. The development of THz technology and nanoscience can complement each other very well and nanomaterials can also be used as THz devices such as THz polarizers, THz modulators, THz emitters, as well as broadband THz anti-reflection coating layers. Finally, we give a perspective on the future development of "TeraNano" field as THz wave is still on the horizon for more applications in new nanomaterials.

15.2 Principles of THz Spectroscopy

15.2.1 THz Experimental Setup

THz–TDS

As shown in Figure 15.2, a femtosecond Ti:Sapphire oscillator laser (central wavelength of 800 nm) or a femtosecond fiber laser (central wavelength of 1,550 nm) with the short pulses (<100 fs) is usually employed for THz–TDS. The femtosecond laser can be divided into a generation beam and a probe beam by a beam splitter (BS). The generation beam is used to excite a THz emitter to generate THz radiation after passing through a delay line. Both electro-optic crystals (such as ZnTe <110>) based on optical rectification effect (Nahata et al. 1996b) and photoconductive antenna (Tani et al. 1997) are used to generate a broadband THz radiation. The THz radiation is collected and focused onto the sample surface by a pair of parabolic mirrors. The transmitted THz wave carrying the sample information is collimated and focused onto a sampling crystal (ZnTe <110>) (Planken et al. 2001) together with the probe beam. In the sampling crystal, the electric field of the THz wave induces birefringence, which would change the polarization of the probe beam based on Pockels effect (Gallot & Grischkowsky 1999). The probe beam out of the sampling crystal is incident on a balance detector, which consists of a quarter wave plate, a Wollaston prism, and two diodes to measure the differential intensity. By varying the time delay line in Figure 15.2, THz time-domain signals can be obtained. The sampling crystal and balance detector can also be replaced by another photoconductive switch to record the time-domain signals (Tani et al. 1997). A mechanical chopper and a lock-in amplifier can be used to improve the signal-to-noise ratio. To minimize the THz absorption or attenuation by water vapor in air, the generated THz beam is enclosed in a vacuum chamber or in the dry nitrogen atmosphere.

OPTP Spectroscopy

OPTP spectroscopy (also called time-resolved THz spectroscopy) is usually used for the carrier dynamic study in nanomaterials. Different from the THz–TDS, another pump beam is needed to excite the materials from ground state to excitation state. As shown in Figure 15.3, a second BS is used to introduce the pump beam, and another mechanical chopper and time delay line are needed to fully acquire

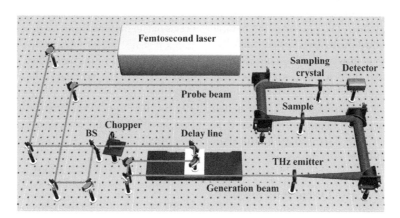

FIGURE 15.2 A typical THz–TDS configuration. THz emitter can be a ZnTe <110> crystal and a photoconductive antenna for generating a broadband THz radiation. BS is the abbreviation for beam splitter.

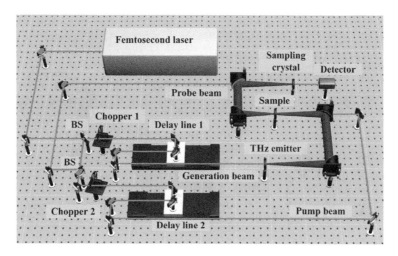

FIGURE 15.3 A typical OPTP spectroscopy setup. Two ZnTe crystals are used as THz emitter and detector.

experimental data. Besides, an amplified femtosecond laser with low repetition rate (kHz) is usually used for OPTP experiment. Therefore, it is appropriate to use ZnTe crystals for THz generation and detection as the high energy is easy to damage the antenna.

In experiment, it is needed to keep the pump spot size larger than the THz spot size to ensure a uniformly excited area on samples (Beard 2000). In terms of data acquirement, OPTP spectroscopy has two delay lines to adjust the time delay between the THz generation beam and the pump beam. By varying "delay line 1" in Figure 15.3, a frequency-dependent photoconductivity can be obtained at different pump delay time. By adjusting "delay line 2" in Figure 15.3, time-resolved dynamics of photoexcited carriers can be estimated. Both delay lines can scan sequentially for two-dimensional data sets (Beard 2000, Beard et al. 2002b).

THz Emission Spectroscopy

THz emission spectroscopy has similar features as THz–TDS. THz–TDS is used for the characterization of THz wave interaction with materials. However, THz emission spectroscopy is mainly used for the THz radiation property of materials after the laser excitation. As shown in Figure 15.4, a femtosecond laser (either oscillator or amplifier) is tightly focused onto the sample after passing through a BS, a chopper, and a delay line. The photoexcited sample can emit THz radiation outward owing to transient polarization or transient photocurrent processes. The emitted THz wave is collected and focused onto a ZnTe <110> crystal and a balanced detector to get the THz time-domain signal by the free space electro-optical sampling. Usually a pair of wire-grid polarizers (WGPs in Figure 15.4) is used to analyze polarization state of the THz radiation from the samples (Obraztsov et al. 2014b).

15.2.2 THz Spectroscopy Data Analysis

In this part, we focus mainly on the transmission configuration as the phase measurement in the reflection configuration is not as accurate as that in the transmission configuration (Hashimshony et al. 2001, Jeon & Grischkowsky 1998, Nashima et al. 2001). This is due to the reason that the reference signal and sample signal are not reflected from the same position when switching the sample and the reference mirror.

Optical Constant Measurement by THz–TDS

Different from the other spectroscopic tools, THz–TDS can record the coherent THz radiation pulses in time domain as shown in Figure 15.5a. After the fast Fourier transform (FFT), the amplitude (Figure 15.5b) and phase information (Figure 15.5c) of the THz pulse can be obtained. Combined with theoretical calculation, the optical physical parameters such as permittivity (dielectric constant), refractive index, and absorption coefficient of nanomaterials can be obtained.

Optical constants of materials are the physical quantities used to characterize macroscopic optical properties of materials, which are the fundamental data for further optical characterization such as transmission coefficient, reflection coefficient, photoconductivity, and so on. The optical constants of materials in THz region are less studied but can be easily extracted by THz–TDS. The extraction calculation by THz–TDS was proposed by Dorney et al. (2001) and Duvillaret et al. (2002). In general, optical properties of the materials can be described by complex refractive index $\tilde{n} = n + i\kappa$ (for engineering representation $\tilde{n} = n - i\kappa$). Here, n is a real refractive index, which describes the dispersion of the sample, while κ is the extinction coefficient, which describes the absorption properties of the sample. Generally, both n and κ are functions of frequency. The extinction coefficient and the absorption coefficient (α) can be described as:

$$\alpha = 2\omega\kappa/c \tag{15.1}$$

Consider a free-standing flat slab, under the p-polarization of THz excitation, the complex transmission can be described as follows (Dorney et al. 2001):

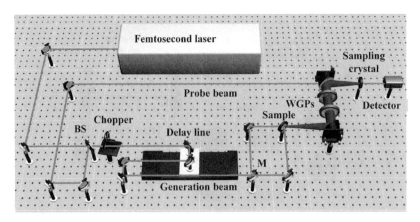

FIGURE 15.4 A typical THz emission spectroscopy in reflection and transmission configurations. M is a removable mirror, which can switch the system between the reflection and transmission configuration. WGPs are a pair of WGPs for THz radiation data analysis.

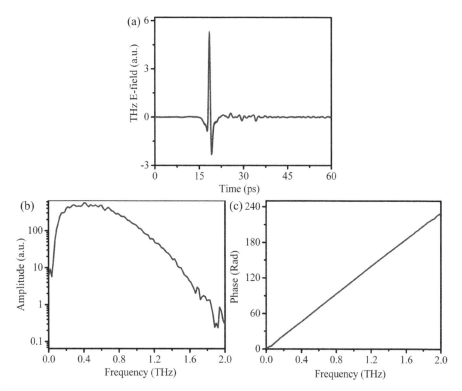

FIGURE 15.5 THz time-domain signal (a) and its corresponding amplitude signal (b) and phase signal (c) after FFT.

$$H(\omega) = \frac{E_{\text{total}}(\omega)}{E_{\text{ref}}(\omega)}$$

$$= \frac{4\tilde{n}_1(\omega)\tilde{n}_2(\omega)\cos\varphi_1\cos\varphi_2}{[\tilde{n}_1(\omega)\cos\varphi_2 + \tilde{n}_2(\omega)\cos\varphi_1]^2}$$

$$\times \left(\exp\left\{\frac{-j[L\tilde{n}_2(\omega) - b\tilde{n}_1(\omega)]\omega}{c}\right\}\right) FP(\omega) \tag{15.2}$$

At normal incidence, $\cos\varphi_1 = \cos\varphi_2 = 1$ and the refractive index of atmosphere $\tilde{n}_1(\omega) = 1$, the complex transmission can be simplified as:

$$H(\omega) = \frac{4\tilde{n}_2(\omega)}{[1 + \tilde{n}_2(\omega)]^2} \times \left(\exp\left\{\frac{-j[d\tilde{n}_2(\omega) - d]\omega}{c}\right\}\right) FP(\omega), \tag{15.3}$$

where d is the thickness of the sample and FP is the Fabry–Pérot factor for the multi-reflections in samples, which can be expressed as:

$$FP(\omega) = \frac{1 - \left[\frac{1-\tilde{n}_2(\omega)}{1+\tilde{n}_2(\omega)}\right]^{2m+2}\exp\left[-j(2m+2)\tilde{n}_2(\omega)\omega d\right]}{1 - \left[\frac{1-\tilde{n}_2(\omega)}{1+\tilde{n}_2(\omega)}\right]^2\exp\left[-2j\tilde{n}_2(\omega)\omega d\right]}, \tag{15.4}$$

Equation (15.3) can be solved by numerical methods and the refractive index (n) and the extinction coefficient (κ) as a function of frequency can be obtained. If the sample is thick enough and all the reflections can be filtered by windowing with only the main pulse left ($m = 0$, $FP = 1$). Equation (15.3) can be simplified as follows:

$$H(\omega) = \frac{4\tilde{n}_2(\omega)}{[1 + \tilde{n}_2(\omega)]^2}\exp\left\{-j\omega\left[\frac{d\tilde{n}_2(\omega) - d}{c}\right]\right\}, \tag{15.5}$$

Considering the complex refractive index of the sample $\tilde{n}_2(\omega) = n_2(\omega) + j\kappa_2(\omega)$ and the complex transmission function as $H(\omega) = \rho(\omega)\exp[-j\Phi(\omega)]$, Eq. (15.5) can be simplified as (Garet et al. 1999):

$$\rho(\omega) = 4\left[n_2^2(\omega) + \kappa_2^2(\omega)\right]^{1/2}$$
$$\times \frac{1}{\left[n_2^2(\omega) + 1\right]^2 + \kappa_2^2(\omega)}\exp\left[-\kappa_2(\omega)d\omega/c\right], \tag{15.6}$$

$$\Phi(\omega) = \frac{[n_2(\omega) - 1]\omega d}{c}$$
$$+ \arctan\left[\frac{\kappa_2(\omega)}{n_2(\omega)[n_2(\omega) + 1] + \kappa_2^2(\omega)}\right] \tag{15.7}$$

Under the weak absorption approximation of the sample $(\kappa_2(\omega)/n_2(\omega) << 1)$, the approximate analytic solution can be expressed as (Garet et al. 1999):

$$n_2(\omega) = \frac{\Phi(\omega)c}{\omega d} + 1 \tag{15.8}$$

$$\kappa_2(\omega) = \frac{-\ln\left(\rho(\omega)\frac{[n_2(\omega)+1]^2}{4n_2(\omega)}\right)c}{\omega d} \tag{15.9}$$

Under the thin film case, the $FP(\omega)$ can also be regarded as a superposition of an infinite number of echoes and the complex transmission function can be rewritten as:

$$H(\omega) = \frac{4\tilde{n}_2(\omega)}{[1+\tilde{n}_2(\omega)]^2} \exp\left\{-j\omega\left[\frac{d\tilde{n}_2(\omega)-d}{c}\right]\right\} \quad (15.10)$$
$$\times \frac{1}{1-\left[\frac{1-\tilde{n}_2(\omega)}{1+\tilde{n}_2(\omega)}\right]^2 \exp\left[-2j\tilde{n}_2(\omega)\omega d/c\right]}$$

Equation (15.10) can be solved by iterative numerical method.

For nanomaterial characterization, the sample is usually on top of a substrate such as silicon with the refractive index 3.418 in THz region (Fattinger et al. 1990). As shown in Figure 15.6, the complex transmission function at normal incidence can be expressed as:

$$H(\omega) = \frac{E_{\text{sam}}(\omega)}{E_{\text{ref}}(\omega)}$$
$$= \frac{t_{01}t_{12}p_f(\omega, L_1)}{\left\{1+r_{01}r_{12}\left[p_f(\omega, L_1)\right]^2\right\}t_{02}p_{\text{air}}(\omega, L_1)} \quad (15.11)$$

where $p_f(\omega, L_1) = \exp\left[-j\delta_1(\omega)\right], p_{\text{air}}(\omega, L_1) = \exp\left[-j\delta_0(\omega)\right]$
$\delta_1(\omega) = \frac{2\pi}{\lambda}\tilde{n}_1(\omega)L_1 = \tilde{n}_1(\omega)\omega L_1/c, \delta_0(\omega) = \frac{2\pi}{\lambda}\tilde{n}_0 L_1$
$= \tilde{n}_0\omega L_1/c,$

And the $\tilde{n}_0(\omega)$, $\tilde{n}_1(\omega)$, and $\tilde{n}_2(\omega)$ are the complex refractive index of the atmosphere, sample, and substrate respectively. L_1 and L_2 are the thickness of the sample and the substrate, respectively. t_{01} and t_{12} are the transmission coefficient at the interface from 0 to 1 and from 1 to 2 in Figure 15.6, respectively. r_{01} and r_{12} are the reflection coefficient at the interface from 0 to 1 and from 1 to 2 in Figure 15.6, respectively. Then the complex transmission function can be expressed as:

$$H(\omega) = \frac{2\tilde{n}_f(\omega)(1+\tilde{n}_s)e^{j\delta_0}}{(1+\tilde{n}_f(\omega))(\tilde{n}_f(\omega)+\tilde{n}_s)e^{j\delta_1}}, \quad (15.12)$$
$$+ (1-\tilde{n}_f(\omega))(\tilde{n}_f(\omega)-\tilde{n}_s)e^{-j\delta_1}$$

where $\tilde{n}_f(\omega)$ is the complex refractive index of the thin film and $\tilde{n}_s$ is the complex refractive index of the substrate. Considering the thickness of the sample is much less than

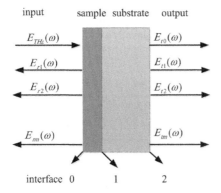

input   sample substrate   output

$E_{THz}(\omega)$ $E_{t0}(\omega)$

$E_{r1}(\omega)$ $E_{t1}(\omega)$

$E_{r2}(\omega)$ $E_{t2}(\omega)$

$E_{rn}(\omega)$ $E_{tn}(\omega)$

interface   0   1   2

FIGURE 15.6 Nanomaterial films on top of a substrate, which can be viewed as THz wave transmission through different layers.

the wavelength of THz wave, then $e^{\pm j\delta} = 1 \pm j\delta$, the complex refractive index of the film can be calculated from Eq. (15.12) by numerical methods.

Other optical properties such as absorption coefficient α, complex dielectric constant $\varepsilon = \varepsilon_r + i\varepsilon_i$, and complex conductivity $\sigma = \sigma_r + i\sigma_i$, electron mobility μ, superconducting material surface resistance $Z_s = R_s + jX_s$ (Kiwa & Tonouchi 2001), can be obtained from the complex refractive index of the materials. For example, the real (ε_r) and imaginary (ε_i) part of the dielectric constant can be expressed as:

$$\varepsilon_r = n^2 - \kappa^2, \varepsilon_i = 2n\kappa, \quad (15.13)$$

The conductivity of materials can be calculated as:

$$\varepsilon = \varepsilon_b + \frac{i\sigma}{\varepsilon_0\omega} \quad (15.14)$$

where ε_b is the background dielectric such as the high frequency (ε_∞) in visible region or the low frequency (ε_s) in infrared region.

Carrier Dynamics in OPTP Spectroscopy

Conventional THz–TDS is far from enough to meet the requirement of material characterization as most optoelectric devices require understanding of the carrier transient process and its associated effects in optoelectric materials. THz–TDS is mainly used to study the static property of materials. However, OPTP spectroscopy is mainly used for the carrier dynamics of materials. This spectroscopic technique has been used to probe the transient conducting and insulating phases in an electron–hole gas successfully (Kaindl et al. 2003).

OPTP measurement is a contact-free and nondestructive method for the photo-carrier lifetimes, mobility of electrons and holes as well as the phononic, excitonic response in THz region. After photoexcitation, electrons transit from the valence band to conduction band and leave holes in valence band. Electrons and holes can experience carrier–carrier scattering for thermalization. During this process, electrons and holes can form excitons or pass the extra energy to the lattice (phonon). Then the carriers can either achieve recombination by radiation or decay by diffusion or be captured by defects. THz wave is sensitive to the dynamics of both local states (Lorentz or Lorentz-like characteristics) and nonlocal state (Drude or Drude-like characteristics). As such, OPTP spectroscopy provides direct evidence and explanation of photoexcited carriers and quasiparticle (exciton and phonon) formation mechanism, many body effects, and decoherent dynamics of many basic physical processes (George et al. 2008, Richter & Schmuttenmaer 2010) in nanomateirals.

In OPTP experiment, the differential THz transmission with and without photoexcitaion can be measured as $\Delta E(t) = E^{\text{on}}(t) - E^{\text{off}}(t)$ in time domain (Cunningham 2013), which can be expressed as:

$$\frac{\Delta E(t)}{E(t)} = \frac{E^{\text{on}}(t) - E^{\text{off}}(t)}{E^{\text{off}}(t)} = \frac{E^{\text{on}}(t)}{E^{\text{off}}(t)} - 1 \quad (15.15)$$

Similar to the analysis by Kužel et al. (2007), Kužel and Němec (2014), and Hannah et al. (2016), the thin layer model in Figure 15.6 can also be used to analyze the differential THz transmission. The differential transmission after FFT can be expressed as:

$$-\frac{\Delta\tilde{E}(\omega)}{\tilde{E}_0(\omega)} = 1 - \frac{n+1}{n+1+Z_0d\tilde{\sigma}(\omega)} = \frac{Z_0d\tilde{\sigma}(\omega)}{n+1+Z_0d\tilde{\sigma}(\omega)}$$ (15.16)

where Z_0 is the impedance of the free space. d is the sample absorption depth at the pump wavelength. n is the THz refractive index of the unexcited sample. The complex photoconductivity can then be expressed as:

$$\tilde{\sigma}(\omega) = \frac{n+1}{Z_0d}\frac{1}{1+\Delta\tilde{E}(\omega)/\tilde{E}_0(\omega)}$$ (15.17)

For samples with relatively small transmission change after excitation, Eq. (15.17) can be further simplified as:

$$\tilde{\sigma}(\omega) = \frac{n+1}{Z_0d}\frac{-\Delta\tilde{E}(\omega)}{\tilde{E}_0(\omega)}$$ (15.18)

This suggests that the photoconductivity is proportional to the THz transmission change after photoexcitation. The real and imaginary photoconductivity dynamics can be evaluated by recording the photo-induced transmission change of THz pulse at pulse peak or zero-crossing positions, respectively, as shown in Figure 15.7a in time-domain scheme. The real conductivity dynamics are proportional to THz peak transmission, while the imaginary conductivity dynamics are proportional to the THz transmission at the cross position, which is due to the phase shift of THz signal as the refractive index change (Cunningham 2013). OPTP spectroscopy can also get the complex photoconductivity in frequency domain.

Models in THz Data Analysis

Usually the measured optical constants can be described well by several models such as Drude, Drude–Smith, and Lorentz models for the carrier dynamics in a view of classical physics. Drude model is a free-electron gas model, which can be used to describe electrons in metals or in highly doped semiconductors or photoexcited free carriers (Hodgson 2012). The complex dielectric constants can be written as:

$$\varepsilon(\omega) = 1 - \frac{\omega_p^2}{\omega^2 + i\Gamma\omega}$$ (15.19)

And the conductivity can be expressed as following:

$$\sigma(\omega) = \frac{Nq^2}{m^*}\frac{\tau}{1-i\omega\tau} = \frac{D_0\tau}{1-i\omega\tau}$$ (15.20)

where $\omega_p^2 = (Nq^2)/(m^*\varepsilon_0)$ is the plasma frequency and m^* is the effective mass. q is the electric charge and $\tau = 1/\Gamma$ is the carrier scattering time (Γ, damping rate). D_0 is the Drude weight. The charge carrier mobility can then be calculated from the Drude model as the following (Parkinson et al. 2012):

$$\mu = \frac{e}{m^*\tau}$$ (15.21)

Lorentz model is usually used to model the response of phonon, exciton, as well as the charge carriers in local response under a restoring force. The dielectric constant considering the background contribution can be expressed as:

$$\varepsilon(\omega) = \varepsilon_b + \frac{S\omega_0^2}{\omega_0^2 - \omega^2 - i\Gamma\omega}$$ (15.22)

where ω_0 is the intrinsic resonant frequency and S is the strength of the oscillator. Considering the photoexcited electrons move in restricted region, the photoconductivity can be obtained as:

$$\tilde{\sigma}(\omega) = \frac{Nq^2}{m^*}\frac{\omega}{\omega\Gamma + i(\omega_0^2 - \omega^2)}$$ (15.23)

If there are multiple oscillators (such as multiple phonon response) as well as free carriers in nanomaterials in THz region, Lorentz–Drude dispersion model would be used, which can be written as:

$$\varepsilon_m(\omega) = \varepsilon_b - \frac{\omega_p^2}{\omega^2 + i\Gamma_p\omega} + \sum_j \frac{S_j\omega_j^2}{\omega_j^2 - \omega^2 - i\Gamma_j\omega}$$ (15.24)

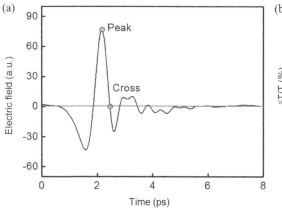

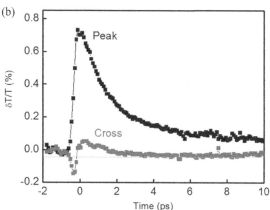

FIGURE 15.7 (a) THz time-domain signal. The peak and zero-crossing positions are shown in the time-domain signal. (b) Photo-induced transmission change of THz at pulse peak (black) or zero-crossing positions.

The first one is the background dielectric function of the sample. The second one corresponds to the Drude response, which takes the nonlocal response into account. The third one corresponds to the Lorentz response, which takes the local response into account.

There are some modified Drude model, such as Drude–Smith model that is used widely to describe the photoexcited carriers, which have taken into account some preferential scattering direction (Smith 2001). The complex conductivity by Drude–Smith model can be written as:

$$\sigma_{DS}(\omega) = \sigma_D(\omega)\left[1 + \sum_{p=1}^{\infty}\frac{C_p}{(1 - i\omega\tau)^p}\right] \quad (15.25)$$

where $\sigma_D(\omega)$ is the Drude conductivity expressed in Eq. (15.20). Here, p stands for the number of scattering event. Where C_p $(-1 \leq C_p \leq 0)$ is the expectation value $\langle\cos\theta\rangle$ for scattering angle θ, and $C_p = -1$ corresponds to complete backscattering response. $C_p = 0$ is fully Drude limit, which is attributed to free-carrier randomized scattering response.

Usually THz–TDS and OPTP spectroscopy measure the average dielectric constant or conductivity of nanomaterials embedded in some matrix such as polymer or air. This composite can be viewed as an effective medium analyzed with the effective medium theory (Niklasson et al. 1981). Two most used effective medium models are Maxwell–Garnett model and Bruggeman model, which can describe the macroscopic properties of composite materials. Maxwell–Garnett model is only suitable for the low volume fraction (f) without percolation pathways. The Maxwell–Garnett model can be written as:

$$f\frac{\varepsilon_{\text{nano}} - \varepsilon_{\text{host}}}{\varepsilon_{\text{nano}} + \eta\varepsilon_{\text{host}}} = \frac{\varepsilon_{MG} - \varepsilon_{\text{host}}}{\varepsilon_{MG} + \eta\varepsilon_{\text{host}}} \quad (15.26)$$

where $\varepsilon_{\text{nano}}$ is the dielectric constant of the nanomaterials. $\varepsilon_{\text{host}}$ is the dielectric constant of the host matrix, and ε_{MG} is the effective dielectric constant of the composite. η is related to the depolarization factor, which is 2 for spherical particles and 1 for infinitely long cylinders (Kužel et al. 2014). Bruggeman model is suitable for the large volume fraction, which can be described as:

$$f\frac{\varepsilon_{\text{nano}} - \varepsilon_{MG}}{\varepsilon_{\text{nano}} + \eta\varepsilon_{MG}} + (1 - f)\frac{\varepsilon_{\text{host}} - \varepsilon_{MG}}{\varepsilon_{\text{host}} + \eta\varepsilon_{MG}} = 0 \quad (15.27)$$

where all the parameters are similar to those in the Maxwell–Garnett model. There are some other seldom used models such as quantum models that can be found in excellent review articles (Lloydhughes 2012).

Analysis of THz Emission Spectroscopy

Interaction of femtosecond laser pulses with materials can generate THz radiation, which promote the development of THz emission spectroscopy. THz emission spectroscopy has been developed to serve as a sensitive and contactless tool for the optoelectronic measurement of semiconductor surfaces and interfaces (Zhang & Auston 1992). When a femtosecond laser beam impinges on the semiconductor surface, photocarriers or photodipoles are excited, which then induce THz radiation with the mechanism of ultrafast photocurrents or optical rectification and so on. This spectroscopy is sensitive to the surfaces and interfaces and the modification of the semiconductor surface could provide a significant strategy for the design and performance evaluation of many electronic and optoelectronic materials for THz applications. Here, we introduce the common mechanism for THz emission spectroscopy.

From Maxwell equations, the mechanism of THz radiation can be ascribed to three terms: dielectric polarization P, free current density J, and magnetization M.

$$E_{\text{THz}} \propto \frac{\partial^2 P}{\partial t^2} + \frac{\partial J}{\partial t} + \frac{\partial}{\partial t}(\nabla \times M) \quad (15.28)$$

We mainly focus on nonmagnetic materials with magnetic permeability $\mu \approx 1$ so that the magnetization terms can be ignored. For magnetic materials, the mechanism of THz emission has been summarized in the review article (Huisman & Rasing 2016).

The first polarization term in Eq. (15.28) may dominate the THz radiation process when the excitation photon energy is lower than the band gap of semiconductors (Côté et al. 2002). The optical rectification effect is a typical example for this case, which can be expressed as:

$$P(\Omega) = \chi^{(2)}E(\omega)E^*(\Omega - \omega) \quad (15.29)$$

where $\chi^{(2)}$ is the second-order nonlinear susceptibility, which can only exist in materials with noncentrosymmetrical point group structures. It is a differential frequency process, which can produce low-frequency electromagnetic wave as well as DC current. A typical example of THz radiation from optical rectification effect can be found in ZnTe and MoS_2 crystals (Huang et al. 2018, Huang et al. 2017).

When the photon energy is larger than the bandgap of semiconductors, the semiconductors absorb photon energy to generate carriers. The second current density term in Eq. (15.28) describes this case, which includes photocurrent by photo-Dember effect, surface depletion field (Zhang et al. 1992), and photon drag effect (PDE) (Zhu et al. 2017) and so on. Photo-Dember effect exists in narrow bandgap semiconductors such as InAs (Gu et al. 2002). The notable different mobility of electrons and holes results in a spatial electrical field, which accelerates the carriers to generate THz radiation. Surface depletion effect takes place in wide-band semiconductors such as InP (Zhang et al. 1990). In this effect, the surface state can bend the surface energy level to form an internal electric field along the surface normal to accelerate carriers, resulting in THz radiation. The surface energy level bends upward for n-type semiconductors, while it bends downward for p-type semiconductors. Therefore, different doped semiconductors can lead to an opposite depletion field, resulting in the opposite polarity of the THz time-domain signals.

The second-order nonlinear current generation such as PDE and photogalvanic effect (PGE) can also generate ultrafast photocurrent under a femtosecond laser excitation. This ultrafast photocurrent density can be written as (Glazov & Ganichev 2014):

$$J_i = \chi_{ijk}E_jE_k^* + \frac{i}{2}T_{ijkl}\left\{E_j\left(\nabla_l E_k^*\right) - \left(\nabla_l E_j\right)E_k^*\right\} \quad (15.30)$$

where χ_{ijk} and T_{ijkl} are three-rank and four-rank nonzero polarizability tensors, which describe PGE and PDE, respectively. Equation (15.30) suggests that PGE can exist only in materials with noncentrosymmatrical crystal structures, while PDE can exist in materials with arbitrary symmetry structures. These nonlinear process can be found in graphene for THz generation (Obraztsov et al. 2014a,b, Zhu et al. 2017), which is one of the novel method to study the nonlinear optical property of nanomaterials.

These THz-generation mechanisms can be used to analyze the nonlinear optical parameters as well as the surface/interface property of the nanomaterials. For example, THz emission spectroscopy can be used to probe the ultrafast photocurrent without electrodes (Braun et al. 2016, Hendry et al. 2004). This could not be achieved by other methods, especially for the nanomaterials as the electrodes will affect the measurement.

15.3 THz Spectroscopy Applications in Nanomaterials

Nanomaterials are the materials with at least one dimension in the nanoscale (0.1–100 nm), which can be classified into 0D, 1D, 2D nanomaterials. To understand the polarization, dielectric constant, conductivity, charge carrier lifetimes, recombination mechanisms as well as mobility are important to optimize the device performance base on nanomaterials. However, conventional electrical transport measurements need electrode contact and it is challenging to make electrodes to nano-sized objects to get the photoconductivity and transport properties. THz spectroscopy can be viewed as one of the wireless method capable of static and dynamic measurement of nanomaterials especially for optimizing nanomaterials for new solar cells (Tiwana et al. 2010). Baxter et al. (2006) directly compared the photoconductivity of ZnO nanoparticles, ZnO nanowires, and ZnO thin films using OPTP spectroscopy. They found that films have the highest mobility and nanoparticles have the lowest mobility, while nanowires are in the middle. Time constants for the carrier decay in thin films are twice longer than those in nanowires and five times longer than those in nanoparticles due to the interfaces and grain boundaries in different morphology of the samples (Baxter et al. 2006). As the 0D and 1D nanomaterials can be viewed as nanomaterials embedded in a dielectric host such as the air (Jeon et al. 2002) or hexene (Bergren et al. 2016), the nanomaterials can be viewed as an effective medium by Maxwell–Garnett model or Bruggman model as discussed in Section 2.2.3.

15.3.1 THz–TDS and OPTP Spectroscopy for 0D Nanoparticles

Nanoparticles are also called quantum dots, which demonstrate quantum effects with the change of particles size. OPTP spectroscopy was first used for the photoconductivity measurement of CdSe nanoparticles by Beard et al. (2002a). They found that the mobility obeys the scaling law with the radius of the nanoparticles (R^4) in the strong confinement regime, while it is governed by a restricted mean free path in the weak confinement regime (Beard et al. 2002a). Wang et al. (2006) further studied the susceptibility of CdSe quantum dots in the strong confinement regime and they pointed out that the measurement of the THz response is from the photo-generated excitons instead of free carriers. This is due to the reason that the radius of quantum dots are less than their exciton Bohr radius and the THz response is from discrete energy level in quantum dots instead of from energy band in bulk materials. As shown in Figure 15.8a, the electric-field waveform of a THz pulse transmitted through an unexcited quantum dot (QD) suspension and the differential THz electric-field transmission due to excitation are depicted. The THz electric-field decay is also shown as an inset in Figure 15.8a. After FFT, the sheet susceptibility $\Delta\chi_s = \Delta\chi_s'(\omega) + i\Delta\chi_s''(\omega)$ can be calculated by the following formula (Wang et al. 2006):

$$\frac{\Delta E(\omega)}{E(\omega)} = i\frac{2\pi\omega}{c}\frac{\Delta\chi_s(\omega)}{\sqrt{\varepsilon}} \quad (15.31)$$

The calculated susceptibility is shown in Figure 15.8b, which is frequency-independent over the whole frequency range. The imaginary part of the susceptibility remains almost unchanged, which approaches zero value. The experimental waveform $\Delta E(t)$ in Figure 15.8a can also be reproduced numerically by $E(t)$ propagating through a frequency-independent real susceptibility as shown in the dashed lines in Figure 15.8b. The exciton polarizability (α) of an individual excited QD can be obtained by (Wang et al. 2006):

$$\Delta\chi_s(\omega) = n_s\frac{9\varepsilon^2}{(\varepsilon_{NP} + 2\varepsilon)^2}\alpha \quad (15.32)$$

where ε is the dielectric constant of the QD suspension and ε_{NP} is the dielectric constant of the unexcited CdSe. n_s is the sheet excitation density. Figure 15.8c demonstrates the exciton polarizability as a function of the QD size. The experimental data follows the scale law (R^4) as shown in Figure 15.8c. The calculation based on a simple parabolic-based effective mass model can give an analytical solution for the exciton polarizability (Dakovski et al. 2007, Wang et al. 2006), which is also shown in Figure 15.8c with $R^{3.6}$. The modest deviation from the simple R^4 scaling could come from the simple parabolic approximation and from the electron–hole interaction. The polarizability of photo-induced excitons of CdSe and PbSe QDs in strongly quantum-confined regime has also been analyzed by Dakovski et al. (2007) as shown in Figure 15.8d. Theoretical results based on a parabolic-band effective mass

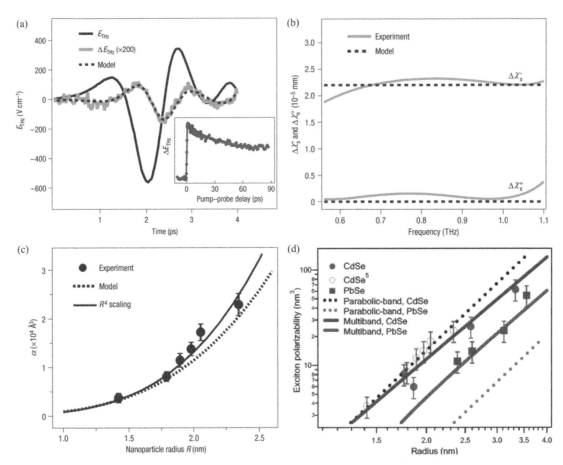

FIGURE 15.8 (a) THz time-domain signal transmitted through an unexcited of CdSe QD suspension and the photo-induced THz signal change at 60 ps after photoexcitation. The decay of the differential THz transmission is plotted as an inset. (b) The real and imaginary part of the photo-induced susceptibility of the CdSe QD as a function of frequency (gray line for the experiment data). The dashed lines demonstrate a frequency-independent and purely real-induced susceptibility that corresponds to the dashed waveform calculated in (a). (c) Polarizability of quantum-confined excitons in photoexcited CdSe QD as a function of QD radius R. The dots with error bars are experimental data and the solid line shows a simple R^4 scaling. The dotted lines are the calculation based on an effective-mass model. (d) Polarizability of photo-induced excitons in CdSe and PbSe QDs as a function of the QD radius. The dots are experimental data. The circular dots for the CdSe QDs are from the reference (Wang et al. 2006). The dotted lines are calculation based on a parabolic-band effective mass model and the solids lines are the calculation based on a multiband effective mass model. (Reprinted with permission Dakovski et al. 2007, Wang et al. 2006. (a–c) Copyright 2006, Springer Nature; (d) Copyright 2007, American Chemical Society.)

model (dotted lines) and a multiband effective mass model (solid line) agree with the experimental results reasonably in Figure 15.8d.

15.3.2 THz–TDS and OPTP Spectroscopy for 1D Nanomaterials

The most important property of 1D nanomaterials such as carbon nanotubes and nanowires is the anisotropic response due to the different response of carriers along the longitudinal and transverse axis of 1D nanomaterials. THz response of carbon nanotube films was first studied by Jeon et al. (2002), and they observed anisotropic absorption and dispersion of carbon nanotube films with THz–TDS. OPTP has also been used to study the dynamic

property of the carbon nanotubes, and the results suggest Auger recombination process occurs at several picoseconds and excitonic-relaxation process happens in hundreds of picosecond (Xu et al. 2009). Furthermore, anisotropic relaxation process by changing the polarization of pump and THz wave has been observed. The results show that the ultrafast exciton is along the longitudinal axis direction (Xu et al. 2010). This further indicates that not only the static response of SWNTs is anisotropic but also the dynamic response is anisotropic. The anisotropy of carbon nanotubes can be induced by the texture without alignment intentionally. This anisotropy can be tuned by mechanical method such as stretching, which increases the anisotropic response as shown in Figure 15.9a (Xu et al. 2015). The anisotropy can be enhanced by the optical pumping as shown in

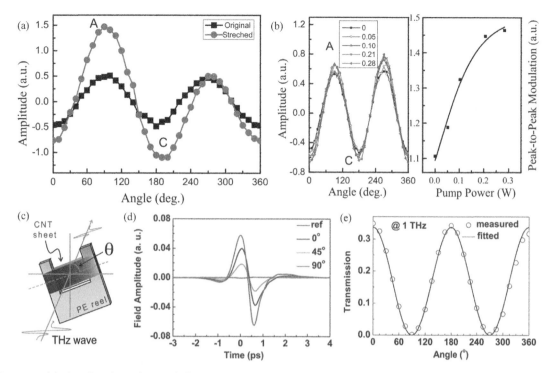

FIGURE 15.9 (a) Angular dependence of the THz pulse peak transmission through original (square-dotted line) and stretched (circle-dotted line) carbon nanotubes. (b) The modulation enhancement of the anisotropic responses by light illuminating with different pump power (unit W). The peak-to-peak modulation versus pump power is shown in the right. The solid line is a fitting result with a hyperbolic tangent function. (c) Experimental scheme for the polarization-state measurements with the incident THz beam polarized along the longitudinal axis of the carbon nanotubes. (d) Time-domain THz signal for different polarization angles. The reference signal is measured without any carbon nanotube polarizer. (e) The transmission spectrum as a function of polarizer angle θ at 1 THz and the corresponding fitting curve to $\cos^2\theta$, which follows Malus's law. (Reprinted with permission Kyoung et al. 2011, Xu et al. 2015. (a–c) Copyright 2015, Elsevier; (d and e) Copyright 2011, American Chemical Society.)

Figure 15.9b. This extra carrier injection enhances the anisotropic response of the carbon nanotubes as the polarization of the pump beam is along the preferentially longitudinal axis (Docherty et al. 2014). The extra carriers injected by photoexcitaiton would also adjust carbon–carbon bonds length and anisotropic deformations for electromagnetic actuators (Gartstein et al. 2002).

The anisotropic response of carbon nanotubes has been used as THz band polarizers. Kyoung et al. (2011) proposed a method by utilizing highly oriented carbon nanotube aerogel sheets winded on a U-shaped polyethylene reel as shown in Figure 15.9c (Kyoung et al. 2011). The incident THz beam is polarized along the longitudinal axis of carbon nanotube sheet and the polarizer can be rotated from 0° to 360°. THz time-domain signal is measured at different polarizer angles (Figure 15.9d) and more than two orders of magnitude extinction are observed between the perpendicular (0°) and parallel (90°) orientation. Figure 15.9e shows the transmission as a function of polarizer angle at 1 THz, which is consistent with the Malus' law $\cos^2\theta$. Further improvement can also be achieved by carbon nanotube fiber as a freestanding THz polarizer with nearly ideal performance (Zubair et al. 2016).

Another important 1D nanomaterials are nanowires such as GaAs, InAs, and InP nanowires, of which, the

carrier lifetimes, carrier mobilities, surface recombination velocities, and donor doping levels can be measured by THz spectroscopy (Joyce et al. 2013). Figure 15.10a depicts photoconductivity lifetimes in four InP nanowire samples with different diameters, which show a rapid rise within 1 ps followed by slow decay (Joyce et al. 2012). The fitting with an exponential function suggests that the time constants (τ) are on the scale of nanoseconds for all diameters. This suggests that the carrier lifetime in InP nanowires is relatively insensitive to surface states. The effective recombination time is closely approximated by the following formula:

$$1/\tau = 1/\tau_{\text{volume}} + 4S/d \qquad (15.33)$$

where d is the diameter and S is the surface recombination velocity and τ_{volume} is the time constant for the recombination in volume. The calculated surface recombination velocity of InP nanowires is approximately 170 cm/s, which is quite low at room temperature (Joyce et al. 2012). Compared with the GaAs nanowires as shown in Figure 15.10b, the GaAs nanowires exhibit a high-surface recombination velocity. Baig et al. (2017) studied the GaAs nanowires embedded in the parylene C, and they observed the photoconductivity response is greater when the pump is polarized parallel to the nanowires, while there is negligible photoconductivity response when the pump is

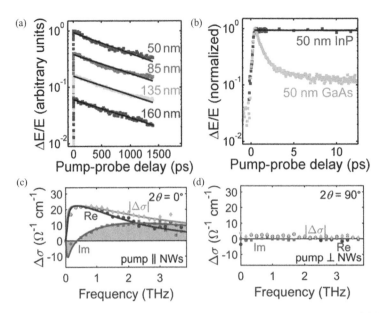

FIGURE 15.10 (a) Photoexcited change in THz waves for InP nanowires with different diameters. (b) Photoconductivity decays for InP nanowires and GaAs nanowires with the same diameter (50 nm) at the same pump fluence (8 μJ/cm^2). The lines are fitting results by monoexponential functions. (c) Frequency-dependent photoconductivity response of GaAs nanowires after photoexcitation with the pump pulse-polarized parallel to the longitudinal axis and (d) perpendicular to the longitudinal axis. The lines are fitting results by a Drude-plasmon model. (Reprinted with permission Baig et al. 2017, Joyce et al. 2012. (a and b) Copyright 2012, American Chemical Society; (c and d) Copyright 2017, American Chemical Society.)

polarized perpendicular to the nanowires as shown in Figure 15.10c,d. The lines in Figure 15.10c are fitting lines by a Drude-plasmon model (similar to Drude–Lorentz model), which reveals a high-electron mobility of 1,800 cm^2 V^{-1} S^{-1}. They also present an ultrafast switchable THz polarization modulator based on these GaAs semiconductor nanowires (Baig et al. 2017).

15.3.3 THz–TDS and OPTP Spectroscopy for 2D Nanomaterials

Two-dimensional nanomaterials are a hot topic in material research since the discovery of graphene in 2004 by Geim et al. (Novoselov et al. 2004). Take graphene as an example. The thickness of one-layer graphene is only 0.345 nm and the mobility is 2×10^5 cm^2/V·s, which is higher than that of silicon (1,500 cm^2/V·s). In the visible region, one-layer graphene only absorb 2.3% of light, which is determined by the fine-structure constant (Nair et al. 2008). In THz region, THz conductivity of graphene can be described by Drude model (Low and Avouris 2014):

$$\sigma(\omega) = -i\frac{e^2 E_F}{\pi\hbar^2(\omega + i\tau)} \qquad (15.34)$$

where E_F is the Fermi level. Unlike traditional Drude model (Section 2.2.3), the Fermi level in Eq. (15.34) can be tuned easily, and it is related to the carrier density (n) by the following formula:

$$|E_F| = \pm\hbar V_F \left(\pi|n|\right)^{1/2} \qquad (15.35)$$

In Eq. (15.35), the positive signal stands for the Fermi level above the Dirac point for electron doping, while the negative signal stands for the Fermi level below the Dirac point for hole doping. V_F is the Fermi velocity with the value 10^6 m/s. It can be seen from Eqs. (15.34) and (15.35) that the THz conductivity is determined by the Fermi level which is related to the carrier density controlled by gate voltage or by chemical doping. Another key parameter is the carrier momentum scattering time (τ), which is related to the synthesis method and doping method, and the optimization of the growth temperature and chemical doping can improve the THz sheet conductivity of graphene significantly (Qi et al. 2014). The reduced graphene oxide films (Zhou et al. 2015) also demonstrate some property like graphene such as Drude response in THz region. The reduced graphene oxide films are tunable and sensitive to the film thicknesses and to the reduction degrees, which can be efficiently controlled by the solution-processable fabrication method. This tunable method can be used for broadband THz antireflection coating based on the impedance matching between the atmosphere and the substrate (Zhou et al. 2014). As shown in Figure 15.11a, the impedance matching condition can be directly observed for the minimum transmission for the second pulse (pulse 2) due to the FP effect from the substrate (Zhou et al. 2014). The well agreement of the experimental data and the theoretical data in Figure 15.11 confirms the broadband impedance matching with graphene layers for THz application. This impedance matching condition can be optimized by doping. After the doping, the Fermi energy drops from ~0.11 to ~0.25 eV and only two layers of graphene are needed to

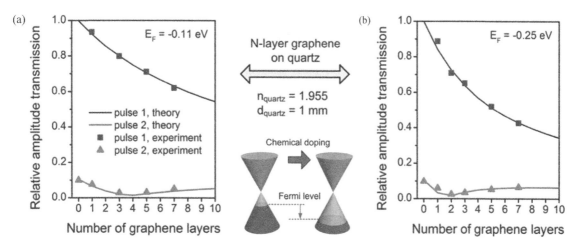

FIGURE 15.11 Relative THz transmission of (a) N-layer graphene on quartz before doping; (b) N-layer graphene on quartz after doping; the inset is the illustration of the decrease in Fermi level after chemical doping. (Reprinted with permission Zhou et al. 2014. Copyright 2014, AIP Publishing.)

achieve impedance matching condition as shown in Figure 15.11b. Broadband impedance matching is needed to minimize the reflections at the interfaces of the THz components, which suppress the unwanted FP effect. This is especially useful for the THz–TDS as the reflections limit the spectral resolution. It is also important for the anti-THz-Radar coatings as THz Radar is on the development.

The photoconductivity of graphene is also sensitive to the environments by the adsorption of atmospheric gases such as nitrogen and oxygen (Docherty et al. 2012). As shown in Figure 15.12, Docherty et al. (2012) demonstrated

THz differential transmission in vacuum, N_2, air, and O_2 at room temperature. In vacuum, THz conductivity increases (photo-induced absorption) within 1 ps followed by fast cooling (within ~0.6 ps) and a longer exponential cooling (~2.5 ps). However, the photoconductivity signal reverses from positive in vacuum to negative in gaseous environment in Figure 15.12a, which suggests bleaching in gaseous environment. Figure 15.12b demonstrated the photoconductivity spectra in different gaseous environment taken after 2 ps photoexcitation. The spectra in the presence of N_2, air, and O_2 have a Lorentzian response (fitting lines in

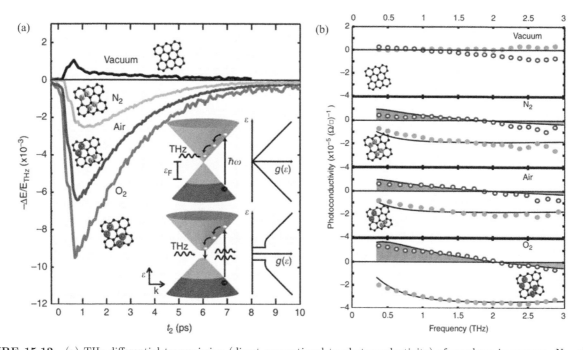

FIGURE 15.12 (a) THz differential transmission (direct proportional to photoconductivity) of graphene in vacuum, N_2, air, and O_2 at room temperature. The carrier processes and density of states of p-doped graphene after the pump are demonstrated in vacuum (top) and in the gases (bottom); (b) THz complex photoconductivity spectrum (at delay time of 2 ps) in vacuum (top) and gases(bottom). The solid lines are Lorentzian model fitting results. (Reprinted with permission Jun et al. 2008. Copyright 2012, Springer Nature.)

Figure 15.12b). The physisorption of O_2 and N_2 on graphene could act as a dopant and lead to the opening of a small band gap, which could lead to a modification of the density of states (inset in Figure 15.12a) (Jun et al. 2008). This further promotes the recombination of photoexcited electrons with holes to generate stimulated emission, resulting in an enhancement of the THz transmission with negative photoconductivity.

Negative photoconductivity has also been observed in other 2D materials such as topological insulator (Bi_2Se_3) (Sim et al. 2014) and transition metal dichalcogenides (MoS_2) (Lui et al. 2014). Even though the negative THz photoconductivity in 2D nanomaterials is a novel property, the mechanism is still on debate and further effort is needed to make the practical applications with the negative photoconductivity of nanomaterials (Lu et al. 2017).

15.3.4 THz Emission Spectroscopy for Nanomaterials

Nanomaterials have the advantages in integrated photonic applications for compact optoelectronic devices. Several progresses have been achieved with nanomaterials for THz

nano-emitters, such as silicon nanowires (Jung et al. 2010), InAs nanowires (Arlauskas et al. 2014, Seletskiy et al. 2011), and GaAs nanowires (Trukhin et al. 2015). Here, we take carbon nanomaterials as examples to explore the potentional applications of THz emission spectroscopy to nanomaterials and the potentional THz nano-emitters based on nanomaterials.

As 1D nanomaterials, THz radiation from carbon nanotubes also demonstrates anisotropic response. To demonstrate anisotropic THz radiation from aligned carbon nanotubes, Titova et al. did the THz generation experiment with 400 nm 100 fs optical pump pulses as shown in Figure 15.13a, in which carbon nanotubes are aligned vertically when $\theta_{SWCNT} = 0°$ (Titova et al. 2015). Figure 15.13b shows the THz time-domain and frequency-domain signals with $\theta_{SWCNT} = 0°$ and $\theta_{pump} = 90°$ (pump polarization along horizontal direction). Figure 15.13c illustrates the THz pulses with different pump polarization and carbon nanotube orientations. Carbon nanotubes can only generate THz emission parallel to the longitudinal axis of the carbon nanotubes, which suggests that ultrafast photocurrent surge excited by the femtosecond laser dominates along the longitudinal axis of the carbon nanotubes. Optical anisotropic

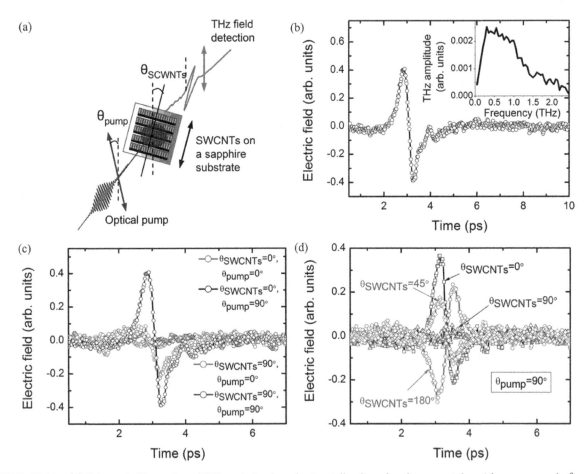

FIGURE 15.13 (a) Schematic illustration of THz emission from horizontally aligned carbon nanotube with a pump angle θ_{pump} and sample orientation angle θ_{SWCNT}. (b) The THz pulse and its frequency spectrum. (c) THz waveforms with different pump polarization and sample orientation angles. (d) The dependence of THz waveform on sample orientation angle with the horizontal pump polarization. (Reprinted with permission Titova et al. 2015. Copyright 2011, American Chemical Society.)

absorption of carbon nanotube can also be observed as the THz radiation intensity for pump polarization $\theta_{\mathrm{pump}} = 90°$ larger than that for pump polarization $\theta_{\mathrm{pump}} = 0°$. The dependence of THz waveform on the orientation angle with pump polarization $\theta_{\mathrm{pump}} = 90°$ is presented Figure 15.13d. The polarity of the THz waveform reverses when the sample orientation angle changed from 0° to 180°. This behavior can be ascribed to top–down anisotropy of the samples tentatively.

Graphene has also been proposed as a THz emitter, which suggests that the dominant THz radiation mechanism is PDE (Maysonnave et al. 2014, Obraztsov et al.

2014a,b). Owing to the weak interaction between pump light and graphene, THz generation from graphene is relatively weak. Plasmonic enhancement by Au film has been proposed to increase the THz radiation from graphene (Bahk et al. 2014). Recently, vertically grown graphene has been proposed to enhance the interaction between pump light and graphene as shown in Figure 15.14a (Zhu et al. 2017). The THz radiation intensity from the vertically grown graphene is ten times larger than that from a single-layer graphene. The dependence of THz p- and s-polarization components on the incidence angle has been shown in Figure 15.14b. A significant feature is that there is no THz signal that

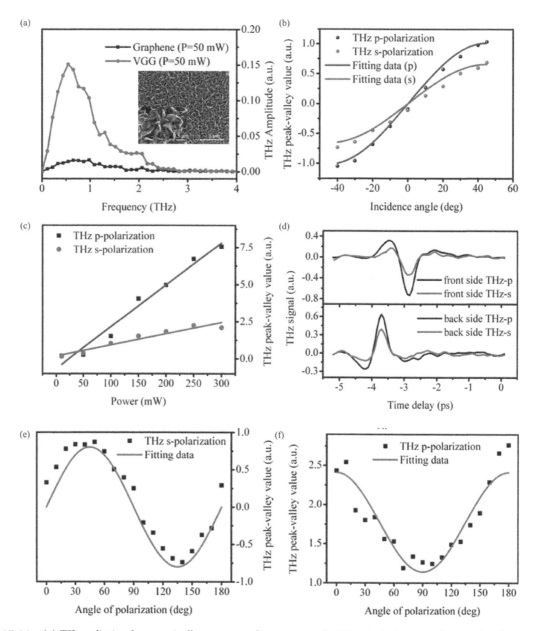

FIGURE 15.14 (a) THz radiation from vertically grown graphene compared with a single-layer graphene under the same conditions. Inset is the SEM image of the vertically grown graphene. (b) THz intensity dependence on the incidence angle. (c) THz intensity dependence on the pump power. (d) THz waveforms taken from the front side and back side of a substrate. (e and f) THz intensity dependence on the pump polarization angle for s- and p-components, respectively. (Reprinted with permission Zhu et al. 2017. Copyright 2017, Royal Society of Chemistry.)

can be detected at normal incidence and a polarity reversal occurred at opposite incidence. Figure 15.14c shows that the THz intensity increases linearly with the increasing average pump power, which indicates that a second-order nonlinear effect dominated the THz radiation process. These results suggest that ultrafast photocurrents from PDE and PGE are the dominant THz radiation mechanism. However, the PGE can be excluded because the graphene flakes grow vertically and show a centrosymmetrical structure. Figure 15.14d illustrates THz waveforms excited from the front and back sides of the substrate, separately. The polarity reversal of THz waveform demonstrates that this generation process is sensitive to the direction, which is the evidence of PDE. The THz intensity dependence on the polarization angle can be fit well with the theory by considering PDE (Figure 15.14e,f), which further proves that PDE dominates the THz radiation from the vertically grown graphene.

15.4 Conclusion and Outlook

In summary, we have presented a comprehensive review on fundamental principles of THz spectroscopy, which includes THz–TDS, OPTP spectroscopy, and THz emission spectroscopy. We analyze the interaction of THz wave with nanomaterials and give the fundamental data analysis with THz–TDS and OPTP spectroscopy. We also give the analysis of THz emission spectroscopy of nanomaterials, from which some nonlinear optical property of nanomaterials can be analyzed. Even though many progresses have been made in "TeraNano" field, THz wave is still on the horizon for practical applications. On one hand, to find nanomaterials with high THz emission efficiency, sensitive detecting and manipulating efficiency is still on the progress. On the other hand, to use the THz spectroscopy to understand the THz properties of new kind of nanomaterials is also very important in "TeraNano" field in the future.

Acknowledgment

This work was supported by National Natural Science Foundation of China (No. 11774288), Key Science and Technology Innovation Team Project of Natural Science Foundation of Shaanxi Province (2017KCT-01, 2019JC-25).

References

Andreeva, V. A., Kosareva, O. G., Panov, N. A. et al. 2016. Ultrabroad terahertz spectrum generation from an air-based filament plasma. *Physical Review Letters* 116(6): 063902.

Arlauskas, A., Treu, J., Saller, K. et al. 2014. Strong terahertz emission and its origin from catalyst-free InAs nanowire arrays. *Nano Letters* 14(3): 1508–1514.

Auston, D. H. and Cheung, K. P. 1985. Coherent time-domain far-infrared spectroscopy. *Journal of the Optical Society of America B* 2(4): 336–339.

Auston, D. H. and Smith, P. R. 1983. Generation and detection of millimeter waves by picosecond photoconductivity. *Applied Physics Letters* 43(7): 631–633.

Bahk, Y.-M., Ramakrishnan, G., Choi, J. et al. 2014. Plasmon enhanced terahertz emission from single layer graphene. *ACS Nano* 8(9): 9089–9096.

Baig, S. A., Boland, J. L., Damry, D. A. et al. 2017. An ultrafast switchable terahertz polarization modulator based on III–V semiconductor nanowires. *Nano Letters* 17(4): 2603–2610.

Baxter, J. B. and Schmuttenmaer, C. A. 2006. Conductivity of ZnO nanowires, nanoparticles, and thin films using time-resolved terahertz spectroscopy. *The Journal of Physical Chemistry B* 110(50): 25229–25239.

Beard, M. C. 2000. Transient photoconductivity in GaAs as measured by time-resolved terahertz spectroscopy. *Physical Review B* 62(23): 15764–15777.

Beard, M. C., Turner, G. M. and Schmuttenmaer, C. A. 2002a. Size-dependent photoconductivity in CdSe nanoparticles as measured by time-resolved terahertz spectroscopy. *Nano Letters* 2(9): 983–987.

Beard, M. C., Turner, G. M. and Schmuttenmaer, C. A. 2002b. Terahertz spectroscopy. *The Journal of Physical Chemistry B* 106(29): 7146–7159.

Bergren, M. R., Palomaki, P. K., Neale, N. R. et al. 2016. Size-dependent exciton formation dynamics in colloidal silicon quantum dots. *ACS Nano* 10(2): 2316–2323.

Braun, L., Mussler, G., Hruban, A. et al. 2016. Ultrafast photocurrents at the surface of the three-dimensional topological insulator Bi$_2$Se$_3$. *Nature Communications* 7: 13259.

Brucherseifer, M., Nagel, M., Bolivar, P. H. et al. 2000. Label-free probing of the binding state of DNA by time-domain terahertz sensing. *Materials Science Forum* 384–385(24): 253–258.

Côté, D., Laman, N. and Van Driel, H. M. 2002. Rectification and shift currents in GaAs. *Applied Physics Letters* 80(6): 905–907.

Cai, Y., Brener, I., Lopata, J. et al. 1998. Coherent terahertz radiation detection: Direct comparison between free-space electro-optic sampling and antenna detection. *Applied Physics Letters* 73(4): 444–446.

Cheville, R. A. and Grischkowsky, D. 1995. Far-infrared terahertz time-domain spectroscopy of flames. *Optics Letters* 20(15): 1646–1648.

Cunningham, P. D. 2013. Accessing terahertz complex conductivity dynamics in the time-domain. *IEEE Transactions on Terahertz Science & Technology* 3(4): 494–498.

Dakovski, G. L., Lan, S., Xia, C. et al. 2007. Terahertz electric polarizability of excitons in PbSe and CdSe quantum dots. *The Journal of Physical Chemistry C* 111(16): 5904–5908.

Docherty, C. J., Lin, C. T., Joyce, H. J. et al. 2012. Extreme sensitivity of graphene photoconductivity to environmental gases. *Nature Communications* 3(6): 1228.

Docherty, C. J., Stranks, S. D., Habisreutinger, S. N. et al. 2014. An ultrafast carbon nanotube terahertz polarisation modulator. *Journal of Applied Physics* 115(20): 203108.

Dorney, T. D., Baraniuk, R. G. and Mittleman, D. M. 2001. Material parameter estimation with terahertz time-domain spectroscopy. *Journal of the Optical Society of America A* 18(7): 1562–1571.

Duvillaret, L., Garet, F. and Coutaz, J. L. 2002. A reliable method for extraction of material parameters in terahertz time-domain spectroscopy. *IEEE Journal of Selected Topics in Quantum Electronics* 2(3): 739–746.

Exter, M., Fattinger, C. and Grischkowsky, D. 1989. Terahertz time-domain spectroscopy of water vapor. *Optics Letters* 14(20): 1128–1130.

Fülöp, J. A., Pálfalvi, L., Almási, G. et al. 2011. High energy THz pulse generation by tilted pulse front excitation and its nonlinear optical applications. *Journal of Infrared Millimeter & Terahertz Waves* 32(5): 553–561.

Fattinger, C. and Grischkowsky, D. 1989. Terahertz beams. *Applied Physics Letters* 54(6): 490–492.

Fattinger, C., Grischkowsky, D., Exter, M. V. et al. 1990. Far-infrared time-domain spectroscopy with terahertz beams of dielectrics and semiconductors. *Journal of the Optical Society of America B* 7(10): 2006–2015.

Fischer, B. M., Walther, M. and Uhd, J. P. 2002. Far-infrared vibrational modes of DNA components studied by terahertz time-domain spectroscopy. *Physics in Medicine & Biology* 47(21): 3807.

Gallot, G. and Grischkowsky, D. 1999. Electro-optic detection of terahertz radiation. *Journal of the Optical Society of America B* 16(8): 1204–1212.

Garet, F., Coutaz, J. L. and Duvillaret, L. 1999. Highly precise determination of optical constants and sample thickness in terahertz time-domain spectroscopy. *Applied Optics* 38(2): 409–415.

Gartstein, Y. N., Zakhidov, A. A. and Baughman, R. H. 2002. Charge-induced anisotropic distortions of semiconducting and metallic carbon nanotubes. *Physical Review Letters* 89(4): 045503.

George, P. A., Strait, J., Dawlaty, J. et al. 2008. Ultrafast optical-pump terahertz-probe spectroscopy of the carrier relaxation and recombination dynamics in epitaxial graphene. *Nano Letters* 8(12): 4248–4251.

Glazov, M. M. and Ganichev, S. D. 2014. High frequency electric field induced nonlinear effects in graphene. *Physics Reports* 535(3): 101–138.

Grischkowsky, D., Ketchen, M. B., Chi, C. C. et al. 1988. Capacitance free generation and detection of subpicosecond electrical pulses on coplanar transmission lines. *IEEE Journal of Quantum Electronics* 24(2): 221–225.

Gu, P., Tani, M., Kono, S. et al. 2002. Study of terahertz radiation from InAs and InSb. *Journal of Applied Physics* 91(9): 5533–5537.

Han, J., Chen, W., Zhang, J. et al. 2008. Terahertz response of bulk and nanostructured ZnO. *Piers Online* 4(3): 391–395.

Han, P. Y., Cho, G. C. and Zhang, X. C. 2000. Time-domain transillumination of biological tissues with terahertz pulses. *Optics Letters* 25(4): 242–244.

Han, P. Y., Tani, M., Usami, M. et al. 2001. A direct comparison between terahertz time-domain spectroscopy and far-infrared Fourier transform spectroscopy. *Journal of Applied Physics* 89(4): 2357–2359.

Han, P. Y. and Zhang, X. C. 2001. Free-space coherent broadband terahertz time-domain spectroscopy. *Measurement Science and Technology* 12(11): 1747.

Hannah, J. J., Jessica, L. B., Christopher, L. D. et al. 2016. A review of the electrical properties of semiconductor nanowires: insights gained from terahertz conductivity spectroscopy. *Semiconductor Science and Technology* 31(10): 103003.

Hashimshony, D., Geltner, I., Cohen, G. et al. 2001. Characterization of the electrical properties and thickness of thin epitaxial semiconductor layers by THz reflection spectroscopy. *Journal of Applied Physics* 90(11): 5778–5781.

Hendry, E., Koeberg, M., Schins, J. M. et al. 2004. Ultrafast charge generation in a semiconducting polymer studied with THz emission spectroscopy. *Physical Review B* 70(3): 033202.

Ho, I. C., Guo, X. and Zhang, X. C. 2010. Design and performance of reflective terahertz air-biased-coherent-detection for time-domain spectroscopy. *Optics Express* 18(3): 2872–2883.

Hodgson J N. Optical absorption and dispersion in solids[M]. Springer Science & Business Media, 2012.

Horiuchi, N. 2012. Terahertz nano-exploration. *Nature Photonics* 6: 82.

Hu, B. B. and Nuss, M. C. 1995. Imaging with terahertz waves. *Optics Letters* 20(16): 1716.

Hu, B. B., Zhang, X. C., Auston, D. H. et al. 1990. Free-space radiation from electro-optic crystals. *Applied Physics Letters* 56(6): 506–508.

Huang, Y., Zhu, L., Yao, Z. et al. 2018. Terahertz surface emission from layered MoS_2 crystal: competition between surface optical rectification and surface photocurrent surge. *The Journal of Physical Chemistry C* 122(1): 481–488.

Huang, Y., Zhu, L., Zhao, Q. et al. 2017. Surface optical rectification from layered MoS_2 crystal by THz time-domain surface emission spectroscopy. *ACS Applied Materials & Interfaces* 9(5): 4956–4965.

Huber, R., Tauser, F., Brodschelm, A. et al. 2001. How many-particle interactions develop after ultrafast excitation of an electron-hole plasma. *Nature* 414(6861): 286–289.

Huisman, T. J. and Rasing, T. 2016. THz emission spectroscopy for THz spintronics. *Journal of the Physical Society of Japan* 86(1): 011009.

Jeon, T. I. and Grischkowsky, D. 1997. Nature of conduction in doped silicon. *Physical Review Letters* 78(6): 1106–1109.

Jeon, T. I. and Grischkowsky, D. 1998. Characterization of optically dense, doped semiconductors by reflection THz time domain spectroscopy. *Applied Physics Letters* 72(23): 3032–3034.

Jeon, T. I., Kim, K. J., Kang, C. et al. 2002. Terahertz conductivity of anisotropic single walled carbon nanotube films. *Applied Physics Letters* 80(18): 3403–3405.

Jeon, T. I., Kim, K. J., Kang, C. et al. 2004. Optical and electrical properties of preferentially anisotropic single-walled carbon-nanotube films in terahertz region. *Journal of Applied Physics* 95(10): 5736–5740.

Jepsen, P. U. and Keiding, S. R. 1995. Radiation patterns from lens-coupled terahertz antennas. *Optics Letters* 20(8): 807–809.

Jepsen, P. U., Winnewisser, C., Schall, M. et al. 1996. Detection of THz pulses by phase retardation in lithium tantalate. *Physical Review E Statistical Physics Plasmas Fluids & Related Interdisciplinary Topics* 53(4): R3052.

Joyce, H. J., Docherty, C. J., Gao, Q. et al. 2013. Electronic properties of GaAs, InAs and InP nanowires studied by terahertz spectroscopy. *Nanotechnology* 24(21): 214006.

Joyce, H. J., Wong-Leung, J., Yong, C. K. et al. 2012. Ultralow surface recombination velocity in InP nanowires probed by terahertz spectroscopy. *Nano Letters* 12(10): 5325–5330.

Jun, I., Jun, N. and Akiko, N. 2008. Semiconducting nature of the oxygen-adsorbed graphene sheet. *Journal of Applied Physics* 103(11): 781.

Jung, G. B., Cho, Y. J., Myung, Y. et al. 2010. Geometry-dependent terahertz emission of silicon nanowires. *Optics Express* 18(16): 16353–16359.

Kaindl, R. A., Carnahan, M. A., Hägele, D. et al. 2003. Ultrafast terahertz probes of transient conducting and insulating phases in an electron-hole gas. *Nature* 423(6941): 734–738.

Katzenellenbogen, N. and Grischkowsky, D. 1992. Electrical characterization to 4 THz of N- and P- type GaAs using THz time-domain spectroscopy. *Applied Physics Letters* 61(7): 840–842.

Kiwa, T. and Tonouchi, M. 2001. High frequency properties of YBCO thin films diagnosed by time-domain terahertz spectroscopy. *Physica C Superconductivity* 362(1–4): 314–318.

Kuzel, P., Kadlec, F. and Němec, H. 2007. Propagation of terahertz pulses in photoexcited media: analytical theory for layered systems. *Journal of Chemical Physics* 127(2): 024506.

Kužel, P. and Němec, H. 2014. Terahertz conductivity in nanoscaled systems: effective medium theory aspects. *Journal of Physics D Applied Physics* 47(37): 374005.

Kyoung, J., Jang, E. Y., Lima, M. D. et al. 2011. A reel-wound carbon nanotube polarizer for terahertz frequencies. *Nano Letters* 11(10): 4227–4231.

Liu, H., Lu, J., Tang, S. H. et al. 2014. Composition-dependent electron transport in CdS_xSe_{1-x} nanobelts: a THz spectroscopy study. *Optics Letters* 39(3): 567–570.

Lloydhughes, J. 2012. A review of the terahertz conductivity of bulk and nano-materials. *Journal of Infrared Millimeter & Terahertz Waves* 33(9): 871–925.

Low, T. and Avouris, P. 2014. Graphene plasmonics for terahertz to mid-infrared applications. *Acs Nano* 8(2): 1086–1101.

Lu, J., Liu, H. and Sun, J. 2017. Negative terahertz photoconductivity in 2D layered materials. *Nanotechnology* 28(46): 464001.

Ludwig, C., Balakirev, F. F., Habermeier, H. U. et al. 1996. Electrodynamics of high-temperature superconductors investigated with coherent terahertz pulse spectroscopy. *Journal of the Optical Society of America B* 13(9): 1979–1993.

Lui, C. H., Frenzel, A. J., Pilon, D. V. et al. 2014. Trion-induced negative photoconductivity in monolayer MoS_2. *Physical Review Letters* 113(16): 166801.

Maysonnave, J., Huppert, S., Wang, F. et al. 2014. Terahertz generation by dynamical photon drag effect in graphene excited by femtosecond optical pulses. *Nano Letters* 14(10): 5797–5802.

Mittleman, D. M., Cunningham, J., Nuss, M. C. et al. 1997. Noncontact semiconductor wafer characterization with the terahertz Hall effect. *Applied Physics Letters* 71(1): 16–18.

Mittleman, D. M., Jacobsen, R. H., Neelamani, R. et al. 1998. Gas sensing using terahertz time-domain spectroscopy. *Applied Physics B* 67(3): 379–390.

Mittleman, D. M., Jacobsen, R. H. and Nuss, M. C. 1996. T-ray imaging. *IEEE Journal of Selected Topics in Quantum Electronics* 2(3): 679–692.

Nahata, A., Auston, D. H., Heinz, T. F. et al. 1996a. Coherent detection of freely propagating terahertz radiation by electro-optic sampling. *Applied Physics Letters* 68(2): 150–152.

Nahata, A., Weling, A. S. and Heinz, T. F. 1996b. A wideband coherent terahertz spectroscopy system using optical rectification and electro-optic sampling. *Applied Physics Letters* 69(16): 2321–2323.

Nair, R. R., Blake, P., Grigorenko, A. N. et al. 2008. Fine structure constant defines visual transparency of graphene. *Science* 320(5881): 1308–1308.

Nashima, S., Morikawa, O., Takata, K. et al. 2001. Measurement of optical properties of highly doped silicon by terahertz time domain reflection spectroscopy. *Applied Physics Letters* 79(24): 3923–3925.

Niklasson, G. A., Granqvist, C. and Hunderi, O. 1981. Effective medium models for the optical properties of inhomogeneous materials. *Applied Optics* 20(1): 26–30.

Novoselov, K. S., Geim, A. K., Morozov, S. V. et al. 2004. Electric field effect in atomically thin carbon films. *Science* 306(5696): 666–669.

Obraztsov, P. A., Kanda, N., Konishi, K. et al. 2014a. Photon-drag-induced terahertz emission from graphene. *Physical Review B* 90(24): 241416.

Obraztsov, P. A., Kaplas, T., Garnov, S. V. et al. 2014b. All-optical control of ultrafast photocurrents in unbiased graphene. *Scientific Reports* 4: 4007.

Parkinson, P., Dodson, C., Joyce, H. J. et al. 2012. Noncontact measurement of charge carrier lifetime and mobility in GaN nanowires. *Nano Letters* 12(9): 4600–4604.

Parkinson, P., Lloydhughes, J., Gao, Q. et al. 2007. Transient terahertz conductivity of GaAs nanowires. *Nano Letters* 7(7): 2162–2165.

Planken, P. C. M., Nienhuys, H. K., Bakker, H. J. et al. 2001. Measurement and calculation of the orientation dependence of terahertz pulse detection in ZnTe. *Journal of the Optical Society of America B* 18(3): 313–317.

Qi, M., Zhou, Y. X., Hu, F. R. et al. 2014. Improving terahertz sheet conductivity of graphene films synthesized by atmospheric pressure chemical vapor deposition with acetylene. *Journal of Physical Chemistry C* 118(27): 15054–15060.

Richter, C. and Schmuttenmaer, C. A. 2010. Exciton-like trap states limit electron mobility in TiO_2 nanotubes. *Nature Nanotechnology* 5(11): 769.

Rudd, J. V., Johnson, J. L. and Mittleman, D. M. 2000. Quadrupole radiation from terahertz dipole antennas. *Optics Letters* 25(20): 1556–1558.

Schall, M., Helm, H. and Keiding, S. R. 1999. Far infrared properties of electro-optic crystals measured by THz time-domain spectroscopy. *International Journal of Infrared & Millimeter Waves* 20(4): 595–604.

Seletskiy, D. V., Hasselbeck, M. P., Cederberg, J. G. et al. 2011. Efficient terahertz emission from InAs nanowires. *Physical Review B* 84(11): 115421.

Sim, S., Brahlek, M., Koirala, N. et al. 2014. Ultrafast terahertz dynamics of hot Dirac-electron surface scattering in the topological insulator Bi_2Se_3. *Physical Review B* 89(16): 4006–4006.

Smith, N. V. 2001. Classical generalization of the Drude formula for the optical conductivity. *Physical Review B Condensed Matter* 64(15): 155106.

Smye, S. W., Chamberlain, J. M., Fitzgerald, A. J. et al. 2001. The interaction between Terahertz radiation and biological tissue. *Physics in Medicine & Biology* 46(9): R101.

Tani, M., Matsuura, S., Sakai, K. et al. 1997. Emission characteristics of photoconductive antennas based on low-temperature-grown GaAs and semi-insulating GaAs. *Applied Optics* 36(30): 7853–7859.

Titova, L. V., Pint, C. L., Zhang, Q. et al. 2015. Generation of terahertz radiation by optical excitation of aligned carbon nanotubes. *Nano Letters* 15(5): 3267–3272.

Tiwana, P., Parkinson, P., Johnston, M. B. et al. 2010. Ultrafast terahertz conductivity dynamics in mesoporous TiO_2: influence of dye sensitization and surface treatment in solid-state dye-sensitized solar cells. *The Journal of Physical Chemistry C* 114(2): 1365–1371.

Trukhin, V. N., Bouravleuv, A. D., Mustafin, I. A. et al. 2015. Generation of terahertz radiation in ordered arrays of GaAs nanowires. *Applied Physics Letters* 106(25): 252104.

Uhd Jepsen, P., Schairer, W., Libon, I. H. et al. 2001. Ultrafast carrier trapping in microcrystalline silicon observed in optical pump–terahertz probe measurements. *Applied Physics Letters* 79(9): 1291–1293.

Wang, F., Shan, J., Islam, M. A. et al. 2006. Exciton polarizability in semiconductor nanocrystals. *Nature Materials* 5(11): 861–864.

Wu, Q., Hewitt, T. D. and Zhang, X.-C. 1996. Two-dimensional electro-optic imaging of THz beams. *Applied Physics Letters* 69(8): 1026–1028.

Wu, Q. and Zhang, X. C. 1996. Electrooptic sampling of freely propagating terahertz fields. *Optical & Quantum Electronics* 28(7): 945–951.

Wu, X., Quan, B., Xu, X. et al. 2013a. Effect of inhomogeneity and plasmons on terahertz radiation from GaAs (100) surface coated with rough Au film. *Applied Surface Science* 285(11): 853–857.

Wu, X., Xu, X., Lu, X. et al. 2013b. Terahertz emission from semi-insulating GaAs with octadecanthiol-passivated surface. *Applied Surface Science* 279(8): 92–96.

Xu, L., Zhang, X. C. and Auston, D. H. 1992. Terahertz beam generation by femtosecond optical pulses in electro-optic materials. *Applied Physics Letters* 61(15): 1784–1786.

Xu, X., Li, S., Shi, Y. et al. 2005. Light-induced dielectric transparency in single-walled carbon nanotube films. *Chemical Physics Letters* 410(4–6): 298–301.

Xu, X., Yao, Z. and Jin, Y. 2015. Texture and light-induced anisotropic terahertz properties of free-standing single-walled carbon nanotube films with random networks. *Materials Chemistry and Physics* 162: 743–747.

Xu, X. L., Chuang, K., Nicholas, R. J. et al. 2009. Terahertz excitonic response of isolated single-walled carbon nanotubes. *Journal of Physical Chemistry C* 113(42): 18106–18109.

Xu, X. L., Parkinson, P., Chuang, K. C. et al. 2010. Dynamic terahertz polarization in single-walled carbon nanotubes. *Physical Review B Condensed Matter* 82(8): 085441.

Zhang, L., Huang, Y., Zhao, Q. et al. 2017. Terahertz surface emission of d-band electrons from a layered tungsten disulfide crystal by the surface field. *Physical Review B* 96(15): 155202.

Zhang, X. C. and Auston, D. H. 1992. Optoelectronic measurement of semiconductor surfaces and interfaces with femtosecond optics. *Journal of Applied Physics* 71(1): 326–338.

Zhang, X. C., Hu, B. B., Darrow, J. T. et al. 1990. Generation of femtosecond electromagnetic pulses from semiconductor surfaces. *Applied Physics Letters* 56(11): 1011.

Zhou, Y., Xu, X., Hu, F. et al. 2014. Graphene as broadband terahertz antireflection coating. *Applied Physics Letters* 104(5): 051106.

Zhou, Y., Yiwen, E., Ren, Z. et al. 2015. Solution-processable reduced graphene oxide films as broadband terahertz wave impedance matching layers. *Journal of Materials Chemistry C* 3(11): 2548–2556.

Zhu, L., Huang, Y., Yao, Z. et al. 2017. Enhanced polarization-sensitive terahertz emission from vertically grown graphene by a dynamical photon drag effect. *Nanoscale* 9(29): 10301–10311.

Zubair, A., Tsentalovich, D. E., Young, C. C. et al. 2016. Carbon nanotube fiber terahertz polarizer. *Applied Physics Letters* 108(14): 141107.

16

TXRF Spectrometry in Conditions of Planar X-ray Waveguide-Resonator Application

V. K. Egorov
IMT RAS

E. V. Egorov
RUDN

16.1 Introduction

Experimental science experience of the last century brought out clearly that the X-ray radiation is the most commonly encountered and very effective means for a material properties study. X-ray diffraction method is best suited to the structural features study of substance [1,2]. The material element composition and some peculiarities of the chemical bond are investigated by X-ray fluorescence (XRF) analysis methods [3–6]. It is known that the X-ray characteristical fluorescence can be excited by hard X-ray radiation, electron and ion beams. The spectrometrical method based on the XRF yield excitation by high-energy ion beams was called the particle-induced X-ray emission (PIXE). The method of XRF yield excitation by electron beams got the name electron-probe microanalysis. But XRF material diagnostics by X-ray beam excitation is the most simple, cheap and widely-distributed experimental procedure for the experimental practice. This procedure is carried out usually in 45°–45° geometry (Figure 16.1a). Similar conventional geometry implies that the exiting radiation beam generated by Mo, Ag, Ru or W X-ray sources incidences on the studied target at an angle $\theta = 45°$ to the target surface, and the characteristical fluorescence yield is collected at the same angle to the surface and at an angle $\psi = 90°$ to the incident beam direction. The XRF yield collection is carried out by PIN or silicon drift detector (SDD). Similar X-ray spectrometry variety has named the method of energy-dispersive analysis. The method collects XRF spectra in the wide-energy range allowing to register the XRF yield for almost all elements

existing in the nature. The method is not effective for the light-element diagnostics owing to a strong absorption of X-ray low-energy radiation in the detector window material. The energy resolution method is defined by detector characteristics and usually near 125 eV. The resolution improvement can be achieved by different diffraction procedures' application for the XRF spectra registration. It is XRF with the wavelength dispersion. The XRF wavelength-dispersive spectrometry allows to decrease the spectrum energy resolution more than one order by the X-ray flux-diffraction effect use. The method is oriented on the separate line's registration and characterized by low-radiation gathering power.

In spite of its merits, XRF spectrometry in conditions of any type of excitation application is characterized by some specific problems and cannot be considered as the quantitative analytical procedure. The first problem is connected with the mutual influence of X-ray characteristical lines excited by an internal impact, which has the name matrix effect. This effect can be compensated partly by correcting introduction on the atomic number (Z), on the radiation absorption (A) and on the second fluorescence (F) by the matrix (ZAF) correction conception use. In addition to matrix problem, there is a need to take into account the fluorescence yield effective depth for different fluorescence lines typical for the material. For example, CaF$_2$ monocrystal excitation will be accompanied by yield of CaK$\alpha\beta$ fluorescence lines with the effective depth 2.5 μm and by yield of FK$\alpha\beta$ lines with the effective depth 0.15 μm, only. The spectrum background is the additional factor defining

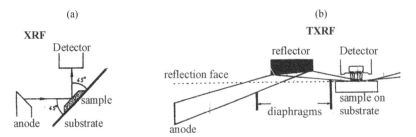

FIGURE 16.1 Comparison geometries of conventional XRF material diagnostics (XRF) (a) and surface element analysis in conditions of total external reflection of hard X-ray radiation flux on the studied surface (TXRF) (b).

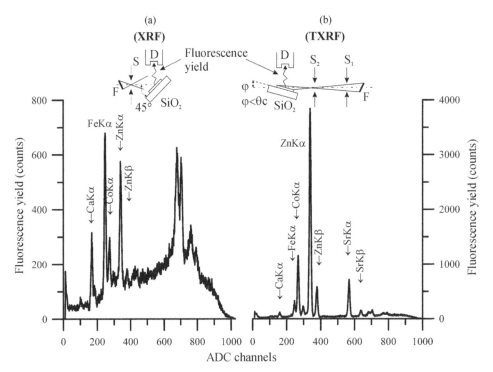

FIGURE 16.2 Patterns of XRF yield for the liquid dry residue contained Ca, Fe, Co, Zn and Zr atoms collected in the conventional XRF measurement geometry (a) and used the sliding excited beam incidence (b) with BSW-24 (Mo) radiation source in regime $U = 25$ keV, $I = 10$ mA, measurement time 300 s. Energy step 26 eV/channel. θ_c – is the total external reflection critical angle for the fluorescence exciting flux of the studied material.

the procedure's analytical quality. It is known that the background deposit in XRF spectrum is directly proportional to the thickness of target exciting layer [7,8]. On the basis of this conception, Yoneda and Horiuchi [9] suggested to decrease the background deposition in the collected spectrum by the incidence angle decreasing of X-ray exciting beam up to magnitudes of its total external reflection values. Figure 16.1b visualizes its idea. At the same time, the exciting beam must be prepared in the filament form. Figure 16.2 shows XRF yield spectra of the solution dry residue containing Ca, Fe, Co, Zn and Zr atoms collected by using the XRF conventional geometry (a) and in conditions of TXRF measurements (b). Thickness of the solution dry residue film was near 100 nm. Spectra were registered at the same regime of BSW-24 (Mo) X-ray source function ($U = 25$ keV, $I = 10$ mA). TXRF spectrum of the sample is distinguished by the background deposit's

insignificant value and X-ray characteristical fluorescence lines' high-intensity magnitudes for atoms of the dry residue film. Moreover, TXRF spectrum of the sample is characterized by a deposit decreasing of the exciting beam parasite scattering intensity. Very high efficiency of the thin film-coating TXRF study is presented by Figure 16.3. There are XRF and TXRF spectra of the Au film with thickness 9 nm deposited on Si substrate. RBS spectrum presented on the insertion shows that the film is free from the island effect. The RBS spectrum of Au/Si target was collected by using the ion-beam analytical complex Sokol-3 functioned in IMT RAS [10]. On base of the direct observation, it is apparent that the TXRF spectrum is more informative in comparison with the XRF spectrum because it allows to identify Ga, Se and Zr atoms' presence in the film. SiKα line belongs to the fluorescence yield of substrate atoms. Similar material investigations executed in different

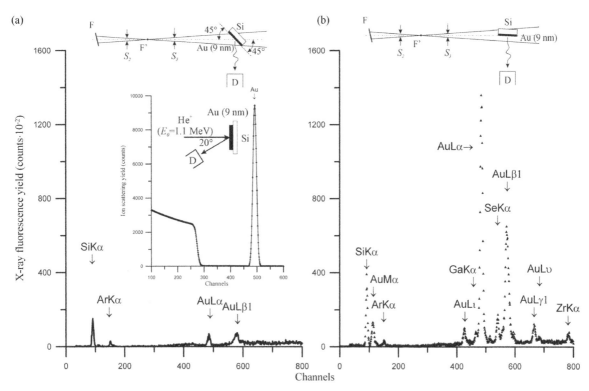

FIGURE 16.3 Patterns of XRF yield for Au (9 nm)/Si film sample collected in the conventional XRF measurement geometry (a) and used in the sliding exciting beam incidence (b). Measurements were carried out with BSW-24 (Mo) radiation source in regime $U = 25$ keV, $I = 10$ mA, measurement time 300 s. Energy step 20 eV/channel. RBS spectrum of the sample is shown on insertion. Energy step 1.9 keV/channel.

laboratories in the wake of Japan scientists give grounds to claim that the XRF analysis at exciting beam total external reflection on the studied surface is today the more effective method for element diagnostic of the material surface layer [11].

16.2   TXRF Material Diagnostics Peculiarities

In result of a great experimental activity of prof. H. Aiginger, P. Wobrauschek, A. Von Bohlen, C. Strely and other scientists [12–15], the TXRF spectrometry turned into the conventional method of trace element analysis of solid surfaces and dry residue of liquids deposited on specially prepared carries. Further methodological development of XRF analysis in the total external reflection conditions led to the specific modification of an element's diagnostics [16]. Methodological realism and appearing of new experimental procedures brought about the necessity to improve and widen the fundamental monograph of prof. R. Klockenkamper by additional data [17].

X-ray optical scheme of TXRF spectrometer built on base of the conventional conception is presented on Figure 16.4. It contains the X-ray radiation source, unit for the X-ray beam formation, system for the XRF yield registration and minigoniometer for the studied target orientation near an exciting beam. X-ray tubes with Mo, Ag, Rh and W

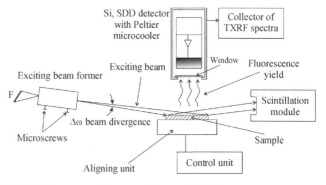

FIGURE 16.4   Schematic design of TXRF spectrometer built on base of the conventional X-ray optics setup.

anodes are usually used as X-ray radiation sources for TXRF measurements. Critical angles of the total external reflection featured for these radiation beams' interaction with the material surface are very small. This angle for MoKα total external reflection on the quartz surface is equal to 0.1°. So, TXRF measurements demand high precision of the studied target spatial positioning. Simplest TXRF facility can be prepared by using the conventional digital goniometer with the use of standard XRF spectrometer. Figure 16.5 demonstrates such example by using the HZG-4 goniometer and X-ray SDD spectrometer X-123 type. Figure 16.6 presents X-ray source types used for TXRF investigations, conventional types of X-ray exciting beam forms and experimental schemes for the TXRF measurements execution. Demands

FIGURE 16.5 Setup of TXRF simplest spectrometer built on the basis of standard digital goniometer HZG-4. 1. BSW-24 (Mo) X-ray source; 2. Holder of X-ray flux former; 3. Studied object on the goniometer attachment; 4. X-123 X-ray spectrometer of Amptec firm with Be window $t = 12~\mu m$; 5. HZG-4 goniometer; 6. Scintillation detector for the aligning procedure.

to X-ray source are not strict. But the beam using for the characteristic fluorescence yield excitation must be satisfied to some limitation. The exciting beam for TXRF analysis

must be prepared in the filament form with minimization of its width and angular divergence. The total external reflection phenomenon imposes some limitations on the studied target surface quality. The surface must characterize by a very small waviness and roughness. This demand is connected with the exciting beam incidence angle's trifle and, as a result, with very small value of the penetration depth parameter in conditions of the beam total reflection. Dependence of this parameter on the incidence angle value θ, presented on Figure 16.7a, can be described by the next analytical form [18]:

$$Z_p(\theta) = \frac{\lambda_0}{4\pi\sqrt{\delta}\sqrt{\sqrt{(x^2-1)^2 + y^2} - \sqrt{x^2-1}}} \qquad (16.1)$$

where λ_0 is the mean wavelength of exciting beam, $x = \theta/\theta_c$, $y = \beta/\delta$, where θ_c is the total reflection critical angle, δ and β parameters are the polarization and the absorption factors in conventional expression for the material refraction coefficient [19]:

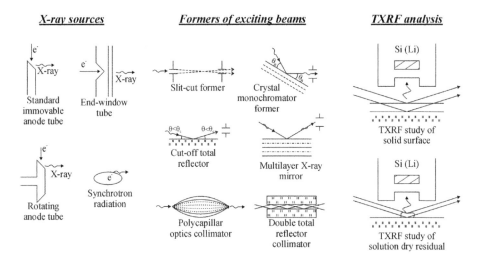

FIGURE 16.6 Schemes show types of X-ray sources, X-ray beam formers and analyzing objects. TXRF spectrometer is built on the basis of various combination of these units.

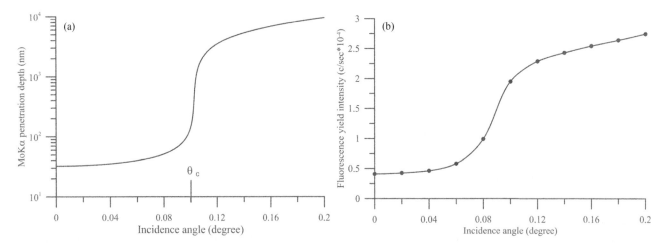

FIGURE 16.7 Depth penetration function of Mo radiation flux (a) and SiKα fluorescence yield intensity on the flux incident angle for the quartz-polished surface.

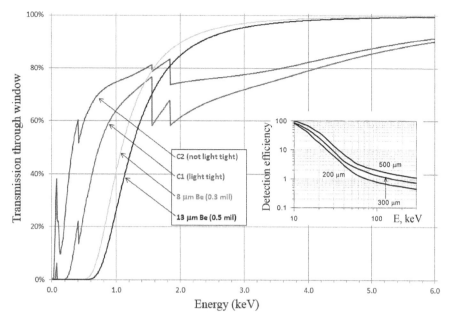

FIGURE 16.8 Functions of X-ray radiation registration efficiency of X-ray SDD equipped by different input windows in the energy area 0–5 keV are shown. The same functions for the high-energy radiation registration are presented in inset [24].

$$n = 1 - \delta - i\beta^1 \qquad (16.2)$$

The penetration depth function characterizes inward distance from the studied target surface corresponding to the exciting beam intensity decreasing on "e" times. At the same time, experimental investigations show that the real thickness of target surface layer being responsible for the XRF yield formation exceeds the penetration depth magnitude on 2 or 3 times, usually. This discrepancy depends on the exciting beam incidence angle and the studied material atomic density [21]. The function outlines of X-ray characteristical fluorescence yield on the incidence angle magnitude is akin to $Z(\theta)$ one in the total reflection angular area (Figure 16.7b). Beyond this range, the XRF yield is described by the expression characteristics for XRF conventional spectrometry [22,23]:

$$J_y = I_0 D \sin\theta \qquad (16.3)$$

where I_0 is the exciting beam intensity, D is the factor defined by material properties and θ is the exciting beam incidence angle. The comparison analysis of XRF and TXRF fluorescence yield shows that the total reflection area is characterized by a more less value of its integral intensity. But the spectral acutance of TXRF measurements is a more high owing to decreasing of a material volume excited by the initial X-ray beam. As a result, TXRF analysis is characterized by the background deposit decreasing in registered fluorescence spectrum and by the abrupt reduction

of element detection limits. But the greatest significance of TXRF measurements is connected with the absence of the matrix effect influence – the effect of mutual influence of XRF lines on its yield intensities. Moreover, line intensities in the TXRF spectrum are connected with atomic concentrations in the exciting material layer by the linear dependence, and the fluorescence yield does not depend on an absorption factor. So, the TXRF spectrometry of solid materials can be treated in distinction to XRF measurements as the quantitative analytical method as much as the host atoms quantity in the sample exciting volume can be evaluated from structural considerations, lightly. Host atoms in similar investigations will play a role of the internal standard. Similar approach can be used for the quantitative element analysis of liquid dry residues by metered addition into the initial solution of definite atoms. But direct comparison of the characteristical line's intensity is true only by taking into account the radiation absorption by the X-ray detector window and cross-section values of the atom's fluorescence excitation. Figure 16.8 presents functions of the X-ray efficiency registration by Si SDD on energy of X-ray quants in conditions of the different input windows' application for energy range 0.2–5 keV [24]. Consideration of these data is very important for the light element diagnostics. At the work with high-energy executed fluorescence yield ($E > 10$ keV), it is a need to introduce a correction on the detector efficiency registration (see insertion on Figure 16.8).

Figure 16.9 shows the experimental dependences for characteristical fluorescence excitation cross sections of Kα and Lα different elements lines by the MoK$\alpha\beta$ radiation beam application. Experimental data were carried out by the use of BSW-24 (Mo) radiation source in regime $U = 25$ keV, $I = 20$ mA. Cross-section excitation magnitudes are very sensitive to the radiation source regime. So, the quantitative

---

[1]Such form of the material refraction factor is usually used in works but it is not legitimate because of X-ray wavelength is commensurable or smaller in comparison with interatomic distances in materials [20].

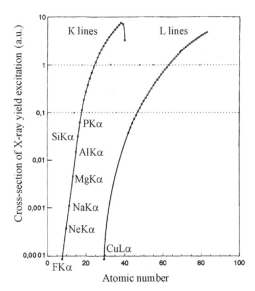

FIGURE 16.9 Cross-section energy dependences of $K\alpha$ and $L\alpha$ different element lines' excitation by the MoK$\alpha\beta$ radiation beam.

investigation must be executed after the specific excitation cross-section measurement for lines observed in the TXRF spectrum. Taking into account the data about lines' excitation cross section, its factor absorption in the detector window material and the host atom quantity N_h in the exciting volume, one can define an admixture concentration N_{ad} by the expression [25]:

$$N_{ad} = N_h \frac{S_{ad}}{S_h} \frac{F_h}{F_{ad}} \frac{\varepsilon_h}{\varepsilon_{ad}} \qquad (16.4)$$

where S_h and S_{ad} are host and admixture lines' intensity yield minus the background ones, F_h and F_{ad} – the atoms lines' excitation cross sections, ε_h and ε_{ad} – the atoms lines detector registration efficiencies. The minimum admixture atom's content of any experimental measurements is regulated by the detection limit's parameter defined by the conventional expression [3]:

$$C_{ad}^{LD} = \frac{3\sqrt{S_b^{ad}}}{S_h - S_b^h} \qquad (16.5)$$

where S_b^{ad} and S_b^h are background intensities in the line positions of host and admixture atoms. In conditions of TXRF spectrometry, the expressions (16.5) must be normalized on the host element concentration [25]:

$$N_{ad}^{L\alpha} = N_h \frac{3\sqrt{S_b^{ad}}}{S_h - S_b^h} \frac{F_h}{F_{ad}} \frac{\varepsilon_h}{\varepsilon_{ad}} \qquad (16.6)$$

TXRF investigations of monocrystalline objects and epitaxial structures must take into account a possiblity of the fluorescence spectrum additional lines appearing connected with the diffraction phenomenon with intensities being commensurable with ones of XRF lines. Moreover, its intensities can be changed at the condition's measurement variation. These peaks' migration and its intensity changing at the experimental conditions variation are beautifully

illustrated in Figure 16.10. Azimuthal angle variation at the TXRF testing of Si [111] monocrystalline target leads to very strong variation of the registered spectrum. Areas of the structural reflections appearing are presented on the top side of the Figure. Absence of Kβ and L$\beta\gamma$ lines is the fundamental criterion for rejection of structural reflections from the TXRF spectrum. In addition to the structural reflections and SiKα line emitted by host atoms, these TXRF spectra contain CaKα, FeKα and ArKα fluorescence lines. Calcium and iron lines characterize a pollution existence on the silicon wafer surface. ArKα line present in the spectrum reflects the fact that these investigations were executed without vacuum processing.

TXRF analytical method being similar to other experimental procedures is characterized by its own critical parameter. Total external reflection conditions of an exciting beam on the planar target with great surface area demand to form the beam in filament form with minimization of its width and angular divergence. Thickness of the target excited layer is limited by the penetration depth parameter. So, there is the geometrical limitation for interaction of the exciting beam with the target surface for any incident angle of the beam. In the result of this limitation, the efficiency of TXRF measurements may be improved by the beam radiation density increasing, only. And it provides strong evidence that the critical parameter of TXRF spectrometry is the radiation density factor of exciting beam emitted by the X-ray radiation former system. The increase in radiation density factor can be achieved either by a primitive augmentation of X-ray source power upto the synchrotron facility application or by the elaboration of specific X-ray beam formers being capable of increasing the X-ray beam radiation concentration [26,27]. This feature is typical for the planar X-ray waveguide resonator (PXWR), and there is good reason to believe that the TXRF spectrometry's further development will be oriented on the application of these devices and its parameter's modification [28].

16.3 TXRF Study of PbWO$_4$

The interest to study PbWO$_4$ scintillation possibilities was connected with existence in its scintillation response to the short duration light pulse (near 10 ns). Similar scintillation material was required for equipment of Large Hadron Collider. It is well known that the lead tungsten is the stoichiometric compound with homogeneity area nearly 0.01% at. At the same time, the Czochrolski method of PbWO$_4$ monocrystal boules preparation produced material with different magnitudes of the short-scintillation component intensity's yields [29]. Results of these measurements are presented on Figure 16.11. Scheme of the scintillation registration is shown on upper part of the figure. Scintillation spectra are presented in the form of light yield intensity for the same dose of gamma radiation. These investigations showed that PbWO$_4$ monocrystal is characterized by three types of the excitation relaxation (A, B and C types). X-ray

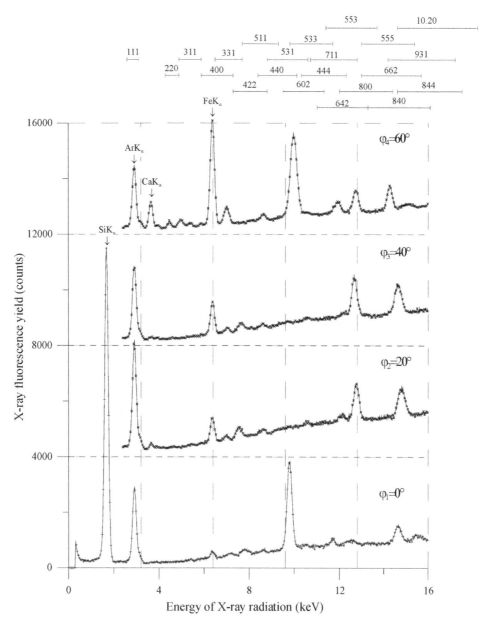

FIGURE 16.10 TXRF spectra collected for four positions of the silicon wafer with [111] surface orientation at different azimuth angles ($\psi = 0°$, $20°$, $40°$, $60°$). Expected area of the structural reflections appearing is shown on top part of figure.

diffraction study did not demonstrate structural anomalies related to stolzite structure, and difference in structural parameters was observed in fifth symbol after comma. Diffraction measurements with similar precision are very difficult task. So, TXRF measurements of different crystalline types were executed. Results of these TXRF investigations are presented on Figure 16.12. Comparison of TXRF spectra featured for crystals of A and C types shows that Pb and W atoms' concentration are different. A type crystal is characterized by more high concentration of lead atoms, and its factor defines enhanced scintillation activity of the crystal. At the same time, the element concentration difference for these crystals is smaller as 0.01% at. These results provided enough reasons to suppose that the homogeneity area of the compound has a composite structure with

three substructure zones. Presented example shows that the TXRF measurements are able to help in solution of structure problems. And a waveguide-resonance device included into the TXRF facility allows to increase the XRF analysis efficiency of material in conditions of the exiting beam total external reflection on its surface.

16.4 MoKαβ Beam Formation by PXWR

The waveguide-resonance propagation conception of X-ray fluxes was suggested by V.K. Egorov and E.V. Egorov [30] owing to impossibility to find a suitable explanation for peculiarities of the flux stream through the supernarrow

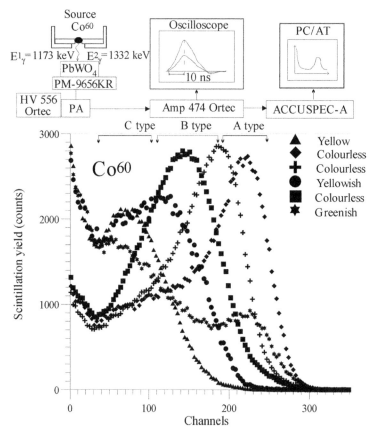

FIGURE 16.11   Scheme for $PbWO_4$ scintillation yield registration and intensities of optical yield registration for $PbWO_4$ crystal of A, B and C types differed by small composition destination.

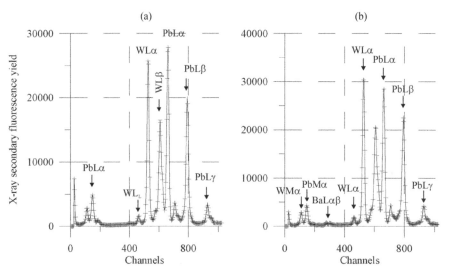

FIGURE 16.12   TXRF spectra, collected for $PbWO_4$ crystals of A-type (a) and C-type (b). Spectra collected by the use of BSW-24 (Mo) X-ray source in regime $U = 25$ keV, $I = 10$ mA during 600 s. Channel step is equal 16 eV/channel.

extended silt clearance on the basis of the flux direct propagation and the multiple external total reflection mechanism's superposition [31–33]. Foundation of this conception is the phenomenon of X-ray standing wave uniform interference field appearing in all air space of the slit clearance formed by planar dielectric reflectors in conditions of X-ray quasimonochromatic flux propagation. It is well known that

the X-ray polycapillary optics functions on the basis of the X-ray flux multiple total external reflection mechanism [34]. Every act of the X-ray quasimonochromatic flux's total reflection on planar material surfaces initiates the local interference field of X-ray standing wave appearing owing to the interaction between incident and reflected fluxes. The model scheme of the X-ray flux total external reflection phenomenon is shown on Figure 16.13a. This field arises in an air space under the material interface [35] and penetrates into the material volume with exponential attenuation [36]. And it is very important to understand that the interference field area in the air space is limited but in the material volume its propagation is not bounded.

Local area size evaluation of the X-ray standing wave interference field presents the specific interest. From optical investigations it is known that the area's longitudinal size is defined by half of the radiation coherence length parameter (L) [37]:

$$l = L/2 = \lambda_0^2/2\Delta\lambda \qquad (16.7)$$

where λ_0 is the radiation mean wavelength and $\Delta\lambda$ is the radiation monochromatization degree. Transverse size of the area is defined by the experimental condition's specificity [38]. At the total external reflection of X-ray quasimonochromatic flux on the material interface, the interference field area transverse size is approximately equal to the longitudinal length [39]. The standing wave period (D) in the air local area is defined by the expression [35]:

$$D = \frac{\lambda_0}{2\sin\theta} \approx \lambda/2\theta \qquad (16.8)$$

where θ is the flux incidence angle on the material-air interface. When we have planar extended slit clearance with distance between reflectors smaller as half coherence length of the radiation transporting by the clearance, we realize the waveguide-resonance propagation of the radiation flux [40]. The waveguide-resonance propagation mechanism is characterized by arising of the X-ray standing wave uniform interference field in all air-slit space, and the similar field with damped amplitude will appear in the material of reflectors.

It is shown on Figure 16.13b. In this case, any variation of the flux falling angle on the slit clearance input does not lead to the interference field reshaping. These variations provoke the standing wave period change, only. By this means, the X-ray planar narrow extended slit clearance is able to transport quasimonochromatic radiation fluxes falling on the device inlet under any angles, which are smaller as the critical angle of total external reflection for the reflector's material.

Our experimental investigations of the CuKαβ radiation beam intensity dependence on the width of planar extended slit clearance formed by quartz reflectors showed that the nature allows to realize the waveguide-resonance mechanism for the X-ray quasimonochromatic flux propagation in nanosize range of the slit clearance width [41]. Peculiarities of CuKαβ radiation fluxes generated by standard X-ray tubes are well known [42]. Comparison of the experimental data and the model conception allowed to conclude that the planar extended slit clearance prepared by quartz reflectors works for the CuKαβ radiation as the waveguide-resonator at the slit width range 0–220 nm. Our further investigations showed that the waveguide-resonance regime can be realized by the reflector's use prepared from monocrystalline materials and materials with amorphous structure. The polycrystalline structure of the reflector material cannot support the waveguide-resonance regime for X-ray quasimonochromatic flux propagation because the interference field of X-ray standing wave losses its uniformity.

In addition to our investigations of the CuKαβ radiation flux transportation by planar extended slit clearance in conditions of its width variation and for the speculations exclusion of the angular experiments interpretation, we carried out the systematic investigations of the

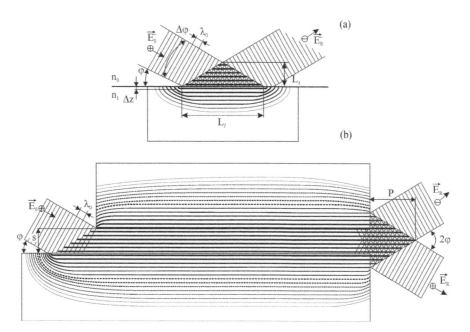

FIGURE 16.13 Scheme of X-ray flux total external reflection on the material interface (a) and its transportation in frame of the waveguide-resonance phenomenon at the flux propagation through planar extended slit clearance when its width is smaller as half of the radiation coherence length transporting by the clearance (b).

MoKαβ radiation beam intensity dependence on width of
the extended slit clearance formed by quartz reflectors in
the wide range of its variation [43]. These investigations
were executed by use of HZG-4 goniometer fabricated by
Carl Zeiss Jena firm. The construction of X-ray extended
slit clearance used for our investigations is presented on
Figure 16.14. Its use allowed to study the spatial intensity distribution and to fix the integral intensity of MoKαβ
radiation beams formed by the slit clearance in wide range
of width magnitudes. All measurements were executed by
use of BSW-24 (Mo) X-ray source in regime $U = 25$ keV,
$I = 10$ mA. Results of these investigations are presented on
Figure 16.15.

Experimental data show that the MoKαβ beam spatial
intensity distribution envelopes are close to Gauss outline.
At the same time, the result's comparison obtained for
CuKαβ and MoKαβ radiation beams engaged out attention to the absence of the distribution parameters variation for the nanosize area of slit clearance width up to the
slitless collimator application. Moreover, the integral intensity of MoKαβ radiation beams formed by the slit clearance
device in the width magnitude range 0–110 nm remained the

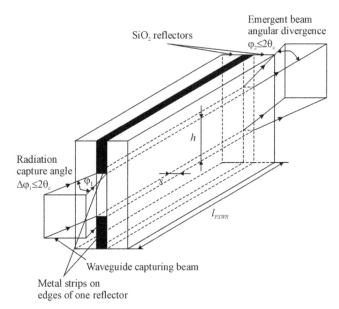

FIGURE 16.14 Scheme of X-ray initial flux capture area and
the emergent beam formation by the planar extended slit clearance
formed by two quartz reflectors.

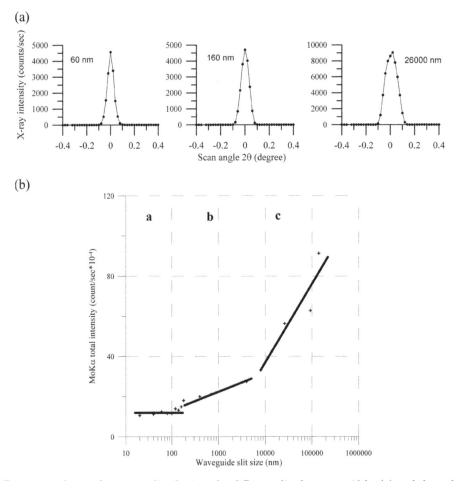

FIGURE 16.15 Experimental spatial intensity distributions for different slit clearance widths (a) and dependence of the MoKαβ
radiation flux total intensity on the slit clearance width formed by device built on scheme shown in Figure 16.14b. First dimension
interval can be interpreted as answering to the waveguide-resonance mechanism of MoKαβ flux propagation.

same. For the experimental data interpretation, it is a need to remember that the coherence length dimension of MoK$\alpha\beta$ radiation is equal to $L = 220$ nm. Similar to the experimental data interpretation obtained for the CuK$\alpha\beta$ radiation, it may be concluded that the first size interval permits the waveguide-resonance propagation of the MoK$\alpha\beta$ radiation beam. The slit clearance structure corresponding to the first size interval was designated as the planar PXWR [41]. The radiation density parameter of X-ray beams formed PXWR structures is of considerable interest to X-ray users, particularly in the light of PXWR possible application for the TXRF spectrometry.

Figure 16.16 shows the experimental dependence of the radiation density parameter of MoK$\alpha\beta$ radiation beams formed by the slit clearance device (a) and the double slit-cut system (b) on the slits width magnitude. Comparison of these dependences shows that the beam radiation density formed by the slit clearance device is more high in comparison with one after characteristics for the slit-cut system. The experimental values presented on the Figure are correspondent to position near the outlet of these beam formers. The radiation density parameter magnitude of MoK$\alpha\beta$ radiation beam formed by the waveguide-resonance unit is approximately 10,000 times higher than expected one in the X-ray beam formed by the conventional slit-cut system. So, the waveguide-resonance formers including into the TXRF X-ray optical schemes for X-ray exciting beams formation is quite reasonability. Its application allows to expect the detection limit decreasing of admixture elements on one – two orders. Moreover, it is known that the formation of

X-ray filament beam with nanosize width by slit-cut or shield systems is not easily soluble task. X-ray filament beam's application with a micrometric width for the TXRF analysis leads to increasing of the background deposit in collected spectra owing to the beam stray scattering excitation. This factor will be minimized in case of the waveguide-resonance former application. Because of this, there is good reason to believe that the waveguide-resonance structures including into the TXRF spectrometer facility will lead to considerable enhancement of the method analytical characteristics.

16.5 PXWR Application for TXRF Spectrometry

In the previous section, we discussed fundamental features of PXWR with the simplest construction. Such structure is built on the basis of two polished quartz reflectors with equivalent lengths and an identical form of its butt ends. PXWR with the simplest structure possesses by series of useful practical properties [44]. It captures the X-ray flux generated by radiation source into its inlet in the angular range, which cannot exceed double magnitude of the total external reflection critical angle for the reflectors material. The PXWR transports X-ray flux by its planar slit clearance almost without attenuation and forms the filament emergent beam. The emergent beam is characterized by very high-radiation density and has nanosize width. But the beam losses its phenomenal advantages already on small

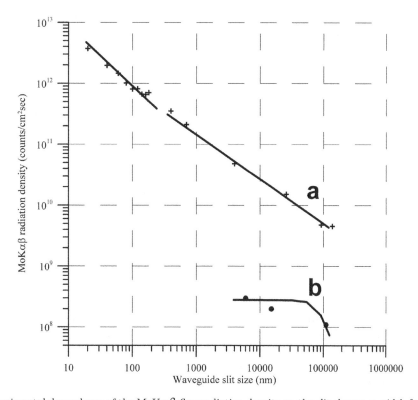

FIGURE 16.16 Experimental dependence of the MoK$\alpha\beta$ flux radiation density on the slit clearance width for emergent beam formed by device shown in Figure 16.10a (a) and similar dependence for X-ray beam formed by double slit-cut system (b).

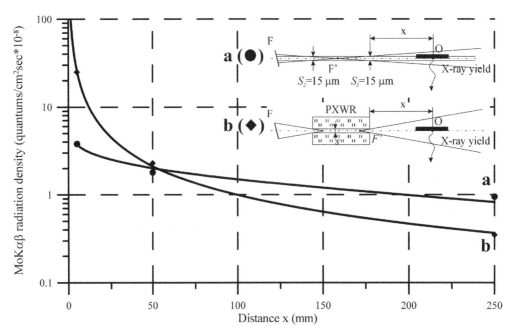

FIGURE 16.17   Experimental dependences of the MoKαβ beams radiation density formed by PXWR simplest design (b) and double slit-cut system (a) on distance from output cuts of these formers. Measurement geometries are presented in inset.

distances ($d \sim$3–5 cm) from the PXWR output owing to the beam angular divergence existence. Figure 16.17 shows dependences of the radiation density diminution in MoKαβ radiation beams formed by PXWR with the simplest design and the double slit-cut system on distance from the formers outlet. Analysis of these dependences allows to affirm that application of the waveguide-resonance former with simplest design for TXRF spectrometry losses the practical sense when the distance between its outlet and the studied sample position will be large, 50 mm. When the distance between outputs of X-ray beam formers and studied target position is near 45 nm, the average radiation density in the exciting beam is higher in one formed by PXWR with simplest design in comparison with the emergent beam formed by double slit-cut system. As a result, the fluorescence yield of X-ray characteristic lines of V-Ta-C sample is higher in case of PXWR application. It is demonstrated by Figure 16.18. At the increasing distance, the waveguide-resonance device with simplest design application for TXRF measurements will not be effective. So, it is clear, that the PXWR device including into TXRF facility will increase the analytical method efficiency either in case of the exciting beam divergence decreasing or by principal reorganization of the X-ray optical spectrometric scheme.

Figure 16.19 presents the principal pattern of the similar X-ray optical scheme being capable of increasing the TXRF measurement efficiency. This scheme provides the direct introduction of studied sample into the planar slit clearance of waveguide-resonance structure. For similar procedure execution, the waveguide-resonance structure is completed by quartz reflectors with holes for the sample and for the XRF yield leading out. This X-ray optical scheme allows to achieve maximum of the exciting beam radiation density

on the studied surface. Moreover, this measurement scheme allows to decrease the parasite scattering yield especially in case of the helium atmosphere introduction into the PXWR slit clearance. TXRF experimental investigations carried out with use of the measurement cell built on the basis of the modified X-ray optical scheme showed very beautiful results [45,46]. The idea of studied sample introduction into the X-ray standing wave interference field space of PXWR structure is the breakthrough step in TXRF spectrometry development. But the further growth in efficiency of this diagnostical method is connected with the upgrading of its own waveguide-resonance structure.

One approach to the waveguide-resonance structure upgrading is search of ways for the efficiency increasing of PXWR input radiation capture without the emergent beam parameters worsening. This approach can be realized on the basis of two technological solutions presented on Figure 16.20. According to the first proposal, the waveguide-resonance structure is equipped by the input skewed radiation concentrator with capture angle, which cannot exceed double value of the total external refraction critical angle for the reflector's material [47]. Length and surface form of the skewed concentrator is the subject of the specific investigation. For the formation of MoKαβ radiation beam by the modified waveguide-resonance device, the skewed concentrator angle must not exceed the magnitude 0.2°. For experimental realization of the approach, we elaborated PXWR with the skewed concentrator for the MoKαβ radiation capture with length $l_3 = 30$ mm at the total slit-clearance length $l_p = 100$ mm and with the width $s = 80$ nm. Figure 16.21 presents measurement schemes for the study of spatial intensity distribution in MoKαβ radiation beams formed by the PXWR with simplest design and by

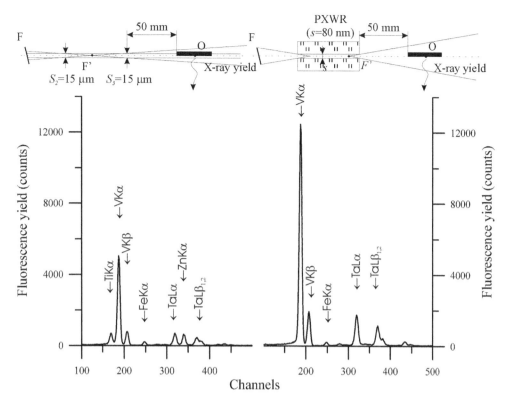

FIGURE 16.18 Geometries of measurements and TXRF spectra of V-Ta-C film collected by using of X-ray excite slit-cut system (a) and PXWR (b).

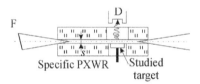

FIGURE 16.19 Alternative X-ray optical scheme of TXRF measurements, which insures the radiation density maximum of exciting flux on studied sample by its direct introduction into the interference field of X-ray standing wave initiated by the exciting flux.

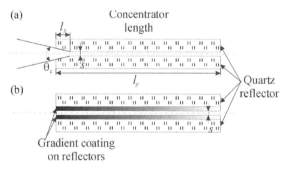

FIGURE 16.20 Principal schemes of X-ray waveguide-resonance structures allowing to increase the radiation capture value by application of the input skewed concentrator (a) and by deposition of the gradient film coating on quartz reflectors surfaces (b). Coatings must supply high structural density on PXWR input and low structural density on its output.

the waveguide-resonator equipped by skewed concentrator. Application of the modified waveguide-resonator did not lead to change of the radiation intensity distribution envelope. But the integral intensity of emergent beam increased approximately five times. It is obvious that the efficiency of waveguide-resonance device application will be higher at suitable selection of the skewed concentrator length and its surface form. The form modification can increase the concentrator efficiency on 1–1.5 order [48].

Second scheme presented on Figure 16.20 is an equally interesting and perspective. Fundamental essence of the scheme lies in the fact that the inlet material of waveguide-resonance structure is characterized by high atomic density, and this parameter is more smaller for its outlet material. It was expected that the waveguide-resonator equipped by the slit-clearance gradient structure can increase the emergent beam integral intensity by three times. But these expectations were destroyed because the solid solution structure of the deposited films demonstrated the polycrystalline construction, and similar slit clearances do not support the waveguide-resonance propagation of MoKαβ radiation flux.

Aside from search of waveguide-resonance constructions oriented on the increasing emergent beam integral intensity, it was a need to solve the task of the emergent beam angular divergence reduction at the retention of its integral intensity. This task can be solved by application of two simple schemes of the waveguide-resonance structure completing. Simplest solution of the problem can be achieved by the nonequivalent length reflector's use for PXWR building.

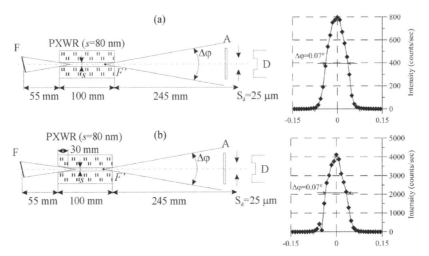

FIGURE 16.21 Experimental schemes for the spatial intensity distribution study in MoKαβ radiation fluxes formed by PXWR with simplest design (a) and waveguide-resonator equipped by skewed radiation concentrator (b) and real intensity distributions collected with these devices' application.

Similar construction suggests that the one half of X-ray flux composing the PXWR emergent beam undergoes total external reflection on the ledge of long reflector and increases the second half. (Because the first and second fluxes are coherent, the reflection conditions must be selected specially.) The experimental scheme for the study of spatial intensity distribution in MoKαβ radiation beam formed by the PXWR simplest design and one built on the basis of nonequivalent length reflectors and collected diagrams of the intensity distribution are presented on Figure 16.22. The spatial intensity diagram of the beam formed by modified unit shows that the emergent beam angular divergence was halved at the integral intensity conservation. But the spatial distribution form changes from the Gauss outline characteristic for beams formed by the simplest PXWR to the half Gauss one. Similar outline form variation is not the principle for XRF analysis but is very negative for X-ray diffractometrical measurements. The TXRF spectrometry

with application of waveguide-resonance structures built on the basis of nonequivalent length reflectors shows evident efficiency. Figure 16.23 presents experimental schemes for TXRF measurements with simplest and modified PXWRs application and XRF spectra of the high temperature superconducting film on the strontium titanate monocrystalline substrate collected with its application. The TXRF spectra comparison shows that the modified PXWR application leads to enhancing of the method efficiency.

Another design godsend allowing to reduce the PXWR emergent beam angular divergence without changing of its integral intensity is the composite planar X-ray waveguide-resonator (CPXWR) [49]. The classical scheme of the composite waveguide-resonator involves two simplest PXWRs, which are installed one after the other and characterized by the mutual alignment. But it is not enough for the CPXWR function. Our investigations showed that the distance magnitude between two simplest PXWRs

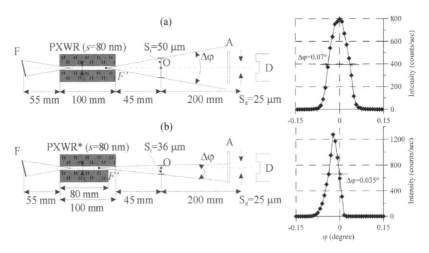

FIGURE 16.22 Experimental scheme of the MoKαβ radiation fluxes formation by PXWR with simplest design (a) and by the waveguide-resonator built on the basis of nonequivalent length quartz reflectors (b) and real intensity distributions collected with the use of these devices. Measurements were carried out with BSW-24 (Mo) radiation source in regime $U = 25$ keV, $I = 10$ mA. Absorption factor of A-attenuator $K = 700$.

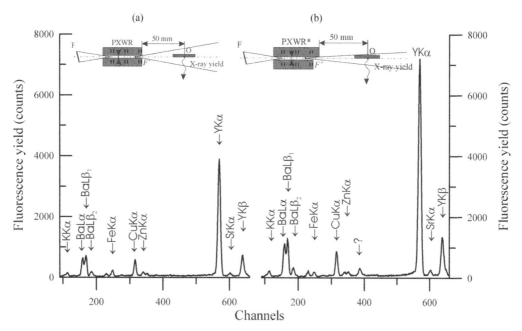

FIGURE 16.23 Experimental schemes for TXRF comparative investigations of $Ba_{1.4}Y_{1.0}Cu_{2.4}O_{7.0}$ epitaxial film on $SrTiO_3$ monocrystal substrate by application of PXWR with simplest design (a) and the waveguide-resonator built on the basis of nonequivalent length reflectors for the MoK$\alpha\beta$ exciting beam formation. (b) Measurements were carried out with BSW-24 (Mo) radiation source $U = 25$ keV, $I = 10$ mA. Incidence angle of the beams was equal to zero. Energy step 20 eV/channel.

has the principal significance for correct work of the device. When the distance is large for some critical value L_p (Figure 16.13b), the double PXWR structure forms emergent beam with very small angular divergence and very small integral intensity. Such device cannot call the composite PXWR. At the same time, when the distance between the simplest PXWRs will be smaller to the L_p critical magnitude defined by the expression [49]:

$$L_p = \lambda_0^3/8\Delta\lambda^2 \tag{16.9}$$

The emergent beam parameters change, cardinally. At the distance between PXWRs being not exceeded the critical parameter magnitude, the integral intensity of the device emergent beam will be correlated with X-ray radiation intensity of the captured flux. Moreover, our investigations showed that in this case, the angular divergence of formed X-ray beam is smaller in comparison with the radiation capture angle, and the emergent beam angular divergence magnitude depends on the value of distance between PXWRs. The analytical connection between the distance magnitude and the degree of emergent beam angular divergence decreasing remains to be received. Chief problems of these experimental investigation are connected with the influence ambiguity of form and treatment quality of reflector's ends on the emergent beam parameters. The investigation data allowed to understand that the effect of emergent beam angular divergence reduction is the result of partial angular tunneling phenomenon of X-ray flux in the gap between PXWRs installed one after the other. The phenomenon of X-ray flux partial angular tunneling is a consequence of the X-ray standing waves interference

fields interaction initiated in the slit clearances of PXWRs composed of the composite waveguide-resonator. The gap critical size between PXWRs is defined by the dimension of X-ray standing waves interference field protrusion (P) (Figure 16.13b). When the interference field protruded from the slit clearance of first PXWR achieves the second PXWR slit clearance, the effect of the interference field's interaction arises. This effect leads to the capture of all radiation flux abandoning first PXWR by the slit clearance of second PXWR. As a result, CPXWR generates X-ray emergent beam with the angular divergence diminution value and the invariable integral intensity. At the same time, accordingly to the Liouville theorem, the variation of statistical ensemble of some parameter must provoke the others parameters changing [50]. Our precise measurement showed that the X-ray beam angular divergence decreasing is accompanied by the radiation monochromatization degree deterioration of CPXWR emergent beam [51]. The model of composite waveguide-resonator function allowed to establish the connection between factor of the monochromatization degree deterioration $\delta\lambda$ and parameter of the CPXWR emergent beam angular divergence decreasing $\Delta\varphi$ by the next expression:

$$\delta\lambda = \varphi_1\lambda_0\Delta\varphi \tag{16.10}$$

where φ_1 is the input capture angle of CPXWR. As a result, the emergent beam will be characterized by $\varphi_2 = \varphi_1 - \Delta\varphi$ angular divergence and its monochromatism degree will be defined by the expression $\Delta\lambda_2 = \Delta\lambda_0 + \delta\lambda$. The factor of emergent beam radiation monochromatism decreasing is not the principle for TXRF measurements, but it is not useful for X-ray diffractometry. Results of the CPXWR application for angular divergence decreasing of its emergent beam are presented on Figure 16.24. In this measurement, we used the composite waveguide-resonator with modified design. The Figure shows experimental schemes for the study of the spatial intensity distribution in MoKαβ radiation beams formed by the simplest PXWR and by the composite waveguide-resonator with original design and distribution diagrams collected by the schemes' application. Emergent beams formed by PXWR and CPXWR are characterized by approximately equal intensities, but they have different angular divergences

and, consequently, different radiation densities. One can expect that the composite waveguide-resonator insertion into the conventional type TXRF facility can improve the measurement analytical efficiency. Figure 16.25 shows justice of this expectation. There is TXRF spectra of the Ti$_2$Zn film coating on Si substrate excited by the MoKαβ radiation beams use formed by the double slit-cut system and the CPXWR original design presented. On the basis of the spectra comparison, one can conclude that the composite waveguide-resonator application allows to improve the TXRF analysis efficiency more than one

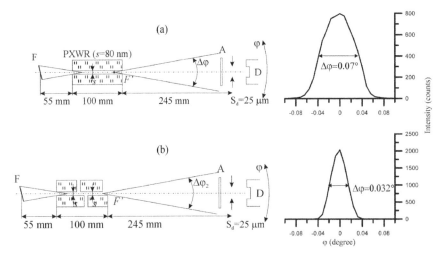

FIGURE 16.24 Experimental schemes of MoKαβ radiation flux formation by PXWR with simplest design (a) and by the composite waveguide-resonator of original construction (b) and real intensity distributions, collected with the use of these devices. Measurements were carried out with BSW-24 (Mo) radiation source in regime $U = 25$ keV, $I = 10$ mA. Absorption factor of A-attenuator $K = 700$.

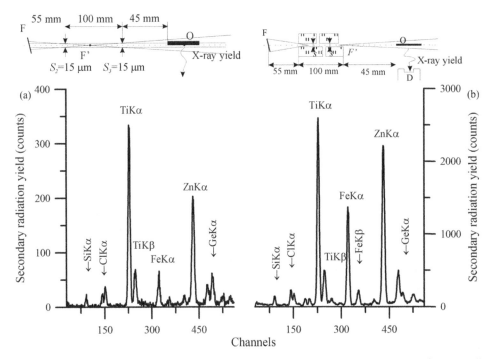

FIGURE 16.25 Experimental schemes for TXRF comparative investigations of Ti$_2$Zn film on silicon substrate collected in conditions of the MoKαβ exciting beam formation by double slit-cut system (a) and the composite waveguide-resonator of original construction (b) and spectra of X-ray characteristic fluorescence yield excited by these fluxes. Measurements were carried out with BSW-24 (Mo) radiation source in regime $U = 25$ keV, $I = 10$ mA. Incidence angle of the beams was equal to zero. Energy step 20 eV/channel.

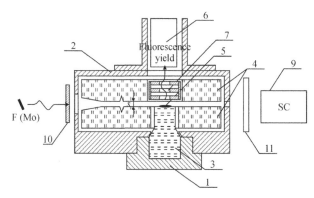

FIGURE 16.26 Schematic construction of the TXRF original cell allowing to introduce the studied sample into the interference field of X-ray standing wave. The cell includes its own design last modifications of PXWR properties. 1 – the screw fixture; 2 – the cell holder; 3 – the sample's substrate; 4 – quartz reflectors; 5 – the studied sample; 6 – SSD detector of TXRF yield; 7 – the collimator; 8 – X-ray source; 9 – scintillation counter for aligning procedure; 10 – inlet filter; and 11 – attenuator.

order. It is necessary to notice that the SiKα line initiated by the substrate material has more high intensity in the spectrum collected at the CPXWR application. Moreover, TXRF spectra presented on the Figure demonstrates the visible distinction for $I(\text{TiK}\alpha)/I(\text{ZnK}\alpha)$ relations calculated from the spectra. Rutherford backscattering study of the sample allowed to define the film thickness ($t = 20$ nm) and its depth element concentration profile. RBS data provide enough reason to interpret the spectra intensity distinction as a result of element's concentration variation on the film's thickness. Ti$_2$Zn stochiometry reflects the average film composition, but the film surface layer is characterized by the titanium excess concentration. As a result, one can conclude that the radiation density magnification leads to increasing of the surface layer thickness defining the X-ray characteristical fluorescence yield. This effect will be absent for homogeneous samples.

Waveguide-resonator modification by the skewed radiation concentrator's application and the construction's elaboration oriented on the emergent beam angular divergence decreasing are directed to the magnification of its radiation density. So, these elaborations were used for the TXRF cell for further modification. Figure 16.26 shows the principal scheme of the supermodified cell including the skewed radiation concentrator and constructive approaches for the X-ray standing wave interference field modification at the studied sample position.

16.6 TXRF Measurements with Compact Cell

Reflectors of the cell have angular slopes providing the PXWR input skewed concentrator preparation, deep scratches for the composite PXWR effect initiation and holes with different diameters for the effect creation of

nonequivalent length reflectors. For the background deposit decreasing the cell is equipped by Ta collimator and possibility of the helium atmosphere introduction. The cell construction is defined by the Russian patent [52]. It is the main unit of TXRF spectrometer functioning in the Saint Petersburg Electrotechnical University (LETI) in frame of the "Polus" group. The TXRF spectrometer is provided by BSW-22 (Mo) and BSW-22 (Ag) X-ray sources. Figure 16.27 demonstrates the cell disposition on the digital goniometer HZG-4. In principle, the cell can be included into setup of any X-ray goniometers and combined with any types of X-ray radiation sources. Near the cell, the additional holder of samples is shown. Compact TXRF cell is dedicated for surface-layer element composition study of dry residue solutions on Be or quartz substrates and solid targets. The total external reflection phenomenon of X-ray flux on material surface is characterized by the penetration depth parameter (see expression 16.1). This expression characterizes the depth for incident flux intensity decreasing on "e" times and allows to estimate the layer thickness responsible for the XRF yield formation. Because the exciting beam formed by PXWR is characterized by some angular divergence, it is important to evaluate the layer thickness answered for TXRF spectrum generation by direct experiment. For the experimental evaluation of this layer thickness, we studied set of thin film model targets with different coating thickness. The set presented Co thin films on Si substrate. Thickness of these films was measured by the Rutherford backscattering method by using Sokol-3 ion beam analytical complex [53]. Film's thickness varied from 3 to 30 nm. Measurement results for the four targets are presented in Figure 16.28a. TXRF spectra were collected at the same regime of BSW-24 (Mo) X-ray source $U = 25$ keV, $I = 5$ mA during 300 s. The measurements were carried out by using SDD detector of X-123 spectrometer. Spectra demonstrate CoKα and CoKβ of the films' host atoms. CaKα, FeKα, ZnKα and PbMα lines characterize main contaminations containing

FIGURE 16.27 The original TXRF cell position on HZG-4 digital goniometer.

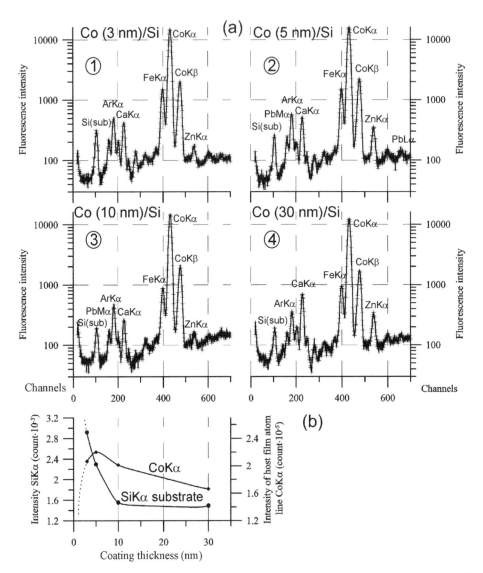

FIGURE 16.28 Experimental pattern of TXRF spectra for Co thin films on Si substrate (a) and curves for CoKα and SiKα line intensities for these targets characterizing the thin film thickness influence on X-ray secondary radiation yield (b). Channel step 16 eV/channel. Spectra collected by Si detector with Be window 12 μm.

in the studied films. ArKα line presence in the spectra shows that the measurements were executed without helium washing of the cell. SiKα line existence in TXRF spectra reflects the fact of substrate atoms excitation by X-ray beam formed by waveguide-resonator structure included into the cell construction. The comparison of CoKα and SiKα line's intensity variation on the coating thickness change allows to evaluate the real thin-film thickness creating X-ray characteristic fluorescence yield in conditions of Co target excitation by X-ray beam formed by the PXWR. These data are presented on Figure 16.28b. The presented diagram demonstrates that the TXRF active layer has the thickness near 5 nm. This magnitude is higher as one expected from the assumption that the fluorescence yield area is defined by the target surface layer whose thickness is corresponded to the penetration depth value of X-ray exciting beam at the half of total external reflection critical angle θ_c for the target material [54]. The expected magnitudes of this thickness

evaluated on the basis of the penetrate depth function calculated for interaction of MoKα flux with Co material surface showed on Figure 16.29 is equal to 3.7 nm. The experimental result obtained on the basis of Co/Si thin film target's investigations allows to conclude that the real value of TXRF active layer thickness in condition of PXWR application can be evaluated as the magnitude of X-ray beam penetration depth corresponding to its effective incident angle $\varphi = (0.7 \div 0.8)\theta_c$. This fact must be taken into account at the contamination content calculation on the basis of the TXRF method when the exciting beam is formed by the PXWR. Similar approach presents the background for TXRF spectrometry use a real quantitative method for the material element analysis.

The approach was applied for the contamination element diagnostics of silicon wafer by the use of our compact TXRF cell built on the basis of the waveguide-resonance system for MoKα exciting beam formation. Pattern of the spectrum is

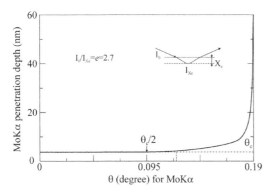

FIGURE 16.29 The function of MoKα radiation depth penetration into Co surface layer on the exciting beam incident angle. At the first approximation of surface layer thickness forming TXRF yield in conditions of the beam formation by the slitless collimator, one was suggested to use the magnitude corresponding to incident angle $\theta_c/2$ (θ_c – critical angle of total reflection).

presented on Figure 16.30. Spectrum consists of the X-ray characteristic fluorescence line set. Ca, Fe, Co, Ni and Zn are the atom's contamination of wafer surface. Silicon is the host of surface-layer exciting volume. Line intensity of silicon can be used as the internal standard datum mark. By using the expression (16.4), one can calculate the element's concentration of pollutions. There are $N(\mathrm{Ca}) = 9.2 \times 10^{11}$ at/cm^2; $N(\mathrm{Fe}) = 4.2 \times 10^{11}$ at/cm^2; $N(\mathrm{Co}) = 0.9 \times 10^{11}$ at/cm^2; $N(\mathrm{Ni}) = 2.1 \times 10^{11}$ at/cm^2; $N(\mathrm{Zn}) = 1.1 \times 10^{11}$ at/cm^2. Studied sample of the wafer is not enough clear for use in the microelectronics technology.

Figure 16.31 demonstrates TXRF spectrum collected for the iron film with thickness of 200 nm on silicon substrate. The film has polycrystalline structure. The spectrum is presented in logarithmic scale for a comfortable visualization

of its background deposit. Spectrum shows big set of XRF lines. Iron atoms are the host of the film and FeKα line intensity after background subtraction can be used as the internal standard datum mark. By use of this magnitude, one can calculate the admixture element concentrations in the film material. As a result, we have: $N(\mathrm{Al}) = 5.0 \times 10^{12}$ at/cm^2; $N(\mathrm{Si}) = 3.0 \times 10^{12}$ at/cm^2; $N(\mathrm{S}) = 6.1 \times 10^{11}$ at/cm^2; $N(\mathrm{Ti}) = 1.1 \times 10^{12}$ at/cm^2; $N(\mathrm{V}) = 1.9 \times 10^{12}$ at/cm^2; $N(\mathrm{Cr}) = 1.2 \times 10^{12}$ at/cm^2; $N(\mathrm{Mn}) = 2.1 \times 10^{12}$ at/cm^2; $N(\mathrm{Cu}) = 1.8 \times 10^{12}$ at/cm^2. Similar quantitative analysis is possible when we are sure that set of light elements are not present in surface of the studied object. Great concentration of light elements can be revealed by Rutherford backscattering measurements or AES method [55]. But its trace content can be detected by PIXE spectrometry [56] in case of use X-ray detectors with C1 or C2 windows (see Figure 16.8). When the host element of structure formation is featured by low-energy fluorescence yield, the TXRF analysis is able to diagnose the contamination presence in the structure, only.

For example, Figure 16.32 shows TXRF spectrum of the Dutch cheese sample. HKα, CKα, NKα, OKα lines are not presented in the spectrum because the spectrometer is equipped by X-ray detector with Be window. The spectrum shows CaKα and ClKα lines' characteristic for additional components of the product. Analogical measurements presented on Figure 16.33 were carried out for the petroleum film deposited on the Be substrate. But in this case, the TXRF method was underpinned by RBS results. RBS of H$^+$ ion beam ($E_0 = 0.953$ MeV) demonstrates the three-step diagram. Height of these steps characterizes the atom quantity in studied material [53]. The energy position of the steps is corresponded to the ion beams scattering energy on the nuclear of material atoms. Computer approximation of the

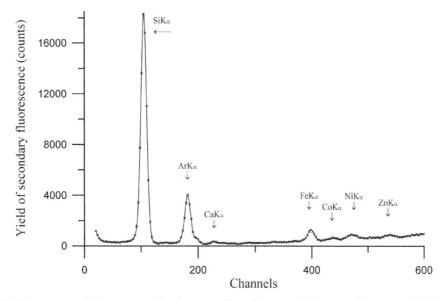

FIGURE 16.30 TXRF spectrum of Si monocrystal substrate collected by use of the original compact TXRF cell built on the basis of the modified PXWR. It was used BSW-24 (Mo) X-ray source in the regime $U = 25$ keV, $I = 10$ mA during 1,000 s. Channel step 16 eV/channel.

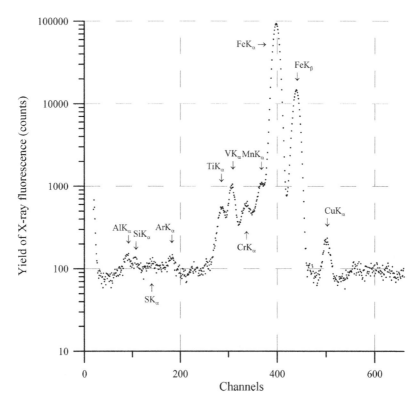

FIGURE 16.31 TXRF spectrum of Fe (200 nm)/Si polycrystal thin film. Regime of X-ray source BSW-24 (Mo) $U = 25$ keV, $I = 10$ mA. Registration time 500 s. Channel step 16 eV/channel. Spectrum collected by Si detector with Be window 12 μm.

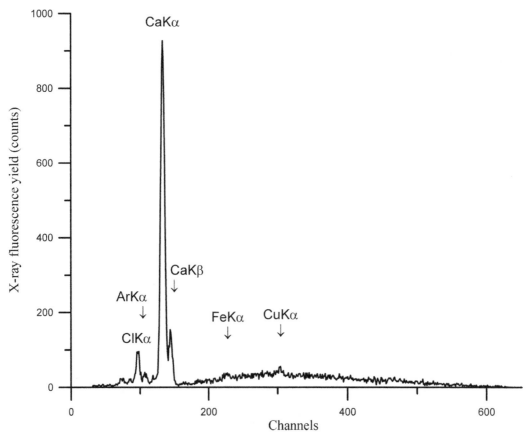

FIGURE 16.32 TXRF spectrum of Dutch cheese sample collected by use of the original TXRF cell built on the basis of modified waveguide-resonance structure and allowed to introduce studied sample into the waveguide-resonance slit clearance. Measurements were carried out with BSW-22 (Mo) radiation source in regime $U = 40$ keV, $I = 3$ mA. Energy step 28 eV/channel.

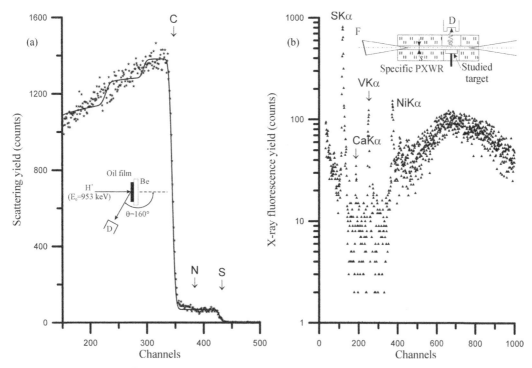

FIGURE 16.33 RBS spectrum of H^+ ion beam ($E_0 = 0.953$ MeV) (a) and TXRF spectrum collected with use the original TXRF cell (b) for the petroleum film deposited on Be substrate. Measurements geometries are shown on insertions. Energy step for RBS spectrum – 1.9 keV/channel, for TXRF spectrum – 20 eV/channel. Petroleum admixture concentration related to S:

S	Na	Cl	Ca	V	Fe	Ni	La
1	0.005	0.005	0.005	0.015	0.001	0.005	0.0005

spectrum allowed to define the sample element composition, which was $C_{0.87}H_{0.1}S_{0.028}N_{0.002}$ for the studied petroleum. Hydrogen content was defined by method of the nuclear recoil [57]. TXRF measurements presented possibilities to display the admixture petroleum composition. For similar investigations, $SK\alpha$ line can offer as the internal standard. TXRF spectrum registered $NaK\alpha$, $ClK\alpha$, $CaK\alpha$, $VK\alpha$, $FeK\alpha$, $NiK\alpha$ and $LaK\alpha$ set of lines. These element contents normalized to the sulfur concentration are presented on the Figure caption. Our petroleum investigations served as the basis for conclusion that the heavy petroleum can be used as real source for the vanadium industrial production and for the experimental extraction of rare-earth elements.

TXRF cell efficiency discussion elaborated on the basis of the modified waveguide-resonance structure is impossible without of the element detection limits evaluation provided by the cell. This evaluation was executed in process of the water study selected from Saint Petersburg water pipe. Figure 16.34 shows TXRF spectrum collected for sample of the water drop dry residue deposited on the quartz substrate. In the measurements, we used the Fe-containing solution with 1.8 mg/L of atom concentration as the internal standard. The detection limit for the iron atoms was evaluated on the basis of the specific expression for this parameter [22]:

$$m_{LD} = \frac{3m_i}{\sqrt{KN\tau}} \qquad (16.11)$$

where m_{LD} is the detection limit of atoms in the mass form, m_i is the total mass of analyzed atoms, N is the characteristical line yield intensity (counts in sec), τ is the measurement period and K is the analytical line contrast parameter defined by the expression:

$$K = \frac{I_i - I_b}{I_b} \qquad (16.12)$$

where I_i is the analytical line total yield and I_b is the background total yield in the analytical line area. The spectrum treatment allowed to calculate the detection limit for iron atoms as 4×10^{-13} g. Ca atoms with concentration 22 mg/L is the main admixture for Saint Petersburg water pipe. Experimental data reported in the work shows that our TXRF cell built on the basis of the supermodified waveguide-resonance structure is the beautiful analytical construction. TXRF measurement's efficiency can be raised by directly subtracting one TXRF spectrum from another. Figure 16.35 presents three TXRF spectra collected for pure substrate excitation by $MoK\alpha$ radiation flux in the compact cell (a), for the dry residue solution sample (b) and result in its subtraction (c). Spectrum of the pure glassy carbon substrate shows that it is characterized by some contaminations existence. It is S, Ca, Fe and Zn atoms. Spectrum of the dry residue solution sample demonstrates additional lines $CrK\alpha$ and $CrK\beta$. Both experimental spectra contain $ArK\alpha$ line. Because both spectra were collected at

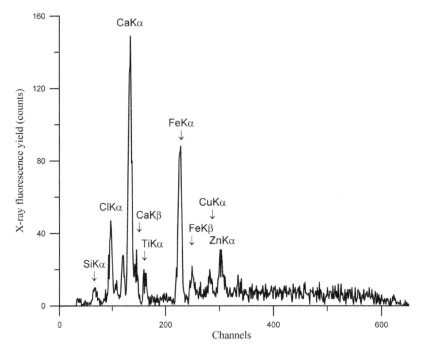

FIGURE 16.34   TXRF spectrum of dry residue water drop extracted from Saint Petersburg water pipe collected by using the original TXRF cell. Measurements were carried out with BSW-22 (Ag) radiation source in regime $U = 40$ keV, $I = 4$ mA, $\tau = 1,000$ s.

the same X-ray source function regimes and collection exposition times, one can expect that the direct subtraction of these spectra will be correct if the ArKα line position will show the background middle intensity. The subtractive spectrum shows presents of SKα, CaKα, CrKα, CrKβ, NiKα and ZnKα. Analysis of the line set allows to propose that SKα, CaKα and ZnKα lines are connected with substrate contamination. But the NiKα line reflects the nickel atom's presence in the initial solution. The spectrum quality is not enough for the quantitative evaluation, but the direct subtraction approach to visualize some lines concealed by background.

TXRF investigations can be having direct practical tendency. In frame of our experimental activity, we studied the low-temperature plasma influence on surface of different fabrics and leather wares. Figure 16.36 shows experimental and theoretical Rutherford backscattering spectra of H^{+} ion beam ($E_0 = 1.3$ MeV) on the leather sample pre (a) and post (b) the plasma treatment. Spectra do not show the visual variation of the host element concentration. At the same time, TXRF measurements shown on Figure 16.37 demonstrate some changes in the contamination composition. It is apparent that the plasma treatment is removed from the leather surface of sulfur atoms and vanadium. Concentration of potassium atom in the surface layer diminished. Cr atom's content in the layer did not change. In the practical sense, the low-temperature plasma treatment led to appearing of the surface glitter and leather ware strength growth. TXRF and RBS testing of leather wares allowed to find the optimal conditions for the low-temperature plasma treatment of fabric and leather wares for achievement of its best consumer properties.

16.7   X-ray Fluorescence Analysis of Gas Mixture

Conventional TXRF spectrometers usually can be used for gas pollution investigations by analyzing the filters used for the absorption of gaseous impurities [58]. This procedure is undirect measurement and is coupled with analytical complications. At the same time, PXWR application for the gas composition diagnostics can ensure the direct measurement of gas component concentrations with a high productivity and an operating convenience. Figure 16.38 presented a schematic design of the XRF spectrometer model for a gas composition analysis in the flowing regime. The model contains X-ray source, a waveguide-resonance radiation concentrator, a microvolume for activation of the secondary X-ray yield, X-ray detection equipment and the collection system for XRF spectra registration. Elaborated in our laboratory model of the spectrometer has the slit with width $s = 43$ nm and height $h = 12$ mm. Both reflectors formed by the slit have round holes opposite to each other. Between holes is the microvolume ($v = 2.6 \times 10^{-6}$ cm^3) for the flowing gas activation. Input and output gas volumes are near holes and restrict other hands by Be windows. For the light element analysis, special windows must be applied. The gas flow across the waveguide-resonator slit is characterized by a small value ($< 1 \times 10^{-2}\%$) about the main stream.

Experimental XRF spectrum of our laboratory gas atmosphere is represented on Figure 16.39. Spectrum collects by using of X-ray tube (Mo) at $U = 25$ keV and $I = 10$ mA in during $\tau = 1,500$ s. Presence of conventional Be windows with thickness $t = 20$ mkm strikes off the appearance of lines

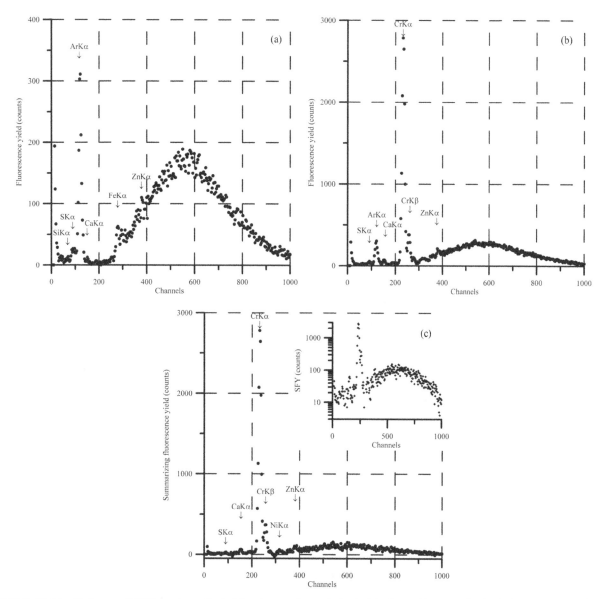

FIGURE 16.35 Patterns of XRF spectra collected for clean carbon substrate (a), dry residue sample of solution containing Cr atoms in concentration 10 μkg/mL (b) and spectrum of the direct subtraction result (c). Spectra were collected in identical conditions. Channel step 23 eV/channel. Resulting spectrum in the logarithmic scale is shown on insertion.

suited to oxygen in nitrogen atoms. The spectrum demonstrates ArK_α line as the main spectral deposit. It is well known that the argon atom's concentration in air is nearly 1% at. So, these atoms can be used as an internal standard. Intensity of an argon peak presented in the spectrum conforms to total quantity of Ar atoms in the activating volume on level 1.1×10^{11} at.

The spectrum analysis showed that most of the peaks did not appertain to gas impurities. It is more likely that the magnesium line reflects the presence of these atoms in the window material. Peak MoL_α is own line of the exciting radiation. Au line's existence can reflect the presence of these atoms on contacts of X-ray semiconductor detector. Presence of peaks connected with Pb atoms can be resulted by a lead collimator installation between tube and PXWR. Elements presented in atmosphere of our laboratory are Cl,

S, Fe and Zn. Contents of Fe and Zn atoms in the exciting volume do not enhance 2×10^8 each. S and Cl atoms' quantities determination are not possible on the basis of the spectrum.

It is interesting to evaluate detection limits for diagnostics of Cl and S atoms as the most prevalent pollutants of the atmosphere. At first, it is needed to notice that the molybdenum radiation is not best activator for the analysis of gas mixtures. The position of MoL_α line is near the position of SK_α one. Moreover, the excitation mode is higher as the energy of exciting radiation nearer to energy of the impurity atomic level is activated. Analysis of the spectrum shown on Figure 16.39 allows to evaluate that the detection limits for S and Cl at exciting the radiation generated by tube with Mo anode are nearly 0.05% at. Application of a more soft radiation (for example CrK_α)

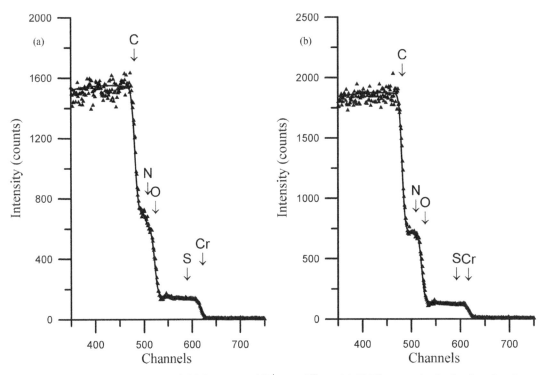

FIGURE 16.36 Experimental and theoretical RBS spectra of H^+ ions ($E_0 = 1.3$ MeV) scattering by leather-dressing sample before (a) and after (b) low temperature plasma treatment. Channel step 1.9 keV/channel.

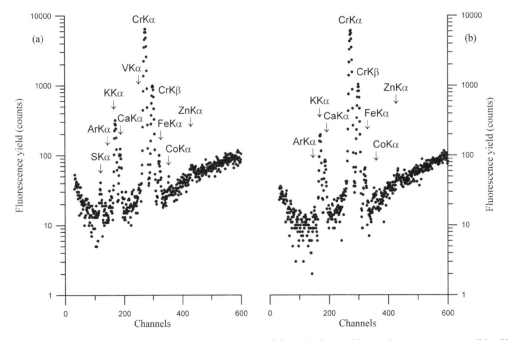

FIGURE 16.37 TXRF spectra of leather-dressing sample before (a) and after cold-ion plasma treatment (b). Channel step 20 eV/channel.

will decrease these detection limits up to 0.001 at.% or better.

This gas analyzer can have some applications in medical practice. The state of health of patients is usually fixed on the basis of instrumental data on urine or blood. The XRF gas analyzer will allow instrumental multielement control of the air exhaled by the patient.

16.8   TXRF Analysis Perspectives

PXWR application for TXRF spectrometry allowed to increase the analysis sensitivity. Elaboration of the compact TXRF cell built on the basis of the specific design waveguide-resonance structure raises the work convenience. The gas analyzer building on the basis of the

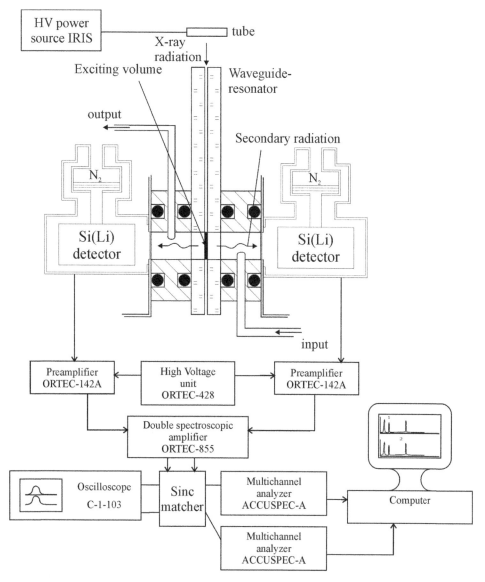

FIGURE 16.38 Schematic design of the XRF gas analyzer equipped by PXWR. The analyzer can provide the flow-type function.

waveguide-resonance unit is the beautiful illustration of XRF analysis development by this device application. But it realizes only part of potential possibilities for the TXRF and gas analysis efficiency increasing embedded in the waveguide-resonance idea.

It is known that the conventional scheme of TXRF measurements can be oriented on the sole exciting beam source, only. The waveguide-resonance formation of X-ray exciting beam for TXRF changes this stereotype. Figure 16.40 demonstrates the specific scheme of X-ray waveguide-resonance structure allowing to work with set from four independent X-ray sources. The structure is built in the form of two quartz disk reflectors with holes for the studied sample placing and for the XRF yield bringing out. For the waveguide-resonance channels creation, the figured deposition of titanium film on one reflector surface is executed. Clean and covered disks' superposition creates the waveguide-resonance cell orienting on the use of set

of independent radiation sources. It is clear that every channel of the waveguide-resonance structure can be modified according to the scheme described early. Realization of the similar constructions can be considered as the next step of the TXRF spectrometry development.

16.9 TXRF Spectrometry in Conditions of Ion Beam Excitation

The conventional TXRF spectrometric X-ray optical scheme is designed for the sliding incidence of X-ray exciting beam on the studied sample surface. At the same time, the alternative approach is possible to TXRF measurements at the perpendicular incidence of X-ray exciting beam on the surface. Figure 16.41 presents principal schemes for the TXRF material diagnostics execution by conventional and

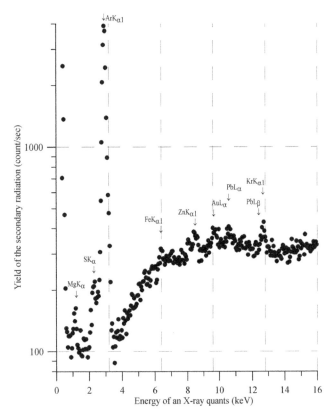

FIGURE 16.39 XRF spectrum of atmosphere in laboratory collected with PXWR using BSW-27 (Mo) tube regime $U = 25$ keV, $I = 10$ mA. Exposition $\tau = 1,500$ s. Channel step 16 eV. Every third channel is shown.

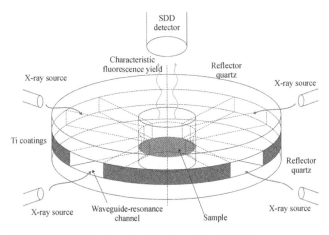

FIGURE 16.40 Principal design of the TXRF spectrometric cell built on the basis of disk waveguide-resonator oriented on using the set of X-ray exciting sources for TXRF investigation of different samples. Sources quantity for such structures is not limited.

alternative manners. The TXRF alternative spectrometry is characterized by specific features connected with excitation of a large material volume. As a result, the method has the matrix effect problem but demonstrates the increasing value of X-ray characteristical fluorescence yield. TXRF spectrum can be selected by the slit-cut systems (Figure 16.41b) or application of the waveguide-resonance selector. But

the best effect for TXRF alternative spectrometry will be obtained in case of the scheme realization presented on Figure 16.41c. In this scheme, the waveguide-resonance configuration appears in the result of narrow extended slit clearance creation between the studied target surface and the Be polished reflector one. Because Be is transparent to the hard X-ray radiation, it excites the target characteristical fluorescence with high efficiency. The waveguide-resonance slit clearance selects from the total XRF yield the specific part of characteristic radiation emitted by the thin surface layer of the studied target, only. This experimental scheme possesses very-high X-ray gathering power, but it is unsuitable for the TXRF element quantitative analysis. At the same time, similar scheme can be used for TXRF measurements in conditions of the PIXE spectrometry [59].

It is well known that the XRF spectrometry using X-ray beam excitation has shown very little promise for the light element diagnostics in materials (Figure 16.9). At the same time, the PIXE spectrometry is very effective for the analysis of these elements. Figure 16.42 presents experimental data for the excitation cross section of Kα XRF yield on the hydrogen ion beam energy for the different chemical elements [60]. Comparison of these data with the excitation cross-section data presented on Figure 16.9 shows that AlKα and YKα fluorescence excitation by the MoK$\alpha\beta$ radiation beam and the hydrogen ion beam with energy $E_0 = 1$ MeV leads to distinction on six orders! But for all its attractiveness, the PIXE spectrometry is not quantitative method of the element diagnostics. Its possibilities are limited by the matrix effect and the factor connected with the absorption coefficient difference for X-ray radiation with different wavelengths. The matrix effect compensation presence is not possible, today. At the same time, there are still rooms for neutralization of the line's absorption difference influence on XRF yield intensity. Figure 16.43 presents the experimental scheme for XRF yield bringing out excited by the high-energy ion beam in conditions of the specific waveguide-resonance structure use. This scheme allows to register the XRF yield corresponding to thin surface layer of the studied target [57]. Hole in the center of polished beryllium reflector was designed for the exciting ion beam penetration to the studied sample. The experimental unit presented on the Figure was positioned on goniometer of the Sokol-3 ion beam analytical complex chamber [53]. Experiments were executed with use of the H$^+$ ion beam with energy 1.25 MeV. The old copper coin with silver coating was the subject of our measurements. The silver coating thickness 170 nm was defined from the Rutherford backscattering spectrum presented on Figure 16.44. The approximation showed that the silver coating has the thin oxidizing layer with thickness 10 nm. Surface carbon film appeared during measurements. RBS spectrometry is the absolute diagnostical method, but it is characterized by very low sensitivity. This methodical shortage is beautifully compensated by parallel PIXE measurements. The coin XRF spectrum obtained in the conventional PIXE geometry is presented on Figure 16.45a. The spectrum characterizes element composition of the coin

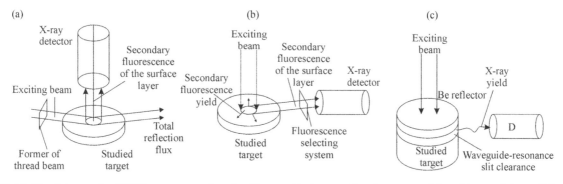

FIGURE 16.41 Schemes for TXRF measurements using X-ray optical conventional geometry (a) and in conditions of X-ray source and detector positions inversion by using the conventional surface fluorescence yield collection (b) and its case of the specific PXWR application (c).

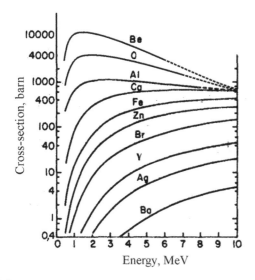

FIGURE 16.42 Functions of XRF excitation cross section for Kα line's different atoms on the proton beam energy in PIXE experiments [60].

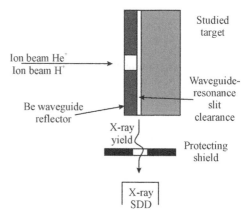

FIGURE 16.43 Experimental scheme for TXRF measurements in which X-ray characteristic excitation yield is initiated by high-energy proton beam. Scheme is realized by creation of the waveguide-resonance channel between Be polished reflector and studied surface [53].

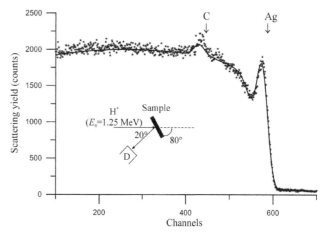

FIGURE 16.44 Experimental and theoretical RBS spectra of H^+ ion beam ($E_0 = 1.25$ MeV) for old copper coin coated by Ag coating. Oxygen and carbon atoms are presented in thin surface film of the coating. Energy step 1.9 keV/channel.

as the spectral ensemble featuring of its coating and volume. The waveguide-resonator application allows to select the XRF spectrum corresponding to the coin coating, only. This spectrum is presented on Figure 16.45b. As a result, one can conclude that the silver coating contains Mg, Al and Fe atoms. The coin volume material contains in addition to copper, Si, Cl, S, Fe and Ni atoms. Owing to the matrix effect existence, the quantitative determination of the admixtures' concentration has some difficulties. The total and selected PIXE spectra comparison demonstrates a visible decreasing of the background deposit in case of the waveguide-resonance application. As a whole, one can establish that the waveguide-resonance structure application for PIXE measurements creates new experimental method for the surface element diagnostics as the TXRF and PIXE hybrid – TXRF PIXE.

Similar investigations were carried out for study of soot element composition formed as a result of jet engine function. Figure 16.46 presents experimental and theoretical RBS spectra of helium and hydrogen ions collected for the soot sample deposited on Be

substrate. Spectra approximation allowed to define the element composition of the soot as $C_{0.43}Al_{0.07}Fe_{0.002}$ $O_{0.25}Cr_{0.004}K_{0.054}Mg_{0.07}B_{0.07}Cl_{0.05}$. Important factor of RBS measurement with H^+ ions is the real diagnostics

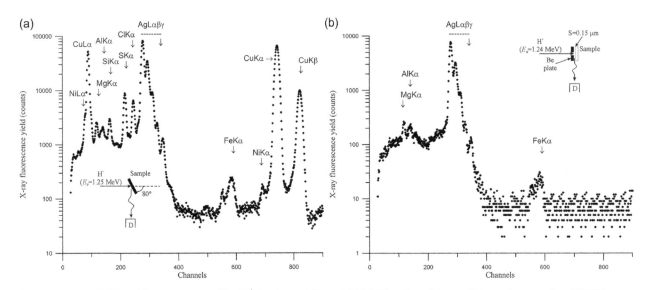

FIGURE 16.45   XRF yield spectra excited by H^+ ion beam ($E_0 = 1.24$ MeV) collected in conditions of conventional PIXE measurements (a) and in the geometry of TXRF material diagnostics (b). Energy step 10 eV/channel.

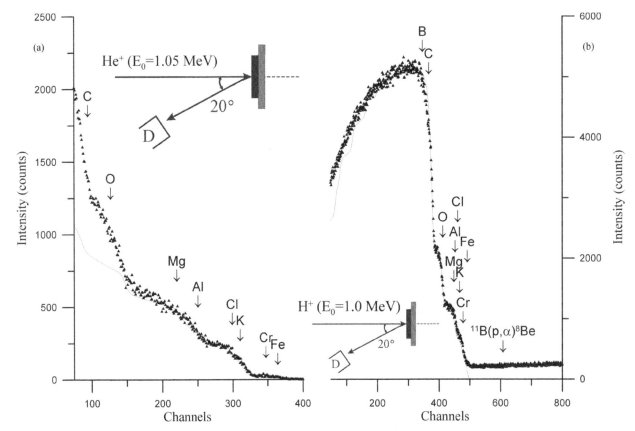

FIGURE 16.46   Experimental and theoretical RBS spectra of He^+ (a) and H^+ ions (b) on soot sample formed as a result of jet engine function deposited on Be substrate.

of B atoms availability in the studied object. Spectrum of helium ions scattering supplies better diagnostics of more heavy atoms. At the same time, detection limits of RBS spectrometry are restricted by magnitude near 0.1% at. So, small admixtures can be detected by TXRF or TXRF-PIXE methods. Figure 16.47 presents these spectra. It shows XRF lines of elements displayed by RBS method and addition lines connected with small admixture atoms.

It is very interesting to compare intensity of lines in TXRF and TXRF PIXE spectra. In any case, the spectra present very useful information about parameters of jet engine function and degree of the fuel combustion.

TXRF PIXE methods is characterized by very small magnitude of a background deposit in comparison with PIXE in its conventional realization. Figure 16.48 shows XRF spectra collected in the conventional PIXE geometry

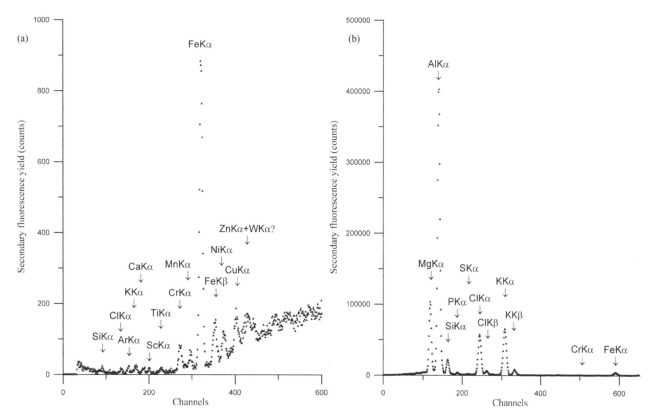

FIGURE 16.47 TXRF (a) and TXRF PIXE spectra (b) collected for the soot sample formed as a result of jet emgine function deposited on Be substrate. Channel step 20 (a) and 10 (b) eV/channel.

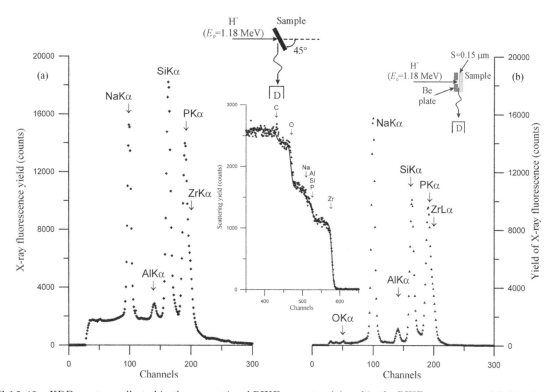

FIGURE 16.48 XRF spectra, collected in the conventional PIXE geometry (a) and in the PIXE geometry modified by the waveguide-resonator application (b). Insets show geometries of the fluorescence spectra collection and RBS H^+ ($E_0 = 1.18$ MeV) ions spectrum. Arrows on fluorescence spectra point energy positions of X-ray characteristical lines and on RBS spectrum of ion-scattering energies on nuclear target surface atoms. Energy step for XRF spectra 10.5 eV/channel, for RBS – 1.9 keV/channel.

and in conditions of TXRF PIXE one for the nature composite compound $Na_3Zr_{1.3}Si_{1.9}Al_{0.1}P_{1.0}O_{12}C_2$. RBS spectrum of the sample is shown on insertion. TXRF PIXE spectrum is characterized the element composition on thin surface layer. As a result, the background deposition is very small. Fluorescence line's intensity relations in these spectra are differed. It connects with different influence of matrix factor on thin surface layer and on material volume forming XRF yield. Owing to the matrix influence existence, TXRF PIXE is the semi-quantitative procedure as the conventional PIXE one, but the waveguide-resonator application increases the method sensitiveness.

16.10   Conclusion

The work presents short characteristic of TXRF and PIXE experimental methods elaborated for the precision material element's diagnostics and its efficiency increasing as a result of the X-ray waveguide-resonance structures including into its experimental facility. There a specific discussion about physics and practical application of the waveguide-resonance structures oriented on the MoKαβ radiation beams is presented.

Acknowledgments

The authors wish to thank Dr. M.S. Afanasiev for help in preparation of PXWR, CPXWR and TXRF cells and prof. Jun Kawai for the interest to works connected with the waveguide-resonance idea. This work was carried out with partial financial support of Russian Fundamental Investigations Foundation (grant in frame of #19-07-00271) and State task #075-00475-19-00. This paper also received partial financial support from the Ministry of Education and Science of the Russian Federation on the program to improve the competitiveness of Peoples' Friendship University of Russia (RUDN University) among the world's leading research and education centers in the 2016–2020.

References

1. H.P. Klug, L.E. Alexander. *X-ray Diffraction Procedures*. New York: Wiley. 1974. 966 p.

2. A. Authier. *Dynamical Theory of X-ray Diffraction*. Oxford: Oxford University Press. 2001. 661 p.

3. E.P. Bertin. *Principles and Practice of X-ray Spectrometric Analysis*. New York: Plenum Press. 1975. 1079 p.

4. *Handbook of X-ray Spectrometry*, Eds. by R. Van Grieken, A.A. Markowicz, 2nd Ed., 2009. 666 p.

5. S.A.E. Johanson, J.L. Campbell, K.G. Malquist. *Principles Induced X-ray Emission Spectrometry (PIXE)*. New York: Wiley. 1995. 451 p.

6. *Quantitative Electron-Probe Microanalysis*, Eds. by V.D. Scott, G. Love. New York: Wiley. 1983. 321 p.

7. R. Jenkins, R.W. Gould, D. Dedcke. *Quantitative X-ray Spectrometry*, 2nd Ed. New York: Dekker. 1995. 484 p.

8. *X-ray Spectroscopy*, Ed. by S.K. Sharma. Croatia: InTech. 2012. 280 p. (ISBN 978-953-307-967-7).

9. Y. Yoneda, T. Horiuchi. Optical flats for use in X-ray spectrochemical microanalysis. *Review of Scientific Instruments*. 42 (1971) 1069–1070.

10. V.K. Egorov, A.P. Zuev, E.V. Egorov. Scintillation response of monocrystal $PbWO_4$ to random and channeled ions. *NIM*. B119 (1996) 418–424.

11. R. Klockenkamper. *Total Reflection X-ray Fluorescence Analysis*. New York: Wiley. 1997. 245 p.

12. H. Aiginger. History development and principles of total reflection X-ray fluorescence analysis (TXRF). *Spectrochimica Acta*. 46B (1991) 1313–1321.

13. P. Wobrauschek. Total reflection X-ray fluorescence analysis (Review). *X-ray Spectrometry*. 36 (2001) 289–300.

14. A. Von Bohlen. Total reflection X-ray fluorescence and grazing incidence X-ray spectrometry – tools for micro- and surface analysis (review). *Spectrochimica Acta*. B64(10) (2009) 821–832.

15. C. Streli, P. Wobrauschek, F. Meirer, G. Perroni. Synchrotron radiation induced TXRF. *Journal of Analytical Atomic Spectrometry*. 23 (2008) 792–798.

16. *X-ray Spectrometry: Recent Technological Advanced*, Eds. by K. Tsuji, J. Injuk, R. Van Grieken. Chichester: Wiley. 2004. 517 p.

17. R. Klockenkamper, A. Von Bohlen. *Total Reflection X-ray Fluorescence Analysis and Related Methods*. New York: Wiley. 2015. 519 p.

18. J.H. Underwood. Glancing incidence optics in astronomy. *Space Science Instrumentation*. V1 (1975) 289–299.

19. M. Blochin. *Physik Runtgenstrahlen*. Berlin: Verlag der Technic. 1957. 535 p.

20. L.D. Landau, E.M. Lifshits. *Electrodynamics of Continuous Medium*. Reading, MA: Addison-Wesley. 1965. 586 p.

21. A. Prange, H. Schwenke. Trace element analysis using total reflection X-ray fluorescence spectrometry. *Advances in X-ray Analysis*. 35 (1992) 899–924.

22. I.F. Losev. *Quantitative X-ray Spectral Fluorescence Analysis*. Moscow: Nauka. 1969. 336 p. (In Russian).

23. *Rontgenfluoreszenzanalyse, Anwendung in Betriebslaboratorien*, Ed. by H. Ehrhard. Leipzig: VEB Deutscher Verlag fur Grundstoffindustrie. 1981. 261 p.

24. X-123SDD Complete X-Ray Spectrometer with SDD. https://www.amptek.com/products/sdd-x-ray-detectors-for-xrf-eds/x-123sdd-complete-x-ray-spectrometer-with-silicon-drift-detector-sdd

25. V.K. Egorov, A.P. Zuev, E.V. Egorov. X-ray fluorescence analysis of ultralow pollution concentration in condition of X-ray beam total external reflection formed by slitless collimator. *Zavodtskaja Laboratoria.* 67(3) (2001) 3–11 (In Russian).

26. K. Yakushiji, S. Ohkawa, A. Yoshinaga, J. Hagara. Origin of spurious peaks of total reflection X-ray fluorescence analysis of Si wafers excited by monochromatic X-ray beam WLβ. *Japanese Journal of Applied Physics.* 33(2 part 1) (1994) 1130–1135.

27. V.K. Egorov, E.V. Egorov. The experimental background and the model description for the waveguide-resonance propagation of X-ray radiation through a planar narrow extended slit (Review). *Spectrochimica Acta.* B59 (2004) 1049–1069.

28. V.K. Egorov, E.V. Egorov. Peculiarities of the planar waveguide-resonator application for TXRF spectrometry (Review). *Advances in X-ray Chemical Analysis (Japan).* 44 (2013) 21–40.

29. V.K. Egorov, A.P. Zuev, E.V. Egorov. Scintillation response of monocrystal $PbWO_4$ to random and channeling ions. *NIM.* B119 (1996) 418–424.

30. V.K. Egorov, E.V. Egorov. Discovery of a planar waveguide for X-ray radiation. The Los Alamos e-print archive: https://arxiv.org/abs/physics/0101059.

31. T. Mingazin, V. Zelenov, V. Lejkin. Slitless collimator for X-ray beams. *Instruments and Experimental Techniques.* 24(1 part 2) (1981) 244–247.

32. V. Lejkin, T. Mingazin, V. Zelenov. Collimating device for X-ray radiation. *Pribori i Technika Ekxperimenta.* 24(3) (1981) 208–211. (In Russian).

33. V. Lejkin, T. Mingazin, V. Zelenov. X-ray beam forming by using of slitless collimator. *Instruments and Experimental Techniques.* 27(6 part 1) (1984) 1333–1336.

34. M. Kumakhov, F. Komarov. Multiple reflection from surface X-ray optics. *Physics Reports.* 191 (1990) 289–352.

35. M. Bedzyk, G. Bommarito, J. Schildkraut. X-ray standing waves at a reflecting mirror surface. *Physical Review Letters.* 69 (1989) 1376–1379.

36. V.K. Egorov, E.V. Egorov. Waveguide-resonance mechanism for X-ray beam propagation: physics and experimental background. *Advances in X-ray Analysis.* 46 (2003) 307–315.

37. M. Born, E. Wolf. *Principle of Optics Electromagnetic Theory of Propagation of Interference and Diffraction of Light.* Oxford: Pergamon Press. 1993. 808 p.

38. L. Mondel, E. Wolf. *Optical Coherence and Quantum Optics.* Cambridge: Cambridge University Press. 1995. 837 p.

39. V.K. Egorov, E.V. Egorov. Background of X-ray nanophotonics based on the planar air waveguide-resonator. *X-ray Spectrometry.* 36 (2007) 381–397.

40. V.K. Egorov, E.V. Egorov. Physics of planar X-ray waveguide. *Proceedings of SPIE.* 4502 (2001) 148–172.

41. V.K. Egorov, E.V. Egorov. Planar waveguide-resonator: a new device for X-ray optics. *X-ray Spectrometry.* 33 (2004) 360–371.

42. L.I. Mirkin. *Handbook of X-ray Analysis of Polycrystal Analysis.* New York: Consultants Bureau. 1964. 920 p.

43. V.K. Egorov, E.V. Egorov, E.M. Loukianchenko. High effective TXRF spectrometry with waveguide-resonance devices applications. *Aspects in Mining & Mineral Science.* 2(4) (2018) 1–23.

44. V.K. Egorov, E.V. Egorov. Planar X-ray waveguide-resonator features. *Trends in Applied Spectroscopy.* 8 (2010) 67–83.

45. V.K. Egorov, E.V. Egorov. The compact TXRF cell on base of the planar X-ray waveguide-resonator. *Advances in X-ray Chemical Analysis (Japan).* 43 (2012) 139–146.

46. E. Lukianchenko, V. Egorov, V. Rudenko, E. Egorov. TXRF spectrometer on base of the waveguide-resonator with specific design. *Journal of Physics: Conference Series.* 729 (2016) 012028.

47. V.K. Egorov, E.V. Egorov. Implementation of the light gathering power enhancement of waveguide-resonator. *Surface Investigation.* 3(1) (2009) 41–47.

48. J.H. Underwood. Glancing incidence optics in X-ray astronomy: a short review. *Space Science Instrumentation.* 1 (1975) 289–304.

49. V.K. Egorov, E.V. Egorov. Composite X-ray waveguide-resonator as a background for a new generation of material testing equipment for films on Si substrates. *Proceedings of MRS.* 716 (2002) 189–195.

50. H.J.W. Muller-Kirsten. *Basics of Statistical Physics.* Singapore: World Scientific. 2013. 248 p.

51. V.K. Egorov, E.V. Egorov. Peculiarities in the formation of X-ray fluxes by waveguide-resonators of different construction. *Optics and Spectroscopy.* 126(6) (2018) 838–849.

52. V.K. Egorov, E.M. Loukianchenko, V.N. Rudenko, E.V. Egorov. The device for X-ray fluorescence analysis of materials equipped by the planar X-ray waveguide-resonator for formation of radiation exciting flux. Russian patent #2555191 with priority from 24 April 2014.

53. V.K. Egorov, E.V. Egorov. Ion beam for material analysis: conventional and advanced approaches. In *Ion Beam Application.* London: InTechOpen. 2018. pp. 37–71. doi: 10.5772/intechopen.76297.

54. V.K. Egorov, A.P. Zuev, B.A. Maljukov. Surface contamination diagnostics of silicon wafers by total

reflection X-ray fluorescence spectrometry. *Izvestija VUZov, Tsvetnaya Metallurgija.* 5 (1997) 54–69. (In Russian).

55. J. Wolstenholme. *Auge Electron Spectroscopy.* New York: Moment press, LLC. 2015. 256 p.

56. S.A.E. Johanson, J.L. Campbell, K.G. Malquist. *Particle Induced X-ray Emission Spectrometry (PIXE).* New York: Wiley. 1995. 451 p.

57. H. Hofsas. *Forward Recoil Spectrometry.* New York: Plenum. 1996. 278 p.

58. L.R. Doolitle. Algorithm for the rapid simulation of Rutherford backscattering spectra. *NIM.* B9 (1985) 344–351.

59. V.K. Egorov, E.V. Egorov, M.S. Afanas'ev. TXRF spectrometry at ion beam excitation. *Journal of Physics*: *Conference Series.* 808 (2017) 012002.

60. T.A. Cahill. Proton microprobs and particle induced X-ray analytical system. *Annual Review of Nuclear and Particle Science.* 30 (1980) 211–256.

17

Nanoscale XPEEM Spectromicroscopy

C. A. F. Vaz, Armin Kleibert,
and Mario El Kazzi
Paul Scherrer Institute

The development of new characterization techniques with enhanced sensitivities, spatial resolutions down to the atomic scale, and temporal resolutions down to the femtosecond regime have been instrumental to the understanding of the physical processes governing the properties of materials and to the discovery of new physical phenomena with potential for applications. In particular, scattering, spectroscopy, and imaging techniques based on *synchrotron X-ray light* have contributed immensely to such progress by taking full advantage of intense, monochromatic, and polarized X-rays available over a large photon energy range, typically from 50 eV to 100 keV [106,186]. These energies cover the typical atomic-binding energies of core electrons, such that, by tuning the X-ray energy, one can probe specific atomic transitions of individual elements of a compound. In X-ray scattering processes, the photon interacts with a core electron via the electric-dipole interaction and excites the electron into a higher energy state, with a transition probability that, according to Fermi's golden rule, depends on the matrix element of the electric-dipole operator between the initial and all other available states, where the atomic states (including the excited states) reflect strongly the local crystal field created by the surrounding atoms. It is this sensitivity to the electronic environment and to the particular electronic state that is at the base of X-ray spectroscopy, such as X-ray absorption (XAS) [37,38,213] and X-ray photoemission spectroscopy (XPS) [224], among others [22,153,183].

When combined with high-resolution imaging, spectroscopy can provide unique insights into nanoscale materials, including composite materials and nanodevice structures. For example, the physical arrangement of different species in a composite material can be identified or local changes in the electronic state of a device structure can be studied through changes in the spectral signal. In addition, properties that depend on the X-ray light polarization (circular and linear dichroic effects) can be explored to obtain maps of that very material property, such as magnetic configuration, molecular orientation, and ferroelectric domains [245]. All these aspects are widely utilized in X-ray microscopy: both the capability of local spectroscopy and the ability to obtain spatial maps that reveal nonuniform distributions of a sample property, be it the elemental composition of composite materials or the nonuniform distribution of the magnetization in magnetic materials.

In general terms, we may divide imaging techniques in two classes, scanning or full field. In scanning techniques, the incoming beam is focused to a small spot size and raster-scanned through the sample to obtain local spectroscopic information [144], as in scanning transmission X-ray microscopy (STXM) [97,169,212,232], scanning photoemission microscopy (SPEM) [3,5,68], and ptychography [42,143,199,234,252], the latter a more recent scattering-based imaging technique that relies on phase-recovery algorithms to retrieve the phase information and the full real-space information about the scatterer (sample). In full-field imaging, the X-ray beam illuminates uniformly the sample and either the transmitted X-ray intensity is directly imaged, as in transmission X-ray microscopy (TXM) [50,97] and X-ray holography [24,46,219], or the photoemitted electron intensity, as in X-ray photoemission electron microscopy (XPEEM) [6,17,18,29,30,35,68,107,109, 123,177,212,214].

In XPEEM, the local intensity of photoelectrons emitted from the sample surface upon X-ray illumination is imaged in full field using high-resolution electron optics [18]. The relatively short escape depth of the photoemitted electrons makes photoemission-based techniques highly surface sensitive, while the high yield of secondary electrons per impinging photon leads to sub-monolayer sensitivity and short acquisition times (high throughput). XPEEM further combines several measurement modes in one instrument (including XAS, XPS, and angle-resolved photoemission spectroscopy, ARPES); it is often also possible to use other excitation sources in the same microscope, such as electrons and UV light, for low-energy electron microscopy (LEEM) and photoemission electron microscopy (PEEM), respectively [16]. This combination of capabilities makes XPEEM a technique of choice for local spectroscopy measurements and imaging in a wide variety of systems. In this chapter, we focus on the ability of XPEEM to provide local spectroscopic information and therefore of probing the electronic structure of materials down to the nm scale. Our goal here is to provide a brief description of the working principles of XPEEM and to illustrate with examples from the literature the gamut of possibilities for studying materials systems in areas ranging from physics, chemistry, biology, and materials science, with an emphasis on local, nanoscale spectromicroscopy. XPEEM has also been extensively used to determine the magnetic domain structure of magnetic materials and magnetic elements using the X-ray magnetic circular and linear dichroism (XMCD and XMLD) effects as contrast mechanisms, a topic that is well covered in the literature [20,82,123,153,189,192,195,213] and which will not be discussed in detail here.

17.1 Introduction to XPEEM

XPEEM relies on the imaging of electrons emitted from the sample upon excitation with X-rays (photoemission). The atomic de-excitation process following photon absorption includes the emission of primary photoelectrons and Auger electrons; these in turn can undergo inelastic electron–electron and electron–plasmon interactions and quasi-elastic electron–phonon interactions to generate a cascade of secondary electrons [211], with kinetic energies in the range from 0 to more than 10 eV. Depending on which type of electrons are probed, different types of measurements are possible in XPEEM. In the following, we will consider the three fundamental components to a PEEM instrument: (i) the light source, (ii) the microscope itself, and (iii) the sample.

17.1.1 Synchrotron X-ray Light Sources

Historically, the development of LEEM/PEEM dates back to the early 1930s, but XPEEM became only possible with the advent of modern electron storage rings providing intense and tunable synchrotron radiation [18]. The latter is produced by forcing a relativistic electron beam (with kinetic energies in the GeV range) to orbit under a magnetic field, whose amplitude and spatial configuration determine the peak energy and polarization of the electromagnetic light produced by the accelerated electrons. The high energy of the electron beam and the high currents in a typical

storage ring (100–400 mA) lead to the creation of very intense X-ray light beams that are strongly focused in the forward direction. One source of polarized photons are the *bending magnets* that are used to make the electrons describe a closed orbit, although more complex *insertion devices* are commonly used, such as *undulators*, consisting of arrays of permanent magnets whose field strength and relative phase are controlled to define the photon energy and polarization [162,241]. The X-ray light thus produced is further monochromatized using line gratings or crystal reflections and directed to the sample using X-ray optical elements. Due the finite size of the source (defined by the entrance slit), the limited resolving power of the monochromator and the finite acceptance angle defined by the exit slit, an energy spread is present in the scattering plane of the beam that limits the ultimate energy resolution that is possible to achieve [44,215,240,241]. An example of a typical configuration of a soft X-ray (photon energies in the range from 100–2000 eV) synchrotron beamline is given in Figure 17.1 showing the disposition of the various optical elements just described and typical physical dimensions of such instrumental installations [52].

Synchrotron X-ray sources are large and technically complex machines that are available mostly in so-called *large-scale user facilities* equipped with many types of measurement techniques, including XPEEM (for a recent table of synchrotron light sources where XPEEM microscopes are installed, we refer the reader to Ref. [83], even if new instruments are constantly being developed or installed and made available for users). The newest development in synchrotron X-ray light sources is that of *diffraction limited* light sources, where the product of the linear source size and the electron beam divergence becomes smaller than that of the photon beam [45,114,241]; in this limit, the x-ray beam becomes fully coherent across its transverse dimension. Another important development are X-ray free electron lasers (FELs), which provide ultrashort X-ray pulses with a length of a few fs and very high intensities [135]. Such sources are particularly interesting for studying ultrafast phenomena and may offer measurements with high spatial and temporal resolution with suitable XPEEM instruments [196]. However, while such instruments are available at UV FELs (and used for biological research [2]), space-charge problems due to the high peak intensities render these developments at present difficult for standard XPEEM microscopes [194].

17.1.2 The Photoemission Electron Microscope

A simplified schematic of a spectroscopic low-energy electron microscope (SPELEEM) instrument suitable for combining XPEEM and LEEM operation is presented in Figure 17.2 [187]. The sample is placed facing the objective lens under a large difference of potential (in the range from 10 to 20 kV) that accelerates the photoemitted electrons towards the electron optics. The small distance between

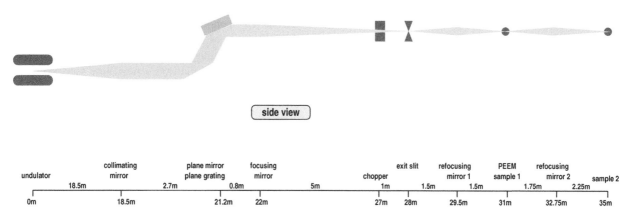

FIGURE 17.1 Simplified optical layout of the Surface/Interface Microscopy (SIM) beamline of the Swiss Light Source, Switzerland. (Reproduced from Ref. [52], with the permission of AIP Publishing.)

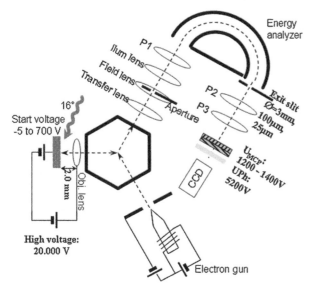

FIGURE 17.2 Schematic diagram of a SPELEEM instrument. (From [70] with permission from Elsevier, Copyright 2012.)

sample and objective lens, of a few mm, constrains the X-ray light to impinge the sample at a low angle of incidence with respect to the sample plane, although normal illumination is also realized in some instruments [249]. The scattering geometry is important for measurements with polarized light, for example, in measurements of dichroic effects such as XMCD/XMLD for probing magnetic phenomena, X-ray natural linear dichroism (XLND) for probing crystal orientation of materials, or when using the "search light"-like effect to probe the direction of chemical bonds of molecular layers by means of polarization-dependent near-edge XAS spectroscopy fine structure (NEXAFS). In combination with a sample rotation stage, full three-dimensional mapping of those properties becomes possible.

The electron optics of a spectroscopic PEEM instrument consists of an imaging column, an energy analyser, and projector optics. Each unit consists of a number of electromagnetic or electrostatic lenses, which enable a fast switching between the different imaging/detection modes and field of views (typically in the range from 1 to 100 μm).

The detector stage typically employs a microchannel plate assembly combined with a phosphorescent screen and a charge-coupled device (CCD) camera [17,18]. A more recent development is the use of hybrid electron detectors for direct recording of the image [145,200,220]. Several apertures are introduced into the electron beam to improve the spatial resolution and to reduce aberrations of the lenses. A contrast aperture can be placed in the (transferred) back focal plane of the objective lens and has the effect of removing those electrons that are ejected from the sample at angles too far from the normal direction (which lead to spherical aberration); an energy slit placed at the exit of the energy analyser allows one to select electrons with a given kinetic energy for detection, improving the image sharpness by reducing the electron energy dispersion, effectively reducing chromatic aberrations. Moreover, another set of pinholes can be introduced in one image plane to select smaller sample areas for averaged spectroscopic investigation (e.g. for operation modes such as micro-XPS and micro-ARPES). The selected area aperture further improves the resolution of the energy analyzer.

Several measurement configurations are possible in XPEEM. The most common imaging mode consists of acquiring a magnified image of the locally photoemitted electrons. These can be secondary electrons, used for measuring XAS spectra when varying the photon energy, or direct photoemitted electrons for acquiring X-ray photoemission spectra by measuring the kinetic energy of the latter; switching between these types of spectroscopy is possible by controlling the pass energy of the energy analyzer, which in some cases can be made spin-sensitive [107]. The first mode takes advantage of the comparably high intensity of secondary electrons and is particularly well suited for imaging. Spatially resolved XAS spectroscopy is possible in this mode since the number of secondary electrons is proportional to the local XAS cross section of the sample. When imaging with secondary electrons, the (material dependent) information depth is limited by the electron escape depth, of typically about 3–5 nm in metals and to about 10 nm in oxides. When probing photoelectrons, the information depth can be varied by selecting the kinetic energy. When using soft X-rays (100–2,000 eV) for sample excitation, photoemission measurements exhibit an enhanced surface sensitivity with a probing depth of typically 1 nm or less, since the corresponding kinetic energies center around the minimum in the universal curve of the inelastic mean free path [18,122,198]. The lateral spatial resolution in both XAS and XPS modes is typically limited to 30–100 nm by chromatic and spherical aberrations and by the available sample contrast and ultimately by space-charge effects [193]. Some instruments are nowadays equipped with aberration-correcting optics typically based on an electron mirror. With such instruments, improved lateral spatial resolution down to 18 nm was reported when imaging with photoelectrons [188]. More recently, also hard X-ray (photon energies above 5 keV) photoemission electron microscopes (HAXPEEMs) have been developed which can detect kinetic energies up to 10 kV [161]. With such instruments, a probing depth larger than 10 nm has been achieved, albeit with a lateral spatial resolution of about 500 nm.

In some instruments, it is possible to measure the energy dispersion of the photoemitted electrons for direct acquisition of X-ray photoemission spectra, typically in a window of 10 eV around the energy set by the energy analyzer. In combination with the selective area aperture, local XPS measurements can be carried out in regions down to 1 μm in size. This mode is often referred to as micro-XPS in the literature. Yet another possibility is to measure the diffraction plane of a selected area of the sample for micro-ARPES. In this mode, images of the k_x-k_y plane at constant energy (with an energy resolution limited to about 200 meV) are obtained [133,136,137]. Spectroscopic PEEM instruments are often additionally equipped with an electron gun, which can be used for imaging with low-energy electrons in LEEM mode or for local structural characterization by means of low-energy electron diffraction (LEED). A UV source can be used also for imaging without X-rays and to provide work function contrast. A more recent development consists of time-of-flight momentum-resolved microscopes, with and without spin filters [55,136,197].

An important aspect of XPEEM imaging relates to the contrast mechanism. For photoemission to occur, the electron must have a kinetic energy in excess of the material work function, which in turn, depends strongly on the condition of the surface, including the presence of adsorbates, surface termination, surface crystal termination, and surface topography [14,173,247,248]. As a consequence, even at constant photon energy, a strong topographic contrast of the sample surface can be obtained. Such contrast contains a wealth of information concerning the surface potential, although extracting quantitative information is challenging. The sensitivity to the surface potential can be used to advantage to image the ferroelectric polarisation of ferroelectrics in LEEM, in so-called mirror imaging mode, where the kinetic energy of the incident electron beam is reduced such that they are reflected from the surface [13,30,172,218]; or in PEEM mode, where the ferroelectric polarization gives rise to a shift in the emission spectra of the secondary

electrons [13]. Topographic contrast in XPEEM can be used to identify areas of interest in the sample but plays no role for spectroscopy, since it remains constant when the photon energy is changed. Finally, we may note that synchrotron radiation is pulsed in nature with a well-defined pulse length, typically in the range from 50 to 100 ps and a repetition rate of a few hundred MHz. Using gateable detectors and locked excitation sources with controllable time delay relative to the synchrotron pulse, time-resolved pump-probe experiments can be carried out with a time resolution given by the X-ray pulse length [70,168].

17.1.3 Sample Requirements

In XPEEM, the sample is the first optical element of the microscope. Due to the high voltage applied between the sample and the objective lens, the photoemitted electrons describe parabolic trajectories when not emitted at right angles from the sample. Also, given the distribution in kinetic energy of the photoemitted electrons, the electrons will be accelerated differently and attain different velocities. This initial angular and kinetic energy distribution of the photoemitted electrons gives rise to spherical and chromatic aberrations, respectively, which largely limit the ultimate spatial resolution possible with XPEEM. Hence, for this technique, it is important that the sample be as flat as possible for best spatial resolution. Since XPEEM relies on measuring photoemitted or secondary electrons, it is necessary that the sample be slightly conducting so that local charging is avoided. For nonconducting samples, charging has the effect of modifying the local electric field of the sample, which then strongly alters the kinetic energy of the photoemitted electrons and the imaging conditions. One solution to achieve sufficient conductivity is to cover the sample with a thin conducting film, for example, gold or carbon. However, even for conducting samples, too high photon densities (particularly, when using isolated, high-intensity synchrotron radiation pulses for time-resolved measurements) can lead to *space-charge* effects due to Coulomb repulsion between photoelectrons emitted from the surface or to the distribution of mirror charges in the sample, which lead to energy shifts in the photoemission spectra [251] and to a degradation of the spatial resolution in XPEEM [124]. In addition, due to the ultrahigh vacuum (UHV) conditions of the XPEEM measurements, the samples need to have very low vapor pressures in UHV; however, exposure to gases for adsorption or catalysis studies is still acceptable as long as the pressure during the measurement remains in the high vacuum range (although recent developments in 'environmental' and liquid cells for XPEEM makes it now possible to study solid–liquid and liquid–gas interfaces [69,148,149]). Despite the above restrictions, the range of material systems and physical phenomena that can be measured with XPEEM is vast, partly made possible by the design of sample holders that permit various kinds of excitations. In addition to control of the temperature [43], local magnetic [107] and electric [26] fields, ultrafast laser excitations [70], electric current pulses [79,174], and r.f. excitations [238] can be applied to the sample, making it possible, for example, to directly image magnetic switching processes [9,75,184,242], the displacement of magnetic domain walls in device structures subject to electric currents [102], the study of magnetoelectric phenomena [25,26,49,182], or the direct imaging of fast magnetization dynamics [168,196].

17.2 Applications of XPEEM for Local Spectroscopy

The need for local spectroscopy arises typically when investigating materials systems that are intrinsically nonhomogeneous, such as in composites, small structures fabricated on a supporting substrate, and nanodevices, but also in nominally homogeneous systems that may contain local inhomogeneities due to defects, phase impurities, or phase segregation when more than one equilibrium state are available to the system. X-ray spectroscopy can then be used to ascertain the local electronic structure, including the chemical or magnetic state, and thereby gain a better understanding of the physical, chemical, and magnetic properties of the system at the electronic level, down to the nanoscale. In this section, we present various examples where XPEEM has been employed to probe the electronic properties of materials, with a focus on the *local* spectroscopic response. The goal is to highlight the capability of this technique to measure a wide variety of systems and to address a wide range of problems in the natural sciences and beyond.

17.2.1 Organic Matter

Biological tissue poses the challenge that it tends not to be compatible with the UHV conditions under which XPEEM measurements are carried out, such that elaborate sample preparation steps are required, for example, involving drying and fixation of the specimens on a substrate [40,207] or ashing the organic material to expose inorganic species for trace element analysis [61]. Also, care has to be taken to prevent or minimize radiation damage to the tissue [1,95,108,131] (a problem which is common to all techniques employing strongly interacting radiation beams). However, some studies have been carried out that take advantage of the high spatial and spectral resolution and chemical sensitivity of XPEEM to detect specific chemical species in different areas of biological tissues, including those of atomic species introduced for therapeutic purposes [204] or as toxic elements [205,206,208]; of the detection of species associated to pathologies [94]; the identification of different types of cells in a tissue [207]; for elemental composition and respective chemical environment identification of individual cells [201]; and to benchmark the adsorption of proteins to nonbiodegradable polymers used for drug delivery or medical implants [28,116–119], for instance. In one example, XPEEM has been use to locally characterize human Glioblastoma cells grown on an Au substrate after

exposure to a Gd-containing molecule (Gd-DTPA). For this study, the tissue samples were first dried in air and then ashed in ozone to remove carbon and nitrogen while preserving all other elements. Local spectroscopy carried out in different parts of the cell showed a higher accumulation of Gd in the cell nuclei compared with the cytoplasm, which is deemed beneficial for neutron irradiation therapy for cancer treatment, given the large cross section of Gd for neutrons [204,209,210].

XPEEM has also been used to study molecular layers on surfaces for applications in *spintronics* [167,243,255], where molecules carrying magnetic moments are used as potential building blocks for magnetic data recording or processing devices [185]. While the properties of molecules in the gas phase or liquid are often well known, they can change drastically when in contact with a solid substrate due to conformational changes or charge transfer processes, for instance. Such effects can affect not only the magnetic properties of the molecules but also their stability, and therefore, characterization of the supported molecules is important. In Ref. [64], the authors used XPEEM to investigate the magnetic and chemical coupling of magnetic molecules to a ferromagnetic substrate through a nonmagnetic layer. For this purpose, the authors deposited a monolayer of manganese(III) tetraphenylporphyrin chloride (MnTPPCl) on a chromium wedge which was prepared on a thin ferromagnetic fcc cobalt layer on a Cu(111) substrate. Using spatially resolved XAS spectroscopy they found that, upon adsorption on both the bare cobalt and on the chromium wedge, the molecules are reactively modified to Mn(II)TPP by loss of the Cl ligand. On the bare cobalt surface, the Mn ions couple ferromagnetically to the Co film, while they become magnetically decoupled from the latter on the chromium wedge. The work showed that the RKKY effect, which is present in thin-film multilayer systems, is too weak to effectively couple the molecular magnetic moments to the ferromagnetic layer through a nonmagnetic interlayer at room temperature.

17.2.2 Mineralogy, Metallurgy, and Composite Structures

One advantage of the surface sensitivity of XPEEM is that one can probe the local properties of bulk materials by looking at a cross section of the material, provided that no significant modifications in the electronic structure of the surface take place. Such an approach has been used to obtain information about the local structure and morphology of composite materials or of variations in local composition and electronic state in minerals and metals, including steels [14,59,163] and clays [225]. Particularly striking are the studies of the local morphology and crystalline orientation of biominerals [62]. Examples include the determination of the calcite orientation of the sea urchin tooth [99,100,127]; of aragonite in nacre in mollusc shells [41,63,139,140,155,156]; the composition and structure of oyster adhesive [142]; and the orientation of aragonite in geological aragonite samples

[141]. In these instances, XPEEM was used to identify the carbonate from the residual organic material (through elemental contrast), the different carbonate species (calcite or aragonite) based on the local spectroscopic characteristics, and the crystal orientation of crystalline aragonite, based on the natural linear dichroic effect associated with the orthorhombic crystal symmetry. These studies illustrate well the capability of XPEEM to investigate the local structure of organic minerals and in providing a better understanding of the biological formation and their structural properties. An example is shown in Figure 17.3 for red abalone nacre, where the layered structure of nacre (separated by layers of organic matter) can be identified on the upper part of the XPEEM images, with a Ca spectrum characteristic of aragonite; the lower part corresponds to the prismatic calcite layer (the outer part of the shell), consisting of calcite and of a disordered, spherulitic aragonite phase in between [63]. The XPEEM image shown in Figure 17.3a is the ratio between the π^*/σ^* images at the carbon K edge and the changes in contrast give a measure of the relative c-axis crystal orientation of aragonite due to X-ray linear natural dichroism [62]. In another instance, local spectroscopy using XPEEM has been used to investigate the influence of ocean acidification on the amorphous calcium carbonate formation in biomineral shells to show that ocean acidification reduces the crystallographic control of shell formation [51].

In the investigation of mineral samples, one is interested in identifying and characterizing mineral intergrowths that can be used to establish geochemical conditions of formation and evolution [60,191,202]. One example is given in Figure 17.4, which shows an XPEEM image measured at the Fe L_3 edge from a sample of the mineral carrolite ($CuCo_2S_4$), confirming the presence of carrolite as well as a vein in the mineral of different Cu-Fe sulphides (bornite and chalcopyrite), whose identification is made

by measuring the local Cu L_3 edge spectra [191]. The study of meteoritical minerals provides another example where the capability of measuring small specimens and the nondestructive nature of XPEEM are used to advantage [190,191].

Composite structures correspond to bulk materials made of two or more components that interact in such a way as to yield properties that differ from the isolated parts. Composite systems can be categorized in terms of their connectivity, which indicates the dimensionality of the separate phases; for example, a mixture of nanoparticles would be of type 0-0, of nanoparticles embedded in a matrix, 0-3, etc. [151]. Due to the multicomponent nature of such materials systems, techniques that are capable of spatially resolving each component can provide unique insights into the individual role of each part to the ensemble properties. One challenge for XPEEM is that the surface of such samples needs to be sufficiently smooth for high-resolution imaging and the feature size needs to be larger than the lateral spatial resolution in order to resolve the individual signal from each component. In many instances, such difficulties can be overcome and XPEEM can then provide important insights into the mechanisms responsible for the interactions within the composite system. One type of composite structures and heterostrutures that has been extensively studied recently consists of a combination of magnetic and ferroelectric components to form so-called multiferroic composite structures, engineered to exhibit a coupling between the magnetic and ferroelectric order parameters, i.e., a *magnetoelectric coupling* [56,67, 85,170,226,229]. Three types of interfacial coupling have been devised for such control, either via strain, charge, or through exchange-bias effects [226,229]. Examples of investigations of magnetoelectric coupling using XPEEM include the study of FeCo/BiFeO₃ heterostructures, where a modification in the magnetic structure of the magnetic FeCo

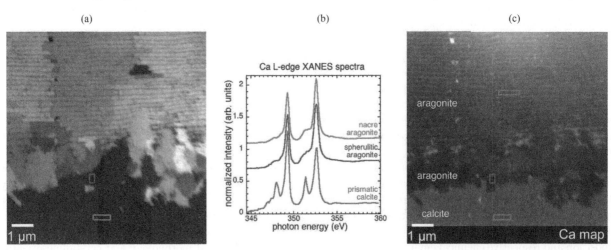

FIGURE 17.3 The nacre-prismatic calcite interface of red abalone. (a) Ratio of XPEEM images taken at energies corresponding to the π^*/σ^* carbon K-edge lines, with linearly polarized light set with polarization vector at 45° from the vertical nacre growth direction. (b) Local XAS spectra over regions marked in the XPEEM images. (c) Calcite elemental contrast image obtained by division of XPEEM images at 351.6 eV (sharp calcite peak) and at the Ca pre-edge energy (345 eV). (Reprinted with permission from Ref. [63]. Copyright 2008, American Chemical Society.)

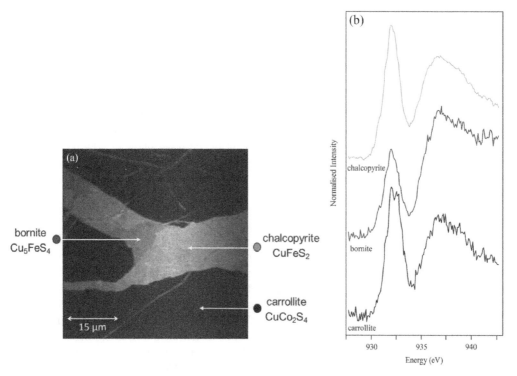

FIGURE 17.4 (a) XPEEM image taken at the Fe L$_3$ edge of a sample of carrolite containing a vein with different Cu-Fe sulphides, whose identification can be made by measuring the local Cu L$_3$ edge, shown in (b). (From [191], with permission from Elsevier, Copyright 2014.)

elements were found upon changes in the ferroelectric state of BiFeO$_3$ films [32]; of La$_{0.7}$Sr$_{0.3}$MnO$_3$ films grown epitaxially on BaTiO$_3$ substrates [31]; of Ni elements fabricated on piezoelectric [Pb(Mg$_{0.33}$Nb$_{0.66}$)O$_3$]$_{0.68}$-[PbTiO$_3$]$_{0.32}$ (PMN-PT) substrates [25,49]; of CoFe$_2$O$_4$-BaTiO$_3$ nanoparticle composites [47]; and of Co films grown on BiFeO$_3$ films [182]. In the latter example, a reproducible switching of the Co magnetic state could be achieved by switching the ferroelectric polarization direction of BiFeO$_3$. These studies mostly take advantage of the magnetic imaging capabilities of XPEEM to determine the changes in the magnetic state with the applied electric field [87], while local spectroscopy has been used to confirm phase purity and chemical state.

17.2.3 Thin Films, 2D Materials, and Heterostructures

When the physical scale of a system is reduced in one dimension down to sizes where the thickness becomes comparable to characteristic lengths of the system, such as correlation and exchange lengths, the materials properties of the system start to deviate from that of the bulk to become either two-dimensional (2D) in character or to start manifesting the role of interface interactions with the supporting medium [228,233]. Both the understanding of physical phenomena that occur in thin films or in 2D materials systems and the search for novel functional properties suitable for device applications have motivated intense research work. Such effort is made possible on the one hand by the development of advanced thin-film deposition

techniques, including molecular beam epitaxy, pulsed laser deposition, and magnetron sputtering that enabled a fine control over the deposition process at the atomic level, and on the other, by the development of suitable characterization techniques, including PEEM. The early stages of film growth have been extensively investigated with LEEM, where a strong topographical contrast and high lateral spacial resolution combine to provide important information concerning the growth nucleation mechanism and its progress up to full coverage of the supporting substrate [16,18,19,73,138]. XPEEM has been used in such studies to determine chemical homogeneity after growth and to determine whether chemical reactions between substrate and film take place through absorption or photoemission spectroscopy, often in combination with LEEM [18,164]. The growth of metals on semiconductor surfaces has been extensively investigated, such as noble metals for the creation of low-dimensional nanowire structures by preferential adsorption to step edges of a vicinal surface [254] or ferromagnetic metals for hybrid ferromagnetic/semiconductor device structures [129]. In the latter case, magnetic contrast images provide information on the presence of magnetism, while local spectroscopy enables the identification of interface reactions or interdiffusion that may blunt the semiconductor/metal interface, which is in general detrimental for applications in spintronics [4,255].

One example of the use of XPEEM to investigate local changes in the electronic structure of materials is the study of the metal-insulator transition, consisting of a change between metallic and insulating states driven by an external field, such as magnetic or electric field, pressure, or temperature. The microscopic mechanisms behind such electronic changes can be manifold, including Mott localization processes associated with strong electron–electron interactions, Anderson localization associated with disorder, from strong electron–phonon interaction (Peirls metal-insulator) [244] or magnetic-order driven (Slater metal-insulator transition) [86]. The onset of a metal-insulator transition is also often accompanied by phase coexistence characteristic of first-order phase transitions, although in strongly correlated materials it may also originate as a consequence of the presence of competing ground states in the system [36,146,221]. Examples include the temperature-driven, metal-insulator phase separation in the mixed valence manganites such as $La_{0.7}Ca_{0.3}MnO_3$ [48], in VO_2 [65,244], and in the nickelates, such as $NdNiO_3$ [134,166]. XPEEM has been employed to investigate the local evolution of the metal-insulator transition in materials such as VO_2 [65] and $NdNiO_3$ [134,166]. As an example, we show in Figure 17.5 the local spectroscopy data taken with XPEEM on a 30 unit cell $NdNiO_3(001)$ film grown on $NdGaO_3$, showing in (a) the spectral evolution of the $NdNiO_3$ with decreasing temperature averaged over the field of view and (c) measured locally in insulating and metallic regions of the sample, where subtle changes in the spectroscopic response can be identified; the difference in the absorption spectra is then explored to obtain images of the metallic and insulating regions across the metal-insulating phase transition, shown in (b) and (d) at two different temperatures, demonstrating the appearance

of elongated insulating patches which are aligned along the terraces of the atomically flat stepped surface, pointing to the role of the surface morphology to the nucleation of the insulating domains [134].

Intrinsic 2D systems have also been investigated extensively with XPEEM, for example, boron nitride [34,253], the transition metal dichalcogenides WS_2 [53,223] and MoS_2 [222], and graphene [93,218,239]. Graphene refers to a single layer of the basal plane of graphite; although studied theoretically in the past as a building block of the 3D graphite structure [235], only more recently, with the controlled growth [157] and mechanical exfoliation [154] of single graphite layer, was there an outburst of interest in the electronic and transport properties of graphene [237]. One interesting feature of graphene is the linear electron energy dispersion with momentum that resembles that of ultra-relativistic particles described by a massless Dirac equation [150]; from analogies with quantum electrodynamics, spin-like quasi-particles and chirality can be inferred [58]. While the properties of graphene are of interest for device applications, including in spintronics [71,98], the challenge of producing graphene over extended areas still remains [21,181,218]. Although high-quality graphene sheets can be obtained with mechanical exfoliation, they tend to be of micrometric size and the creation of multilayer graphene patches is difficult to avoid [159,160]; in epitaxially grown graphene, for example, obtained by thermal decomposition of SiC(0001) substrates, by migration of carbon dissolved

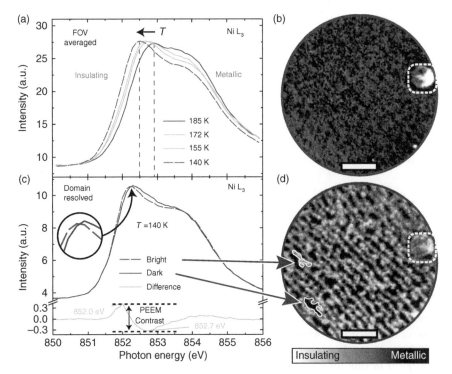

FIGURE 17.5 XPEEM results showing the changes in the spectroscopic response of $NiNdO_3$ across the metal-insulator transition, (a) averaged over the whole field of view, and (c) taken at metallic and insulating regions of the sample. Contrast images of the sample in the metallic (b) and at the phase separated state (d) correspond to differences of images taken at 852.0 and 852.7 eV (scale bar is 1 μm). (From [134], with permission.)

into the bulk of a metal substrate to the surface, or by hydrocarbon vapor deposition, extended films can be obtained, although the growth of regions with more than one layer are difficult to prevent [18,39,81,92,128,218], making area-averaging measurements problematic. In this context, local probes such as LEEM/PEEM have proved very valuable [18, 27,128,218] and the full gamut of capabilities of those techniques have been exploited, including in the identification of the layer thickness in LEEM by using beam interference effects [39,81,92], changes in the K-edge absorption spectra with layer thickness in XAS [133,159], and the direct measurement of the band structure using micro-ARPES [54,92,132,133,158,175,217,218]. An example of the latter is given in Figure 17.6, where the local angle-resolved photoemission spectra of single and multilayer graphene films grown on Ru(0001) are shown, where the thickness identification was first carried out with LEEM [218]. The measurement of the energy dispersion via ARPES in one-, two-, and three-layer graphene shows that the expected linear energy dispersion develops only at the bilayer graphene, while the single layer displays rather broad energy bands [217,218]. The latter effect is attributed to hybridization between graphene and Ru that quenches the 2D properties of graphene [217,218]. On the other hand, at three layers, the energy bands broaden, signaling a transition to the bulk graphite behavior [175,217,218]. Noteworthy in this context is the fact that the theoretical description of graphene usually considers a lattice motive of *two* different carbon atoms, a legacy of the two different perpendicular bonds of the hexagonal close packing of graphite [33], although for graphene no such break in symmetry should be present—suggesting that, despite the 2D character of graphene, out-of-plane interactions that break the lattice symmetry still occur. In another study, suspended graphene

was obtained by laying sheets of graphene onto Si substrates where trenches had been lithographically defined and where the XAS spectra could be measured for graphene flakes with different thicknesses [159]; for the case of the single-layer graphene, a special signature in the XAS spectra was identified that was associated to the presence of an *interlayer* state [159] ascribed to a theoretically predicted additional conduction band of three-dimensional character [165].

17.2.4 Surface Studies and Catalysis

Catalytically relevant processes have been intensively investigated using PEEM with UV light excitation, in particular for studying dynamic pattern formation during chemical reactions at surfaces [89,179]. Many such studies focused on CO oxidation, water formation ($H_2 + O_2$), and the reaction between NO and H_2 on the surfaces of catalytically active metals, such as Pt, Rh, or Pd, and make use of changes in the work function that occur during the chemical reaction and that result in a well detectable intensity change in UV PEEM images [88,90,180,216]. This approach led, for instance, to the discovery of spatiotemporal chemical patterns in the oxidation of CO on low-index surfaces of platinum crystals [80], a phenomenon that was later also discovered in other reactions [18,89,179]. The work function modifications detected in UV PEEM were also correlated quantitatively by the simultaneous detection of reactants in the gas phase using mass spectroscopy, resulting in a "kinetics by imaging" approach to study the CO oxidation in Pd or Pt surfaces [180].

Although UV PEEM can provide a wealth of information on the surface reaction processes, it lacks the spectroscopic capabilities of XPEEM, which can be important to gain insights into the chemistry of the reaction [89]. For

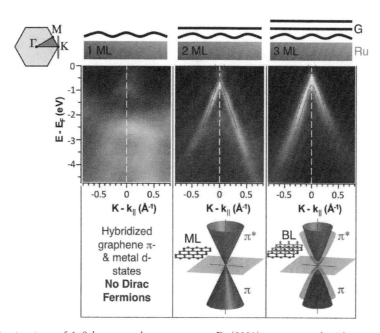

FIGURE 17.6 Electronic structure of 1–3 layer graphene grown on Ru(0001) as measured with micro-ARPES in XPEEM. (From [218], with permission.)

instance, by correlating UV PEEM with chemical imaging (XAS) and micro-XPS, the microscopic origin of the work function decrease in the pattern-forming reaction systems, such as the chemical wave patterns found in the CO oxidation on Pt and Rh surfaces, could be investigated [76]. XPEEM was also used to study water formation reactions *in situ* on Rh(110) partially covered with Au and Pd. The micro-XPS mode revealed that Au and Pd form alloy stripes separated by oxygen-covered Rh regions during this reaction [125]. In another XPEEM work, the reduction of Ce^{4+} to Ce^{3+} in nanosized CeO_x islands grown on Ru upon exposure to molecular hydrogen could be studied in detail [84].

More recently, *in situ* spectroscopic imaging using XPEEM was combined with advanced sample preparation techniques in order to gain insight into the hydrogen spillover effect on alumina and titania substrates [96], cf. Figure 17.7. Using electron beam lithography, arrays of iron oxide and platinum dots with well-defined separations from 0 to 45 nm were fabricated as schematically shown in Figure 17.7b. At 343 K, molecular hydrogen splits on the platinum surface and the atomic hydrogen spills across the sample and reaches the iron oxide dots, reducing the iron oxide. The chemical state of the latter is ascertained by acquiring XAS spectra at the Fe L_3 edge, as shown in Figure 17.7a. The experiments revealed that, on alumina, the hydrogen reduces only iron oxide dots at distances below or close to 15 nm to the platinum dots. In contrast, on the titania substrate, the reduction of the iron oxide dots was observed over the full range of investigated separation distances. XAS spectra recorded at the Ti $L_{2,3}$ revealed further that the titania surface itself is reduced, even at distances over 1 μm from the nearest platinum dot, demonstrating the long diffusion length of hydrogen

on this substrate. Reference measurements on bare titania surfaces under similar condition revealed that no reduction is observed without the presence of the platinum dots. Finally, theoretical investigations showed that the efficiency of hydrogen spillover is determined by both the mobility and the desorption of hydrogen atoms on the different surfaces. In the case of the alumina surface, hydrogen is easily desorbed from the surface, which limits the surface diffusion and reduction of iron oxide to the closest-lying dot pairs, while the high-mobility and low-desorption rate on titania lead to a nearly infinite diffusion of hydrogen and reduction of the iron oxide.

17.2.5 Nanodevices, Nanostructures, and Nanoparticles

The high lateral spatial resolution of XPEEM is particularly well suited for the investigation of nanodevices or their building blocks, which might be composed of semiconducting, metallic, or oxide materials. Similar to thin films, numerous studies have been performed on magnetic nano- or submicrometer structures which form the basis of magnetic nanodevices. In the case of isolated patterned elements, one seeks to understand the role of the different magnetic energy terms in determining the low-energy configuration of the system, in particular the magnetostatic energy contribution, which depends sensitively on the shape of the element; at the micrometer and submicrometer scale, the number of magnetic domains and of equilibrium states that are energetically favored reduce in number, resulting in much simpler magnetic configurations as compared to extended systems [103,230,231]. XPEEM was also used for the investigation of arrays of interacting nanomagnets [78], which can either represent metamaterials with tunable properties or

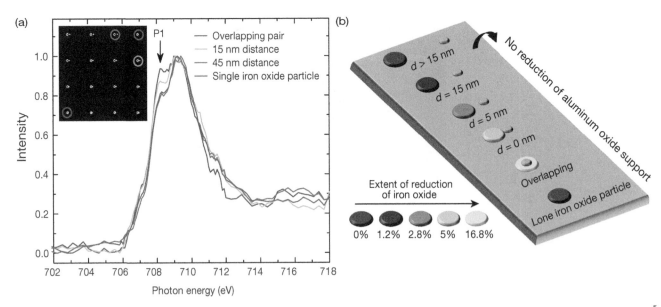

FIGURE 17.7   (a) XAS spectra of individual iron oxide dots recorded at the Fe L_3 edge while dosing molecular hydrogen at 1×10^{-5} mbar for the Pt-FeO_x dots shown in the scanning electron micrographs (inset). (b) Sketch illustrating the observed reduction of the iron nanoparticles as a function of the distance from the Pt nanoparticles on alumina. (From [96], with permission.)

serve as a proxy for understanding the microscopic properties of real materials such as spin ices [152]. Similarly, XPEEM has been used to study the operation of nanomagnetic logic devices [7,11]. In another work, current-induced switching of antiferromagnetic domains in CuMnAs cross devices was imaged [66]. However, in most of these cases, the XPEEM measurements are carried out at fixed photon energies and the spectroscopic response tends to be of secondary importance.

Spectroscopic XPEEM investigations have been performed on systems with relevance for the development of electronic devices such as transistors or diodes. For instance, two-dimensional p- and n-doped lines with variable spacing and width in the range from 0.1 to 100 μm implanted into silicon were studied [15]. The authors used a double hemispherical analyzer for spectroscopic imaging using photoelectrons and extracted the local band alignments by taking account of doping, band bending, and surface photovoltage. *Operando* imaging as a function of applied bias voltage was also demonstrated [12]. XPEEM was further used to investigate redox-based memristive (ReRAM) metal-oxide devices. In such devices, the resistance can be switched by voltage pulses, a process which is being considered for computation or data storage applications. Resistive switching effects were discovered, for instance, in transition metal oxides and were attributed to the motion of mobile donor-type defects, such as oxygen vacancies, and the corresponding valence change in the transition metal cation. XPEEM allowed one to validate such processes qualitatively and quantitatively by means of spatially resolved XAS spectroscopy. For instance, a chemical reduction at the Ti sites was found in Au/Fe:STO/Nb:STO devices upon electrical discharge due to the large electrical potential gradient combined with Joule heating [115]. *Operando* measurements employing a photoelectron transparent graphene top electrode made possible a direct visualization of conductivity filaments [8]. Local XAS spectroscopy allowed the authors to quantitatively determine the changes in the donor concentration at the electrode-oxide interfaces responsible for the change in the resistivity. Simulations indicate that the latter occurs due to a modulation of the effective Schottky barrier. HAXPEEM was used to probe deeply buried components in such devices and revealed also a reduction of the oxidizable bottom Ta electrode in the low-resistance state of ZrO_x cells with respect to the high-resistive state [101].

Nanostructures such as nanowires or epitaxially grown islands have been characterized as well. For instance, XPEEM was used to clarify the growth mechanism of InAs nanowires on SiO_x-covered InAs(111) substrates [130]. By means of spectroscopic imaging with photoelectrons and micro-XPS, it was shown that the role of SiO_x is to immobilize In nanodroplets which form at the surface due to diffusion through the SiO_x at elevated temperatures and to act as nucleation centers for the self-catalyzed growth of the InAs nanowires in metal-organic vapor phase epitaxy. XPEEM was also used to identify surface transport processes in the

epitaxial growth of Ge(Si) islands on Si(111) substrates by quantitative analysis of images recorded with Ge 3d and Si 2p core-level electrons. The measurements revealed Si-richer edges with respect to the centers of the islands with lateral dimensions a few hundred nanometers and thus highlight the impact of surface transport processes for nanoalloying [171]. XPEEM was also used to reveal the role of the local chemical composition of mesoscopic permalloy nanotubes on their low-temperature magnetic properties [23]. The nanotubes were grown as a 30 nm thick polycrystalline shell around GaAs nanowires grown with a length of 10–20 μm long and with an hexagonal cross-section of 150 ± 20 nm. Spatially resolved XAS spectroscopy performed at the Fe L_3 and Ni L_3 edges revealed the presence of an inhomogeneous native oxide layer on top of the permalloy shell, which seemingly exhibits antiferromagnetic order at low temperatures, giving rise to a sizeable exchange-bias effect in the magnetization reversal behavior of the permalloy nanotube.

Nanoparticle systems have diameters typically below about 100 nm, and one could expect XPEEM not to be the most suitable technique to study such systems, since the particle size becomes comparable to the lateral resolution limit of XPEEM. Indeed, while this is expected for nanoparticle agglomerates, where it may no longer be possible to identify individual nanoparticles, for the case where the particles are sufficiently well separated, one retains the capability of probing each individual nanoparticle, although the particle size will appear larger than the actual size (the image then becomes a convolution between the point-like particle size and the microscope resolution). This was for instance demonstrated for InAs islands with a lateral size of about 50 nm and a height of about 20 nm grown on a Se-terminated GaAs(100) surface. The samples were studied by correlating images recorded using Se 3d, Ga 3d, and In 4d core-level photoelectrons [77]. The experiments revealed a complex redistribution of chemical species on the sample surface when evaporating InAs, resulting eventually in InAs islands which covered with a layer of $(In_xGa_{1-x})_2Se_3$. Such information is crucial for instance for understanding and controlling the growth of quantum dots. First successful chemical characterization of particles with much smaller diameters was demonstrated in 10 nm Fe_2O_3 particles [176] and on 8 nm oxidized Co nanoparticles [178].

Using the XMCD effect with circularly polarized X-rays on *in situ* prepared nanoparticle deposits, obtained by means of an UHV compatible arc cluster ion source (ACIS) [105], made possible the magnetic characterization of individual nanoparticles. For example, in metallic Fe and Co nanoparticles 10–20 nm in size, it is found that particles with both stable and fluctuating (superparamagnetic) magnetism coexist, showing the unexpected presence of strong magnetic anisotropies in a significant number of nanoparticles, an effect which is ascribed to the presence of internal structural defects [9,10,104]. This is illustrated in Figure 17.8c showing an elemental contrast image of Fe nanoparticles taken at the L_3 edge, and where the bright spots correspond

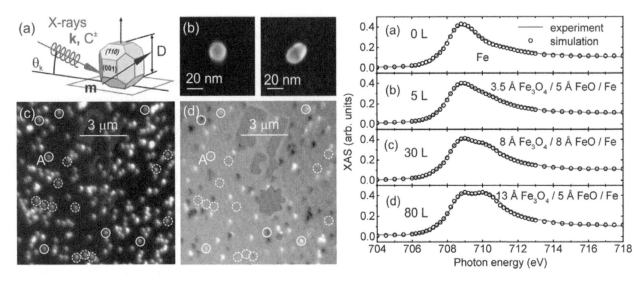

FIGURE 17.8 Left: (a) Experimental geometry for the XPEEM experiment and (b) *ex situ* scanning electron microscopy images of two particles recorded after the XPEEM experiments. (c) Fe elemental and (d) magnetic contrast map of nanoparticles on a Si substrate. Full and dashed circles denote examples of particles with and without magnetic contrast, respectively. Right: XAS spectra at the Fe L_3 edge as a function of O_2 dosage (full lines) and simulated data (circles). The corresponding layer thicknesses used in the simulations are as denoted. (Reproduced from Ref. [227] with permission from the PCCP Owner Societies.)

to nanoparticles with a size range from 8 to 22 nm, as measured *ex situ* with scanning electron and atomic force microscopies [227]; the magnetic XMCD contrast image of the very same nanoparticles, Figure 17.8d, shows that many particles are in a magnetically blocked state, unexpected for particles in this size range with bulk Fe properties [104]. The capability of probing individual nanoparticles has also been explored to investigate the initial oxidation stages of Fe nanoparticles at the nanoparticle level [227], as illustrated in Figure 17.8 (right panel), showing the XAS spectra of the same particles as a function of molecular oxygen exposure. From fits to the data using reference spectra, one can obtain quantitative information regarding the type of oxide formed and its layer structure. The data show that the oxidation starts with the formation of an FeO shell, which progressively oxidizes to Fe_3O_4 to form a $Fe_3O_4/FeO/Fe$ core-shell structure up to 80 L oxygen [227].

17.2.6　Electrochemistry and Li-Ion Batteries

Several approaches have been envisaged when using XPEEM to study the electrochemical properties of materials, including in *operando* conditions using solid-state electrolytes or by using specially designed cells for containing liquids or in a *post mortem* approach, where the cell is disassembled at specific points of the voltammogram cycle. Examples of the first approach include the PEEM study of the electrocatalysis in Pt/YSZ (yttria-stabilized zirconia), where changes in the surface work function were used as a proxy for detecting oxidation and reduction fronts at the Pt surface [91,126]. In this instance, Pt electrodes were patterned lithographically on a YSZ substrate and where

additional electrical contacts were established to connect to a potentiostat. The use of cells for liquids in XPEEM is a more recent development; one demonstration of that capability was carried out for the copper-electroplating reaction using a solution of $CuCO_4$, where the evolution of the Cu and oxygen XAS signal could be followed with the voltammogram cycle [148,149]. A simpler approach to using XPEEM in electrochemistry is to measure the status of the system at discrete steps of the voltammogram cycle, to determine the evolution in the oxidation state of the different species, or to determine changes in composition. This approach has been used to determine the electrode composition in Pt/YSZ [147] and for detailed studies of the evolution of the electrode and electrolyte interfaces of Li-ion batteries [110–113,236,250].

Li-ion batteries are a key contender for energy storage for electric mobility and backup power supply connected to homes or intermittent sustainable power plant's grids [72]. Among the most important requirements for their large-scale commercialization are the long autonomy range, fast charging, stable battery life cycle, low cost per kW, and safety [74,120]. To ensure the continuous increase of their electrochemical performance and safety, a good understanding of the physics and chemistry of the various component parts of the battery and in particular of the working electrodes is required [246]. However, while *post mortem* and *operando* techniques have helped tremendously to gain fundamental understanding of the structural and chemical evolution of the bulk of electrode materials [121], a full understanding of the processes taking place at the surface and at the electrolyte–electrode interface still remains illusive [57], a direct consequence of the complex chemistry and morphology of Li-ion battery electrodes. These complexities are due to various aspects: (i) the high roughness, inhomogeneity, and porosity of the electrode surface, which is composed of a mixture of active particles, conductive carbon, and organic binder; (ii) the complex nature of the electrolyte–electrode interface that is formed during (de-)lithiation processes, which result in the formation of organic and inorganic species with preferential surface reaction on the different particles; and (iii) the surface structural degradation, in particular, of the cathode at high voltage, where leaching of transition metals from the active material and of fluorinated species from the binder occur, leading to contamination of the counter electrode [74]. XPEEM, due to its high spatial resolution and surface sensitivity, is an ideal tool to study the evolution of the interface between electrode and electrolyte in Li-ion batteries [111–113,250].

A first example of the power of XPEEM to investigate the interface of Li-ion batteries concerns the elucidation of the controversial surface reaction of cycled $Li_4Ti_5O_{12}$ (LTO) electrode, in particular, whether it can induce reduction of the commonly used aprotic electrolyte and develop a stable surface layer at 1.55 V vs. Li^+/Li [203]. In Figure 17.9 we show the results of a local spectroscopic investigation of

this electrode vs. cycling. The local XAS spectra acquired at the C K-edge on carbon (Figure 17.9e) and LTO particles (Figure 17.9f) show that reduction of the electrolyte occurs only on the latter particles and exclusively during lithiation (charge). A second reaction is detected during the delithiation (discharge), leading to a partial dissolution of the organic–inorganic species covering the LTO particles. This behavior correlates strongly with the Li^+ insertion/de-insertion and the thermodynamic stability of the electrolyte. In combination with DFT calculations, it is found that the electrolyte reduction is caused by the solvents adsorbed on the LTO outer planes driven by the Li-ion insertion. The adsorption of solvents leads to a shift of their LUMO to energies below the operating potential of LTO, opening the way to a favorable electron transfer process towards their reduction.

In another study, XPEEM was used to identify the surface degradation on high-voltage cathode materials and their impact on the counter electrode, namely, in high-energy Li-rich NCM ($Li_{1.17}(Ni_{0.22}Co_{0.12}Mn_{0.66})_{0.83}O_2$) cycled in aprotic electrolyte against $Li_4Ti_5O_{12}$ (LTO) [112]. Thanks to the excellent lateral resolution of the XPEEM, it

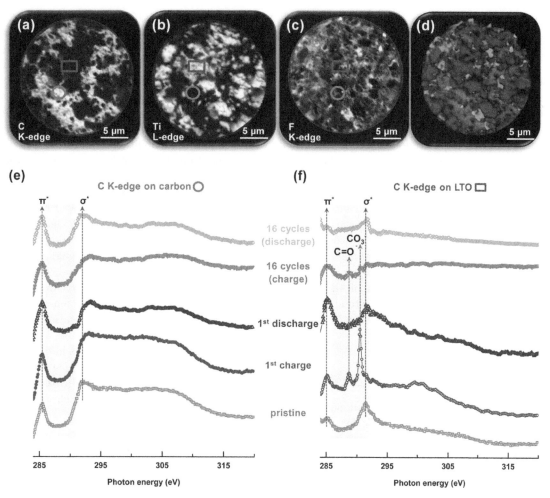

FIGURE 17.9 XPEEM elemental contrast images acquired on LTO pristine electrodes (a) at the C K-edge, (b) Ti L-edge, and (c) F K-edge. In (d) a compositional map is obtained by the overlaying of (a), (b), and (c) images. XAS measurements carried out at the C K-edge on single particles of (e) carbon and (f) LTO. Reproduced from Ref. [111] with permission from The Royal Society of Chemistry.

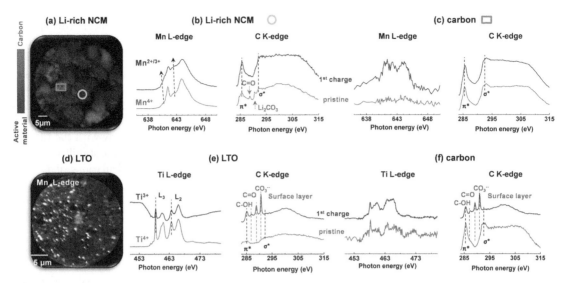

FIGURE 17.10 Element-specific XPEEM contrast images performed on (a) Li-rich NCM electrode at the Mn L-edge and C K-edge and (d) LTO electrode at the Mn L-edge, both after 30 cycles. Local XAS evolution upon Li-rich NCM electrode delithiation (at 5.1 V vs. Li$^+$/Li) performed at the Mn-L edge and C K edge acquired on (b) Li-rich NCM particles and (c) carbon particles. Local XAS evolution on the LTO counter electrode upon lithiation at the Ti-L edge and C-K edge acquired on (e) LTO particles and (f) carbon particles. Adapted from Ref. [112].

is found that, despite the highly oxidative potential of 5.1 V vs. Li$^+$/Li, there is no formation of oxidized electrolyte by-products layer on any of the cathode particles, Figure 17.10b,c; instead, a homogeneous organic–inorganic layer builds up across the particles of the LTO anode arising from the electrolyte and PVDF (polyvinylidene fluoride) binder decomposition on Li-rich NCM, Figure 17.10e, f. In addition, such layer incorporates, already from the first charge, micrometer-sized agglomerates of transition metals, Figure 17.10d. The presence of the latter on the anode is explained by the instability of the reduced Mn, Co, and Ni at the surface of Li-rich NCM formed mainly during delithiation, Figure 17.10b. Reduced transition metals are unstable and prone to be transported to the LTO, where they are further reduced to metallic-like clusters. These results demonstrate that a dual reaction takes place at the Li-rich NCM-electrolyte interface if subject to a high potential, namely, degradation of the surface structure and decomposition of the electrolyte and binder, affecting directly the anode surface through migration-diffusion processes [112].

17.3   Conclusions

Far from exhausting the range of topics and possibilities for the use of XPEEM as a research tool for the study of the properties of materials, the present overview simply aimed at providing a glance at some of the current topics that have benefited from the spectromicroscopic capabilities of this technique. Despite the stringent and demanding setups required for the installation and running of such an instrument, the pay-off in terms of electronic, chemical, and magnetic local information afforded by XPEEM

remain tremendous. One can anticipate that with the ongong technical developments in both instrumentation and in X-ray source characteristics, new possibilities for the characterization of materials with new excitation probes will become possible, for example, for *operando* measurements, where the response of the system to an excitation is measured in real time. Our goal here is to stimulate new research possibilities, rather than circumscribing the range of measurement capabilities of XPEEM. In particular, we have focused on the use of PEEM as a spectromicroscopy technique using X-ray light as the excitation source for spatially resolved XAS and XPS characterization, although we have mentioned throughout that PEEM systems can be used in a versatile fashion that allows different operation modes, including UV PEEM, LEEM, micro-LEED, micro-XPS, and micro-ARPES; although the energy resolution in some of these spectroscopy modes is lower than that of the conventional area-averaging techniques, the gain in spatial information more than compensates for that loss. The range of materials systems is also vast and can be extended even into cases which would not be the normal purview of a UHV technique, such as in liquid electrochemistry and catalysis studies. We anticipate that XPEEM, in all its variants, will continue to be a technique of choice whenever access to local, nanoscale electronic information is required.

Acknowledgments

We are very grateful to the authors that kindly allowed reproduction of their work here. CAFV acknowledges support by the Swiss National Science Foundation (SNF) (Grant No. 200021_153540).

Bibliography

1. H. Ade and A. P. Hitchcock. *Polymer*, 49:643, 2008.

2. H. Ade, W. Yang, S. L. English, J. Hartman, R. F. Davis, R. J. Nemanich, V. N. Litvinenko, I. V. Pinayev, Y. Wu, and J. M. J. Madey. *Sur. Rev. Lett.*, 5:1257, 1998.

3. W. H. Ade. *Nucl. Instrum. Methods Phys. Res. A*, 319:311, 1992.

4. C. Adelmann, J. Q. Xie, C. J. Palmstrøm, J. Strand, X. Lou, J. Wang, and P. A. Crowell. *J. Vac. Sci. Technol. B*, 23:1747, 2005.

5. M. Amati, M. Dalmiglio, M. K. Abyaneh, and L. Gregoratti. *J. Electron Spectrosc. Relat. Phenom.*, 189S:30, 2013.

6. S. Anders, H. A. Padmore, R. M. Duarte, T. Renner, T. Stammler, A. Scholl, M. R. Scheinfein, J. Stöhr, L. Séve, and B. Sinkovic. *Rev. Sci. Instrum.*, 70:3973, 1999.

7. H. Arava, P. M. Derlet, J. Vijayakumar, J. Cui, N. S. Bingham, A. Kleibert, and L. J. Heyderman. *Nanotechnology*, 29:265205, 2018.

8. C. Baeumer, C. Schmitz, A. Marchewka, D. N. Mueller, R. Valenta, J. Hackl, N. Raab, S. P. Rogers, M. I. Khan, S. Nemsak, M. Shim, S. Menzel, C. M. Schneider, R. Waser, and R. Dittmann. *Nat. Commun.*, 7:12398, 2016.

9. A. Balan, P. M. Derlet, A. F. Rodríguez, J. Bansmann, R. Yanes, U. Nowak, A. Kleibert, and F. Nolting. *Phys. Rev. Lett.*, 112:107201, 2014.

10. A. Balan, A. F. Rodríguez, C. A. F. Vaz, A. Kleibert, and F. Nolting. *Ultramicroscopy*, 159:513, 2015.

11. A. D. Bang, F. K. Olsen, S. D. Slöetjes, A. Scholl, S. T. Retterer, C. A. F. Vaz, T. Tybell, E. Folven, and J. K. Grepstad. *Appl. Phys. Lett.*, 113:132402, 2018.

12. N. Barrett, D. M. Gottlob, C. Mathieu, C. Lubin, J. Passicousset, O. Renault, and E. Martinez. *Rev. Sci. Instrum.*, 87:053703, 2016.

13. N. Barrett, J. E. Rault, J. L. Wang, C. Mathieu, A. Locatelli, T. O. Mentes, M. A. Niño, S. Fusil, M. Bibes, A. Barthélémy, D. Sando, W. Ren, S. Prosandeev, L. Bellaiche, B. Vilquin, A. Petraru, I. P. Krug, and C. M. Schneider. *J. Appl. Phys.*, 113:187217, 2013.

14. N. Barrett, O. Renault, H. Lemaître, P. Bonnaillie, F. Barcelo, F. Miserque, M. Wang, and C. Corbel. *J. Electron Spectrosc. Relat. Phenom.*, 195:117, 2014.

15. N Barrett, L F Zagonel, O Renault, and A Bailly. *J. Phys.: Condens. Matter*, 21:314015, 2009.

16. E. Bauer. *Rep. Prog. Phys.*, 57:895, 1994.

17. E. Bauer. *J. Phys.: Condens. Matter*, 13:11391, 2001.

18. E. Bauer. *Surface Microscopy with Low Energy Electrons*. Springer, New York, 2014.

19. E. G. Bauer. *Microsc. Microanal.*, 12:347, 2006.

20. E. Beaurepaire, H. Bulou, F. Scheurer, and J.-P. Kappler, editors. *Magnetism: A Synchrotron Radiation Approach*. Lecture Notes in Physics 697. Springer, Berlin Heidelberg, 2006.

21. C. Berger, Z. Song, X. Li, X. Wu, N. Brown, C. Naud, D. Mayou, T. Li, J. Hass, A. N. Marchenkov, E. H. Conrad, P. N. First, and W. A. de Heer. *Science*, 312:1191, 2006.

22. P. M. Bertsch and D. B. Hunter. *Chem. Rev.*, 101:1809, 2001.

23. A. Buchter, R. Wölbing, M. Wyss, O. F. Kieler, T. Weimann, J. Kohlmann, A. B. Zorin, D. Rüffer, F. Matteini, G. Tütüuncüoglu, F. Heimbach, A. Kleibert, A. Fontcuberta i Morral, D. Grundler, R. Kleiner, D. Koelle, and M. Poggio. *Phys. Rev. B*, 92:214432, 2015.

24. F. Buettner, M. Schneider, C. M. Guenther, C. A. F. Vaz, B. Laegel, D. Berger, S. Selve, M. Klaeui, and S. Eisebitt. *Opt. Express*, 21:30563, 2013.

25. M. Buzzi, R. V. Chopdekar, J. L. Hockel, A. Bur, T. Wu, N. Pilet, P. Warnicke, G. P. Carman, L. J. Heyderman, and F. Nolting. *Phys. Rev. Lett.*, 111:027204, 2013.

26. M. Buzzi, C. A. F. Vaz, J. Raabe, and F. Nolting. *Rev. Sci. Instrum.*, 86:083702, 2015.

27. M. Cattelan and N. A. Fox. *Nanomaterials*, 8:284, 2018.

28. G. Ceccone, B. O. Leung, M. J. Perez-Roldan, A. Valsesia, P. Colpo, F. Rossi, A. P. Hitchcock, and A. Scholl. *Surf. Interface Anal.*, 42:830, 2010.

29. X. M. Cheng and D. J. Keavney. *Rep. Prog. Phys.*, 75:026501, 2012.

30. S. Cherifi, R. Hertel, S. Fusil, H. Béa, K. Bouzehouane, J. Allibe, M. Bibes, and A. Barthélémy. *Phys. Status Solidi RRL*, 4:22, 2010.

31. R. V. Chopdekar, J. Heidler, C. Piamonteze, Y. Takamura, A. Scholl, S. Rusponi, H. Brune, L. J. Heyderman, and F. Nolting. *Eur. Phys. J. B*, 86:241, 2013.

32. Y.-H. Chu, L. W. Martin, M. B. Holcomb, M. Gajek, S.-J. Han, Q. He, N. Balke, C.-H. Yang, D. Lee, W. Hu, Q. Zhan, P.-L. Yang, A. Fraile-Rodríguez, A. Scholl, S. X. Wang, and R. Ramesh. *Nat. Mater.*, 7:478, 2008.

33. D. D. L. Chung. *J. Mater. Sci.*, 37:1475, 2002.

34. H. Cun, A. Hemmi, E. Miniussi, C. Bernard, B. Probst, K. Liu, D. T. L. Alexander, A. Kleibert, G. Mette, M. Weinl, M. Schreck, J. Osterwalder, A. Radenovic, and T. Greber. *Nano Lett.*, 18:1205, 2018.

35. S. Czekaj, F. Nolting, L. J. Heyderman, P. R. Willmott, and G. van der Laan. *Phys. Rev. B*, 73:020401, 2006.

36. E. Dagotto, T. Hotta, and A. Moreo. *Phys. Rep.*, 344:1, 2001.

37. F. de Groot and J. Vogel. Fundamentals of X-ray absorption and dichroism: the multiple approach. In F. Hippert, E. Geissler, J. L. Hodeau,

E. Lelièvre-Berna, and J.-R. Regnard, editors, *Neutron and X-ray spectroscopy*, page 3. Springer-Verlag, Dordrecht, 2006.

38. F. M. F. de Groot. *J. Electron Spectrosc. Relat. Phenom.*, 67:529, 1994.

39. T. A. de Jong, E. E. Krasovskii, C. Ott, R. M. Tromp, S. J. van der Molen, and J. Jobst. *Phys. Rev. Mater.*, 2:104005, 2018.

40. G. de Stasio, C. Capasso, W. Ng, A. K. Ray-Chaudhuri, S. H. Liang, R. K. Cole, Z. Y. Guo, J. Wallace, F. Cerrina, G. Margaritondo, J. Underwood, R. Perera, J. Kortright, D. Mercanti, M. T. Ciotti, and A. Stecchi. *Eusrophys. Lett.*, 16:411, 1991.

41. R. T. DeVol, C.-Y. Sun, M. A. Marcus, S. N. Coppersmith, S C. B. Myneni, , and P. U. P. A. Gilbert. *J. Am. Chem. Soc.*, 137:13325, 2015.

42. M. Dierolf, A. Menzel, P. Thibault, P. Schneider, C. M. Kewish, R. Wepf, O. Bunk, and F. Pfeiffer. *Nature*, 467:436, 2010.

43. A. Doran, M. Church, T. Miller, G. Morrison, A. T. Young, and A. Scholl. *J. Electron Spectrosc. Relat. Phenom.*, 185:340, 2012.

44. J. Dvorak, I. Jarrige, V. Bisogni, S. Coburn, and W. Leonhardt. *Rev. Sci. Instrum.*, 87:115109, 2016.

45. W. Eberhardt. *J. Electron Spectr. Rel. Phenom.*, 200:31, 2015.

46. S. Eisebitt, J. Lüning, W. F. Schlotter, M. Lörgen, O. Hellwig, W. Eberhardt, and J. Stöhr. *Nature*, 432:885, 2004.

47. D. Erdem, N. S. Bingham, F. J. Heiligtag, N. Pilet, P. Warnicke, C. A. F. Vaz, Y. Shi, M. Buzzi, J. L.M. Rupp, L. J. Heyderman, and M. Niederberger. *ACS Nano*, 10:9840, 2016.

48. M. Fäth, S. Freisem, A. A. Menovsky, Y. Tomioka, J. Aarts, and J. A. Mydosh. *Science*, 285:1540, 1999.

49. S. Finizio, M. Foerster, M. Buzzi, B. Krüger, M. Jourdan, C. A. F. Vaz, J. Hockel, T. Miyawaki, A. Tkach, S. Valencia, F. Kronast, G. P. Carman, F. Nolting, and M. Kläui. *Phys. Rev. Appl.*, 1:021001, 2014.

50. P. Fischer, T. Eimüller, G. Schütz, G. Denbeaux, A. Pearson, L. Johnson, D. Attwood, S. Tsunashima, M. Kumazawa, N. Takagi, M. Khler, and G. Bayreuther. *Rev. Sci. Instrum.*, 72:2322, 2001.

51. S. C. Fitzer, P. Chung, F. Caccherozzi, S. S. Dhesi, N. A. Kamenos, V. R. Phoenix, and M. Cusack. *Sci. Rep.*, 6:21076, 2016.

52. U. Flechsig, F. Nolting, A. Fraile Rodríguez, J. Krempaský, C. Quitmann, T. Schmidt, S. Spielmann, and D. Zimoch. *AIP Conf. Proc.*, 1234:705, 2010.

53. S. Forti, A. Rossi, H. Büch, T. Cavallucci, F. Bisio, A. Sala, T. O. Mentes, A. Locatelli, M. Magnozzi, M. Canepa, K. Müller, S. Link, U. Starke, V. Tozzini, and C. Coletti. *Nanoscale*, 9:16412, 2017.

54. S. Forti and U. Starke. *J. Phys. D: Appl. Phys.*, 47:094013, 2014.

55. G. Schönhense, K. Medjanik, C. Tusche, M. de Loos, B. van der Geer, M. Scholz, F. Hieke, N. Gerken, J. Kirschner, and W. Wurth. *Ultramicroscopy*, 159:488, 2015.

56. V. Garcia, M. Bibes, and A. Barthélémy. *C. R. Physique*, 16:168, 2015.

57. M. Gauthier, T. J. Carney, A. Grimaud, L. Giordano, N. Pour, H.-H. Chang, D. P. Fenning, S. F. Lux, O. Paschos, C. Bauer, F. Maglia, S. Lupart, P. Lamp, and Y. Shao-Horn. *J. Phys. Chem. Lett.*, 6:4653, 2015.

58. A. K. Geim and K. S. Novoselov. *Nat. Mater.*, 6:183, 2007.

59. H. M. Ghomi, U. D. Lanke, and A. G. Odeshi. *Can. Metall. Q.*, 21:202, 2012.

60. B. Gilbert, B. H. Frazer, F. Naab, J. Fournelle, J. W. Valley, and G. de Stasio. *Am. Mineral.*, 88:763, 2003.

61. B. Gilbert, L. Perfetti, R. Hansen, D. Mercanti, M. T. Ciotti, P. Casalbore, R. Andres, P. Perfetti, G. Margaritondo, and G. De Stasio. *Front. Biosci.*, 5:10, 2000.

62. P. U. P. A. Gilbert. *J. Electron Spectrosc. Relat. Phenom.*, 185:395, 2012.

63. P. U. P. A. Gilbert, R. A. Metzler, D. Zhou, A. Scholl, A. Doran, A. Young, M. Kunz, N. Tamura, and S. N. Coppersmith. *J. Am. Chem. Soc.*, 130:17519, 2008.

64. J. Girovsky, M. Buzzi, C. Waeckerlin, D. Siewert, J. Nowakowski P. M. Oppeneerand F. Nolting, T. A. Jung, A. Kleibert, and N. Ballav. *Chem. Commun.*, 50:5190, 2014.

65. A. X. Gray, J. Jeong, N. P. Aetukuri, P. Granitzka, Z. Chen, R. Krukeja, D. Higley, T. Chase, A. H. Reid, H. Ohldag, M. A. Marcus, A. Scholl, A. T. Young, A. Doran, C. A. Jenkins, P. Shafer, E. Arenholz, M. G. Samant, S. S. P. Parkin, and H. A. Dürr. *Phys. Rev. Lett.*, 116:116403, 2016.

66. M. J. Grzybowski, P. Wadley, K. W. Edmonds, R. Beardsley, V. Hills, R. P. Campion, B. L. Gallagher, V. Novak J. S. Chauhan, T. Jungwirth, F. Maccherozzi, and S. S. Dhesi. *Phys. Rev. Lett.*, 118:057701, 2017.

67. M. Guennou, M. Viret, and J. Kreisel. *C. R. Physique*, 16:182, 2015.

68. S. Günther, B. Kaulich, L. Gregoratti, and M. Kiskinova. *Prog. Surf. Sci.*, 70:187, 2002.

69. H. Guo, E. Strelcov, A. Yulaev, J. Wang, N. Appathurai, S. Urquhart, J. Vinson, S. Sahu, M. Zwolak, and A. Kolmakov. *Nano Lett.*, 17:1034, 2017.

70. L. Le Guyader, A. Kleibert, A. F. Rodríguez, S. El Moussaoui, A. Balan, M. Buzzi, J. Raabe, and F. Nolting. *J. Electron Spectrosc. Relat. Phenom.*, 185:371, 2012.

71. W. Han, R. K. Kawakami, M. Gmitra, and J. Fabian. *Nat. Nanotechnol.*, 9:794, 2014.

72. M. A. Hannan, Md. M. Hoque, A. Hussain, Y. Yusof, and P. Jern Ker.IEEE Access, 6:19362, 2018.

73. J. B. Hannon and R. M. Tromp. *Annu. Rev. Mater. Res.*, 33:263, 2003.

74. T. R. Hawkins, B. Singh, G. Majeau-Bettez, and A. H. Strømman. *J. Ind. Ecol.*, 83:3912, 2013.

75. J. Heidler, J. Rhensius, C. A. F. Vaz, P. Wohlhüter, H. S. Körner, A. Bisig, S. Schweitzer, A. Farhan, L. Méchin, L. Le Guyader, F. Nolting, A. Locatelli, T. O. Menteş, M. A. Niño, F. Kronast, L. J. Heyderman, and M. Kläui. *J. Appl. Phys.*, 112:103921, 2012.

76. M. Hesse, S. Günther, A. Locatelli, T. O. Menteş, B. Santos, and R. Imbihl. *J. Phys. Chem. C*, 120:26864, 2016.

77. S. Heun, Y. Watanabe, B. Ressel, D. Bottomley, Th. Schmidt, and K. C. Prince. *Phys. Rev. B*, 63:125335, 2001.

78. L. J. Heyderman and R. L. Stamps. *J. Phys.: Condens. Matter*, 25:363201, 2013.

79. L. Heyne, M. Kläui, J. Rhensius, L. Le Guyader, and F. Nolting. *Rev. Sci. Instrum.*, 81:113707, 2010.

80. H. H. Rotermund, W. Engel, M. Kordesch, and G. Ertl. *Nature*, 343:355, 1990.

81. H. Hibino, H. Kageshima, M. Kotsugi, F. Maeda, F.-Z. Guo, and Y. Watanabe. *Phys. Rev. B*, 79:125437, 2009.

82. W. Kuch, R. Schäfe, P. Fischer, and F. U. Hillebrecht. *Magnetic Microscopy of Layered Structures*. Springer Series in Surface Sciences 57. Springer, Heidelberg, 2016.

83. A. P. Hitchcock. *J. Electron Spectrosc. Relat. Phenom.*, 200:49, 2015.

84. J. Höcker, T. O. Mentes, A. Sala, A. Locatelli, Th. Schmidt, J. Falta, S. D. Senanayake, and J. I. Flege. *Adv. Mater. Interfaces*, 2:1500314, 2015.

85. M. B. Holcomb, S. Polisetty, A. Fraile Rodríguez, V. Gopalan, and R. Ramesh. *Int. J. Mod. Phys. B*, 26:1230004, 2012.

86. M. A. Hossain, B. Bohnenbuck, Y. D. Chuang, M. W. Haverkort, I. S. Elfimov, A. Tanaka, A. G. Cruz Gonzalez, Z. Hu, H.-J. Lin, C. T. Chen, R. Mathieu, Y. Tokura, Y. Yoshida, L. H. Tjeng, Z. Hussain, B. Keimer, G. A. Sawatzky, and A. Damascelli. *Phys. Rev. B*, 86:041102(R), 2012.

87. M.-A. Husanu and C. A. F. Vaz. Spectroscopic characterisation of multiferroic interfaces. In C. Cancellieri and V. N. Strocov, editors, *Spectroscopy of Complex Oxide Interfaces*, volume 266 of *Springer Series in Materials Science*, page 245. Springer, Cham, 2018.

88. R. Imbihl. *Surf. Sci.*, 603:1671, 2009.

89. R. Imbihl. *J. Electron Spectrosc. Relat. Phenom.*, 185:347, 2012.

90. R. Imbihl and G. Ertl. *Chem. Rev.*, 95:697, 1995.

91. R. Imbihl and J. Janek. *Solid State Ionics*, 136-137:699, 2000.

92. L. I. Johansson and C. Virojanadara. *J. Mater. Res.*, 29:426, 2014.

93. L. I. Johansson, C. Xia, J. U. Hassan, T. Iakimov, A. A. Zakharov, S. Watcharinyanon, R. Yakimova, E. Janzén, and C. Virojanadara. *Crystals*, 3:1, 2013.

94. C. J. Johnson, P.U.P.A. Gilbert, M. Abrecht, K. L. Baldwin, R. E. Russell, J. A. Pedersen, J. M. Aiken, and D. McKenzie. *Viruses*, 5:654, 2013.

95. A. Kade, K. Kummer, D. V. Vyalikh, S. Danzenbächer, A. Blüher, M. Mertig, A. Lanzara, A. Scholl, A. Doran, and S. L. Molodtsov. *J. Phys. Chem. B*, 114:8284, 2010.

96. W. Karim, C. Spreafico, A. Kleibert, J. Gobrecht, J. VandeVondele, Y. Ekinci, and J. A. van Bokhoven. *Nature*, 541:68, 2017.

97. B. Kaulich, P. Thibault, A. Gianoncelli, and M. Kiskinova. *J. Phys.: Condens. Matter*, 23:083002, 2011.

98. D. Khokhriakov, A. W. Cummings, K. Song, M. Vila, B. Karpiak, A. Dankert, S. Roche, and S. P. Dash. *Sci. Adv.*, 4:eaat9349, 2018.

99. C. E. Killian, R. A. Metzler, Y. U. T. Gong, T. H. Churchill, Ian C. Olson, V. Trubetskoy, M. B. Christensen, J. H. Fournelle, F. De Carlo, S. Cohen, J. Mahamid, A. Scholl, A. Young, A. Doran, F. H. Wilt, S. N. Coppersmith, and Pupa U. P. A. Gilbert. *Adv. Funct. Mater.*, 21:682, 2011.

100. C. E. Killian, R. A. Metzler, Y. U. T. Gong, I. C. Olson, J. Aizenberg, Y. Politi, F. H. Wilt, A. Scholl, A. Young, A. Doran, M. Kunz, N. Tamura, S. N. Coppersmith, and P. U. P. A. Gilbert. *J. Am. Chem. Soc.*, 131:18404, 2009.

101. A. Kindsmüller, C. Schmitz, C. Wiemann, K. Skaja, D. J. Wouters, R. Waser, C. M. Schneider, and R. Dittmann. *APL Mater.*, 6:046106, 2018.

102. M. Kläui, M. Laufenberg, L. Heyne, D. Backes, U. Rüdiger, C. A. F. Vaz, J. A. C. Bland, L. J. Heyderman, S. Cherifi, A. Locatelli, T. O. Mentes, and L. Aballe. *Appl. Phys. Lett.*, 88:232507, 2006.

103. M. Kläui and C. A. F. Vaz. Magnetisation configurations and reversal in small magnetic elements. In H. Kronmüller and S. Parkin, editors, *Handbook of Magnetism and Advanced Magnetic Materials*, volume 2, page 879. John Wiley & Sons, Hoboken, NJ, 2007.

104. A. Kleibert, A. Balan, R. Yanes, P. M. Derlet, C. A. F. Vaz, M. Timm, A. F. Rodríguez, A. Béché, J. Verbeeck, R. S. Dhaka, M. Radovic, U. Nowak, and F. Nolting. *Phys. Rev. B*, 95:195404, 2017.

105. A. Kleibert, J. Passig, K.-H. Meiwes-Broer, M. Getzlaff, and J. Bansmann. *J. Appl. Phys.*, 101:114318, 2007.

106. J. B. Kortright, D. D. Awschalom, J. Stöhr, S. D. Bader, Y. U. Idzerda, S. S. P. Parkin, I. K. Schuller, and H.-C. Siegmann. *J. Magn. Magn. Mater.*, 207:7, 1999.

107. F. Kronast, J. Schlichting, F. Radu, S. K. Mishra, T. Noll, and H. A. Dürr. *Surf. Interface Anal.*, 42:1532, 2010.

108. A. Kade, D. V. Vyalikh, S. Danzenbächer, K. Kummer, A. Blüher, M. Mertig, A. Lanzara, A. Scholl, A. Doran, and S. L. Molodtsov. *J. Phys. Chem. B*, 111:13491, 2007.

109. I. Krug, N. Barrett, A. Petraru, A. Locatelli, T. O. Mentes, M. A. Nio, K. Rahmanizadeh, G. Bihlmayer, and C. M. Schneider. *Appl. Phys. Lett.*, 97:222903, 2010.

110. D. Leanza, M. Mirolo, C. A. F. Vaz, P. Novák, and M. El Kazzi. *Batteries & Supercaps*, 2:482492, 2019.

111. D. Leanza, C. A. F. Vaz, I. Czekaj, P. Novák, and M. El Kazzi. *J. Mater. Chem. A*, 6:3534, 2018.

112. D. Leanza, C. A. F. Vaz, G. Melinte, X. Mu, P. Novák, and M. El Kazzi. *ACS Appl. Mater. Interfaces*, 11:6054, 2019.

113. D. Leanza, C. A. F. Vaz, P. Novak, and M. El Kazzi. *Annual reports of electrochemistry Laboratory*, page 70, 2015.

114. S. C. Leemann, M. Sjöström, and Å. Andersson. *Nucl. Instrum. Methods Phys. Res., A*, 883:33, 2018.

115. C. Lenser, M. Patt, S. Menzel, A. Köhl, C. Wiemann, C. M. Schneider, R. Waser, and R. Dittmann. *Adv. Funct. Mater.*, 24:4466, 2014.

116. B. O. Leung, A. P. Hitchcock, J. L. Brash, A. Scholl, A. Doran, P. Henklein, J. Overhage, K. Hilpert, J. D. Hale, and R. E. W. Hancock. *Biointerphases*, 3:FB27, 2008.

117. B. O. Leung, A. P. Hitchcock, R. Cornelius, J. L. Brash, A. Scholl, and A. Doran. *Biomacromolecules*, 10:1838, 2009.

118. B. O. Leung, A. P. Hitchcock, R. Cornelius, J. L. Brash, A. Scholl, and A. Doran. *J. Electron Spectrosc. Relat. Phenom.*, 185:406, 2012.

119. L. Li, A. P. Hitchcock, R. Cornelius, J. L. Brash, A. Scholl, and A. Doran. *J. Phys. Chem. B*, 112:2150, 2008.

120. W. Li, B. Song, and A. Manthiram. *Chem. Soc. Rev.*, 46:3006, 2017.

121. F. Lin, Y. Liu, X. Yu, L. Cheng, A. Singer, O. G. Shpyrko, H. L. Xin, N.Tamura, C. Tian, T.-C. Weng, X.-Q. Yang, Y. S. Meng, D. Nordlund, W. Yang, and M. M. Doeff. *Chem. Rev.*, 117:13123, 2017.

122. I. Lindau and W. E. Spicer. *J. Electron Spectrosc. Relat. Phenom.*, 3:409, 1974.

123. A. Locatelli and E. Bauer. *J. Phys.: Condens. Matter*, 20:093002, 2008.

124. A. Locatelli, T. O. Mentes, M. A. Niño, and E. Bauer. *Ultramicroscopy*, 111:1447, 2011.

125. A. Locatelli, T. O. Mentes, L. Aballe, A. Mikhailov, and M. Kiskinova. *J. Phys. Chem. B*, 110:19108, 2006.

126. B. Luerssen, J. Janek, and R. Imbihl. *Solid State Ionics*, 141-142:701, 2001.

127. Y. Ma, B. Aichmayer, O. Paris, P. Fratzl, A. Meibom, R. A. Metzler, Y. Politi, L. Addadi, P. U. P. A. Gilbert, and S. Weiner. *PNAS*, 106:6048, 2009.

128. K. L. Man and M. S. Altman. *J. Phys.: Condens. Matter*, 24:314209, 2012.

129. K. L. Man, A. Pavlovska, E. Bauer, A. Locatelli, T. O. Mentes, M. A. Niño, G. K. L. Wong, I. K. Sou, and M. S. Altman. *J. Phys.: Condens. Matter*, 26:315006, 2014.

130. B. Mandl, J. Stangl, E. Hilner, A. A. Zakharov, K. Hillerich, A. W. Dey, L. Samuelson, G. Bauer, K. Deppert, and A. Mikkelsen. *Nano Lett.*, 10:4443, 2010.

131. G. Margaritondo and G. De Stasio. *Int J Imaging Syst Technol*, 8:188, 1997.

132. C. Mathieu, N. Barrett, J. Rault, Y. Y. Mi, B. Zhang, W. A. de Heer, C. Berger, E. H. Conrad, and O. Renault. *Phys. Rev. B*, 83:235436, 2011.

133. C. Mathieu, E. H. Conrad, F. Wang, J. E. Rault, V. Feyer, C. M. Schneider, O. Renault, and N. Barrett. *Surf. Interface Anal.*, 1268:46, 2014.

134. G. Mattoni, P. Zubko, F. Maccherozzi, A. J. H. an der Torren, D. B. Boltje, M. Hadjimichael, N. Manca, S. Catalano, M. Gibert, Y. Liu, J. Aarts, J.-M. Triscone, S. S. Dhesi, and A. D. Caviglia. *Nat. Commun.*, 7:13141, 2016.

135. B. W. J. McNeil and N. R. Thompson. *Nat. Photonics*, 4:814, 2010.

136. K. Medjanik, O. Fedchenko, S. Chernov, D. Kutnyakhov, M. Ellguth, A. Oelsner, B. Schönhense, T. R. F. Peixoto, P. Lutz, C.-H. Min, F. Reinert, S. Däster, Y. Acremann, J. Viefhaus, W. Wurth, H. J. Elmers, and G. Schönhense. *Nat. Mater.*, 16:615, 2017.

137. T. O. Menteş and A. Locatelli. *J. Electron Spectrosc. Relat. Phenom.*, 185:323, 2012.

138. T. O. Menteş, G. Zamborlini, A. Sala, and A. Locatelli. *Beilstein J. Nanotechnol.*, 5:1873, 2014.

139. R. A. Metzler, M. Abrecht, R. M. Olabisi, D. Ariosa, C. J. Johnson, B. H. Frazer, S. N. Coppersmith, and P. U. P. A. Gilbert. *Phys. Rev. Lett.*, 98:268102, 2007.

140. R. A. Metzler, J. S. Evans, C. E. Killian, D. Zhou, T. H. Churchill, N. P. Appathurai, S. N. Coppersmith, and P. U. P. A. Gilbert. *J. Am. Chem. Soc.*, 132:6329, 2010.

141. R. A. Metzler and P. Rez. *J. Phys. Chem. B*, 118:6758, 2014.

142. R. A. Metzler, R. Rist, E. Alberts, P. Kenny, and J. J. Wilker. *Adv. Funct. Mater.*, 26:6814, 2016.

143. J. Miao, P. Charalambous, J. Kirz, and D. Sayre. *Nature*, 400:342, 1999.

144. L. Mino, E. Borfecchia, J. Segura-Ruiz, C. Giannini, G. Martinez-Criado, and C. Lamberti. *Rev. Mod. Phys.*, 90:025007, 2018.

145. G. Moldovan, J. Matheson, G. Derbyshire, and A. Kirkland. *Nucl. Instrum. Methods Phys. Res. A*, 596:402, 2008.

146. A. Moreo, S. Yunoki, and E. Dagotto. *Science*, 283:2034, 1999.

147. E. Mutoro, B. Luerssen, S. Günther, and J. Janek. *Solid State Ionics*, 180:1019, 2009.

148. S. Nemšák, E. Strelcov, T. Duchoš, H. Guo, J. Hackl, A. Yulaev, I. Vlassiouk, D. N. Mueller, C. M. Schneider, and A. Kolmakov. *J. Am. Chem. Soc.*, 139:18138, 2017.

149. S. Nemšák, E. Strelcov, H. Guo, B. D. Hoskins, T. Duchoš, D. N. Mueller, A. Yulaev, I. Vlassiouk, A. Tselev, C. M. Schneider, and A. Kolmakov. *Top. Catal.*, 61:2195, 2018.

150. A. H. C. Neto, F. Guinea, N. M. R. Peres, K. S. Novoselov, and A. K. Geim. *Rev. Mod. Phys.*, 81:109, 2009.

151. R. E. Newnham, D. P. Skinner, and L. E. Cross. *Mat. Res. Bull.*, 13:525, 1978.

152. C. Nisoli, R. Moessner, and P. Schiffer. *Rev. Mod. Phys.*, 85:1473, 2013.

153. F. Nolting. Magnetic imaging with X-rays. In E. Beaurepaire et al., editor, *Magnetism and Synchrotron Radiation*, volume 133, of *Proceedings in Physics*, page 345. Springer-Verlag, Berlin, 2010.

154. K. S. Novoselov, A. K. Geim, S. V. Morozov, D. Jiang, Y. Zhang, S. V. Dubonos, I. V. Grigorieva, and A. A. Firsov. *Science*, 306:666, 2004.

155. I. C. Olson, A. Z. Blonsky, N. Tamura, M. Kunz, B. Pokroy, C. P. Romao, M. A. White, and P. U. P. A. Gilbert. *J. Struct. Biol.*, 184:454, 2013.

156. I. C. Olson, R. Kozdon, J. W. Valley, and P. U. P. A. Gilbert. *J. Am. Chem. Soc.*, 134:7351, 2012.

157. C. Oshima and A. Nagashima. *J. Phys.: Condens. Matter*, 9:1, 1997.

158. A. Ouerghi, M. Ridene, C. Mathieu, N. Gogneau, and R. Belkhou. *Appl. Phys. Lett.*, 102:253108, 2013.

159. D. Pacilé, M. Papagno, A. F. Rodríguez, M. Grioni, L. Papagno, Ç. Ö. Girit, J. C. Meyer, G. E. Begtrup, and A. Zettl. *Phys. Rev. Lett.*, 101:066806, 2008.

160. M. Papagno, A. F. Rodríguez, Ç. Ö. Girit, J. C. Meyer, A. Zettl, and D. Pacilé. *Chem. Phys. Lett.*, 475:269, 2009.

161. M. Patt, C. Wiemann, N. Weber, M. Escher, A. Gloskovskii, W. Drube, M. Merkel, and C. M. Schneider. *Rev. Sci. Instrum.*, 85:113704, 2014.

162. B. D. Patterson. *Am. J. Phys.*, 79:1046, 2011.

163. G. Pereira, A. Lachenwitzer, D. Munoz-Paniagua, M. Kasrai, P. R. Norton, M. Abrecht, and P. U. P. A. Gilbert. *Tribol. Lett.*, 23:109, 2006.

164. C. Pettenkofer, A. Hofmann, W. Bremsteller, C. Lehmann, and F. Kelleter. *Ultramicroscopy*, 119:102, 2012.

165. M. Posternak, A. Baldereschi, A. J. Freeman, E. Wimmer, and M. Weinert. *Phys. Rev. Lett.*, 50:761, 1983.

166. D. Preziosi, L. Lopez-Mir, X. Li, T. Cornelissen, J. H. Lee, F. Trier, K. Bouzehouane, S. Valencia, A. Gloter, A. Barthélémy, and M. Bibes. *Nano Lett.*, 18:2226, 2018.

167. G. A. Prinz. *Phys. Today*, April:58, 1995.

168. J. Raabe, C. Quitmann, C. H. Back, F. Nolting, S. Johnson, and C. Buehler. *Phys. Rev. Lett.*, 94:217204, 2005.

169. J. Raabe, G. Tzvetkov, U. Flechsig, M. Böge, A. Jaggi, B. Sarafimov, M. G. C. Vernooij, T. Huthwelker, H. Ade, D. Kilcoyne, T. Tyliszczak, R. H. Fink, and C. Quitmann. *Rev. Sci. Instrum.*, 79:113704, 2008.

170. R. Ramesh. *Phil. Trans. R. Soc. A*, 372:20120437, 2014.

171. F. Ratto, A. Locatelli, S. Fontana, S. Kharrazi, S. Ashtaputre, S. K. Kulkarni, S. Heun, and F. Rosei. *small*, 2:401, 2006.

172. J. E. Rault, T. O. Menteş, A. Locatelli, and N. Barret. *Sci. Rep.*, 4:6792, 2014.

173. O. Renault, A. M. Pascon, H. Rotella, K. Kaja, C. Mathieu, J. E. Rault, P. Blaise, T. Poiroux, N. Barrett, and L. R. C. Fonseca. *J. Phys. D: Appl. Phys.*, 47:295303, 2014.

174. J. Rhensius, L. Heyne, D. Backes, S. Krzyk, L. J. Heyderman, L. Joly, F. Nolting, and M. Kläui. *Phys. Rev. Lett.*, 104:067201, 2010.

175. C. Riedl, A. A. Zakharov, and U. Starke. *Appl. Phys. Lett.*, 93:033106, 2008.

176. J. Rockenberger, F. Nolting, J. Lüning, J. Hu, and A. P. Alivisatos. *J. Chem. Phys.*, 116:6322, 2002.

177. A. F. Rodríguez, A. Kleibert, J. Bansmann, A. Voitkans, L. J. Heyderman, and F. Nolting. *Phys. Rev. Lett.*, 104:127201, 2010.

178. A. F. Rodríguez, F. Nolting, A. Kleibert J. Bansmannb, and L. J. Heyderman. *J. Magn. Magn. Mater.*, 316:426, 2007.

179. H. H. Rotermund, M. Pollmann, and I. G. Kevrekidis. *Chaos*, 12:157, 2002.

180. H. H. Rotermund. *Surf. Sci.*, 603:1662, 2009.

181. G. M. Rutter, J. N. Crain, N. P. Guisinger, T. Li, P. N. First, and J. A. Stroscio. *Science*, 317:219, 2007.

182. W. Saenrang, B. A. Davidson, F. Maccherozzi, J. P. Podkaminer, J. Irwin, R. D. Johnson, J. W. Freeland, J. Íñiguez, J. L. Schad, K. Reierson, J. C. Frederick, C. A. F. Vaz, L. Howald, T. H. Kim, S. Ryu, M. v. Veenendaal, P. G. Radaelli, S. S. Dhesi, M. S. Rzchowski, and C. B. Eom. *Nat. Commun.*, 8:1583, 2017.

183. M. Salluzzo and G. Ghiringhelli. Oxides and their heterostructures studied with X-ray absorption spectroscopy and resonant inelastic X-ray scattering in the "soft" energy range. In C. Cancellieri and V. N. Strocov, editors, *Spectroscopy of Complex Oxide Interfaces: Photoemission and Related Spectroscopies*, page 283. Springer International Publishing, Cham, 2018.

184. O. Sandig, J. Herrero-Albillos, F. M. Römer, N. Friedenberger, J. Kurde, T. Noll, M. Farle, and F. Kronast. *J. Electron Spectrosc. Relat. Phenom.*, 185:365, 2012.

185. S. Sanvito. *Nat. Phys.*, 6:562, 2010.

186. A. S. Schlachter and F. J. Wuilleumier, editors. *New Directions in Research with Third-Generation Soft X-Ray Synchrotron Radiation Sources.* Springer-Science+Business Media, Dordrecht, 1992.

187. Th. Schmidt, S. Heun, J. Slezak, J. Diaz, K. C. Prince, G. Lilienkamp, and E. Bauer. *Surf. Rev. Lett.*, 5:1287, 1998.

188. Th. Schmidt, A. Sala, H. Marchetto, E. Umbach, and H.-J. Freund. *Ultramicroscopy*, 126:23, 2013.

189. C. M. Schneider. *J. Magn. Magn. Mater.*, 160:9517, 1997.

190. P. F. Schofield, A. D. Smith, J. F. W. Mosselmans, H. Ohldag, A. Scholl, S. Raoux, G. Cressey, B. A. Cressey, P. D. Quinn, C. A. Kirk, and S. C. Hogg. *Geostand. Geoanal. Res.*, 34:145, 2010.

191. P. F. Schofield, A. D. Smith, A. Scholl, A. Doran, S. J. Covey-Crump, A. T. Young, and H. Ohldag. *Coord. Chem. Rev.*, 277-278:31, 2014.

192. A. Scholl. *Curr. Opin. Solid State Mater. Sci.*, 7:59, 2003.

193. A. Scholl, M. A. Marcus, A. Doran, J. R. Nasiatka, A. T. Young, A. A. MacDowell, R. Streubel, N. Kent, J. Feng, W. Wan, and H. A. Padmore. *Ultramicroscopy*, 188:77, 2018.

194. B. Schönhense, K. Medjanik, O. Fedchenko, S. Chernov, M. Ellguth, D. Vasilyev, A. Oelsner, J. Viefhaus, D. Kutnyakhov, W. Wurth, H. J. Elmers, and G. Schönhense. *New J. Phys.*, 20:033004, 2018.

195. G. Schönhense. *J. Phys.: Condens. Matter*, 11:9517, 1999.

196. G. Schönhense, H. J. Elmers, S. A. Nepijko, and C. M. Schneider. *Adv. Imaging Electron Phys.*, 142:159, 2006.

197. G. Schönhense, K. Medjanik, and H.-J. Elmers. *J. Electron Spectrosc. Relat. Phenom.*, 200:94, 2015.

198. M. P. Seah. *Surf. Interface Anal.*, 44:1353, 2012.

199. D. A. Shapiro, Y.-S. Yu, T. Tyliszczak, J. Cabana, R. Celestre, W. Chao, K. Kaznatcheev, A. L. D. Kilcoyne, F. Maia, S. Marchesini, Y. S. Meng, T. Warwick, L. L. Yang, and H. A. Padmore. *Nat. Photonics*, 8:765, 2014.

200. I. Sikharulidze, R.van Gastel, S. Schramm, J. P. Abrahams, B. Poelsema, R. M. Tromp, and S. J.van der Molen. *Nucl. Instrum. Methods Phys. Res. A*, 633:S239, 2011.

201. A. Skallberg, C. Brommesson, and K. Uvdal. *Biointerphases*, 12:02C408, 2017.

202. A. D. Smith, G. Cressey, P. F. Schofield, and B. A. Cressey. *J. Synchrotron Rad.*, 5:1108, 1998.

203. M. S. Song, R. H. Kim, S. W. Baek, K. S. Lee, K. Park, and A. Benayad. *J. Mater. Chem. A*, 2:631, 2014.

204. G. De Stasio, P. Casalbore, R. Pallini, B. Gilbert, F. Sanità, M. T. Ciotti, G. Rosi, A. Festinesi, L. M. Larocca, A. Rinelli, D. Perret, D. W. Mogk, P. Perfetti, M. P. Hehta, and D. Mercanti. *Cancer Res.*, 61:4272, 2001.

205. G. De Stasio, D. Dunham, B. P. Tonner, D. Mercanti, M. T. Ciotti, P. Perfetti, and G. Margaritondo. *J. Synchrotron Rad.*, 2:106, 1995.

206. G. De Stasio, S. Hardcastle, S. F. Koranda, B. P. Tonner, D. Mercanti, M. T. Ciotti, P. Perfetti, and G. Margaritondo. *Phys. Rev. E*, 47:2117, 1993.

207. G. De Stasio, S. F. Koranda, P . Tonner, G. R. Harp, D. Mercant, M. T. Ciotti, and G. Margaritondo. *Europhys. Lett.*, 19:655, 1992.

208. G. De Stasio, D. Mercanti, M. T. Ciotti, T. C. Droubay, P. Perfetti, G. Margaritondo, and B. P. Tonner. *J. Phys. D: Appl. Phys.*, 29:259, 1996.

209. G. De Stasio, D. Rajesh, P. Casalbore, M. J. Daniels, R. J. Erhardt, B. H. Frazer, L. M. Wiese, K. L. Richter, B. R. Sonderegger, B. Gilbert, S. Schaub, R. J. Cannara, J. F. Crawford, M. K. Gilles, T. Tyliszczak, J. F. Fowler, L. M. Larocca, S. P. Howard, D. Mercanti, M. P. Mehta, and R. Pallini. *Neurol. Res.*, 27:387, 2005.

210. G. De Stasio, D. Rajesh, J. M. Ford, M. J. Daniels, R. J. Erhardt, B. H. Frazer, T. Tyliszczak, M. K. Gilles, R. L. Conhaim, S. P. Howard, J. F. Fowler, F. Estève, and M. P. Mehta. *Clin. Cancer Res.*, 12:206, 2006.

211. J. Stöhr. *NEXAFS Spectroscopy.* Springer-Verlag, Heidelberg, 1996.

212. J. Stöhr, H. A. Padmore, S. Anders, T. Stammler, and M. R. Scheinfein. *Surface Rev. Lett.*, 5:1297, 1998.

213. J. Stöhr and H. C. Siegmann. *Magnetism.* Springer-Verlag, Berlin, 2006.

214. J. Stöhr, Y. Wu, B. D. Hermsmeier, M. G. Samant, G. R. Harp, S. Koranda, D. Dunham, and B. P. Tonner. *Science*, 259:658, 1993.

215. V. N. Strocov, T. Schmitt, U. Flechsig, T. Schmidt, A. Imhof, Q. Chen, J. Raabe, R. Betemps, D. Zimoch, J. Krempasky, X. Wang, M. Grioni, A. Piazzalunga, and L. Patthey. *J. Synchrotron Rad.*, 17:631, 2010.

216. Y. Suchorski and G. Rupprechter. *Surf. Sci.*, 643:52, 2016.

217. P. Sutter, M. S. Hybertsen, J. T. Sadowski, and E. Sutter. *Nano Lett.*, 9:2655, 2009.

218. P. Sutter and E. Sutter. *Adv. Funct. Mater.*, 23:2617, 2013.

219. C. Tieg, R. Frömter, D. Stickler, S. Hankemeier, A. Kobs, S. Streit-Nierobisch, C. Gutt, G. Grübel, and H. P. Oepen. *Opt. Express*, 18:27251, 2010.

220. G. Tinti, H. Marchetto, C. A. F. Vaz, A. Kleibert, M. Andrä, R. Barten, A. Bergamaschi, M. Brückner, S. Cartier, R. Dinapoli, E. Fröjdh, D. Greiffenberg, C. Lopez-Cuenca, D. Mezza, A. Mozzanica,

F. Nolting, M. Ramilli, S. Redford, M. Ruat, Ch. Ruder, L. Schädler, Th. Schmidt, B. Schmitt, F. Schütz, X. Shi, D. Thattil, S. Vetter, and J. Zhang. *J. Synchrotron Radiat.*, 24:963, 2017.

221. Y. Tokura and Y. Tomioka. *J. Magn. Magn. Mater.*, 200:1, 1999.

222. D. J. Trainer, A. V. Putilov, C. Di Giorgio, T. Saari, B. Wang, M. Wolak, R. U. Chandrasena, C. Lane, T.-R. Chang, H.-T. Jeng, H. Lin, F. Kronast, A. X. Gray, Xiaoxing Xi, J. Nieminen, A. Bansil, and M. Iavarone. *Sci. Rep.*, 7:40559, 2017.

223. S. Ulstrup, J. Katoch, R. J. Koch, D. Schwarz, S. Singh, K. M. McCreary, H. K. Yoo, J. Xu, B. T. Jonker, R. K. Kawakami, A. Bostwick, E. Rotenberg, and C Jozwiak. *ACS Nano*, 10:10058, 2016.

224. P. van der Heide. *X-ray Photoelectron Spectroscopy.* John Wiley & Sons, Hoboken, NJ, 1912.

225. D. Vantelon, R. Belkhou, I. Bihannic, L. J. Michot, E. Montargès-Pelletier, and J.-L. Robert. *Phys. Chem. Minerals*, 36:593, 2009.

226. C. A. F. Vaz. *J. Phys.: Condens. Matter*, 24:333201, 2012.

227. C. A. F. Vaz, A. Balan, F. Nolting, and A. Kleibert. *Phys. Chem. Chem. Phys.*, 16:26624, 2014.

228. C. A. F. Vaz, J. A. C. Bland, and G. Lauhoff. *Rep. Prog. Phys.*, 71:056501, 2008.

229. C. A. F. Vaz, J. Hoffman, C. H. Ahn, and R. Ramesh. *Adv. Mater.*, 22:2900, 2010.

230. C. A. F. Vaz, M. Kläui, J. A. C. Bland, L. J. Heyderman, C. David, and F. Nolting. *Nucl. Instrum. Methods. Phys. Res. B*, 246:13, 2006.

231. C. A. F. Vaz, M. Kläui, L. J. Heyderman, C. David, F. Nolting, and J. A. C. Bland. *Phys. Rev. B*, 72:224426, 2005.

232. C. A. F. Vaz, C. Moutafis, C. Quitmann, and J. Raabe. *Appl. Phys. Lett.*, 101:083114, 2012.

233. C. A. F. Vaz, F. J. Walker, C. H. Ahn, and S. S. Ismail-Beigi. *J. Phys.: Condens. Matter*, 27:123001, 2015.

234. D. J. Vine, G. J. Williams, B. Abbey, M. A. Pfeifer, J. N. Clark, M. D. de Jonge, I. McNulty, A. G. Peele, and K. A. Nugent. *Phys. Rev. A*, 80:063823, 2009.

235. P. R. Wallace. *Phys. Rev.*, 71:622, 1947.

236. J. Wang, Y. Ji, N. Appathurai, J. Zhou, and Y. Yang. *Chem. Commun.*, 53:8581, 2017.

237. J. H. Warner, F. Schäffel, A. Bachmatiuk, and M. H. Rümmeli. *Graphene: Fundamentals and Emergent Applications.* Elsevier, Amsterdam, 1st edition, 2013.

238. A. Wartelle, C. Thirion, R. Afid, S. Jamet, S. Da Col, L. Cagnon, J.-C. Toussaint, J. Bachmann, S. Bochmann, A. Locatelli, T. O. Mentes, and O. Fruchart. *IEEE Trans. Magn.*, 11:4300403, 2015.

239. S. Watcharinyanon, C. Virojanadara, J. R. Osiecki, A. A. Zakharov, R. Yakimova, R. I. G. Uhrberg, and L. I. Johansson. *Surf. Sci.*, 605:1662, 2011.

240. J. B. West and H. A. Padmore. Optical engineering. In G. V. Marr, editor, *Handbook on Synchrotron Radiation*, volume 2, page 21. North-Holland Publishing Company, Amsterdam, 1987.

241. P. R. Willmott. *An Introduction to Synchrotron Radiation.* John Wiley and Sons, Ltd, 2011.

242. P. Wohlhüter, J. Rhensius, C. A. F. Vaz, J. Heidler, H. S. Körner, A. Bisig, M. Foerster, L. Méchin, F. Gaucher, A. Locatelli, M. A. Niño, S. El Moussaoui, F. Nolting, E. Goering, L. J. Heyderman, and M. Kläui. *J. Phys.: Condens. Matter*, 25:176004, 2013.

243. S. A. Wolf, D. D. Awschalom, R. A. Buhrman, J. M. Daughton, S. von Molnár, M. L. Roukes, A. Y. Chtchelkanova, and D. M. Treger. *Science*, 294:1488, 2001.

244. Z. Yang, C. Ko, and S. Ramanathan. *Annu. Rev. Mater. Res.*, 41:337, 2011.

245. T. Yokoyama, T. Nakagawa, and Y. Takagi. *Int. Rev. Phys. Chem.*, 27:449, 2008.

246. X. Yu and A. Manthiram. *Energy Environ. Sci.*, 11:527, 2018.

247. L. F. Zagonel, N. Barrett, O. Renault, A. Bailly, M. Bäurer, M. Hoffmann, S.-J. Shih, and D. Cockayne. *Surf. Interface Anal.*, 40:1709, 2008.

248. L. F. Zagonel, M. Bäurer, A. Bailly, O. Renault, M. Hoffmann, S.-J. Shih, D. Cockayne, and N. Barrett. *J. Phys.: Condens. Matter*, 21:344013, 2009.

249. A. A. Zakharov, A. Mikkelsen, and J. N. Andersen. *J. Electron Spectrosc. Relat. Phenom.*, 185:417, 2012.

250. J. Zhou, J. Wang, Y. Hu, and M. Lu. *ACS Appl. Mater. Interfaces*, 9:39336, 2017.

251. X. J. Zhou, B. Wannberg, W. L. Yang, V. Brouet, Z. Sun, J. F. Douglas, D. Dessau, Z. Hussain, and Z.-X. Shen. *J. Electron Spectrosc. Relat. Phenom.*, 142:27, 2005.

252. X. Zhu, A. P. Hitchcock, D. A. Bazylinski, P. Denes, J. Joseph, U. Lins, H.-W. Shiu S. Marchesini and, T. Tyliszczak, and D. A. Shapiro. *PNAS*, 113:E8219, 2016.

253. S. Zihlmann, P. Makk, C. A. F. Vaz, and C. Schönenberger. *2D Mater.*, 3:011008, 2016.

254. F.-J. Meyer zu Heringdorf, R. Hild, P. Zahl, Th. Schmidt, B. Ressel, S. Heun, E. Bauer, and M. Horn von Hoegen. *Surf. Sci.*, 480:103, 2001.

255. I. Žutić, J. Fabian, and S. Das Sarma. *Rev. Mod. Phys.*, 76:323, 2004.

Integrating Cavities and Ring-Down Spectroscopy

Edward S. Fry and John Mason
Texas A&M University

18.1 Introduction

When electromagnetic radiation passes through a medium, some of its energy may be absorbed. The fraction of the energy that is absorbed will depend on the frequency of the electromagnetic radiation and the composition of the medium. The process of measuring the portion of radiation absorbed when it interacts with a given medium is called absorption spectroscopy. This is an analytical chemistry tool used to determine the presence of a chemical compound or element in a medium. If the electromagnetic absorption properties are known for a particular compound or element, absorption spectroscopy can be used to quantify the amount present in the medium. This chapter describes a new approach that makes it possible to obtain the absorption spectra for optically complex samples in which the scattering is much larger than the absorption; in these samples, the absorption spectra were previously unobtainable.

The classical approach to absorption spectroscopy is to measure the absorption via the fraction of the electromagnetic radiation that is transmitted through a medium as a function of the frequency of that radiation – the fraction transmitted decreases as the absorption increases. However, this simplistic approach can be prone to errors; *e.g.* (i) Any scattering in the medium must be negligible (much smaller than the absorption) or must be measured and used to correct the absorption data. (ii) For very weak absorption, a small attenuation in the transmitted light must be measured on top of a large background. Consequently, other important experimental approaches to absorption spectroscopy have been developed; two of these are cavity ring-down absorption spectroscopy [1] and integrating cavity absorption spectroscopy [2,3]. These two approaches will be reviewed. Then, their combination, integrating cavity ring-down spectroscopy (ICRDS), will be

described; it has produced an exceptionally powerful new approach to absorption spectroscopy, even in the presence of severe scattering [4].

18.2 Cavity Ring-Down Spectroscopy (CRDS)

Cavity ring-down spectroscopy (CRDS) was first introduced by O'Keefe and Deacon in 1988 [1]. Figure 18.1 shows a schematic of the CRDS concept. Briefly, a short laser pulse is incident on one of the mirrors of a two-mirror optical cavity (the two mirrors must have a very high reflectivity). A fraction of the laser pulse energy enters the cavity during the time duration of the pulse, but after the pulse ends, no more energy enters the cavity. That fraction of pulse energy that entered the cavity through the first mirror bounces back and forth between the mirrors. Since a small fixed fraction of it is lost at each reflection, the total energy inside the cavity decays exponentially. The energy that leaks through the second mirror of the cavity is directly proportional to the exponentially decaying energy inside the cavity and is recorded by a photodetector. When the cavity is filled with an absorbing sample, the exponential decay is faster because energy is also lost due to the absorption during the passage through the sample after each reflection from a mirror. The increase in the rate of the exponential decay gives the absorption coefficient.

CRDS drastically improves the measurement sensitivity for absorption spectra by generating an extremely long path length through the absorbing sample and removing background fluctuations from the laser amplitude. The long effective path length is due to the fact that the laser energy bounces back and forth between the mirrors (and hence through the sample) many times. For example, assume the reflectivity of each mirror is $\rho = 0.9999$. If the distance

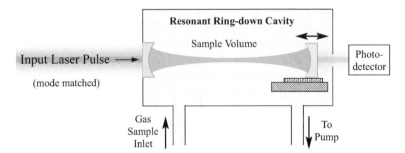

FIGURE 18.1 Schematic of the CRDS concept.

between the mirrors is $d = 10$ cm, then the light energy in an empty cavity will be reduced to $1/e$ of its initial value after the light has traveled an effective path length D given by,

$$D = \frac{d}{1-\rho} = 1.0 \text{ km.} \quad (18.1)$$

When an absorbing sample is placed in the cavity, the exponential decay rate of the light energy in the cavity will be faster and the effective path length will be shorter. For weakly absorbing samples, the path length through the sample will be greater, and the corresponding increase in the exponential decay time means there will be more data points in the exponential decay. Hence, there will be a better measurement accuracy for samples with a smaller absorption.

The fact that CRDS absorption measurements are independent of laser amplitude fluctuations/variations is a consequence of the fact that the sample absorption is determined by the rate of decay of the pulse energy in the cavity rather than the magnitude of the energy in the cavity. Basically, the rate of decay is not affected by the amplitude or temporal shape of the laser pulse (so long as all the input pulse energy is supplied to the cavity in a time short compared to the decay time).

Another important advantage of CRDS is the capability to measure an absolute absorption coefficient without requiring a reference sample of known absorption for calibration. Specifically, let E_0 be the irradiance on the mirror at one end of the cavity at time $t = 0$. At a later time t, the irradiance on that mirror will be

$$E(t) = E_0 e^{-t/\tau}, \quad (18.2)$$

where τ is the time constant for the cavity (*i.e.* the time for the intensity to decrease by a factor $1/e$). If the reflectivity of each mirror is ρ and $n_{\tau e}$ is the number of round trips in the empty cavity during a decay time τ_{empty}, then

$$E\left(\tau_{\text{empty}}\right) = \rho^{2n_{\tau e}} E_0 = E_0 e^{2n_{\tau e} \ln \rho}. \quad (18.3)$$

Note, the coefficient is $\rho^{2n_{\tau e}}$, because there is a reflection from two mirrors (ρ^2) during each complete round trip. From Eq. (18.2), when $t = \tau$, the exponent is –1; hence, from Eq. (18.3), the number of round trips $n_{\tau e}$ is given by

$$2n_{\tau e} \ln \rho = -1 \text{ or } n_{\tau e} = \frac{1}{2 \ln \rho} \quad (18.4)$$

If the distance between the cavity mirrors is d, then the travel time for one round trip in an empty cavity is $2d/c$ where c is the vacuum speed of light. Since there are n_τ round trips in a decay time τ, the decay time τ_{empty} when the cavity is empty is

$$\tau_{\text{empty}} = n_{\tau e} \frac{2d}{c} = -\frac{1}{\ln \rho} \frac{d}{c}. \quad (18.5)$$

If the cavity is filled with an absorbing sample (absorption coefficient α), and if the modes and linewidths are such that the Bouguer–Lambert–Beer law applies, [5] then Eq. (18.3) becomes

$$E\left(\tau_{\text{sample}}\right) = \rho^{2n_\tau} e^{-2n_{\tau s}\alpha d} E_0 = E_0 e^{2n_{\tau s} \ln \rho - 2n_{\tau s}\alpha d}, \quad (18.6)$$

where $n_{\tau s}$ is the number of round trips through the sample during the decay time τ_{sample}, and is given by

$$2n_{\tau s} \ln \rho - 2n_{\tau s}\alpha d = -1 \text{ or } n_{\tau s} = -\frac{1}{2(\ln \rho - \alpha d)}. \quad (18.7)$$

The travel time for one round trip through a sample that fills the cavity is $2n_s d/c$ where c is the vacuum speed of light and n_s is the index of refraction of the sample medium. Consequently, the decay time τ_{sample} can be written,

$$\tau_{\text{sample}} = n_{\tau s} \frac{2n_s d}{c} = -\frac{1}{\ln \rho - \alpha d} \frac{n_s d}{c}. \quad (18.8)$$

Equations 18.5 and 18.8 give

$$\frac{1}{\tau_{\text{sample}}} - \frac{1}{n_s \tau_{\text{empty}}} = \frac{c}{d} \frac{(-\ln \rho + \alpha d)}{n_s} + \frac{c}{d} \frac{\ln \rho}{n_s} = \frac{c\alpha}{n_s}, \quad (18.9)$$

and solving for α gives

$$\alpha = \frac{1}{c} \left(\frac{n_s}{\tau_{\text{sample}}} - \frac{1}{\tau_{\text{empty}}} \right). \quad (18.10)$$

Thus, Eq. (18.10) shows that, given the index of refraction of the sample medium, the measurements of the decay time with the sample filling the cavity and with the cavity empty provide an accurate measurement of the absorption coefficient without any need for calibration with a standard reference absorber.

18.3 Integrating Cavity Spectroscopy (ICS)

The integrating cavity is a closed container with a diffuse reflecting wall that has a very high reflectivity. Such cavities have a long history beginning with Sumpner's 1892 discussion [6] of the increased brightness of the walls in a space with diffuse reflecting walls, and Ulbricht's 1900 discussion [7] of a closed sphere with a diffuse reflecting wall, the "Ulbricht sphere". But, applications and discussions about integrating cavities were mainly directed towards photometric applications until 1970 when Elterman demonstrated the use of an integrating cavity to measure absorption [2].

The basic concept of Elterman's scheme is shown in Figure 18.2. Light enters the integrating cavity through a small window and diffusely reflects in all directions when it strikes the wall; all the reflected rays are again diffusely reflected each time they strike the cavity wall. The diffuse

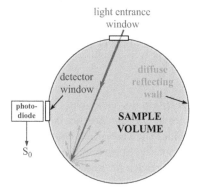

FIGURE 18.2 Elterman's integrating cavity design. An example of a light ray entering through the window and incident on the diffusing surface is shown.

reflection results in two important characteristics: (i) the sample volume is uniformly illuminated with light traveling in all directions so scattering within the sample volume does not affect the uniformity of illumination (unless absorption or scattering are too extreme); and (ii) light makes many transits through the absorbing sample, so there is a long path length and hence greater sensitivity even for a small sample volume. Elterman's introduction of the use of an integrating cavity to measure optical absorption was a truly major contribution because of those two characteristics. But his theoretical analysis left some concerns, and his experimental design was problematic due to the directionality of the incident light entering the cavity. More recent integrating cavity innovations have worked to improve the theoretical framework and remove directionality.

Figure 18.3 shows an integrating cavity design that does eliminate the directionality problem and provides an isotropic illumination of the sample. The design consists of integrating cavity enclosed in a second integrating cavity. The sample volume is formed by the inner wall **wI** which is enclosed by wall **wII** with an air gap between the walls. Both walls are made from a diffuse reflecting material, *e.g.* Spectralon [8]. Electromagnetic energy (*e.g.* visible light) is injected into the system through the outer wall **wII** and diffusely reflects back and forth between the walls. Although the reflectivity might be greater than 99% at each reflection, it is incident multiple times and some of it diffuses all the way through wall **wI** and into the sample volume. In fact, if there were no absorption losses, the steady-state electromagnetic energy density in the sample volume and in the air gap between the walls would be equal; but as absorption in the sample volume increases, the electromagnetic energy density in the sample volume decreases.

After many reflections, the radiation that diffuses through wall **wI** enters the sample volume with a Lambertian

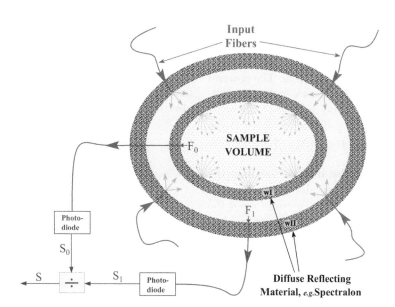

FIGURE 18.3 Double-wall integrating cavity provides isotropic illumination of the sample. The inner and outer diffuse reflecting cavity walls are labeled **wI** and **wII**; they are separated by an air gap.

distribution and has the same radiant flux at every point on the wall; hence, the sample is isotropically illuminated.

The irradiances F_0 and F_1 (Figure 18.3) are directly proportional to the radiant energy densities U_0 in the sample volume and U_1 in the air gap. U_0 depends on the absorption in the sample volume; it decreases as the absorption increases. U_1 provides a measure of the amount of electromagnetic energy being supplied to the sample volume. The measured signal voltages S_0 and S_1 are directly proportional to F_0 and F_1 and have been shown to be simply related to the absorption coefficient α of the sample, [3].

$$\alpha = C\frac{S_1}{S_0} - D, \qquad (18.11)$$

where C and D are instrument calibration factors. Both C and D have a very strong dependence on the magnitude of the diffuse reflectivity of the wall of the sample volume. They can be determined by measuring S_0 and S_1 for two calibration samples with known absorption coefficients α and then using the two resulting equations corresponding to Eq. (18.11) to solve for C and D. Elterman proposed determining the calibration factors in an absolute sense by introducing absorbing areas of known size into the wall (**wI**) of the sample cavity. In practice, the accuracy of this approach was found to be poor because of errors associated with determining the actual loss produced by the absorbing area.

An important application of an integrating cavity would be accurate measurements of the spectral absorption of pure water. Extensive references on this subject are available [9,10], and they indicate wide variability in observed values and considerable uncertainty as to the actual values. A major problem in the observed values is the scattering correction required for transmission measurements. Consequently, an integrating cavity in which the data is unaffected by scattering in the sample was introduced to measure the absorption coefficient for pure water in the visible [11]. The double walls of this integrating cavity were made from that diffuse reflecting material known as Spectralon [8] that has a reflectivity of ∼99.2% in the visible spectrum. The spectral absorption data for pure water are shown in Figure 18.4; data in the dark gray and light gray regions of the spectrum were in agreement with the generally accepted values at the time, but data in the light gray lines (∼420 nm) were a factor of 4–5 lower.

It would be good to extend this data further into the ultraviolet (UV) because pure water absorption down to 250 nm is very important, and there are concerns about the available data. Scattering continues to be the major problem when measuring the absorption spectra of pure water for UV wavelengths. Specifically, pure water absorption in the UV is so weak that even scattering by molecules and density fluctuations requires significant corrections. Unfortunately, around 400 nm, the absorption of Spectralon has begun to become appreciable compared to the absorption of pure water, and as one goes further into the UV, the

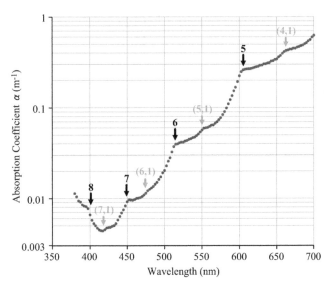

FIGURE 18.4 Spectral absorption data for pure water using a double-wall integrating cavity made from Spectralon. The black arrows with an integer **n** indicate the predicted positions of the **n**th harmonic of the O–H stretch; the arrows in light gray with mode assignments (j,1) indicate the predicted positions of the combination of the jth harmonic of the O–H stretch with the fundamental of the scissor's mode.

pure water absorption decreases and the Spectralon absorption increases. Since the integrating cavity will always give data that combines the absorption contributions of the pure water with that of the Spectralon, one cannot use a Spectralon integrating cavity for absorption studies of water in the UV.

Consequently, a new diffuse reflecting material is required. It is important to understand that when light reflects from a diffuse reflecting material such as Spectralon, it does not just reflect from the surface. Instead, it penetrates the surface and scatters around inside the material. When light exits the surface of the material, it does so with a Lambertian distribution; see Figure 18.5. If the reflecting material has a significant absorption cross section, then an appreciable fraction of light will be lost as the light travels inside the material. Thus, an effective UV diffuse reflector must not absorb in the UV. One such material is quartz.

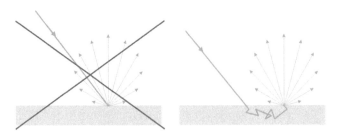

FIGURE 18.5 Light incident on a diffuse reflecting material such as Spectralon does not just reflect from the surface; it enters the material and exits in a Lambertian distribution after multiple scatterings within the material.

As an abstract model, consider a thin quartz plate with light at normal incidence on it as shown in Figure 18.6 [12].

Eventually, only surface reflections will be of interest, so interference effects will be ignored; also absorption/scattering in the quartz plate will be assumed negligible. Then, the three simultaneous equations relating i_0, i_1, i_2 and i_3 in Figure 18.6 are easily written as,

$$i_1 = \rho i_0 + (1 - \rho) i_3$$
$$i_2 = \rho i_3 + (1 - \rho) i_0,$$
$$i_3 = \rho i_2 \qquad (18.12)$$

where ρ is the fraction of the irradiance that is reflected at normal incidence on a quartz–air interface. If the indices of refraction of air and of quartz are 1.0 and n_q, respectively, then the fraction ρ of the irradiance reflected at normal incidence is

$$\rho = \left(\frac{n_q - 1}{n_q + 1} \right)^2. \qquad (18.13)$$

The relations (Eq. 18.12) can be solved to obtain the reflectance R_1 of one plate,

$$R_1 = \frac{i_1}{i_0} = \frac{2\rho}{1 + \rho}. \qquad (18.14)$$

Now consider two plates with a small air gap between them as in Figure 18.7.

There are now seven simultaneous equations that relate the irradiances i_1 through i_7; these equations are analogous to those in Eq. (18.12). They can be solved for the overall reflectance R_2 due to two plates,

$$R_2 = \frac{i_1}{i_0} = \frac{4\rho}{1 + 3\rho}. \qquad (18.15)$$

$\Uparrow i_4$

$\Uparrow i_2$ $\Downarrow i_3$	Plate 1

$\Uparrow i_0$ $\Downarrow i_1$

FIGURE 18.6 The irradiance on the bottom surface of quartz Plate 1, is i_0. From within Plate 1, the irradiance on the upper surface is i_2 and on the lower surface it is i_3. The upward radiosity from the top surface of Plate 1 is i_4, and the downward radiosity from the bottom surface of Plate 1 is i_1.

$\Uparrow i_8$

$\Uparrow i_6$ $\Downarrow i_7$	Plate 2

$\Uparrow i_4$ $\Downarrow i_5$

$\Uparrow i_2$ $\Downarrow i_3$	Plate 1

$\Uparrow i_0$ $\Downarrow i_1$

FIGURE 18.7 Two quartz plates and the labels for the upward and downward irradiances within and outside each plate.

Finally, this concept can be extended to N plates for which there will now be $4N-1$ simultaneous equations analogous to Eq. (18.12); their solution gives

$$R_N = \frac{i_1}{i_0} = \frac{2N\rho}{1 + (2N - 1)\rho}. \qquad (18.16)$$

This same result was already obtained in 1862 by George Stokes [13].

First, note that as $N \to \infty$, the overall reflectivity R approaches 100%. For visible light at normal incidence on a quartz plate, the reflectivity ρ is ~4%. Then from Eq. (18.16), achieving a reflectivity of $R = 0.9995$ (99.95%) would require

$$N = \frac{R(1 - \rho)}{2\rho(1 - R)} = 23,988 \text{ plates}. \qquad (18.17)$$

Of course, such a large number of plates is not a viable option for an integrating cavity. But the important feature is the quartz–air interfaces, and these could be obtained using quartz particles rather than plates. With randomly shaped and oriented quartz particles, the reflections (scattering) would be in all directions, but for a diffuse reflector, this is exactly what is desired. Quartz powder is readily available commercially; it is mixed with waxes (*e.g.* lipstick), tars, etc. as a flow enhancer. A typical quartz powder consists of ~40 nm size particles that tend to aggregate into submicron clumps. Figure 18.8 shows an electron microscope picture of typical quartz powder aggregate. A stack of 24,000 quartz particles with dimensions of 400 nm would be <1 cm high.

To make a diffuse reflecting material for an integrating cavity, the quartz powder is baked under vacuum at 1,000°C and back filled with high-purity Argon. The powder is then compressed at a pressure of about 100 psi and becomes a semisolid because the irregular-shaped particles get physically entangled with each other. In fact, with considerable care, the compressed powder can actually be machined into various shapes. The choice of 100 psi is based on experimental testing that shows the total reflectivity R increases as

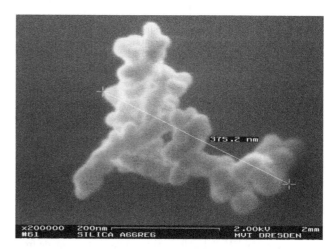

FIGURE 18.8 An electron microscope picture of a typical quartz powder aggregate.

the pressure increases up to about 100 psi, but then R starts decreasing as the pressure is increased further. At higher pressures, the optical contact between particles improves transmission and decreases reflectivity. The measured reflectivity of this compressed quartz powder is typically 99.89%, but measurements as high as 99.92% have been observed in the visible (for comparison, the maximum reflectivity of Spectralon in the visible is 99.3%). But most importantly, in the UV at 250 nm, the reflectivity of this compressed quartz powder is 99.6%.

In a major breakthrough, this new quartz powder diffuse reflector has been used to make the first reliable measurements of the spectral absorption of pure water at wavelengths down to 250 nm [14]. A vertical cross section through the center of the instrument that was designed for these measurements is shown in Figure 18.9. The instrument consists of two concentric integrating cavities centered around a quartz glass sample cell. The integrating cavity walls (**wI** and **wII**) are made from the compressed quartz powder described above. Quartz fibers are used to inject light into the outer cavity (air gap); the light then diffuses through the inner cavity wall (**wI**) and isotropically illuminates the sample region. The quartz sample cell has a volume of 1.538 L; and it has inlet and outlet ports located at the bottom and top, respectively, of the assembly. The irradiances from the inner cavity wall (S_0) and from the outer cavity wall (S_1) were sampled through two optically polished 6-mm- and 5-mm-diameter quartz rods. A larger quartz rod was chosen for the inner cavity to increase the signal from the weaker intensity light. The signal from each

rod was passed through a separate optical chopper wheel, one at a frequency of 1.552 kHz and the other at 1.207 kHz. Both chopped light signals are then delivered to a second fumed silica-integrating cavity that is monitored by a single quartz-windowed photomultiplier tube whose output is sent to two lock-in amplifiers. The frequency from each chopper wheel was provided as a reference to one of the lock-ins whose output provided the corresponding S_0 or S_1. By referencing the signal from the inner cavity to the outer cavity with a single detector, intensity fluctuations from the light source as well as thermal fluctuations from the PMT were eliminated as sources of noise.

The results of these measurements are shown in Figure 18.10. Also shown are previous measurements using other techniques. The severity of the disarray in previous data is clear. The black diamonds in this region in Figure 18.10 are not data points; these are extrapolations by Morel et al. who were attempting to provide some kind of data to fill in this gap. The references to this data are: Lee et al. [15], Morel et al. [16], Pope and Fry [11], Mason et al. [14], Quickenden and Irvin [17], Cruz et al. [18], Kröckel and Schmidt [19].

18.4 Integrating Cavity Ring-Down Spectroscopy (ICRDS)

Both integrating cavity spectroscopy (ICS) and CRDS offer different and very powerful approaches to absorption spectroscopy. Even though combining them offers extraordinary capabilities, they have not been combined previously because commercially available diffuse reflectors have not had a sufficiently high reflectivity to do ring-down spectroscopy. However, the quartz diffuse reflector described above has a sufficiently high reflectivity to make the combination a possibility. This is illustrated in Figure 18.11 for 5.0 cm diameter spherical cavities made of Spectralon and of compressed quartz powder. In this example, the input pulse (dark gray solid line) was shown with a slight "after pulse" to further demonstrate the importance of the higher reflectivity. Because of the greater energy lost on each reflection, the pulse in the Spectralon cavity (filled gray lines) decays more quickly than the pulse in the powdered quartz cavity (dark gray straight line).

The "after pulse" produces a bump in the powdered quartz cavity signal (dark gray straight line), but that signal then becomes a long perfect exponential decay with many data points that facilitate the determination of the decay constant. In the Spectralon cavity, the decay is so fast that fewer data points are available for an accurate determination of the decay constant. In fact, the decay closely follows the tail of the input pulse. In this case, the small after pulse disturbs the decay so that it is no longer a simple, one parameter exponential decay. Spectralon, which was once considered to have the highest known diffuse reflectivity (99.1%) of any material, is simply not very promising for ICRDS applications; that is why it has not been previously

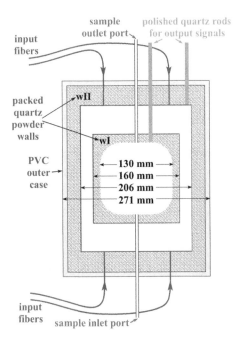

FIGURE 18.9 Cross section through the middle of the integrating cavity device used to measure the optical absorption of pure water in the UV. All walls in the system that are made from quartz are shown in dark gray straight lines to simplify the identification of their locations.

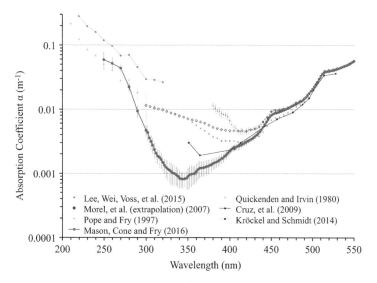

FIGURE 18.10 Absorption spectra of pure water. The data obtained in an integrating cavity using the new compressed quartz powder are shown in black.

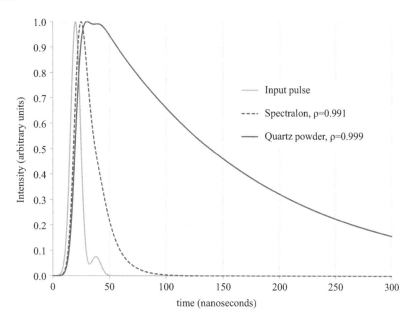

FIGURE 18.11 Decay of the energy in spherical (5.0 cm diameter) integrating cavities made of Spectralon (dashed line) and compressed quartz powder (line).

implemented. But, the high reflectivity (99.92%) available with the compressed quartz powder has now made ICRDS a reality.

Consider the temporal dynamics in an integrating cavity. Let $E(t)$ be the irradiance on the wall of the cavity at time t; and let τ be the time constant of the cavity (*i.e.* the time for $E(t)$ to decrease by a factor $1/e$). Then $E(t)$ has the form

$$E(t) = E_0 e^{-t/\tau}, \qquad (18.18)$$

where E_0 is the irradiance at some arbitrary time during the exponential decay of E that is chosen as the time $t = 0$. Now, let n be the number of reflections from the cavity wall in a decay time τ. If ρ is the reflectivity at each reflection from the Lambertian surface of the cavity, then after n reflections, the remaining irradiance will be $\rho^n E_0$ and

$$E(\tau) = \rho^n E_0 = E_0 e^{n \ln \rho}. \qquad (18.19)$$

But, from Eq. (18.18), $E(\tau) = E_0 \, e^{-1}$. Combining this with Eq. (18.19) gives

$$n = -\frac{1}{\ln \rho} \approx \frac{1}{1 - \rho}. \qquad (18.20)$$

For Spectralon with $\rho = 0.991$, n will be ~ 111, but with compressed quartz powder with $\rho = 0.999$, n will be $\sim 1,000$. Clearly, seemingly small changes in ρ make a significant difference in the number of reflections before the irradiance drops by a factor $1/e$.

Let $\bar{d}$ be the average distance a photon travels between successive reflections at the cavity wall, and let $\bar{t}$ be the average time between reflections at the cavity wall. If c is the speed of light in the cavity, then, $\overline{d} = c\overline{t}$, and since there are n reflections in time τ,

$$\tau = n\overline{t} = -\frac{1}{\ln \rho}\frac{\overline{d}}{c}, \qquad (18.21)$$

If the cavity is filled with a sample whose absorption coefficient is α, Eq. (18.19) becomes

$$E(\tau) = \rho^n e^{-n\alpha\overline{d}}E_0 = E_0 e^{n \ln \rho - n\alpha\overline{d}}. \qquad (18.22)$$

As in the derivation of Eq. (18.20), the number of reflections n_{sample} with a sample present is

$$n_{\text{sample}} = \frac{1}{-\ln \rho + \alpha\overline{d}}, \qquad (18.23)$$

and

$$\tau_{\text{sample}} = \frac{m\overline{d}}{c\left(-\ln \rho + \alpha\overline{d}\right)}, \qquad (18.24)$$

where c is the speed of light in vacuum and m is the index of refraction of the sample medium filling the cavity. When the cavity is empty, τ_{empty} is given by τ in Eq. (18.21),

$$\tau_{\text{empty}} = \frac{\overline{d}}{-c \ln \rho}. \qquad (18.25)$$

Solving for α using Eqs. (18.24) and (18.25) gives

$$\alpha = \frac{1}{c}\left(\frac{m}{\tau_{\text{sample}}} - \frac{1}{\tau_{\text{empty}}}\right). \qquad (18.26)$$

This is a powerful result. If the index of refraction m of the sample medium is known, one only has to measure two decay times: (i) the decay time τ_{empty} with the cavity empty, and (ii) the decay time τ_{sample} with the cavity completely filled with the sample. These two measurements are then sufficient to obtain an accurate value of the absolute absorption coefficient for almost any sample without need for a calibration with a reference absorber.

However, there is another phenomenon that must also be kept under consideration. As shown in Figure 18.5, when light reflects from a diffusing surface, it spends some time inside the surface. Consequently, the time τ given by Eq. (18.21) should instead be written

$$\tau = -\frac{1}{\ln \rho}\left(\frac{\overline{d}}{c} + \delta t\right), \qquad (18.27)$$

where $\overline{d}/c$ is the average time of a transit across the cavity and δt is the average time spent in the wall after each transit. A measurement of δt in quartz powder cavities was obtained by measuring τ for spherical cavities with different diameters D (note, $\overline{d} = 2D/3$). From Eq. (18.27), a plot of τ versus $\overline{d}$ is a straight line with slope $\mathbf{m}$ and y-intercept $\mathbf{b}$,

$$\mathbf{m} = -\frac{1}{c \ln \rho}, \quad \mathbf{b} = -\frac{\delta t}{\ln \rho}. \qquad (18.28)$$

Solving these for $\ln \rho$ and δt gives

$$\ln \rho = -\frac{1}{\mathbf{m}c}, \quad \delta t = \frac{\mathbf{b}}{\mathbf{m}c}. \qquad (18.29)$$

Spherical cavities were made from the quartz powder by compressing the powder in two solid blocks and then machining a hemispherical depression in each one. The two blocks were then held together to make the spherical cavity and the decay constant τ was measured. The two blocks were then separated, and the hemispherical cavity was machined to a larger diameter and τ was measured again. This process was repeated eight times to collect the set of data shown in Figure 18.12.

A least-squares fit to the straight-line data in Figure 18.12 gives the equation shown in the figure with $\mathbf{m} = 3.453$ ns/mm and $\mathbf{b} = 4.001$ ns. The corresponding results for ρ and δt are

$$\rho = 0.99904 \pm 0.00003, \quad \delta t = 3.9 \times 10^{-3}\text{ns} \pm 66\%. \quad (18.30)$$

This time δt spent in the wall will impact absolute absorption measurements described by Eq. (18.26). To estimate the size of the effect, consider a sample whose index of refraction can be approximated as 1.0. By replacing the average transit time $\overline{d}/c$ in Eqs. (18.24) and (18.25) with the total time $\overline{d}/c + \delta t$, the equation for α is

$$\alpha = \frac{1}{c}\left(\frac{1}{\tau_{\text{sample}}} - \frac{1}{\tau_{\text{empty}}}\right)\left(1 + \frac{c}{\overline{d}}\delta t\right). \qquad (18.31)$$

The magnitude of the error produced by δt is the magnitude of $c\,\delta t/\overline{d}$ compared to 1.0. Take $\delta t = 3.9 \times 10^{-3}$ ns and $c = 30$ cm/ns, then if $\overline{d} = 3$ cm, the error, $c\,\delta t/\overline{d}$, is $0.039 \approx 4\%$; if $\overline{d} = 30$ cm, the error is only $0.0039 \approx 0.4\%$. Obviously, δt has less impact as the cavity increases in size and the average transit time increases.

Clearly, a key factor in the temporal response of an integrating cavity is that average distance $\overline{d}$ that a photon travels between reflections from the cavity wall. To obtain a general result for $\overline{d}$ [20], consider an integrating cavity of arbitrary shape whose volume is V and surface area is S. In an integrating cavity that contains N photons per unit volume, the photons will be uniformly distributed both throughout the volume and in all directions of propagation.

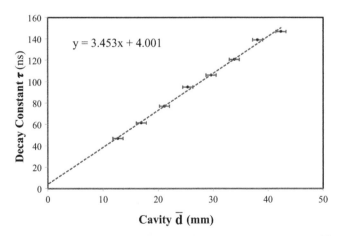

FIGURE 18.12 Measured decay constants τ as a function of $\overline{d}$ for spherical cavities made from compressed quartz powder.

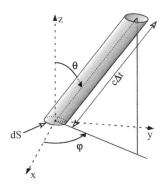

FIGURE 18.13 Schematic for the photon flux incident on the wall of a cavity.

Consider an infinitesimal element of area dS on the wall of the cavity and an infinitesimal cylindrical volume as shown in Figure 18.13.

With all photons in the cavity uniformly distributed throughout the cavity, the number of photons in the cylinder of length $c\Delta t$ in Figure 18.13 is $N\,(c\Delta t)\,(\cos\theta\,dS)$, where c is the speed of light in the cavity and θ, dS are defined in Figure 18.13. Since at any point in the volume the photon propagation directions are uniformly distributed over a 4π solid angle, the fraction of photons propagating in the direction θ, φ in the solid angle $d\Omega = \sin\theta\,d\theta\,d\varphi$ is $d\Omega/4\pi$. Hence, the total number of photons in the cylinder of Figure 18.13 that are traveling in the direction θ, φ in the solid angle $d\Omega = \sin\theta\,d\theta\,d\varphi$ is

$$dN_{\text{uniform}} = N\,(c\,\Delta t)\,(\cos\theta\,dS)\left(\frac{1}{4\pi}\sin\theta\,d\theta\,d\varphi\right)$$

$$= \left(\frac{N\,c}{4\pi}\right)\cos\theta\sin\theta\,d\theta\,d\varphi\,dS\,\Delta t. \quad (18.32)$$

Since there is no photon–photon scattering within the volume, all these photons must be supplied by reflection from the cavity surface.

Now, consider the reflection of photons from the cavity surface. The total number of photons in the cavity is N, V and all of these photons will, on average, be incident on the wall in a time $\bar{d}/c$. Thus, the number of photons incident on the cavity wall per unit area per unit time is

$$\left(\frac{NV}{S}\right)\bigg/\left(\frac{\bar{d}}{c}\right) = \frac{Nc}{\bar{d}}\frac{V}{S}, \quad (18.33)$$

and, the total number of photons incident on area dS in time Δt is

$$\frac{Nc}{\bar{d}}\frac{V}{S}dS\,\Delta t, \quad (18.34)$$

Assuming the diffusing cavity surface to be Lambertian, the probability that photons will be reflected from the surface at an angle θ to the normal is proportional to $\cos\theta$, independent of the angle of incidence. The probability p that photons will be reflected at angles θ, φ into an infinitesimal solid angle $d\Omega = \sin\theta\,d\theta\,d\varphi$ is then

$$p = \frac{1}{\pi}\cos\theta\sin\theta\,d\theta\,d\varphi, \quad (18.35)$$

where the factor $1/\pi$ provides normalization for the integral of p over 2π steradians. Thus, the number of photons, dN_{surf}, contained in that cylinder shown in Figure 18.13 is the number incident on the element of cavity surface area dS in time Δt and reflected at angles θ, φ into the solid angle $d\Omega = \sin\theta\,d\theta\,d\varphi$, *i.e.* dN_{surf} is just the product of Eqs. (18.34) and (18.35),

$$dN_{\text{surf}} = \left(\frac{Nc}{\bar{d}}\frac{V}{S}dS\,\Delta t\right)\left(\frac{1}{\pi}\cos\theta\sin\theta\,d\theta\,d\varphi\right)$$

$$= \left(\frac{Nc}{\pi\bar{d}}\frac{V}{S}\right)(\cos\theta\sin\theta\,d\theta\,d\varphi)\,dS\,\Delta t. \quad (18.36)$$

Equation 18.32 gives the number of photons in that infinitesimal cylinder and within that infinitesimal solid angle in Figure 18.13 based on the uniform distribution of both position and directions of propagation for photons throughout the cavity. Equation 18.36 gives the number of photons in that cylinder and within that infinitesimal solid angle based on the fact that they can only meet those conditions via reflection from the surface, *i.e.* photon–photon scattering is negligible. Thus, $dN_{\text{uniform}} = dN_{\text{surf}}$, and since the θ, φ dependences in dN_{uniform} and dN_{surf} are identical in Eqs. (18.32) and (18.36), the lead coefficients are equal,

$$\left(\frac{Nc}{\pi\bar{d}}\frac{V}{S}\right) = \left(\frac{Nc}{4\pi}\right), \quad (18.37)$$

or

$$\bar{d} = \frac{4V}{S}. \quad (18.38)$$

This is a completely general result for $\bar{d}$; it is independent of cavity shape and depends only on the volume-to-surface ratio of the closed cavity. An identical result but with more restrictive assumptions has also been obtained from neutron transport theory [21]. For a sphere of diameter D, the parameter $\bar{d}$ is

$$\bar{d} = \frac{2D}{3}. \quad (18.39)$$

For a cylinder of arbitrary height H and diameter D, the parameter $\bar{d}$ is

$$\bar{d} = \frac{2DH}{2H + D}. \quad (18.40)$$

If $D = H$, this result is identical to that of a sphere of diameter D. If $H \gg D$, then Eq. (18.40) gives $\bar{d} = D$. Another interesting cavity is that of a torus; it is both a non-simply connected and a non-convex region. But, previous derivations have assumed the region was both simply connected and convex [21]; hence, this sets the above derivation apart from previous results. For a torus, the volume to surface ratio is $D/4$, where D is the diameter of the ring cross section, *i.e.* $\bar{d}$ is totally independent of the overall diameter of the torus. Using Eq. (18.33), $\bar{d} = D$ for the torus, a result that is identical to $\bar{d}$ for a long cylinder. This makes physical sense because as a cylinder gets longer the relative contribution from the caps on the ends becomes less significant and it will begin to "look" like a torus that has been cut and stretched into a cylinder. Other results for $\bar{d}$ are available in the literature [20].

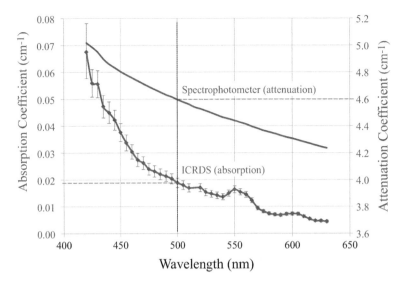

FIGURE 18.14   Absorption and attenuation coefficients for 6×10^7 human retinal pigmented epithelium (RPE) cells.

One of the first applications of ICRDS was a determination of spectral absorption coefficients of human pigmented epithelium cells with the pigmentation removed [4]. The data in Figure 18.14 shows both the absorption coefficient using ICRDS and the attenuation coefficient (absorption plus scattering) using a spectrophotometer. This data illustrates the special capability of ICRDS; *e.g.* the attenuation is 4.6 cm^{-1} at the wavelength of 500 nm, but the absorption is only $\sim$0.02 cm^{-1}. Absorption is a factor of over 200 smaller than attenuation and could not have been measured prior to ICRDS.

18.5   Summary

ICRDS is an extremely powerful new technique. It has all the advantages of both conventional CRDS and ICS and avoids their disadvantages. Advantages of ICRDS include the following:

1. It illuminates the sample isotropically.
2. It is independent of scattering effects in the sample.
3. It can provide effective path lengths of a hundred meters or more.
4. It provides absorption data for scattering particulates in a medium as well as the absorption of the medium.
5. It measures the rate of absorption rather than magnitude and thereby provides intrinsic insensitivity to source intensity variations.
6. It has no cavity mode structure or mode matching requirements.
7. It operates over a very broad spectral bandwidth because of the very broad bandwidth of the diffuse reflectivity.

References

1. O'Keefe A, and Deacon DAG, *Rev. Sci. Instrum.* **59**, 2544–2551 (1988).
2. Elterman P, *Appl. Opt.* **9**, 2140–2142 (1970).
3. Fry ES, Kattawar GW, and Pope RM, *Appl. Opt.* **31**, 2055–2065 (1992).
4. Cone MT, Mason JD, Figueroa E, Hokr BH, Bixler JN, Castellanos CC, Wigle JC, Noojin GD, Rockwell BA, Yakovlev VV, and Fry ES, *Optica* **2**, 162–168 (2015).
5. Zalicki P, and Zare RN, *J. Chem. Phys.* **102**, 2708–2717 (1995).
6. Sumpner WE, *Proc. Phys. Soc. London* **12**, 10–29 (1892).
7. Ulbricht VR, *Elektrotechnische Zetschrift* **21**, 595–597 (1900).
8. Labsphere, Spectralon® Diffuse Reflectance Material. www.labsphere.com/labsphere-products-solutions/materials-coatings-2/coatings-materials/spectralon/.
9. Buiteveld H, Hakvoort JHM, and Donze M, in *Ocean Optics XII*, in Proc. Soc. Photo-Opt. Instrum. Eng. **2258**, edited by J. S. Jaffe, Bergen, Norway, 1994, pp. 174–183.
10. Querry MR, Wieliczka DM, and Segelstein DJ, in *Handbook of Optical Constants of Solids II*, edited by E. D. Palik, (Academic Press, Inc.: Cambridge, MA), 1991, pp. 1059–1077.
11. Pope RM, and Fry ES, *Appl. Opt.* **36**, 8710–8723 (1997).
12. Cone MT, Musser JA, Figueroa E, Mason JD, and Fry ES, *Appl. Opt.* **54**, 334–346 (2015).
13. Stokes GG, *Proc. R. Soc. London* **11**, 545–556 (1862).
14. Mason JD, Cone MT, and Fry ES, *Appl. Opt.* **55**, 7163–7172 (2016).

15. Lee Z, Wei J, Voss K, Lewis M, Bricaud A, and Huot Y, *Appl. Opt.* **54**, 546–558 (2015).

16. Morel A, Gentil B, Claustre H, Babin M, Bricaud A, Ras J, and F.Tieche, *Limnol. Oceanogr.* **52**, 217–229 (2007).

17. Quickenden TI, and Irvin JA, *J. Chem. Phys.* **72**, 4416–4428 (1980).

18. Cruz RA, Marcano A, Jacinto C, and Catunda T, *Opt. Lett.* **34**, 1882–1884 (2009).

19. Kröckel L, and Schmidt MA, *Opt. Mater. Express* **4**, 1932–1942 (2014).

20. Fry ES, Musser J, Kattawar GW, and Zhai P-W, *Appl. Opt.* **45**, 9053–9065 (2006).

21. Case KM, and Zweifel PF, *Linear Transport Theory*, (Addison-Wesley: Boston, MA), 1967, p. 56.

19

Inductively Coupled Plasma Mass Spectrometry for Nanomaterial Analysis

Francisco Laborda,
Eduardo Bolea, and
Maria S. Jimenez
University of Zaragoza

19.1 Introduction

Inductively coupled plasma mass spectrometry (ICP-MS) is an atomic spectrometry technique that provides information for most elements in the periodic table (noble gases, H, N, O, F and C are excluded), with very low detection limits. Thus, ICP-MS is one of the techniques of choice for the analysis of inorganic engineered nanomaterials (ENMs). However, when considering the analytical chemistry related to nanomaterials, it must bear in mind that the analyses are not just focused on the characterization of pristine nanomaterials, as those synthesized in the laboratory or manufactured by the industry, but also on samples containing such nanomaterials. Whereas the analysis of pristine nanomaterials involves their characterization at different levels (ISO TC 229 2012, Tantra 2016), when a nanomaterial is part of a sample, firstly the presence of the nanomaterial must be confirmed, followed by its characterization and/or quantification, which are hindered by the complexity of the sample matrix and the concentration of the nanomaterial itself (Baalousha & Lead 2015, Laborda et al. 2016a). These samples include industrial or consumer products (e.g., cosmetics, textiles, polymers, foods), as well as any kind of biological or environmental sample, such as those produced under laboratory conditions to assess the release, fate, behavior, and (eco)toxicity of nanomaterials, or collected from "real world" compartments related to the life cycle of nanomaterials (e.g., waters, soils, organisms, tissues, cells).

At first glance, ICP-MS can not only be used for chemical characterization, providing information about the elemental composition of inorganic nanomaterials, but also for the determination of elemental impurities present in any type of nanomaterial. In the case of samples containing nanomaterials, the presence and the total mass content of the element/s monitored are determined in a more or less straightforward way. Selective information about the presence of a nanomaterial containing a specific element in a sample can be achieved by previous isolation of the nanomaterial by a batch separation technique (e.g., cloud point extraction, solid phase extraction) followed by its elemental detection and quantification. Alternatively, the dissolved fraction of the element can be separated (by dialysis, ultracentrifugation or ultrafiltration) and determined by ICP-MS. In any case mentioned, ICP-MS behaves as an ensemble technique (like dynamic light scattering [DLS]), providing bulk elemental information from a large number of nanoparticles measured simultaneously. However, ICP-MS can also be used as a counting technique (like electron microscopy) providing information of nanoparticles in a particle-by-particle basis when used in the so-called single particle mode (SP-ICP-MS). Finally, ICP-MS can also be used as an element-specific detector, both working in standard and single particle modes, in combination with different separation techniques, like field-flow fractionation, hydrodynamic chromatography (HDC) or electrophoresis. These techniques separate polydisperse suspensions into monodisperse fractions prior to their detection/quantification. Figure 19.1 shows an overview of the different types of analytical information achievable by using ICP-MS in the different modes and ways, in relation with the analysis of pristine nanomaterials and samples containing nanomaterials, which will be discussed along the chapter.

The chapter has been organized to provide the basic principles of ICP-MS (Section 19.2) and SP-ICP-MS

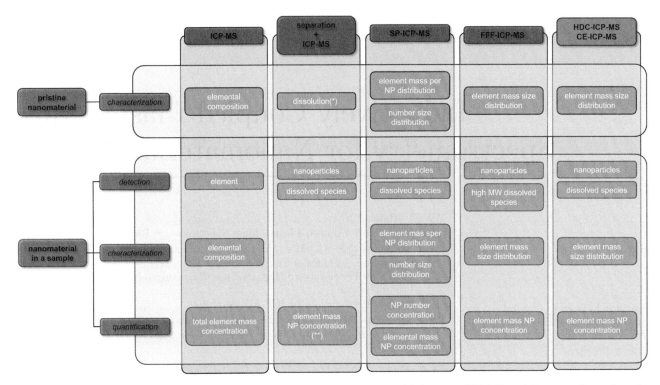

FIGURE 19.1 Overview of the different types of analytical information achievable by using ICP-MS in different modes and configurations. (*) In combination with ultrafiltration, dialysis or ultracentrifugation; (**) in combination with cloud point extraction.

(Section 19.3). Section 19.4 focuses on different continuous separation techniques commonly coupled to ICP-MS. Sample preparation methods usually applied in ICP-MS analysis are presented in Section 19.5. Section 19.6 discusses reliable approaches for the analysis of nanomaterials and nanomaterial containing samples by ICP-MS in combination with other techniques. Finally, Section 19.7 includes two paradigmatic case studies of nanomaterial analysis.

19.2   Inductively Coupled Plasma Mass Spectrometry: Basic Principles

ICP-MS is considered one of the most versatile atomic spectrometric techniques, providing rapid multielement analysis, from major to ultra-trace components, with detection limits down to the part-per-trillion level, for a large range of samples on a routine basis. Although detailed information about the technique can be found in several reference works (Thomas 2013, Becker 2007, Montaser 1998), here we will provide a brief overview.

In ICP-MS, elements are ionized in an inductively coupled plasma and then separated and detected in a mass spectrometer. The extraction of ions from the plasma source, which is at atmospheric pressure and 5,000–10,000 K, into the mass spectrometer involves an interface consisting of a series of apertures and chambers held at progressively lower pressures (down to 10^{-6} torr) and ambient temperature. A schematic diagram of a typical ICP-MS instrument is shown in Figure 19.2. Although samples can be analyzed in any aggregation state, most commonly, analytes are introduced in solutions or suspensions using a nebulization system. It consists of a nebulizer and a spray

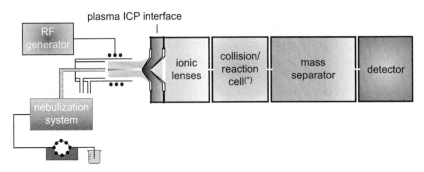

FIGURE 19.2 Schematic diagram of a typical ICP-MS instrument. (*) Collision/reaction cells just present in quadrupole instruments.

chamber, which produces an aerosol of small droplets. Once the droplets, containing the analyte, pass into the plasma, solvent evaporates, forming solid particles, which in turn are vaporized and their elements atomized and subsequently ionized. Analyte ions are confined in the central channel of the plasma, which is positioned horizontally in front of the mass spectrometer interface, being sampled along with the plasma gases through the apertures of the sampling and skimmer cones. Positive ions are focalized and separated from neutral species by an ion optic system, consisting of one or more electrostatically controlled lens components. Next, ions are introduced into the mass separator, where they are separated according to their mass-to-charge ratio (m/z). Quadrupoles, double-focusing sector field or time-of-flight separators are used in commercial ICP-MS instruments. Finally, detection is performed using electron multipliers or Faraday cups. Electron multipliers working in pulse counting mode are used for ultimate sensitivity. Quadrupole instruments are available with one or more additional multipoles (quadrupoles and/or octopoles), placed between the ion optics and the mass separator, which work in combination with a gas (He, O_2, H_2, CH_4, NH_3) as collision or reaction cells, reducing or eliminating the occurrence of spectroscopic interferences.

The high temperature of the argon plasma sources facilitates the volatilization and dissociation of the sample into its constituent atoms, resulting in a large proportion of singly charged atomic ions. Standard measurements in ICP-MS provide a mass spectrum whereby the intensity of the ion peaks at the corresponding m/z is plotted. These mass spectra are very simple, comprising mostly isotopic lines, although polyatomic ions formed from the plasma gas and/or the sample matrix can be present, particularly below 80 m/z. The occurrence of the ion peaks along with the knowledge of the isotopic pattern of the elements allow to get qualitative information about the elemental composition, whereas elemental quantitative information is obtained from the ion peak intensity of the isotope/s selected for each element. Quantification involves the calibration with dissolved element standards or the use of isotope dilution. Ideally, the response of an element in the ICP-MS is similar when it is introduced as particles or dissolved species, although it may be not the case for very refractory nanomaterials or large particles.

19.3 Single Particle ICP-MS

Whereas conventional ICP-MS just provides information about elemental composition and element mass concentration, SP-ICP-MS allows to obtain information about: (i) the presence of particulate and/or dissolved forms of elements; (ii) the mass of element/s per nanoparticle, which can be converted into nanoparticle size as long as information about the composition, shape and density of the nanoparticles is known or assumed; (iii) nanoparticle number concentrations

and (iv) mass concentrations of the dissolved and/or particulate forms (Laborda et al. 2016b, Montaño et al. 2016).

The transformation of conventional ICP-MS into a counting technique is based on the different behavior of dissolved and particulate forms of an element in the plasma, along with the use of very fast data acquisition. The key point is that soluble forms of an element are distributed homogenously within a solution, whilst in a nanoparticle suspension, the element is distributed heterogeneously, being present just in the discrete nanoparticles. This means that a solution produces a constant flux of element and a steady signal, whereas a suspension of nanoparticles produces packs of ions from each single nanoparticle (Figure 19.3a,b). If the suspension is sufficiently dilute and the data acquisition sufficiently fast, these packs of ions can be detected as individual signals. Figure 19.3 shows these time-resolved signals, with a duration of about 400–800 μs, recorded at different acquisition frequencies. By using fast data acquisition (e.g., 10^4 Hz, i.e. using reading times of 100 μs), the profile of the transient signal produced by each particle can be recorded (Figure 19.3b). Working at lower frequencies (i.e. at reading times in the milliseconds range), the packs of ions are measured as individual pulses, as it is shown in Figure 19.3c,d. SP-ICP-MS signals are recorded during several minutes, obtaining time scans as shown in Figure 19.4a, which consist of a number of events, recorded as transient signals or pulses, depending on the selected acquisition frequency, above a steady baseline. Whereas the intensity of each event is due to the ions detected from each particle, the baseline is due to the background at the mass recorded or to the presence of dissolved forms of the element

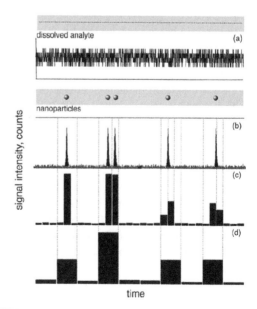

FIGURE 19.3 Simulated time-resolved ICP-MS signals from a solution (a) and a nanoparticle suspension (b–d) of the same element at different data acquisition frequencies (reading times). (b) 10,000 Hz (100 μs), (c) 300 Hz (3 ms), (d) 100 Hz (10 ms). Not to scale. (Reproduced with permission from Laborda et al. 2014. Copyright 2014, American Chemical Society.)

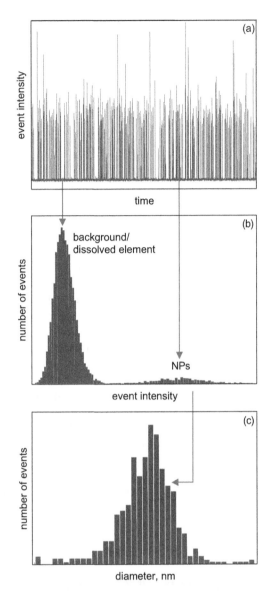

FIGURE 19.4 (a) Time scan of a nanoparticle suspension containing dissolved forms of the element present in the nanoparticles. (b) Event intensity frequency histogram of data from (a). (c) Number size distribution of spherical nanoparticles calculated from the second intensity distribution in (b). (Reproduced with permission from Laborda et al. 2016b. Copyright 2016, Elsevier.)

measured. Raw time scans can be processed by plotting the event intensity vs. the event intensity frequency, obtaining histograms as shown in Figure 19.4b, where the first distribution is due to the background and/or the presence of dissolved forms of the element measured and the second to the particles themselves.

The basic assumption behind SP-ICP-MS is that each recorded event represents a single nanoparticle. Under such conditions, the number of events counted during a fixed acquisition time is directly related to the number concentration of nanoparticles. On the other hand, the intensity of each event is proportional to the number of atoms of the element monitored in each detected nanoparticle and, hence, to the mass of element per nanoparticle. The mass

of the element can be related to the size of the nanoparticle if the composition, shape and density of the particle are known, as in Figure 19.4c, where the corresponding number size distribution is shown.

The detection of nanoparticles is associated to the capability of identifying the nanoparticle events over the baseline produced by the background (Figure 19.4a). The basic approach for identification of these events consists in applying a threshold criterion based on the standard deviation of the baseline (3σ or 5σ, where σ is the standard deviation of the baseline). In a similar way, size-detection limits are usually calculated by using a 3σ criterion. Metal nanoparticles above 5–10 nm can be detected by SP-ICP-MS, whereas the sizes increase for oxides up to several hundreds of nanometers, depending on their stoichiometry and composition. With respect to mass, limits of detection down to attograms per particle can be achieved (Lee et al. 2014, Laborda et al. 2016b). Double-focusing instruments, with better ion transmission and hence higher detection efficiency, as well as lower background signals, provide the best size-detection limits. With quadrupole instruments, the use of dwell times in the microsecond range and collision/reaction cells for isotopes subjected to polyatomic background interferences also can contribute to lower attainable detection limits. Number concentration detection limits are directly related to the number of particles that can be delivered to the plasma and, hence, to the nebulization efficiency and the sample flow rate but also to the time spent counting particles, i.e. the acquisition time. In practice, number concentration detection limits in the range of 10^6 L^{-1} are reported with current ICP-MS instruments and acquisition times around 1 min (Laborda et al. 2014).

19.4 Separation Techniques Coupled to ICP-MS

Different continuous separation techniques have already proven suitable for the separation of nanoparticles based on different properties. Here attention is paid to field-flow fractionation, HDC and capillary electrophoresis (CE). Their coupling to ICP-MS involves the transformation of ICP-MS from a bulk element-specific technique into a hyphenated technique selective to element species, providing two-dimension information (as fractograms, chromatograms or electropherograms) by relating the element mass concentration with the separation property, which is linked to the size of the nanoparticles mostly. When the separation technique is coupled to the ICP-MS working in single particle mode, a third dimension is added by considering the mass of element per nanoparticle.

19.4.1 Field-Flow Fractionation

Field-flow fractionation (FFF) is a family of flow-based, versatile fractionation techniques that allows the separation and size characterization of nanoparticles and their agglomerates/aggregates in the range from 1 nm to several μm (Schimpf et al. 2000). FFF operation entails the injection of a narrow sample band into a liquid stream and a subsequent focusing step inside the channel to produce a narrow sample band (Figure 19.5a). Separation takes place in a thin, elongated channel with a high-dimensional aspect ratio without a stationary phase. Retention is caused by the action of an externally generated field, applied perpendicularly to the laminar flow governed by a parabolic flow velocity profile inside the channel. This field drives the species to different positions between the channel walls according to the diffusive forces that counteract the field. Because of the velocity profile, as can be seen in Figure 19.5, the species are eluted at different times. There are two main elution modes in FFF, the normal or Brownian mode (Figure 19.5b), which is the principal elution mode for particles in the nanoscale size range (and up to 1 μm), and the steric mode (Figure 19.5c), for large particles for which the action of diffusion becomes negligible and they are driven by the field directly to the bottom wall of the channel. In normal mode, smaller particles (with higher diffusion coefficients) tend to elute first, whereas in steric mode, the order of elution is reversed, and larger particles elute first, since the elution is determined by the extent of penetration towards the center of the parabolic flow velocity profile.

The nature of the field applied (thermal, sedimentation, flow, electrical or magnetic) defines the type of FFF subtechnique. Until now only sedimentation FFF, in which a centrifugal force is applied, and flow FFF, where a perpendicular flow (crossflow) is applied, have been coupled to ICP-MS for the analysis of nanomaterials. Flow FFF separate sample components as a function of their diffusion coefficient and therefore to their hydrodynamic diameter (the diameter of a hypothetical hard sphere that diffuses in the same way that the particle under examination, including solvent and shape effects), whereas in sedimentation FFF, separation is based on the effective or buoyant mass of the particles, which includes a correction for the solvent density (Schimpf et al. 2000). Sedimentation FFF, also known as centrifugal FFF (CFFF) is especially suitable for high-density particles (e.g., metallic particles) in the range of 20 nm–20 μm, although the lower size is limited by the rotation speed attainable by the instrument and the density of the particle. On the other hand, flow FFF is the most universal of all FFF techniques since the crossflow applied can displace to all the species inside the channel. In flow FFF, the field consists of a second independent flow that permeates through the bottom

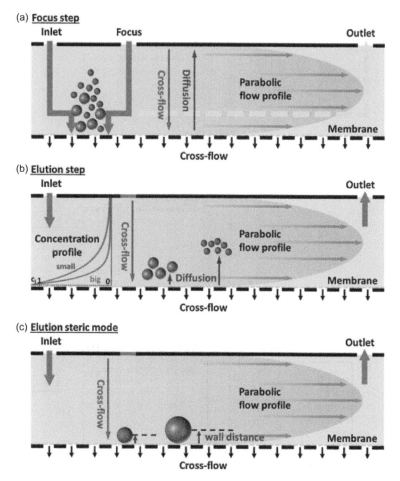

FIGURE 19.5 Schematic representation of asymmetrical flow FFF separation. (a) Focus step, (b) elution in normal mode, (c) elution in steric mode. (Reproduced with permission from Wagner et al. 2014. Copyright 2014, American Chemical Society.)

channel wall equipped with a permeation membrane. The pore size of this membrane (typically 1–10 kDa) limits the lower size attainable for this technique; size range that goes from 1–5 nm to 50 μm. There are two instrumental configurations commercially available: asymmetrical (AF4), which is the most used currently in the analysis of ENMs and hollow fiber (HF5). These configurations differ in terms of the geometrical channel shape and the way the cross-flow is applied.

FFF is commonly coupled to ICP-MS as online elemental detector (Meermann 2015, Pornwilard & Siripinyanond 2014), with element detection limits down to 1–10 μg L^{-1} for the hyphenated technique (Laborda et al. 2016a), and the ability to provide rapid multielemental and isotopic information. The interfacing of FFF and ICP-MS is straightforward, by connecting the outlet of the FFF system directly to the nebulizer of the ICP-MS with a tube or also including a T-piece to deliver simultaneously an internal standard to compensate for signal drift (Dubascoux et al. 2010). The typical instrumental configuration of an AF4 system coupled to an ICP-MS is shown in Figure 19.6a. FFF outflow range (from 0.5 to 1.0 mL min^{-1}) is compatible with those used

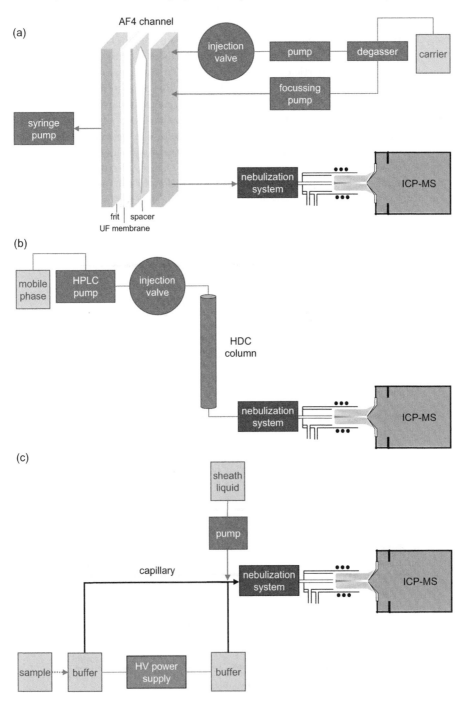

FIGURE 19.6 Schematic instrumental configuration of online couplings of ICP-MS with different separation techniques. (a) Asymmetrical flow FFF, (b) HDC, (c) CE.

by nebulizers. The coupling to ICP-MS working in single particle mode has also been described (Huynh et al. 2016, Hetzer et al. 2017), so information of number concentration and element mass per nanoparticle or core diameter of nanoparticles along the fractogram is obtained in addition to hydrodynamic diameters, as can be seen in Figure 19.7a. Prior to ICP-MS, nondestructive detectors such as UV-Vis absorption and fluorescence, DLS or multi-angle light scattering (MALS) can also be coupled to the exit of the FFF prior to the ICP-MS to provide complementary information. UV-Vis can inform about the nature of the species eluted (such as metallic nanoparticles and quantum dots through their plasmon resonance or fluorescence, respectively). An independent measurement of the dimensions of nanoparticles eluted can be obtained by detectors based on light scattering (the hydrodynamic diameter in the case of DLS and the radius of gyration for MALS). In general terms, the lower sensitivity of these detectors can limit their use for some applications.

The absence of stationary phase inside the channel, which reduces the interaction of the species during the separation step, the possibility of applying different field programs to separate particles largely different in size or the capability of injecting large volumes of diluted samples, allowing the preconcentration of species inside the channel, are some of the main advantages of FFF that usually are pointed out respect to other separation techniques (Meermann 2015, Contado 2017). Moreover, FFF systems are compatible with aqueous carriers and many organic solvents, which allows keeping, in most cases, the ENMs in the native conditions of the sample, minimizing the risks of altering their original size distribution. However, all these advantages should be put into context when the characterization of ENMs in complex matrixes is considered, where time-consuming optimization of the separation conditions (e.g., as carrier composition, type of permeation membrane, field applied) is required for the different ENMs and samples studied in order to obtain suitable channel recoveries and peak resolution (Gigault et al. 2014).

Although resolution in FFF is hindered by the inherent dispersity of most nanoparticle populations, it resembles that of size-exclusion chromatography, which is lower that other high-performance liquid chromatography techniques, as can be seen in Figure 19.8a for AF4-ICP-MS. Analysis time is another variable that must be considered, with fractograms in the range of 20–60 min, although times up to 90 min may be required depending on the grade of complexity of the sample.

19.4.2 Hydrodynamic Chromatography

HDC is a liquid chromatography separation technique in which the sample is injected into an open capillary tube or a column packed with solid beads, although porous beads with pore size smaller than the size of the particles can also be used (Striegel & Brewer 2012). In packed-column HDC, which is the configuration mostly used in

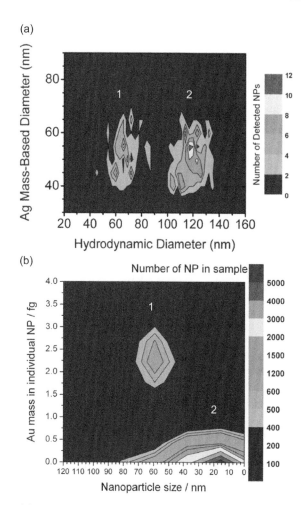

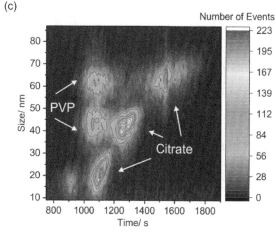

FIGURE 19.7 (a) Contour plot result of an AF4-SP-ICP-MS analysis of a mixture containing 60 nm Ag nanoparticles (1) and Ag-SiO$_2$ nanoparticles (2). (Reproduced with permission from Huynh et al. 2016. Copyright 2016, American Chemical Society.) (b) Contour plot of an HDC-SP-ICP-MS analysis of a drinking water sample spiked with Au nanoparticles having nominal sizes of 60 (1) and 30 nm (2). (Reproduced with permission from Pergantis et al. 2012. Copyright 2012, American Chemical Society.) (c) CE-SP-ICP-MS two-dimensional map acquired from a complex five-component mixture of different PVP- and citrate-coated Ag nanoparticles. (Reproduced with permission from Mozhayeva and Engelhard 2017. Copyright 2017, American Chemical Society.)

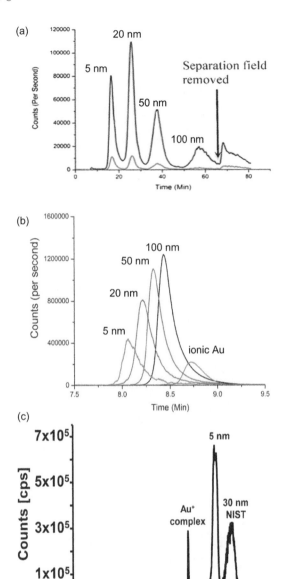

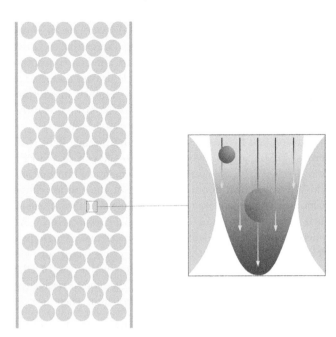

FIGURE 19.9   Schematic representation of HDC separation.

FIGURE 19.8   (a) AF4-ICP-MS fractogram of Au nanoparticles of different sizes. (Reproduced with permission from Gray et al. 2012. Copyright 2012, Royal Society of Chemistry.) (b) HDC-ICP-MS chromatogram of Au nanoparticles of different sizes and ionic gold. (Reproduced with permission from Gray et al. 2012. Copyright 2012, Royal Society of Chemistry.) (c) CE-ICP-MS electropherogram of gold nanoparticles of different sizes and ionic gold. (Reproduced with permission from Franze & Engelhard 2014. Copyright 2014, American Chemical Society.)

combination with ICP-MS, the beads should be inert, i.e., made of a material that minimizes non-HDC interactions between the stationary phase and the species under separation. Such non-HDC effects can be minimized through the addition of salts and/or surfactants to the mobile phase to screen electrostatic van der Waals interactions, which are especially common in aqueous media. Separation in HDC (Figure 19.9) arises from the parabolic flow velocity profile,

which develops under laminar flow conditions in the interstitial medium of the packed column, where the fastest streamlines of flow are in the middle of the interstitial medium and the slowest near the packing particles. Since larger nanoparticles cannot approach the walls of the packing beads, they remain near the center of the flow profile, where they preferentially experience faster velocities along the packed column. On the contrary, smaller nanoparticles, in addition to experiencing these faster velocities, also experience the slower ones, traveling through the packed column with a slower average velocity than that of their larger-sized counterparts. Therefore, larger nanoparticles in the sample elute from the packed column earlier than smaller ones. Because diffusion is the driving separation mechanism in HDC, nanoparticles are separated according to their hydrodynamic diameter, like in AF4.

HDC uses the same instrumental configuration that any other high-performance liquid chromatography technique. At present, two columns packed with solid beads are commercially available (800 mm length and 7.5 mm diameter), covering the size ranges of 5–300 and 20–1,200 nm. Mobile phases with low ionic strength containing anionic surfactants are recommended for such columns. Typical compositions include millimolar concentrations of phosphate salts, sodium dodecyl sulfate, Triton X-100 and formaldehyde, adjusting the pH to 7–8. Both mobile phase composition as well as column flow rates (1–2 mL min^{-1}) are compatible with the use of ICP-MS as elemental detector, which is coupled in a similar way than FFF and used in combination with other nondestructive detectors (Figure 19.6b).

Although HDC resembles AF4 in some ways, HDC has distinctive features which makes it appealing (Gray et al. 2012, Chang et al. 2017). HDC separations take place in a shorter time, with analysis times below 10 min. Moreover,

HDC is considered more robust than AF4, with higher averaged recoveries, including low molecular-weight dissolved species, which are not lost through the channel membrane (Roman et al. 2016). On the other hand, the resolving power of HDC is lower than AF4, as it can be seen comparing Figures 19.8a,b. When coupled to ICP-MS, both techniques provide the same information as element mass-based size (hydrodynamic diameter) distributions, with similar mass concentration detection limits. HDC can also be coupled to ICP-MS in single particle mode (HDC-SP-ICP-MS) (Pergantis et al. 2012, Proulx et al. 2014, Roman et al. 2016), which provides information about the element mass per nanoparticle along the chromatogram (Figure 19.7b), as described previously for AF4.

19.4.3 Capillary Electrophoresis

Electromigration separation techniques have demonstrated to be powerful analytical tools for the separation and characterization of ENMs (Trapiella-Alonso et al. 2016). CE, the most employed format of electromigration separations, provides a high-resolution separation of any type of compounds, based on their intrinsic properties such as the charge and the hydrodynamic diameter or the viscosity and temperature of the separation medium. CE employs narrow-bore capillaries, where species are separated due to their different electrophoretic mobility within an electroosmotic flow of an electrolyte solution under a high voltage (Ban et al. 2015). Capillary zone electrophoresis (CZE) is the simplest mode of CE, it is based on the use of separation media with relatively low viscosity where the species move from one end of the capillary to the other according to the balance between their electrophoretic mobility and the electroosmotic flow. Separations can be improved by adding specific additives to the separation medium, which are the base of other CE modes. In micellar electrokinetic chromatography (MEKC), a suitable charged surfactant, such as sodium dodecylsulfate, is added in a concentration sufficiently high to allow the formation of micelles, so that species are separated according to their different partition between the micelle, acting as a pseudostationary phase, and the separation medium. In capillary gel electrophoresis (CGE) the capillary is filled with a polymeric gel, the electroosmotic flow is eliminated and separation is based just on the electrophoretic mobility and size of the species. In capillary electrochromatography (CEC), the capillary is packed with microparticles coated with a bonded stationary phase and species are separate according to their partition between this stationary phase and the separation medium, which is moving as a result of the electroosmotic flow.

A typical CE instrument consists of a high-voltage power supply, a capillary and a detector (Figure 19.6c). Most capillary tubes are made of fused silica (inner diameter 25–75 μm) externally coated with a thin layer of polyimide to give it mechanical strength. Sample injection is different than in HDC or FFF, instead of using six-port valves for microliter delivery, volumes in the nanoliter range are introduced into the capillary by hydrodynamic or electrokinetic injection, applying a pressure or a potential during a short period of time, respectively. UV-visible absorption and fluorescence detection is performed on capillary by using the capillary itself as detection cell. However, the coupling of external detectors, that is the case of ICP-MS and other mass spectrometers, requires a specialized interface. Because the flow rate of the liquid emerging from the capillary is very small, the interface provides a sheath flow to get a flow rate compatible with the nebulizer of the ICP-MS, as well as a connection to the power supply of the CE system, as can be seen in Figure 19.6c.

CE is applicable to nanoparticles because they exhibit surface charge depending on the medium and their surface composition (López-Lorente et al. 2011). However, the addition of surfactants and the use of MEKC appears to be the most convenient mode for separation of metallic nanoparticles (Franze and Engelhard 2014, Qu et al. 2014). In addition, separation of ionic species and nanoparticles can be performed in the same run by addition of a suitable complexing agent, as can be seen in Figure 19.8c.

CE-ICP-MS provides information in the form of an electropherogram (Figure 19.8c), which resembles the chromatograms obtained in high-performance liquid chromatography. When compared to FFF and HDC, CE offers ultimate resolution because of the low dispersion of the sample moving along the capillary under electroosmotic flow, which shows a flat velocity profile, instead of the parabolic one of laminar flows. This lower dispersion also implies better detection limits than FFF or HDC-ICP-MS, below 1 μg L^{-1}. In spite of the lower flow rates of CE, separations take place in less than 10 min. CE has also been coupled to ICP-MS in single particle mode (Franze et al. 2017, Mozhayeva & Engelhard 2017, Mozhayeva et al. 2017). In this case, SP-ICP-MS provides information about the number concentration and core diameter of those particles separated by CE. Given that separation in CE is based on electrophoretic mobility, differences on surface coatings can be used to separate nanoparticles with the same nominal diameter (Mozhayeva & Engelhard 2017), as can be seen in Figure 19.7c.

19.5 Sample Preparation for ICP-MS Analysis

Although the bulk analysis of solid samples by ICP-MS can be carried out in combination with laser ablation (Russo et al. 2002), most of the routine procedures involve the digestion/decomposition of the solid samples to get the analyte transformed into soluble species for its introduction in solution into the ICP-MS. This is also the case for nanomaterials or solid samples containing nanomaterials, which are analyzed after a suitable digestion step. For samples consisting of aqueous suspensions of nanoparticles, although they can also be analyzed after digestion, direct analysis is a convenient option because it preserves the original characteristics of the nanoparticles. In any case, matrix removal or isolation and/or concentration of nanomaterials may require additional sample preparation steps apart from digestion when analyzing samples containing nanomaterials. Additionally, when ICP-MS is used in single particle mode or coupled to a continuous separation technique, the preservation of most of the properties of the nanomaterial is mandatory. Most frequently applied pretreatments in the analysis of nanomaterials by ICP-MS are presented in brief.

Digestion. Digestion of solid samples containing inorganic nanomaterials can involve the dissolution of the nanomaterial, the degradation of the sample matrix or both. For the decomposition of organic matrices, concentrated oxidizing acids like nitric acid, alone or in combination with hydrogen peroxide or hydrochloric acid, are commonly used in combination with conventional heating systems at atmospheric pressure or under pressure with microwave-assisted techniques. Under acidic conditions, nanomaterials like metallic silver or copper and copper or zinc oxides get dissolved, whereas others require additional reagents (e.g., aqua regia for gold, hydrogen peroxide for CeO_2, or hydrofluoric acid for TiO_2). In any case, these acid-based digestions are oriented to get information of the total element content in the nanomaterial or in the sample.

Alternatively, organic matrices can be degraded by using alkaline reagents, like tetramethylammonium hydroxide, or enzymes (proteases or pectinases) to solubilize biological materials (Loeschner et al. 2014, Jiménez-Lamana et al. 2016) while preserving the core of inorganic nanoparticles. These strategies allow the direct detection, quantification and size characterization of the nanomaterials by using SP-ICP-MS or ICP-MS in combination with a separation technique.

Dissolved/particulate fractionation. Ultracentrifugation, ultrafiltration and dialysis have been used to isolate dissolved species from nanoparticles, followed by the quantification of the dissolved fraction; hence, they have been extensively applied to study the dissolution of nanomaterials (Misra et al. 2012). By centrifugation, species are fractionated according to their size and density, whereas in dialysis and ultrafiltration, fractionation is based on the use of nanoporous membranes of different materials and nominal molecular-weight cut-offs. Because dialysis is based on pure diffusion, it takes longer times to achieving equilibrium; hence, ultrafiltration is preferred to speed up the separation process, because species below the cut-off are forced to cross the membrane using a centrifugal force. Centrifugation is not considered so efficient because of the harsh ultracentrifugation conditions needed for removal of nanoparticles.

Liquid phase extraction. Nanomaterials can be extracted from solid and liquid samples by using water or organic solvents while preserving some of their properties. This is the case of clean-up procedures based on the use of hexane for defatting sunscreen samples containing TiO_2 (Nischwitz & Goenaga-Infante 2012).

Cloud point extraction is also used for the isolation and concentration of nanoparticles, preserving their core size and morphology (Chao et al. 2011). Cloud point extraction involves the addition of a nonionic surfactant (e.g., Triton X114) at concentrations over the critical micellar concentration, the incorporation of the nanoparticles in the micellar aggregates and the separation of the surfactant phase from the aqueous one by mild heating (ca. 40°C). By adding a complexing agent, selective extraction of the nanoparticles in the presence of the corresponding cations can be achieved. This strategy has been combined with the element determinations to obtain information about the total element content in the sample and in the particulate fraction, as well as in the dissolved fraction by difference.

19.6 Analytical Approaches for the Analysis of Nanomaterials by ICP-MS

As we have summarized in Figure 19.1, the contribution of ICP-MS for the characterization of inorganic pristine nanomaterials lies in their bulk chemical characterization, through the determination of their elemental composition, as well as their behavioral characterization in relation with their solubility in specific media, combining ICP-MS with the isolation of the dissolved fraction under specified conditions. When ICP-MS is used in single particle mode, the technique is capable of providing reliable information about the number-based size distribution of the nanomaterial. In this respect, SP-ICP-MS is being considered on similar terms than other well-established counting techniques, like electron microscopy (Lövestam et al. 2010, ISO TC 229 2017). Separation techniques coupled to ICP-MS can also provide size information as mass-based size distributions.

The situation is far more complex when analyzing samples containing nanomaterials. The simplest approach is based on the use of ICP-MS for monitoring specific element/s present in the nanomaterial, both for detection and total element quantification, directly or by prior digestion of the sample. This approach may be sufficient when tackling with consumer and industrial products, as well as with samples from environmental and (eco)toxicological studies, where control samples with no added nanomaterials are available for comparison, like in bioaccumulation assays (Krystek et al. 2014). Although the determination of the total element content may provide the first evidence of the presence of a nanomaterial in a sample, this information can be inconclusive in many cases and more specific information may be required, involving more elaborated approaches that must include other analytical techniques apart from those based in ICP-MS (e.g., Lombi et al. 2012).

Table 19.1 shows different analytical approaches that have been applied to the analysis of nanomaterials and samples containing nanomaterials. Although not explicitly stated, most of the approaches includes the determination of the total element content. In general terms, liquid consumer products (disinfectants, dietary supplements...) as well as water samples can be analyzed directly or after a suitable

TABLE 19.1 Selected Analytical Approaches for the Analysis of Nanomaterials and Samples Containing Nanomaterials

Sample	ENM	Sample Preparation	ICP-MS	SP-ICP-MS	FFF-ICP-MS	HDC-ICP-MS	CE-ICP-MS	Other Techniques	Analytical Information	Refs.
Biological tissues	TiO_2	Acid digestion	●	–	–	–	–	–	Total element mass conc.	(Krystek et al. 2014)
Waste water	ZnO	Acid digestion	●	–	–	–	–	XAS	Total element mass conc.	(Lombi et al. 2012)
Sewage sludge	Ag	Ultrafiltration	●	–	–	–	–	–	NP dissolution	(Caballero-Díaz 2013)
Functionalized NPs	Ag	Filtration; Ultracentrifugation	●	–	–	–	–	–	Dissolved/NP element mass concentration	(Unrine et al. 2012)
Microcosm water	Ag	Filtration	●	–	–	–	–	TEM	Element size fractionation; NP characterization	(Mitrano et al. 2014a)
Wash water from textiles	Ag	Ultrafiltration	●	–	–	–	–	–	Dissolved/NP element mass conc.	(Chao et al. 2011)
Antibacterial products	Ag	Cloud point extraction	●	–	–	–	–	–	Dissolved/NP element mass conc.	(Mitrano et al. 2014b)
Waters	Ag	–	–	●	–	–	–	–	NP number size distribution; NP dissolution	(Dan et al. 2015)
Sunscreen	TiO_2	Dilution (surfactant)	–	●	–	–	–	–	NP number size distribution	(Jiménez-Lamana et al. 2016)
Plant tissue	Au	Enzymatic digestion	–	●	–	–	–	–	NP number concentration	(Praetorius et al. 2017)
Soil	CeO_2	Aqueous extraction	–	●	–	–	–	–	NP number size distribution; NP number concentration; Natural/engineered NP	(Praetorius et al. 2017)
Chicken meat	Ag	Enzymatic digestion	–	●	● (AF4)	–	–	TEM	NP number size distribution	(Loeschner et al. 2013)
Waste water	Ag	–	–	–	● (AF4)	–	–	–	NP mass size distribution; NP mass concentration	(Hoque et al. 2012)
Sunscreen	TiO_2	Clean-up (hexane)	–	–	● (AF4)	–	–	–	–	(Nischwitz and Goenaga-Infante 2012)
Sunscreen	TiO_2	Clean-up (hexane)	–	–	● (AF4)	–	–	–	–	(Samontha et al. 2011)
Coffee creamer	SiO_2	Clean-up (hexane)	–	–	● (C)	–	–	TEM	NP mass size distribution	(Heroult et al. 2014)
Sunscreen	TiO_2 ZnO	Dilution (surfactant)	–	–	● (AF4)	●	–	–	NP mass concentration; NP detection; NP size	(Philippe and Schaumann 2014)
Sewage sludge / Tap water	Ag	–	–	–	–	●	–	TEM	NP size	(Tiede et al. 2010)
Waste water	Ag	–	–	–	–	(HDC-SP-ICP-MS)	–	–	NP detection; NP Size	(Proulx et al. 2016)
Blood / Plasma	Ag	Dilution (mobile phase)	–	–	–	(HDC-SP-ICP-MS)	–	–	Dissolved/NP detection; NP Size	(Roman et al. 2016)
Cytosol simulants	Au	Ultrafiltration	–	–	–	–	●	–	Characterization of NPs conjugates	(Legat et al. 2018)
Consumer products	Ag	Acid digestion	●	–	–	–	–	TEM, UV-Vis, DLS, ISE	Total element mass conc.	(Tulve et al. 2015)
Coating	Ag	Water release	●	●	(AF4)	–	–	–	NP characterization; Total element mass conc.; Dissolved/NP element mass conc.; NP size	(Abad-Alvaro 2017)

pretreatment depending on the required information. When ICP-MS is used in combination with cloud point extraction, detailed information about the fraction of element present as nanoparticles is obtained (Chao et al. 2011). In a similar way, when it is combined with dialysis, ultrafiltration or ultracentrifugation, the dissolved fraction of the element can be determined directly (Caballero-Díaz 2013).

In the case of solid consumer products, including foods, personal care products, sunscreens or textiles, the grade of complexity of the matrix requires some kind of sample preparation. For instance, TiO_2 nanoparticles have been size-characterized in sunscreens by SP-ICP-MS (Dan et al. 2015) and HDC-ICP-MS (Philippe & Schaumann 2014) just after dilution with a surfactant solution, whereas defatting with hexane was needed prior to the analysis by AF4 (Nischwitz & Goenaga-Infante 2012) or CFFF (Samontha et al. 2011) coupled to ICP-MS. Similarly, Ag nanoparticles added to chicken meat have been detected and size-characterized by SP-ICP-MS and AF4-ICP-MS after an enzymatic digestion of the sample (Loeschner et al. 2013). The characterization of SiO_2 nanoparticles in coffee creamer by separation techniques such as AF4 required the use of ICP-MS as elemental detector, as well as a MALS detector, and transmission electron microscopy (TEM) with energy-dispersive X-ray (EDX) analysis offline (Heroult et al. 2014).

Environmental samples including waters, soils and sludges have been also analyzed to detect the presence of different nanomaterials, although most of the applications reported correspond to natural samples spiked with these ENMs. Using online AF4-ICP-MS, Ag nanoparticles previously added were detected and quantified in untreated wastewaters (Hoque et al. 2012). Also fate studies in different scenarios, such as natural water mesocosms or sewage sludges have been carried out, although the use of ICP-MS has been limited to the determination of nanoparticle mass concentrations and nanoparticle/dissolved ratios (Mitrano et al. 2014b). Tiede et al. (2010) studied the behavior of Ag nanoparticles in sewage sludge supernatant by ICP-MS, HDC-ICP-MS and TEM. Philippe and Schaumann (2014) demonstrated the effectiveness of HDC separation combined with UV-visible, fluorescence and ICP-MS detectors for determining the size and concentration of silver- and gold-based colloids in synthetic surface waters containing salts and natural organic matter. The presence of nanoparticles in wastewaters and river waters has been confirmed by the use of HDC-ICP-MS and HDC coupled to ICP-MS in single particle mode (Proulx et al. 2016). More recently, an approach based on the use of SP-ICP time-of-flight MS, capable of simultaneous multielement analysis in individual particles, coupled with a machine learning data treatment, has allowed to distinguish CeO_2-engineered nanoparticles from natural Ce-bearing nanoparticles based on their individual multielement fingerprints (Praetorius et al. 2017).

In relation with biological samples, HDC-SP-ICP-MS has been used for the determination of dissolved Ag and the distribution of Ag nanoparticles in terms of hydrodynamic diameter, mass-derived diameter, number and mass

concentration in plasma and blood (Roman et al. 2016). CE-ICP-MS has been used to separate and characterize Au nanoparticles and their protein conjugates in cytosols (Legat et al. 2018).

19.7   Analysis of Nanomaterials by ICP-MS: Case Studies

Two illustrative case studies are presented to highlight the capabilities and limitations of ICP-MS in relation with the analysis of nanomaterials and the need of multimethod approaches for solving analytical problems related to the analysis of nanomaterials. The first one makes use of ICP-MS in a more conventional way, just for the determination of total element contents, complementing this information with that obtained from other techniques, whereas in the second case, different analytical methods solely based on ICP-MS are used.

Case Study 1

Analysis of consumer products containing silver nanoparticles by ICP-MS and other analytical techniques. Tulve et al. (2015) developed a tiered approach for the analysis of consumer products (textile, plastic and liquid samples) containing silver nanoparticles. Total silver content was determined by ICP-MS, directly or after acid digestion. Visualization and sizing of nanoparticles were performed by scanning or TEM in combination with energy-dispersive X-ray spectroscopy to confirm the composition of the nanoparticles. For liquid samples, UV-visible absorption spectrometry was used as a supplementary technique to confirm the presence of metallic silver nanoparticles, based on their surface plasmon resonance absorption around 400 nm, DLS to provide information about hydrodynamic diameters and ion-selective electrode potentiometry to measure free ionic silver. ICP-MS could be used in combination with the rest of techniques because total silver contents in the samples were in the mg L^{-1}/mg kg^{-1} range or higher. The situation would be far more difficult if the expected concentrations were in the range of ng L^{-1} or ng kg^{-1}, where just ICP-MS could provide useful information because of its higher sensitivity.

Case Study 2

Release of silver species from a silver nanocoating by ICP-MS-based methods. In the approach developed by Abad-Alvaro et al. (2017), the sensitivity and element-specific response of conventional ICP-MS were complemented with the use of the technique in single particle mode and in combination with ultrafiltration and asymmetrical flow FFF. The release studies were performed by placing the coated samples in contact with water for 24 h followed by the analysis of the resulting suspensions. Total silver content of the coatings was determined by ICP-MS after acid digestion, whereas total silver released was determined directly without any treatment. Dissolved (ionic) silver in the suspensions was isolated by ultrafiltration using membranes of 3 kDa cut-off (equivalent to 2 nm hydrodynamic diameter) and determined by ICP-MS. SP-ICP-MS was used to detect the presence of dissolved and particulate forms of the element. It was not possible to obtain more information from this method due to the small size of the nanoparticles released. Finally, AF4-ICP-MS allowed to obtain information about the size of the nanoparticles and their mass concentration, which was in good agreement with the quantitative information obtained by ultrafiltration and total analysis. By using the combination of the ICP-MS methods described, information about the release of both dissolved and particulate forms of silver, as well as the size of the nanoparticles, could be obtained at concentrations down to 0.1 μg L^{-1} and nanoparticle diameters of 5 nm.

References

Abad-Alvaro, I.; Bolea, E.; Laborda, F.; Castillo, J.R., 2017. An ICP-MS-based platform for release studies on silver-based nanomaterials. *J. Anal. At. Spectrom.* 32, 1101–8.

Baalousha, M.; Lead, J.R. 2015. *Characterization of Nanomaterials in Complex Environmental and Biological Media.* Amsterdam: Elsevier.

Ban, E.; Yoo, Y.S.; Song, E.J. 2015. Analysis and applications of nanoparticles in capillary electrophoresis. *Talanta* 141: 15–20

Becker, J.S. 2007. *Inorganic Mass Spectrometry: Principles and Applications.* Chichester: Wiley.

Caballero-Díaz, E.; Pfeiffer, C.; Kastl, L. et al. 2013. The toxicity of silver nanoparticles depends on their uptake by cells and thus on their surface chemistry. *Part. Part. Syst. Charact.* 30: 1079–85.

Chang, Y.; Shih, Y.; Su, C. et al. 2017. Comparison of three analytical methods to measure the size of silver nanoparticles in real in environmental water and wastewater samples. *J. Hazard. Mater.* 322: 95–104.

Chao, J.; Liu, J.; Yu, S. et al. 2011. Speciation analysis of silver nanoparticles and silver ions in antibacterial products and environmental waters via cloud point extraction-based separation. *Anal. Chem.* 83: 6875–82.

Contado, C. 2017. Field flow fractionation techniques to explore the "nano-world". *Anal. Bioanal. Chem.* 409: 2501–18.

Dan, Y.; Shi, H.; Stephan, C.; Liang, X. 2015. Rapid analysis of titanium dioxide nanoparticles in sunscreens using single particle inductively coupled plasma–mass spectrometry. *Microchem. J.* 122: 119–26.

Dubascoux, S.; Le Hécho, I.; Hassellöv, M.; Von Der Kammer, F.; Potin Gautier, M.; Lespes, G. 2010. Field-flow fractionation and inductively coupled plasma mass spectrometer coupling: History, development and applications. *J. Anal. At. Spectrom.* 25: 613–23.

Franze, B.; Engelhard, C. 2014. Fast separation, characterization and speciation of gold and silver nanoparticles and their ionic counterparts with micellar electrokinetic chromatography coupled to ICP-MS. *Anal. Chem.* 86: 5713–20.

Franze, B., Strenge, I., Engelhard, C. 2017. Separation and detection of gold nanoparticles with capillary electrophoresis and ICP-MS in single particle mode (CE-SP-ICP-MS). *J. Anal. At. Spectrom.* 32, 1481–89.

Gigault, J.; Pettibone, J.M.; Schmitt, C.; Hackley, V.A. 2014. Rational strategy for characterization of nanoscale particles by asymmetric-flow field flow fractionation: a tutorial. *Anal. Chim. Acta* 809: 9–24.

Gray, E.P.; Bruton, T. A.; Higgins, C.P. et al. 2012. Analysis of gold nanoparticle mixtures: a comparison of hydrodynamic chromatography and asymmetric flow field flow fractionation coupled to ICP-MS. *J. Anal. At. Spectrom.* 27: 1532–39.

Heroult, J.; Nischwitz, V.; Bartczak, D.; Goenaga-Infante, H. 2014. The potential of asymmetric flow field-flow fractionation hyphenated to multiple detectors for the quantification and size estimation of silica nanoparticles in a food matrix. *Anal. Bioanal. Chem.* 406: 3919–27.

Hetzer, B.; Burcza, A.; Gräf, V. et al. 2017. Online-coupling of AF4and single particle-ICP-MS as an analytical approach for the selective detection of nanosilver release from model food packaging films into food simulants. *Food Control* 80: 113–24.

Hoque, M.E.; Khosravi, K.; Newman, K.; Metcalfe, C.D. 2012. Detection and characterization of silver nanoparticles in aqueous matrices using asymmetric-flow field flow fractionation with inductively coupled plasma mass spectrometry. *J. Chromatogr. A* 1233: 109–15.

Huynh, K.A.; Siska, E.; Heithmar, E. et al. 2016. Detection and quantification of silver nanoparticles at environmentally relevant concentrations using asymmetric flow field-flow fractionation online with single particle inductively coupled plasma mass spectrometry. *Anal. Chem.* 88: 4909–16.

ISO TC 229. 2012. *ISO/TR 13014:2012 Nanotechnologies - Guidance on Physico-Chemical Characterization of Engineered Nanoscale Materials for Toxicologic Assessment.* Geneve: International Organization for Standardization.

ISO TC 229. 2017. *ISO/TS 19590:2017 Nanotechnologies – Size Distribution and Concentration of Inorganic Nanoparticles in Aqueous Media via Single Particle*

Inductively Coupled Plasma Mass Spectrometry. Geneve: International Organization for Standardization.

Jiménez-Lamana, J.; Wojcieszek, J.; Jakubiak, M.; Asztemborska, M.; Szpunar, J. 2016. Single particle ICP-MS characterization of platinum nanoparticles uptake and bioaccumulation by Lepidium sativum and Sinapis alba plants. *J. Anal. At. Spectrom.* 31: 2321–29.

Krystek, P.; Tentschert, J.; Nia, Y. et al. 2014. Method development and inter-laboratory comparison about the determination of titanium from titanium dioxide nanoparticles in tissues by inductively coupled plasma mass spectrometry. *Anal. Bioanal. Chem.* 406: 3853–61.

Laborda, F.; Bolea, E.; Jiménez-Lamana, J. 2014. Single particle inductively coupled plasma mass spectrometry: a powerful tool for nanoanalysis. *Anal. Chem.* 86: 2270–8.

Laborda, F.; Bolea, E.; Cepriá, G. et al. 2016a. Detection, characterization and quantification of inorganic engineered nanomaterials: a review of techniques and methodological approaches for the analysis of complex samples. *Anal. Chim. Acta* 904: 10–32.

Laborda, F.; Bolea, E.; Jiménez-Lamana, J. 2016b. Single particle inductively coupled plasma mass spectrometry for the analysis of inorganic engineered nanoparticles in environmental samples. *Trends Environ. Anal. Chem.* 9: 15–23.

Lee, S.; Bi, X.; Reed, R.B.; Ranville, J.F.; Herckes, P.; Westerhoff, P. 2014. Nanoparticle size detection limits by single particle ICP-MS for 40 elements. *Environ. Sci. Technol.* 48: 10291–300.

Legat, J.; Matczuk, M.; Timerbaev, A. et al. 2018. Cellular processing of gold nanoparticles: CE-ICP-MS evidence for the speciation changes in human cytosol. *Anal. Bioanal. Chem.* 410: 1151–56.

Loeschner, K.; Navratilova, J.; Købler, C. et al. 2013. Detection and characterization of silver nanoparticles in chicken meat by asymmetric flow field flow fractionation with detection by conventional or single particle ICP-MS. *Anal. Bioanal. Chem.* 405: 8185–95.

Loeschner, K.; Brabrand, M.S.J.; Sloth, J.J.; Larsen, E.H. 2014. Use of alkaline or enzymatic sample pretreatment prior to characterization of gold nanoparticles in animal tissue by single-particle ICPMS. *Anal. Bioanal. Chem.* 406: 3845–51.

Lombi, E.; Donner, E.; Tavakkoli, E. et al. 2012. Fate of zinc oxide nanoparticles during anaerobic digestion of wastewater and post-treatment processing of sewage sludge. *Environ. Sci. Technol.* 46: 9089–96.

López-Lorente, A.I.; Simonet, B.M.; Valcarcel, M. 2011. Electrophoretic methods for the analysis of nanoparticles *Trends Anal. Chem.* 30: 58–71.

Lövestam, G.; Rauscher, H.; Roebben, G. et al. 2010. *Considerations on the Definition of Nanomaterial for Regulatory Purposes.* Luxembourg: Publication Office of the European Union.

Meermann, B. 2015. Field-flow fractionation coupled to ICP-MS: separation at the nanoscale, previous and recent application trends. *Anal. Bioanal. Chem.* 407: 2665–74.

Misra, S.K.; Dybowska, A.; Berhanu, D.; Luoma, S.N.; Valsami-Jones, E. 2012. The complexity of nanoparticle dissolution and its importance in nanotoxicological studies. *Sci. Total Environ.* 438: 225–32.

Mitrano, D.M.; Rimmele, E.; Wichser, A.; Erni, R.; Height, M.; Nowack, B. 2014a. Presence of nanoparticles in wash water from conventional silver and nano-silver textiles. *ACS Nano* 8: 7208–19.

Mitrano, D.M.; Ranville, J.F.; Bednar, A.; Kazor, K.; Hering, A. S.; Higgins, C.P. 2014b. Tracking dissolution of silver nanoparticles at environmentally relevant concentrations in laboratory, natural, and processed waters using single particle ICP-MS (spICP-MS). *Environ. Sci. Nano* 1: 248–59.

Montaño, M.D.; Olesik, J.W.; Barber, A.G.; Challis, K.; Ranville, J.F. 2016. Single particle ICP-MS: advances toward routine analysis of nanomaterials. *Anal. Bioanal. Chem.* 408: 5053–74.

Montaser, A. 1998. *Inductively Coupled Plasma Mass Spectrometry.* Chichester: Wiley.

Mozhayeva, D.; Engelhard, C. 2017. Separation of silver nanoparticles with different coatings by capillary electrophoresis coupled to ICP-MS in single particle mode. *Anal. Chem.* 89: 9767–74.

Mozhayeva, D.; Strenge, I.; Engelhard, C. 2017. Implementation of online preconcentration and microsecond time resolution to capillary electrophoresis single particle inductively coupled plasma mass spectrometry (CE-SP-ICP-MS) and its application in silver nanoparticle analysis. *Anal. Chem.* 89: 7152–59.

Nischwitz, V.; Goenaga-Infante, H. 2012. Improved sample preparation and quality control for the characterisation of titanium dioxide nanoparticles in sunscreens using flow field flow fractionation on-line with inductively coupled plasma mass spectrometry. *J. Anal. At. Spectrom.* 27: 1084–92.

Pergantis, S.A.; Jones-Lepp, T.L.; Heithmar, E.M. 2012. HDC with SP-ICP-MS for ultratrace detection of metal-containing nanoparticles. *Anal. Chem.* 84: 6454–62.

Philippe, A.; Schaumann, G.E. 2014. Evaluation of hydrodynamic chromatography coupled with UV-visible, fluorescence and inductively coupled plasma mass spectrometry detectors for sizing and quantifying colloids in environmental media. *PLoS One* 9: e90559.

Pornwilard, M.-M.; Siripinyanond, A. 2014. Field-flow fractionation with inductively coupled plasma mass spectrometry: past, present, and future. *J. Anal. At. Spectrom.* 29: 1739–52.

Praetorius, A.; Gundlach-Graham, A.; Goldberg, E. et al. 2017. Single-particle multi-element fingerprinting (spMEF) using inductively-coupled plasma time-of-flight mass spectrometry (ICP-TOFMS) to identify engineered nanoparticles against the elevated natural background in soils. *Environ. Sci. Nano* 4: 307–14.

Proulx, K.; Hadioui, M.; Wilkinson, K.J. 2016. Separation, detection and characterization of nanomaterials in municipal wastewaters using hydrodynamic chromatography coupled to ICPMS and single particle ICPMS. *Anal. Bioanal. Chem.* 408, 5147–55.

Qu, H.; Mudalige, T.K.; Linder, S.W. 2014. Capillary electrophoresis/inductively-coupled plasma-mass spectrometry: development and optimization of a high resolution analytical tool for the size-based characterization of nanomaterials in dietary supplements. *Anal. Chem.* 86: 11620–7.

Roman, M.; Rigo, C.; Castillo-Michel, H. et al. 2016. HDC coupled to SP-ICP-MS for the simultaneous characterization of AgNPs and determination of disssolved Ag in plasma and blood of burn patients. *Anal. Bioanal. Chem.* 408: 5109–24.

Russo, R.E.; Mao, X.; Liu, H.; Gonzalez, J.; Mao, S.S. 2002. Laser ablation in analytical chemistry – a review. *Talanta* 57: 425–51.

Samontha, A.; Shiowatana, J.; Siripinyanond, A. 2011. Particle size characterization of titanium dioxide in sunscreen products using sedimentation field-flow fractionation-inductively coupled plasma-mass spectrometry. *Anal. Bioanal. Chem.* 399: 973–78.

Schimpf, M.E.; Caldwell, K.; Giddings, J.C. 2000. *Field-Flow Fractionation Handbook.* Chichester: Wiley.

Striegel, A.M.; Brewer, A.K. 2012. Hydrodynamic chromatography. *Ann. Rev. Anal. Chem.* 5: 15–34.

Tantra, R. 2016. *Nanomaterial Characterization: An Introduction.* Chichester: Wiley.

Trapiella-Alonso, L.; Ramírez-García, G.; d'Orlyé, F. et al. 2016. Electromigration separation methodologies for the characterization of nanoparticles and the evaluation of their behaviour in biological systems. *Trends in Anal. Chem.* 84: 121–30.

Thomas, R. 2013. *Practical Guide to ICP-MS: A Tutorial for Beginners.* Boca Raton, FL: CRC Press.

Tiede, K.; Boxall, A.B.A.; Wang, X. et al. 2010. Application of hydrodynamic chromatography-ICP-MS to investigate the fate of silver nanoparticles in activated sludge. *J. Anal. At. Spectrom.* 25: 1149–54.

Tulve, N.S.; Stefaniak, A.B.; Vance, M.E. et al. 2015. Characterization of silver nanoparticles in selected consumer products and its relevance for predicting children's potential exposures. *Int. J. Hyg. Environ. Health* 218: 345–57.

Unrine, J.M.; Colman, B.P.; Bone, A.J.; Gondikas, A.P.; Matson, C.W. 2012. Biotic and abiotic interactions in aquatic microcosms determine fate and toxicity of Ag nanoparticles. Part 1. Aggregation and dissolution. *Environ. Sci. Technol.* 46: 6915–24.

Wagner, M.; Holzschuh, S.; Traeger, A.; Fahr, A.; Schubert, U.S. 2014. Asymmetric flow field-flow fractionation in the field of nanomedicine. *Anal. Chem.* 86: 5201–10.

20

Determination of Nanomaterial Electronic Structure via Variable-Temperature Variable-Field Magnetic Circular Photoluminescence (VTVH-MCPL) Spectroscopy

Patrick J. Herbert and
Kenneth L. Knappenberger Jr.
The Pennsylvania State University

20.1 Electronic Structure Characterization of Nanoscale Materials

The electronic properties for atoms and some molecular systems can be understood through computational and experimental quantum mechanical approaches that analyze discrete electronic transitions between orbitals with well-defined spin and orbital angular momentum quantum numbers (King 1964). Many of these approaches can be adapted to understand quantum-confined nanostructures. For semiconductor nanoparticles, the electronic properties are determined, in part, by material band structure. For these systems, photon absorption generates a bound electron–hole pair (i.e. exciton). An example of a well-understood excitonic system is a zero-dimensional CdSe nanocrystal (i.e. quantum dot). Figure 20.1 shows a comparison of the electronic state diagrams for a quantum dot relative to a molecular system. For a molecular system, a state is defined by its principle quantum numbers (n, l, m_l, and m_s). For CdSe quantum dots, only the total angular momentum (J) quantum number is valid (Efros & Rosen 2000). The total angular momentum of an excitonic state is dependent of electron–hole interactions and crystal field splitting. Furthermore, these electron–hole and crystal field interactions can generate energy separations, known as zero-field splittings (ZFS), between states on the order sub-to-hundreds of meV. Electronic relaxation and subsequent optical properties of excitonic systems are dependent

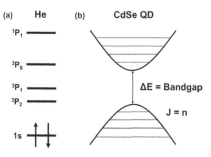

FIGURE 20.1 Comparison of atomic (a) and semiconductor (b) electronic structure.

on the relative total angular momenta (J) and the ZFS energies separating the states. As such, it is necessary to characterize these electronic structure parameters in order to understand the exciton dynamics and optical properties of nanomaterials. An established experimental method that has been widely used to characterize the electronic structure of molecules is been variable-temperature variable-field magnetic circular dichroism (VTV$\vec{H}$-MCD) spectroscopy (Stephens et al. 1975, Neese & Solomon 1999, Piepho & Schatz 1983, Mason 2007). However, more recently, it has been shown that VTV$\vec{H}$-MCD spectroscopy can also be employed to study nanoscale excitonic materials (Kuno et al. 1998, Ando et al. 1993, Archer et al. 2007). Of specific interest to this chapter is the emission analogue to VTV$\vec{H}$-MCD, magnetic circular photoluminescence (MCPL), which has proven useful in transient electronic structure characterization of sub-nanometer ligand-protected metal nanoclusters and 0-D and 1-D semiconducting nanocrystals (Herbert et al. 2017, Green et al. 2016, Zhao et al. 2018, Blumling et al. 2011, Blumling et al. 2013). Throughout this chapter, several examples from our research on metal nanoclusters are described as illustrations of the information content available from VTVH-MCPL. These nanocluster systems are of note because while dimensionally being on the order of other nanoscale materials, the electronic structure is modeled as a jellium superatom. As such, these systems serve as a prototypical transition point between atomic and nanoparticle systems.

In the remainder of this chapter, we will use recently published work studying the transient electronic structure of metal nanoclusters as a case study for the experimental method and information content available through VTVH-MCPL spectroscopy. In doing so, the fundamental background, experimental observables and data analysis/interpretation will be discussed in such a way that this method can be utilized for electronic structure characterization for a variety of nanoscale material systems. The remainder of this chapter is as follows: (i) an introduction to magnetic field effects on atomic systems; (ii) an overview of the electronic structure information provided through magneto-optical spectroscopy; (iii) a description of the experimental observables observed for the VTVH-MCPL method and data analysis procedures; and (iv) a nanoscale material case study utilizing the derived VTVH experimental formulism.

20.1.1 Magnetic Field Effects

The Normal Zeeman Effect

Although many magnetic-field interactions exist, a primary focus of the VTVH-MCPL method involves the Zeeman effect. The Zeeman effect arises from inherent magnetic moments present in molecular systems due to the acceleration and spin of electrons. The application of an external magnetic field on a quantum system results in a force applied to the magnetic moments (μ_H) of electronic transitions. The effect of this force depends on the magnitude and symmetry of each magnetic moment. These Zeeman interactions are especially important for spin-orbit-coupled systems, such as the metal nanoclusters described here. For atomic orbitals, the magnetic moment associated with each orbital is dependent on the orbital angular momentum (l) and orbital magnetic (m_l) quantum numbers. When an external magnetic field ($\vec{H}$) is applied, the field will act on the orbital magnetic moment, $\vec{\mu}_z$, along the axis of field propagation. Depending on relative orientation and magnitude of $\vec{H}$ and $\vec{\mu}_z$, the field will have a stabilizing or destabilizing effect on the orbital. For example, Figure 20.2 depicts the Zeeman effect for s and p orbitals. For an s-orbital, the orbital magnetic moment, $m_j = 0$. As a result, the application of a magnetic field has no influence on the relative orbital energy. For a p-orbital, the magnetic field lifts the degeneracy of the $m_j = +1$ and

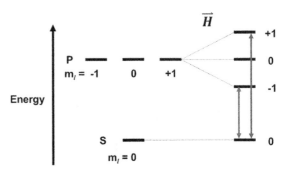

FIGURE 20.2 An illustration of the normal Zeeman effect for s and p orbitals.

-1. For an unoccupied orbital, the energy splitting (ΔE) is given as

$$\Delta E = \mu_z \vec{H} \qquad (20.1)$$

Here, μ_z is related to the orbital magnetic quantum number and the Bohr magneton.

$$\mu_z = -m_l \mu_B \quad (m_l = l, l-1, \ldots, -l) \qquad (20.2)$$

The Bohr magneton is the inherent magnetic moment of an electron based on its mass and charge. Equation (20.2) gives the formal expression for μ_B.

$$\mu_B = \frac{e\hbar}{2m_e} \qquad (20.3)$$

Here, e is the elementary electron charge, m_e is the mass of an electron and $\hbar$ is Plank's constant. The value of μ_B equals 5.788×10^{-5} eV/Tesla. For weak magnetic fields on the order of several Tesla (T), Eqs. (20.1)–(20.3) can be used to predict the Zeeman effect for unoccupied orbitals or diamagnetic systems (i.e. $S = 0$). For partially occupied or paramagnetic systems (i.e. $S \geq 1/2$), an additional contribution to the effective magnetic moment via the spin of an electron must be accounted for. This effect is known as spin-orbit coupling and will be discussed in the next section.

The Anomalous Zeeman Effect

For an occupied orbital, the magnetic moment is given by the total angular momentum, J. The total angular momentum is determined by the sum of orbital angular momentum (L) and the spin of an electron (S).

$$J = L + S \qquad (20.4)$$

For a single electron system, the spin of an electron (m_s) can equal $\pm 1/2$. For a multielectron system, the value S is defined as the sum of all individual m_s values. For example, a two-electron system where $s = \pm 1/2$ for each electron can have $S = 1$ or 0 depending on if the spins are aligned ($\uparrow\uparrow$) or misaligned ($\uparrow\downarrow$). The multiplicity of values for S is given as ($2s + 1$). In a similar fashion, the spin of an electron can couple to the orbital angular momentum depending on the relative orientation of their angular momenta projections. This is known as spin-orbit coupling. Figure 20.3 shows the

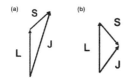

FIGURE 20.3 General schematic of spin-orbit coupling resulting from the vector addition of spin and orbital angular momentum where L and S projections are aligned (a) and misaligned (b).

vector addition of angular moment contributions from the electron spin (S) and the orbital angular momentum (L). As shown in Figure 20.3, the coupling between L and S can be additive (Figure 20.3a) or subtractive (Figure 20.3b) depending on the relative alignment of L and S angular momenta projections. The result of this vector addition is the total angular momentum of the system, J.

For nanoscale materials, excitonic states are defined by the value of J. The promotion/relaxation of an electron from one state to another via the absorption/emission of a photon is subject to spectroscopic selections rules. In general, a photon carries an angular momentum value of 1. Due to the law of conservation of angular momentum, the transition of an electron from one state to another via a photon must be accompanied by a change in the angular momentum by a value of 1. This is known as the LaPorte selection rule and is one of many spectroscopic selection rules. In general, for an allowed transition, $\Delta J = 0, \pm 1$, but not $J = 0$ to $J = 0$. Other selection rules apply to the allowed changes in spin ($\Delta S = 0$) and orbital angular momentum ($\Delta L = 0, \pm 1$) values for a spectroscopic transition. Because of these spectroscopic selection rules, it is convenient to define a state's total angular momentum, along with the spin and orbital angular momenta. To do this, each state is assigned a spectroscopic term symbol (Eq. 20.5)

$$^{2S+1}L_J \qquad (20.5)$$

From spectroscopic term symbols, the spin character of a state can be defined, and allowed electronic transitions can be predicted. Magneto-optical spectroscopy can be used to determine these spectroscopic term symbols. For spin-orbit-coupled systems, the magnetic moment of the state is defined by the total angular momentum, J. As with normal Zeeman splitting, the application of an external magnetic field will apply a force on the M_J vector. To account for the influence of spin-orbit coupling, we must adjust Eq. (20.1) by including the Landé g-factor (g_L).

$$\Delta E = g\mu_B \vec{H} \qquad (20.6)$$

From Eq. (20.6), we can define the Landé g-factor as the quantitative representation for spin-orbit coupling of an electronic state. Furthermore, the Landé g-factor can be related to spectroscopic term symbols by applying a spin-orbit coupling model such as Russell-Saunders (RS) or j–j coupling.

$$g_L = 1 + \frac{J(J+1) + S(S+1) - L(L+1)}{2J(J+1)} \qquad (20.7)$$

In this way, the Landé g-factor can be experimentally characterized, and hence, the angular and spin momenta of the state can be determined. In addition, through various magnetic field-dependent analysis, other influences of spin-orbit coupling such as ZFS can be quantified. In doing so, we can characterize the electronic structure of a system. These electronic structure parameters provide valuable information into understanding a system's optical properties. This includes the determination of allowed and forbidden electronic transitions, the quantification of energy barriers between electronic relaxation processes and an understanding of electronic and vibrational coupling between states. The remainder of this chapter will focus on the experimental aspects of magneto-optical spectroscopy. We will develop the underlying formalisms for using variable-temperature and variable-field analyses to extract electronic structure information including Landé g-factors, ZFS and relative Faraday terms. This will conclude with a case study for utilizing VTV$\vec{H}$ magneto-optical spectroscopy to characterize the transient electronic structure of a nanoscale system.

20.1.2 Magneto-Optical Spectroscopy

The experimental determination of Landé g-factors relies on the field-induced differential absorption or emission of circularly polarized light. From Eq. (20.6), the Zeeman splitting of a state where $g = 2$ and $\vec{H}= 10$ T is approximately 1 meV. For molecular and nanoscale systems, absorption line widths can be on the order of tens to hundreds of meV. As a result, direct spectral resolution of the energy separation between Zeeman split electronic transitions is not a sufficient experimental observable. To overcome this experimental limitation, MCD and MCPL spectroscopies rely on inherent spectroscopic selection rules which arise from external field effects. To illustrate this, a hypothetical electronic transition from p to s orbitals is considered. Figure 20.4 depicts the promotion of an electron from the p to s orbital upon the absorption of a photon. Upon the application of a magnetic field, the p-orbital undergoes Zeeman splitting. This splitting has the effect

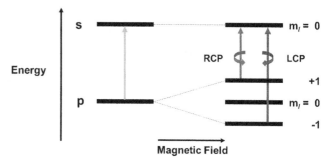

FIGURE 20.4 Illustration of electronic absorption with and without an applied external magnetic field.

of stabilization and destabilization of the $m_l = -1$ and $m_l = +1$ subshells, respectively. Because $m_l = 0$ for an s orbital, no magnetic field interaction occurs. Due to this splitting, two electronic transitions, from the $m_l = \pm 1$ subshells, are observed at lower and higher energy relative to the zero-field transitions. The orbital magnetic quantum number for Zeeman split states adds an additional spectroscopic selection rule requirement. As previously mentioned, a photon carries an angular momentum of 1. For linear polarized light, the angular momentum of the photon is a superposition of angular momentum values of ± 1. When a quarter-wave phase shift is introduced to linear polarized light, the photon becomes circularly polarized. We denote left- and right-circularly polarized angular momenta values of +1 or −1, respectively. Experimentally, this means, we can selectively probe Zeeman split states by controlling or monitoring the polarization state of the incident/emitted light.

MCD spectroscopy relies on the differential absorption of left- and right-circularly polarized light to characterize steady-state electronic transitions. For transient electronic characterization, MCPL can be used. By determining the differential response of left- and right-circularly polarized light as a function of applied field strength, information about the spin-orbit coupling, namely the Landé g-factor, can be obtained. Similarly, the differential response can be monitored as a function of sample temperature. By utilizing Boltzmann statistics, electronic structure information such as ZFS and relative Faraday term contributions can be determined. In the next section, the determination of these electronic structure parameters through MCD and MCPL spectroscopic analysis are derived.

Variable-Field MO Spectroscopy

Circular dichroism can occur naturally, i.e. without the application of a magnetic field, due to the spatial distribution of electric charge in a system. We can begin by relating circular dichroism to Beer's Law for a system of known molar concentration, c, and pathlength of light, l, using Eq. (20.8).

$$\Delta A = \text{LCP} - \text{RCP} = \Delta\varepsilon_M cl \qquad (20.8)$$

Here, the differential absorption, ΔA, of left- and right-circularly polarized light is dependent on the differential molar absorptivity coefficient, $\Delta\varepsilon_M$. For magnetically induced circular dichroism, the term $\vec{H}$ is included to account for the applied field strength (Eq. 20.9)

$$\Delta A_H = \text{LCP} - \text{RCP} = \Delta\varepsilon_M cl\vec{H} \qquad (20.9)$$

From Eq. (20.9), we can predict the differential absorption response should be linear with respect to $\vec{H}$. At room temperature, a linear response is observed as each Zeeman split state is equally populated. However, when sample temperatures are lowered well below the energy separation induced from Zeeman splitting (i.e. $\Delta E < kT$), a Boltzmann population distribution between Zeeman split states leads to a deviation from linear behavior. Consider a diamagnetic system ($g = 1$) under the influence of a 10 T applied field. Using Eq. (20.6), the resultant Zeeman splitting would be approximately 0.58 meV. If we consider a Boltzmann's distribution where $\Delta E = k_B T$, the energy separation from Zeeman splitting is equal to ΔE at approximately 7 K. This means at temperatures lower than 7 K, an unequal population distribution between Zeeman states will persist. Figure 20.5a demonstrates temperature-dependent population redistribution for Zeeman split states. At sample temperatures where Zeeman splitting is on the order of $k_B T$, an increase in the energy separation as a function of applied field strength will result in an increased thermal redistribution of electrons to the lowest energy state. At a critical applied field strength, ΔE greatly exceeds $k_B T$. At this limit, further increases to the magnetic field strength has no effect on the population distribution and a saturation behavior is observed. Figure 20.5b shows this field-dependent saturation behavior at various sample temperatures. At sample temperatures where $\Delta E_{\text{Zeeman}} < k_B T$ (20 K and 10 K traces), a linear behavior is observed as predicted by Eq. (20.9). For sample temperatures where $\Delta E_{\text{Zeeman}} > k_B T$ (4.5 K and 1.7 K traces), deviations from linear behavior at high fields are observed. Experimentally, this nonlinear field- and temperature-dependent response can be modeled using Eq. (20.10).

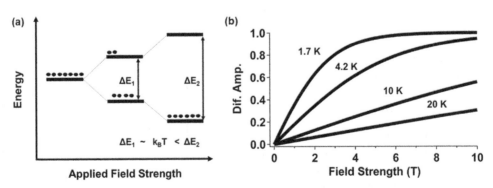

FIGURE 20.5 (a) Schematic of field-dependent Boltzmann populations of Zeeman states at cryogenic temperatures (i.e. ∼4.2 K). (b) Saturation behavior of field-dependent differential spectral amplitude at various sample temperatures.

$$\Delta I = A_{\text{sat}} \tanh\left(\frac{g\mu_B \vec{H}}{2k_B T}\right) \qquad (20.10)$$

Here the differential absorption, ΔI, is modeled using a hyperbolic function where A_{sat} is the saturation intensity, g is the Landé g-factor, $\vec{H}$ is the applied field strength and T is the sample temperature. It worth reinforcing that from inspection of Eq. (20.10), the only variable that is system specific is the Landé g-factor. The other variables in the Eq. (20.10) are either physical constants (μ_B, k_B) or experimental parameters ($\vec{H}$, T). The value of the term A_{sat}, is dependent on the signal-to-noise ratio of the measurement. As a result, from Eq. (20.10), we can directly determine the Landé g-factor for an electronic state. Further experimental considerations pertaining to Eq. (20.10) are that the differential response intensity is directly proportional to $\vec{H}$ and indirectly proportional to T. This would suggest that experimental conditions should be performed where the highest magnetic field and lowest sample temperatures are obtainable.

Multistate System: The Faraday A, B and C-Terms

Up to this point, we have considered the differential response resulting from Zeeman splitting of a single state. However, for nanoscale, and molecular, systems with high density of states (DOS), energy separation between individual electronic states is commonly on the order of ΔE_{Zeeman}. Just as thermal population redistribution between Zeeman states must be considered, temperature-dependent thermalization processes can occur between individual states. Furthermore, we must consider the effect of field-induced mixing between energetically nearly degenerate states. In order to expand our description of the magnetically induced differential response derived in the last section, we must incorporate Faraday A, B and C terms to our model. In short, each Faraday term represents one of three possible field-induced processes on a system. Faraday A and C terms denote Zeeman splitting of unoccupied and occupied states, respectively. Faraday B-terms denote field-induced mixing processes between nearly degenerate states. In this section, we will introduce the full Faraday term-included expression for MCD and MCPL acquired differential signal. Next, we will describe the physical mechanism and experimental observable of each term individually. Finally, we will derive the analytical formulism required to extract electronic structure information from experimental data.

The differential signal acquired from VTV$\vec{H}$-MCD or -MCPL methods is given by Eq. (20.11).

$$\frac{\Delta A}{E} = \left(cg\mu_B \vec{H}\right)\left[A_1\left(\frac{-\delta f(E)}{\delta E}\right) + \left(B_0 + \frac{C_0}{k_B T}\right)f(E)\right] \qquad (20.11)$$

Here, ΔA is the field-dependent difference between the absorption/emission of left- and right-circularly polarized light, $E = h\nu$, c is a constant that accounts for sample concentration and sample pathlength, g is the Landé g-factor, μ_B is the Bohr magneton, H is the applied magnetic field, $f(E)$ is the absorption bandshape and $\delta f(E)/\delta(E)$ is the first derivative. A_1, B_0 and C_0, are the Faraday A, B, and C terms, respectively. In total, Eq. (20.11) accounts for all contributions to the differential signal that can occur for a given system. By analyzing the differential signal response as a function of applied field strength and sample temperature, system-specific electronic structure information can be extracted. In particular, determination of relative Faraday A-, B- and C-terms can provide information on spin-orbit coupling interactions, orbital occupation and energy separation between states. To properly analyze differential MCD and MCPL signal as a function of Eq. (20.11), the experimental observables and data processing methods must be understood. Figure 20.6 depicts the physical mechanism for each Faraday term and the line shape profile observed in the differential spectral signal. We will now discuss each term in detail using Figure 20.6 as a reference.

A-Term

An A-term denotes differential absorption or emission into unoccupied Zeeman split states. Figure 20.6a shows the differential absorption of left- and right-circularly polarized light into unoccupied states and the resulting derivative line shape in the differential spectrum. For a Gaussian transition profile, absorption or emission peak energies are shifted depending on the Zeeman splitting energy. Assuming a similar transition dipole value, the magnitude of left- and right-circularly polarized absorption or emission is equal. For MCD and MCPL, a rigid shift (RS) approximation is used to predict differential signal line shape. The RS approximation states that while the center peak energy will shift in energy as a function of applied field strength, the spectral line profile will be unchanged. Subtraction of equal Gaussian profile's offset in energy results in a symmetric derivative line shape. This symmetric derivative line shape is a specific experimental signature of A-terms. Because the differential response arises from Zeeman splitting of unoccupied states, A-terms are not subject to Boltzmann thermal distributions and are accordingly temperature independent.

B-Term

Unlike A- and C-terms, B-terms arise not from Zeeman splitting but rather from field-induced mixing of nearly degenerate states. Field-induced mixing describes the interaction between magnetic moments of two or more separate spin-orbit coupled states. At sufficiently high fields, the magnetic moments of individual states will begin to vectorially add. This vector addition results in a transfer of each state's magnetic moment character to each other. Figure 20.6b shows this mixing process and resultant differential response. In the differential spectra, a symmetric Gaussian line shape is observed. The magnitude of B-terms is directly proportional to applied field strength and inversely proportional to the energy separation between mixing states. As with A-terms, B-terms are temperature independent. As a result, experimentally differentiating between A- and B-terms relies on the inspection of the differential line shape.

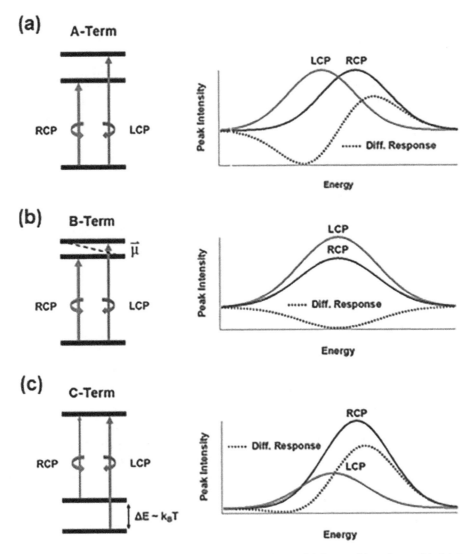

FIGURE 20.6   (a) Schematic of A-term behavior and contributions to observed differential line shape. (b) Origin of B-terms resulting from field-induced mixing interactions. (c) Temperature-dependent C-term behavior.

C-Term

Contrary to A-terms, C-terms arise from Zeeman splitting of occupied states. For MCD, this manifests as Zeeman splitting of ground states, while for MCPL C-terms commonly, but not exclusively, arise from Zeeman splitting of excited states. Figure 20.6c shows the differential response for C-terms. In comparison to A-terms, C-terms exhibit an asymmetric differential line shape. As previously mentioned, the energy separation between states resulting from Zeeman splitting is on the order of 1 meV. At low temperatures (<4.2 K), the differential optical response will be subject to the relative population of Zeeman split states. As shown in Figure 20.5a, the relative population of Zeeman split states is subject to the sample temperature and described by a Boltzmann distribution. As a result, the differential signal amplitude for a C-term at a fixed field will be inversely proportional to the sample temperature. This temperature-dependence provides an experimental approach to differentiating C-term behavior from A- and B-terms.

The temperature-dependent response may also result from thermal population redistribution between a manifold of Zeeman split states. Figure 20.7 shows the relative energy diagram for two nearly degenerate states undergoing

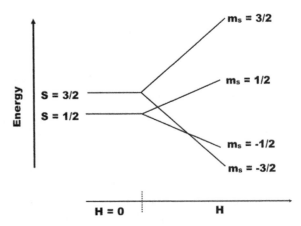

FIGURE 20.7   Zeeman splitting of zero-field split states.

Zeeman splitting. The energy between $S = 3/2$ and $S = 1/2$ arises from spin-orbit coupling and is denoted as the ZFS energy. Due to the higher angular momentum (g-factor) of the $S = 3/2$ state, the energy separation induced by Zeeman splitting is more sensitive to field strength than the lower angular momentum $S = 1/2$ state. As shown in Figure 20.7, at high fields, the $m_S = -3/2$ state resides at the lowest energy. At low temperatures, the $m_S = -3/2$ state will be preferentially populated due to Boltzmann statistics. As the sample temperature increases, the higher energy $m_s = -1/2, 1/2$ and $3/2$ states will begin to be populated dependent on the energy separation between states. This temperature-dependent thermal population will be reflected in the differential optical response as a C-term. We can also observe that at low field strengths, the relative ordering of the Zeeman split states changes. As a result, the temperature-dependent differential optical response will change relative to high fields. By acquiring a series of temperature-dependent measurements at fixed field strengths, we can extrapolate the ZFS energies of the unperturbed states. As we will see in the next section, quantifying these ZFS energies allows us to define spin-orbit coupling interactions and quantify relevant energy barriers that modify the optical properties of a system.

Experimental determination of A-, B- and C-terms requires spectral line shape and sample temperature-dependent analysis. Identification of A-terms is achieved by simple observation of a symmetric derivative line shape in the differential spectrum. Further confirmation of A-term assignment is achieved by an independence of the line shape with changing sample temperature. Determination of B- and C-term contributions is less straight forward. Due to the high DOS for nanoscale systems, it is ubiquitous for both B- and C-terms to contribute to the differential optical response. Quantifying the relative contributions of B- and C-terms to the differential response can provide information about state degeneracy and potential electronic coupling phenomena. Experimentally, the differential response is analyzed as a function of sample temperature. The temperature-dependent signal amplitude is fit to Eq. (20.12).

$$\text{Amplitude} = \sum_i \left(\frac{C_i}{k_B T} \alpha_i H + B_i \alpha_i H \right);$$

$$\alpha_1 = \frac{1}{1 + e^{\frac{-\Delta}{k_B T}}}; \alpha_2 = \frac{e^{\frac{-\Delta}{k_B T}}}{1 + e^{\frac{-\Delta}{k_B T}}} \quad (20.12)$$

Here, C_i and B_i denote relative amplitudes for the C- and B-terms. Each α represents the Boltzmann population weighting factors specific to B- and C-terms. For multicomponent systems, α can be expanded by inclusion of additional Boltzmann population terms. Finally, Δ is the energy barrier between states or ZFS value. For the remainder of this chapter, we will utilize a case study to show the application of the formalisms derived in this section. This will include the experimental design, data analysis and interpretation.

20.1.3 Case Study: VTVH-MCPL of $Au_{25}(PET)_{18}$

Here, we will apply the formalisms developed in the previous sections to a model nanoscale system. Our model system is the $Au_{25}(PET)_{18}$, where PET stands for phenylethanethiol, monolayer-protected nanocluster (MPC). The $Au_{25}(PET)_{18}$ nanocluster is structurally defined by three domains: (i) a 13 metal atom icosahedral core; (ii) an inorganic layer of alternating -S-Au-S- semiring units and (iii) a passivating layer of organic ligands (PET) which serve to stabilize the nanocluster in solution (Zhu et al. 2008). As previously mentioned, the electronic structure of the $Au_{25}(PET)_{18}$ nanocluster is defined by the jellium superatom model (Walter et al. 2008). Here, the valence electrons of each metal atom (6s) are taken to be delocalized over the metal core of the nanocluster. Each electron occupies a superatomic orbital denoted by an orbital angular momentum value. Total electron counts which equate to closed orbital shells (i.e. $n = 2, 8, 18, \ldots$) are expected to be stable. For MPCs with electron-withdrawing ligands, the total electron count must be adjusted. For the neutral $Au_{25}(PET)_{18}$ nanocluster, the total electron count is 7 and the electron configuration is expressed as S^2P^5. This predicted paramagnetic superatom electron configuration has been verified through electron paramagnetic resonance studies (Zhu et al. 2009). Furthermore, theoretical calculations using this superatom theory in party have developed an understood model for the electronic structure of this system. In general, low-energy electronic transitions arise from transitions between superatom P- (HOMO) and D- (LUMO) orbitals (Aikens 2010). Higher energy transitions arise from the promotion of electrons from the nanocluster ligand band to core-based superatom orbitals. Through VTVH-MCPL spectroscopy measurements, carrier relaxation proceeding through these two distinct energy-dependent pathways are characterized.

For VTVH-MCPL study, samples of $Au_{25}(PET)_{18}$ were dispersed in a 17% w/w polystyrene/toluene solution which was then dried as a film onto a microscopic slide. Before the sample slide is loaded into the magnetic, it must be affixed with a polarization film. The polarization film is composed of a quarter-wave plate (QWP) and a linear polarizer (LP) (Figure 20.8b inset). Unlike MCD where the circular polarization state of the incident light is controlled experimentally, emission from Zeeman split states occurs spontaneously (Figure 20.8a). As a result, for MCPL spectroscopy, the isolation of emission from Zeeman split states must occur through polarization optics after the sample. Circularly polarized light transmitting the QWP experiences a quarter-wave phase shift and becomes linearly polarized. For left- and right-circularly polarized light, this phase shift will result in photons of linear polarization state axes rotated 90° from each other. Next, a linear polarizer is fixed to a set transmission axis, allowing the transmission of one polarization state and extinguishing the other. The transmitted photon is then collected using a fiber

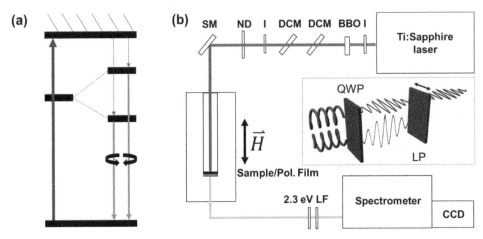

FIGURE 20.8 (a) Schematic of polarization-dependent emission from Zeeman split states. (b) Experimental design of MCPL spectroscopy. (Reproduced with permission from Herbert et al. 2017. Copyright 2017, Optical Society of America.)

optic, passed through spectral filters to extinguish light from the excitation source and spectrally dispersed onto a detector using a spectrometer. In order to observe emission from the extinguished Zeeman state, the direction of the applied field is inverted 180° while maintaining constant field strength, $\vec{H}$. This field direction inversion has the effect of inverting the order of Zeeman splitting. As a result, the previously extinguished emission will exhibit a polarization state enabling transmission through the polarization optics. In this way, MCPL differential spectra are generated by subtracting the PL spectra collected at parallel and antiparallel field directions.

Once the polarization film is affixed, the sample slide and film are mounted onto a nonconductive, fiber optic containing probe. The probe is then lowered, under vacuum, into the bore of a superconducting magnet. The magnet is composed of a superconducting solenoid coil dispersed in liquid helium. The sample probe is centered within the solenoid coil. The application of an external electric voltage drives a current through the solenoid coil generating a magnetic field along the direction to the laser excitation source. The field direction can be oriented parallel or antiparallel to the direction of the laser excitation source based on the direction of the electric current. The sample temperature was controlled via a conductive heater attached to the sample probe. Excitation of the sample was achieved by the frequency doubled output of a 1 kHz regeneratively amplified Ti:Sapphire laser system generating pulses centered at 1.55 eV. The frequency doubled 3.1 eV photons were directed onto the sample. The resulting PL was collected using a fiber optic and sent to a spectrometer. Residual 3.1 eV photons were extinguished before the slit of the spectrometer using 2.3 eV long-pass spectral filters. The PL was then dispersed using a spectral grating onto a liquid nitrogen-cooled charge coupled 1,024 linear pixel array.

Results: Variable-Field Analysis

Figure 20.9a shows the PL spectrum (black) of $Au_{25}(PET)_{18}$ taken at 4.5 K upon excitation of 3.1 eV. The PL spectrum

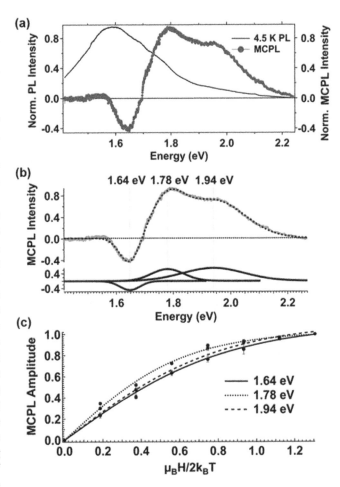

FIGURE 20.9 (a) $Au_{25}(SC_8H_9)_{18}$ photoluminescence spectrum collected at 4.5 K (black), following 3.1-eV excitation. Representative differential MCPL spectrum generated from PL collected at ±17.5 T applied field strength (gray). (b) Three component fitting scheme for differential spectra. (c) Differential peak amplitudes monitored as a function of applied field strength, at constant sample temperature of 4.5 K; the data correspond as follows: 1.64 eV (solid), 1.78 eV (dotted) and 1.98 eV (dashed). (Reproduced with permission from Herbert et al. 2017. Copyright 2017, Optical Society of America.)

is characterized by a broad spectral range. Also shown in Figure 20.9a is the differential PL spectrum (gray) measured at 17.5 T. The differential spectrum exhibits three discrete components centered at 1.64, 1.78 and 1.94 eV. This illustrates the advantage of MCPL spectroscopy in that spectrally congested PL components can be resolved by polarization-dependent detection. For VTVH analysis, each component is fit to a Gaussian function. Figure 20.9b shows the three Gaussian components and resultant fit (dashed) to the $Au_{25}(PET)_{18}$ differential spectrum. In Section 20.1.2.1, we derived how to determine the Landé g-factor from field-dependent analysis of the differential peak amplitudes. Figure 20.9c shows the field-dependent amplitudes for each component measured at 4.5 K. The solid lines show the best fits for the field-dependent data using Eq. (20.10). The g-factors for the 1.64, 1.78 and 1.94 eV components are 1.1 ± 0.1, 1.7 ± 0.1 and 1.05 ± 0.04, respectively. As discussed in Section 20.1.1.1, we can relate the g-factors to spectroscopic term symbols using a spin-orbit coupling model (Eq. 20.7). From the g-factors, the spectroscopic term symbols for the 1.64 and 1.94 eV components agree with a D-Doublet. The high g-factor 1.78 eV component is assigned as a P-Quartet.

Results: Variable-Temperature Analysis

As discussed in Section 20.1.2.2, field-dependent MCPL analysis is acquired at the lowest sample temperatures achievable to avoid influences of Boltzmann thermal population of higher energy states. However, by analyzing the differential response as a function of sample temperature and at a set field strength, we can determine relative Faraday term contributions to the differential response and quantify ZFS energies that arise from spin-orbit coupling interactions or crystal field splitting.

Figure 20.10a shows the differential MCPL spectra for $Au_{25}(PET)_{18}$ MPCs acquired at sample temperatures ranging from 4.5 to 40 K. Inspection of the three differential components as a function of sample temperature reveals two different trends. For the two higher energy components, 1.78 and 1.92 eV, a gradual signal amplitude decrease with increasing sample temperature is observed. The lowest energy, 1.64 eV, component conversely exhibits an increase in signal amplitude with increasing sample temperature. Figure 20.10b shows the temperature-dependent differential amplitudes of each component plotted versus reciprocal temperature. As shown in Eq. (20.12), variable-temperature amplitudes are inversely proportional to sample temperature. As a result, it is standard practice to report amplitudes as reciprocal temperatures. To this point, experimental considerations should be made as to the temperature steps sampled. The resolution of your energy gap is dependent on the temperature steps taken; smaller steps lead to finer energy gaps. The lower limit of any achievable energy gap quantification is given by the lowest sample energy achievable. As the differential signal amplitude is inversely proportional to sample temperature, it is imperative to take finer temperature steps at lower temperatures.

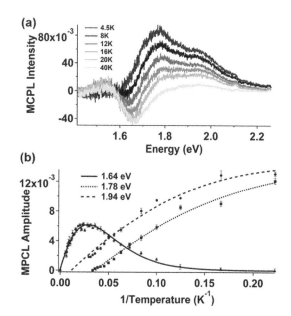

FIGURE 20.10 (a) Sample temperature-dependent differential spectra for $Au_{25}(PET)_{18}$. (b) Temperature-dependent differential amplitudes for 1.64 eV (solid), 1.78 eV (dotted) and 1.98 eV (dashed) components. (Reproduced with permission from Herbert et al. 2017. Copyright 2017, Optical Society of America.)

Due to the absence of a symmetric derivative line shape, contributions from Faraday A-terms can be neglected. As a result, we can use Eq. (20.12) to fit the temperature-dependent differential amplitudes for each component and determine the ZFS, Δ, energies and relative B- and C-term contributions. The results for 1.78 and 1.94 eV components yielded Δ values of 420 ± 20 μeV and 320 ± 50, respectively. These several hundred μeV Δ energies agree with expected energy separations due to ligand field splitting. For the 1.64 eV component, analysis of temperature-dependent differential amplitudes yielded a Δ value of 3.8 ± 0.1 meV. This value was interpreted as ZFS splitting of core-based superatomic orbitals due to strong spin-orbit coupling. These Δ values suggest that carrier relaxation of the higher energy 1.78 and 1.94 eV components proceed through a distinct set of states, namely the ligand band, relative to the core-based 1.64 eV PL component.

Analysis of temperature-dependent differential amplitude was also used to determine relative Faraday B- and C-terms. The results of the 1.78 and 1.94 eV components yielded relative B-term contributions of 0.17 ± 0.02 and 0.06 ± 0.01. These B-term weighting factor values indicate that field-induced mixing processes are present. The magnitude of B-term contributions is dependent on the energy separation between mixing states. For Δ energies on the order of several hundred μeV, strong mixing is expected at 17.5 T ($\vec{H}\mu_B = 1$ meV). Unlike the 1.78 and 1.94 eV components, the 1.64 eV component amplitude exhibited temperature-dependent behavior characteristic of pure C-terms as previously observed in VTVH-MCD studies. Correspondingly, analysis of the 1.64 eV component yielded a C-term weighting factor of 0.97 ± 0.01. Previous studies on the $Au_{25}(PET)_{18}$ revealed a similar temperature-dependent behavior to the global PL yield as observed from VTVH-MCPL analysis of the 1.64 eV component (Green et al. 2014). Furthermore, analysis of temperature-dependent global PL intensity was performed to quantify an activation barrier between bright and dark emission pathways in agreement with the 1.64 eV Δ value. These results suggest that VTVH-MCPL can correlate state-specific energy barriers to relaxation processes which determine the global optical properties of a system.

VTVH-MCPL can be used in conjunction with zero-field temperature-dependent PL analysis to extract electronic-vibrational coupling information (Green et al. 2014). By monitoring the peak energy of individual PL components, vibrational coupling strengths and energies can be quantified (O'Donnell & Chen 1991). Quantification of vibrational coupling strengths can be used to predict state-specific PL yields, while determination of coupled vibrational energies can correlate PL relaxation pathways to structure-specific motifs. For the $Au_{25}(PET)_{18}$ MPC, vibrational coupling analysis for the low-energy 1.64 eV component exhibited low vibrational coupling strength while the 1.78 eV component exhibited high vibrational coupling. These vibrational coupling strengths can be related to VTVH analysis. Inspection of the vibrational-coupling constants reveals that the relative magnitude of coupling for each component follows as the relative B-term weighting factors determined from VTVH analysis. Additionally, the g-factors determined for the 1.64 and 1.78 eV components are indicative of low- and high-angular momenta configurations, respectively. Spectroscopic selection rules, in accordance to the law of conservation of angular momenta, dictate that relaxation from high-spin electronic configuration must involve the partitioning of angular momenta to available vibrational degrees of freedom. As a result, the 1.78 eV component couples more strongly to vibrational modes resulting in lower PL yield. This suggests that VTVH analysis can be utilized to predict relative electronic-vibrational coupling for state-specific relaxation channels.

Vibrational energies determined for each component agree well with Au-Au stretching for the lower energy 1.64 eV component and Au-S stretching mode for both 1.78 and 1.94 eV components. This suggests the 1.64 eV emission channels relaxes through superatomic states localized to the metal core. Conversely, the Au-S stretching mode determined for the 1.72 and 1.96 eV components is associated with relaxation through the inorganic semirings. This assignment agrees with the interpretation of Δ values as resulting from either ligand field splitting (1.78 and 1.94 eV components) or spin-orbital coupling interactions (1.64 eV component). Furthermore, the ^{2}D term symbol assignment for the 1.64 eV component from field-dependent MCPL analysis agrees with theoretical calculations, which predict the low-energy emission channel to originate from core-based superatom D orbitals. Taken together, we can develop

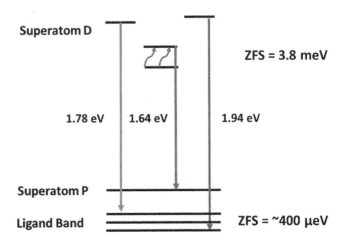

FIGURE 20.11 State-resolved mechanism for emission pathways of $Au_{25}(PET)_{18}$.

a state-resolved mechanism for which describes radiative- and non-radiative relaxation channels for the $Au_{25}(PET)_{18}$ MPC (Figure 20.11).

20.1.4 Summary and Outlook

To summarize, several examples of how VTVH magneto-optical methods can be used to obtain detailed understanding of the electronic properties of nanoparticles were given. These insights arise from the experimental ability to manipulate state-specific magnetic moments with a combination of sample temperature variation and applied magnetic fields. Transient signals from photo-excited states are isolated spectroscopically using polarization-resolved detection schemes. From these variable-temperature and field-dependent analyses, Landé g-factors can be quantized electronic transitions. Determination of g-factors leads to the assignment of spectroscopic term symbols using a spin-orbit coupling model. These spectroscopic term symbols can be used to understand the spin character of an electronic state and predict spectroscopic transitions. VTVH analysis enables the determination of relative Faraday terms and the quantification of energy separation between states arising from spin-orbital coupling interactions such as ZFS. As demonstrated here, this information provides valuable insight into dissection electronic relaxation pathways and subsequent optical properties of a system. In particular, Landé g-factor analysis enables the characterization of specific molecular orbitals and how the spin character of a state influences carrier dynamics. From ZFS analysis, different manifold of states can be identified by relative energy splitting values. For the study $Au_{25}(PET)_{18}$, this ZFS resulted in the discrimination of core- and ligand-based relaxation pathways. Furthermore, absolute values of ZFS energies can have direct thermodynamic implications on the global optical properties of a system. This includes ZFS values, which serve as the energy barrier between bright and dark emission pathways. When VTVH analysis is used in parallel with temperature-dependent PL analysis, information about electronic-vibrational coupling mechanisms can be determined. Finally, this work demonstrates that the steady-state electronic information content of MCD spectroscopy is directly transferable to MCPL spectroscopy for transient electronic structure characterization. Moving forward, exciting opportunities exist for the development VTVH magneto-optical spectroscopies with high time resolution, such as ultrafast multidimensional spectroscopy. Such an advance could pave the way for examination of spin coherences and spin-state-resolved energy transfer mechanisms.

Acknowledgments

Different aspects of the work summarized here have been supported by the Air Force Office of Scientific Research, grant number FA-9550-18-1-0347, and the National Science Foundation through awards CHE-1801829, CHE-1806222, and CHE-1807999.

References

Aikens, C. M. 2010. Geometric and electronic structure of $Au_{25}(SPhX)_{18}^{-}$ (X = H, F, Cl, Br, CH_3, and OCH_3). *J. Phys. Chem. Lett.* 1: 2594–2599.

Ando, K., Yamada, Y., Shakin, V. A. 1993. Magneto-optical study of quantum confinement in Cd(S,Se) quantum dots. *Phys. Rev. B* 47: 13462–13465.

Archer, P. I., Santangelo, S. A., Gamelin, D. R. 2007. Direct observation of sp-d exchange interactions in colloidal Mn^{2+} - and Co^{2+} -doped CdSe quantum dots. *Nano Lett.* 7: 1037–1043.

Blumling, D. E., McGill, S., Knappenberger, K. L. 2013. The influence of applied magnetic fields on the optical properties of zero- and one-dimensional CdSe nanocrystals. *Nanoscale* 5: 9049–9056.

Blumling, D. E., Tokumoto, T., McGill, S. et al. 2011 Magneto-photoluminescence properties of colloidal CdSe nanocrystal aggregates. *J. Phys. Chem. C* 115: 14517–14525.

Efros, A. L., Rosen, M. 2000. The electronic structure of semiconductor nanocrystals. *Annu. Rev. Mater. Sci.* 30: 475–521.

Green, T. D., Herbert, P. J., Yi, C. et al. 2016. Characterization of emissive states for structurally precise $Au_{25}(SC_8H_9)_{18}^{0}$ monolayer-protected gold nanoclusters using magnetophotoluminescence spectroscopy. *J. Phys. Chem. C* 120: 17784–17790.

Green, T. D., Yi, C., Zeng, C. et al. 2014 Temperature-dependent photoluminescence of structurally-precise quantum-confined $Au_{25}(SC_8H_9)_{18}^{0}$ and $Au_{38}(SC_{12}H_{25})_{24}$ metal nanoparticles. *J. Phys. Chem. A* 118: 10611–10621.

Herbert, P. J., Mitra, U., Knappenberger, K. L. 2017. Variable-temperature variable-field magnetic circular photoluminescence (VTVH-MCPL) spectroscopy for electronic-structure determination in nanoscale chemical systems. *Opt. Lett.* 42: 4833–4836.

King, G. W. 1964. *Spectroscopy and Molecular Structure.* New York: Holt, Rinehart and Winston, Inc.

Kuno, M., Nirmal, M., Bawendi, M. G. et al. 1998. Magnetic circular dichroism study of CdSe quantum dots. *J. Chem. Phys.* 108: 4242–4247.

Mason, W. R. 2007. *Magnetic Circular Dichroism Spectroscopy.* Hoboken: John Wiley and Sons.

Neese F., Solomon, E. I. 1999. MCD C-term signs, saturation behavior, and determination of band polarizations in randomly oriented systems with spin $S \geq 1/2$. applications to $S = 1/2$ and $S = 5/2$. *Inorg. Chem.* 38: 1847−1865.

O'Donnell, K. P., Chen, X. 1991. Temperature dependence of semiconductor band gaps. *Appl. Phys. Lett.* 58: 2924–2926.

Piepho, S. B., Schatz, P. N. 1983. *Group Theory in Spectroscopy.* Hoboken: John Wiley and Sons.

Stephens, P. J. 1975. Magnetic circular dichroism. *Adv. Chem. Phys.* 35: 197–264.

Walter, M., Akola, J., Lopez-Acevedo, O. et al. 2008. A unified view of ligand-protected gold clusters as superatom complexes. *Proc. Natl. Acad. Sci. USA* 105: 9157–9162.

Zhao, T., Herbert, P. J., Zheng, H. et al. 2018. State-resolved metal nanoparticle dynamics viewed through the combined lenses of ultrafast and magneto-optical spectroscopies. *Acc. Chem. Res.* 51: 1433–1442.

Zhu, M., Aikens, C. M., Hendrich, M. P. et al. 2009. Reversible switching of magnetism in thiolate-protected Au_{25} superatoms. *J. Am. Chem. Soc.* 131: 2490–2492.

Zhu, M., Aikens, C. M., Hollander, F. J. et al. 2008. Correlating the crystal structure of a thiol-protected Au_{25} cluster and optical properties. *J. Am. Chem. Soc.* 130: 5883–5885.

Photoluminescence Spectroscopy of Single Semiconductor Quantum Dots

Kateřina Kůsová
Institute of Physics of AS CR

Luminescence spectroscopy is a powerful tool to investigate the properties of semiconductor quantum dots (QDs). The interpretation of such measurements, however, is not straightforward. Individual QDs inevitably vary in their properties, which causes variations in the corresponding luminescence characteristics. Thus, a luminescence spectrum, or decay, of an ensemble of such QDs is a sum of a very large number of different signals emitted by individual QDs and there is simply no way to decompose the ensemble signal back into its individual components.

Consequently, single-QD luminescence spectroscopy is the ultimate tool to study the optical properties of these nanostructures, because it allows the characterization of (preferably a large number of) signals emitted by individual quantum emitters with well-defined properties. It unveils the differences between the individual QDs in an ensemble caused by size, even slightly different shape, surface composition, or the distribution of charges surrounding the QD. Thus, these experiments yield unique comparisons of the experimentally obtained and theoretically predicted behavior of quantum emitters and provide invaluable additional information to ensemble photoluminescence (PL) methods, which, on the other hand, probe the "average" optical properties.

Clearly, the main experimental obstacles are distinguishing the individual QDs and the inherently low levels of the emitted light, the detection of which is, however, feasible with modern sensitive detectors. Although microscopic PL studies of many types of nanostructures bring invaluable information, here, we will focus on the studies of single colloidal semiconductor QDs, which are a very interesting class of optically active materials suitable for a wide variety of applications (Kovalenko et al., 2015). In colloidal QDs, the type II–VI semiconductor QDs are usually in the spotlight because many cutting-edge measurements are carried out using those QDs due to the availability of well-defined samples as well as the excellent level of understanding of the behavior of these materials. Nevertheless, here, we want to focus on other types of QDs as well, as they offer interesting alternative to CdSe-based QDs in terms of material availability or toxicity.

This chapter is organized as follows: in the first two sections Section 21.1 and 21.2, the general considerations, providing a basis for the understanding of the rest of the text, and a historical background of single-QD experiments are given. Then, the general methods used in single-QD experiments are briefly described in Section 21.3. The chapter closes with the descriptions of what can be and has been found out about the PL of colloidal QDs using single-QD PL methods in two sections describing spectrally and temporally resolved experiments, Section 21.4 and 21.5, respectively. Further information can be found in various reviews, focusing on the PL spectroscopy of single QDs (Fernée et al., 2013, 2014; Sychugov et al., 2017) or temporally resolved measurements (Gómez et al., 2006; Frantsuzov et al., 2008; Cichos et al., 2007; Cordones and Leone, 2013; Efros and Nesbitt, 2016) and signal analysis (Verberk and Orrit, 2003; Lippitz et al., 2005; Cui et al., 2014).

21.1 General Considerations

21.1.1 Electronic Properties of Colloidal QD

Colloidal QDs are tiny crystals with sizes close to the Bohr exciton radius

$$a_B = 4\pi\epsilon \frac{\hbar^2}{e^2} \left(\frac{1}{m_e} + \frac{1}{m_h} \right), \qquad (21.1)$$

ϵ being the permittivity of the material, $\hbar$ the reduced Planck's constant, e the charge of an electron, and m_e and m_h the effective masses of an electron and a hole, respectively (Pelant and Valenta, 2012). The Bohr exciton radius corresponds to the distance of an electron and a hole in an exciton in a given material and therefore, when the sizes comparable to the Bohr exciton radius are reached in a crystal, the electron–hole pair starts to "feel" the confining potential of the crystal, giving rise to the effects of quantum confinement.

The Bohr exciton radius usually ranges between 1 and 5 nm depending on the material constants in Eq. (21.1).

The sizes of QD then vary approximately between 1 and 10 nm, implying that they comprise hundreds to thousands of atoms. As such, they lie somewhere between the scale of atoms or molecules and that of bulk, macroscopic semiconductors exhibiting collective behavior. Being a little bit too big for a molecule, but a bit too small for a crystal, their electronic states are described using different approaches, based on which starting point one chooses to use. Within the bulk semiconductor terminology, or a top-down approach, the size-related quantum-confinement effect adds up to effects of the crystal structure. Whereas quantum confinement gives rise to the well-known size-dependent emission energies, see Figure 21.1a, the contribution of the underlying crystal structure, usually expressed in terms of the effective mass, determines the number of states which comprise the band edge, and the structural anisotropy terms remove some of the degeneracies (Fernée et al., 2014). An opposite, bottom-up approach treats a QD as a large "molecule" and the corresponding molecular orbitals are computed (or estimated) (Fernée et al., 2014). The highest occupied molecular orbital and the lowest unoccupied molecular orbital are then treated as the edges of the valence and conduction bands, respectively. This approach inherently includes the quantum confinement effects and is often utilized in theoretical simulations. It is already well developed for the case of type II–VI semiconductor QDs (Gómez et al., 2006).

An interesting transition between these two approaches was suggested in Si QDs, where molecular orbitals computed

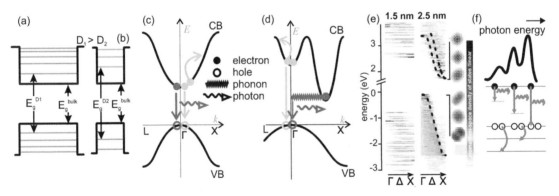

FIGURE 21.1 Overview of electronic properties of QDs. (a,b) A QD seen as a zero-dimensional quantum well. Electronic levels become discretized (the gray lines) and the QD's energy gap E_g widens with respect to the bandgap of the bulk semiconductor E_g^{bulk}. The energy levels also depend on the size of the QD $D_{1/2}$. (c,d) The bandstructure of a direct-bandgap (c) and an indirect-bandgap (d) bulk semiconductor. The radiative recombination of an electron–hole pair (the process in darker gray) is much more efficient in a direct-bandgap semiconductor, because in an indirect-bandgap semiconductor it requires the participation of a third quasi-particle, a phonon, to conserve the quasi-momentum $\vec{k}$. On the other hand, the non-radiative Auger recombination of a trion (the process in light gray) is about the same magnitude in both types of material. (e) A transition between the level-based and band-based description. The computed real-space molecular orbitals (density of space) of two sizes of hydrogen-terminated Si QDs were projected into the reciprocal space, the real-space localizations of orbitals of the larger QD are also shown for a few energy levels close to the bandgap as the circular objects next to the reciprocal-space density of states. Above a certain size threshold, the maxima of the reciprocal density of space very well copy the bandstructure of bulk Si, proving that the bandstructure concept is still valid, with some limitations, in such Si QDs. After (Hapala et al., 2013). (f) A scheme of the radiative recombination of a positively charged trion shown in the real-space representation. In contrast to a non-radiative Auger process from panels (c) and (d), the trion can recombine radiatively, transferring its excess energy to a photon. Alternatively, only a part of the energy can be transformed into an emitted photon, whereas the rest is consumed by the extra hole, which is then excited to a higher level. This process gives rise to replicas in the photoluminescence spectrum as shown at the top of the panel.

using the density functional theory approach were mapped into the reciprocal space (Hapala et al., 2013). Thus, the reciprocal-space bandstructure of these QDs was produced without *a priori* assuming any particular shape of bands. Above a certain size limit, this nano-bandstructure very well copied the shape of bulk silicon's bands, as shown in Figure 21.1e. Depending on the symmetry of the investigated crystal, band-edge states became discretized, in agreement with the existence of discrete energy levels at band edges.

No matter which approach is used to describe the evolution of electronic states (especially close to the band edges, which are of special interest for optical studies), the processes of radiative or non-radiative recombination and charge relaxation are then treated analogically to other systems (for an overview of the dynamics of the excited states in semiconductor colloidal QD, see (Rabouw and de Mello Donegá, 2016)). In short, the act of excitation gives rise to an electron–hole pair, which loses excess energy during a relaxation process, after which it forms an exciton. This exciton can then, as sketched in Figure 21.1c,d for a direct- and indirect-bandgap semiconductor, respectively, recombine either radiatively or non-radiatively, depending on the material and the particularities of the QD (see Table 21.1 for radiative exciton lifetimes τ_{rad} in different materials). An important difference between direct- and indirect-bandgap materials lies in the fact that whereas the non-radiative processes have comparable lifetimes regardless of the nature of the bandgap, the characteristic lifetimes of the radiative recombination are several orders of magnitude different. This difference implies that the indirect-bandgap materials are usually weak light emitters, because the excited level becomes depleted by non-radiative recombination before the act of radiative recombination can occur.

Under higher excitation photon fluxes (high excitation intensities), multiple excitons can form. Out of these quasi-particles, a biexciton and a singly charged exciton, or a trion, are the most important ones. In bulk semiconductors,

non-radiative Auger recombination is an efficient route causing quenching of the emission from multiple (or charged) excitons (Pelant and Valenta, 2012). However, the scenario of efficient Auger recombination in some QDs has been recently challenged based on the results of single-QD studies (see Section 21.4.3 and 21.5.3).

An important difference between PL studies of bulk semiconductors and QDs is the fact that, unlike in a bulk semiconductor, any excited state inside a QD is isolated from the rest of the QDs. This implies that under a certain excitation-intensity threshold, roughly corresponding to one absorbed photon per a QD and per its exciton lifetime, no multiexcitonic effects usually appear because the impinging photons are simply absorbed by the remaining, as of then non-excited QDs. Therefore, it is very useful to estimate the average number of absorbed photons (per a unit temporal interval or per the temporal difference between two consecutive excitation pulses) $\langle N \rangle$. On the other hand, in analogy to bulk semiconductors, the intensity of the emission from multi-excitonic features changes with the intensity of excitation and certain types of such power dependences are characteristic of a particular process, based on kinetic equations governing the temporal evolution of the system (Pelant and Valenta, 2012). Importantly, due to the temperature dependence of τ_{rad}, one also has to take into account the measurement temperature when trying to exclude saturation-related effects (Hartel et al., 2012).

21.1.2 Expected Signal in single-QD Experiments

Typical signal levels in optical experiments involving single QDs as light sources can be estimated using simple back-of-the-envelope calculations. First, it is useful to express the excitation laser power P_{laser} in terms of a photon flux

$$P_{\mathrm{laser}}\,(\mathrm{photons/cm^2 s}) = \frac{P_{\mathrm{laser}}\,(\mathrm{W/cm^2})}{E_{1\mathrm{photon}}} = \frac{P_{\mathrm{laser}}\,(\mathrm{W/cm^2})}{hc/\lambda_{\mathrm{laser}}}$$
$$= 5.03 \times 10^{15}\, P_{\mathrm{laser}}\,(\mathrm{W/cm^2})\lambda(\mathrm{nm}), \qquad (21.2)$$

TABLE 21.1 Comparison of Estimation of Number of Emitted Photons N_{emit} in Single Experiments for $P_{\mathrm{laser}} = 100\ \mathrm{W/cm^2}$, $\lambda_{\mathrm{laser}} = 400$ nm (633 nm for InAs) and Continuous Excitation.

Material	CdSe	InAs	Si	CsPbI$_3$	InP	PbS
Citation	(Leatherdale et al., 2002)	(Yu et al., 2005)	(Valenta et al., 2016)	(Becker et al., 2018)	(Chandrasekaran et al., 2017)	(Chen et al., 2018)
Size (nm)	3.4	3.4	3.3	9.3	3.0	4.5
σ_{abs} (cm^2)	1.5×10^{-14}	10^{-15}	10^{-15}	8×10^{-14}	2×10^{-14}	2×10^{-16}
N_{abs} (ph/s)	3×10^6	3×10^5	2×10^5	2×10^7	4×10^6	4×10^4
QY	0.5	0.5	0.05	0.7	0.6	0.7
N_{emit} (ph/s)	$\mathbf{1.5 \times 10^6}$	$\mathbf{1.6 \times 10^5}$	10^4	$\mathbf{10^7}$	2×10^6	$\mathbf{3 \times 10^4}$
τ_{rad}	20 ns	200 ns	300 μs	1 ns	50 ns	2 μs
R_{rad} (ph/s)	5×10^7	5×10^6	$\mathbf{3.3 \times 10^3}$	10^9	2×10^7	5×10^5
100 counts	6 ms	60 ms	1 s	1 ms	4 ms	0.5 s
50,000 counts	3 s	30 s	8 min	0.5 s	2 s	3 min

Rows with gray background highlight values which were estimated from the input white-background rows using Eqs. (21.3) and (21.4). Detectivity of the detection system can be around 0.1 CCD counts/emitted photon in excellent setups (Pevere et al., 2015), but the value 0.01 CCD counts/emitted photon is used. Please note that in very slowly emitting systems (Si), the signal might be limited by the slow radiative lifetime. The two bottommost rows show acquisition times necessary to achieve intensity levels of 100 counts (binning for a blinking intensity–time trace, see Section 21.5.5) and 50,000 counts (a single-QD PL spectrum, see Section 21.4). However, even in a single material, these values can vary from one sample to another.

where h is the Planck's constant and c the speed of light in vacuum. Using this expression and considering the absorption cross-section σ_{abs} of the studied QDs, the number of absorbed photons N_{abs} per second can be easily calculated as

$$N_{\mathrm{abs}} = P_{\mathrm{laser}} \sigma_{\mathrm{abs}}. \tag{21.3}$$

Since σ_{abs} of QDs is typically $10^{-17} - 10^{-14}$ cm^2 (Pelant and Valenta, 2012), one can see from Table 21.1 that in efficiently emitting QDs, the detected signal can easily be of the order of 10^4 counts per second (considering a simplified situation when only a single QD is present in the excited spot and its QY can be characterized using ensemble data) even below the $1^{\,\mathrm{exciton}}/_{\mathrm{QD}}$ excitation level. However, it is important to realize that the σ_{abs} values are typically deduced from macroscopic ensemble measurements, where more than a single QD absorbs light. On a single-QD level, especially under continuous excitation, where the excitation is not limited by the laser repetition rate, the maximal possible emitted signal might also be limited by the rate at which this QD can absorb/emit light (see Table 21.1). This situation can occur in QDs characterized by long radiative exciton lifetimes τ_{rad}, which imply slow emission rates R_{rad}, i.e. a small number of emitted photons per second

$$R_{\mathrm{rad}} = 1/\tau_{\mathrm{rad}}. \tag{21.4}$$

Apart from the signal-level itself, systems under laser illumination can be unstable and their emission can severely deteriorate with time. This phenomenon of photobleaching is definitely an issue in bio-imaging studies utilizing various types of molecules as light-emitting markers (Bruchez et al., 1998). Interestingly, semiconductor QDs have proven to have much better photostability (Bruchez et al., 1998), with Si QDs even excelling in stability over CdSe or CdTe QDs (Zhong et al., 2015; Bruhn et al., 2014b).

21.1.3 Heavy-Tailed Distributions

The behavior and treatment of heavy-tailed statistical distributions definitely deserve a more in-depth discussion since these distributions play an important role in some of the single-QD phenomena. Mathematically speaking, heavy-tailed distributions are those which have their moments undefined. For example, taking a probability density function $y(x)$ of a (continuous) statistical distribution, its first moment $\int_{-\infty}^{+\infty} x y(x)\,\mathrm{d}x$ represents the mean of the distribution. However, the integral might diverge and in that case the mean of such a distribution is undefined. (The second moment then represents variance, or deviation, the broadness of the distribution.)

In practice, the divergence of the moment-integral is caused by the much-too-slow decrease of $y(x)$ towards infinity. This causes a random variable obeying that distribution to assume values ranging over many orders of magnitude, which does not happen at "well-behaved" distributions. For example, the distribution of heights of adults,

whose mean is around 180 cm, is clearly not heavy-tailed, since one cannot find adults 5 mm, or 10 m tall.

An example of a heavy-tailed distribution occurring in single-QD spectroscopy is the Lorentzian distribution[1] (LD)

$$y_{\mathrm{Lorentz}}(x; \mu, \gamma) = \frac{1}{\pi\gamma \left[1 + \left(\frac{x-\mu}{\gamma} \right)^2 \right]}, \quad \mathrm{FWHM} = 2\gamma \tag{21.5}$$

characterized by the scale parameter (describing the broadness of the distribution) γ and median μ. It has both its mean and variance undefined. Its heavy tail can be nicely illustrated on a comparison with the standard normal (Gaussian, GD) distribution with median μ and standard deviation σ defined as

$$y_{\mathrm{Gauss}}(x; \mu, \sigma) = \frac{1}{\sigma\sqrt{2\pi}} \exp\left\{ -\frac{(x-\mu)^2}{2\sigma^2} \right\},$$
$$\mathrm{FWHM} = 2\sigma\sqrt{\ln(4)} = 2.355\sigma. \tag{21.6}$$

If two sets of e.g. 10,000 random data each following one of these distributions are generated, the histograms of these two datasets can be directly compared. Figure 21.2a shows a zoom-in on the central parts of such histograms, where the tails of the LD are cut off (the whole GD is shown). While the two central parts do not look much different at first sight, one should realize that the LD is smaller because the rest of its data is contained in its tails, which becomes apparent when the two datasets are plotted in logarithmic scales (separately for the positive and negative sides) as in Figure 21.2b. Here, the LD continues three orders of magnitude further than the GD on both the positive and negative side, implying a six-order-of-magnitude difference altogether, and this difference gets even larger for larger datasets.

Perhaps even more counterintuitive is a second heavy-tailed distribution which is often encountered in the spectroscopy of single-QD, namely the discrete power-law distribution[2] (PLD) with exponent α starting from $x_{\min}$

$$y_{\mathrm{PL}}(x; \alpha, x_{\min}) = \frac{x^{-\alpha}}{\zeta(\alpha, x_{\min})} \sim x^{-\alpha}, \tag{21.7}$$

where $\zeta()$ is the generalized or Hurwitz zeta function. A power law has a well-defined mean over $x \in [1, \infty)$ only if $\alpha > 2$, and it has a finite variance only if $\alpha > 3$. Once again,

---

[1] Lorentzian distribution is sometimes also referred to as a Cauchy distribution. Random numbers distributed according to the Lorentzian distribution can be generated using uniformly distributed random numbers on the $[0, 1]$ interval $\mathrm{Rnd}_{[0,1]}$ as: $\mathrm{LorentzRnd} = \gamma \tan(\pi(\mathrm{Rnd}_{[0,1]} - 0.5)) + \mu$.

[2] Other common names for a PLD include Zipf law or the Pareto distribution (Newman, 2005). Discrete PLD random data PLRnd can be generated (Clauset et al., 2009) using a set of uniform random numbers $\mathrm{Rnd}_{[0,1)}$ in the range $0 \le \mathrm{Rnd} < 1$ as: $\mathrm{PLRnd} = \lfloor (x_{\min} - 1/2)(1 - \mathrm{Rnd}_{[0,1)})^{-\frac{1}{\alpha-1}} + 1/2 \rfloor$ for $\alpha > 0$, where the symbols $\lfloor \ldots \rfloor$ denote rounding to the nearest integers.

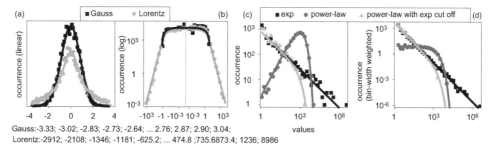

FIGURE 21.2 Heavy-tailed distributions. (a,b) Comparison of a Lorentzian from Eq. (21.5) and Gaussian from Eq. (21.6) distributions with the same FWHM (2) and number of data (10,000). In panel (a), some data points of the LD are left out to fit it on the same scale as the GD. In (b), the whole distributions are plotted in logarithmic scales (these histograms are bin-width weighted). The first and last five data points from each distribution are listed below the panels. (c and d) Comparison of a three discrete random datasets all starting from $x_{min} = 1$ containing 5,000 data points: exponential from Eq. (21.8) ($x_0 = 1,000$), power-law from Eq. (21.7) ($\alpha = 3/2$), and power-law with an exponential tail from Eq. (21.10) ($x_0 = 1,000$, $\alpha = 3/2$). Panel (c) plots the classical occurrence-based histogram (the number of occurrences per a bin), please note how the uneven (in this case logarithmic) bin widths influence the apparent shape of the distributions. For the case of logarithmic bin widths, the functions appear to be multiplied by the x scale. Panel (d) depicts histograms from (c) divided by bin widths, which yields the expected shape of the distributions.

a "common" distribution for the PLD to be compared to is the discrete exponential distribution[3] (expD) described by x_0 starting from x_{min}

$$y_{\exp}(x; x_0, x_{min}) = \left(1 - \exp\left\{-\frac{1}{x_0}\right\}\right) \exp\left\{\frac{x_{min}}{x_0}\right\}$$
$$\times \exp\left\{-\frac{x}{x_0}\right\}. \quad (21.8)$$

A comparison of the shape of these two distributions is shown in Figure 21.2c,d with the expD falling off much more steeply, the difference is again around three orders of magnitude for datasets containing 5,000 points.

A random process obeying an expD (say in the continuous form) is easily understandable from a physicist's point of view in terms of a process proceeding on a given frequency $F = 1/x_0$. To see how much more complex a PLD is, we can borrow two equalities from the theory of the Laplace transform

$$\int_0^{+\infty} \exp\{-Fx\}\, dx = \frac{1}{F} = F^{-1} \quad \text{and}$$
$$\int_0^{+\infty} x \exp\{-Fx\}\, dx = \frac{1}{F^2} = F^{-2}, \quad (21.9)$$

which imply that basically an infinite number of exponentials, and consequently also an infinite number of frequencies, are needed to give rise to a PLD. This means that there is no characteristic frequency for a PLD process (all frequencies in the range where the PLD holds are represented, albeit with different weights depending on the PLD exponent). This property is also sometimes referred to as self-similarity

or scale invariance, because the observed behavior will be the same no matter at what scale one chooses to look at it. The same definitely does apply to an expD: the behavior of an expD process far from its characteristic frequency is characterized by the absence of any observable events, in agreement with the much steeper decrease from Figure 21.2c,d.

Although numerous processes in nature (and in social sciences) are governed by a PLD (Newman, 2005; Clauset et al., 2009), such as the intensities of earthquakes, the population of cities or citations of scientific articles, in reality, there always have to be a long-scale cut-off in addition to the x_{min}, simply because nothing can go on *ad infinitum*. In such cases, when a process is observed on a scale where the PLD starts to cease to exist, (discrete) power law with an exponential cutoff [4]

$$y_{PL-\exp}(x; \alpha, x_0) \approx x^{-\alpha} \exp\left\{-\frac{x}{x_0}\right\} \quad (21.10)$$

can be useful.

While analyzing observed data, it is often beneficial to generate an artificial dataset, such as those in Figure 21.2 and study how the changes of some of the parameters influence the observed result. Especially in the case of a dataset obeying a discrete PLD, however, extreme care needs to be taken in order to correctly replicate the PLD's heavy tail in the simulation. Simple rounding off does not work with PLDs, the heavy tail manifests itself as scarce events scattered over long scales (see Figure 21.2c). Useful tips for generating randomly distributed datasets are listed as footnotes at the corresponding distributions.

---

[3]Using the same symbols as for a PLD, discrete random data drawn from an expD ExpRnd can be generated (Clauset et al., 2009) as: ExpRnd = $\lfloor x_{min} - x_0 \ln(1 - \text{Rnd}_{[0,1)})\rfloor$.

---

[4]For the case of the power law with cutoff there is no closed-form expression, but one can generate exponentially distribute random numbers and then accept them with probability $p = (x/x_{min}) - \alpha$ (Clauset et al., 2009).

One more important note to PLDs is the estimation of the exponent α an experimentalist usually uses to characterize the observed data. Perhaps the most common procedure is to preform a least-squares linear regression on the histogram plotted in logarithmic scales, where a PLD should transform to a straight line with its slope yielding the desired parameter: $\ln(y(x)) = \alpha \ln(x) + \text{constant}$. This procedure dates back to the early 19th century works on PLDs (Clauset et al., 2009). Unfortunately, such ancient methods generate significant systematic errors under relatively common conditions (Clauset et al., 2009) and should be avoided once modern computers can do the task. Instead, to accurately describe the observed data, one should use the maximum likelihood estimators of the exponent α (Clauset et al., 2009; Hoogenboom et al., 2006) (although they can be a bit more complicated for the discrete case (Clauset et al., 2009)) and bounds of the power-law behavior should be tested (Clauset et al., 2009; Hoogenboom et al., 2006).

Linewidth

The homogeneous, lifetime-limited linewidth of a transition is given by the Heisenberg's uncertainty principle

$$\Delta E \Delta t \geq \hbar/2; \qquad \Delta E(\Delta t = 1 \text{ ns}) \approx 0.3 \text{ μeV}, \quad (21.11)$$

However, in semiconductors, the transitions are thermally broadened with linewidths

$$\Delta E_{\text{thermal}}^{\text{3D}} = k_B T; \quad \Delta E_{\text{thermal}}^{\text{3D}}(300 \text{ K}) = 25.85\text{meV},$$
$$\Delta E_{\text{thermal}}^{\text{3D}}(10 \text{ K}) = 0.862 \text{ meV}, \quad (21.12)$$

k_B being the Boltzmann's constant. In quantum-confined systems, this thermal broadening decreases due to a different shape of the density of states and, in the simplest approximation, should completely disappear for a zero-dimensional system, whose density of states is ideally a set of delta functions (Arzberger and Amann, 2000).

21.1.4 Diffraction Limit

When studying object microscopically by optical means, the resolution of the system becomes limited due to the wave properties of light. Considering light a wave characterized by a certain wavelength λ, it will, after passing through an aperture, due to diffraction, form a spot, or a blur, with a size given by the so-called Airy disk. Imaging two (very small) objects will then result in a combination of two such diffraction-limited spots, which are, however, resolvable only if their centers are far apart enough from each other with respect to their sizes. Thus, diffraction sets a fundamental limitation on achievable resolution D given by

$$D = \text{const.}\frac{\lambda}{\text{NA}} \approx \frac{1}{2}\frac{\lambda}{\text{NA}}, \quad (21.13)$$

NA being the numerical aperture of the imaging system. Depending on what is deemed "resolvable", the const. in Eq. (21.13) slightly vary, being 0.5 for the so-called Abbe diffraction limit and 0.61 for the Rayleigh criterion, but the difference is in practice negligible. In modern optics, NA can reach 1.5, yielding the diffraction limit roughly in the form $D = \lambda/2$ or around 250 nm. In practice, all optical elements of a system have their own optical aberrations, and thus, the fundamental diffraction-limited imaging is achievable only in confocal microscopes (see Section 21.3).

The diffraction limit is small when compared to the sizes of, for example, certain biological objects such as individual cells or bacteria to be imaged, but still very large for optical imaging of single-cell organelles, proteins, let alone single QDs.

21.2 Historical Background

Microscopy in general offers a powerful insight into the properties on small scales. A crucial breakthrough in microscopic techniques was the discovery of the possibility of studying single quantum systems on (or as a part of) surfaces using probe-based techniques. First such technique was scanning tunneling microscopy (Binnig et al., 1982, 1983). Its discovery in 1981 was awarded the Nobel Prize as early as in 1986, only a few years after this method was first reported. Other derived microscopic techniques, including atomic force microscopy or scanning near-field optical microscopy (SNOM), soon followed. All these probe-based techniques have by now matured into routine and incredibly powerful tools.

The study of light emission of single nano-objects with sizes much smaller than the diffraction limit described in Section 21.1.4, such as molecules or quantum wells, is a direct extension of probe-based microscopy. The first studies of optical spectra of a single molecule were carried out on a model of pentacene molecules in a p-terphenyl crystal at 1.5 K. The very first optical spectrum was recorded using a sophisticated zero-scattering-background absorption technique (Moerner and Kador, 1989) in 1989. Soon after that, much simpler fluorescence excitation proved to produce superior signal (Orrit and Bernard, 1990) and many subsequent experiments thus relied on fluorescence excitation. The possibility of detecting the spectra of single molecules at room temperature was first demonstrated in 1993 (Betzig and Chichester, 1993) using SNOM, a probe-based method where a low-aperture optical fiber is scanned over the studied surface in the near-field regime. After this boost, optical studies of single dye molecules using far-field fluorescence techniques, the interpretation of which is far less complex because of the absence of a metal-coated tip in close proximity of the studied object, soon followed (Nie et al., 1994; Macklin et al., 1996). For early reviews on this topic, see (Moerner, 1994; Moerner and Orrit, 1999).

In parallel to optical studies of single molecules, low-dimensional quantum structures were also under close scrutiny. With regards to "zero"-dimensional structures, two different classes of materials were found. In structures based on epitaxially grown layers, narrow low-temperature PL spectra were observed (Brunner et al., 1992; Leon et al., 1995), in agreement with the theoretical predictions of the discrete density of states in zero-dimensional structures from Section 21.1.3. However, in colloidally prepared QDs, PL excitation and PL narrowing techniques had not seemed to have been able to reach this goal for some time (Norris and Bawendi, 1996). Such a contrast might have seemed strange, especially considering the fact that epitaxially grown low-dimensional structures typically based on the group III–V semiconductors (GaAs/AlGaAs) are generally much larger than colloidal QDs typically based on the group II–VI semiconductors (CdS/CdSe). This size difference is not negligible: whereas epitaxial dots may have tens (Leon et al., 1995) but also hundreds (Brunner et al., 1992) of nms, QDs typically have 1–10 nm in diameter and should thus be closer to a "truly zero-dimensional" object.

This problem was overcome in 1996, when far-field epiflurescence optical microscopy at 10 K of ZnS-overcoated CdSe QDs finally produced an optical spectrum with the FWHM of 120 µeV (Empedocles et al., 1996). In this study, the authors utilized far-field fluorescence spectroscopy, the very same technique which had successfully yielded luminescence spectra of single molecules (Moerner and Orrit, (1999)) and they compared this techniques to PL narrowing spectroscopy, which had been the technique of choice in previous studies of single low-dimensional quantum structures. In this comparison, the simpler fluorescence spectroscopy proved superior. After this starting point, not only spectrally but also temporally resolved PL experiments investigating single QDs has become an important tool verifying various theoretical predictions of QD behavior.

An interesting continuation of the development of high-resolution microscopy techniques is described later in Section 21.5.7.

21.3 General Methods

21.3.1 The Experiment

Even though the optical investigation of single QDs seemingly requires that the diffraction limit from Eq. (21.13) is surpassed, it is actually possible to circumvent rather than surpass it. A single-QD experiment breaks down into three deceptively simple steps: (i) ensuring that only a single object of interest is present in the investigated optical spot, which ensures that the diffraction limit does not interfere with the measurement, (ii) keeping the level of undesirable background light smaller than the signal of the investigated QD and (iii) collecting as much light as possible.

When colloidal QDs are to be investigated on a microscopic level, they clearly need to be deposited on a substrate, which provides a simple solution to the first problem. When a colloid diluted to very low concentrations is deposited, a single QD can be resolved using an optical microscope. The areal density of QDs after the deposition should amount approximately to 1 QD per 500×500 µm^2 with preferably homogeneous coverage. Higher areal densities lead to

too many QDs in the optically investigated spot, whereas lower densities might mean that the emitted optical signal is too low to allow for the efficient alignment of the optical setup and finding an emitting QD can become difficult. When depositing from a stable colloid, simple drop- or spin-casting will usually do, albeit the necessity of fine-tuning the correct concentration. In some non-colloidal QDs with sizes still comparable to the Bohr radius, much more sophisticated alternative preparation methods had been applied with tremendous success (see Figure 21.10).

As for the second background-related problem, there are several possible sources of a background signal. Apart from the obvious dark counts of the detector, residual PL can sometimes arise from the optical parts of the collection system, but in a modern setup (e.g. using interference instead of color glass optical filters) PL of impurities from the solvent and/or the substrate (or matrix) can be much more detrimental. Therefore, it is of utmost importance to excite (or collect light from) as small a volume as possible in order to expose as little non-studied material to the excitation light as possible, to use a substrate of very high optical quality (e.g. crystalline quartz) and to thoroughly check the PL signal of the substrate and the solvent (Wang et al., 2009). Also, although the excitation wavelength clearly needs to be chosen so that efficient excitation of the sample is achieved, it is generally beneficial to excite at longer wavelengths, since more energetic excitation photons at shorter wavelengths substantially increase the possibility of exciting undesirable impurities.[5]

Finally, efficient collection of light requires a sensitive detector (e.g. single photon-counting systems) and as wide a collection angle as possible.

Single-QD PL can be collected in various modes, depending on whether the signal is resolved spectrally, temporally, with regards to light polarization, etc. The generally lower levels of signal do not usually allow for a combination of the above. For example, acquiring a spectrum using a CCD camera might require the collection times of seconds to minutes (or even tens of minutes, depending on the signal as shown in Table 21.1), implying that one cannot study the temporal evolution of such a spectrum at timescales smaller than this collection time (unless specialized, in their nature non-single-QD methods are utilized, see Section 21.4.4).

In reality, an optical microscope is usually a heart of single-QD studies. It can be used in the epifluorescence configuration, i.e. using the microscope objective to both excite the sample and subsequently collect the emitted light. Taking the abovementioned measurement of a single-QD spectrum as an example illustrating how a single-QD

---

[5]In QDs, the spectrum of wavelengths which excite PL efficiently is broader than in molecules and, consequently, some excitation wavelength tuning is possible. Also, there is generally no need to collect a spectrum very close to the excitation line.

measurement works, the optical microscope can be used to image the investigated spot as sketched in Figure 21.3d. The collected light passes through a spectrometer set into the direct imaging mode (e.g. using a mirror instead of a dispersive element) and is acquired using a CCD camera, as shown in Figure 21.3a. This image features mainly agglomerates of the investigated material, which might not, however, be detrimental to the measurement because these agglomerates allow for finding the signal of interest and efficient focusing the collection system. On closer inspection, small emitting spots are also noticeable. Then, the spectrograph is switched to the spectrum-acquiring mode and a narrow slit is placed to its entrance to choose the investigated spot, yielding a set of PL spectra each corresponding to an emitting spot in the image (compare Figure 21.3a,b).

In the example above, the setup was based on wide-field imaging, meaning that the excitation light is not focused directly on the sample but illuminates a wider area (Figure 21.3a). In such a setup, it is important to use a high magnification objective and a tube lens combination in order to generate sufficient magnification for resolving individual QDs. Also, the pixel size of the CCD needs to be taken into account to optimally map a single QD (i.e. a diffraction-limited spot) onto approximately a pixel-sized area.

A different possibility is to employ confocal laser microscopy for sample imaging. In a confocal microscope, the excitation laser is focused onto the sample into an ideal diffraction-limited spot and any light away from this focal point is filtered out using a diaphragm iris. This means that only a diffraction-limited spot is imaged at a time and an image analogous to that in Figure 21.3a can be generated only by scanning (either the laser beam

or the sample). On the other hand, the signal from a single QD studied in this mode can be passed directly to different devices, e.g. either a CCD camera to acquire a PL spectrum or an avalanche photodiode to achieve high temporal resolution. Consequently, a confocal setup is generally more versatile, even if wide-field imaging provides an easier way of mapping the investigated area in inhomogeneous samples.

A photon counting avalanche photodiode is often utilized in temporally resolved experiments because of its low dark counts and extremely high sensitivity. However, avalanche photodiodes do feature the so-called dead time, during which the photodiode cannot detect any photon. Modern commercial avalanche photodiodes working in the visible spectrum can reach as low dead times as 35 ns; however, this value might still be too high for some experiments (see Section 21.5.3). In such cases, the techniques of time-correlated single photon counting (TCSPC), such as multiple detectors, can be utilized to circumvent the dead-time problem. A popular setup is the Hanbury Brown and Twiss correlation setup (see Figure 21.4a), which employs two avalanche photodiodes and a beamsplitter. When one of the detectors records a photon, it is held off and only the second detector is waiting for the next photon, which allows for the detection of arrival times of PL photons with respect to the arrivals of the excitation photons with temporal resolution far exceeding the dead time. The data recorded during such a measurement are then made up by a series of temporal intervals between the excitation and detection; a histogram constructed from such intervals represents the PL (exciton) lifetime, as shown in Figure 21.4d. In addition to the measurements of single-QD lifetimes (see

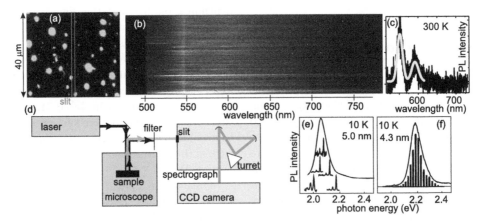

FIGURE 21.3 Basic single-QD PL spectra. (a) Image showing PL of agglomerates and single QDs from an optical microscope. PL spectra were acquired from the area inside the slit. (b) After a dispersion element is inserted, the image of PL spectra of QDs inside the slit can be collected by a CCD camera. (c) A representative single-QD PL spectrum from panel (b) plotted as a function of wavelength. Raw data are in black, smoothed spectrum is shown in gray. (d) A schematic of an experiment employing an optical microscope in the epifluorescence configuration for the characterization of single-QD PL spectra. (e) Representative examples of single-QD PL spectra measured at cryogenic temperatures (the smaller curves) and the corresponding ensemble PL spectrum (the larger curve). The single-QD spectra clearly exhibit phonon replicas, whereas the ensemble spectrum remains featureless. (f) A histogram of the positions of peaks from 513 single-QD PL spectra compared to the corresponding ensemble spectrum, again at low temperatures. Panels (a–c) showing PL of surface-engineered direct-bandgap Si QDs are taken from (Kůsová et al., 2010). Panels (e) and (f) were one of the very first published single-QD spectra and correspond to CdSe (Empedocles et al., 1996). (Reproduced with permission.)

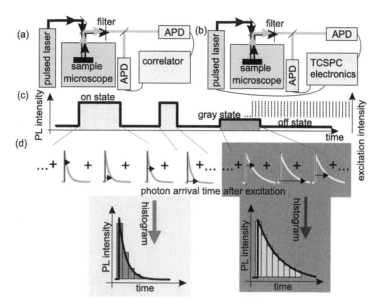

FIGURE 21.4 Basics of photon counting experiments. (a) The scheme of the Hanbury Brown and Twiss correlation setup. (b) The scheme of a time-tagged time-resolved single photon counting setup. (c) An idealized model intensity–time trace on the left and pulsed excitation on the right. This particular $I(t)$ trace features three different states (on, gray, off) characterized by different intensity levels (for more information, see Section 21.5.5). (d) The recorded photon arrival times after the excitation pulse are designated using the short black arrows. The PL photons will arrive at different delays after the excitation pulses. Despite being random, the arrival times sample the single-QD PL decay, shown as gray curves for illustration. The histogram of the photon arrival times after excitation then yields the single-QD PL decay curve (the bottom panel). In TCSPC experiments, the photon arrival times within the whole measurement window are recorded, which allows for the assignment of decay curves from (d) only to selected temporal intervals of the $I(t)$ trace from (c). In the example above, decay curves are produced for the intervals corresponding to two of the different states (on and gray).

Section 21.5.4), such a setup can also be exploited for a measurement of basic antibunching (see Section 21.5.3).

However, in other cases, more information than just the delay of the emitted photon with respect to the timing of the excitation photon is necessary. In order to record or use all the information contained in the measured PL dynamics, the arrival times of all emitted photons are recorded relatively to the beginning of the experiment (the so-called time tag) in addition to the (picosecond) TCSPC timing relative to the excitation pulses. Figure 21.4b depicts a time-tagged time-resolved single photon counting setup and Figure 21.4c illustrates an application of such a setup to a model long-term dynamics of a single QD exhibiting intermitency (see Section 21.5.5). Using this setup, the PL decay curve histograms of photon arrival times can be constructed from temporal intervals corresponding to the chosen emitted intensity levels, thus allowing for the correlation of an intensity level with a lifetime. Another possible application of a time-tagged time-resolved single photon counting is the determination of the ratio of integral intensities of the zero delay and the first side peaks in an antibunching experiment under pulsed excitation, indicative of biexciton quantum yield, see Section 21.4.3. Excellent summaries of single TCSPC techniques are available from the websites of Picoquant, under the section Technical notes (Picoquant, 2018).

Moreover, measuring any type of signal, be it a single-QD spectrum or dynamics, at cryogenic temperatures is a demanding task and would deserve a chapter for itself.

Generally speaking, a small cryostat only cooling the sample with an external microscope can be employed, or a PL setup can be built within the cryostat. The latter system has many advantages over the former one in terms of improved collection efficiency and increased sample space; however, they come at a cost of much higher complexity to build the setup. More details about experimental setups for single-QD experiments can be found e.g. in (Pelant and Valenta, 2012; Sychugov et al., 2017) or (Fernée et al., 2014), which also focuses on cryogenic operation.

21.3.2 Signatures of single-QD Emission

When measuring the optical signal on a single-QD level, it is usually relatively difficult to ascertain that only a single QD is present in the investigated spot, which is, however, of paramount importance for the experiment. Therefore, the experimentalist typically relies on various single-QD PL signal signatures, which are either predicted theoretically or known to occur from other experiments, and which are characteristic for single QDs. Typical signatures include theoretically predicted line narrowing (see Section 21.4.4) and antibunching (see Section 21.5.3), or experimentally discovered spectral diffusion (see Section 21.4.1) and blinking (see Section 21.5.5). Each of these signatures has their merits as well as limitations, as described in the corresponding sections. The typical effects of spectral diffusion and blinking on single-QD PL spectra are illustrated in Figure 21.5.

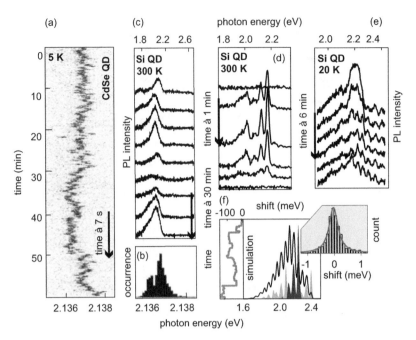

FIGURE 21.5 Spectral diffusion and blinking in single-QD spectra. (a) Shifts of 512 single-QD spectra of CdSe. (b) A histogram of spectral positions of peaks from panel (a). Both after (Fernée et al., 2010). (c) Single-QD spectra of a single Si QD exhibit both blinking and spectral diffusion, please note the long acquisition time of each spectrum. After (Valenta et al., 2008). (d and e) Single-QD spectra of a surface-engineered direct-bandgap Si QD. Panel (d) exhibits blinking and relative spectral stability, panel (e) illustrates an atypical influence of spectral diffusion on the spectral shape at longer acquisition times. (f) Simulation showing how a spectral shape such as that in panel (e) can evolve. Ideal, diffusion-free spectrum is indicated by the gray shaded area, the darker part highlights a single exciton and its phonon replicas. This diffusion-free spectrum undergoes a sequence of 1,000 spectral shifts shown on the left, which were generated as random data following a Lorentzian distribution depicted in the inset. The resulting spectrum in black very well copies the shape of the exprimentally acquired bottommost spectrum from panel (e). (d–f) after (Kůsová et al., 2015). (All panels reproduced with permission.)

21.3.3 Selectivity

One more word of caution to be said is that optical single-QD studies are selective in nature. Although they can help gain invaluable insight into the fundamental optical properties of semiconductor QDs and can elucidate the variability of these properties between individual QDs, they lead the experimentalist to focus on a particular studied object. Unavoidably, the selected QDs will be the brightest ones or those exhibiting the expected properties (such as narrow spectra or blinking, see Section 21.4.4 and 21.5.5). Therefore, the analysis of results has to be performed cautiously, in parallel to other experimental methods.

21.4 Spectrally and Polarization-Resolved single-QD Experiments

21.4.1 Spectral Diffusion

Already the first single-QD experiments (Empedocles et al., 1996) revealed the presence of a phenomenon called spectral diffusion (SD) or random shifts of the whole single-QD PL spectrum. SD results from the interaction of a QD with its surroundings and had previously been observed in PL studies involving single molecules (albeit the magnitude of

the shifts, which is several orders of magnitude higher in QDs than in molecules (Empedocles et al., 1996)). Be as it may, this phenomenon was unexpected in QDs at first experiments, because the exciton inside the QD was assumed to be well isolated from the surrounding matrix or a substrate (Empedocles et al., 1996).

The very first insights into the SD process revealed, using electro-optical spectroscopy, a correlation between the fluctuations of the local electrical fields around the studied QDs and the spectral peak position, implying that SD is caused be these local fields (a local Stark effect) (Empedocles, 1997). It also exhibits strong dependence on both the pump power and excitation wavelength, being caused by the dissipation of excess energy via hot carrier relaxation (Empedocles and Bawendi, 1999). SD characteristics are compatible with such changes as the re-arrangement of surface ligands or even atoms (Fernée et al., 2014). As for the SD spectral shifts, they were observed to follow a Lorentzian distribution (Fernée et al., 2010), whose heavy tail (see Section 21.1.3) implies the occurrence of very large spectral jumps (as much as 0.6 meV at 5 K during a 50-min measurement (Fernée et al., 2010)).

In addition to the above "standard", slow SD, which is observed on timescales of seconds and more, a fast SD process manifesting itself on the timescale from microseconds to seconds was also identified (Coolen et al., 2008;

Fernée et al., 2013) using specialized techniques. Unlike slow SD, the fast process is not laser driven, and more studies are still needed to shed more light on it.

The main manifestation of SD in a measurement is the fact that it leads to the broadening of the observed spectrum, which increases with longer acquisition times. Such a shift of single-QD PL spectra detected throughout a 1-h detection window is shown in Figure 21.5a. Thus, as the PL spectrum emitted by a QD shifts with time due to SD, a CCD camera collecting a signal will sum over all the shifted spectra within the acquisition time, inevitably smearing out or skewing any present spectral features, and producing a much broader PL spectrum. This can be a problem especially at samples with low emission, where long acquisition times are a necessity, and for prospective applications of QD, where SD puts the limit on their spectral stability.

A second type of instability of the emission single QDs exhibit is the so-called blinking, i.e. changes in intensity levels (see Section 21.5.5). A combined effect of these two phenomena is documented in Figure 21.5.

21.4.2 Band-Edge Exciton

A spectrally resolved single-QD measurement can reveal the spectral fingerprints of various processes in contrast to ensemble measurements, which are basically a sum of a very large number of spectra emitted by individual QDs. However, the individual spectra are not exact copies of each other, but differ from one another due to differences in the QDs. For example, differences in size lead to spectral shifts. Even if the PL spectra of individual QDs were well-resolved with narrow spectral lines, they would be shifted with respect to each other due to size fluctuations of QDs and an ensemble PL measurement would sum over these shifted spectra, inevitably producing a broad featureless one. This process is illustrated in Figure 21.3e.

The very first single-QD PL spectra (Empedocles et al., 1996) carried out at 10 K on CdSe QD revealed, in addition to SD (see Section 21.4.1), spectral narrowing with respect to the ensemble measurements and longitudinal optical phonon replicas (see Figure 21.3d,e). Moreover, a sum of multiple PL spectra from individual QDs were shown to have corresponded to the macroscopic ensemble PL

measurement, in agreement with the concept of an ensemble measurement being a sum over all the single-QD PL spectra. Later, cryogenic studies going down to 2 K carried out on commercial CdSe/ZnS QDs with much better photostability (and lowered SD) even closely mapped the fine structure of the band-edge exciton, including its splitting into a dark and bright states (Biadala et al., 2009; Fernée et al., 2009). The identification of the lines with the split excitonic states was confirmed by polarization-resolved (Fernée et al., 2009) and even magneto-optical (Biadala et al., 2010) studies. Thus, single-QD spectroscopy at low temperatures allows the testing of long-standing theoretical predictions with unprecedented detail.

Although the cutting-edge studies are usually performed with II–VI semiconductor QDs because of the in-depth theoretical understanding of how these materials work and because of the availability of high-quality samples, spectroscopic single-QD studies were performed also in other materials. For example, several types of silicon QDs, most of which feature much less efficient (see Table 21.1), indirect-bandgap phonon-mediated emission (Hybertsen, 1994), had already been studied thoroughly (Sychugov et al., 2017). One body of work constitutes silicon QDs designed especially for single-QD studies, fabricated by a top-down combination of lithography and self-limiting oxidation of silicon wafers. This method yields unique arrays of stably positioned QDs (and other nanostructures), however, only QDs with oxidized surface are accessible, since oxidation is a mandatory step. Such samples enabled thorough measurements of the evolution of single-QD PL spectra of oxidized Si QDs including phonon replicas present at some spectra (Sychugov et al., 2005, 2017), as shown in Figure 21.6d.

Just to mention a few other materials, as shown in Figure 21.6, exciton fine structure has also been investigated in perovskite $CsPbI_3$ QDs (Yin et al., 2017) and, more generally, in cesium lead halide perovskite QDs, the lowest excitonic state has even very recently been shown to involve a highly emissive state, unlike in other semiconductor QDs, where the lowest energy sub-level is usually dark (Becker et al., 2018). Moreover, room-temperature single-QD PL spectroscopy has been applied to carbon QDs to test the hypothesis that the emission tunability observed previously

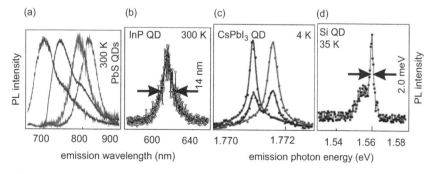

FIGURE 21.6 Selected single-QD spectra from non-CdSe colloidal semiconductor QDs. (a) After (Peterson and Krauss, 2006). (b) After (Chandrasekaran et al., 2017). (c) After (Yin et al., 2017). (d) After (Sychugov et al., 2005). (All panels reproduced with permission.)

in ensembles was a result of selective excitation of different subsets of carbon QDs within the ensemble (van Dam et al., 2017). Contrary to this hypothesis, single-QD PL spectroscopy produced tunable emission spectra from individual QDs using several excitation wavelengths, proving the incorrectness of the assumed model. Various other materials exhibit different degrees of spectral narrowing (Sychugov et al., 2017).

21.4.3 Biexcitons and Trions

Once again contrary to expectations (see Section 21.1.1), efficient light emission from multiple or charged exciton has been detected in some types of QDs using single-QD studies. The first hints were given by the spectroscopic studies at 4.2 K performed on very small core-shell CdTe/CdSe QDs with the average core size of 1.6 nm, which clearly revealed several multiexcitonic bands (Osovsky et al., 2009). Further studies were carried on *elongated* commercial CdSe/ZnS QDs at 2 K. A biexcitonic line, as confirmed also by magneto-optical experiments, appeared (Louyer et al., 2011) in the elongated QDs, whereas QDs in the sphere-like regime were observed not to have produced detectable biexcitonic emission. A rate-equation fit to the power dependence in these QDs of both the exciton and biexciton emission areas has even led to the conclusion that a unity biexciton quantum efficiency can be obtained for some QDs (Louyer et al., 2011). Optical properties of these QDs are summarized in Figure 21.7. Another example of efficient multiexcitonic emission is that

of core–"giant" shell CdSe/CdS QDs at 4 K again confirming efficient multiexcitonic (and charged-exciton-related) bands (Htoon et al., 2010).

Apart from simple multiexcitonic effects, a QD, especially under high excitation fluence and on a substrate, QDs can become charged, with the first excited state forming a trion (a negatively or positively charged exciton, see Figure 21.1f). Similarly to multiexcitons, efficient emission from trions has been observed, for example, in single CdSe–ZnS QDs at low temperatures, where both the excitonic emission and its trionic replica showed the same polarization (Fernée et al., 2009). Other types of QDs with efficient trionic emission evidenced by single-QD PL spectroscopy include perovskite CsPbI$_3$ QDs (Hu et al., 2016b, Yin et al., 2017) or surface-engineered direct-bandap Si QDs (Kůsová et al., 2015), see Figure 21.5d–f.

All these results suggest that it is actually the shape and the shell of the QD which determines its multiexcitonic PL efficiency, which is basically limited by the rate of the most important competitive non-radiative process, the Auger recombination as described in Section 21.1.1. The efficiency of multiexcitons (and, conversely, the Auger recombination) can be further evidenced by photon-statistics measurements described later in Section 21.5.3.

21.4.4 The Ultimate Linewidth

The optical resonance linewidth of a single quantum emitter provides a direct measurement of decoherence in the

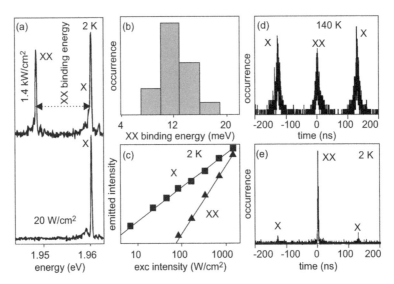

FIGURE 21.7 Single-QD study of biexcitons at low temperatures. (a) Two single-QD PL spectra of an individual elongated CdSe/ZnS QD under two different excitation intensities. At higher excitation intensity, a clear biexcitonic (*XX*) line appears in addition to the exciton recombination (*X*). (b) Distribution of biexciton binding energies derived from single-QD spectra such as the one in panel (a) for 14 individual QDs. (c) The emitted PL intensity vs. the excitation intensity dependence for the two spectral lines. (d) Photon correlation under pulsed excitation shows no signs of antibunching, confirming the high efficiency of biexcitonic emission. The excitation intensity corresponds to the mean number of excitons per pulse of 0.2. The ratio of integrated intensities of the zero-delay peak and the side peaks is 0.8. (e) The same experiment as in (d) but at lower temperature indicates strong bunching because of the exciton residing in a long-lived state (see text for more information). (All panels after (Louyer et al., 2011), reproduced with permission, copyright the American Chemical Society.)

system, which is a quantity of interest to certain quantum technological applications and ultimately sets a fundamental limitation for these materials. Therefore, the question of how narrow a spectral line can be produced using QD-based emitters is a fundamental one.

Generally speaking, in QDs, the linewidth can be broadened, with respect to the lifetime limit from Eq. (21.11), by several processes, namely the interaction with various types of phonons, the fine structure of the ground state and spectral diffusion (SD, see Section 21.4.1) (Cui et al., 2016). The quantification of these influences in various samples is still an active area of research, as is excellently discussed in (Sychugov et al., 2014). In contrast to simple expectation (Section 21.1.3), the linewidth of a single QD PL spectrum depends on temperature (Arzberger and Amann, 2000; Sychugov et al., 2014) via coupling with acoustic phonons, whose population can be reduced at lower temperatures (Fernée et al., 2008). (Nevertheless, narrow linewidth below the $k_B T$ limit from Eq. (21.12) is usually considered the first signature of single-QD emission (Sychugov et al., 2017).)

Therefore, the search for the ultimate, homogeneous linewidth is carried out at cryogenic temperatures. Here, what seems to complicate the experiments the most is SD or, more precisely, the fast type of SD (see Section 21.4.1). Using specialized experimental techniques, the lowest detected linewidth is as narrow as < 2 µeV in CdSe QDs (Fernée et al., 2013), which is still about an order of magnitude larger than the lifetime limit from Eq. (21.11). Other measurements aiming at the narrowest linewidth possible are discussed in (Fernée et al., 2014; Sychugov et al., 2014). Interestingly, in indirect-bandgap Si QDs with phonon-assisted radiative transitions, strong exciton–phonon coupling was theoretically predicted to significantly broaden the homogeneous linewidth (Delerue et al., 2001). However, this prediction turned out to be incorrect and (resolution-limited) linewidths of 200 µeV, comparable with the linewidth of QD of direct-bandgap materials, were achieved (Sychugov et al., 2014).

An opposite approach to the selective measurement where QDs with as narrow a linewidth as possible are chosen to be characterized (see Section 21.3.3) is solution photon-correlation Fourier spectroscopy. It measures a spectral analogy to the autocorrelation function from Section 21.5.2, the spectral correlation function, yielding the spectral information of single QDs in their native environment at fast timescales with short exposure times, without the experimentalist selection bias and with ensemble-level statistics (Cui et al., 2016). Unlike true single-QD spectral experiments, it produces an "average" single-QD linewidth for the whole sample. These measurements document how varying size, shell composition, and shell thickness influence the single linewidth (at room temperature) and yield values around 50–100 meV for various CdSe QDs (Cui et al., 2016).

The same technique has also been applied to the study of the zero phonon line in a trion in 3-nm-core-10-nm-shell CdSe/CdS QDs at 4K, under which conditions these QDs are permanently negatively charged (Biadala et al., 2015). The characteristic linewidth at 4 K was of the order of 50 µeV, at a timescale as short as 250 ns, which is a broader linewidth than the one observed for an exciton. This broadening was shown not to be a result of SD and was proposed to be due to the addition of the third charge, which weakens the coherence of the excited state through its stronger coupling to the environment.

21.5 Temporally Resolved single-QD Measurements

21.5.1 Timescales

When investigating a temporally resolved PL signal, it is important to take into account the wealth of processes which one can encounter. Figure 21.8 illustrates the broad range of timescales which can in principle be covered. It is useful to express a particular timescale in terms of the number of exciton lifetimes which can fit into it, because this value

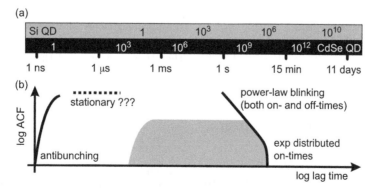

FIGURE 21.8 Timescales accessible to temporally resolved single-QD measurements. (a) The upper line shows the timescale in terms of the exciton lifetimes of an indirect-bandgap single-QD emitter such as Si QDs, the bottom line in terms of a direct-bandgap single-QD emitter, such as CdSe QDs. (b) A scheme of an overall PL dynamics expressed in terms of the autocorrelation function, black curve for a direct-bandgap emitter, the gray area for an indirect-bandgap emitter. Depending the on experimentally accessible timescale, different regimes can be observed.

gives a rough estimate of how many emission events might need to occur in order to reach this timescale.[6] The values of approximately 10^{10} emission events occurring on a less than a minute timescale imply that even relatively improbable processes can dominate the behavior on such long timescales, provided that their impact is high enough.

21.5.2 Photon Correlation

In addition to traditional techniques, the analysis of temporally resolved single-QD data sometimes relies on the use of a specialized statistical tool, namely the (normalized second-order) (auto)correlation function $g^{(2)}(\tau)$ (ACF)

$$g^{(2)}(\tau) \equiv \frac{\langle I(t)I(t+\tau)\rangle}{\langle I(t)\rangle^2}, \qquad (21.14)$$

where $I(t)$ is the intensity–time trace and τ is sometimes referred to as the lag time. The meaning of the ACF can be expressed as "the probability to find any other photon in time $t + \tau$ if one was observed at time t" (Verberk and Orrit, 2003). The ACF can be directly measured using a hardware correlator, or computed from intensity–time traces of a single-QD emitter, either as an autocorrelation of one intensity-time trace with itself or a cross-correlation of two different intensity–time traces. An efficient way of calculating the non-normalized ACF $G^{(2)}(\tau)$ relies on the use of the Fourier transform FT{} and its inverse form iFT{}

$$G^{(2)} = i\text{FT}\{|\text{FT}\{I(t)\}|^2\}. \qquad (21.15)$$

A specialized Hanbury Brown Twiss correlation setup, which reaches higher temporal resolution, can also be employed, see Section 21.3 and Figure 21.4.

Generally speaking, the ACF-based signal analysis can be applied to any investigated timescale, the only limitation being the underlying measured data. It is a strong statistical tool, which allows for the identification of even very weak signals hidden in random, non-correlating background, and it does not involve any experimentalist bias which can be present in other types of analyses (such as thresholding data); however, its application can be limited in random data following heavy-tailed distribution (see Section 21.5.5). More information on the analysis of a stream of photons can be found in reviews, e.g. (Verberk and Orrit, 2003; Lippitz et al., 2005; Cui et al., 2014).

21.5.3 Antibunching and Photon Statistics: Less Than the Exciton Lifetime Timescale

Short, sub-lifetime timescales are usually characterized using the autocorrelation function of the emitted photon

stream (ACF), see Section 21.5.2. In a source of classical, non-quantized electromagnetic field, the ACF for very short lag times would approach a non-zero constant C

$$\lim_{\tau \to 0} g^{(2)}_{\text{classical}}(\tau) = \frac{\langle I(t)^2\rangle}{\langle I(t)\rangle^2} = C. \qquad (21.16)$$

However, in a two-level quantum-mechanical source emitting single photons, the situation is completely different and the detection of two photons simultaneously is practically zero (Lounis and Orrit, 2005; Fernée et al., 2013; Cui et al., 2014)

$$\lim_{\tau \to 0} g^{(2)}_{\text{quantized}}(\tau) = 0. \qquad (21.17)$$

The concept of a two-level quantum-mechanical emitter can be realized in a semiconductor QD (under low excitation). Such a QD can emit only a single photon after the excitation event and it is the exciton lifetime which puts the lower limit on the delay between two consecutive emitted photons. Thus, under continuous illumination, an ACF corresponding to a two-level quantum-mechanical source exhibits the so-called antibunching dip with the onset of the signal rising exponentially characterized by the exciton lifetime (Lounis et al., 2000). Under pulsed illumination, the ACF correlation comprises a series of peaks with the frequency given by the laser repetition rate due to the periodic coincidences between PL photons emitted after different excitation pulses (compare Figure 21.9a and b). The weight of the zero-delay peak reflects the probability that two photons were emitted under the same excitation pulse, which approaches zero in a two-level quantum-mechanical emitter (Lounis and Orrit, 2005). These antibunching dips have long been considered the ultimate proof of the identification of a single photon emitter (Fernée et al., 2013; Chandrasekaran et al., 2017, Hu et al., 2016b) and were utilized to exclude the possible presence of more emitting QDs in the investigated optical spot.

However, a QD is only a good representation of a two-level quantum system in the limit of high Auger recombination and low biexciton efficiency (Lounis et al., 2000; Fernée et al., 2013). If the biexciton emission is efficient, which does happen at some QDs as explained in Section 21.4.3, the temporal signature of a biexciton–exciton radiative cascade under pulsed excitation shows up in the zero-delay peak and the integrated areas of the zero-delay and side peaks $g^{(2)}_0$ is quantitatively related to the ratio of biexciton XX and exciton X quantum yields η (Nair et al., 2011) assuming zero background noise level

$$g^{(2)}_0 \approx \frac{\eta_{XX}}{\eta_X}. \qquad (21.18)$$

(A similar quantity can be used to quantify the depth of the antibunching dip.)

Importantly, the existence of a zero-delay peak in the ACF signal does not automatically imply efficient biexcitonic emission, because this non-zero signal can also arise due to the presence of multiple emitters in the investigated optical spot. Thus, the zero-delay ACF peak signal is an

---

[6]Naturally, these values are valid under sufficiently high continuous excitation, under pulsed excitation the number of excitation events is determined by the repetition rate of the laser.

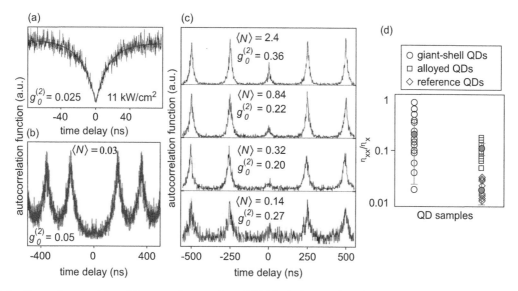

FIGURE 21.9 Examples of antibunching experiments. (a and b) Typical examples of antibunched emission from single QDs under continuous excitation for CdSe QDs (a) and under pulsed excitation for CsPbI₃ QDs (b). $\langle N \rangle$ designates the excitation intensity level with respect to the 1 exciton per a QD threshold. In panel (a), the exciton lifetime fit of 16 ns is shown by the smooth black curves. (a) After (Lounis et al., 2000) and (b) after (Hu et al., 2016b). (c) Autocorrelation measurements of giant-shell CdSe QDs under pulsed excitation for different excitation intensities exhibiting reduction in antibunching. Interestingly, the zero-delay signal persists even under the 1 exciton per QD threshold. After (Park et al., 2011). (d) Spread of biexciton quantum yields η_{XX} for different CdSe QDs. Giant-shell QDs have higher maximal values than alloyed QDs but also a much larger spread of values. Non-engineered, reference CdSe QDs are also plotted for comparison. After (Park et al., 2014). (All panels reproduced with permission.)

indicator of biexcitonic emission *if and only if* a single QD is ensured to be present within the investigated area.

Early examples of antibunching measurements include those performed on CdSe/ZnS QDs with the average core size of 1.8 nm under continuous excitation at room temperature (Lounis et al., 2000). Deep antibunching dips were observed with the minimum-to-maximum ratio of 0.025 (see Figure 21.9a), implying no biexcitonic emission and rapid Auger rates. The dependence on excitation intensity of the antibunched emission even yielded a single-QD absorption cross-section (Lounis et al., 2000). A similar trend of low $g_0^{(2)}$ is also typically observed under pulsed excitation, as illustrated in Figure 21.9b.

However, later measurements showed that some QDs can exhibit a completely different type of behavior, reduced auntibunching, which, in agreement with the spectral measurements mentioned in Section 21.4.3, implies slower Auger rates. For example, CdSe/CdZnS QDs at 140 K under pulsed excitation exhibited higher biexciton efficiencies, which varied from one QD to another between 2% and 12% (Nair et al., 2011). Elongated commercial CdSe/ZnS QDs mentioned previously (see Section 21.4.3) also produced significantly reduced auntibunching at 140 K ($g_0^{(2)} = 0.8$), with surprisingly high bunching at 2 K, where the exciton resides in a long-lived ground state and thermal mixing with the higher shorter-lived state does not occur any more due to the low temperatures (Louyer et al., 2011). This behavior is documented in Figure 21.7d,e. An increase of biexciton quantum yield was also observed when CdSeTe/ZnS QDs were enclosed in an ITO layer (Li et al., 2017).

Two other very important studies were aimed at revealing the relationship between slower Auger rates and the type and the thickness of the QD's shell (Park et al., 2011, 2014). In these studies, the increased thickness of the shell was observed to lead to slower Auger rates in CdSe/CdS QDs with a "traditional", step-like transition between the core and the shell materials and a large shell thickness (the so-called giant-shell QDs). Interestingly, reduced antibunching persisted even for excitation intensities below the 1 exciton per QD threshold, see Figure 21.9c. A second type of sample was specially prepared CdSe/CdS QDs with an alloyed shell, where the transition between the core and shell materials is more gradual. Again, Auger rates were reduced, but this time they did not significantly decrease with the shell thickness. The resulting biexciton PL efficiencies are compared in Figure 21.9d. Overall, biexciton efficiencies in alloyed-shell CdSe QDs (Park et al., 2014) were lower than the highest ones in thick-shell CdSe QDs (Park et al., 2011), however, the overall spread in efficiencies was lower in the alloyed-shell samples. These results clearly indicate that the intentional fabrication of QDs exhibiting slower Auger rates, or "interface" or "Auger engineering", is possible through fine-tuning the overlap of electron(s) and hole(s) wavefunctions comprising the trion. However, the Auger rates are clearly enormously sensitive to the defects in the shell and within the core–shell interface, the presence of which immediately makes the Auger rates much more rapid.

Last but not least, to the best of our knowledge, antibunching has not yet been reported in slowly emitting QDs associated with indirect-bandgap materials such as silicon, apparently due to the corresponding low emission rates (see Table 21.1).

21.5.4 PL Decay of Single QDs: Nanoseconds to Microseconds

The PL decay of a single QD can be determined either from antibunching measurements, see Section 21.5.3, or using single-photon counting techniques as described in Section 21.3. The single-QD PL decay usually provides additional information to another characterization technique. Recently, single-QD lifetimes have become a popular tool to study the details of long-timescale intermittency dynamics, as described in Section 21.5.5. Examples of single-QD decays include e.g. giant-shell CdSe/CdS QDs at 4 K mentioned previously, which show the emergence of faster PL decay components with increasing excitation intensity, in accord with the emergence of multiexcionic spectral bands (Htoon et al., 2010). The biexciton decay time in these QDs was measured to be 11 ns. Temperature-dependent single-QD PL lifetimes were also used to ascertain the identification of the observed lines of the exciton fine structure in the single-QD PL spectra in CdSe/ZnS QDs (Biadala et al., 2009).

Single-QD measurements which deserve to be mentioned separately include those performed on oxidized Si QDs, see Figure 21.10. In one of these experiments, single-QD PL decays were found to exhibit a single-exponential form with the decay time in the order of units to tens of µs, while a sum of such decays measured on several individual QDs well copied the stretched-exponential shape typically observed in ensembles (Sangghaleh et al., 2013). This observation shows which one of the several proposed scenarios explaining the stretched-exponential shape of the ensemble PL decay curve is valid. A next experiment probing the

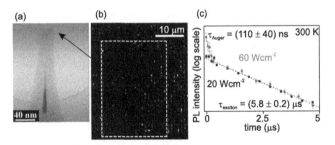

FIGURE 21.10 Si QDs fabricated specially for single-QD measurements by a combination of lithography and oxidation. (a) A typical TEM image of a nanopillar with a single Si QD at the top, as indicated by the arrow. After (Sychugov et al., 2011). (b) A PL image of a regular array of such QDs. After (Bruhn et al., 2011). (c) Normalized single-QD PL decay curves of these Si QDs under two excitation intensities. At higher excitation intensity, a faster biexcitonic emission appears. However, this biexcitonic emission decays with the lifetime characteristic of Auger rates, because the Auger rates are in this case much faster than the biexcitonic emission. After (Pevere et al., 2015). Please note the microsecond timescale. (All panels reproduced with permission.)

dependence on excitation intensity revealed the presence of a faster component at higher excitation intensities, which was interpreted in terms of the emission of a biexciton, however, decaying with the decay rate of a non-radiative Auger process, which is in this case much faster than the radiative recombination of a biexciton (Pevere et al., 2015). However, no particular emission line of a biexciton was reported. Variations in Auger lifetimes between individual Si QDs were observed, ranging from 80 to as much as 300 ns. Importantly, such long lifetimes were obtained without any intentional engineering, however, the selectivity of single-QD measurements mentioned in Section 21.3.3 plausibly plays a very important role here.

21.5.5 Intermittency: The Timescale Up to Tens of Minutes

Single QDs do exhibit random switching off and back on of their emission even under continuous illumination. This phenomenon is known as emission intermittency or more colloquially blinking. Its first report in QDs (Nirmal et al., 1996) came out nearly at the same time as the very first single-QD PL study (Empedocles et al., 1996), where the QD intermittency was also observed. Blinking was a bit unexpected. Although it had been detected in single molecules, QDs were expected to isolate the exciton inside of them and any external influence switching the emission off was supposed not to exist.

When studying this long-timescale PL dynamics of QDs, one can simply record an intensity–time trace $I(t)$ of a single QD using experimental tools described in Section 21.3. At first studies, QDs were observed to switch their emission simply between two levels of signal: an on state, corresponding to an emitting QD, and an off state with no emission. Such an $I(t)$ trace is shown in Figure 21.11. Probably the most common analysis of such a signal lies in constructing the so-called probability distributions ($P_{on/off}$) of the on- and off-times ($T_{on/off}$). In order to construct these distributions, the measured $I(t)$ trace is thresholded to determine the intensity level separating the on and the off state and then transformed into a "digital" off or on signal. Next, the temporal intervals which the QD spends in the off and the on states, respectively, are extracted and their histograms are plotted, giving rise to the $P_{on/off}$. This analysis is illustrated in Figure 21.11.

Although the first studies treated the $P_{on/off}$ as exponential and reported a characteristic timescale, analogically to blinking dynamics in common single molecules, QDs were relatively soon proven to exhibit $P_{on/off}$ following a power-law dependence (PLD) (Kuno et al., 2000), as shown in Figure 21.12. A PLD is a non-trivial dependence, because it looks reasonably single exponential over a limited experimental temporal window, and unlike an exponential distribution, it lacks a characteristic timescale (frequency), basically being a sum of exponential distributions over all the timescales on which it applies. (A PLD and an exponential distribution are compared in more detail in

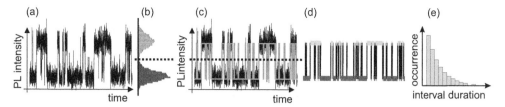

FIGURE 21.11 Analysis of an intensity–time trace using the approach of the distributions of on and off times. The $I(t)$ trace in (a) is used to construct the histogram of intensity levels in the measurement (b). If a clear threshold between an on and an off state is found (the dotted line), the $I(t)$ signal can be "digitalized" into a signal fluctuating between two levels (c). The durations of the on-times (light gray) and off times (dark gray) are then extracted (d) and used to construct the on-times (e) and off-times histograms. The $I(t)$ curve after (Bruhn et al., 2014a).

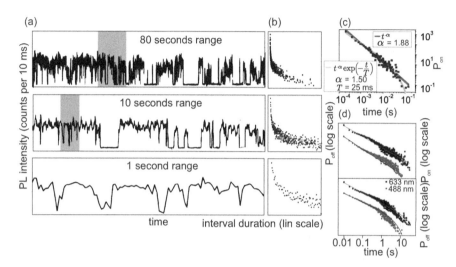

FIGURE 21.12 Self-similarity of power-law blinking. (a) Intensity–time traces of a single CdSe QD, three different time ranges are shown. The gray rectangles denote temporal intervals which are zoomed-in in the plots below. (b) The semilogarithmic distributions of off-times for the three time ranges deduced from the blinking traces next to them. Despite the nearly order-of-magnitude difference in timescales, the distributions appear to have a very similar shape. Both after (Kuno et al., 2000). (c) A log–log plot of the distribution of on times of a single CdSe QD following a power-law distribution over more than three decades in time. After (Kuno et al., 2001). However, the difference between the fit with a power law and an exponentially cut-off power law using the least-square fitting procedure is very small. The second fit after (Kůsová et al., 2016). (d) The distribution of on- and off-times for PbS/CdS QDs for 15 individual single QDs under two different excitation wavelengths. Unlike in other materials, in these QDs, it is the off-times that exhibit an exponential cut-off to the power-law blinking dynamics on the timescale of seconds. After (Chen et al., 2018). (All panels reproduced with permission.)

Section 21.1.3.) These self-similar, fractal characteristics as illustrated in Figure 21.12a,b imply that a process responsible for such a phenomenon also must take place over many decades of time, making it challenging to find a plausible theoretical explanation.

The robustness of two-level PLD blinking can be illustrated by the fact that it has been observed in many direct-bandgap II–VI QDs (Cichos et al., 2007), but also in indirect-bandgap Si QDs (Sychugov et al., 2017) as well as in other materials (InP/ZnSe (Chandrasekaran et al., 2017; Reid et al., 2018), PbS (Peterson and Krauss, 2006; Chen et al., 2018), etc.). Generally speaking, the exponents $\alpha_{\mathrm{on/off}}$ of the PLD $P_{\mathrm{on/off}}$ from Eq. (21.7) are often close to 3/2 (Frantsuzov et al., 2008; Cichos et al., 2007), generally varying between the values of 1 and 2.

The simplest and historically the oldest theoretical explanation of why the emission suddenly turns off is based on a transition between a neutral (on-) and a charged (off-) state (Cook and Kimble, 1985), in which the suppression of the emission is a direct result of Auger recombination. However, a simple model of a carrier escaping to a trap state to render the QD ionized and then back to make it neutral again would yield exponentially distributed $P_{\mathrm{on/off}}$. Therefore, a fluctuation in a property (e.g. a tunneling barrier (Kuno et al., 2003)) or a process needs to be postulated to obtain a PLD of *both* switching on and off of the emission. For example, an electron tunneling to trap states approximately 1–2 nm away from the QD would reflect in blinking dynamics at the timescales between 10^{-4} and 10^{2} s, illustrating that due to the exponential dependence of the tunneling barrier on the distance to the trap site a relatively small spread in the distances can map into a large interval of timescales in blinking dynamics (Kuno et al., 2003). However, there is still no scientific consensus on the particular mechanisms of the ionization and reneutralizaion of QDs. Detailed information on the mechanisms of charge trapping can be found e.g. in review (Cordones and Leone, 2013).

Pure PLD $P_{on/off}$ blinking dynamics can persist to the timescales of about seconds or even longer. Then, an exponential cut-off following Eq. (21.10) appears in the P_{on} and on much longer timescales, the on-times are distributed exponentially, see Figure 21.8. Hints to the existence of an exponential cut-off also in the P_{off} on much longer timescales (tens of minutes or more) have also been put forward (Wang et al., 2008; Issac et al., 2012; Krasselt et al., 2011) and recently, relatively short-timescale (seconds) exponential tail has been observed in the blinking dynamics of PbS QDs (Chen et al., 2018), as shown in Figure 21.12d. The long-timescale P_{on} cut-off can be a result of a process such as the recombination of a biexciton, which has low probability on the timescales comparable with the exciton lifetime, but which, in accord with Figure 21.8, becomes much more probable to occur at least once on the timescales of seconds involving a large number of recombination event (Peterson and Nesbitt, 2009).

The analysis of a long-scale single-QD PL dynamics using the $P_{on/off}$ approach might bring about several problems. First, the experimentalist has to choose a threshold intensity level between an on- and off-state, see Figure 21.11b,c as well as a binning time for the $I(t)$ trace, which would be an easy task in an ideal, noise-free signal. However, the inevitable presence of noise in a signal emitted by a single QD severely complicates this analysis and different choices of threshold or binning time might yield different results (Crouch et al., 2010). Also, PLD data require specialized statistical tools. The frequently used fit of a PLD as linear in a log–log scale might yield completely erroneous results (Clauset et al., 2009), see Section 21.1.3. Consequently, some of the spread in the reported differences in power-law exponents might be simply a result of an incorrect analysis.

A possible alternative description of measured blinking dynamics employs the correlation function from Eq. (21.14) in Section 21.5.2 (Verberk and Orrit, 2003; Houel et al., 2015; Kůsová et al., 2016). Unlike the $P_{on/off}$, the application of an ACF is free of an experimentalist's bias and, moreover, it allows one to characterize a broader timescale. However, "pure" power laws, as heavy-tailed distributions, complicate the description of the measured data with an ACF because the shape of the ACF depends more on the particular shape of the heavy tail (i.e. on which of the possible long on times from the random distribution really occur) than on other parameters of the distribution. This means that if both the $P_{on/off}$ are power-law distributed, statistically the same data lead to a different ACF and the ACF depends on the measurement time (the so-called ergodicity breaking) (Margolin et al., 2006; Messin et al., 2001). Clearly, the ACF is not a suitable tool for the characterization of such data. On the other hand, this problem disappears on the timescales where the exponential cut-offs of the P_{on} become relevant (Margolin et al., 2006; Kůsová et al., 2016).

Generally speaking, if the switching is random (which usually applies for blinking), the ACF and $P_{on/off}$ descriptions are directly related (Verberk and Orrit, 2003). Starting from an $I(t)$ trace (or the corresponding TCSPC data as explained in Section 21.3), the data can be analyzed in terms of both $P_{on/off}$ or the ACF. However, the correspondence between the ACF and the $P_{on/off}$ without knowing the underlying dataset is not that straightforward (Verberk and Orrit, 2003; Kůsová et al., 2016), implying that it might become difficult to compare dataset characterized by different approaches in the literature.

The upper limit in terms of timescales of the power-law blinking dynamics has already been described. As for the lower limit, it has not yet been studied so thoroughly, probably due to the difficulties with lower signal levels at shorter timescales and several a bit conflicting studies have been published. Experiments performed in both bare CdSe and CdSe/ZnS core-shell QDs suggest that in these QDs the power-law persists to the timescales comparable with the excited-state lifetime (Sher et al., 2008). A different study of uncapped colloidal CdSe reported a power-law dependence over five decades in time when combined from several individual QDs (200 μs–100 s) (Kuno et al., 2001). On the other hand, a change in the power-law blinking dynamics has been observed using the ACF description, where the ACF changed its slope on the timescale of about units to tens of milliseconds in both CdSe/ZnS core-shell and surface-engineered direct-bandgap Si QDs (Pelton et al., 2007; Kůsová et al., 2016), as illustrated in Figure 21.13a. Figure 21.13b then depicts a completely different dynamics of an uncapped CdSe QD (Verberk et al., 2002), where a power-law ACF was observed over six decades of time (1 μs–1 s), suggesting PLD $P_{on/off}$ around the μs timescale but an exponential cut-off of the on-times probably at as short timescales as somewhere around tens of milliseconds. Clearly, the extent of the PLD blinking dynamics depends on the type of the QD and probably even more on its shell. As a result of the low number of studies to date, much more information awaits to be uncovered.

Obviously, several temporal regimes exist. The most common ones are those with PLD of both $P_{on/off}$ and the longer timescales with a PLD P_{off} but exponentially decreasing P_{on}. Moreover, the P_{off} also possibly exhibits an exponential tail at even longer timescale and there might be a change in the PLD exponent at shorter timescales, see Figure 21.8. Clearly, the temporal boundaries between these different regimes vary with the type of QD and they very likely depend more on the shell of the QD than on the number of recombination events, as the typical on-time exponential tail was reported also in indirect-bandgap Si QDs with the exciton lifetime longer than microseconds (Valenta et al., 2008). The limits of these regimes are clearly very important parameters, describing how much the surrounding environment influences a QD emitting for a given period of time and how stable the emission actually is.

Importantly, the basic above-described scenario with two-level blinking does not always apply. Some QDs in the off

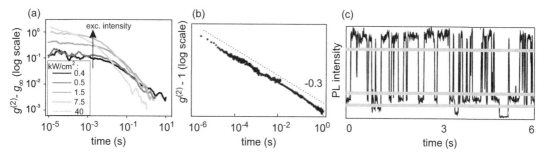

FIGURE 21.13 Atypical examples of blinking studies. (a) An autocorrelation function documenting blinking of surface-engineered direct-bandgap Si QDs. A change in dynamics is observed around the millisecond timescale, the dynamics also depends on the excitation intensity from a certain threshold. After (Kůsová et al., 2016). (b) An autocorrelation function documenting blinking of an uncapped CdS QD over six decades of time. A power-law correlation function with an exponent of 0.3 (the dotted line) translates into power-law distributed off-times and exponentially distributed on-times. After (Verberk et al., 2002). (c) An intensity–time trace of a single CdSe/CdS QD exhibiting three distinct levels of intensity. The gray line denotes the intensity thresholds used to discriminate the intensity levels. After (Gómez et al., 2009). (All panels reproduced with permission.)

state can continue to photoluminesce, albeit weakly. In the first blinking experiments, the PL intensity of a single QD during these off periods was significantly smaller (ten times) than during an adjacent on PL period. However, in later studies, some groups reported that the off state does not have to be completely dark (Spinicelli et al., 2009), or the observation of a so-called "gray" state with the PL intensities intermediate between an on and an off state (Gómez et al., 2009), as shown in Figure 21.13c. Thus, more relaxation channels are clearly possible, in accord with the non-negligible contributions of multiexcitons and trions to the emission of some QDs (Zhao et al., 2010), as also mentioned in Section 21.4.3 and 21.5.2.

An immediate prediction of Auger-switched-off model is that the decrease of PL intensity (PL blinking) should be accompanied by a proportional shortening of the PL lifetime (lifetime blinking). Indeed, a direct correlation between the PL intensity and lifetime (dubbed A-type blinking) has been clearly demonstrated in core–thick-shell CdSe/CdS QDs using time-tagged time-resolved single-photon counting techniques (see Section 21.3) (Galland et al., 2011). Interestingly, however, experiments on core–shell CdSe/CdS QDs with moderate shell thicknesses revealed another type of blinking behavior, where the PL lifetimes are uncorrelated with the PL intensity (Galland et al., 2011) (the so-called B-type blinking). This phenomenon has been interpreted in terms of hot electrons from the photoexcitation event being sufficiently energetic to tunnel rapidly to a surface state, which can then undergo fast non-radiative recombination with the hole state. To explain the switching of the emission off and on, the activation and deactivation of surface trap states was suggested and tested by spectroelectrochemistry on single QDs. In that particular study (Galland et al., 2011), B-type blinking was found to be prevalent in QDs with an intermediately thick shell and was characterized by a PLD of both on and off states. In QDs with a thicker shell, however, a shift towards either A-type blinking or even completely suppressed blinking was observed; the A-type blinking had a short exponential cut-off of 70 ms. Apart

from CdSe QDs, A- and B-type blinking was reported also in short-wave-infrared-emitting InAs/CdZnS QDs (Bischof et al., 2014).

These later results indicate that the picture of long-term blinking dynamics is not so simple and more types of phenomena can occur in different types of QDs. It should be noted that the above "traditional" model of an Auger-switched-off QD is not the only one explaining blinking dynamics and many other plausible models were proposed, see e.g. (Efros and Nesbitt, 2016) and references therein. One of the most discussed "competing" models is the diffusion-controlled electron transfer model, which relies on the assumption of spectral diffusion of both the QD and the "trap" state level (Tang and Marcus, 2005). This model successfully predicts the existence of the long-timescale exponential cut-off, as well as a shorter-timescale regime with a different PLD slope, which was observed experimentally (Pelton et al., 2007; Kůsová et al., 2016).

An interesting set of experiments utilizing blinking as a tool for a direct investigation of the fundamental properties of QDs was carried out using TCSPC methods, see Section 21.3. Specifically, as depicted in Figure 21.14a–d, in phase-pure zinc-blende core-shell 4.2-nm-in-diameter CdSe/CdS QDs, the identification of certain intensity levels in blinking time traces with the neutral and charged states of the QDs, respectively, allowed for the quantification of the lifetimes of both the radiative and non-radiative (Auger) processes of an exciton and both the negatively and positively charged trion. Moreover, the excitation-intensity dependence yielded the quantum yields and the absorption cross section for these states as well as that of the neutral, positive, and negative biexciton states on a single-QD level (Xu et al., 2017). The Auger lifetimes of all the states were around 1 ns, but they (and all other investigated parameters) varied with the charge state of the QD in qualitatively the same manner for several individual QDs (Xu et al., 2017). At a different experiment, the still slightly controversial phenomenon of carrier multiplication, i.e. the excitation of more than one exciton in a QD with a single excitation

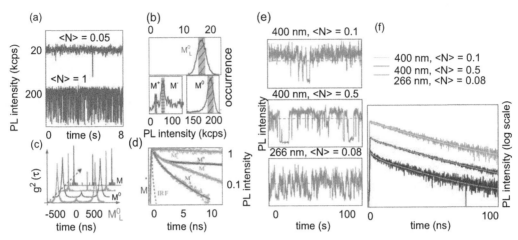

FIGURE 21.14 Time-tagged time-resolved single-photon counting, a tool for the study of fundamental properties. (a) Blinking $I(t)$ traces of CdSe/CdS QDs under low and high excitation. (b) Intensity-level histograms for the $I(t)$ traces from (a) document the existence of several states. By analyzing only the temporal intervals characterized by these intensity levels, the antibunching signal (c) and decay curved (d) corresponding to the individual states can be quantified. Based on the analysis, these states can be identified with a monoexciton M_L^0, biexciton M^0, and positively M^+ and negatively M^- charged trions. (a–d) After (Xu et al., 2017). (e) Blinking $I(t)$ traces of a single CdSe QD at different excitation intensities and excitation wavelengths. (f) The single-QD PL decay curves from the on-states from panel (e) reveal the contributions of a radiative single-exciton decay and faster non-radiative Auger recombination. The corresponding lifetimes allows one to directly quantify the efficiency of carrier multiplication, which is around 10 % for this particular QD. (e,f) After (Hu et al., 2016a). (All panels reproduced with permission.)

photon provided that the energy of the excitation photon is high enough ($E_{\mathrm{photon}} > 2E_g$), was also documented on a single-QD level for CdSe/ZnS QDs (Hu et al., 2016a), as shown in Figure 21.14e,f. Again, the analysis of the single-exciton and Auger lifetimes of single QDs under different excitation intensities and with the excitation photon energy below and above the $2E_g$ threshold directly yielded the efficiency of carrier multiplication of single QDs, being about 13 % for 50 QD on average.

21.5.6 Suppression of Blinking

Blinking is usually perceived as an undesired phenomenon, because it decreases the signal a QD can emit. It negatively influences quantum yield, because even though the excitation is absorbed, the corresponding emission is switched off for a non-negligible period of time in an unpredictable manner. Therefore, intense research has been focused on the fabrication of QDs where this phenomenon is suppressed.

As was explained in the previous Section 21.5.5, switching off of the emission is connected with the escape of photoexcited carriers, which subsequently causes a strong enhancement of non-radiative recombination rates. Thus, there are two possible ways of suppressing blinking: either the carriers are prevented from escaping the QD (preventing ionization) or the QDs are fabricated in such a way that they are on even when one of the carriers has already escaped (suppression of Auger recombination).

The most intuitive way of fabricating QDs which do not ionize is using a thick shell. Indeed, strong suppression of blinking was achieved in core–thick-shell CdSe/CdS QDs

(Chen et al., 2008; Mahler et al., 2008; Vela et al.,2010), where a significant fraction of the fabricated QDs (> 50 %) exhibited either blinking-free, or at least nearly blinking-free behavior. Naturally, the fraction of blinking-free QDs was observed to increase with increasing shell thickness (Mahler et al., 2008). Interestingly, even the thick-shell QDs possessing suppressed blinking in terms of PL intensity fluctuations were found to exhibit B-type blinking, i.e. the fluctuation of lifetimes in their long-term dynamics (Galland et al., 2012). The lifetime blinking was explained by the efficient emission of a trion (in the charged state), as confirmed by measurements in an electrochemical cell, i.e. when electrochemically controlling the degree of QD's charging (Galland et al., 2012). Thus, some degree of ionization clearly occurs even at thick-shell QDs and curbing Auger recombination becomes a necessity.

As already discussed in Section 21.5.3, a thick shell of a QD helps suppress the Auger rates. However, probably even a better option is a slowly changing confinement potential with the core and shell materials gradually transiting into one another (Efros and Nesbitt, 2016). Moreover, the thick-shell QDs were actually engineered for an efficient emission of a *negatively* charged trion. However, from time to time the QD can become positively charged, in which state the trion emission is quenched more than in the negative case (Park et al., 2014; Xu et al., 2017). Thus, the most successful suppression of blinking was reported in "core-alloyed-"bulky"-shell CdSe/CdSe$_{1-x}$S$_x$ QDs (the overall size > 40 nm) with a radially dependent composition resulting in a soft confinement potential effective for both electrons and holes (Nasilowski et al., 2015), where complete suppression

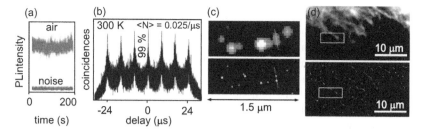

FIGURE 21.15 Suppression and utilization of blinking. (a) Blinking $I(t)$ trace of core-alloyed-"bulky"-shell CdSe/CdSe$_{1-x}$S$_x$ QDs (the overall size > 40 nm) in air at $\langle N \rangle = 0.5/$ µs, no off-states are observed. (b) The ACF measurements of the QDs from (a) documenting 100% biexcitonic quantum yield. (a,b) after (Nasilowski et al., 2015). (c,d) Blinking super-resolution achieved with carbon QDs. The top panels show conventional diffraction-limited images from a confocal microscope of (c) carbon QDs on a coverslip and (d) microtubulus inside a cell labeled with carbon QDs. The bottom panels depict the corresponding super-resolved images. (c,d) after (He et al., 2017). (All panels reproduced with permission, copyright the Americal Chemical Society.)

of blinking and 100% biexciton quantum yield at room temperature were documented, as shown in Figure 21.15a,b. More generally, core–shell engineering of II–IV and IV–IV QDs is summarized in a thorough review (Jang et al., 2017).

Once again, far from the perfection of CdSe QDs, suppression of blinking was observed also in other materials. Nearly blinking-free behavior was achieved in 1.6-nm-core-0.6-nm-shell CdTe/CdSe QDs at 4 K (Osovsky et al., 2009). A non-blinking regime appeared in 9.3-nm-in-diameter CsPbI$_3$ QDs under low-excitation $\langle N \rangle = 0.03$ with the emergence of a gray and an off state at higher excitation levels $\langle N \rangle = 1.7$ at room temperature; these QDs also exhibited narrow resolution-limited linewidth of 200 µeV at cryogenic temperature and negligible spectral diffusion (Hu et al., 2016b). Suppressed blinking was also observed in InP/ZnSe QDs (Chandrasekaran et al., 2017; Reid et al., 2018), where the on-state fraction was found to be 95.5% on average for 25 studied QDs (Chandrasekaran et al., 2017). Interestingly, these QDs were shown to exhibit traditionally fast Auger rates of 70 ps (when compared to the PL decay time of ≈ 20 ns (Chandrasekaran et al., 2017)), suggesting efficient suppression of ionization.

21.5.7 Super-Resolution

As undesired as blinking is usually perceived to be, it has, somewhat surprisingly, already found its use. The whole field of single-molecule spectroscopy gave rise to a class of novel microscopy techniques enabling optical imaging with resolution surpassing the fundamental diffraction limit from Eq. (21.13). This so-called super-resolution microscopy came to use in 2006 and its development was awarded the 2014 Nobel Prize in Chemistry, see Weiss (2014) and references therein.

Nowadays, the term super-resolution microscopy encompasses a class of methods with many acronyms derived from the two original wide-field methods of photo-activated localization microscopy, or PALM, and stochastic optical reconstruction microscopy, or STORM. The trick leading to "breaking" the diffraction limit lies in resolving the spatially

unresolvable objects in time. A series of images, or frames, of the set of objects under study is taken, and in each frame, only some of these objects emit light, whereas the rest are off. Dedicated algorithms are then used to track the individual studied objects in the separate frames, and their position is determined with sub-diffraction-limit precision using a fitting procedure. Thus, as long as the areal density of the momentarily emitting objects is kept low, a super-resolved image can be superposed from the individual frames with spatial resolution typically in tens of nanometers, see Figure 21.15c,d.

Originally, photobleaching was exploited in PALM to render some emitting objects dark, whereas reversible switching between an on- and a dark off-state was employed in STORM. However, blinking of molecules or QDs also involves changes in the emitted intensity, and the same type of procedure can thus be applied to the intermittent emission. Recently, various types of nanoparticles including semiconductor QDs has started to be considered as the light-emitting species for super-resolution microscopy imaging (Lin et al., 2018). In contrast to the suppression of blinking, however, in super-resolution microscopy, it is useful to apply QDs exhibiting only short on-times characterized by high emitted intensity and separated by long off-times to obtain the ideal low-areal densities of emitting objects in the individual frame.

21.6 Summary

PL spectroscopy of single colloidal QDs is a powerful tool providing a unique way of verifying theoretical predictions or scenarios used to describe the optical behavior of QDs. It has made possible a direct investigation of the fine structure of the lowest excitonic states in QDs and was utilized to document carrier multiplication on a single-QD level. However, single-QD PL experiments do not have to be viewed as investigations focused purely on the fundamental research, because the naturally occurring variations between the individual QDs observed by these methods can be further exploited for engineering QDs with a particular application in mind. Specifically, light emission under high excitation revealed the possibility of curbing the omnipresent competing non-radiative processes, yielding much more efficiently emitting QDs with non-radiative recombination rates lowered to values unparalleled in bulk semiconductors. Moreover, the investigations of the process of blinking, or the random switching off of the emission, which was discovered at the microscopic level but which still adversely influences the macroscopic emission efficiency, is now understood well enough to allow for the fabrication of non-blinking QDs. Despite these successes, there are still blank spaces to fill in and blinking of QDs in particular is still a very active area of research.

As for the characterization techniques, the next step might lie in the fusion of single-QD PL spectroscopy with probe-based microscopy techniques such as atomic force microscopy, which definitely has the potential to yield novel insights and uncover new ways to tune the already known processes. With regards to new functionalities, in addition to "simple" QDs, also various nano-heterostructures comprising different materials are under close scrutiny now. Such nano-heterostructures, whose design is based on what is already known about QDs, can be engineered to meet particular demands and thus push the boundaries of what semiconductor QD can do even further.

Acknowledgments

The Czech Science Foundation funding, Grant No. 18-05552S is gratefully acknowledged. Prof. I. Pelant in acknowledged for going through the text.

Bibliography

Arzberger M and Amann MC (2000) Homogeneous line broadening in individual semiconductor quantum dots by temperature fluctuations. *Phys. Rev. B*, 62:11029–11037.

Becker MA et al. (2018) Bright triplet excitons in caesium lead halide perovskites. *Nature*, 553(7687):189–193.

Betzig E and Chichester RJ (1993) Single molecules observed by near-field scanning optical microscopy. *Science*, 262(5138):1422–1425.

Biadala L, Louyer Y, Tamarat P and Lounis B (2009) Direct observation of the two lowest exciton zero-phonon lines in single CdSe/ZnS nanocrystals. *Phys. Rev. Lett.*, 103:037404.

Biadala L, Louyer Y, Tamarat P and Lounis B (2010) Band-edge exciton fine structure of single CdSe/ZnS nanocrystals in external magnetic fields. *Phys. Rev. Lett.*, 105:157402.

Biadala L et al. (2015) Photon-correlation Fourier spectroscopy of the trion fluorescence in thick-shell CdSe/CdS nanocrystals. *Phys. Rev. B*, 91:085416.

Binnig G, Rőhrer H, Gerber C and Weibel E (1982) Tunneling through a controllable vacuum gap. *Appl. Phys. Lett.*, 40:178–180.

Binnig G, Rőhrer H, Gerber C and Weibel E (1983) 7 × 7 Reconstruction on Si(111) Resolved in Real Space. *Phys. Rev. Lett.*, 50(2):120–123.

Bischof TS, Correa RE, Rosenberg D, Dauler EA and Bawendi MG (2014) Measurement of emission lifetime dynamics and biexciton emission quantum yield of individual InAs colloidal nanocrystals. *Nano Lett.*, 14(12):6787–6791.

Bruchez M Jr, Moronne M, Gin P, Weiss S and Alivisatos AP (1998) Semiconductor nanocrystals as fluorescent biological labels. *Science*, 281(5385):2013–2016.

Bruhn B, Qejvanaj F, Gregorkiewicz T and Linnros J (2014a) Temporal correlation of blinking events in CdSe/ZnS and Si/SiO$_2$ nanocrystals. *Phys. B Condens. Matter*, 453:63–67.

Bruhn B, Qejvanaj F, Sychugov I and Linnros J (2014b) Blinking statistics and excitation-dependent luminescence yield in Si and CdSe nanocrystals. *J. Phys. Chem. C*, 118:2202–2208.

Bruhn B, Valenta J, Sangghaleh F and Linnros J (2011) Blinking statistics of silicon quantum dots. *Nano Lett.*, 11:5574.

Brunner K et al. (1992) Photoluminescence from a Single GaAs/AlGaAs Quantum Dot. *Phys. Rev. Lett.*, 69(22):3216–3219.

Chandrasekaran V, Tessier MD, Dupont D, Geiregat P, Hens Z and Brainis E (2017) Nearly blinking-free, high-purity single-photon emission by colloidal InP/ZnSe quantum dots. *Nano Lett.*, 17(10):6104–6109.

Chen JS, Zang H, Li M and Cotlet M (2018) Hot excitons are responsible for increasing photoluminescence blinking activity in single lead sulfide/cadmium sulfide nanocrystals. *Chem. Commun.*, 54:495–498.

Chen Y et al. (2008) "Giant" multishell CdSe nanocrystal quantum dots with suppressed blinking. *J. Am. Chem. Soc.*, 130(15):5026–5027.

Cichos F, von Borczyskowski C and Orrit M (2007) Power-law intermittency of single emitters. *Curr. Opin. Colloid Interface Sci.*, 12(6):272–284. ISSN 1359-0294.

Clauset A, Shalizi CR and Newman MEJ (2009) Power-law distributions in empirical data. *SIAM Rev.*, 51(4):661–703.

Cook RJ and Kimble HJ (1985) Possibility of direct observation of quantum jumps. *Phys. Rev. Lett.*, 54:1023–1026.

Coolen L, Brokmann X, Spinicelli P and Hermier JP (2008) Emission characterization of a single CdSe-ZnS nanocrystal with high temporal and spectral resolution by photon-correlation Fourier spectroscopy. *Phys. Rev. Lett.*, 100:027403.

Cordones AA and Leone SR (2013) Mechanisms for charge trapping in single semiconductor nanocrystals probed by fluorescence blinking. *Chem. Soc. Rev.*, 42:3209–3221.

Crouch CH et al. (2010) Facts and artifacts in the blinking statistics of semiconductor nanocrystals. *Nano Lett.*, 10(5):1692–1698.

Cui J, Beyler AP, Bischof TS, Wilson MWB and Bawendi MG (2014) Deconstructing the photon stream from single nanocrystals: from binning to correlation. *Chem. Soc. Rev.*, 43:1287–1310.

Cui J et al. (2016) Evolution of the single-nanocrystal photoluminescence linewidth with size and shell: Implications for exciton-phonon coupling and the optimization of spectral linewidths. *Nano Lett.*, 16(1): 289–296.

van Dam B et al. (2017) Excitation-dependent photoluminescence from single-carbon dots. *Small*, 13(48):1702098. ISSN 1613-6829.

Delerue C, Allan G and Lannoo M (2001) Electron-phonon coupling and optical transitions for indirect-gap semiconductor nanocrystals. *Phys. Rev. B*, 64(19):193402.

Efros AL and Nesbitt DJ (2016) Origin and control of blinking in quantum dots. *Nat. Nanotechnol.*, 11(8):661–71.

Empedocles Sa (1997) Quantum-confined stark effect in single CdSe nanocrystallite quantum dots. *Science*, 278(5346):2114–2117.

Empedocles SA and Bawendi MG (1999) Influence of spectral diffusion on the line shapes of single CdSe nanocrystallite quantum dots. *J. Phys. Chem. B*, 103(11):1826–1830.

Empedocles SA, Norris DJ and Bawendi MG (1996) Photoluminescence spectroscopy of single CdSe nanocrystallite quantum dots. *Phys. Rev. Lett.*, 77(18):3873–3876.

Fernée M, Littleton B, Cooper S, Rubinsztein-Dunlop H, Gómez DE and Mulvaney P (2008) Acoustic Phonon Contributions to the Emission Spectrum of Single CdSe Nanocrystals. *J. Phys. Chem. C*, 112(6):1878–1884.

Fernée MJ, Littleton B, Plakhotnik T, Rubinsztein-Dunlop H, Gómez DE and Mulvaney P (2010) Charge hopping revealed by jitter correlations in the photoluminescence spectra of single CdSe nanocrystals. *Phys. Rev. B*, 81:155307.

Fernée MJ, Littleton BN, Rubinsztein-Dunlop H and Ferne MJ (2009) Detection of bright trion states using the fine structure emission of single CdSe/ZnS colloidal quantum dots. *ACS Nano*, 3(11):3762–8. ISSN 1936-086X.

Fernée MJ, Sinito C, Louyer Y, Tamarat P and Lounis B (2013) The ultimate limit to the emission linewidth of single nanocrystals. *Nanotechnology*, 24(46):465703. ISSN 1361-6528.

Fernée MJ, Tamarat P and Lounis B (2013) Cryogenic single-nanocrystal spectroscopy: Reading the spectral fingerprint of individual CdSe quantum dots. *J. Phys. Chem. Lett.*, 4(4):609–618. ISSN 1948-7185.

Fernée MJ, Tamarat P and Lounis B (2014) Spectroscopy of single nanocrystals. *Chem. Soc. Rev.*, 43(4): 1311–37.

Frantsuzov P, Kuno M, Jankó B and Marcus RA (2008) Universal emission intermittency in quantum dots, nanorods and nanowires. *Nat. Phys.*, 4:519–522.

Galland C, Ghosh Y, Steinbrück A, Hollingsworth JA, Htoon H and Klimov VI (2012) Lifetime blinking in nonblinking nanocrystal quantum dots. *Nat. Commun.*, 3(May):908.

Galland C et al. (2011) Two types of luminescence blinking revealed by spectroelectrochemistry of single quantum dots. *Nature*, 479(7372):203–207.

Gómez DE, Califano M and Mulvaney P (2006) Optical properties of single semiconductor nanocrystals. *Phys. Chem. Chem. Phys.*, 8:4989–5011.

Gómez DE, van Embden J, Mulvaney P, Fernée MJ and Rubinsztein-Dunlop H (2009) Exciton–trion transitions in single CdSe-CdS core-shell nanocrystals. *ACS Nano*, 3(8):2281–2287.

Hapala P, Kůsová K, Pelant I and Jelínek P (2013) Theoretical analysis of electronic band structure of 2- to 3-nm Si nanocrystals. *Phys. Rev. B*, 87(19):195420.

Hartel AM, Gutsch S, Hiller D and Zacharias M (2012) Fundamental temperature-dependent properties of the Si nanocrystal band gap. *Phys. Rev. B*, 85(16): 165306.

He H et al. (2017) High-density super-resolution localization imaging with blinking carbon dots. *Anal. Chem.*, 89(21):11831–11838.

Hoogenboom JP, den Otter WK and Offerhaus HL (2006) Accurate and unbiased estimation of power-law exponents from single-emitter blinking data. *J. Chem. Phys.*, 125(20):204713.

Houel J et al. (2015) Autocorrelation analysis for the unbiased determination of power-law exponents in single-quantum-dot blinking. *ACS Nano*, 9(1):886–893.

Htoon H et al. (2010) Highly emissive multiexcitons in steady-state photoluminescence of individual "giant" CdSe/CdS core/shell nanocrystals. *Nano Lett.*, 10(7):2401–2407.

Hu F et al. (2016a) Carrier multiplication in a single semiconductor nanocrystal. *Phys. Rev. Lett.*, 116:106404.

Hu F et al. (2016b) Slow Auger recombination of charged excitons in nonblinking perovskite nanocrystals without spectral diffusion. *Nano Lett.*, 16(10):6425–6430.

Hybertsen MS (1994) Absorption and emission of light in nanoscale silicon structures. *Phys. Rev. Lett.*, 72(10):1514–1517.

Issac A, Krasselt C, Cichos F and von Borczyskowski C (2012) Influence of the dielectric environment on the photoluminescence intermittency of CdSe quantum dots. *ChemPhysChem*, 13(13):3223–3230.

Jang Y et al. (2017) Interface control of electronic and optical properties in IV-VI and II-VI core/shell colloidal quantum dots: A review. *Chem. Commun.*, 53:1002–1024.

Kovalenko MV et al. (2015) Prospects of nanoscience with nanocrystals. *ACS Nano*, 9(2):1012–1057.

Krasselt C, Schuster J and von Borczyskowski C (2011) Photoinduced hole trapping in single semiconductor quantum dots at specific sites at silicon oxide interfaces. *Phys. Chem. Chem. Phys.*, 13:17084–17092.

Kuno M, Fromm DP, Hamann HF, Gallagher A and Nesbitt DJ (2000) Nonexponential blinking kinetics of single CdSe quantum dots: A universal power law behavior. *J. Chem. Phys.*, 112(7):3117–3120.

Kuno M, Fromm DP, Hamann HF, Gallagher A and Nesbitt DJ (2001) "On"/"off" fluorescence intermittency of single semiconductor quantum dots. *J. Chem. Phys.*, 115:1028.

Kuno M, Fromm DP, Johnson ST, Gallagher A and Nesbitt DJ (2003) Modeling distributed kinetics in isolated semiconductor quantum dots. *Phys. Rev. B*, 67(12):125304.

Kůsová K, Pelant I, Humpolíčková J and Hof M (2016) Comprehensive description of blinking-dynamics regimes in single direct-band-gap silicon nanocrystals. *Phys. Rev. B*, 93(3):035412.

Kůsová K, Pelant I and Valenta J (2015) Bright trions in direct-bandgap silicon nanocrystals revealed by low-temperature single-nanocrystal spectroscopy. *Light Sci. Appl.*, 4:e336.

Kůsová K et al. (2010) Brightly luminescent organically capped silicon nanocrystals fabricated at room temperature and atmospheric pressure. *ACS Nano*, 4(8):4495–4504.

Leatherdale CA, Woo WK, Mikulec FV and Bawendi MG (2002) On the absorption cross section of CdSe nanocrystal quantum dots. *J. Phys. Chem. B*, 106(31):7619–7622.

Leon R, Petroff PM, Leonard D and Fafard S (1995) Spatially resolved visible luminescence of self-assembled semiconductor quantum dots. *Science*, 267(5206):1966–1968.

Li Z et al. (2017) Enhanced biexciton emission from single quantum dots encased in N-type semiconductor nanoparticles. *Appl. Phys. Lett.*, 111(15):153106.

Lin Y, Nienhaus K and Nienhaus GU (2018) Nanoparticle probes for super-resolution fluorescence microscopy. *ChemNanoMat*, 4(3):253–264.

Lippitz M, Kulzer F and Orrit M (2005) Statistical evaluation of single nano-object fluorescence. *Chemphyschem: Eur. J. Chem. Phys. Phys. Chem.*, 6(5):770–789.

Lounis B, Bechtel H, Gerion D, Alivisatos P and Moerner W (2000) Photon antibunching in single CdSe/ZnS quantum dot fluorescence. *Chem. Phys. Lett.*, 329(5):399–404. ISSN 0009-2614.

Lounis B and Orrit M (2005) Single-photon sources. *Rep. Prog. Phys.*, 68(5):1129.

Louyer Y, Biadala L, Trebbia JB, Fernée MJ, Tamarat P and Lounis B (2011) Efficient biexciton emission in elongated CdSe/ZnS nanocrystals. *Nano Lett.*, 11(10):4370–4375.

Macklin JJ, Trautman JK, Harris TD and Brus LE (1996) Imaging and time-resolved spectroscopy of single molecules at an interface. *Science*, 272(5259):255–258.

Mahler B, Spinicelli P, Buil S, Quelin X, Hermier Jp and Dubertret B (2008) Towards non-blinking colloidal quantum dots. *Nat. Mat.*, 7(8):659–664. ISSN 1476-1122.

Margolin G, Protasenko V, Kuno M and Barkai E (2006) Power-law blinking quantum dots: Stochastic and physical models. *Fractals, Diffus. Relax. Disord. Complex Syst. Adv. Chem. Physics, Part A*, 133: 327–356.

Messin G, Hermier JP, Giacobino E, Desbiolles P and Dahan M (2001) Bunching and antibunching in the fluorescence of semiconductor nanocrystals. *Opt. Lett.*, 26(23):1891–1893.

Moerner WE (1994) Examining nanoenvironments in solids on the scale of a single, isolated impurity molecule. *Science*, 265(5168):46–53.

Moerner WE and Kador L (1989) Optical detection and spectroscopy of single molecules in a solid. *Phys. Rev. Lett.*, 62:2535–2538.

Moerner WE and Orrit M (1999) Illuminating single molecules in condensed matter. *Science*, 283(5408):1670–1676.

Nair G, Zhao J and Bawendi MG (2011) Biexciton quantum yield of single semiconductor nanocrystals from photon statistics. *Nano Lett.*, 11(3):1136–1140.

Nasilowski M, Spinicelli P, Patriarche G and Dubertret B (2015) Gradient CdSe/CdS quantum dots with room temperature biexciton unity quantum yield. *Nano Lett.*, 15(6):3953–3958.

Newman MEJ (2005) Power laws, Pareto distributions and Zipf's law. *Contem. Phys.*, 46(5):323–351.

Nie S, Chiu D and Zare R (1994) Probing individual molecules with confocal fluorescence microscopy. *Science*, 266(5187):1018–1021. ISSN 0036-8075.

Nirmal M et al. (1996) Fluorescence intermittency in single cadmium selenide nanocrystals. *Nature*, 383:802–804.

Norris DJ and Bawendi MG (1996) Measurement and assignment of the size-dependent optical spectrum in CdSe quantum dots. *Phys. Rev. B*, 53:16338–16346.

Orrit M and Bernard J (1990) Single pentacene molecules detected by fluorescence excitation in a p-terphenyl crystal. *Phys. Rev. Lett.*, 65:2716–2719.

Osovsky R, Cheskis D, Kloper V, Sashchiuk A, Kroner M and Lifshitz E (2009) Continuous-Wave Pumping of Multiexciton Bands in the Photoluminescence Spectrum of a Single CdTe–CdSe Core-Shell Colloidal Quantum Dot. *Phys. Rev. Lett.*, 102:197401.

Park YS, Bae WK, Padilha LA, Pietryga JM and Klimov VI (2014) Effect of the core/shell interface on Auger recombination evaluated by single-quantum-dot spectroscopy. *Nano Lett.*, 14(2):396–402.

Park YS et al. (2011) Near-unity quantum yields of biexciton emission from CdSe/CdS nanocrystals measured using single-particle spectroscopy. *Phys. Rev. Lett.*, 106:187401.

Pelant I and Valenta J (2012) *Luminescence Spectroscopy of Semiconductors*. Oxford University Press, Oxford.

Pelton M, Smith G, Scherer NF and Marcus RA (2007) Evidence for a diffusion-controlled mechanism for fluorescence blinking of colloidal quantum dots. *P. Natl. Acad. Sci. USA*, 104(36):14249–14254.

Peterson JJ and Krauss TD (2006) Fluorescence Spectroscopy of Single Lead Sulfide Quantum Dots. *Nano Lett.*, 6(3):510–514.

Peterson JJ and Nesbitt DJ (2009) Modified power law behavior in quantum dot blinking: A novel role for biexcitons and Auger ionization. *Nano Lett.*, 9(1): 338–345.

Pevere F, Sychugov I, Sangghaleh F, Fucikova A and Linnros J (2015) Biexciton Emission as a probe of Auger recombination in individual silicon nanocrystals. *J. Phys. Chem. C*, 119(13):7499–7505.

Picoquant (2018) www.picoquant.com, web pages of the Picoquant company, accessed 2018.

Rabouw FT and de Mello Donegá C (2016) Excited-state dynamics in colloidal semiconductor nanocrystals. *Top. Curr. Chem.*, 374(5):58.

Reid KR, McBride JR, Freymeyer NJ, Thal LB and Rosenthal SJ (2018) Chemical structure, ensemble and single-particle spectroscopy of thick-shell InP–ZnSe quantum dots. *Nano Lett.*, 18(2):709–716.

Sangghaleh F, Bruhn B, Schmidt T and Linnros J (2013) Exciton lifetime measurements on single silicon quantum dots. *Nanotechnology*, 24(22):225204. ISSN 1361-6528.

Sher PH et al. (2008) Power law carrier dynamics in semiconductor nanocrystals at nanosecond timescales. *Appl. Phys. Lett.*, 92(2008):101111.

Spinicelli P, Buil S, Quelin X, Mahler B, Dubertret B and Hermier JP (2009) Bright and grey states in CdSe-CdS nanocrystals exhibiting strongly reduced blinking. *Phys. Rev. Lett.*, 102:136801.

Sychugov I, Fucikova A, Pevere F, Yang Z, Veinot JGC and Linnros J (2014) Ultranarrow luminescence linewidth of silicon nanocrystals and influence of matrix. *ACS Photonics*, 1(10):998–1005.

Sychugov I, Juhasz R, Valenta J and Linnros J (2005) Narrow luminescence linewidth of a silicon quantum dot. *Phys. Rev. Lett.*, 94:087405.

Sychugov I, Valenta J and Linnros J (2017) Probing silicon quantum dots by single-dot techniques. *Nanotechnology*, 28(7):072002.

Sychugov I, Valenta J, Mitsuishi K, Fujii M and Linnros J (2011) Photoluminescence measurements of zero-phonon optical transitions in silicon nanocrystals. *Phys. Rev. B*, 84(12):125326.

Tang J and Marcus RA (2005) Diffusion-controlled electron transfer processes and power-law statistics of fluorescence intermittency of nanoparticles. *Phys. Rev. Lett.*, 95:107401.

Valenta J, Greben M, Remeš Z, Gutsch S, Hiller D and Zacharias M (2016) Determination of absorption cross-section of Si nanocrystals by two independent methods based on either absorption or luminescence. *Appl. Phys. Lett.*, 108(2):023102.

Valenta J et al. (2008) Light-emission performance of silicon nanocrystals deduced from single quantum dot spectroscopy. *Adv. Funct. Mater.*, 18(18):2666–2672.

Vela J et al. (2010) Effect of shell thickness and composition on blinking suppression and the blinking mechanism in "giant" CdSe/CdS nanocrystal quantum dots. 3:706–17.

Verberk R, van Oijen AM and Orrit M (2002) Simple model for the power-law blinking of single semiconductor nanocrystals. *Phys. Rev. B*, 66(23):233202.

Verberk R and Orrit M (2003) Photon statistics in the fluorescence of single molecules and nanocrystals: Correlation functions versus distributions of on- and off-times. *J. Chem. Phys.*, 119(4):2214–2222.

Wang S et al. (2008) Blinking statistics correlated with nanoparticle number. *Nano Lett.*, 8(11):4020–4026.

Wang X et al. (2009) Non-blinking semiconductor nanocrystals. *Nature*, 459(7247):686–689, retracted (2015): *Nature*, 527(7579):544.

Weiss PS (2014) Nobel prizes for super-resolution imaging. *ACS Nano*, 8(10):9689–9690.

Xu W et al. (2017) Deciphering charging status, absolute quantum efficiency, and absorption cross section of multi-carrier states in single colloidal quantum dots. *Nano Lett.*, 17(12):7487–7493.

Yin C et al. (2017) Bright-exciton fine-structure splittings in single perovskite nanocrystals. *Phys. Rev. Lett.*, 119:026401.

Yu P et al. (2005) Absorption cross-section and related optical properties of colloidal InAs quantum dots. *J. Phys. Chem. B*, 109(15):7084–7087.

Zhao J, Nair G, Fisher BR and Bawendi MG (2010) Challenge to the charging model of semiconductor-nanocrystal fluorescence intermittency from off-state quantum yields and multiexciton blinking. *Phys. Rev. Lett.*, 104:157403.

Zhong Y et al. (2015) Facile, large-quantity synthesis of stable, tunable-color silicon nanoparticles and their application for long-term cellular imaging. *ACS Nano*, 9(6):5958–5967.

Scanning Electrochemical Microscopy and Its Potential in Nanomaterial Characterization

Christine Kranz
Ulm University

22.1 Introduction

Nanoscience and its applications have revolutionized not only research but also our daily life, e.g., in consumer products, medical diagnostics, and energy-related topics. A key aspect of nanomaterials is that their functionality and (re)activity are directly related to size, shape, structure, chemical composition, and also concentration (Mourdikoudis et al., 2018). Therefore, characterization methods are required that provide high-resolution information on these features down to the nanometer level. Next to the characterization of the nanostructured materials, the main challenges are also associated with gaining insight and detailed understanding during *operando* or *in situ* conditions of surface and interface processes at such nanomaterials, e.g., nanoparticles, including phenomena such as aging, changes of morphology, and accompanying changes in properties (Kleijn et al., 2014). Along with the revolution in nanoscience, it is evident that the introduction of scanning probe microscopy (SPM) techniques with the breakthrough initiated by the scanning tunneling microscopy (STM) by Binnig and Rohrer (Binnig et al., 1982; Binnig and Rohrer, 1982) has revolutionized the characterization possibilities, as high-resolution nanometer-scale imaging became available in vacuum, at ambient condition and even in solution. Derived from electrochemical (ec)-STM (Sonnenfeld and Hansma, 1986), an in situ version of STM performed in electrolyte solution at a biased sample surface termed 'scanning electrochemical microscopy' (SECM) was introduced by Bard and co-workers (Bard et al., 1991; Liu et al., 1986) and developed into an attractive *in situ* surface-characterization technique providing information on the (electro)activity and topography of a sample. This led to applications in multidisciplinary research areas including biomedicine, materials

science, corrosion, catalysis, photocatalysis, fuel cells, and battery research. It should be noted that initial SECM-type experiments were also performed by Engstrom and co-workers using microelectrodes to probe the concentration profiles at macroscopic electrode surfaces (Engstrom et al., 1987, 1986). However, the term 'scanning electrochemical microscopy' and the major instrumental developments have been introduced by the group of Bard (Bard et al., 1989). With improving achievable spatial resolution in SECM, recently the correlation of reactivity with size and structure of nano-sized objects became a focus in SECM research. All scanning probe techniques have in common that information is obtained by the localized interaction of the SPM tip—frequently termed 'probe'—and the sample surface. For example, in atomic force microscopy (AFM), local forces acting on a sharp tip at the end of a cantilever are detected. If the tip is in close proximity or in contact with the sample, AFM provides information on the sample topography (Binnig et al., 1986) or quantitative information from force spectroscopy (Marti, 2001). In SECM, localized electrochemical properties of the sample such as mapping chemical fluxes and interfacial reactivity are determined by a micro- or nanoelectrode (mostly disc-shaped electrode), which is positioned in close proximity to the sample surface (the distance is dependent on the electroactive area of the SECM tip and is several radii of the electroactive surface). The recorded signal (e.g., potentiometric or voltammetric in nature) at the micro- or nanoelectrode is recorded in dependence of the position. The SECM signal is hereby dependent on the nature and reactivity of the sample, the properties of the SECM tip, and the distance between the electrode and sample surface. SECM is an electroanalytical SPM technique, as in fact any electroanalytical technique such as potentiometry (Janotta et al., 2004; Wei et al., 1995; Wipf et al., 2000), voltammetry including square wave voltammetry (Lima et al., 2012; Rudolph et al., 2016), stripping polarography at mercury-coated microelectrodes (Barton and Rodríguez-López, 2017, 2014a; Souto et al., 2012), and electrochemical impedance spectroscopy (Ballesteros et al., 2002; Bandarenka et al., 2014; Keddam et al., 2009) can be performed in SECM experiments. In addition, theoretical models have been developed for the quantitative treatment of SECM data (Amphlett and Denuault, 1998; Lefrou and Cornut, 2010; Mirkin et al., 1992; Zoski and Mirkin, 2002). SECM belongs to the group of so-called non-invasive SPM techniques, as the SECM tip is not in contact with the sample surface in most applications. However, the distance between the sample and the SECM electrode plays not only a pivotal role in terms of achievable resolution but may be also a source of erroneous data, in particular, if sub-micrometer-sized SECM tips are used in voltammetric experiments.

In recent years, significant developments have been made in distance-controlled—usually current independent—positioning of the SECM tip, which allows independent recording the electrochemical reactivity and surface topography. In Section 22.2.3, recent advances in distance-controlled methodologies will be discussed. A major advantage related to the constant distance-controlled SECM is that nanometer-sized electrodes can be positioned with high accuracy, and convolution of the electrochemical information with sample morphology is almost entirely circumvented. This is a critical factor, if the topological features of the sample are of the dimensions of the used SECM tip and nano-sized objects are the focus of the investigation. Besides, novel developments in respect to probes, instrumental improvements, and novel application fields contributed to the fact that SECM has become a valuable SPM technique. Several companies offer nowadays commercial SECM instrumentation. Within recent years, several excellent reviews and monographs have been published on SECM and its various fields of application (Amemiya et al., 2008; Holzinger et al., 2016; Izquierdo et al., 2018; Kai et al., 2018; Kranz, 2014; Mirkin and Bard, 2012; Polcari et al., 2016; Takahashi et al., 2017; Zoski, 2016). Hence, this chapter will give an overview on basic concepts, recent instrumental and probe developments, and then highlight only some of the applications related to nanomaterials and especially to energy-related SECM studies, which are in these days and age of prevalent interest.

22.2 Instrumentation and Operation Modes

22.2.1 Hardware

The hardware of SECM instruments comprises the following main components shown in Figure 22.1a. A positioning system, which allows the movement of SECM tip and/or sample in x,y,z direction. In contrast to AFM, and STM where piezoelectric tube scanners are used, SECM instruments rely frequently only stepper motors when micro-sized electrodes are used as SECM tips or

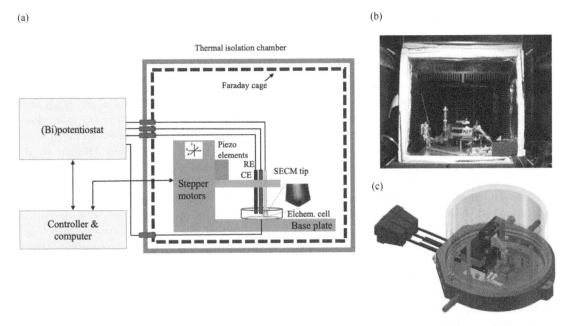

FIGURE 22.1   (a) Schematic view of SECM instrument located in a Faraday cage and a thermal isolation chamber; the 3-electrode or 4-electrode setup comprises the SECM tip as working electrode, optional the sample as working electrode 2, the CE and the RE. (b) Photograph of an SECM system placed in an isothermal chamber using vacuum insulated panels and aluminum heat sinks for drift compensation. (Reprinted with permission from Kim et al., 2016a, Copyright (2012) American Chemical Society.) (c) SECM scan head and electrochemical cell placed in an Argon-filled glove box. (Reprinted with permission from Bülter et al., 2015.)

a combination of stepper motors for large movements and piezoelectric positioning elements for accurate fine positioning for sub-micrometer probes. For nanopipette-based techniques, piezoelectric scanners are preferred. To control electrochemical experiments, typically a high-end (bi)potentiostat is employed, which provides current ranges down to the sub-pico-ampere regime. An electrochemical cell consists of a three- or four-electrode (i.e., in case a potential should be applied to the sample), comprising the sample at the bottom of the cell, the SECM tip as working electrode (WE), a counter electrode (CE) and a reference electrode (RE), respectively. Positioning and bi(potentiostat) are controlled via an acquisition system equipped with the appropriate software. The SECM may be established on top of an inverted light or fluorescence microscope in case of biological/biomedical studies. Specialized electrochemical cells that allow controlling physical parameters such as temperature (Hirano et al., 2008; Schäfer et al., 2013) and/or the surrounding atmosphere, e.g., fixed CO_2 content or oxygen-reduced atmospheres (Nogala et al., 2010) have been implemented in SECM. Controlled atmospheres are a prerequisite for applications in, e.g., battery research (Ventosa and Schuhmann, 2015) but also for live cell studies (incubator conditions). For example, Wittstock and co-workers introduced a custom-designed Ar-fillable SECM chamber including ports for the necessary electrical connections, Ar-inlet and -outlet for battery research (Bülter et al., 2015, 2014).

SECM equipment, placed in a glove box for battery-related research, has also been reported (Zampardi et al., 2013). For nano-sized SECM probes, improvements in instrumentation are required related to thermal drift suppression, as the distance between the sample and the nano-sized SECM probe is on the order of few nanometers (Kim et al., 2012). Potential damage by electrostatic discharge (Nioradze et al., 2013) and vibrations are also an issue. Hence, proper grounding and modifications of the (bi)potentiostats have recently been reported and are a prerequisite to avoid damaging nano-sized SECM probes (Kim et al., 2016a).

22.2.2 SECM Probes

Ultramicroelectrodes

To date, the majority of SECM experiments are still conducted with disc-shaped microelectrodes typically sealed in glass with radii of the electroactive disc in the range of 1–25 μm (see inset in Figure 22.1a). Ultramicroelectrodes (UMEs) can be readily fabricated on the lab bench and characterized by conventional optical microscopy and electrochemical methods such as cyclic voltammetry and SECM approach curves (see Section *Feedback Mode*). The voltammetric behavior of electrodes is related to the timescale of the experiment. For micro-/ultramicroelectrodes with disc or spherical geometry, the thickness of the diffusion layer $\delta = \sqrt{\pi D t}$ (D: diffusion coefficient and t: time) exceeds the radius of the electrode at relatively short times, resulting in an enhanced mass transport and a steady-state current. Besides the enhanced mass transport, they show decreased double-layer charging currents due to their reduced surface area and reduced ohmic potential (iR) drop, which is associated with the small Faradaic currents in the nA and pA region. Soft microelectrodes have been introduced by the groups of Wittstock and Girault, which can be mass fabricated using laser ablation and ink jet printing (Cortes-Salazar et al., 2009). A microchannel is cut into polyethylene terephthalate (PET) or polyimide (Kapton HN®), respectively, which is then filled with electroactive material such as conductive carbon ink or nanoparticulate gold ink (Lesch et al., 2012a). After lamination with a polymer film (e.g., Parylene C), the electrode area is exposed by simple cutting with a scalpel. This fabrication procedure has been demonstrated for microelectrode arrays (Lesch et al., 2012b) and push-pull probes (Momotenko et al., 2012), which allow localized delivery of agents or combination with mass spectrometry. The advantage of arrayed probes is that relatively large areas can be imaged within reasonable periods of times. In addition, a constant distance between the electroactive area can be maintained, if the insulation layer is brought into direct contact with the sample surface. The soft electrode bends slightly, and if the microelectrode is dragged in contact across the sample surface, the electroactive area keeps largely the distance to the sample surface. Although, this is a viable approach in case large scans are required, the trend in SECM probe design clearly focuses on decreasing the SECM probes to nanoscale dimensions, which is required for high-resolution investigation of nano-sized objects or features. Nanoelectrodes not only provide superior spatial resolution but allow also to perform experiments with high temporal resolution.

Nanoelectrodes

So-called nanoelectrodes have already been fabricated and described in detail in the 1990s based on electrochemical etching procedures of microwires and consecutive insulation (Mirkin et al., 1992; Penner et al., 1990; Slevin et al., 1999) similar to the fabrication of ec-STM probes. Nowadays, solid nanoelectrodes for SECM experiments are mostly fabricated using a laser-assisted pipette puller. Nanoelectrodes with diameters ranging from 2 to 500 nm by controlling local heating of the glass and the pulling force have been reported (Li et al., 2009; Shao et al., 1997) (e.g., see Figure 22.2a). In order to expose the electroactive area, the pulled enclosed nanowire is exposed by etching in 40% hydrofluoric acid (HF) or by micropolishing, which however may limit the reproducibility of the shape and size of the obtained electrodes. A more detailed description with optimized conditions was published by Schuhmann and co-workers for the fabrication of tight-sealed, disc-shaped nanoelectrodes (Katemann and Schuhmann, 2002). These procedures are nowadays widely adopted to fabricate platinum, gold, carbon, and silver nano-sized electrodes for

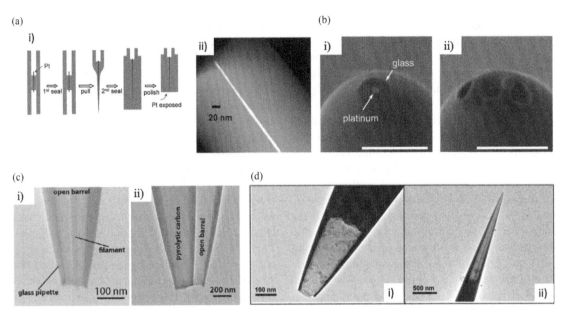

FIGURE 22.2 Fabrication scheme and scanning electron microscopy (SEM) images of nanoscopic scanning electrochemical probes. (a) (i) Schematic view of the laser-assisted pulling process and exposure of the nanowire by polishing. (ii) Transmission electron microscopy (TEM) image of a 3 nm radius Pt nanoelectrode sealed in SiO_2. (Reprinted with permission from Li et al., 2009, Copyright (2009) American Chemical Society.) (b) SEM images of a $\sim$110 nm diameter Pt nanoelectrode (i) before and (ii) after damage through electrostatic discharge. Scale bars are 1 µm. (Reprinted with permission from Nioradze et al., 2013, Copyright (2013) American Chemical Society.) (c) TEM images of (i) double-barrel nanopipette, (ii) double-barrel nanopipette for hybrid SICM-SECM with one barrel filled with pyrolytic carbon and the second barrel open. (Reprinted with permission from Kang et al., 2016, Copyright (2016) American Chemical Society. (d) TEM images of carbon-coated nanopipettes with different cavity depths. (i) $r = 33$ nm, $\theta = 8°$, $H = 12$; (ii) $r = 10$ nm, $\theta = 5°$, $H = 190$. (Reprinted with permission from Yu et al., 2014, Copyright (2013) American Chemical Society.)

SECM experiments (Clausmeyer and Schuhmann, 2016; Cox and Zhang, 2012; Hua et al., 2018; Li et al., 2009; Ying et al., 2017 and references in there).

Nanopipette-Based Electrodes

Nanopipettes, which are commonly used in scanning ion conductance microscopy (SICM) (Chen et al., 2012; Hansma et al., 1989; Novak et al., 2009), have also gained significant interest for the fabrication of electrochemical SPM probes and carbon-based nanoelectrodes, as reviewed recently (Clausmeyer and Schuhmann, 2016). Nanopipettes can be readily pulled using a laser-pipette puller with sufficient reproducibility and characterized in respect to their orifice and geometrical features via transmission electron microscopy (TEM) as recently described in detail (Wilde et al., 2018). Such pipettes can then be filled with carbon material by pyrolysis using a carbon source, such as methane, acetylene, and butane (McKelvey et al., 2013; McNally and Wong, 2001; Takahashi et al., 2011). However, such fabricated carbon electrodes frequently show sluggish electron behavior due to the high content of amorphous carbon and mainly outer-sphere electron transfer mediators, e.g., water-soluble ferrocene derivates are used for their characterization and for imaging experiments. In case redox species with inner-sphere electron transfer characteristics should be investigated, the nanosized-carbon electrodes need to be modified by electrochemical deposition of Pt (O'Connell and Wain, 2014) or as shown for potentiometric

measurements with iridium oxide (Nadappuram et al., 2013). However, such modifications may change the overall geometry, which adds to the complexity in data interpretation. Alternatively, theta capillaries may be pulled into two channel nanopipettes (see Figure 22.2c(i)), which are unique as they may serve as bifunctional SPM probes. One channel is filled with electrolyte to position the probe via the ion current (SICM), and the second channel is filled with electrode material for Faradaic current measurements (as shown Figure 22.2c(ii)) (Takahashi et al., 2011). In addition, such dual-channel probes are used in scanning electrochemical cell microscopy (SECCM), as discussed in Section *AFM-SECM*. In general, the electrochemical performance of nano-sized electrodes is extremely sensitive to the smallest changes in electrode geometry (e.g., recessed electrode or protruding rather than disc). Hence, a thorough characterization of nanoelectrodes is a requirement that remains an ongoing challenge. For visualization, scanning electron microscopy (SEM) is used for the characterization for submicroelectrodes (i.e., down to approx. 100 nm of the active electrode radius), yet, is increasingly difficult and insufficient, if the active electrode radius is reduced to tens of nanometers. Recently, TEM became the method of choice for the characterization of nanopipettes and nanoelectrodes. However, routine characterization before and after usage of the nanosized probes may be difficult due to limited access and costs to high-resolution TEM instruments. Bulk electrochemical characterization, e.g., cyclic voltammetry may assist in estimating the active electrode size but does not provide information on the precise geometry. SECM approach curves—as discussed below—are also employed to characterize the geometry of nanoelectrodes (Mirkin et al., 1992; Shao et al., 1997). The Faradaic current measured at the nanoelectrode in dependence of the distance when approaching a pure conductor is not only dependent on the conductivity of the sample but also on the shape and insulation quality of the nanoelectrode. Evaluating such approach curves via numerically fitted data provides information on the geometric features of the electroactive area.

Although tremendous improvements have been reported in nanoscale electrochemical imaging, providing quantitative SECM data, sufficient reproducibility, and robustness of nanoscale probes remain challenging. In addition, appropriate handling of nanoelectrodes is crucial (e.g., electrostatic charging, see Figure 22.2b); likewise, solutions and chemicals may potentially block the nanoscale electrode area due to contaminations.

22.2.3 Imaging and Positioning Modes

This section is divided into imaging modes (Figure 22.3), modes for surface modifications, and positioning modes (Figure 22.4). Conventional positioning using SECM approach curves, which is basically a feedback mode experiment will be discussed in Section *Feedback Mode*. Positioning modes independent of the recorded Faradaic current will be addressed below. In respect to imaging modes, generation-collection mode (G/C) (Lee et al., 1991) and feedback (FB) mode (Kwak and Bard, 1989) are still the most widely used operation modes in SECM. The redox-competition (RC)

mode introduced by Schuhmann and co-workers (Eckhard et al., 2006) is frequently applied when imaging living cells, as well as in electrocatalysis studies.

Imaging Modes

Generation-Collection Mode

Generation-collection (G/C) mode experiments were among the first SECM-type experiments (Engstrom et al., 1986). G/C experiments can be subdivided into tip generation-substrate collection (TG/SC) mode (Figure 22.3a(i)) or substrate generation-tip collection (SG/TC) mode (Figure 22.3a(ii)). In SG/TC, the SECM probe detects species that are generated at or released from the sample surface. For TG/SC, the tip and the sample are biased at a potential, where a specific electroactive species is produced at the SECM tip and detected at the sample tip, as shown in Figure 22.3a(i). In this mode, the collection efficiencies given by the ratio of substrate current to tip current i_S/i_T is 100%, if the tip is close to the substrate surface (i.e., at a distance of several radii of the electroactive tip area). In case the reactive active species undergoes a homogenous reaction, i_S/i_T becomes smaller and the rate constant of the homogeneous reaction can be determined (Unwin, 2012). Although the GC mode usually suffers from limited lateral resolution in imaging experiments compared to feedback mode, it has gained importance in catalysis, e.g., for screening experiments (Minguzzi et al., 2015), for investigating nanoparticles in respect to oxygen reduction reaction (ORR) (Sánchez-Sánchez et al., 2010) and hydrogen evolution reaction (HER) (Sun et al., 2014), and for mapping biological samples (e.g., investigating enzyme activity, membrane transport, release

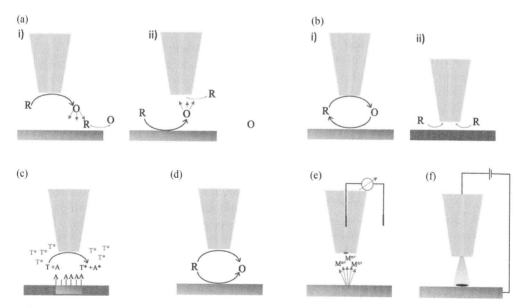

FIGURE 22.3 Schematic of SECM imaging modes. (a) Generation-collection mode; (i) TG/SC mode; (ii) SG/TC mode. (b) Feedback mode (i) over a conducting substrate (positive feedback) and (ii) over an inert substrate (negative feedback). (c) Surface interrogation (SI) mode; a chemically or electrochemically adsorbed species A reacts with a titrant T generated at the tip, resulting in a positive feedback current. (d) RC mode; substrate and SECM tip are competing for the same redox active species. (e) Potentiometric mode with an ion-selective microelectrode, where M^{n+} is a metal ion in solution with charge n (n = integer). (f) Direct mode where the SECM tip is the CE and the sample the working electrode.

of signaling molecules at single cells or respiratory activity (oxygen consumption of living cells) (Holzinger et al., 2016). GC experiments have been performed via amperometric probes (Pierce et al., 1992; Wittstock and Schuhmann, 1997), microbiosensors (Hecht et al., 2013; Horrocks et al., 1993b; Kueng et al., 2005), UMEs modified with electrocatalytic layers (Voronin et al., 2012), and potentiometric probes (Horrocks et al., 1993a). Potentiometric probes have the advantage that the substrate diffusion is negligibly disturbed in contrast to amperometric probes. Quantitative evaluation of SG/TC experiments can be obtained at steady-state conditions, which require that the imaged structures (i.e., micrometer-sized domains) have to be on the order of magnitude of the UME (Sánchez-Sánchez et al., 2008).

Feedback Mode

In feedback mode, an artificial redox mediator with fast electron transfer characteristic is added to the solution. A comprehensive review of useful redox mediators was provided in a recent review (Polcari et al., 2016). As long as the tip is in bulk solution and the SECM tip is biased at a potential where the mediator is oxidized or reduced, and the mass transport process is purely diffusional, a steady-state current due to hemispherical diffusion towards the UME is obtained, which is given as $i_{T\infty} = 4nFDcr$ for a disc microelectrode, and as $i_{T\infty} = 4nFDcr(1 + qH^p)$ for a conical microelectrode (Zoski and Mirkin, 2002) with $i_{T\infty}$: steady-state current, n: number of transferred electrons, F: Faraday constant, D and c: diffusion coefficient and concentration of the redox mediator, and r: radius of the disc microelectrode. For conical electrodes, the parameters are modified as $q = 0.3661$, $p = 1.14466$, and $H = h/r$ (h = height), respectively.

If the UME is approached to the sample surface, the nature of the sample and the distance will influence the concentration profile at the UME leading either to a reduced current (a.k.a., negative feedback effect) or increased current due to the regeneration of the redox mediator at the sample surface (a.k.a., positive feedback effect) (Kwak and Bard, 1989) as shown in Figure 22.3b. The approach curve towards an insulator additionally depends on the so-called RG value, described by the ratio of the radius of the insulating sheath to the radius of the electroactive embedded wire (RG = rg/r) (Mirkin et al., 1992). In conventional SECM experiments, the feedback mode is also used to position the SECM tip. The current is dependent on both the tip-to-substrate distance (i.e., diffusional process via pure positive or negative feedback) and the apparent electron transfer kinetics. Analytical approximations have been developed to describe the theoretical feedback current for pure diffusional mass transport and for processes where the electron transfer is limiting the overall recorded current. Lefrou and Cornut summarized these analytical expressions in a review (Cornut and Lefrou, 2008). For imaging topography and/or electroactivity, the feedback mode is the most frequently used mode and has been employed in almost all fields

of application of SECM ranging from corrosion research to biomedical studies, surface modifications, and energy-related topics.

Surface Interrogation Mode

Surface interrogation (SI) mode (Rodríguez-López et al., 2008) reflects a transient SECM feedback mode experiment that allows the direct detection and quantification of adsorbed species at a conductive surface. In SI mode, the dimensions of the substrate and the SECM tip have to be on the same order of magnitude. Figure 22.3c shows the SI mode where a reactive species A that is adsorbed at the sample surface reacts with a species T in a titration-like reaction with T produced at the SECM from T*, which was added to solution. The reaction of A with the locally produced T regenerates the species T*, thereby resulting in a positive feedback effect (see Figure 22.3c). As the adsorbed species A is depleted by the reaction with T, the regeneration of T*, and hence, the increased concentration of T* in the gap between the SECM tip and the surface decreases leading to a negative feedback current during the time course of the reaction. The transient feedback signal then allows the quantification of adsorbed species. The adsorbed species may be generated by applying a potential (i.e., potential scan or potential pulse), which adsorbs at the surface while the SECM tip is kept at open circuit potential. The redox active species added to solution for the "surface titration experiment" has to be stable at these conditions and should not interfere with the formation of the adsorbed species. The SI mode has been employed to study water splitting induced by electrocatalysts or photo-induced processes, whereby intermediate adsorbed species are determining the overall reaction (Barforoush et al., 2018; Krumov et al., 2018; Zigah et al., 2012).

Redox-Competition Mode

The RC mode was first demonstrated to investigate local catalytic activity (Eckhard et al., 2006). In RC mode, the SECM tip and the substrate are competing for the same redox species, when the tip is in close proximity to the sample surface as shown in Figure 22.3d. The advantage of this mode for screening catalytic activity (i.e., heterogeneous oxygen reduction) is that high-background signals are avoided, and hence, the sensitivity is increased. In this mode, the current response is no longer dependent on the sample dimensions (i.e., in comparison to SG-TC mode) but is dominated by the probe radius. The RC mode is also highly suitable in corrosion science (González-García et al., 2011; Santana et al., 2010), in catalysis (Nagaiah et al., 2013; Nebel et al., 2014; Zeradjanin et al., 2011), and for mapping the respiratory activity of single cells (Nebel et al., 2013). Changes in the oxygen concentration within the gap between probe and cell are influenced not only by depletion but also changes of the sample-to-probe distance.

Potentiometric Mode

In potentiometric mode, the SECM tip is a miniaturized ion-selective electrode (ISE) such as solid-state pH microelectrodes, e.g., antimony/antimony oxide or iridium/iridium oxide microelectrodes (Horrocks et al., 1993a; Wipf et al., 2000) or ISE for monovalent and divalent cations (Figure 22.3e). Reproducible fabrication procedures are available for micro-ISEs (Filotás et al., 2017; Izquierdo et al., 2013), which allow the local selective detection of various ions. Several issues have to be taken into account though. In contrast to amperometric microelectrodes, potentiometric probes cannot be positioned via approach curves. Shear-force mode SECM in combination with micro-ISEs has been reported based on ion-selective micropipettes (Hengstenberg et al., 2000). Recently, a carbon-based Ca^{2+}-sensitive micro-ISE was reported, which can be switched between amperometric and potentiometric mode (Ummadi et al., 2016). Hence, an approach curve can be recorded for positioning before the electrode is switched to potentiometric mode. The usually slow response time of potentiometric measurements may be an issue depending on the time constant of the measuring circuit and the thickness of the diffusion layer. In addition to the slow scan speeds required for micro-ISEs also increased noise issues have been reported. Nonetheless, potentiometric mode plays a significant role in corrosion science, where not only local pH changes but also changes in local concentrations of ions play an important role and during the investigation of biofilm formation.

Direct Mode

SECM and related techniques such as nanopipette-based techniques are also suitable for surface modifications generating 2D and 3D micro- and nanosized structures *in situ* (Mandler, 2012). Surface modifications have been demonstrated shortly after the introduction of the SECM via deposition of metal and polymer structures and etching of conducting and semiconducting materials. Such surface modification can be achieved using generation-collection mode and feedback mode SECM (Mandler et al., 1996). Moreover, surface modification has also been demonstrated using the so-called 'direct mode' with the sample serving as WE and the microelectrode as CE as shown in Figure 22.3f. If the counter microelectrode is positioned close to the sample surface, the electrical field lines are confined. Once, e.g., a deposition process is induced, structures with approximately the dimensions of the SECM tip are obtained. Combining shear-force regulation with direct mode, 3D micro-sized objects can be obtained (Kranz et al., 1996). Direct mode in combination with pulse-deposition techniques have been used for microstructured polymer deposits (Kranz et al., 1995). In recent years, SECM was employed for the localized deposition of nanoparticles (Danieli and Mandler, 2013; Fedorov and Mandler, 2013).

Current-Independent Positioning Strategies

Shear-Force Mode

One drawback of conventional SECM may be associated with that the SECM tip is positioned via recording an approach curve and then scanned in a fixed height across the sample surface, which is frequently termed "constant height mode". In case the surface morphology is in the range of the radius of the electroactive area of the SECM tip, convoluted signals related to changing reactivity of the sample surface and topographic surface features may lead to erroneous data (Hengstenberg et al., 2000), due to distance dependency of SECM signals. In addition, correlation of topographical and electrochemical information is required to understand structure–activity relationships, which is a general challenge in nanoscience. First reports on using shear-force-based distance regulation, which allows the positioning of the SECM tip independently from the electrochemical response have been published by Ludwig et al. (1995) and James et al. (1998). Hydrodynamic forces dampen the vibration amplitude of a fiber-shaped micro- or nanoelectrode in close proximity to the sample surface. The amplitude signal of the horizontal excitation is used as feedback signal maintaining a constant distance between the tip and the sample surface. First shear-force-based SECM systems used optical shear-force detection (Figure 22.4a(i)) (Ludwig et al., 1995), which are nowadays replaced by tuning fork-type resonators (Figure 22.4a(ii)) first applied in nearfield scanning optical microscopy (NSOM) (Brunner et al., 1997; Karrai and Grober, 1995) and piezo-based actuator-detector systems (Figure 22.4a(iii)) (Ballesteros Katemann et al., 2003) due to less-demanding instrumental aspects. Shear-force positioning is routinely applied for current-independent positioning of microelectrodes and nanoelectrodes (Etienne et al., 2012; Nebel et al., 2014, 2010; Takahashi et al., 2009). The lateral topographical resolution achieved in shear-force mode SECM is not comparable to other scanning probe techniques such as AFM or SICM, as the size of the SECM probe (i.e., electroactive area and insulating sheath) determines the lateral resolution in this noncontact mode operation. Nonetheless, modified SECM tips such as miniaturized biosensors or potentiometric probes (e.g. ISEs) can be reliably positioned and used for imaging.

Alternative modes for positioning microelectrodes have been introduced over the years. One early approach first described by Wipf et al. (Wipf et al., 1993; Wipf and Bard, 1992) is based on recording the distance dependence of the impedance signal when an AC voltage is superimposed onto a DC voltage. This impedance-related signal has been used for topographical maps of, e.g., PC12 cells (Kurulugama et al., 2005). Schuhmann and co-workers developed a distance-dependent SECM mode based on AC signals, which they termed "alternating current" SECM (Ballesteros et al., 2002). Although, the measurements reflect electrochemical impedance spectroscopy (EIS) measurements, it is different from localized EIS techniques as no bias is applied to the sample. The AC signal has been used in feedback signal for the constant-distance imaging of nonconductive samples (Etienne et al., 2004). The measurements are conducted in a redox-mediator free, low-conductivity electrolyte solution, where the solution resistance dominates the overall impedance signal (Z) (Figure 22.4b).

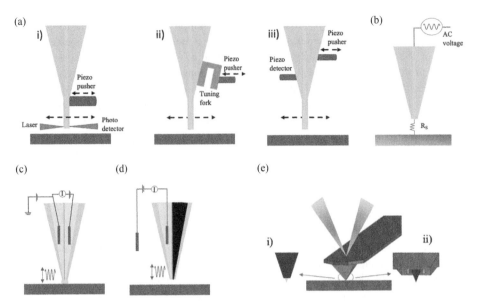

FIGURE 22.4 Schematic of constant distance positioning modes. (a) Shear-force mode: (i) optical readout; (ii) tuning fork-based detection; (iii) piezo-based actuator–detector systems. (b) AC mode with an AC voltage signal superimposed to a small DC voltage at the SECM tip; magnitude of the current and phase-angle response are recorded. (c) Double-barrel probe for SECCM; when the tip is moved to the sample, the electrolyte meniscus is brought into contact with the surface. (d) Hybrid SICM–SECM based on a dual-barrel nanopipette with one barrel filled with pyrolytic carbon as nanoelectrode and the second barrel as SICM probe. (e) Hybrid AFM–SECM probe with (i) AFM tip with conical electrode located at the tip apex and (ii) microfabricated cantilever with tip-integrated electrode recessed from the AFM tip.

Changing the electrolyte to high ionic strength, also constant-distance SECM experiments could be obtained on conductive samples as reviewed by Eckhard and Schuhmann (2008).

With soft probes and soft probe arrays, constant distance imaging can be achieved in contact mode, where the polymer insulation of the soft microelectrode or microelectrode array bend when in contact with the sample surface. With a tilted end of the electrode, the electrode or electrode array is "dragged" over the surface and keeps by this a constant distance. Further modes, such as intermittent-contact SECM (IC-SECM) (McKelvey et al., 2010), also allow a Faraday current independent positioning. However, as this mode is not widely used in the SECM community, it will be not discussed here.

Nanopipette-Based Modes

Within recent years, substantial improvements in nanoscale imaging have been achieved by using modified nanopipettes (Bentley et al., 2019) and by combining SECM with complementary scanning probe techniques such as AFM (Kranz et al., 2001; Macpherson and Unwin, 2000) and SICM (Comstock et al., 2010; Takahashi et al., 2010). Unwin and co-workers introduced the term 'scanning electrochemical probe microscopy (SEPM)', which refers to SPM techniques using a sub-micrometer-to-nanoscale electrode probe to map chemical fluxes and interfacial reactivity. Such a probe may reflect a classical nanoelectrode (sealed in glass), a nanopipette-based electrochemical SPM probe as shown in Figure 22.4c or an electrode implemented in the tip of an

AFM cantilever (Figure 22.4e). The recent progress particular for nanoscale electrochemical imaging may be associated with the relative ease in fabricating nanopipettes using a laser-pipette puller, improved instrumentation and characterization techniques, and the access to commercial AFM–SECM probes.

Based on theta capillaries, bifunctional probes have been fabricated, which use one channel as SICM probe to position the dual-pipette based on the change in ion current of the electrolyte-filled channel, which contains a reference electrode (Takahashi et al., 2011). A second reference electrode in the sample solution and a small potential difference applied between the two reference electrodes inside the nanopipette and in solution measures the ion current in dependence of the pipette position analogous to SICM. The second barrel is filled with carbonaceous material by pyrolytic decomposition (Figure 22.4d), as described in Section 2.2.3. The recorded ion current is governed by the overall resistance dominated by the pipette resistance and the access resistance between the pipette tip and the sample surface, which is significantly changed once the pipette is in close proximity to the sample surface. Several imaging modes have been developed, which have been recently reviewed. (Chen et al., 2012) Alternatively, also ring electrodes around pulled pipettes have been reported for micro-sized probes (Walsh et al., 2005) and for ring electrodes with dimensions around 300 nm, respectively (Comstock et al., 2010; Takahashi et al., 2010; Thakar et al., 2013). Within recent year, such nanopipettes have been employed for biomedically relevant research such as mapping immobilized enzymes with unprecedented resolution, which was obtained in hopping mode SICM and GC and FB mode SECM (Takahashi et al., 2010) but also for (electro)active material-related topics and nanoparticle characterization (O'Connell et al., 2015).

Scanning Electrochemical Cell Microscopy

SECCM is a nanopipette-derived technique, which was introduced for high-resolution electrochemical imaging by Unwin and co-workers (Lai et al., 2011; Snowden et al., 2012) and recently reviewed in depth by the same group (Bentley et al., 2019; Ebejer et al., 2013; Kang et al., 2016). Derived from the scanning droplet cells, which was used for corrosion science in the 1990s, Unwin and co-workers consequently translated the concept to the microscale introducing the 'scanning micropipette contact method (SMCM)' (Williams et al., 2009). While initially, micropipette openings in the range of 300–1,000 nm were used, now single-barrel pipettes and double-barreled (theta) pipette having orifices down to 30–50 nm are reported. The basic principle is shown in Figure 22.4c. The single or double barrels contain electrolyte solution and a quasi-reference electrode or a combined reference-and-CE (i.e., quasi-reference counter electrode; QRCE). If the single or theta nanopipette is moved towards the sample surface, the formed meniscus makes contact with the sample surface due to attractive capillary forces and a miniaturized electrochemical cell is formed, which in addition changes the recorded ion current. Analog the ac feedback mode in SICM (Li et al., 2008), the pipette can be oscillated (i.e., sinusoidal waveform) providing an AC component of the ionic current as the meniscus is periodically altered. This ac component of the ion current strongly depends on the meniscus height and can be used as signal for a feedback loop allowing constant distance operation. The technique offers several advantages. For reactive samples, immersion of the whole surface area in electrolyte solution is prevented, thereby suppressing unwanted reaction at the sample surface. In addition, simultaneously multiple information on topography, electroactivity, and conductivity can be obtained (Lai et al., 2011). By changing the format how the tip is scanned across the sample surface using an Archimedes spiral scanning pattern, high-speed and high-resolution image sequences with thousands of pixels at rates as fast as approximately 4 s per frame have been demonstrated (Momotenko et al., 2015), as recently reviewed by the same group (Kang et al., 2016).

AFM–SECM

AFM–SECM provides the direct correlation of topological information with electro chemical surface activity, both at an excellent lateral resolution and with the possibility obtaining local quantitative physical parameters. Basically, the AFM cantilever—more precisely, the sharp tip of the AFM probe—is modified in a way that it comprises an electroactive area typically with sub-micrometer dimensions. Due to the inherent accurate positioning of the AFM probe using force interactions, which induces the bending of the

cantilever (with force constants ranging from 100 down to 0.01 N/m) that is used as signal in a feedback circuit for regulating the distance between the tip and the sample at nanoscale accuracy, the AFM tip-integrated electrode is simultaneously positioned at constant distance with the same accuracy. Different designs and fabrication schemes have been introduced for batch-fabricated and lab-bench made AFM–SECM probes (Figure 22.4e) (Kranz, 2014). Derived from etched wires, probes with conical or spherical electrodes located at the apex of the tip have been individually fabricated (Abbou et al., 2004; Macpherson and Unwin, 2000). Batch-fabricated AFM probes with conical electrodes also constituting the AFM tip (Dobson et al., 2006; Frederix et al., 2005; Nellist et al., 2017; Wain et al., 2014; Zigah et al., 2012) and with recessed ring-shaped electrodes (Shin et al., 2007) or disc-shaped electrodes (Salomo et al., 2010) have been described in literature. Modification of commercially available silicon-nitride probes with a conductive layer and insulation layer followed by exposure of an electrode at the base of the pyramidal AFM tip maintain a nonconductive AFM tip (Eifert et al., 2012; Kranz et al., 2001; Lee et al., 2013). Carbon nanotube-modified probes (Burt et al., 2005; Patil et al., 2010) have also been used in combined AFM–SECM. The advantage of probes with tip-recessed integration of the electrode in comparison to the electrode at the AFM tip apex is evident, as direct contact between the electrode and sample surface is avoided even in contact-mode AFM imaging. Figure 22.4e gives an overview of the two principle AFM–SECM probe designs. Recently, a combined AFM–SECM probe with a conductive polymer tip for potential electrical stimulation and force spectroscopy under potential control has been introduced (Knittel et al., 2014). The high force sensitivity ranging from 10 to 10^6 pN (Helenius et al., 2008) is the basis for recording force–distance (FD) curves. Information on tip–sample interaction such as adhesion, elasticity, and energy dissipation can be derived from such force-spectroscopic measurements (Helenius et al., 2008). Quantitative nanomechanical mapping marketed by the Bruker Corporation as PeakForce Tapping™ (PFT) mode (Pittenger et al., 2011) in combination with SECM gives access to a broad range of physical and chemical high-resolution parameters. The combination STM-SECM will not be covered here, as only a very limited number of studies has been published on this topic within the last decades.

22.3 Highlighted Applications

SECM has been used to study a wide variety of different samples and scenarios across disciplines. For example, SECM has been extensively employed for biological and biomedical applications including investigations at living cells (e.g., respiratory activity), biofilms, and immobilized biomacromolecules such as enzymes, antibodies, and DNA fragments. The interested reader is directed to recent reviews focusing on biological imaging with SECM

(Beaulieu et al., 2011; Conzuelo et al., 2018; Filice and Ding, 2019; Holzinger et al., 2016), as this topic will not be addressed within the present chapter. Also, SECM is an interesting technique addressing corrosion phenomena, whereby changes in morphology are correlated to microelectrochemical processes (see Payne et al., 2017; Thomas et al., 2015).

The following section will highlight selected applications of SECM and hybrid SECM techniques related to nanoscience and nanomaterials. The correlation of structural features and structural changes during operation with functionality or activity is challenging as high-resolution *in situ* and *operando* methods are required providing information on structural features such as size, facets, and grain boundaries, but at the same time high-resolution information on reactivity ideally correlated in space and in time. For example, nanostructured materials have gained importance in fundamental and applied electrochemistry (e.g., electrocatalysis and sensing), which requires understanding on how the (electro)activity is related to structures at the nanoscale. There is an ever-increasing number of applications of SECM and derived *in situ* electrochemical SPM techniques using nanoscopic electrochemical probes to understand interfacial processes. In recent years, there is a pronounced emphasis on single entity studies in electrochemistry, which is documented, e.g., in recent Faraday Discussions Meetings in 2016 and 2018, respectively.

22.3.1 Nanoparticles

The common aspect of nanomaterials including nanoparticles, nanowires, nanorods, and nanosheets is that the physical and chemical properties are distinctly different from the related bulk materials. The application of nanoparticles in biomedical research, in catalysis and sensing requires high-resolution studies beyond vacuum techniques to correlate activity to physical properties. For example, tailoring the properties, e.g., optimization of performance in electrocatalysis including the HER and ORR, requires SEPM techniques that allow *in situ* studies on single entities to correlate physical properties such as size, shape, orientation at the surface, and crystallographic domains (i.e., facets) with the electrochemical properties. The recent instrumental and probe advancements in SECM allow nowadays the *in situ* investigation of single nanoparticles. Several microelectrochemical approaches have been employed to study single nanoparticles. Exemplarily, nanoparticles have been directly attached at a microelectrode; e.g., Bard and co-workers studied a single Pt nanoparticle attached to a carbon fiber microelectrode (Xiao and Bard, 2007). However, linker molecules are required to attach the NP to the electrode surface. As such, the procedure is labor intensive. Nanoparticle collision experiments using micro- and nanoelectrodes (Gao et al., 2018; McKelvey et al., 2018; Robinson et al., 2017), nanopipettes (Mirkin et al., 2016) and SECCM (Chen et al., 2015a) have been used to study various nanoparticles including metal NPs but also biomedically relevant vesicles

(Li et al., 2015). When a NP attaches stochastically to a small disc electrode, current spikes are observed associated with, e.g., a catalytic reaction at the NP. Alternatively, the microelectrode can be biased at a potential where NP oxidation or reduction can take place. Both approaches provide quantitative information on size, concentration, aggregation, and catalytic reactivity of the investigated NPs. For example, the electrochemical detection of single iridium oxide nanoparticle (IrO_X NP) collisions at a Pt UME have been investigated by SECM (Kwon et al., 2010) and with a carbon-modified quartz nanopipette (Mirkin et al., 2016).

The electroactivity of NPs such as HER and ORR has been intensively studied with SECM. Hydrogen evolution of single Pd nanoparticles on Au <111> electrodes, which exhibited different kinetics from bulk Pd was studied in a SECM-type experiment using an ec-STM tip for localized electrochemical measurements (Meier et al., 2002). The electrocatalytic activity towards the ORR at Au NP spot arrays has been investigated by conventional SECM imaging (Sánchez-Sánchez et al., 2010; Wain, 2013). For example, the influence of shapes of Au NPs (cubic, sphere, and rods) was studied in respect the activity towards oxygen reduction (Sánchez-Sánchez et al., 2010), revealing that cubic Au NPs were more active towards ORR in basic solution compared to spherical Au NPs, and gold nanorods, which is related to a higher ratio of Au <111>.

Individual nanoparticles were also imaged without distance regulation (i.e., constant height mode) using SECM tips with radii as small as several nanometers. Mirkin and co-workers among others have published a series of papers on studying single-metal nanoparticles using constant height mode SECM. Single Pd nanocubes (Blanchard et al., 2016), Pt nanoparticles (Kim et al., 2016b), and gold NPs deposited on polyphenylene-modified highly ordered pyrolytic graphite (HOPG) (i.e., as small as 10–20 nm) (Sun et al., 2018, 2014) were studied. Glass-sealed Pt nanoelectrodes with radii as small as ≥ 3 nm were used to record approach curve at Au NPs obtaining good agreement with theoretical approach curves (Sun et al., 2014). SECM images recorded in feedback mode revealed positive feedback over a single-particle Au NP or Pd cubic NP as shown in Figure 22.5a(iii) using nanoelectrodes with 14 and 10 nm, respectively. The study was intended to investigate the blockage of electron transfer by electrografted thiophenol diazonium films on HOPG (Figure 22.5a(ii)) as schematically shown in Figure 22.5a(i). In addition, anchoring of Pd nanocubes by thiol groups allowed imaging of the nonspherical nanoparticles.

The Au NPs were investigated in respect to the HER in generation/collection experiment (Sun et al., 2014). Although high electrochemical resolution could be achieved, the images did not provide detailed structural information, as obtainable by STM. In a recent paper, the same group studied the transition from Faradaic feedback response to electron tunneling between the nanoscopic SECM tips with radii of 14 and 42 nm, respectively and an Au NP ($r = 50$ nm) immobilized on an insulating surface (Sun

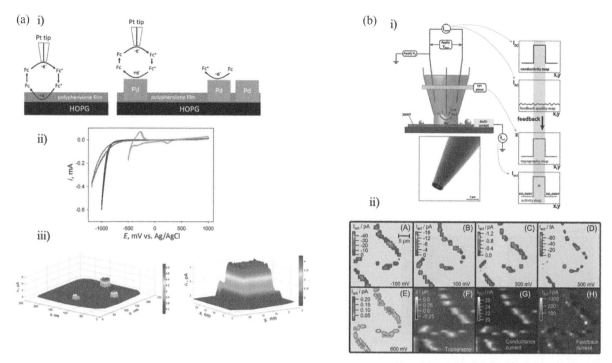

FIGURE 22.5 SECM investigation of nanoparticles. (a) (i) Schematic representation of the feedback mode SECM experiment at HOPG modified with a partially blocking aryldiazonium film modified with Pd nanocubes. (i) Cyclic voltammograms obtained in 1 M $HClO_4$ at bare HOPG (black), HOPG modified with thiophenol diazonium (TPD) (gray), and the same electrode with attached Pd nanocubes (light gray). (iii) SECM images of HOPG with Pd nanocubes attached to the electrografted TPD film: (left) 200×200 nm^2 image (60 lines, 60 pixels per line) obtained with a 13 nm radius Pt tip and (right) 25×25 nm^2 image (25 lines, 50 pixels per line) of a Pd nanocube obtained with a 10 nm radius Pt tip. Images were recorded in 1 mM solution of Fc with $E_S = -0.1$ V, and $E_T = 0.4$ V. (Reprinted with permission from Blanchard et al., 2016, Copyright (2016) American Chemical Society.) (b) (i) Schematic of the SECCM for simultaneously recording topography, conductivity, and activity maps. (ii) Pt NPs deposited at carbon nanotube, (A–E) electrochemical response at different potentials; (F) surface topography; (G) DC conductance current, and (H) AC component of the conductance current. (F–H) were recorded at a sample potential of 300 mV. (Reprinted with permission from Lai et al., 2011, Copyright (2011) American Chemical Society.)

et al., 2018). Electron tunneling was observed at distances smaller than 3 nm. From the recorded current distance curves, the heterogenous rate constant and the tunneling constant were evaluated. Bard and co-workers studied individual Pt NPs, which were directly electrodeposited onto HOPG via nucleation and growth avoiding any capping agents or anchoring molecules. These experiments were also performed in constant height mode. The Pt NPs were investigated in respect to the hydrogen oxidation reaction (HOR) in feedback mode using nanoscopic SECM tips (Kim et al., 2016b). The difference in resolution of the NPs, which could not fully be resolved, was associated with the different heights the images were recorded. Also, quantitative data was obtained with individual Pt NPs showing a lower limit of the heterogeneous rate constant $k°$ of 2 cm/s for the HOR reaction.

High-resolution imaging of individual nanoparticles can also be achieved via SICM—which however goes beyond the scope of this chapter—using a nanopipette with an opening as small as 30 nm. The electrocatalytic oxidation of borohydride at Au NPs deposited on a carbon fiber support was mapped in alkaline media (Kang et al., 2017). Due to the reaction, hydroxide ions are depleted

and water is released during the reaction, which leads to a change in the ionic composition around the NPs dependent on the activity and morphology of the individual particles. Unwin and co-workers used SECCM to record redox activity of individual platinum nanoparticles deposited onto carbon nanotubes along with the topography and conductivity (Lai et al., 2011). These studies revealed that the catalytic properties of the individual Pt NPs towards ORR were dependent on the morphology of the NPs as shown in Figure 22.5b.

Also, hybrid techniques such as SICM-SECM and AFM–SECM have be employed to study single nanoparticles. For example, Demaille and co-workers used AFM–SECM to investigate individual Au NPs (Huang et al., 2013). To achieve high-resolution electrochemical mapping, the redox active molecules were tethered via a flexible polyethylene glycol (PEG) chain to the apex of the conical AFM–SECM probe (electrode at the tip apex), which suppressed the diffusional dispersion of the redox active species. Simultaneously, the topography of the Au NPs and the electrochemical response (i.e., positive feedback) was recorded with a resolution of the electrochemical signal in the order of 20 nm as shown in Figure 22.6a.

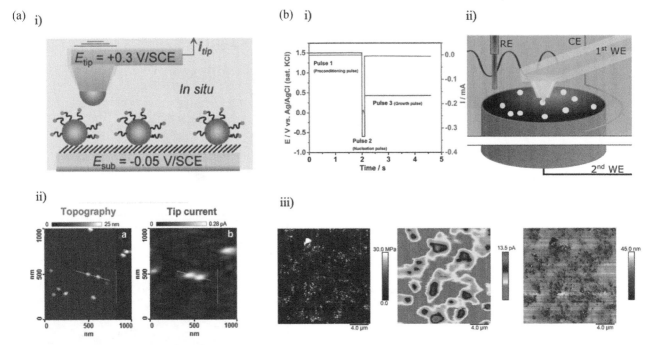

FIGURE 22.6 High-resolution AFM–SECM images of gold nanoparticles. (a) (i) Measurement scheme of molecular-touching (Mt)-AFM–SECM with a cantilever-shaped spherical gold electrode; (ii) topography (tapping mode) and current image of Fc-PEGylated gold nanoparticles. Tip and substrate potential: E_{tip} = +0.30 V/SCE, E_{sub} = −0.05 V/SCE. The probe is oscillated at its fundamental flexural frequency of 2.41 kHz, ∼15% damping, imaging rate 0.4 Hz, in aqueous 0.1 M citrate buffer pH 6. (Reprinted with permission from Huang et al., 2013, Copyright (2013) American Chemical Society.) (b) AFM–SECM imaging of stabilizer-free AuNPs deposited on conductive carbon-doped poly(dimethylsiloxane) (c-PDMS); (i) potential profile (gray curve) and current profile (gray curve) of the potentiostatic triple pulse (PTP) electrosynthesis of bare AuNPs; (ii) measurement scheme; (iii) simultaneous PFT mode AFM–SECM images: (left) map of Young's modulus, (middle) SECM feedback current image and (right) deformation map; images recorded in 5 mM Fc(MeOH)$_2$/(0.1 M KCl; AFM–SECM tip potential of 0.5 V with a peak force of 15 nN, an amplitude of 60 nm and a frequency of 1 kHz. (Reprinted with permission of Guin et al., 2017.)

Although not comparable with the approach by Demaille, in respect to resolution of the electrochemical response, PFT AFM–SECM measurements have the advantage that quantitative physical parameters such as adhesion, elasticity, and energy dispersion can be simultaneously recorded. Combined PFT-SECM measurement have been recorded for Au NPs. Single Au nanostars (Knittel et al., 2016) on poly(dimethylsiloxane) (PDMS)—a quite challenging sample—and nonstabilized spherical Au NPs deposited onto carbon-doped PDMS (c-PDMS) by a nucleation growth process (Guin et al., 2017) have been imaged in feedback mode SECM providing simultaneously physical properties such as the Young's modulus and deformation maps. Using a potentiostatic triple-pulse strategy (Figure 22.6b(i)), spherical, nonstabilized AuNPs with a narrow size distribution of 76 ± 5 nm (diameter) could be deposited within 5 s onto hydrophobic c-PDMS substrate in the absence of any structure directing or stabilizing agent. It could be shown that the Au NPs are preferably deposited at the non-uniformly distributed areas with high carbon content. Increased stiffness due to the particles at or close to the surface was observed, which in combination with the feedback current and deformation profile clearly reveals the presence of AuNPs on C-PDMS (Figure 22.6b(iii)).

22.3.2   Carbon (Nano)Materials

The allotropes of carbon (among those, graphite, HOPG (sp^2), boron-doped diamond (BDD) (sp^3)) and nanostructured carbon such as nanodiamond, graphene, graphene oxide, carbon nanotubes, and carbon nanodots play a significant role in sensing, energy conversion and storage, and applied electrochemical research. The electronic structure, the present defects, the heterogeneity in conductivity, and contaminations of the carbon surface determine the heterogeneous ET kinetics at the different carbon materials, which is still subject of substantial research and debate. Due to a wide variety of surface bonds and groups, graphite and carbon sp^2 nanomaterials are preferred for inner-sphere redox processes, which require the adsorption of the species or the reaction product at the electrode surface. In contrast, for outer-sphere redox processes, the redox species does not require adsorption at the electrode surface and bonds do not have to be broken as described by the Marcus–Hush–Chidsey theory. For example, the electrochemical conversion of catecholamines (i.e., neurotransmitters), electrocatalytic water splitting, and ORR are examples of inner-sphere redox processes, which are strongly dependent on the surface state of the carbonaceous material. Changes in overpotentials over several orders of magnitude in dependence of the electrochemical conditions and pretreatment of the carbon electrodes have been reported (McCreery, 2008). Electrochemical SPM techniques such as SECCM and SECM have been used to study electron transfer kinetics at graphene, CNT, HOPG, and BDD. Single-walled, multi-walled carbon nanotubes and graphene have been extensively studied via SECM using micro-sized and nanosized electrodes and via SECCM using nanopipettes by Amemiya and co-workers, and Unwin and co-workers, as recently reviewed by both groups (Amemiya et al., 2016; Unwin et al., 2016). Amemiya and co-workers used predominately nanogap voltammetry SECM experiments where they observed, e.g., extremely high rate constants for polystyrene (PS) supported chemical vapor deposited (CVD) graphene with $k°$ values ≥ 25 cm/s for ferrocene methanol (Chen et al., 2015b). Interestingly, both groups obtained quite different kinetic values for outer-sphere redox processes, where defects should play a minor role. If experimental parameters such as handling of the samples to avoid organic contaminations, or specific experimental differences, possible adsorption of the redox species on carbon (HOPG surface) and on the insulating glass sheet, attribute to the differing results have not yet been fully resolved. Unwin and co-workers demonstrated that for some inner-sphere processes such as ORR, similar kinetics at the sidewalls of SWNTs comparable to gold electrocatalysts are prevalent with significant enhancements in activity at kink sites (Byers et al., 2014).

Similar to CNTs, also for graphene, the localized electrochemical experiments gave different results. The electrochemistry of graphene—either obtained by chemical vapor deposition, reduction of graphene oxide or exfoliation—has been subject of intensive electrochemical studies due to its unique properties and multitude of applications. Related to the different fabrication routes and associated quality of the graphene sheets, the obtained data are however difficult to compare. The effect of contamination was addressed by taking significant care in sample preparation to avoid contamination of the surface with airborne hydrocarbons of exfoliated graphene (Chen et al., 2015b). Standard rate constants of $k° \geq 12$ cm/s were determined. The obtained $k°$ values are much higher compared to values, obtained for the same redox couple ($FcTMA^{2+/+}$) using SECCM where values of $k° = 0.1 - 1$ cm/s for the $FcTMA^{2+/+}$ couple at the HOPG surface were obtained. Consistency in electrochemical data across length and timescales still remains challenging in nanoscale electrochemistry. Figure 22.7 shows SECCM measurements along with AFM measurements (height information), which clearly reveal differences in the electrochemical electron transfer characteristic at the step edges of graphene flakes, obtained via exfoliation, in dependence of the redox mediator. Two outer-sphere redox couples, $FcTMA^{2+/+}$ and $[Ru(NH_3)_6]^{3+}$ were used to investigate differences in step edges within the same sample. ET kinetics of $FcTMA^{2+/+}$ at the investigated

single layer and multilayer graphene samples were found to be fast and reversible in contrast to $[Ru(NH_3)_6]^{3+}$, which standard potential is close to the intrinsic Fermi level of graphene/graphite (Kneten and McCreery, 1992). A strong dependence of the electron transfer kinetics was observed on the number of graphene layers as shown in Figure 22.7b(iv). Enhanced activity was observed at some step edges but not at all of them (Figure 22.7b(v)).

Boron-Doped Diamond

Doping of diamond with boron concentrations of 10^{20}–10^{21} cm^{-3} leads to semi-metallic characteristics with resistivities <0.001 Ω/cm (Lagrange et al., 1998). The level of boron uptake in the crystal lattice during growth depends on the crystal face orientation and results in areas with a higher boron concentration. The electrochemical behavior of polycrystalline BDD strongly depends on the doping level (Patten et al., 2012a), structural defects (Kraft, 2007), non-diamond carbon impurities (mainly sp^2 hybridized carbon) (Bennett et al., 2004), crystallographic orientation (Pleskov, 2011), surface termination (Vanhove et al., 2007), fraction of grain boundaries (Williams, 2011), and surface modifications (Hoffmann et al., 2011), resulting in a rather heterogeneous electrode material. Nonetheless, BDD electrodes have gained substantial interest in electroanalytical applications due to the robustness and chemical inertness, low background currents, and a broad electrochemical window in aqueous solution related to the overpotential for oxygen and hydrogen evolution (Angus, 2011; Luong et al., 2009).

As-grown diamond is hydrogen-terminated and hydrophobic in nature, whereas oxygen-terminated BDD is hydrophilic. The surface termination has an impact on the kinetics of certain redox mediators (Hutton et al., 2013; Yagi et al., 1999). H-termination is less stable as it is partially changing to O-termination via oxidation (slow) in air or via anodic treatment in aqueous solution (Hoffmann et al., 2010). The local heterogeneity of polycrystalline BDD electrodes has been studied with SPM techniques such as conductive AFM (Colley et al., 2006), SECM (Dincer et al., 2012; Holt et al., 2004; Tomlinson et al., 2016), and SECCM (Liu et al., 2018; Maddar et al., 2016; Patten et al., 2012a, b). A focus of these studies was the heterogeneity in boron doping in dependence on the crystal faces and how this varying boron content and the surface termination influences the potential window and kinetic of the HER. The potential window is influenced by the content of sp^2 carbon content (Bennett et al., 2004), which catalyzes water electrolysis evident in the anodic window. Differences in HET related to the facet-dependent boron content were observed in SECCM experiments (Patten et al., 2012a, b). The <111> facets may show an—up to 10 times higher—boron content compared to <100> facets (Janssen et al., 1992). A recent study investigated the influence of surface termination and doping level on the electrochemical solvent window of O-terminated BDD and H-terminated polycrystalline BDD in aqueous KCl (Liu et al., 2018). In combination with micro-Raman

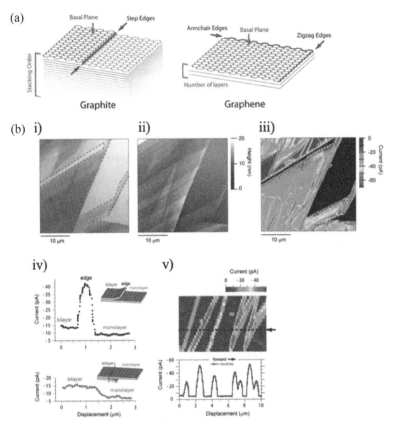

FIGURE 22.7 (a) A schematic of sp^2 carbon materials (graphite, graphene) showing the key intrinsic structural motifs. (b) (i) Optical microscopy image; (ii) AFM image, and (iii) SECCM electroactivity map of the reduction of $Ru(NH_3)_6{}^{3+}$ for the same area (as shown in (ii)) of an exfoliated graphene sample on a silicon/silicon oxide substrate. (iv) SECCM current scan profiles of two characteristics over step edges: electrochemically active (top) and nonactive (bottom); activity is dependent on, if the step edge being exposed or buried. (v) SECCM electroactivity map of step edges of different overall heights. (Reprinted with permission from Unwin, Gell and Zhang, 2016, Copyright (2016) American Chemical Society.)

imaging, grains were visualized, which were correlated to the electrochemical data. It was shown that lower dopant levels resulted in an expanded potential window of 450 mV for O-terminated BDD indicating that the higher doped grains are more (electro)catalytically active. Surface termination has a strong influence on the potential window, with O-terminated BDD exhibiting an expanded potential window in comparison to H-terminated BDD.

22.3.3 Catalysis and Energy-Related Material

Research into renewable and sustainable energy sources has gained tremendous importance, as fossil sources begin to become scarce resources in addition to the global threat of CO_2-induced climate change based on burning fossil energy resources. Fuel cells, photocatalysis, and novel battery concepts require microscopic and high-resolution methods to characterize and understand complex processes at solid/liquid interfaces. Within the last two decades, SECM and SECM-related techniques have been employed to locally study the efficiency of materials in dependence of the nature and structure as electrocatalytic reactions for

inner-sphere redox processes, which strongly depend on the surface structure of the electrode material. SECM studies on electrocatalytic materials have been performed in generation collection mode, RC mode, feedback mode, and surface interrogation mode.

Electrocatalysts have been mainly screened for HOR, ORR, methanol oxidation, oxygen evolution reaction (OER), and hydrogen reduction reaction. Electrocatalysts for methanol oxidation were investigated by a technique termed 'scanning differential electrochemical mass spectrometry' (Jambunathan and Hillier, 2003) via scanning of a capillary, which transfers locally generated products to a mass spectrometer for further analysis. Bard and co-workers published a series of papers employing G/C mode to screen the activity of binary and ternary metallic micro-sized spotted arrays of the catalysts for ORR (Fernández et al., 2005; Jung et al., 2009; Minguzzi et al., 2015). Schuhmann and co-workers investigated binary co-deposits of Pd-Pt and Pd-Au in respect to the electrocatalytic activity towards H_2O_2 in respect to a varying metal ratio (Nagaiah et al., 2013). In RC mode, oxygen was produced at the microelectrode positioned in close proximity to the sample surface by applying an anodic pulse for oxygen evolution from the aqueous solvent. Then, a potential cathodic pulse of -0.55 V vs. Ag/AgCl was applied for oxygen reduction. At this potential, the microelectrode is competing for oxygen, which is available in the gap formed by the microelectrode and substrate. The reduction current measured at the SECM tip is dependent on the activity of the sample for ORR. Most of these experiments were conducted with microelectrodes, which limits the achievable resolution to the sub-micrometer level. SECCM maps and movies were recorded for high-resolution topographical and voltammetric data with a spatial resolution of 50 nm for the HER at molybdenite (MoS_2) (Bentley et al., 2017). MoS_2 is intensively studied as an abundant and low-cost alternative electrocatalyst to Pt for hydrogen evolution. It was shown that the basal plane of bulk MoS_2 (2H crystal phase) possesses uniform activity, but the HER is enhanced at the edge plane (steps, edges, or crevices), which was also reported for MoS_2 investigated via macroscopic electrochemical experiments. The authors report for the basal plane of bulk MoS_2 a Tafel slope and J_o of approx. 120 mV per decade and 2.5×10^{-6} A/cm^2, respectively, while the edge plane has a similar Tafel slope ($\sim$130 mV per decade) but an estimated J_o of approx. 1×10^{-4} A/cm^2. Besides the improved resolution, the SECCM technique avoids issues with aging and degradation, as the sample surface is not immersed in solution and only local electrochemical cells are formed with the meniscus touching the surface. In addition, issues with ohmic drop are also avoided in this configuration. The importance of local heterogeneities could also be shown using SECCM of single-crystal iron nickel sulfides ($Fe_{4.5}Ni_{4.5}S_8$), which are highly efficient HER electrocatalysts (Bentley et al., 2018). SECCM data suggest apparently lower activity from the <111> planes of $Fe_{4.5}Ni_{4.5}S_8$ compared to the macroscopic activity. This observation suggests that local heterogeneities, such as defects (or compositional differences) are largely responsible for the observed macroscopic activity.

Photoelectrochemical Studies

SECM has also been employed to study photoelectrochemical processes during light-driven water splitting generating hydrogen and oxygen. There is a pronounced demand for novel electrocatalysts replacing Pt as the most effective but cost-intensive material. Transition metal phosphides such as cobalt phosphide are attractive materials, as they show only a slight increased overpotentials (80–100 mV) compared to Pt for the HER. The stability of cobalt phosphide at mild experimental condition has been studied via SECM (Ahn and Bard, 2017), indicating that even under mild conditions further modification of the CoP catalyst would be necessary to increase stability as the material degrades fast. Bismuth vanadate—when doped with W and Mo—is an excellent photocatalyst for water splitting (Tokunaga et al., 2001). $BiVO_4$ photoanodes are attractive due to their relative low costs and the fact that they show high solar-to-hydrogen conversion efficiency of 5.2% (highest for metal-oxide photo-electrode) (Han et al., 2014). Initial

SECM studies for characterization of semiconductor materials were performed under irradiation of the entire sample (Zigah et al., 2012) resulting in a macroscopically generated photocurrent, which impedes the investigation of local changes of photoelectrocatalytic activity. Local illumination was achieved with an optical fiber with a diameter $\sim$400 μm modified with a ring electrode, which was scanned across the sample surface, while recording the photocurrent of semiconductor materials (see Figure 22.8a(i)) (Lee et al., 2008; Liu et al., 2010). For example, screening of arrays of catalysts drop-coated onto a $BiVO_4$-modified FTO substrate (Liu et al., 2010; Ye et al., 2010) and analysis of electrocatalyst-modified Ta_3N_5 nanotube arrays were performed with this approach (Cong et al., 2012). The activity of photocatalyst arrays such as bimetallic oxide combinations (Fe–Pd, Fe–Eu and Fe– Ru) and trimetallic Bi–V–Zn oxide compounds on $BiVO_4$-modified FTO have been studied in the presence of a sacrificial Na_2SO_3 electron donor. Small additions of W (5%–10%) to $BiVO_4$ lead to an enhanced photocurrent (Lee et al., 2008).

Figure 22.7a(ii) shows the effect of addition of europium and rubidium to iron oxide, which also produced an enhancement in the observed photocurrent. For example, the Fe 90/Eu 10 spot showed a 10% increase in photocurrent. However, higher concentrations of Eu resulted in an adverse effect as the photocurrent dropped at the Fe 80/Eu 20 spot and spots with higher EU ratio. Rb showed a higher effect compared to Eu. All spots containing 10%–40% Rb resulted in a higher photocurrent in comparison to the pure Fe_2O_3.

An increased lateral resolution should result by directly coupling light into the glass sheath of the microelectrode (see Figure 22.8b(i)), which does no longer limit the size of the electroactive microelectrode by the diameter of the optical fiber (Conzuelo et al., 2017). The improved performance was demonstrated by analysis of thin $BiVO_4$ films deposited on an FTO substrates, which also contained regions of Mo-doping and CoO_x modification. As visible in Figure 22.8b(iii), a clear change in photocurrent is observed when the UME is scanned across the unmodified FTO, where a minimum cathodic current at the tip is recorded (see Figure 22.8b(iii) top), in comparison to the Mo-doped $BiVO_4$ stripe where higher cathodic O_2 collection currents were recorded upon illumination. Increased current was also observed at the edge a Mo-doped $BiVO_4$ layer (Figure 22.8b(ii) (middle)). The sample also contained areas with a CoO_x layer, a water oxidation catalyst, where a maximum of cathodic current was recorded as these areas contained both Mo-doped $BiVO_4$ and CoO_x (Figure 22.8b(iii) (bottom)).

22.3.4 Battery Materials

Various types of fuel cells, such as biofuel cells, polymer electrolyte fuel cells, and proton exchange membrane fuel cells biofuel cells (PEMFC) have been studied, which is summarized in several reviews (Bertoncello, 2010; Ramasamy, 2014;

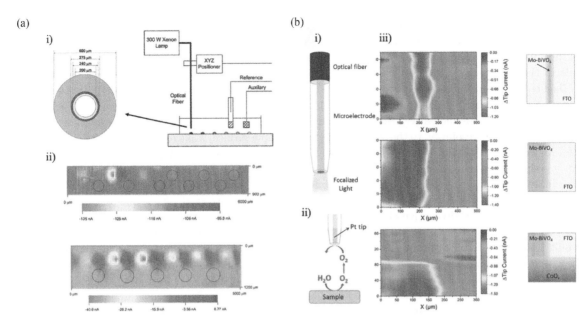

FIGURE 22.8 Photoelectrochemical (PEC) SECM studies with localized illumination. (a) (i) Schematic diagram of the setup for SECM measurements of PEC reactions under illumination with the bottom view of the schematic of the Au ring optical fiber (left); (ii) SECM images of (top) Fe/Eu photocatalyst and (bottom) Fe/Rb photocatalyst at an applied potential of 0.2 V vs. Ag/AgCl in 1 M KOH under UV-Vis light illumination. (Adapted and reprinted with permission from Lee et al., 2008, Copyright (2008) American Chemical Society.) (b) (i) Schematic of a microelectrode for simultaneously local illumination (through the glass sheath) and as sensing probe. (ii) Detection scheme for the water splitting under illumination. The evolved oxygen is collected at a Pt microelectrode polarized at an adequate potential for ORR. (iii) Difference in tip current recorded under light and dark conditions for different semiconductor samples deposited on a FTO substrate. (Top) Mo-doped stripe of about 100 μm width deposited on BiVO$_4$; (middle) Mo-doped BiVO$_4$ deposited on the left region of the substrate; (bottom) Mo-doped BiVO$_4$ deposited on the left region of the substrate; CoO$_x$ catalyst is additionally deposited on top of the bottom half of the sample. The schemes on the right side illustrate the composition of the modified surfaces. Sample polarized at +200 mV vs. Ag/AgCl/3 M KCl. Pt tip (diam. 25 μm) polarized at −600 mV vs. Ag/AgCl/3 M KCl for ORR. Scans were recorded at constant distance of the Pt UME of 20 μm from the surface. Electrolyte: 0.1 M phosphate buffer, pH 8.5. (Adapted and reprinted with permission from Conzuelo et al., 2017, Copyright (2017) American Chemical Society.)

Schuhmann and Bron, 2012). Lately, SECM and hybrid SECM techniques have sparked some attention for studying battery materials and processes, as recently reviewed by several research groups active in battery research (Barton and Rodríguez-López, 2016; Schwager et al., 2016; Ventosa and Schuhmann, 2015). The ability to perform *operando* studies on heterogeneities, local changes of the interface, surface modifications due to degeneration and the influence on the electrochemical performance of electrode materials and fluxes towards and from electrodes with high resolution may contribute to address some of the major scientific challenges in Li-ion batteries (LIBs), and in particular, in post-Li battery research. This includes phenomena such as low ionic mobility, degradation of active electrode materials, and electrolytes and insufficient reversibility in the charge and discharge processes. SECM studies relating to battery research are to date predominantly focused on LIBs. For LIBs, the solid-electrolyte interphase (SEI), which is formed during the first cycle through reductive decomposition of the electrolyte has been a topic of intensive studies and its structure is still not fully clarified (Verma et al., 2010). In fact, the SEI formation plays the essential role in LIBs as it allows Li$^+$ transport but blocks electrons and thus

prevents further electrolyte decomposition. The structure, compactness, morphology and composition of the SEI layer strongly influences the performance of the LIBs (Aurbach, 2000; Verma et al., 2010). Instrumental modifications are required for SECM studies of batteries to tightly exclude oxygen and humidity due to the highly reactive materials. SECM instrumentation in a glovebox (Ventosa and Schuhmann, 2015), or a specially designed cover chamber for the scan head and electrochemical cell (Bülter et al., 2015) has been used to study battery materials. Although, generation-collection mode SECM for studying the SEI layer has been reported (Xu et al., 2012), feedback mode SECM experiments are preferred to investigate the electrochemical reactivity of the electrode surface, which is directly related to the electric properties of the SEI. Different redox mediators have been employed for such studies. Schuhmann and co-workers performed feedback mode experiments using ferrocene as redox species, whereby the reduction of Fc$^+$ is inhibited on various electrode materials such as TiO$_2$ paste (Zampardi et al., 2013, 2015c), glassy carbon (Zampardi et al., 2015a), and graphite composite electrodes (Zampardi et al., 2015b) by the SEI formation. Wittstock and co-workers preferred 2,5-Di-tert-butyl-1,4-dimethoxy benzene

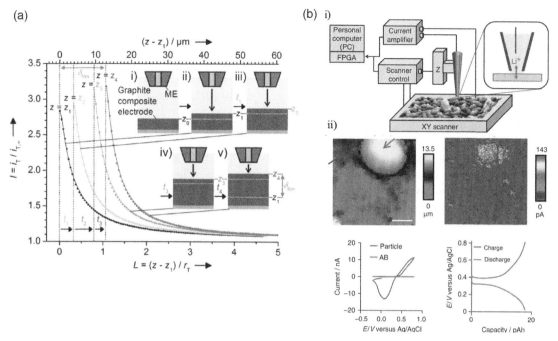

FIGURE 22.9 Approaches to LIBs via SECM. (a) Concept for the in situ characterization of the swelling behavior of a graphite composite electrode during cycling via fitted approach curves. (Reprinted with permission from Schwager et al., 2016.) (b) Topography and current activity of a LiFePO$_4$ electrode. (i) (top) Simultaneous SECCM topography (left) and current (right) images. Scan ranges are 20 × 20 μm. The substrate potential was +0.65 V vs. Ag/AgCl QRCE (Li$^+$ deintercalation; scale bar, 5 μm). (ii) (bottom, left) Cyclic voltammograms recorded at different points on a LiFePO$_4$ electrode surface (location corresponds to the arrows of the topography image). Scan rate is 0.1 V/s. (bottom, right) Local charge (deintercalation) and discharge (intercalation) characteristics applying current magnitudes of 200 pA in each case via SECCM. (Reprinted with permission from Takahashi et al., 2014.)

(DBDMB) as redox mediator (Huang et al., 2015) due to its stable diffusion-controlled steady-state current at the Pt SECM tip at potentials of 4.1 V vs. Li/Li$^+$, which allowed long-term studies (Schwager et al., 2016). This redox species was used, e.g., to study the spatio-temporal changes of SEI on graphite composite electrodes (Bülter et al., 2014). The authors associated the observed spontaneous spatiotemporal changes of SEI to volume changes during charging/discharging, dissolution of SEI components, or gas formation and mechanical disruption of the SEI inside the porous composite material. This was attributed to small movements of particles as a result of volume expansion during lithiation, swelling of polymeric binder, and/or relaxation of mechanical stress associated to the preparation of the material. Volume changes could be studied via approach curves recorded during cycling as shown in Figure 22.9a.

Rodríguez-López and co-workers used RC mode experiments with a Hg-capped Pt SECM tip that had a linear response to the Li$^+$ from 20 μM to around 5 mM as determined by anodic stripping voltammetry (Barton and Rodríguez-López, 2014). Model samples such as a macroscopic and partially PFTE-covered Au electrode and a microsized gold electrode (120 μm) competed locally with the Hg microelectrode for the uptake of Li-ions. An increase in the Li$^+$ consumption at the Au–Li-alloy sample was determined by the decrease in current at the Hg microelectrode.

The demonstrated limited lateral resolution and the challenge mapping rough battery electrodes with a growing cap electrode may limit the practicability of the approach. Also, first imaging attempts with SECCM using a single-barrel nanopipette with an orifice as small as 50 nm have been presented to study battery-related problems with unprecedented resolution down to 100 nm (see Figure 22.9b).

The spatial heterogeneity of a LiFePO$_4$ electrode was mapped in LiCl solution and could be correlated to topographical features (Takahashi et al., 2014). In order to improve the cycle durability of LIBs, cathodes can be coated with metal-oxide thin layers, which is a focus in current lithium ion battery research. Recently, ZrO$_2$-coated LiCoO$_2$ thin-film electrodes were investigated using SECCM performing local cyclic voltammetry with nanopipettes to study the influence of inhomogeneous metal-oxide coating (Inomata et al., 2019). So far, all SECCM studies related to battery research have been made in aqueous solutions. However, in order to address real-world scenarios, such studies need to be performed in organic electrolytes. Also, the roughness of the electrodes may be a challenge for this high-resolution electrochemical SPM techniques. Nonetheless, overcoming such experimental challenges would establish a family of electrochemical scanning probe methods, highly suitable for the correlation of topographical changes with reactivity, required for *operando* studies in battery research.

22.4　Conclusions and Outlook

Within the last two decades, substantial instrumental improvements and the development of innovative imaging modalities have been reported rendering SECM and derivative techniques an indispensable tool in nanoscience. With novel techniques such as nanopipette-based approaches or combined SPM techniques such as AFM–SECM and SICM-SECM, topography may be simultaneously recorded with the electrochemical information at high spatial and temporal resolution. Hence, the correlation of structural features to (electrochemical) reactivity is enabled, which is a key requirement for the characterization of nanomaterials. As each individual technique has its strengths and weaknesses, hybrid approaches overcome inherent limitations enabling to study complex real-world problems. The opportunity to perform the investigations in solution finally opens the route to *operando* studies and in combination with other spectroscopic or microscopic techniques will build the future basis to understand complex processes at the solid/liquid interface. This is of particular relevance for energy-related studies but also in biomedical and corrosion research, which is not discussed within this chapter. More than 25 years have passed since the introduction of SECM. Among the scanning probe techniques, the wide range of developments in SECM has demonstrated the need and the challenges when characterizing interfacial processes, in particular at heterogeneous materials. Importantly, the development of nanoscopic SECM probes and nanopipettes has driven the developments described within this chapter and enabled the investigation even for individual nanoparticles. A remaining challenge is the fact that many of the interfacial processes, e.g., in corrosion, biology, and energy-related aspects also require high temporal resolution of the recorded signals, which is currently a limitation in SECM-related techniques. This issue was most recently addressed by Unwin and coworkers, who introduced spiral scanning for imaging up to 1,000 times faster than typical speeds in SECM (Momotenko et al., 2015). Another issue that needs to be addressed in future is related to the quality control procedures, proficiency tests, and standardization of SECM imaging experiments. Validation includes the entire procedure from sampling or sample preparation to publication of the results, as well as reporting the associated uncertainties; the latter is not yet fully addressed in SECM and frequently makes the comparison of data in nanoscience applications difficult or impossible. Hence, future efforts in SECM may focus on improving the temporal resolution, as well as the reliability and comparability of data by the introduction of standardized reference samples and methods.

References

Abbou, J., Anne, A., Demaille, C., 2004. Probing the structure and dynamics of end-grafted flexible polymer chain layers by combined atomic force-electrochemical microscopy. Cyclic voltammetry within nanometer-thick macromolecular poly(ethylene glycol) layers. *J. Am. Chem. Soc.* 126, 10095–10108. doi:10.1021/ja0493502

Ahn, H.S., Bard, A.J., 2017. Assessment of the stability and operability of cobalt phosphide electrocatalyst for hydrogen evolution. *Anal. Chem.* 89, 8574–8579. doi:10.1021/acs.analchem.7b02799

Amemiya, S., Bard, A.J., Fan, F.-R.F., Mirkin, M.V, Unwin, P.R., 2008. Scanning electrochemical microscopy. *Annu. Rev. Anal. Chem.* 1, 95–131. doi:10.1146/annurev.anchem.1.031207.112938

Amemiya, S., Chen, R., Nioradze, N., Kim, J., 2016. Scanning electrochemical microscopy of carbon nanomaterials and graphite. *Acc. Chem. Res.* 49, 2007–2014. doi:10.1021/acs.accounts.6b00323

Amphlett, J.L., Denuault, G., 1998. Scanning Electrochemical Microscopy (SECM): An investigation of the effects of tip geometry on amperometric tip response. *J. Phys. Chem. B* 102, 9946–9951. doi:10.1021/jp982829u

Angus, J.C., 2011. *Synthetic Diamond Films.* John Wiley & Sons, Inc., Hoboken, NJ. doi:10.1002/9781118062364

Aurbach, D., 2000. Review of selected electrode–solution interactions which determine the performance of Li and Li ion batteries. *J. Power Sources* 89, 206–218. doi:10.1016/S0378-7753(00)00431-6

Ballesteros, B., Schulte, A., Calvo, E.J., Koudelka-Hep, M., Schuhmann, W., 2002. Localised electrochemical impedance spectroscopy with high lateral resolution by means of alternating current scanning electrochemical microscopy. *Electrochem. Commun.* 4, 134–138. doi:10.1016/S1388-2481(01)00294-6

Ballesteros Katemann, B., Schulte, A., Schuhmann, W., 2003. Constant-distance mode scanning electrochemical microscopy (SECM)—Part I: Adaptation of a non-optical shear-force-based positioning mode for SECM tips. *Chem. Eur. J.* 9, 2025–2033. doi:10.1002/chem.200204267

Bandarenka, A.S., Maljusch, A., Kuznetsov, V., Eckhard, K., Schuhmann, W., 2014. Localized impedance measurements for electrochemical surface science. *J. Phys. Chem. C* 118, 8952–8959. doi:10.1021/jp412505p

Bard, A.J., Fan, F.R., Pierce, D.T., Unwin, P.R., Wipf, D.O., Zhou, F., 1991. Chemical imaging of surfaces with the scanning electrochemical microscope. *Science* 254, 68–74. doi:10.1126/science.254.5028.68

Bard, A.J., Fu-Ren, F.F., Kwak, J., Lev, O., 1989. Scanning electrochemical microscopy. Introduction and principles. *Anal. Chem.* 61, 132–138. doi:10.1021/ac00177a011

Barforoush, J.M., Seuferling, T.E., Jantz, D.T., Song, K.R., Leonard, K.C., 2018. Insights into the active electrocatalytic areas of layered double hydroxide and amorphous nickel–iron oxide oxygen evolution electrocatalysts. *ACS Appl. Energy Mater.* 1, 1415–1423. doi:10.1021/acsaem.8b00190

Barton, Z.J., Rodríguez-López, J., 2014a. Lithium ion quantification using mercury amalgams as in situ electrochemical probes in nonaqueous media. *Anal. Chem.* 86, 10660–10667. doi:10.1021/ac502517b

Barton, Z.J., Rodríguez-López, J., 2016. Emerging scanning probe approaches to the measurement of ionic reactivity at energy storage materials. *Anal. Bioanal. Chem.* 408, 2707–2715. doi:10.1007/s00216-016-9373-7

Barton, Z.J., Rodríguez-López, J., 2017. Fabrication and demonstration of mercury disc-well probes for stripping-based cyclic voltammetry scanning electrochemical microscopy. *Anal. Chem.* 89, 2716–2723. doi:10.1021/acs.analchem.6b04022

Beaulieu, I., Kuss, S., Mauzeroll, J., Geissler, M., 2011. Biological scanning electrochemical microscopy and its application to live cell studies. *Anal. Chem.* 83, 1485–1492. doi:10.1021/ac101906a

Bennett, J.A., Wang, J., Show, Y., Swain, G.M., 2004. Effect of sp[sup 2]-bonded nondiamond carbon impurity on the response of boron-doped polycrystalline diamond thin-film electrodes. *J. Electrochem. Soc.* 151, E306. doi:10.1149/1.1780111

Bentley, C.L., Andronescu, C., Smialkowski, M., Kang, M., Tarnev, T., Marler, B., Unwin, P.R., Apfel, U.-P., Schuhmann, W., 2018. Local surface structure and composition control the hydrogen evolution reaction on iron nickel sulfides. *Angew. Chemie Int. Ed.* 57, 4093–4097. doi:10.1002/anie.201712679

Bentley, C.L., Edmondson, J., Meloni, G.N., Perry, D., Shkirskiy, V., Unwin, P.R., 2019. Nanoscale electrochemical mapping. *Anal. Chem.* 91, 84–108. doi:10.1021/acs.analchem.8b05235

Bentley, C.L., Kang, M., Maddar, F.M., Li, F., Walker, M., Zhang, J., Unwin, P.R., 2017. Electrochemical maps and movies of the hydrogen evolution reaction on natural crystals of molybdenite (MoS$_2$): Basal vs. edge plane activity. *Chem. Sci.* 8, 6583–6593. doi:10.1039/C7SC02545A

Bertoncello, P., 2010. Advances on scanning electrochemical microscopy (SECM) for energy. *Energy Environ. Sci.* 3, 1620. doi:10.1039/c0ee00046a

Binnig, G., Quate, C.F., Gerber, C., 1986. Atomic force microscope. *Phys. Rev. Lett.* 56, 930–933. doi:10.1103/PhysRevLett.56.930

Binnig, G., Rohrer, H., 1982. Scanning tunneling microscopy. *Helv. Phys. Acta* 55, 726–735. doi:10.5169/seals-115309

Binnig, G., Rohrer, H., Gerber, C., Weibel, E., 1982. Tunneling through a controllable vacuum gap. *Appl. Phys. Lett.* 40, 178–179. doi:10.1063/1.92999

Blanchard, P.-Y., Sun, T., Yu, Y., Wei, Z., Matsui, H., Mirkin, M. V., 2016. Scanning electrochemical microscopy study of permeability of a thiolated aryl multilayer and imaging of single nanocubes anchored to it. *Langmuir* 32, 2500–2508. doi:10.1021/acs.langmuir.5b03858

Brunner, R., Bietsch, A., Hollricher, O., Marti, O., 1997. Distance control in near-field optical microscopy with piezoelectrical shear-force detection suitable for imaging in liquids. *Rev. Sci. Instrum.* 68, 1769–1772. doi:10.1063/1.1147990

Bülter, H., Peters, F., Schwenzel, J., Wittstock, G., 2014. Spatiotemporal changes of the solid electrolyte interphase in lithium-ion batteries detected by scanning electrochemical microscopy. *Angew. Chemie Int. Ed.* 53, 10531–10535. doi:10.1002/anie.201403935

Bülter, H., Peters, F., Schwenzel, J., Wittstock, G., 2015. Comparison of electron transfer properties of the SEI on graphite composite and metallic lithium electrodes by SECM at OCP. *J. Electrochem. Soc.* 162, A7024–A7036. doi:10.1149/2.0031513jes

Burt, D.P., Wilson, N.R., Weaver, J.M.R., Dobson, P.S., Macpherson, J. V., 2005. Nanowire probes for high resolution combined scanning electrochemical microscopy—Atomic force microscopy. *Nano Lett.* 5, 639–643. doi:10.1021/nl050018d

Byers, J.C., Güell, A.G., Unwin, P.R., 2014. Nanoscale electrocatalysis: Visualizing oxygen reduction at pristine, kinked, and oxidized sites on individual carbon nanotubes. *J. Am. Chem. Soc.* 136, 11252–11255. doi:10.1021/ja505708y

Chen, C.-C., Zhou, Y., Baker, L.A., 2012. Scanning ion conductance microscopy. *Annu. Rev. Anal. Chem.* 5, 207–228. doi:10.1146/annurev-anchem-062011-143203

Chen, C.-H., Ravenhill, E.R., Momotenko, D., Kim, Y.-R., Lai, S.C.S., Unwin, P.R., 2015a. Impact of surface chemistry on nanoparticle–electrode interactions in the electrochemical detection of nanoparticle collisions. *Langmuir* 31, 11932–11942. doi:10.1021/acs.langmuir.5b03033

Chen, R., Nioradze, N., Santhosh, P., Li, Z., Surwade, S.P., Shenoy, G.J., Parobek, D.G., Kim, M.A., Liu, H., Amemiya, S., 2015b. Ultrafast electron transfer kinetics of graphene grown by chemical vapor deposition. *Angew. Chemie Int. Ed.* 54, 15134–15137. doi:10.1002/anie.201507005

Clausmeyer, J., Schuhmann, W., 2016. Nanoelectrodes: Applications in electrocatalysis, single-cell analysis and high-resolution electrochemical imaging. *TrAC—Trends Anal. Chem.* 79, 46–59. doi:10.1016/j.trac.2016.01.018

Colley, A.L., Williams, C.G., Johansson, U.D.H., Newton, M.E., Unwin, P.R., Wilson, N.R., Macpherson, J.V., 2006. Examination of the spatially heterogeneous electroactivity of boron-doped diamond microarray electrodes. *Anal. Chem.* 78, 2539–2548. doi:10.1021/ac0520994

Comstock, D.J., Elam, J.W., Pellin, M.J., Hersam, M.C., 2010. Integrated ultramicroelectrode—Nanopipet probe for concurrent scanning electrochemical microscopy and scanning ion conductance microscopy. *Anal. Chem.* 82, 1270–1276. doi:10.1021/ac902224q

Cong, Y., Park, H.S., Wang, S., Dang, H.X., Fan, F.-R.F., Mullins, C.B., Bard, A.J., 2012. Synthesis of Ta 3 N 5 nanotube arrays modified with electrocatalysts for photoelectrochemical water oxidation. *J. Phys. Chem. C* 116, 14541–14550. doi:10.1021/jp304340a

Conzuelo, F., Schulte, A., Schuhmann, W., 2018. Biological imaging with scanning electrochemical microscopy.

Proc. R. Soc. A Math. Phys. Eng. Sci. 474. doi:10.1098/rspa.2018.0409

Conzuelo, F., Sliozberg, K., Gutkowski, R., Grützke, S., Nebel, M., Schuhmann, W., 2017. High-resolution analysis of photoanodes for water splitting by means of scanning photoelectrochemical microscopy. *Anal. Chem.* 89, 1222–1228. doi:10.1021/acs.analchem.6b03706

Cornut, R., Lefrou, C., 2008. New analytical approximation of feedback approach curves with a microdisk SECM tip and irreversible kinetic reaction at the substrate. *J. Electroanal. Chem.* 621, 178–184. doi:10.1016/j.jelechem.2007.09.021

Cortes-Salazar, F., Trauble, M., Li, F., Busnel, J.-M., Gassner, A.-L., Hojeij, M., Wittstock, G., Girault, H.H., 2009. Soft stylus probes for scanning electrochemical microscopy. *Anal. Chem.* 81, 6889–6896. doi:10.1021/ac900887u

Cox, J.T., Zhang, B., 2012. Nanoelectrodes: Recent advances and new directions. *Annu. Rev. Anal. Chem.* 5, 253–272. doi:10.1146/annurev-anchem-062011-143124

Danieli, T., Mandler, D., 2013. Local surface patterning by chitosan-stabilized gold nanoparticles using the direct mode of scanning electrochemical microscopy (SECM). *J. Solid State Electrochem.* 17, 2989–2997. doi:10.1007/s10008-013-2194-0

Dincer, C., Laubender, E., Hees, J., Nebel, C.E., Urban, G., Heinze, J., 2012. SECM detection of single boron doped diamond nanodes and nanoelectrode arrays using phase-operated shear force technique. *Electrochem. Commun.* 24, 123–127. doi:10.1016/j.elecom.2012.08.005

Dobson, P.S., Weaver, J.M.R., Burt, D.P., Holder, M.N., Wilson, N.R., Unwin, P.R., Macpherson, J. V, 2006. Electron beam lithographically-defined scanning electrochemical-atomic force microscopy probes: Fabrication method and application to high resolution imaging on heterogeneously active surfaces. *Phys. Chem. Chem. Phys.* 8, 3909. doi:10.1039/b605828k

Ebejer, N., Güell, A.G., Lai, S.C.S., McKelvey, K., Snowden, M.E., Unwin, P.R., 2013. Scanning electrochemical cell microscopy: A versatile technique for nanoscale electrochemistry and functional imaging. *Annu. Rev. Anal. Chem.* 6, 329–351. doi:10.1146/annurev-anchem-062012-092650

Eckhard, K., Chen, X., Turcu, F., Schuhmann, W., 2006. Redox competition mode of scanning electrochemical microscopy (RC-SECM) for visualisation of local catalytic activity. *Phys. Chem. Chem. Phys.* 8, 5359–5365. doi:10.1039/b609511a

Eckhard, K., Schuhmann, W., 2008. Alternating current techniques in scanning electrochemical microscopy (AC-SECM). *Analyst* 133, 1486–1497.

Eifert, A., Smirnov, W., Frittmann, S., Nebel, C., Mizaikoff, B., Kranz, C., 2012. Atomic force microscopy probes with integrated boron doped diamond electrodes: Fabrication and application. *Electrochem. Commun.* 25, 30–34. doi:10.1016/j.elecom.2012.09.011

Engstrom, R.C., Meaney, T., Tople, R., Wightman, R.M., 1987. Spatiotemporal description of the diffusion layer with a microelectrode probe. *Anal. Chem.* 59, 2005–2010. doi:10.1021/ac00142a024

Engstrom, R.C., Weber, M., Wunder, D.J., Burgess, R., Winquist, S., 1986. Measurements within the diffusion layer using a microelectrode probe. *Anal. Chem.* 58, 844–848. doi:10.1021/ac00295a044

Etienne, M., Layoussifi, B., Giornelli, T., Jacquet, D., 2012. SECM-based automate equipped with a shear-force detection for the characterization of large and complex samples. *Electrochem. Commun.* 15, 70–73. doi:10.1016/j.elecom.2011.11.028

Etienne, M., Schulte, A., Schuhmann, W., 2004. High resolution constant-distance mode alternating current scanning electrochemical microscopy (AC-SECM). *Electrochem. Commun.* 6, 288–293. doi:10.1016/j.elecom.2004.01.006

Fedorov, R.G., Mandler, D., 2013. Local deposition of anisotropic nanoparticles using scanning electrochemical microscopy (SECM). *Phys. Chem. Chem. Phys.* 15, 2725. doi:10.1039/c2cp42823g

Fernández, J.L., Walsh, D.A., Bard, A.J., 2005. Thermodynamic guidelines for the design of bimetallic catalysts for oxygen electroreduction and rapid screening by scanning electrochemical microscopy. M−Co (M: Pd, Ag, Au). *J. Am. Chem. Soc.* 127, 357–365. doi:10.1021/ja0449729

Filice, F.P., Ding, Z., 2019. Analysing single live cells by scanning electrochemical microscopy. *Analyst.* doi:10.1039/C8AN01490F

Filotás, D., Fernández-Pérez, B.M., Izquierdo, J., Kiss, A., Nagy, L., Nagy, G., Souto, R.M., 2017. Improved potentiometric SECM imaging of galvanic corrosion reactions. *Corros. Sci.* doi:10.1016/j.corsci.2017.10.006

Frederix, P.L.T.M., Gullo, M.R., Akiyama, T., Tonin, A., Rooij, N.F. De, Staufer, U., Engel, A., 2005. Assessment of insulated conductive cantilevers for biology and electrochemistry. *Nanotechnology* 16, 997–1005. doi:10.1088/0957-4484/16/8/001

Gao, R., Ying, Y.-L., Li, Y.-J., Hu, Y.-X., Yu, R.-J., Lin, Y., Long, Y.-T., 2018. A 30 nm nanopore electrode: Facile fabrication and direct insights into the intrinsic feature of single nanoparticle collisions. *Angew. Chemie Int. Ed.* 57, 1011–1015. doi:10.1002/anie.201710201

González-García, Y., García, S.J., Hughes, A.E., Mol, J.M.C., 2011. A combined redox-competition and negative-feedback SECM study of self-healing anticorrosive coatings. *Electrochem. Commun.* 13, 1094–1097. doi:10.1016/j.elecom.2011.07.009

Guin, S.K., Knittel, P., Daboss, S., Breusow, A., Kranz, C., 2017. Template- and additive-free electrosynthesis and characterization of spherical gold nanoparticles on hydrophobic conducting polydimethylsiloxane. *Chem. Asian J.* 12, 1615–1624. doi:10.1002/asia.201700444

Han, L., Abdi, F.F., van de Krol, R., Liu, R., Huang, Z., Lewerenz, H.-J., Dam, B., Zeman, M., Smets, A.H.M., 2014. Inside cover: Efficient water-splitting device based on a bismuth vanadate photoanode and thin-film silicon solar cells (*ChemSusChem* 10/2014). *ChemSusChem* 7, 2758–2758. doi:10.1002/cssc.201402901

Hansma, P.K., Drake, B., Marti, O., Gould, S.A.C., Prater, C.B., 1989. Thes canning ion-conductance microscope. *Science* 243, 641–643. doi:10.1126/science.2464851

Hecht, E., Liedert, A., Ignatius, A., Mizaikoff, B., Kranz, C., 2013. Local detection of mechanically induced ATP release from bone cells with ATP microbiosensors. *Biosens. Bioelectron.* 44, 27–33. doi:10.1016/j.bios.2013.01.008

Helenius, J., Heisenberg, C.-P., Gaub, H.E., Muller, D.J., 2008. Single-cell force spectroscopy. *J. Cell Sci.* 121, 1785–1791. doi:10.1242/jcs.030999

Hengstenberg, A., Kranz, C., Schuhmann, W., 2000. Facilitated tip-positioning and applications of non-electrode tips in scanning electrochemical microscopy using a shear force based constant-distance mode. *Chemistry* Eur. J. 6, 1547–1554. doi:10.1002/(SICI)1521-3765(20000502)6:9<1547::AID-CHEM1547>3.0.CO;2-C

Hirano, Y., Nishimiya, Y., Kowata, K., Mizutani, F., Tsuda, S., Komatsu, Y., 2008. Construction of time-lapse scanning electrochemical microscopy with temperature control and its application to evaluate the preservation effects of antifreeze proteins on living cells. *Anal. Chem.* 80, 9349–9354. doi:10.1021/ac8018334

Hoffmann, R., Kriele, A., Obloh, H., Hees, J., Wolfer, M., Smirnov, W., Yang, N., Nebel, C.E., 2010. Electrochemical hydrogen termination of boron-doped diamond. *Appl. Phys. Lett.* 97, 052103.

Hoffmann, R., Kriele, A., Obloh, H., Tokuda, N., Smirnov, W., Yang, N., Nebel, C.E., 2011. The creation of a biomimetic interface between boron-doped diamond and immobilized proteins. *Biomaterials* 32, 7325–7332. doi:10.1016/j.biomaterials.2011.06.052

Holt, K.B., Bard, A.J., Show, Y., Swain, G.M., 2004. Scanning electrochemical microscopy and conductive probe atomic force microscopy studies of hydrogen-terminated boron-doped diamond electrodes with different doping levels. *J. Phys. Chem. B* 108, 15117–15127. doi:10.1021/jp048222x

Holzinger, A., Steinbach, C., Kranz, C., 2016. Chapter 4. Scanning Electrochemical Microscopy (SECM): Fundamentals and applications in life sciences, in: Arrigan, D.W.M. (Ed.), *Electrochemical Strategies in Detection Science*. The Royal Society of Chemistry, Cambridge, UK, pp. 125–169. doi:10.1039/9781782622529-00125

Horrocks, B.R., Mirkin, M.V., Pierce, D., Bard, A.J., Nagy, G.,Toth, K., 1993a. Scanning electrochemical microscopy. 19. Ion-selective potentiometric microscopy. *Anal. Chem.* 65, 1213–1224. doi:10.1021/ac00057a019

Horrocks, B.R., Schmidtke, D., Heller, A., Bard, A.J., 1993b. Scanning electrochemical microscopy. 24. Enzyme ultramicroelectrodes for the measurement of hydrogen peroxide at surfaces. *Anal. Chem.* 65, 3605–3614. doi:10.1021/ac00072a013

Hua, H., Liu, Y., Wang, D., Li, Y., 2018. Size-dependent voltammetry at single silver nanoelectrodes. *Anal. Chem.* 90, 9677–9681. doi:10.1021/acs.analchem.8b02644

Huang, J., Shkrob, I.A., Wang, P., Cheng, L., Pan, B., He, M., Liao, C., Zhang, Z., Curtiss, L.A., Zhang, L., 2015. 1,4-Bis(trimethylsilyl)-2,5-dimethoxybenzene: A novel redox shuttle additive for overcharge protection in lithium-ion batteries that doubles as a mechanistic chemical probe. *J. Mater. Chem. A* 3, 7332–7337. doi:10.1039/C5TA00899A

Huang, K., Anne, A., Bahri, M.A., Demaille, C., 2013. Probing individual redox PEGylated gold nanoparticles by electrochemical-atomic force microscopy. *ACS Nano* 7, 4151–4163. doi:10.1021/nn400527u

Hutton, L.A., Iacobini, J.G., Bitziou, E., Channon, R.B., Newton, M.E., Macpherson, J. V., 2013. Examination of the factors affecting the electrochemical performance of oxygen-terminated polycrystalline boron-doped diamond electrodes. *Anal. Chem.* 85, 7230–7240. doi:10.1021/ac401042t

Inomata, H., Takahashi, Y., Takamatsu, D., Kumatani, A., Ida, H., Shiku, H., Matsue, T., 2019. Visualization of inhomogeneous current distribution on ZrO 2 -coated LiCoO 2 thin-film electrodes using scanning electrochemical cell microscopy. *Chem. Commun.* 55, 545–548. doi:10.1039/C8CC08916G

Izquierdo, J., Knittel, P., Kranz, C., 2018. Scanning electrochemical microscopy: An analytical perspective. *Anal. Bioanal. Chem.* 410, 307–324. doi:10.1007/s00216-017-0742-7

Izquierdo, J., Nagy, L., Bitter, I., Souto, R.M., Nagy, G., 2013. Potentiometric scanning electrochemical microscopy for the local characterization of the electrochemical behaviour of magnesium-based materials. *Electrochim. Acta* 87, 283–293. doi:10.1016/j.electacta.2012.09.029

Jambunathan, K., Hillier, A.C., 2003. Measuring electrocatalytic activity on a local scale with scanning differential electrochemical mass spectrometry. *J. Electrochem. Soc.* 150, E312. doi:10.1149/1.1570823

James, P.I., Garfias-Mesias, L.F., Moyer, P.J., Smyrl, W.H., 1998. Scanning electrochemical microscopy with simultaneous independent topography. *J. Electrochem. Soc.* 145, L64–L66. doi:10.1149/1.1838417

Janotta, M., Rudolph, D., Kueng, A., Kranz, C., Voraberger, H.-S., Waldhauser, W., Mizaikoff, B., 2004. Analysis of corrosion processes at the surface of diamond-like carbon protected zinc selenide waveguides. *Langmuir* 20, 8634–8640. doi:10.1021/la049042h

Janssen, G., van Enckevort, W.J.P., Vollenberg, W., Giling, L.J., 1992. Characterization of single-crystal diamond grown by chemical vapour deposition processes. *Diam. Relat. Mater.* 1, 789–800. doi:10.1016/0925-9635(92)90102-T

Jung, C., Sánchez-Sánchez, C.M., Lin, C.-L., Rodríguez-López, J., Bard, A.J., 2009. Electrocatalytic activity of

Pd—Co bimetallic mixtures for formic acid oxidation studied by scanning electrochemical microscopy. *Anal. Chem.* 81, 7003–7008. doi:10.1021/ac901096h

Kai, T., Zoski, C.G., Bard, A.J., 2018. Scanning electrochemical microscopy at the nanometer level. *Chem. Commun.* 54, 1934–1947. doi:10.1039/C7CC09777H

Kang, M., Momotenko, D., Page, A., Perry, D., Unwin, P.R., 2016. Frontiers in nanoscale electrochemical imaging: Faster, multifunctional, and ultrasensitive. *Langmuir* 32, 7993–8008. doi:10.1021/acs.langmuir.6b01932

Kang, M., Perry, D., Bentley, C.L., West, G., Page, A., Unwin, P.R., 2017. Simultaneous topography and reaction flux mapping at and around electrocatalytic nanoparticles. *ACS Nano* 11, 9525–9535. doi:10.1021/acsnano.7b05435

Karrai, K., Grober, R.D., 1995. Piezoelectric tip-sample distance control for near field optical microscopes. *Appl. Phys. Lett.* 66, 1842–1844. doi:10.1063/1.113340

Katemann, B.B., Schuhmann, W., 2002. Fabrication and characterization of needle-type Pt-disk nanoelectrodes. *Electroanalysis* 14, 22–28. doi:10.1002/1521-4109(200201)14:1<22::AID-ELAN22>3.0.CO;2-F

Keddam, M., Portail, N., Trinh, D., Vivier, V., 2009. Progress in scanning electrochemical microscopy by coupling with electrochemical impedance and quartz crystal microbalance. *ChemPhysChem* 10, 3175–3182. doi:10.1002/cphc.200900506

Kim, J., Renault, C., Nioradze, N., Arroyo-Currás, N., Leonard, K.C., Bard, A.J., 2016a. Nanometer scale scanning electrochemical microscopy instrumentation. *Anal. Chem.* 88, 10284–10289. doi:10.1021/acs.analchem.6b03024

Kim, J., Renault, C., Nioradze, N., Arroyo-Currás, N., Leonard, K.C., Bard, A.J., 2016b. Electrocatalytic activity of individual Pt nanoparticles studied by nanoscale scanning electrochemical microscopy. *J. Am. Chem. Soc.* 138, 8560–8568. doi:10.1021/jacs.6b03980

Kim, J., Shen, M., Nioradze, N., Amemiya, S., 2012. Stabilizing nanometer scale tip-to-substrate gaps in scanning electrochemical microscopy using an isothermal chamber for thermal drift suppression. *Anal. Chem.* 84, 3489–3492. doi:10.1021/ac300564g

Kleijn, S.E.F., Lai, S.C.S., Koper, M.T.M., Unwin, P.R., 2014. Electrochemistry of nanoparticles. *Angew. Chemie Int. Ed.* 53, 3558–3586. doi:10.1002/anie.201306828

Kneten, K.R., McCreery, R.L., 1992. Effects of redox system structure on electron-transfer kinetics at ordered graphite and glassy carbon electrodes. *Anal. Chem.* 64, 2518–2524. doi:10.1021/ac00045a011

Knittel, P., Bibikova, O., Kranz, C., 2016. Challenges in nanoelectrochemical and nanomechanical studies of individual anisotropic gold nanoparticles. *Faraday Discuss.* 193, 353–369. doi:10.1039/C6FD00128A

Knittel, P., Higgins, M.J., Kranz, C., 2014. Nanoscopic polypyrrole AFM-SECM probes enabling force measurements under potential control. *Nanoscale* 6, 2255–2260. doi:10.1039/c3nr05086f

Kraft, A., 2007. Doped diamond: A compact review on a new, versatile electrode material. *Int. J. Electrochem. Sci.* 2, 355–385.

Kranz, C., 2014. Recent advancements in nanoelectrodes and nanopipettes used in combined scanning electrochemical microscopy techniques. *Analyst* 139. doi:10.1039/c3an01651j

Kranz, C., Friedbacher, G., Mizaikoff, B., Lugstein, A., Smoliner, J., Bertagnolli, E., 2001. Integrating an ultramicroelectrode in an AFM cantilever: Combined technology for enhanced information. *Anal. Chem.* 73, 2491–2500. doi:10.1021/ac001099v

Kranz, C., Gaub, H.E., Schuhmann, W., 1996. Polypyrrole towers grown with the scanning electrochemical microscope. *Adv. Mater.* 8, 634–637. doi:10.1002/adma.19960080805

Kranz, C., Ludwig, M., Gaub, H.E., Schuhmann, W., 1995. Lateral deposition of polypyrrole lines by means of the scanning electrochemical microscope. *Adv. Mater.* 7, 38–40. doi:10.1002/adma.19950070106

Krumov, M.R., Simpson, B.H., Counihan, M.J., Rodríguez-López, J., 2018. In situ quantification of surface intermediates and correlation to discharge products on hematite photoanodes using a combined scanning electrochemical microscopy approach. *Anal. Chem.* 90, 3050–3057. doi:10.1021/acs.analchem.7b04896

Kueng, A., Kranz, C., Mizaikoff, B., 2005. Imaging of ATP membrane transport with dual micro-disk electrodes and scanning electrochemical microscopy. *Biosens. Bioelectron.* 21, 346–353. doi:10.1016/j.bios.2004.10.020

Kurulugama, R.T., Wipf, D.O., Takacs, S.A., Pongmayteegul, S., Garris, P.A., Baur, J.E., 2005. Scanning electrochemical microscopy of model neurons: Constant distance imaging. *Anal. Chem.* 77, 1111–1117. doi:10.1021/ac048571n

Kwak, J., Bard, A.J., 1989. Scanning electrochemical microscopy. Theory of the feedback mode. *Anal. Chem.* 61, 1221–1227. doi:10.1021/ac00186a009

Kwon, S.J., Fan, F.-R.F., Bard, A.J., 2010. Observing iridium oxide (IrO x) single nanoparticle collisions at ultramicroelectrodes. *J. Am. Chem. Soc.* 132, 13165–13167. doi:10.1021/ja106054c

Lagrange, J.-P., Deneuville, A., Gheeraert, E., 1998. Activation energy in low compensated homoepitaxial boron-doped diamond films. *Diam. Relat. Mater.* 7, 1390–1393. doi:10.1016/S0925-9635(98)00225-8

Lai, S.C.S., Dudin, P. V., Macpherson, J. V., Unwin, P.R., 2011. Visualizing zeptomole (electro)catalysis at single nanoparticles within an ensemble. *J. Am. Chem. Soc.* 133, 10744–10747. doi:10.1021/ja203955b

Lee, C., Kwak, J., Anson, F.C., 1991. Application of scanning electrochemical microscopy to generation/collection experiments with high collection efficiency. *Anal. Chem.* 63(14), 1501–1504. doi:10.1021/ac00014a030

Lee, E., Kim, M., Seong, J., Shin, H., Lim, G., 2013. An L-shaped nanoprobe for scanning electrochemical microscopy-atomic force microscopy. *Phys. Status Solidi-R.* 7, 406–409. doi:10.1002/pssr.201307120

Lee, J., Ye, H., Pan, S., Bard, A.J., 2008. Screening of photocatalysts by scanning electrochemical microscopy. *Anal. Chem.* 80, 7445–7450. doi:10.1021/ac801142g

Lefrou, C., Cornut, R., 2010. Analytical expressions for quantitative scanning electrochemical microscopy (SECM). *ChemPhysChem* 11, 547–556. doi:10.1002/cphc.200900600

Lesch, A., Momotenko, D., Cortes-Salazar, F., Wirth, I., Tefashe, U.M., Meiners, F., Vaske, B., Girault, H.H., Wittstock, G., 2012a. Fabrication of soft gold microelectrode arrays as probes for scanning electrochemical microscopy. *J. Electroanal. Chem.* 666, 52–61. doi:10.1016/j.jelechem.2011.12.005

Lesch, A., Vaske, B., Meiners, F., Momotenko, D., Cortes-Salazar, F., Girault, H.H., Wittstock, G., 2012b. Parallel imaging and template-free patterning of self-assembled monolayers with soft linear microelectrode arrays. *Angew. Chemie Int. Ed.* 51, 10413–10416, S10413/1–S10413/23. doi:10.1002/anie.201205347

Li, C., Johnson, N., Ostanin, V., Shevchuk, A., Ying, L., Korchev, Y., Klenerman, D., 2008. High resolution imaging using scanning ion conductance microscopy with improved distance feedback control. *Prog. Nat. Sci.* 18, 671–677. doi:10.1016/j.pnsc.2008.01.011

Li, X., Majdi, S., Dunevall, J., Fathali, H., Ewing, A.G., 2015. Quantitative measurement of transmitters in individual vesicles in the cytoplasm of single cells with nanotip electrodes. *Angew. Chemie Int. Ed.* 54, 11978–11982. doi:10.1002/anie.201504839

Li, Y., Bergman, D., Zhang, B., 2009. Preparation and electrochemical response of 1−3 nm Pt disk electrodes. *Anal. Chem.* 81, 5496–5502. doi:10.1021/ac900777n

Lima, A.S., Salles, M.O., Ferreira, T.L., Paixo, T.R.L.C., Bertotti, M., 2012. Scanning electrochemical microscopy investigation of nitrate reduction at activated copper cathodes in acidic medium. *Electrochim. Acta* 78, 446–451. doi:10.1016/j.electacta.2012.06.075

Liu, D.-Q., Chen, C.-H., Perry, D., West, G., Cobb, S.J., Macpherson, J. V., Unwin, P.R., 2018. Facet-resolved electrochemistry of polycrystalline boron-doped diamond electrodes: Microscopic factors determining the solvent window in aqueous potassium chloride solutions. *ChemElectroChem* 5, 3028–3035. doi:10.1002/celc.201800770

Liu, G., Liu, C., Bard, A.J., 2010. Rapid synthesis and screening of Zn x Cd 1− x S y Se 1− y photocatalysts by scanning electrochemical microscopy. *J. Phys. Chem. C* 114, 20997–21002. doi:10.1021/jp1058116

Liu, H.-Y., Fan, F.R.F., Lin, C.W., Bard, A.J., 1986. Scanning electrochemical and tunneling ultramicroelectrode microscope for high-resolution examination of electrode surfaces in solution. *J. Am. Chem. Soc.* 108, 3838–3839. doi:10.1021/ja00273a054

Ludwig, M., Kranz, C., Schuhmann, W., Gaub, H.E., 1995. Topography feedback mechanism for the scanning electrochemical microscope based on hydrodynamic forces between tip and sample. *Rev. Sci. Instrum.* 66. doi:10.1063/1.1145568

Luong, J.H.T., Male, K.B., Glennon, J.D., 2009. Boron-doped diamond electrode: Synthesis, characterization, functionalization and analytical applications. *Analyst* 134, 1965–1979. doi:10.1039/b910206j

Macpherson, J.V, Unwin, P.R., 2000. Combined scanning electrochemical—Atomic force microscopy. *Anal. Chem.* 72, 276–285. doi:10.1021/ac990921w

Maddar, F.M., Lazenby, R.A., Patel, A.N., Unwin, P.R., 2016. Electrochemical oxidation of dihydronicotinamide adenine dinucleotide (NADH): Comparison of highly oriented pyrolytic graphite (HOPG) and polycrystalline boron-doped diamond (pBDD) electrodes. *Phys. Chem. Chem. Phys.* 18, 26404–26411. doi:10.1039/C6CP05394G

Mandler, D., 2012. Micro- and nanopatterning using scanning electrochemical microscopy, in: Mirkin, M.V, Bard, A.J. (Eds.), *Scanning Electrochemical Microscopy*. CRC Press, Boca Raton, FL, pp. 490–521.

Mandler, D., Meltzer, S., Shohat, I., 1996. Microelectrochemistry on Surfaces with the Scanning Electrochemical Microscope (SECM). *Isr. J. Chem.* 36, 73–80. doi:10.1002/ijch.199600010

Marti, O., 2001. Measurements of adhesion and pull-off forces with the AFM. *Handb. Mod. Tribol.* 617–640. doi:10.1201/9780849377877.ch17"n10.1201/9780849377877.ch17

McCreery, R.L., 2008. Advanced carbon electrode materials for molecular electrochemistry. *Chem. Rev.* 108, 2646–2687. doi:10.1021/cr068076m

McKelvey, K., Edwards, M.A., Unwin, P.R., 2010. Intermittent Contact-Scanning Electrochemical Microscopy (IC-SECM): A new approach for tip positioning and simultaneous imaging of interfacial topography and activity. *Anal. Chem.* 82, 6334–6337. doi:10.1021/ac101099e

McKelvey, K., Nadappuram, B.P., Actis, P., Takahashi, Y., Korchev, Y.E., Matsue, T., Robinson, C., Unwin, P.R., 2013. Fabrication, characterization, and functionalization of dual carbon electrodes as probes for Scanning Electrochemical Microscopy (SECM). *Anal. Chem.* 85, 7519–7526. doi:10.1021/ac401476z

McKelvey, K., Robinson, D.A., Vitti, N.J., Edwards, M.A., White, H.S., 2018. Single Ag nanoparticle collisions within a dual-electrode micro-gap cell. *Faraday Discuss.* 210, 189–200. doi:10.1039/C8FD00014J

McNally, M., Wong, D.K.Y., 2001. An in vivo probe based on mechanically strong but structurally small carbon electrodes with an appreciable surface area. *Anal. Chem.* 73, 4793–4800. doi:10.1021/ac0104532

Meier, J., Friedrich, K.A., Stimming, U., 2002. Novel method for the investigation of single nanoparticle reactivity. *Faraday Discuss.* 365–372; discussion 441–462. doi:10.1039/b200014h

Minguzzi, A., Battistel, D., Rodríguez-López, J., Vertova, A., Rondinini, S., Bard, A.J., Daniele, S., 2015. Rapid characterization of oxygen-evolving electrocatalyst spot arrays by the substrate generation/tip collection mode of scanning electrochemical microscopy with decreased O2 diffusion layer overlap. *J. Phys. Chem. C* 119, 2941–2947. doi:10.1021/jp510651f

Mirkin, M.V., Bard, A.J. (Eds.), 2012. *Scanning Electrochemical Microscopy*, 2nd Edition. CRC Press, Boca Raton, FL.

Mirkin, V., Fan, F.-R.F., Bard, A.J., 1992. Scanning electrochemical Part 13 . Evaluation microelectrodes microscopy size of the tip shapes of nanometer. *J. Electroanal. Chem.* 328, 47–62. doi:10.1016/0022-0728(92)80169-5

Mirkin, M. V., Sun, T., Yu, Y., Zhou, M., 2016. Electrochemistry at one nanoparticle. *Acc. Chem. Res.* 49, 2328–2335. doi:10.1021/acs.accounts.6b00294

Momotenko, D., Byers, J.C., McKelvey, K., Kang, M., Unwin, P.R., 2015. High-speed electrochemical imaging. *ACS Nano* 9, 8942–8952. doi:10.1021/acsnano.5b02792

Momotenko, D., Qiao, L., Cortes-Salazar, F., Lesch, A., Wittstock, G., Girault, H.H., 2012. Electrochemical push-pull scanner with mass spectrometry detection. *Anal. Chem.* 84, 6630–6637. doi:10.1021/ac300999v

Mourdikoudis, S., Pallares, R.M., Thanh, N.T.K., 2018. Characterization techniques for nanoparticles: Comparison and complementarity upon studying nanoparticle properties. *Nanoscale* 10, 12871–12934. doi:10.1039/c8nr02278j

Nadappuram, B.P., Mckelvey, K., Al Botros, R., Colburn, A.W., Unwin, P.R., 2013. Fabrication and characterization of dual function nanoscale pH-scanning ion conductance microscopy (SICM) probes for high resolution pH mapping. *Anal. Chem.* 85, 8070–8074. doi:10.1021/ac401883n

Nagaiah, T.C., Schäfer, D., Schuhmann, W., Dimcheva, N., 2013. Electrochemically deposited Pd–Pt and Pd–Au codeposits on graphite electrodes for electrocatalytic H$_2$ O$_2$ reduction. *Anal. Chem.* 85, 7897–7903. doi:10.1021/ac401317y

Nebel, M., Eckhard, K., Erichsen, T., Schulte, A., Schuhmann, W., 2010. 4D shearforce-based constant-distance mode scanning electrochemical microscopy. *Anal. Chem.* 82, 7842–7848. doi:10.1021/ac1008805

Nebel, M., Erichsen, T., Schuhmann, W., 2014. Constant-distance mode SECM as a tool to visualize local electrocatalytic activity of oxygen reduction catalysts. *Beilstein J. Nanotechnol.* 5, 141–151. doi:10.3762/bjnano.5.14

Nebel, M., Grützke, S., Diab, N., Schulte, A., Schuhmann, W., 2013. Microelectrochemical visualization of oxygen consumption of single living cells. *Faraday Discuss.* 164, 19. doi:10.1039/c3fd00011g

Nellist, M.R., Chen, Y., Mark, A., Gödrich, S., Stelling, C., Jiang, J., Poddar, R., Li, C., Kumar, R., Papastavrou, G., Retsch, M., Brunschwig, B.S., Huang, Z., Xiang, C., Boettcher, S.W., 2017. Atomic force microscopy with

nanoelectrode tips for high resolution electrochemical, nanoadhesion and nanoelectrical imaging. *Nanotechnology* 28. doi:10.1088/1361-6528/aa5839

Nioradze, N., Chen, R., Kim, J., Shen, M., Santhosh, P., Amemiya, S., 2013. Origins of nanoscale damage to glass-sealed platinum electrodes with submicrometer and nanometer size. *Anal. Chem.* 85, 6198–6202. doi:10.1021/ac401316n

Nogala, W., Szot, K., Burchardt, M., Roelfs, F., Rogalski, J., Opallo, M., Wittstock, G., 2010. Feedback mode SECM study of laccase and bilirubin oxidase immobilised in a sol-gel processed silicate film. *Analyst* 135, 2051–2058. doi:10.1039/c0an00068j

Novak, P., Li, C., Shevchuk, A.I., Stepanyan, R., Caldwell, M., Hughes, S., Smart, T.G., Gorelik, J., Ostanin, V.P., Lab, M.J., Moss, G.W.J., Frolenkov, G.I., Klenerman, D., Korchev, Y.E., 2009. Nanoscale live-cell imaging using hopping probe ion conductance microscopy. *Nat. Meth.* 6, 279–281. doi:10.1038/nmeth.1306

O'Connell, M.A., Lewis, J.R., Wain, A.J., 2015. Electrochemical imaging of hydrogen peroxide generation at individual gold nanoparticles. *Chem. Commun.* 51, 10314–10317. doi:10.1039/c5cc01640a

O'Connell, M.A., Wain, A.J., 2014. Mapping electroactivity at individual catalytic nanostructures using high-resolution scanning electrochemical—Scanning ion conductance micrococopy. *Anal. Chem.* 86, 12100–12107. doi:10.1021/ac502946q

Patil, A.V., Beker, A.F., Wiertz, F.G.M., Heering, H.A., Coslovich, G., Vlijm, R., Oosterkamp, T.H., 2010. Fabrication and characterization of polymer insulated carbon nanotube modified electrochemical nanoprobes. *Nanoscale* 2, 734. doi:10.1039/b9nr00281b

Patten, H.V, Lai, S.C.S., Macpherson, J. V, Unwin, P.R., 2012a. Active sites for outer-sphere, inner-sphere, and complex multistage electrochemical reactions at Polycrystalline Boron-Doped Diamond Electrodes (pBDD) revealed with Scanning Electrochemical Cell Microscopy (SECCM). *Anal. Chem.* 84, 5427–5432. doi:10.1021/ac3010555

Patten, H.V., Meadows, K.E., Hutton, L.A., Iacobini, J.G., Battistel, D., McKelvey, K., Colburn, A.W., Newton, M.E., MacPherson, J. V, Unwin, P.R., 2012b. Electrochemical mapping reveals direct correlation between heterogeneous electron-transfer kinetics and local density of states in diamond electrodes. *Angew. Chemie Int. Ed.* 51, 7002–7006, S7002/1–S7002/15. doi:10.1002/anie.201203057

Payne, N.A., Stephens, L.I., Mauzeroll, J., 2017. The application of scanning electrochemical microscopy to corrosion research. *Corrosion* 73, 759–780. doi:10.5006/2354

Penner, R.M., Heben, M.J., Longin, T.L., Lewis, N.S., 1990. Fabrication and use of nanometer-sized electrodes in electrochemistry. *Science* 250, 1118–1121. doi:10.1126/science.250.4984.1118

Pierce, D.T., Unwin, P.R., Bard, A.J., 1992. Scanning electrochemical microscopy. 17. Studies of enzyme-mediator kinetics for membrane- and surface-immobilized glucose oxidase. *Anal. Chem.* 64, 1795-1804. doi:10.1021/ac00041a011

Pittenger, B., Erina, N., Su, C., 2011. Quantitative Mechanical Property Mapping at the Nanoscale with PeakForce QNM. Bruker Application Note #128.

Pleskov, Y., 2011. Electrochemistry of diamond, in: E. Brillas, C.A.M.-H. (Ed.), *Synthetic Diamond Films.* John Wiley & Sons, Inc., Hoboken, NJ, p. 79.

Polcari, D., Dauphin-Ducharme, P., Mauzeroll, J., 2016. Scanning electrochemical microscopy: A comprehensive review of experimental parameters from 1989 to 2015. *Chem. Rev.* 116, 13234–13278. doi:10.1021/acs.chemrev.6b00067

Ramasamy, R.P., 2014. Scanning electrochemical microscopy for biological fuel cell characterization, in: Luckarift, H.R., Atanassov, P.B., Johnson, G.R. (Eds.) *Enzymatic Fuel Cells.* John Wiley & Sons, Inc., Hoboken, NJ, pp. 273–303. doi:10.1002/9781118869796.ch14

Robinson, D.A., Liu, Y., Edwards, M.A., Vitti, N.J., Oja, S.M., Zhang, B., White, H.S., 2017. Collision dynamics during the electrooxidation of individual silver nanoparticles. *J. Am. Chem. Soc.* 139, 16923–16931. doi:10.1021/jacs.7b09842

Rodríguez-López, J., Alpuche-Avilés, M.A., Bard, A.J., 2008. Interrogation of surfaces for the quantification of adsorbed species on electrodes: Oxygen on gold and platinum in neutral media. *J. Am. Chem. Soc.* 130, 16985–16995. doi:10.1021/ja8050553

Rudolph, D., Bates, D., DiChristina, T.J., Mizaikoff, B., Kranz, C., 2016. Detection of metal-reducing enzyme complexes by scanning electrochemical microscopy. *Electroanalysis* 28, 2459–2465. doi:10.1002/elan.201600333

Salomo, M., Pust, S.E., Wittstock, G., Oesterschulze, E., 2010. Integrated cantilever probes for SECM/AFM characterization of surfaces. *Microelectr. Eng.* 87, 1537–1539. doi:10.1016/j.mee.2009.11.032

Sánchez-Sánchez, C.M., Rodríguez-López, J., Bard, A.J., 2008. Quantitative calibration of the SECM substrate generation/tip collection mode and its use for the study of the oxygen reduction mechanism. *Anal. Chem.* 80, 3254–3260. doi:10.1021/ac702453n

Sánchez-Sánchez, C.M., Solla-Gullón, J., Vidal-Iglesias, F.J., Aldaz, A., Montiel, V., Herrero, E., 2010. Imaging structure sensitive catalysis on different shape-controlled platinum nanoparticles. *J. Am. Chem. Soc.* 132, 5622–5624. doi:10.1021/ja100922h

Santana, J.J., González-Guzmán, J., Fernández-Mérida, L., Gonzlez, S., Souto, R.M., 2010. Visualization of local degradation processes in coated metals by means of scanning electrochemical microscopy in the redox competition mode. *Electrochim. Acta* 55, 4488–4494. doi:10.1016/j.electacta.2010.02.091

Schäfer, D., Puschhof, A., Schuhmann, W., 2013. Scanning electrochemical microscopy at variable temperatures. *Phys. Chem. Chem. Phys.* 15, 5215. doi:10.1039/c3cp43520b

Schuhmann, W., Bron, M., 2012. Scanning electrochemical microscopy (SECM) in proton exchange membrane fuel cell research and development, in: Hartnig, C., Roth, C. (Eds.), *Polymer Electrolyte Membrane and Direct Methanol Fuel Cell Technology.* Woodhead Publishing, Swaston, UK, pp. 399–424.

Schwager, P., Bülter, H., Plettenberg, I., Wittstock, G., 2016. Review of local in situ probing techniques for the interfaces of lithium-ion and lithium–oxygen batteries. *Energy Technol.* 4, 1472–1485. doi:10.1002/ente.201600141

Shao, Y., Mirkin, M.V., Fish, G., Kokotov, S., Palanker, D., Lewis, A., 1997. Nanometer-sized electrochemical sensors. *Anal. Chem.* 69, 1627–1634. doi:10.1021/ac960887a

Shin, H., Hesketh, P.J., Mizaikoff, B., Kranz, C., 2007. Batch fabrication of atomic force microscopy probes with recessed integrated ring microelectrodes at a wafer level. *Anal. Chem.* 79, 4769–4777. doi:10.1021/ac070598u

Slevin, C.J., Gray, N.J., Macpherson, J.V., Webb, M.A., Unwin, P.R., 1999. Fabrication and characterization of nanometer-sized platinum electrodes for voltammetric analysis and imaging. *Electrochem. Commun.* 1, 282–288. doi:10.1016/S1388-2481(99)00059-4

Snowden, M.E., Güell, A.G., Lai, S.C.S., McKelvey, K., Ebejer, N., O'Connell, M.A, Colburn, A.W., Unwin, P.R., 2012. Scanning electrochemical cell microscopy: Theory and experiment for quantitative high resolution spatially-resolved voltammetry and simultaneous ion-conductance measurements. *Anal. Chem.* 84, 2483–2491. doi:10.1021/ac203195h

Sonnenfeld, R., Hansma, P.K., 1986. Atomic-resolution microscopy in water. *Science* 232, 211–213. doi:10.1126/science.232.4747.211

Souto, R.M., González-García, Y., Battistel, D., Daniele, S., 2012. In situ scanning electrochemical microscopy (SECM) detection of metal dissolution during zinc corrosion by means of mercury sphere-cap microelectrode tips. *Chemistry* 18, 230–236. doi:10.1002/chem.201102325

Sun, T., Wang, D., Mirkin, M. V., 2018. Tunneling mode of scanning electrochemical microscopy: Probing electrochemical processes at single nanoparticles. *Angew. Chemie Int. Ed.* 57, 7463–7467. doi:10.1002/anie.201801115

Sun, T., Yu, Y., Zacher, B.J., Mirkin, M. V., 2014. Scanning electrochemical microscopy of individual catalytic nanoparticles. *Angew. Chemie Int. Ed.* 53, 14120–14123. doi:10.1002/anie.201408408

Takahashi, Y., Kumatani, A., Munakata, H., Inomata, H., Ito, K., Ino, K., Shiku, H., Unwin, P.R., Korchev, Y.E., Kanamura, K., Matsue, T., 2014. Nanoscale visualization of redox activity at lithium-ion battery cathodes. *Nat. Commun.* 5, 1–7. doi:10.1038/ncomms6450

Takahashi, Y., Kumatani, A., Shiku, H., Matsue, T., 2017. Scanning probe microscopy for nanoscale electrochemical imaging. *Anal. Chem.* 89, 342–357. doi:10.1021/acs.analchem.6b04355

Takahashi, Y., Shevchuk, A.I., Novak, P., Murakami, Y., Shiku, H., Korchev, Y.E., Matsue, T., 2010. Simultaneous noncontact topography and electrochemical imaging by SECM/SICM featuring ion current feedback regulation. *J. Am. Chem. Soc.* 132, 10118–10126.

Takahashi, Y., Shevchuk, A.I., Novak, P., Zhang, Y., Ebejer, N., MacPherson, J.V, Unwin, P.R., Pollard, A.J., Roy, D., Clifford, C.A., Shiku, H., Matsue, T., Klenerman, D., Korchev, Y.E., 2011. Multifunctional nanoprobes for nanoscale chemical imaging and localized chemical delivery at surfaces and interfaces. *Angew. Chemie Int. Ed.* 50, 9638–9642, S9638/1–S9638/5. doi:10.1002/anie.201102796

Takahashi, Y., Shiku, H., Murata, T., Yasukawa, T., Matsue, T., 2009. Transfected single-cell imaging by scanning electrochemical optical microscopy with shear force feedback regulation. *Anal. Chem.* 81, 9674–9681. doi:10.1021/ac901796r

Thakar, R., Weber, A.E., Morris, C.A., Baker, L.A., 2013. Multifunctional carbon nanoelectrodes fabricated by focused ion beam milling. *Analyst* 138, 5973–5982. doi:10.1039/c3an01216f

Thomas, S., Izquierdo, J., Birbilis, N., Souto, R.M., 2015. Possibilities and limitations of scanning electrochemical microscopy of Mg and Mg alloys. *Corrosion* 71, 171–183. doi:10.5006/1483

Tokunaga, S., Kato, H., Kudo, A., 2001. Selective preparation of monoclinic and tetragonal BiVO 4 with scheelite structure and their photocatalytic properties. *Chem. Mater.* 13, 4624–4628. doi:10.1021/cm0103390

Tomlinson, L.I., Patten, H. V., Green, B.L., Iacobini, J., Meadows, K.E., McKelvey, K., Unwin, P.R., Newton, M.E., Macpherson, J. V., 2016. Intermittent-contact Scanning Electrochemical Microscopy (IC-SECM) as a quantitative probe of defects in single crystal boron doped diamond electrodes. *Electroanalysis* 28, 2297–2302. doi:10.1002/elan.201600291

Ummadi, J.G., Downs, C.J., Joshi, V.S., Ferracane, J.L., Koley, D., 2016. Carbon-based solid-state calcium ion-selective microelectrode and scanning electrochemical microscopy: A quantitative study of pH-dependent release of calcium ions from bioactive glass. *Anal. Chem.* 88, 3218–3226. doi:10.1021/acs.analchem.5b04614

Unwin, P.R., Güell, A.G., Zhang, G., 2016. Nanoscale electrochemistry of sp 2 carbon materials: From graphite and graphene to carbon nanotubes. *Acc. Chem. Res.* 49, 2041–2048. doi:10.1021/acs.accounts.6b00301

Unwin, P.R., 2012. Visualizing and quantifying homogeneous chemical reactions in electrochemical processes, in: Mirkin, M.V, Bard, A.J. (Eds.), *Scanning Electrochemical Microscopy.* CRC Press, Boca Raton, FL, p. 157.

Vanhove, E., de Sanoit, J., Arnault, J.C., Saada, S., Mer, C., Mailley, P., Bergonzo, P., Nesladek, M., 2007. Stability of H-terminated BDD electrodes: An insight into the influence of the surface preparation. *Phys. Status Solidi* 204, 2931–2939. doi:10.1002/pssa.200776340

Ventosa, E., Schuhmann, W., 2015. Scanning electrochemical microscopy of Li-ion batteries. *Phys. Chem. Chem. Phys.* 17, 28441–28450. doi:10.1039/c5cp02268a

Verma, P., Maire, P., Novák, P., 2010. A review of the features and analyses of the solid electrolyte interphase in Li-ion batteries. *Electrochim. Acta* 55, 6332–6341. doi:10.1016/j.electacta.2010.05.072

Voronin, O.G., Hartmann, A., Steinbach, C., Karyakin, A.A., Khokhlov, A.R., Kranz, C., 2012. Prussian Blue-modified ultramicroelectrodes for mapping hydrogen peroxide in scanning electrochemical microscopy (SECM). *Electrochem. Commun.* 23, 102–105. doi:10.1016/j.elecom.2012.07.017

Wain, A.J., 2013. Imaging size effects on the electrocatalytic activity of gold nanoparticles using scanning electrochemical microscopy. *Electrochim. Acta* 92, 383–391. doi:10.1016/j.electacta.2013.01.074

Wain, A.J., Pollard, A.J., Richter, C., 2014. High-resolution electrochemical and topographical imaging using batch-fabricated cantilever probes. *Anal. Chem.* 86, 5143–5149. doi:10.1021/ac500946v

Walsh, D.A., Ferna, L., Mauzeroll, J., Bard, A.J., 2005. Fabrication and characterization of micropipet probes. *Anal. Chem.* 77, 5182–5188. doi:10.1021/ac0505122

Wei, C., Bard, A.J., Nagy, G., Toth, K., 1995. Scanning electrochemical microscopy. 28. Ion-selective neutral carrier-based microelectrode potentiometry. *Anal. Chem.* 67, 1346–1356. doi:10.1021/ac00104a008

Wilde, P., Quast, T., Aiyappa, H.B., Chen, Y.-T., Botz, A., Tarnev, T., Marquitan, M., Feldhege, S., Lindner, A., Andronescu, C., Schuhmann, W., 2018. Towards reproducible fabrication of nanometre-sized carbon electrodes: Optimisation of automated nanoelectrode fabrication by means of transmission electron microscopy. *ChemElectroChem* 5, 3083–3088. doi:10.1002/celc.2018 00600

Williams, C.G., Edwards, M.A., Colley, A.L., Macpherson, J. V., Unwin, P.R., 2009. Scanning micropipet contact method for high-resolution imaging of electrode surface redox activity. *Anal. Chem.* 81, 2486–2495. doi:10.1021/ac802114r

Williams, O.A., 2011. Nanocrystalline diamond. *Diam. Relat. Mater.* 20, 621–640. doi:10.1016/j.diamond.2011. 02.015

Wipf, D.O., Bard, A.J., 1992. Scanning electrochemical microscopy. 15. Improvements in imaging via tip-position modulation and lock-in detection. *Anal. Chem.* 64, 1362–1367. doi:10.1021/ac00037a011

Wipf, D.O., Bard, A.J., Tallman, D.E., 1993. Scanning electrochemical microscopy. 21. Constant-current imaging with an autoswitching controller. *Anal. Chem.* 65, 1373–1377. doi:10.1021/ac00058a013

Wipf, D.O., Ge, F., Spaine, T.W., Baur, J.E., 2000. Microscopic measurement of pH with iridium oxide microelectrodes. *Anal. Chem.* 72, 4921–4927. doi:10.1021/ac000383j

Wittstock, G., Schuhmann, W., 1997. Formation and imaging of microscopic enzymically active spots on an alkanethiolate-covered gold electrode by scanning electrochemical microscopy. *Anal. Chem.* 69, 5059–5066. doi:10.1021/AC970504O

Xiao, X., Bard, A.J., 2007. Observing single nanoparticle collisions at an ultramicroelectrode by electrocatalytic amplification. *J. Am. Chem. Soc.* 129, 9610–9612. doi:10.1021/ja072344w

Xu, F., Beak, B., Jung, C., 2012. In situ electrochemical studies for Li+ ions dissociation from the LiCoO2 electrode by the substrate-generation/tip-collection mode in SECM. *J. Solid State Electrochem.* 16, 305–311. doi:10.1007/s10008-011-1325-8

Yagi, I., Notsu, H., Kondo, T., Tryk, D.A., Fujishima, A., 1999. Electrochemical selectivity for redox systems at oxygen-terminated diamond electrodes. *J. Electroanal. Chem.* 473, 173–178. doi:10.1016/S0022-0728(99)00027-3

Ye, H., Lee, J., Jang, J.S., Bard, A.J., 2010. Rapid screening of BiVO 4 -based photocatalysts by Scanning Electrochemical Microscopy (SECM) and studies of their photoelectrochemical properties. *J. Phys. Chem. C* 114, 13322–13328. doi:10.1021/jp104343b

Ying, Y.L., Ding, Z., Zhan, D., Long, Y.T., 2017. Advanced electroanalytical chemistry at nanoelectrodes. *Chem. Sci.* 8, 3338–3348. doi:10.1039/c7sc00433h

Yu, Y., Noël, J.-M., Mirkin, M.V., Gao, Y., Mashtalir, O., Friedman, G.,Gogotsi, Y., 2014. Carbon Pipette-Based Electrochemical Nanosampler. *Anal. Chem.* 86, 3365-3372. doi:10.1021/ac403547b

Zampardi, G., La Mantia, F., Schuhmann, W., 2015a. Determination of the formation and range of stability of the SEI on glassy carbon by local electrochemistry. *RSC Adv.* 5, 31166–31171. doi:10.1039/C5RA02940F

Zampardi, G., La Mantia, F., Schuhmann, W., 2015b. In-operando evaluation of the effect of vinylene carbonate on the insulating character of the solid electrolyte interphase. *Electrochem. Commun.* 58, 1–5. doi:10.1016/j.elecom.2015.05.013

Zampardi, G., Ventosa, E., La Mantia, F., Schuhmann, W., 2015c. Scanning electrochemical microscopy applied to the investigation of lithium (De-)insertion in TiO2. *Electroanalysis* 27, 1017–1025. doi:10.1002/elan.201400613

Zampardi, G., Ventosa, E., La Mantia, F., Schuhmann, W., 2013. In situ visualization of Li-ion intercalation and formation of the solid electrolyte interphase on TiO2 based paste electrodes using scanning electrochemical microscopy. *Chem. Commun.* 49, 9347. doi:10.1039/c3cc44576c

Zeradjanin, A.R., Schilling, T., Seisel, S., Bron, M., Schuhmann, W., 2011. Visualization of chlorine evolution at dimensionally stable anodes by means of scanning electrochemical microscopy. *Anal. Chem.* 83, 7645–7650. doi:10.1021/ac200677g

Zigah, D., Rodríguez-López, J., Bard, A.J., 2012. Quantification of photoelectrogenerated hydroxyl radical on TiO2 by surface interrogation scanning electrochemical microscopy. *Phys. Chem. Chem. Phys.* 14, 12764. doi:10.1039/c2cp40907k

Zoski, C.G., 2016. Review—Advances in Scanning Electrochemical Microscopy (SECM). *J. Electrochem. Soc.* 163, H3088–H3100. doi:10.1149/2.0141604jes

Zoski, C.G., Mirkin, M. V., 2002. Steady-state limiting currents at finite conical microelectrodes. *Anal. Chem.* 74, 1986–1992. doi:10.1021/ac015669i

23

Nanostructured Materials Obtained by Electrochemical Methods: From Fabrication to Application in Sensing, Energy Conversion, and Storage

Cristina Cocchiara,
Bernardo Patella,
Fabrizio Ganci,
Maria Grazia Insinga,
Salvatore Piazza,
Carmelo Sunseri, and
Rosalinda Inguanta
Università di Palermo

23.1 Introduction

Nanotechnology is referred to as the design, characterization, production, and application of structures, devices by controlling shape and size at nanoscale through manipulation and modeling of atoms, molecules, and macro-molecules for enhancing properties and functionalities. Nanotechnologies are currently envisaged to revolutionize medicine, manufacturing, energy production, and other fundamental features of everyday life in the 21st century.

The attention toward nanoscience began as far back as several years ago with the talk *There's Plenty of Room at the Bottom* that Richard Feynmann gave on December 29th 1959 at the annual meeting of the American Physical Society at the California Institute of Technology (Feynmann 1960). The talk is considered to have inspired and informed the start of the nanotechnology. A milestone in the diffusion of the nanoscience knowledge is the book by Drexler who provided an excellent introduction to the field of molecular nanoscience through a scientifically detailed description of developments that will revolutionize most of the industrial processes and products currently in use (Drexler 1992). The analysis by Serrano and co-workers focused on the impact of

the nanotechnology on energy production, storage, and use for the development of sustainable energy system (Serrano et al. 2009).

The rapid growth over the years of the nanotechnology can be marked through many indicators such as (i) estimated rise of market volumes for nantechnology-based products, (ii) worldwide growth of public and private funding for nanotechnology research, (iii) number of patent applications, and so on. A very interesting review on the worldwide industrial application of nanoscience was recently provided by Emashova and co-workers who described the nanotechnology growth from the birth to the present day with a particular attention to three principal cluster such as nanomaterials, nano(opto)electronics, and nano-medicine (Emashova et al. 2016). A reliable indicator describing the worldwide impact and trends of the nanotechnology is the number of scientific publication produced over the years. For instance, 6,360 documents were present in SCOPUS database under the heading nanoscience starting from 1991 to February 2018. At least 1,390 of these documents have been published from 2015. This finding is of great relevance because it indicates the real great interest of the scientific world in the nanoscience, which represents the precursor stage for the nanotechnology development. The greater and greater miniaturization of mechanical, optical, electronic products and devices is the visible consequence of the rapid growth of the nanotechnology as periodically predicted by Gordon Moore, co-founder of Fairchild Semiconductor and then INTEL, who predicted that the number of transistors in an integrated circuit doubles approximately every 1 year (Moore 1965). Ten years later, looking ahead to the successive 10 years, he updated his estimate to a doubling every 18–24 months (Moore 1975), while in 2015 foresaw that the rate of progress would reach saturation in the next decade (Courtland 2015). In any case, the periodical forecasts by Moore strongly influenced the high-tech industry because its development plans were based on such projections being the miniaturization of the products as the major innovation objective.

Even if electronic industry is still the prevalent end user of the nanoscience advancements, other technological fields such as catalysts and pigments have drawn new impulse from the nanoscience achievements. Just the intimate relation between the nanoscience advancements and catalysis and pigment industry progress needs a deep analysis about the meaning and correct use of the "nano-" prefix.

According to SI (International System), the prefix "nano-" indicates size of the order of 10^{-9} m. In the field of the nanoscience, the most correct use of "nano-" should be for the structures with a dimension (height, width, or depth) less than 100 nm (Sanjay & Pandey 2017), even if a smaller dimension (of the order of 10 nms) is sometime suggested. Nevertheless, 100 nm size is now extensively shared as the superior bound for a nanostructure, it is very frequent to find structures named as nano- despite none dimension is less than 100 nm. This apparent contradiction is due to the typical nanostructure features. One of the principal characteristics of the very low-sized materials is the very large surface that is evidenced through the aspect ratio value (width to height ratio) for one dimensional (1D) materials. The surface enhancement frequently plays a decisive role also at a scale higher than 100 nm. Materials with high aspect ratio have so high surface-to-volume ratio that their properties become strongly dependent (controllable) on the surface, differently from massive materials. How the large surface area-to-volume ratio increases the activity of the nanostructured materials is clearly evidenced in Figure 23.1 showing that almost 50% of the material is situated at the surface of a 1 nm diameter sphere. For achieving these properties, it is necessary to develop specific preparation methods, the principal of which is shown in Figure 23.2. In general, materials with high surface-to-volume ratios can be more easily obtained through a bottom-up, rather than top-down procedure. The bottom-up procedure allows a best control of the final sizing, because it is based on addition of either single molecules or atoms. On the contrary, the top-down technique can be limited in some cases by dimensional stability problems of the starting material so that it is not possible to reach dimension below a certain value. Therefore, the predominance of the bottom-up technique for fabricating high aspect ratio materials justifies the use

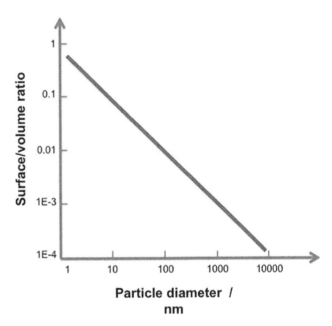

FIGURE 23.1 Typical surface/volume ratio for spherical particles as a function of the particle diameter.

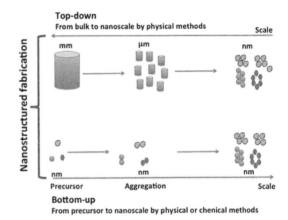

FIGURE 23.2 Schematic diagram of nanostructure fabrication methods.

of the term "nano-" also for materials with none size less than 100 nm.

For better evidencing the role of the size in characterizing the behavior of the materials, it must be considered that several novel properties appear as a dimension goes down in addition to the surface enhancement. For instance, materials with a dimension less than 100 nm show also dramatic change of the chemical reactivity due to change in density and distribution of electrons in the outermost energy level. These changes lead to novel optical, electrical, and magnetic thermal properties. Besides, a high percentage of atoms on surface introduces many size-dependent phenomena because the finite size of the particle confines the spatial distribution of the electrons, leading to the quantized energy levels. The shape of nanocrystals produces surface stress resulting in lattice relaxation (expansion or contraction) and change in lattice constant influencing the electron energy band structure and the bandgap that are sensitive to lattice constant. This quantum confinement like a "particle in a box" creates new energy states and results in a modification of the optoelectronic properties of the semiconductors.

In general, the synthesis of nanomaterials includes control of size, shape, and structure. Assembling the nanostructures into ordered arrays often becomes necessary for rendering them functional and operational. Chemical synthesis permits manipulation of the matter at molecular level, so that a better control of the particle size, shape, and size distribution can be achieved. But there are potential difficulties in chemical processing, because some preparations are complex and hazardous. For instance, entrapment of impurities in the final product needs to be avoided or minimized to obtain the desired properties.

On this basis, it can be concluded that novel properties related to the surface enhancement can be found also at size higher than 100 nm, such as for catalysts and pigments. Therefore, the use of "nano-" only based on the dimension is strongly limitative.

The activity of the materials with high aspect ratio is primarily dependent on the surface properties. For this reason, catalysts and pigments remain among the most important industrial products from the nanoscience. While the electronic industry now slowly advances in the miniaturization, as predicted by Moore in 2015 (Courtland 2015), other sectors continue to grow, especially in the field of sensing and bio-sensing, which in practice concern the catalysis.

In addition to electronics, semiconductors, and chemical industries, several other industries are now nanotechnology-based, such as:

- Aerospace;
- Automobiles;
- Biotechnology;
- Food;
- Healthcare;
- ICT;
- Military;
- Pharmaceuticals;
- Textile.

This industry classification strongly evidences (i) how the market has evolved significantly owing to the incessant development and integration of technologies and (ii) that the global nanotechnology market is segmented on the basis of types, applications, and also geographic area.

The current most usual types of nanostructured materials of industrial interest are:

- Nanoceramics;
- Nanoclays;
- Nanocomposites;
- Nanofibers;

- Nanomagnetics;
- Nanoparticles;
- Nanotubes.

In practice, the different types of nanostructured materials have driven the innovation of the industry in the 21st century, strongly influencing the lifestyle of the worldwide people. Therefore, the technological applications are becoming the real qualification of the nanostructured materials independently of their morphology and size. The low-sized materials could be named as nano when specific preparation methods must be developed and their specific properties cannot be found at higher scale. That is why a classification based only on the size is misleading.

Consequently, in the following, the attention will be paid to the fabrication and characterization of nanostructured materials through template deposition based on redox, such as electrochemical, galvanic, and electroless reactions. Besides, the performances for possible application in the field of sensing, bio-sensing, solar cells, and electrochemical storage and conversion of energy will be detailed, showing the improvements that can be achieved by decreasing the size scale in a simple and inexpensive way based on the use of a support acting as a template.

23.2 Template Synthesis

Template synthesis has been depicted many years ago as a powerful technique for growing regular and uniform nanostructured materials (Hulteen & Martin 1997). It requires the use of a nanostructured material acting as a host into or onto which different types of materials can be conformal grown. One of the advantages of this method is the morphological uniformity of the desired materials like straightforward alignment of NWs when a template having columnar parallel pores is used. Another advantage is the wide flexibility in deposition methods, including either metal- or metal oxide-catalyzed growth (Mader et al. 2014; Wang et al. 2018b), catalyst-free growth from vapor or liquid phases or plasma (Kokai et al. 2008; Le Borgne et al. 2017), dislocation driven (Jin et al. 2010), electrochemical (Shi et al. 2018; Wang et al. 2018c), and solution-gelation (sol-gel) growth (Dorval Courchesne et al. 2015; Yu et al. 2016).

23.3 Membrane as a Template

The most popular and simple template synthesis of nanostructure is based on the use of nano-porous membranes (Chakarvarti & Vetter 1998; Inguanta et al. 2007a, 2009a), which give morphologically mono-disperse materials since their pores are nano-cylinders with the same size. The key requirement of the template is its easy removal at the end of the deposition in order to expose the nanostructure that are nanowires (NWs) or nanotubes (NTs) when anodic alumina or polycarbonate track-etched membranes are used.

Anodic alumina membrane can be easily dissolved through immersion in a concentrated NaOH aqueous solution. The potential drawback of this procedure is the chemical attack of the nanostructures by NaOH. For avoiding this risk, it is necessary to use other porous templates such as polycarbonate membrane (PCM) that can be easily removed by chemical dissolution in an organic solvent such as chloroform. Besides, the polycarbonate in the exhausted organic solution can be easily recovered by solvent evaporation. The drawback of using track edge membrane is the interconnection of the channels giving nanostructures not vertically aligned like anodic alumina (Inguanta et al. 2007b).

Porous anodic alumina was extensively investigated as a typical highly ordered self-assembling structure (Diggle et al. 1968, 1970). Then, the interest was in decorative coatings owing to its chemical resistance against corrosive attack and easy coloring (Thompson et al. 1978; Henley 1982; Thorne et al. 1986). Successively, Forneaux and co-workers found a simple electrochemical method for controlling pore size and detaching the porous mass from the underlying aluminum, opening the way to successive technological developments for the use of the porous anodic membrane in nano-filtration (Furneaux et al. 1989). Over the years, this application has attracted growing interest owing to the alumina chemical stability and possible control of the morphology by adjusting anodizing voltage, nature, and composition of the solution.

Porous alumina is obtained by anodization of aluminum in suitable solutions that determine the size of the porous structure (Thompson & Wood 1981). Typically, about 20 nm diameter pores are formed in sulfuric acid (Wood & O'Sullivan 1970), 40 nm in oxalic acid (Li et al. 2000), and 180–200 nm in phosphoric acid (Shawaqfeh & Baltus 1998; Inguanta et al. 2007a). The final anodizing potential strongly influences the porous morphology in dependence on the anodizing ratio (nm/Volt) that determines the barrier film thickness, i.e. the thickness of the alumina compact film underlying the pores. Figure 23.3 shows a schematic diagram of a porous anodic alumina layer after aluminum anodizing. For using it as a template, the barrier layer is usually removed by chemical dissolution, then a conductive film, usually gold, must be sputtered on one side of the membrane for making it electrically conductive. Figure 23.4a and b show the top and cross-sectional views, respectively, of an anodic alumina porous layer formed by aluminum anodizing in 0.4M H_3PO_4 at 160 V. Porous anodic alumina were also proposed as a photonic crystal for applications in electronics and telecommunications, owing to its high morphological regularity. A two-step process was investigated in order to enhance the order of the structure and avoid formation of defects (Masuda & Fukuda 1995; Masuda 2005). The same procedure has been extensively used for obtaining a template with extremely regular pores vertically aligned.

Track-etched membrane, which is also used as a template, is produced by irradiating polymer films with highly ionizing particles such as either fission fragments or accelerated ions

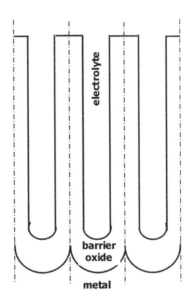

FIGURE 23.3 Schematic diagram of a porous anodic alumina layer after aluminum anodizing.

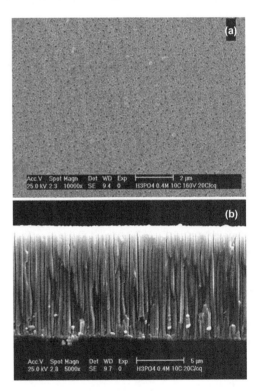

FIGURE 23.4 Top (a) and cross-sectional (b) views of an anodic alumina porous layer formed by aluminum anodizing in 0.6M H_3PO_4 at 160 V.

to form the so-called latent tracks, which are converted into hollow channels by chemical etching. The precise control of the structure distinguishes these membranes from the most conventional polymeric ones because their pore size, shape, and density can be varied in a controllable manner to obtain the required porous morphology. Cylindrical, conical, funnel-like, cigar-like, and other shapes of track-etched pores can be fabricated by different etching methods. For instance, Apel and co-workers detailed the role of surfactant

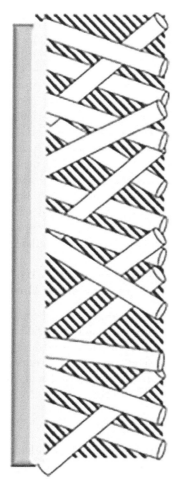

FIGURE 23.5 Schematic diagram of a PCM after deposition of a current collector on the gold sputtered film.

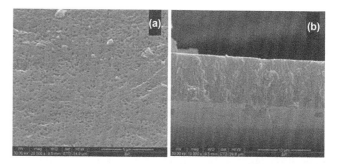

FIGURE 23.6 SEM pictures of top (a) and cross-sectional view (b) of a PCM used as template for the electrochemical growth of nanostructures.

23.4.1 Electroless Deposition

Electroless deposition occurs when a proper reductant is added to the solution containing the desired precursor. It is difficult to grow a deposit by this way, because the redox reaction prevalently occurs when just the reductant is added to the solution containing the precursor. As a consequence, the reaction advances in the bulk of the solution with precipitation of the reduced species, which can occlude the pore entrance, so that the reductant diffusion into the membrane pores is strongly hindered. Under these conditions, irregular deposits are obtained with some pores empty and other ones partially filled up. For this reason, the electroless deposition must be conducted with extreme care as templated nanostructured materials are desired, while it is largely preferred for depositing thin coatings because the process is easy and fast to be conducted (Volpe et al. 2006; Inguanta et al. 2006, 2007c).

23.4.2 Electrochemical Deposition

Electrodeposition is the most widespread technique for filling up the membrane nano-channels in order to fabricate nanostructured materials, as shown in an extensive review dealing with the peculiar advantages of the track-etched membranes (Chakarvarti 2006). The filling up of the porous mass advances through electrochemical reduction of the precursor ions inside the membrane channels. The challenge of the method is related to the confined ambience where the reduction reaction occurs. The low diameter of the channels primarily influences the mass transport because the frequent collisions of the diffusing species with the pore walls greatly reduce the transport rate. A broad quantitative evaluation of this behavior leads to a diffusion coefficient two orders of magnitude less (10^{-7} cm^2/s) than that throughout the bulk of a liquid. Therefore, the diffusion coefficient throughout a liquid filling the membrane channels is more close to that one throughout a solid. The morphology of the growing deposit into the pores is greatly influenced by the electrical parameters (Inguanta et al. 2009b) and also by the parasitic reactions simultaneously occurring with the precursor reduction. As an aqueous solution is used, occurrence of gas evolution

molecules in determining cigar-like pore channels (Apel et al. 2006). Polycarbonate, polyethylene terephthalate, and polyimide are the most used polymers to fabricate track-etched membranes. Polycarbonate ones are largely preferred as templates because can be easily removed by chemical dissolution without damaging metals, alloys, or oxides deposited inside the pores. Figure 23.5 schematically shows a typical morphology diagram of a PCM after deposition of a current collector on the gold sputtered film. Figure 23.6 shows the SEM pictures of top and cross-sectional views of PCMs used as template for the electrochemical growth of nanostructures.

23.4 Deposition into Template by Redox Reaction

Electroless, electrochemical, and galvanic deposition are electrochemistry-based methods for fabricating nanostructured materials through a membrane acting as a template. These techniques can be classified as electrochemistry-based because for all cases, redox reactions are involved like in the electrochemistry, even if the procedures are completely different for each of them.

is highly probable in dependence on the electrode potential and solution pH because water can be electrochemically splitted in hydrogen (at cathode) and oxygen (at anode). As a consequence, hydrogen evolution can interfere with the nanomaterial growth. In dependence on the relative rate of gas production and bubble detachment from the electrode surface, gas bubbles may remain either adhering to the growing surface or trapping into the channels. In both cases, the ohmic drop significantly increases hindering the deposition process. In addition, adhesion of gas bubble to the growing deposit can determine morphology modification because deposition of the desired material is confined to the gap between the hemispheric bubble surface and pore wall. In such conditions, NT morphology is obtained as inferred by Inguanta et al. (2008a). The change of the morphology is also related to the pulsed power supply, evidencing how the deposit morphology can be controlled by adjusting the waveform (Inguanta et al. 2008a).

The possibility to obtain uniform deposits also depends on the distribution of the gold at the pore bottom of the template. Figure 23.7 shows a diagram of a template, which has been schematized with aligned channels for an easier understandability. The figure evidences that every channel is an electrochemical reactor whose behavior is strongly influenced by the gold film covering the bottom. Figure 23.7a shows the differences in the gold distribution, while Figure 23.7b shows the initial phase of the deposition when the desired materials start to cover the gold film by permeating the gold deposit. Then, the growth of the materials continues vertically up to heights which can be slightly different in dependence on the gold distribution at the bottom of the pores. Owing to a nonuniform gold distribution at the pore bottom, ohmic drop throughout every channel can be different, determining significant differences in the height of the nanostructures. Of course, the deposition better advances inside the channels with lower ohmic drop; therefore, nanostructures with different height can be obtained if the regularity of the gold deposit is not carefully considered.

Due to the slow mass transport in the confined ambient where the cathodic reaction occurs, it is necessary to find the proper either constant or pulsed power supply, which together with the solution composition lead to the desired morphology, without dendrite or other morphological defects. Really, deposition with various

imperfections can be obtained if electrodeposition is not carefully controlled.

Overall, these aspects (morphological regularity of the template and mass transport rate) highlight that the critical step to successfully conducting electrodeposition into a membrane is to find the optimum deposition conditions. Of course, such an activity is hard and time consuming but is of great value because after optimization of the process, electrodepositing a nanostructured material becomes very easy. In addition, the electrochemical deposition is practically inexpensive and offers high flexibility because also electric parameters can be adjusted for obtaining high-quality deposits, in addition to the chemical ones. By carefully comparing advantages and difficulties, electrochemical deposition remains the largely preferred method.

23.4.3 Templated Materials by Electrogeneration of Bases

An indirect and valuable electrochemical method for growing nanostructured materials inside the pores of a membrane is based on the electrogeneration of base, which occurs when a cathodic reaction determines a pH increase at the electrode/solution interface. For instance, reduction of NO_3^- to NO_2^-

$$NO_3^- + H_2O + 2e = NO_2^- + OH^- \qquad (23.1)$$

is a typical reaction causing a local pH increase, like also the reduction of perchlorate.

Identically, water and oxygen reductions produce the same effect, according to:

$$2H_2O + 2e = H_2 + 2OH^- \qquad (23.2)$$

$$O_2 + 2H_2O + 4e = 4OH^- \qquad (23.3)$$

Reactions (23.2) and (23.3) are usual when an aqueous solution is used as an electrolyte, unless the cell voltage is carefully monitored. Really, reaction (23.3) is far less effective than (23.2) because the oxygen concentration in water is very low, of the order of 0.0012 mol/kg at 298°K and $P = 1$ bar (Geng & Duan 2010).

As a consequence of the local pH increase, an acid–base reaction can occur involving the precursor cations dissolved in solution and electrogenerated OH^-, with formation of

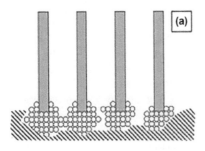 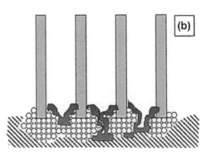

FIGURE 23.7 Schematic diagram of gold distribution at the pore bottom of a template overlying a current collector: (a) pore empty; (b) onset of the deposition on the gold film.

oxygenated species that precipitate inside the template pores as their solubility bound is attained. The electrogeneration of base is a simple method for the electrochemical synthesis of oxygenated multi-element compounds, which, alternatively, could be fabricated by other methods, such as gel-sol that are less easy to be conducted, more expensive, and time consuming.

On the contrary, the only challenge to be faced in the case of the electrogeneration of base is the detection of the suitable solution composition containing the precursor, whose chemical behavior in alkaline ambient leads to the precipitation of the desired compound.

23.4.4 Galvanic Deposition

Another valuable method for synthesizing nanostructured materials by deposition in template is based on establishing a proper galvanic contact (Inguanta et al. 2007d, 2008b). It exists when two materials with different standard electrochemical potentials are electrically brought into contact and both are immersed in an electrolyte, which can be different for each material. In this case, it is mandatory to establish an ionic conductivity between the electrolytes, usually through a salt bridge. Figure 23.8 shows two schemes of galvanic connection when both materials are in contact with one (Figure 23.8a) or two different electrolytes (Figure 23.8b). The less noble material, usually either Zn or Al-Mg alloy, behaves as a sacrificial anode. The process is driven by the electromotive force due to the difference of the standard electrochemical potential of the coupled materials. Therefore, external power supply is not necessary for conducting the deposition. The principal advantage of such a method is the very low deposition rate that cannot be achieved with the conventional electrochemical deposition methods, so that uniform deposits without morphological defects can be obtained. The only way to control the deposition rate is by adjusting the cathode-to-anode surface ratio.

For deposition advancement, the electrochemical standard potential difference must exist between sacrificial anode and the growing material. Since the pore bottom is usually covered by a gold film, as shown in Figure 23.7, it is easy to find a material with higher electrochemical standard potential that behaves as a sacrificial anode. But this is not enough, because the gold film (see Figure 23.7b) is covered by the depositing material after the initial instants; therefore, the electromotive force must

exist between sacrificial anode and depositing material, otherwise the process does not continue. Of course, the deposited materials must be electrically conductive or alternatively must allow the permeation of the electrolyte to an electric conductor, so that a galvanic connection is always operative. This is the case of oxygenated compound deposition through electrogeneration of base, because usually they are poor or not conductors. Nevertheless this difficulty, oxides such as LnO/OH (Ln = La, Ce, Sm, Er) were successfully deposited (Inguanta et al. 2007e, 2012a) with different morphologies in dependence on deposition conditions such as at $-6°C$ or $60°C$ (Inguanta et al. 2012b). At low temperature, high NTs or NWs mixed with NTs are formed, because the electrolyte permeates throughout the tubular structures, so that they can grow. On the contrary, when the deposition rate is enhanced by the temperature increase to $60°C$, only very short NWs are formed, because after the initial phase of deposition, the more compact structure inhibits the permeation of the electrolyte with consequent arrest in the growth of the nanostructures, which consequently are far less high.

23.5 Electrochemical Fabrication of Nanostructured Materials for Sensing

A sensor is a device, module, or subsystem that detects and responds to some type of input from the physical environment. The input for the most common sensors can be light, heat, motion, moisture, pressure, or any one of other environmental parameters. The output is generally a signal that is sent to a computer processor that converts it to human-readable information electronically transmitted to either the local sensor display or to a network for further either processing or remote reading. Over the years, there has been an increasing trend for using sensors as analytical tools for detecting pollutants (Li et al. 2018a), monitoring human vital parameters (Hua et al. 2018) or making early diagnosis of human pathologies (Wang et al. 2018d). A sensor essentially consists of two parts: a detecting element (sensing) for revealing the parameter of interest which is converted into an electrical signal sent to a transducer (the second part of the sensor) which handles and displays it in a readable form for the given application. The transducer is an electronic circuit, while the sensing is a material sensible to revealing the parameter of interest. The fundamental characteristic of the electronic circuit is a very low signal-to-noise ratio, because the output signal from the sensing is extremely weak, as usual. In turn, a valuable sensing must show a linear dynamic response (LDR) with a limit of detection (LOD) as low as possible. In addition, selectivity and reliability are further mandatory characteristics of a sensing element for its successful use. Also the lifetime should be considered, even if disposable devices are increasing everywhere (Bujes-Garrido et al. 2018; Orzari et al. 2018), especially in the field of the human health (Du et al. 2016; Guo &

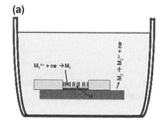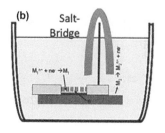

FIGURE 23.8 Schematic configuration of galvanic contact through one (a) or two (b) electrolytes.

Ma 2017; He et al. 2018). About the operational features of a sensor, a very interesting distinction between specificity and sensitivity has been recently proposed in the literature for bio-medical application (Peveler et al. 2016).

Many, if not the most, sensors are based on the electrochemical reaction occurring at the interface between the sensing material and medium containing the analyte. The preference for this type of sensors is due to the electrical signal coupled with the advancement of a redox reaction. Therefore, the electrochemical techniques applied for sensing are those typical of the electro-analysis. The key element of these techniques is the possibility to reveal ultra-traces of the analytes in addition to the simplicity of the operation. Common electrochemical sensing techniques are the voltammetric ones, such as:

- Linear Sweep Voltammetry (LSV);
- Differential Pulse Voltammetry (DPV);
- Square Wave Voltammetry (SWV);
- Normal Pulse Voltammetry (NPV);
- AC Voltammetry (ACV);
- Cyclic Voltammetry (CV).

Other techniques extensively applied for sensing are:

- Chronoamperometry (CA);
- Pulsed Amperometric Detection (PAD);
- Multiple Pulse Amperometry (MPAD);
- Fast amperometry (FAM);
- Chronopotentiometry (CP);
- Open Circuit Potentiometry (OCP);
- Multistep Amperometry (MA);
- Multistep Potentiometry (MP);
- Mixed Mode (MM);
- Impedance spectroscopy/EIS.

All these techniques strongly depend on the true surface area of the sensing electrode and its electrocatalytic feature. The last one requires a proper choice of the material in dependence on the analyte to be revealed, while the surface area can be greatly enhanced through employment of nanostructured materials. Therefore, the use of nanostructures as a sensible element is actually a big help for improving sensor performance. The big challenge in applying these sensing techniques is the miniaturization of the sensor, as a portable device is desired. In particular, the major challenge is the miniaturization of the sensing element as the potential of the electrode involved in the redox reaction of interest (working electrode) must be controlled. In this case, a three-electrode configuration must be designed, one of which behaves as a reference. In addition, the sensing must be electrically powered through a button battery located in the case of the device. For satisfying all these requirements, a miniaturized screen-printed three-electrode device or other similar configurations sensing have been proposed (Hayat & Marty 2014; Cinti & Arduini 2017; Brazey et al. 2018; Ghazizadeh et al. 2018; Wang et al. 2018a).

The role of both electrocatalysis and true sensing surface area has been well evidenced in the literature for the case of H_2O_2 detection (Patella et al. 2016; Sunseri et al. 2016; Patella et al. 2017a, b). Figure 23.8 shows the performance improvement for sensing H_2O_2 when Pd NW array was used in place of copper one. Both copper and palladium nanostructures were fabricated through galvanic deposition into a PCM acting as a template and using an aluminum tube as a sacrificial anode; 5.0% ± 10% μm long Pd NWs showed a sensing linear range from 52 to 4,401 μM with an LOD of 13.5 μM and a sensitivity of 0.37 μA/μM cm²; while 3% ± 10% μm long copper NWs showed a sensing linear range from 62 to 3,708 μM with an LOD of 13.7 μM and a sensitivity of −0.51 μA/μM cm². Of course, a reliable comparison between the two materials must take into account the different height of the NWs. In addition, also the cost of the materials must be taken into account even if the employed amounts are very low. All things considered, copper NW array-based sensor for detecting H_2O_2 can be evaluated as satisfying so that its use can be suggested if not exceptional performances are requested. On the contrary, either Pd NW array (Patella et al. 2017a, b) or more sophisticated Pt-based sensing elements must be used for harder analytical determinations (Leonardi et al. 2014).

The key role of the surface area in determining the sensing performance is shown in Figure 23.9 where dynamic response and LOD of H_2O_2 trough a copper thin film and 3 μm ± 10% long copper NWs are compared. The advantage of the nanostructured sensing array is evident. This finding is confirmed in Figure 23.10 showing an improvement of sensing performance as the NW height increases. Of course, the improvement is due to an enhancement of the wetted surface. Just the wettability can be a severe drawback because the porous mass must be well permeated by the solution containing the analyte for taking advantage of the nanostructured morphology with the best utilization of the sensing element. Unfortunately, metallic porous mass is not hydrophilic; therefore, the permeation of an aqueous

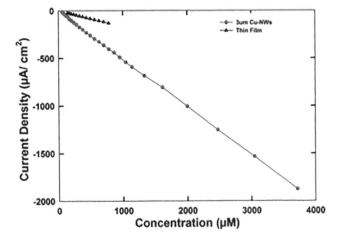

FIGURE 23.9 Dynamic response and LOD of H_2O_2 through a copper sheet (triangles) and 3 μm ± 10% long copper NWs (points).

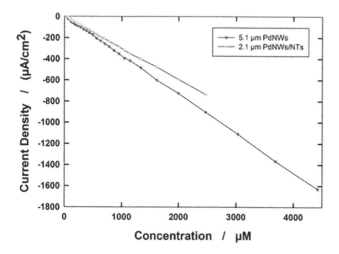

FIGURE 23.10 Improvement of sensing performance as the NW height increases.

solution can be scarce. In the case of detection of H_2O_2 in aqueous solution, the difficulty has been overcome by addition of ethylic alcohol as well shown in Figure 23.11 where LDR and LOD are compared when 5.0 μm ± 10% long palladium NW array is used as a sensing element in the presence and absence of ethylic alcohol. It is evident that the key role of the wetting agent which improves sensor performance owing to the increase of the wetted surface, without any interference, i.e. the selectivity of the sensing is not altered by the addition of ethylic alcohol.

An innovative procedure for analytical determination of mercury ions has been recently proposed (Patella et al. 2017a, b). Square-wave anodic stripping voltammetry (SWASV) is the electrochemical technique for driving the sensing element, and it is considered extremely effective for ultra-trace determinations in comparison to conventional techniques (Kovacs et al. 1995). The main features of this technique are the high sensitivity and reproducibility (standard deviation lower than 5%), besides, the LOD

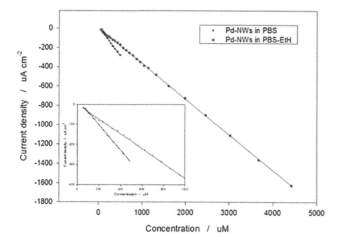

FIGURE 23.11 Influence of the nanostructure wettability. LDR and limit of H_2O_2 detection by 5.0 μm ± 10% long palladium NWs with (point) and without (square) ethyl alcohol in solution acting as a wetting agent.

at ppb level is comparable with standard techniques such as Graphite Furnace Atomic Absorption Spectroscopy (GFAAS), Inductively Coupled Plasma Optical Emission Spectroscopy (ICP-OES), and so on, which provide highly reliable results with high sensitivity. Unfortunately, these techniques are very expensive in terms of both equipment and operating costs; in addition, they are time consuming, require highly skilled personnel, and cannot be used in situ, therefore, real-time results are not available (Ratner & Mandler, 2015).

The procedure for revealing Hg^{2+} or other heavy ions consists of two successive electrochemical reactions. A differential square-wave potential pulse powers the working electrode. Initially, the sensing element is polarized as a cathode in potentiostatic mode, and metallic Hg is deposited (enrichment step). Then, the sensing electrode polarity is inverted and the dissolution current (stripping step) of Hg occurs. The dissolution current peak through a calibration curve reveals the Hg^{2+} ion concentration in solution. NiO thin film (about 900 nm thick) thermally grown on Ni was used as a sensing element, where Ni guarantees only the electrical conduction, while NiO acts as a sensing because it exhibits an excellent catalytic behavior toward the reduction of Hg^{2+} and is very cheap (Wu et al. 2012). In this case, therefore, a two-dimensional material was used in place of the 1D array. The difficulty in using this last type of morphology is due to the proper control of the NiO thickness. Since SWASV is the technique driving the sensing element, it behaves sequentially as a cathode (enrichment step) and anode (stripping step). NiO is a p-type semiconductor, therefore, behaves as an electric rectifier blocking the cathodic current. Therefore, for guaranteeing a current sufficient to reducing Hg^{2+} ions, the NiO thickness has to be adequately thin. Contemporarily, it must be sufficiently thick because it is dissolving into solution especially during the anodic stripping. Therefore, NiO cannot be anodically grown on Ni because an excessively thin film is formed, which dissolves rapidly. The only alternative is the thermal oxidation in air, but both time and temperature must be adjusted for satisfying the opposite requirements. About 2 h of Ni annealing at 773°K were found as the best compromise for fabricating the sensing element (Patella et al. 2017a, b). All the SWASV parameters, like deposition potential, deposition time, pulse amplitude, frequency, and scan rate were optimized in order to enhance the response of the electrode. After that optimization process, SWASV was carried out varying the solution concentration in order to find the sensor characteristics. The effect of the solution pH was also studied using different buffer solutions. It has been found that NiO/Ni sensors have a detection limit of 4 ppb with a linear calibration from 15 ppb to 1.8 ppm (Patella et al. 2017a, b). The sensing performance is shown in Figure 23.12 where the intensity of the stripping current is reported for different Hg^{2+} concentrations. The further challenge is to develop a 1-D nanostructured array for decreasing the LOD value below 2 that is the bound established by EPA for Hg concentration in water.

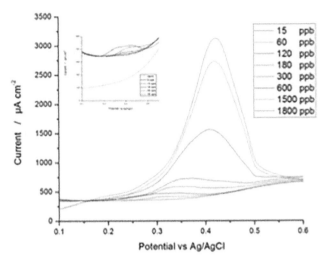

FIGURE 23.12   Intensity of the stripping current for different Hg^{2+} concentrations.

23.6   Electrochemical Fabrication of Nanostructured Semiconductors for Photovoltaic Applications

Over the years, solar cells based on semiconductor thin films have progressively emerged as alternative to the conventional first-generation crystalline silicon solar cell (c-Si), that uses wafers up to 200 μm thick (Deb 1996; Lee & Ebong 2017; Khattak et al. 2018; Reddy et al. 2018).

However, the solar cell giving the highest efficiency (Copper Indium Gallium di-Selenide-CIGS/CdS) contains toxic (Cd) and rare (In) elements that should be substituted with other abundant and nontoxic elements. Compounds of copper, zinc, tin, and selenium (CZTSe) are potentially promising materials, thanks to their capacity to maintain useful properties as absorbers also in compositions different

from the stoichiometric one (Lai et al. 2017; Yao et al. 2017; Taskesen et al. 2018).

In the following, some results will be presented about extensive investigations aimed to find suitable conditions to grow CZTSe nanostructures by one-step electrodeposition into the channels of PCM. A ZnS thin film acting as a buffer was deposited on these nanostructures by chemical bath deposition. Besides, ZnO and ZnO:Al were also deposited on the nanostructured electrodes by RF Magnetron Sputtering to obtain a complete solar cells.

The nanostructure array was obtained by template electrochemical deposition from a bath containing $CuSO_4$, $ZnSO_4$, $SnCl_4$, and H_2SeO_3 at different pH and copper concentration with lactic acid as a complexing agent and Na_2SO_4 as a support electrolyte. The electrodeposition was conducted for 60 min in N_2 atmosphere under pulsed current between 0 and -0.00153 A/cm^2. A PCM was used as a template, one side of which was coated by a Ni layer acting as a current collector. The challenge for the electrochemical deposition of CZTSe is the great difference in the standard electrochemical potentials of the species to be deposited. Therefore, a key role is played by the complexing agent, together with a careful control of the solution composition and pH in order to obtain a Cu/(Zn+Sn) value close to $0.7 \div 0.9$, that is optimal for solar-cell applications (Katagiri et al. 2009; Yu & Carter 2016). Figure 23.13 shows the current density waveform and electrode potential response for the electrodeposition of CZTSe nanostructures. This waveform is the best compromise between deposition rate and inhibition of hydrogen evolution, which is undesired for its negative effect on deposit morphology and ohmic drop enhancement. The morphology is strongly influenced by the addition of Na_2SO_4 as a support electrolyte because it favors the formation of NWs instead of NTs. This is likely due to the higher solution conductivity that prevents an excessive rising of the electrode potential up with consequent plentiful hydrogen evolution leading to the formation of NTs.

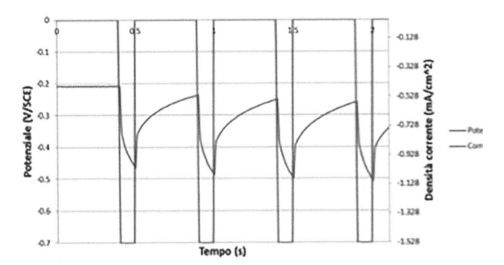

FIGURE 23.13   Current density waveform and potential response for template electrodeposition of CZTS.

Prior to depositing the nanostructured array, a thin film of Mo is deposited on a surface of the template to make the back-contact of the absorber. Then, a Ni layer was deposited on the Mo film, by 2-h long potentiostatic deposition (−1.25 V(SCE) from Watt's bath. This layer plays a key role because it acts as both a current collector and mechanical support for the nanostructures. Then, CZTSe was deposited inside the channel template. The typical morphology of the deposit is shown in Figure 23.14 after the template dissolution. It can be observed that the open space between the NWs could be permeating by the ZnS (acting as buffer) up to contacting the Mo film, with the risk of short circuit. For avoiding such an event, a thin film of CZTSe was potentiostatically deposited prior to ZnS. The porous mass was pretreated with a 0.1M sodium benzenesulfonate solution for guaranteeing the complete permeation by the electrolyte used for ZnS deposition. Interestingly, Figure 23.15a shows that the porous structure is

FIGURE 23.14 SEM pictures at different magnifications (a) low magnification, (b) high magnification of CZTSe electrodeposition in polycarbonate template.

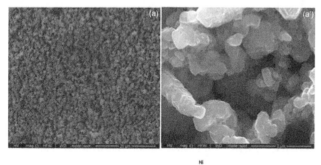

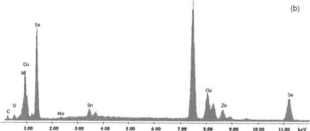

FIGURE 23.15 Morphology of the CZTSe porous mass after potentiostatic deposition of a thin film of the same material (a), and its elemental composition (b), where the Ni peak is due to the layer sustaining the porous mass covered at bottom by a Mo thin film.

FIGURE 23.16 Top views at different magnifications (a) low magnification, (b) high magnification of a Ni/Mo/CZTSe/ZnS/AZO solar cell.

preserved nevertheless deposition of a CZTSe film, even if the morphology is modified, while Figure 23.15b shows the EDS spectrum revealing the elements present in the porous mass. Both Ni and Mo peaks from the back contact are visible in the spectrum. The ZnS buffer layer was obtained by chemical deposition from a properly formulated bath. The deposition of the buffer layer did not mask the porous structure. The chemical composition of the CZTSe/ZnS junction was determined through RAMAN analysis, which revealed also the presence of secondary phases such as $CuSe_2$ and Cu_3Se_2. The fabrication of the solar cell was completed with RF magnetron sputtering of an AZO (Al:ZnO) film which preserved the porous morphology, so that a nanostructured CZTSe-based solar cell was fabricated via a full electrochemical procedure controlling the absorber composition at useful values for photovoltaic application (Figure 23.16).

Also CIGS (copper, indium, gallium, and selenium) NWs were electrochemically deposited into the channels of an anodic alumina membrane by one-step potentiostatic power supply at different applied potentials and ambient temperature (Inguanta et al. 2010a). In particular, when the template was powered at −0.905 V/NHE in potentiostatic mode, the electrodeposition from a bath having a Ga/(Cu+In+Se) molar ratio of 2.4 led to a $CuIn_{0.75}Ga_{0.64}Se_{2.52}$ stoichiometry close to the optimal value of $CuInGaSe_2$, while the photoelectrochemical characterization revealed that the nanostructured mass was a p-type semiconductor with a bandgap of 1.55 eV.

23.7 Electrochemical Fabrication of Nanostructured Active Materials for Energy Storage

The growing technological interest in energy storage is essentially driven by the progressive diminishing of the fossil fuels and simultaneous extensive use of the renewable sources, whose power supply is highly unpredictable. In general, the attention toward energy storage is primarily devoted to establishing a sustainable development model, which requires friendly human actions with the whole ecosystem (Hannoura et al. 2006). In addition to the batteries, also hydrogen production must be considered as an electrochemical way for the energy storage, because it can be produced via water electrolysis by solar cells and used as needed.

23.7.1   Energy Storage through Hydrogen

Hydrogen is a valuable energy carrier that is currently made via steam-methane reforming which is far cheaper than water electrolysis. This is the real alternative process owing to the greenhouse gases produced by the steam-methane reforming, which, consequently, cannot be considered an environmental-friendly process. A report by BOSTON CONSULTING GROUP (BCG 2014) estimated energy storage market growth from 6 to 26 billions of Euros over the next 15 years after 2015. The major increase is expected for hydrogen, which will cover 19% of the worldwide storage market in the year 2030. Therefore, water splitting driven by renewable sources is currently the only way for satisfying the opposite requirements for a clean and simultaneously abundant production of hydrogen (Tee et al. 2017; Trompoukis et al. 2018; Vilanova et al. 2018). Therefore, innovative electrochemical reactors must be developed for reducing the energy requested per unit mass of hydrogen, starting from the thermodynamic value of 79.3 Wh per mol of H_2 (including the entropy increase). At this aim, the two major dissipative contributions to be considered are reaction overvoltage and ohmic drop. The first one depends on the electrode materials, while the second one can be minimized through a proper engineering of the reactor. About electrode materials, the nature of the electrolyte must be primarily considered, because, for instance, precious metals or their alloys are requested for water splitting in acidic aqueous solutions, while cheaper materials, such as Ni and its alloys, can be used as a cathode in alkaline electrolytes. Also the anodic process must be taken into account together with hydrogen evolution reaction (HER). If water electrolysis is the hydrogen source, the highly irreversible oxygen evolution reaction (OER) has to be considered. Therefore, any investigation focused on improving HER must be coupled with proper OER study to optimize the cell performance. A valuable review on the emerging technologies for electrochemical water splitting has been recently published with attention toward fuel-cell application, being hydrogen cold combustion with oxygen the most effective way for recovering the stored energy (Ogawa et al. 2018).

Nanostructured anode and cathode have been extensively investigated for improving the electrode catalytic activity (Ding et al. 2017; Jing et al. 2018; Mollamahale et al. 2018), because reaction overvoltage strongly depends on electrode surface area. In this context, the template electrochemical methods play a relevant role for nanostructured electrode fabrication. Composite Ni-IrO$_2$ electrodes for OER were synthesized through electrodeposition of Ni-NWs in PCM followed by deposition of IrO$_2$ nanoparticles on Ni-NWs (Battaglia et al. 2014). Nickel films on both carbon paper (Ni-CP) and PCMs (Ni-PCM) were also prepared for comparison. Iridium oxide was deposited electrochemically onto the different substrates in three different ways: potentiostatically on Ni-CP, galvanostatically on Ni-PCM, and by cyclic voltammetry on Ni-NWs. Figure 23.17 shows SEM picture of composite Ni-IrO$_2$ NWs in comparison with thin films deposited on both CP and PCM. The advantage of the nanonostructured morphology is evident in terms of enhanced surface area, whose positive effect is further shown in Figure 23.18 where quasi-steady-state polarization curves of current density vs. overpotential are reported for Ni-CP + IrO$_2$, Ni-PCM + IrO$_2$, and Ni-NWs + IrO$_2$ in 1M KOH at room temperature. The beneficial influence of IrO$_2$ addition for improving the electrocatalytic behavior of nanostructured Ni-based electrodes for OER is shown in Figure 23.19

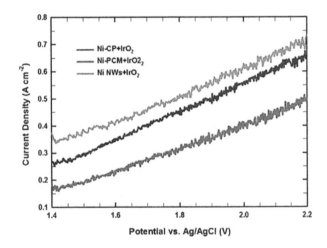

FIGURE 23.18 Quasi-steady-state polarization curves of current density vs. potential for Ni-CP + IrO$_2$, Ni-PCM + IrO$_2$, and Ni-NWs + IrO$_2$ electrodes in 1M KOH at room temperature.

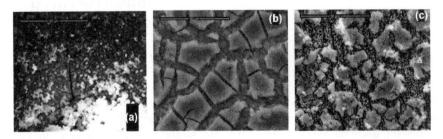

FIGURE 23.17 SEM pictures of composite Ni-IrO$_2$ electrodes where IrO$_2$ nanoparticles are deposited on (a) Ni film supported by carbon paper; (b) Ni film supported by PCMs; (c) Ni-NWs electrochemically grown in template.

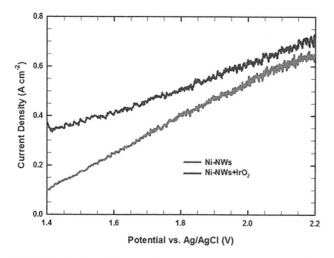

FIGURE 23.19 Current density vs. overpotential plots of Ni-NWs and Ni-NWs + IrO$_2$ anodes for OER in 1M KOH at room temperature.

where current density vs. overpotential plots are reported for Ni-NWs and Ni-NWs + IrO$_2$ anodes in 1M KOH at room temperature.

Also nanostructured electrodes fabricated through electrochemical and galvanic deposition in template have been investigated for HER (Ganci et al. 2017, 2018). Ni and Pd NW cathodes were compared in 30% w/w KOH aqueous solutions. The first ones were fabricated through electrochemical deposition into the pores of a PCM from Watt's bath, while the second ones were fabricated through galvanic deposition into the same template from a bath of 3.5mM Pd(NH$_3$)$_4$(NO$_3$)$_2$ at pH 2. Pd was selected for the sake of comparison because precious metals are good electrocatalysts for HER. Figure 23.20 shows the quasi-steady-state polarization curves for Ni and Pd NWs in 30% w/w KOH solution at room temperature. The better Ni performance than Pd for HER was attributed to the large

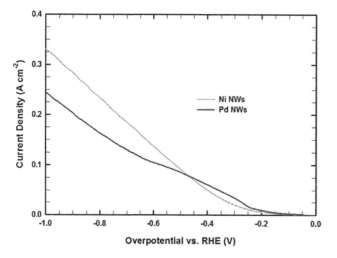

FIGURE 23.20 Quasi-steady-state polarization curves of Ni and Pd NWs for HER in 30% w/w KOH solution at room temperature.

amount of hydrogen adsorption by palladium, which behaves like a sponge for this gas. Initially, Pd cathode shows lower reaction overpotential than Ni, but the progressive adsorption of hydrogen makes palladium NWs less effective than nickel ones. It has been also found that the electrode decay on time is less pronounced for Ni than Pd NWs (Ganci et al. 2017). Since reaction overvoltage and ohmic drop are the principal causes of dissipation in an electrochemical reactor for water splitting, an analysis of their extent has been comparatively conducted. Ohmic drop has been found to weigh on the total dissipation for about 60% (Ganci et al. 2018). This result can be improved by enhancing up to 60°C the electrolysis temperature and modifying the liquid flow rate throughout the cell in order to favoring the gas bubble removal so eliminating any screen effect.

23.7.2 Energy Storage through Secondary Battery

Secondary batteries are the rechargeable ones, differently from the primary ones which cannot be recharged, such as alkaline batteries. Secondary batteries are becoming more and more decisive for energy storage and conversion, because chemical energy is one of the most effective methods of storing and managing energy for various applications such as sustainable mobility, renewable sources, and smart grids (Laurischkat & Jandt 2018). One of the strong advantages of the rechargeable batteries is the thermodynamic efficiency of chemical energy conversion to electrical one and vice versa because it is not subjected to Carnot's constrains. Efficiencies close to 100% can be reached or also higher in dependence on the entropy change of the cell reaction, like in the case of the hypothetical cold combustion of carbon in a fuel cell, according to:

$$C + 0.5O_2 = CO \qquad (23.4)$$

A theoretical efficiency of 124% has been evaluated in s.c. (Standard conditions) at $T = 298.15°K$ for this reaction (Bokris & Srinivasan 1969). Also the fuel cells are classified as secondary batteries even if they can be inverted for recharging only in theory. Therefore, in the following, the attention will be devoted only to those devices which cyclically operate for charging and discharging, neglecting both fuel cells and flow redox batteries, even if also these devices would take advantage from employment of nanostructured electrodes. Here, they are not taken into consideration, because the benefit of nanostructured electrodes is identical to that of a conventional electrochemical reactor in reducing the dissipative contribution owing to the large surface area.

For many years, a dominant position has been held by lead-acid batteries because they are safe, easy to produce, cheap, more than 90% recyclable, and raw materials for their fabrication are abundant. Their prevalent applications are in the automotive for SLI (Starting Light and Ignition), UPS (Uninterruptible Power Supply), and load leveling for renewable energy source. Both science and technology of the lead-acid battery are presented and discussed in detail by Pavlov (2011). The advancement of secondary batteries has been exhaustively described by Goodenough starting from the sodium-sulfur battery discovery by Kummer and Weber in 1967 (Goodenough 2013). He describes also the progressive evolution through which Li-ion batteries successfully penetrated the market, becoming now the most versatile devices for electrochemical storage and conversion of energy. Strong impulse to improving Li-ion battery performance came from the use of $LiMPO_4$ (M = Fe, Mn, Co or Ni) with the olivine structure as a cathode (Padhi et al. 1997), whose major drawback was the intrinsically low ionic and electron conductivity. Such limitation was overcome at nanoscale, by porous $LiFePO_4$ microsphere prepared according to the method proposed by Goodenough and co-workers (Sun et al. 2011). Over the years, also materials for anode have been extensively investigated to replace carbonaceous materials which intercalate Li with equivalent reversible capacity that is of only 372 mAh/g for graphite (Goriparti et al. 2014). Besides, the Li diffusion coefficient throughout these materials is less than 10^{-6} up to 10^{-12} cm^2/s so that batteries result with low power density. Therefore, one of the major current challenge is the selection of anode materials with high specific capacity and Li diffusion coefficient in order to improve the energy and power density, which must be accompanied by long operative life for having quality-reliable battery. For the anode, like for the cathode, a possible solution has been envisaged in the nanoscale morphology of the materials in order to employ anodes with high surface-to-volume ratio in order to fully enhance lithium-ion flux across electrode/electrolyte interface. In this context, various materials have been proposed (Nowak 2018; Qi et al. 2018), some of which are of extreme relevance for the favorable cost-to-specific energy ratio such as graphene-like nanosheets from peanut skin (Li et al. 2018b).

The dimensional stability under cycling is one of the most challenging issue in the selection of the anode materials. The volume changes coupled to Li intercalation/de-intercalation favor the electrode pulverization with consequent shortening of the lifetime. Electrochemical deposition in template of 1-D nanostructures has been proposed for overcoming this drawback, because the voids between the nanostructures can accommodate the mechanical stress. Sn-Co alloys were deposited in potentiostatic mode at -1 V (SCE) into the channels of an alumina membrane with different compositions in dependence on electrodeposition time and concentration of salts dissolved into the electrolytic bath (Ferrara et al. 2010). The same procedure was followed for Sn-Co deposition inside the PCM channels. The comparison between the two arrays as anodes in Li-ion cell revealed that nanostructures grown in polycarbonate template were mechanically more stable over cycling, likely owing to the more free space between NWs (Ferrara et al. 2011). In particular, Sn_2Co_3 NW arrays were found to be excellent anodes of Li-ion battery, because they showed 80% capacity retention after 200 cycles at C/2 and 30°C. Besides, such anodes

showed high charge and discharge rate capability at C-rates from C/3 (0.33 A/g) to 10C (10 A/g) at 30°C and 10°C (Ferrara et al. 2012, 2014). Also amorphous silicon NTs for possible applications in Li-ion battery were fabricated by galvanic deposition into PCM (Battaglia et al. 2013). This material is highly desired as an anode owing to many theoretical considerations (Armstrong et al. 2014), starting from its theoretical capacity of 4,212 mAh/g (Szczech & Jin 2011). In the case of the templated nanostructures, amorphous nature and NT morphology should improve the mechanical stability over cycling that is the current major obstacle to large diffusion of this anode (Battaglia et al. 2013).

As for lead-acid battery, significant performance improvement has been found when nanostructured PbO_2 and Pb are used as positive and negative electrodes, respectively. α-, β-PbO_2, and their mixtures can be selectively deposited in template via anodic potentiostatic polarization (Inguanta et al. 2008c). α-PbO_2 NWs, which are of major interest for battery application, are more regular and compact when deposited at 60°C (Inguanta et al. 2010b). Their performances were tested in a cell simulating a lead-acid battery, where a commercial battery drew the counter electrode in order to investigate only the PbO_2 behavior (Moncada et al. 2014). The nanostructured α-PbO_2 delivered an almost constant capacity of about 190 mAh/g, close to the theoretical value of 224 mAh/g at 1C rate, with 85% of active material utilization for more than 1,000 cycles (Moncada et al. 2014). Further tests showed that α-PbO_2 nanostructured electrode fabricated via template electrodeposition can be cycled at a constant rate from 2 C to 10 C, with a cut-off potential of 1.2 V and discharge depth up to 90% of the gravimetric charge without fading for more than 1,000 cycles (Moncada et al. 2015).

In order to simulate a lead-acid battery with both nanostructured electrodes, also template electrodeposition of lead was investigated, that is severely challenging owing to the possible formation of dendrites, with consequent breakdown of the template. For avoiding this risk, lead was fabricated via electrochemical reduction of PbO_2 NWs after their electrodeposition (Inguanta et al. 2012c). Unfortunately, the lead NWs show an irregular shape owing to the higher density of the metal (11,340 kg/m^3) than oxide (9,400 kg/m^3). Additionally, the molar volume changing during reduction accumulates mechanical stress so that the dimensional stability of the lead NWs is scarce. By deeply investigating electrolyte composition and power supply, it has been found that lead NWs can be successfully deposited through pulsed current from bath containing $Pb(BF_4)_2$ as a precursor, and leveling agents such as HBO_3^+ and lignosulfonic acid (Inguanta et al. 2013). Pb nanostructured electrodes were assembled in a zero gap configuration using a commercial plate of PbO_2 as a counter electrode and an Absorbent Glass Material (AGM) separator. The tests were conducted in 5M H_2SO_4 aqueous solution at room temperature and 1C-rate, with a cut-off potential of 1.2 V. The total capacity of the commercial plate was far higher than lead electrode in order to test only its performance. In these conditions, Pb nanostructured worked at 1C rate without fading for over 1,200 cycles and discharge efficiency around 90% (Insinga et al. 2017).

The performances displayed by both PbO$_2$ and Pb nanostructured electrodes support the conclusion that lead-acid batteries can find new life, because their specific energy can be significantly enhanced by the increase of the active materials utilization degree. Different from Li-ion battery, lead-acid electrodes are subjected to conversion reactions of the active paste during cycling. Since the molar volume of PbSO$_4$ is greater (48.21 mole/cm^3) than Pb (18.27 mole/cm^3) and PbO$_2$ (25.45 mole/cm^3), plates expand on discharging. Therefore, their porosity is diminishing with consequent electrolytic continuity interruption between external and interior parts of the plate. Therefore, the inner active material does not convert unless the discharging rate is as low as possible (0.2C, at least). This limitation is absent for nanostructured electrodes that can work at high C-rate without any fading.

23.8 Conclusions

Nanostructured materials are driving the societal development for the next years because they allow the fabrication of ever more performing devices for a better life style. The challenge is to find easy and cheap preparation methods able to support the technological advancements. Here, it has been shown that template electrochemical deposition is a valuable tool because it is very flexible, cheap, and environmentally friendly. In dependence on the material nature, deposition can be conducted by a galvanic connection, which represents a great advancement because it occurs without external energy supply. In any case, the most conventional electrochemical method where the single template channel behaves like a cell is equally effective. The materials fabricated through these methods can be successfully applied in different fields of technological interest such as electrochemical sensing that is of major interest for environment and human health sensors. The electrochemical template nanostructures can play a key role also in the field of energy storage where one can find application in both hydrogen production or batteries with improved performance, for facing the new challenges coming from the progressive abandon of the fossil fuels.

References

Apel, P.Yu., Blonskaya, I.V., Dmitriev, S.N., Orelovitch, O.L., Sartowska, B. 2006. Structure of polycarbonate track-etch membranes: Origin of the "paradoxical" pore shape. *Journal of Membrane Science* 282:393–400.

Armstrong, M.J., O'Dwyer, C., Macklin, W.J., Holmes, J.D. 2014. Evaluating the performance of nanostructured materials as lithium-ion battery electrodes. *Nano Research* 7:1–62.

Battaglia, M., Piazza, S., Sunseri, C., Inguanta, R. 2013. Amorphous silicon nanotubes via galvanic displacement deposition. *Electrochemistry Communications* 34:134–37.

Battaglia, M., Inguanta, R., Piazza, S., Sunseri, C. 2014. Fabrication and characterization of nanostructured Ni-IrO$_2$ electrodes for water electrolysis. *International Journal of Hydrogen Energy* 39:16797–805.

Bokris, J.O.M., Srinivasan, S. (1969). *Fuel Cells: Their Electrochemistry.* New York: McGraw-Hill.

Boston Consulting Group 2014. Global Energy Summit. www.slideshare.net/globalenergysummit/lbs-ges-2014-keynote-3-holger-rubel-bcg

Brazey, B., Cottet, J., Bolopion, A. et al. 2018. Impedance-based real-time position sensor for lab-on-a-chip devices. *Lab on a Chip* 18:818–31.

Bujes-Garrido, J., Izquierdo-Bote, D., Heras, A., Colina, A., Arcos-Martínez, M.J. 2018. Determination of halides using Ag nanoparticles-modified disposable electrodes. A first approach to a wearable sensor for quantification of chloride ions. *Analytica Chimica Acta* 1012:42–48.

Chakarvarti, S.K., Vetter, J. 1998. Template synthesis—A membrane based technology for generation of nano-/micro materials: A review. *Radiation Measurements* 29:149–59.

Chakarvarti, S.K. 2006. Science and art of synthesis and crafting of nano/microstructures and devices using ion-crafted templates: A review. *Proceedings Volume 6172, Smart Structures and Materials 2006: Smart Electronics, MEMS, BioMEMS, and Nanotechnology.* 61720G, doi: 10.1117/12.640311.

Cinti, S., Arduini, F, 2016. Graphene-based screen-printed electrochemical (bio)sensors and their applications: Efforts and criticisms. *Biosensors and Bioelectronics* 89:107–22.

Courtland, R. 2015. Gordon Moore: The Man Whose Name Means Progress. *IEEE Spectrum: Special Report: 50 Years of Moore's Law.*

Deb, S.K. 1996. Thin-film solar cells: An overview. *Renewable Energy* 8:375–79.

Diggle, J.W., Downie, T.C., Goulding, C.W. 1968. Anodic oxide films on aluminum. *Chemical Reviews* 69:365–405.

Diggle, J.W., Downie, T.C., Goulding, C.W. 1970. The dissolution of porous oxide films on aluminium. *Electrochimica Acta* 15:1079–93.

Ding, R., Cui, S., Lin, J., Sun, Z., Du, P. Chen, C. 2017. Improving the water splitting performance of nickel electrodes by optimizing their pore structure using a phase inversion method. *Catalysis Science and Technology* 7:3056–64.

Dorval Courchesne, N.-M., Steiner, S.A., Cantú, V.J., Hammond, P.T., Belcher, A.M. 2015. Biotemplated silica and silicon materials as building blocks for micro- to nanostructures. *Chemistry of Materials* 27:5361–70.

Drexler, K.E. 1992. *Nanosystems: Molecular Machinery, Manufacturing, and Computation.* Hoboken, NJ: Wiley Interscience.

Du, Y., Zhang, W., Wang, M.L. 2016. An on-chip disposable salivary glucose sensor for diabetes control. *Journal of Diabetes Science and Technology* 10:1344–52.

Emashova, N.A., Kudryashov, V.E., Sorkina, T.A., et al. 2016. Quo vadis, worldwide nanoindustry? *Nanotechnologies in Russia* 11:117–27.

Ferrara, G., Inguanta, R., Piazza, S., Sunseri, C. 2010. Electro-synthesis of Sn–Co nanowires in alumina membranes. *Journal of Nanoscience and Nanotechnology* 10:8328–35.

Ferrara, G., Damen, L., Arbizzani, C. et al. 2011. SnCo nanowire array as negative electrode for lithium-ion batteries. *Journal of Power Source* 196:1469–73.

Ferrara, G., Arbizzani, C., Damen, L. et al. 2012. High-performing Sn-Co nanowire electrodes as anodes for lithium-ion batteries. *Journal of Power Sources* 211:103–107.

Ferrara, G., Arbizzani, C., Damen, L. et al. 2014. Toward tin-based high-capacity anode for lithium-ion battery. *ECS Transactions* 48:153–62.

Feynmann R. 1960. There's plenty of room at the bottom. *Caltech Engineering and Science* 23:22–36. www.zyvex.com/nanotech/feynman.html

Furneaux, R.C., Rigby, W.R., Davidson, A.P. 1989. The formation of controlled-porosity membranes from anodically oxidized aluminium. *Nature* 337:147–49.

Ganci, F., Inguanta, R., Piazza, S., Sunseri, C., Lombardo, S. 2017. Fabrication and characterization of nanostructured Ni and Pd electrodes for hydrogen evolution reaction (HER) in water-alkaline electrolyzer. *Chemical Engineering Transactions* 57:1591–96.

Ganci, F., Lombardo, S., Sunseri, C., Inguanta, R. 2018. Nanostructured electrodes for hydrogen production in alkaline electrolyzer. *Renewable Energy* 123:117–24.

Geng, M., Duan, Z. 2010. Prediction of oxygen solubility in pure water and brines up to high temperatures and pressures. *Geochimica et Cosmochimica Acta* 74:5631–40.

Ghazizadeh, E., Oskuee, R.K., Jaafari, M.R., Hosseinkhani, S. 2018. Electrochemical sensor for detection of miRs based on the differential effect of competitive structures in the p19 function. *Scientific Reports* 8:3786.

Goodenough, J.B. 2013. Evolution of strategies for modern rechargeable batteries. *Accounts of Chemical Research* 46:1053–61.

Goriparti, S., Miele, E., De Angelis, F., Enzo Di Fabrizio, E., Proietti Zaccaria, R., Capiglia, C. 2014. Review on recent progress of nanostructured anode materials for Li-ion batteries. *Journal of Power Sources* 257:421–43.

Guo, J., Ma, X. 2017. Simultaneous monitoring of glucose and uric acid on a single test strip with dual channels. *Biosensors and Bioelectronics* 94:415–19.

Hannoura, A.P., Cothren, G.M., Khairy, W.M. 2006. The development of a sustainable development model framework. *Energy* 31:2269–75.

Hayat, A., Marty, J.L. 2014. Disposable screen printed electrochemical sensors: Tools for environmental monitoring. *Sensors* 14:10432–53.

He, Z., Elbaz, A., Gao, B., Zhang, J., Su, E., Gu, Z. 2018. Disposable morpho menelaus based flexible microfluidic and electronic sensor for the diagnosis of neurodegenerative disease. *Advanced Healthcare Materials* 7:1701306.

Henley V.F. 1982. *Anodic Oxidation of Aluminium and Its Alloys.* Oxford: Pergamon Press.

Hua, Q., Sun, J., Liu, H., et al. 2018. Skin-inspired highly stretchable and conformable matrix networks for multifunctional sensing. *Nature Communications* 9:244.

Hulteen, J.C., Martin, C.R. 1997. A general template-based method for the preparation of nanomaterials. *Journal of Materials Chemistry* 7:1075–87.

Inguanta, R., Amodeo, M., D'Agostino, F., Volpe, M., Piazza, S., Sunseri, C. 2006. Developing a procedure to optimize electroless deposition of thin palladium layer on anodic alumina membranes. *Desalination* 199:352–54.

Inguanta, R., Butera, M., Sunseri, C., Piazza, S. 2007a. Fabrication of metal nano-structures using anodic alumina membranes grown in phosphoric acid solution: Tailoring template morphology. *Applied Surface Science* 253:5447–56.

Inguanta, R., Sunseri, C., Piazza, S. 2007b. Photoelectrochemical characterization of Cu_2O -Nanowire arrays electrodeposited into anodic alumina membranes. *Electrochemical and Solid-State Letters* 10:K63–66.

Inguanta, R., Amodeo, M., D'Agostino, F., Volpe, M., Piazza, S., Sunseri, C. 2007c. Preparation of Pd-coated anodic alumina membranes for gas separation media. *Journal of the Electrochemical Society* 154:D188–94.

Inguanta, R., Piazza, S., Sunseri, C. 2007d. *Italian Patent VI2007A000275*, Octber 12.

Inguanta, R., Piazza, S., Sunseri, C. 2007e. Template electrosynthesis of CeO2 nanotubes. *Nanotechnology* 18(48):485605.

Inguanta, R., Piazza, S., Sunseri, C. 2008a. Influence of electrodeposition techniques on Ni nanostructures. *Electrochimica Acta* 53:5766–73.

Inguanta, R., Piazza, S., Sunseri, C. 2008b. Novel procedure for the template synthesis of metal nanostructures. *Electrochemistry Communications* 10:506–509.

Inguanta, R., Piazza, S., Sunseri, C. 2008c. Growth and characterization of ordered PbO2 nanowire arrays. *Journal of the Electrochemical Society* 155: K205–10.

Inguanta, R., Ferrara, G., Piazza, S., Sunseri, C. 2009a. Nanostructures fabrication by template deposition into anodic alumina membranes. *Chemical Engineering Transactions* 17:957–62.

Inguanta, R., Piazza, S., Sunseri, C. 2009b. Influence of the electrical parameters on the fabrication of copper nanowires into anodic alumina templates. *Applied Surface Science* 255:8816–23.

Inguanta, R., Livreri, P., Piazza, S., Sunseri, C. 2010a. Fabrication and photoelectrochemical behavior of ordered CIGS nanowire arrays for application in solar cells. *Electrochemical and Solid-State Letters* 13:K22–25.

Inguanta, R., Vergottini, F., Ferrara, G., Piazza, S., Sunseri, C. 2010b. Effect of temperature on the growth of α-PbO_2 nanostructures. *Electrochimica Acta* 55:8556–62.

Inguanta, R., Ferrara, G., Piazza, S., Sunseri, C. 2012a. A new route to grow oxide nanostructures based on metal displacement deposition. Lanthanides oxy/hydroxides growth. *Electrochimica Acta* 76:77–87.

Inguanta, R., Piazza, S., Sunseri, C. 2012b. A route to grow oxide nanostructures based on metal. displacement deposition: lanthanides oxy/hydroxides characterization. *Journal of the Electrochemical Society* 159:D493–500.

Inguanta, R., Rinaldo, E., Piazza, S., Sunseri, C. 2012c. Formation of lead by reduction of electrodeposited $PbO2$: Comparison between bulk films and nanowires fabrication. *Journal of Solid State Electrochemistry* 16:3939–46.

Inguanta, R., Randazzo, S., Moncada, A., Mistretta, M.C., Piazza, S., Sunseri, C. 2013. Growth and electrochemical performance of lead and lead oxide nanowire arrays as electrodes for lead-acid batteries. *Chemical Engineering Transactions* 32:2227–32.

Insinga, M.G., Moncada, A., Oliveri, R.L. et al. 2017. Nanostructured Pb electrode for innovative lead-acid battery. *Chemical Engineering Transactions* 60:49–54.

Jin, S., Bierman, M.J., Morin, S.A. 2010. A new twist on nanowire formation: Screw-dislocation-driven growth of nanowires and nanotubes. *Journal of Physical Chemistry Letters* 1:1472–80.

Jing, S., Zhang, L., Luo, L. et al. 2018. N-doped porous molybdenum carbide nanobelts as efficient catalysts for hydrogen evolution reaction. *Applied Catalysis B: Environmental* 224:533–40.

Katagiri, H., Jimbo, K., Maw, W.S. et al. 2009. Development of CZTS-based thin film solar cells. *Thin Solid Films* 517:2455–60.

Kokai, F., Koshio, A., Kobayashi, K., Deno, H. 2008. Formation of nanocarbon and composite materials by laser vaporization of graphite and eleven metals. *Proceedings of SPIE—The International Society for Optical Engineering* 6879, doi: 10.1117/12.761065.

Kovacs, G.T.A., Storment, C.W., Kounaves, S.P. 1995. Microfabricated heavy metal ion sensor. *Sensors and Actuators, B: Chemical* 23:41–47.

Lai, F.-I, Yang, J.-F., Chen, W.-C., Kuo, S.-Y. 2017. $Cu_2ZnSnSe_4$ Thin film solar cell with depth gradient composition prepared by selenization of sputtered novel precursors *ACS Applied Materials & Interfaces* 9: 40224–34.

Laurischkat, K., Jandt, D. 2018. Techno-economic analysis of sustainable mobility and energy solutions consisting of electric vehicles, photovoltaic systems and battery storages. *Journal of Cleaner Production* 179:642–61.

Le Borgne, V., Agati, M., Boninelli, S., et al. 2017. Structural and photoluminescence properties of silicon nanowires extracted by means of a centrifugation process from plasma torch synthesized silicon nanopowder. *Nanotechnology* 28:285702–15.

Lee, T.D., Ebong, A.U. 2017. A review of thin film solar cell technologies and challenges. *Renewable and Sustainable Energy Reviews* 70:1286–97.

Leonardi, S.G., Aloisio, D., Donato, N. et al. 2014. Amperometric sensing of H2O2 using Pt-TiO2/reduced graphene oxide nanocomposites. *ChemElectroChem* 1: 617–24.

Li, A.P., Müller, F., Gösele, U. 2000. Polycrystalline and monocrystalline pore arrays with large interpore distance in anodic alumina. *Electrochemical and Solid-State Letters* 3:131–34.

Li, H., Chen, Q., Hassan, M.M. et al. 2018a. AuNS@Ag core-shell nanocubes grafted with rhodamine for concurrent metal-enhanced fluorescence and surfaced enhanced Raman determination of mercury ions. *Analytica Chimica Acta* 1018:94–103.

Li, J., Qi, H., Wang, Q. et al. 2018b. Constructing graphene-like nanosheets on porous carbon framework for promoted rate performance of Li-ion and Na-ion storage. *Electrochimica Acta* 271:92–102.

Khattak, Y.H., Baig, F., Ullah, S., Marí, B., Beg, S., Ullah, H. 2018. Numerical modeling baseline for high efficiency ($Cu2FeSnS4$) CFTS based thin film kesterite solar cell. *Optik* 164:547–55.

Mader, W., Simon, H., Krekeler, T., Schaan, G. 2014. Metal-catalyzed growth of ZnO nanowires. *Ceramic Engineering and Science Proceedings* 34:51–66.

Masuda, H., Fukuda, K. 1995. Ordered metal nanohole arrays made by a two-step replication of honeycomb structures of anodic alumina. *Science* 268:1466–68.

Masuda, H. 2005. Highly ordered nanohole arrays in anodic porous alumina. In *Ordered Porous Nanostructures and Applications*, ed. R.B. Wehrspohn, 37–56. New York: Springer.

Mollamahale, Y.B., Jafari, N., Hosseini, D. 2018. Electrodeposited Ni-W nanoparticles: Enhanced catalytic activity toward hydrogen evolution reaction in acidic media. *Materials Letters* 213:15–18.

Moncada, A., Mistretta, M.C., Randazzo, S., Piazza, S., Sunseri, C., Inguanta, R. 2014. High-performance of PbO_2 nanowire electrodes for lead-acid battery. *Journal of Power Sources* 256:72–79.

Moncada, A., Piazza, S., Sunseri, C., Inguanta, R. 2015. Recent improvements in PbO_2 nanowire electrodes for lead-acid battery. *Journal of Power Source* 275: 181–88.

Moore, G.E. 1965. Cramming more components onto integrated circuits. *Electronics* 38:1–4.

Moore, G.E. 1975. Progress in digital integrated electronics. *International Electron Devices Meeting, IEEE:* 11–13.

Nowak, A.P. 2018. Composites of tin oxide and different carbonaceous materials as negative electrodes in lithium-ion batteries. *Journal of Solid State Electrochemistry*, doi: 10.1007/s10008-018-3942-y.

Ogawa, T., Takeuchi, M., Kajikawa, Y. 2018. Analysis of trends and emerging technologies in water electrolysis research based on a computational method: A comparison with fuel cell research. *Sustainability (Switzerland)* 10:478.

Orzari, L.O., de Araujo Andreotti, I.A., Bergamini, M.F., Marcolino, L.H., Janegitz, B.C. 2018. Disposable electrode obtained by pencil drawing on corrugated fiberboard substrate. *Sensors and Actuators, B: Chemical* 264:20–26.

Padhi, A.K., Nanjundaswamy, K.S., Goodenough, J.B. 1997. Phospho-olivines as positive-electrode materials for rechargeable lithium batteries. *Journal of the Electrochemical Society* 144:1188–94.

Patella, B., Inguanta, R., Piazza, S., Sunseri, C. 2016. Nanowire ordered arrays for electrochemical sensing of H_2O_2. *Chemical Engineering Transactions* 47:19–24.

Patella, B., Inguanta, R., Piazza, S., Sunseri, C. 2017a. A nanostructured sensor of hydrogen peroxide. *Sensors and Actuators, B: Chemical* 245:44–54.

Patella, B., Piazza, S., Sunseri, C., Inguanta, R. 2017b. NiO Thin film for mercury detection in water by square wave anodic stripping voltammetry. *Chemical Engineering Transactions* 60:1–6.

Pavlov, D. 2011. *Lead-Acid Batteries.* Amsterdam: Elsevier B.V.

Peveler, W.J., Yazdani, M., Rotello, V.M. 2016. Selectivity and specificity: Pros and cons in sensing. *ACS Sensors* 1:1282–85.

Qi, M., Zhong, Y., Chen, M., Dai, Y., Xia, X. 2018. Hollow nickel microtube/carbon nanospheres Core–Shell arrays as electrode material for rechargeable Li-ion batteries. *Journal of Alloys and Compounds* 750:715–20.

Ratner, N., Mandler, D., 2015, Electrochemical detection of low concentrations of mercury in water using gold nanoparticles. *Analytical Chemistry* 87:5148–55.

Reddy, B.P., Sekhar, M.C., Vattikuti, S.V.P., Suh, Y., Park, S.-H. 2018. Solution-based spin-coated tin sulfide thin films for photovoltaic and supercapacitor applications. *Materials Research Bulletin* 103:13–18.

Sanjay, S.S., Pandey, A.C. 2017. A brief manifestation of nanotechnology. In *Advanced Structured Materials,* ed. A. Shukla, 62:47–63. New Delhi: Springer.

Serrano, E., Rus, G., García-Martínez, J. 2009. Nanotechnology for sustainable energy. *Renewable and Sustainable Energy Reviews* 13:2373–84.

Shawaqfeh, A.T., Baltus, R.E. 1998. Growth kinetics and morphology of porous anodic alumina films formed using phosphoric acid. *Journal of the Electrochemical Society* 145:2699–706.

Shi, X., Zeng, Z., Liao, C. et al. 2018. Flexible, planar integratable and all-solid-state micro-supercapacitors based on nanoporous gold/manganese oxide hybrid electrodes via template plasma etching method. *Journal of Alloys and Compounds* 739:979–86.

Sun, C., Rajasekhara, S., Goodenough, J.B., Zhou, F. 2011. Monodisperse porous LiFePO4 microspheres for a high power Li-Ion battery cathode. *Journal of the American Chemical Society* 133:2132–35.

Sunseri, C., Cocchiara, C., Ganci, F., et al. 2016. Nanostructured electrochemical devices for sensing, energy conversion and storage. *Chemical Engineering Transactions* 47:43–48.

Szczech, J.R., Jin, S. 2011. Nanostructured silicon for high capacity lithium battery anodes. *Energy & Environmental Science* 4:56–72.

Taskesen, T., Neerken, J., Schoneberg, J. et al. 2018. Device characteristics of an 11.4% CZTSe solar cell fabricated from sputtered precursors. *Advanced Energy Materials* 1703295:1–6.

Tee, S.Y., Win, K.Y., Teo, W.S. et al. 2017. Recent progress in energy-driven water splitting. *Advanced Science News* 4:1600337.

Thompson, G.E., Furneaux, R.C., Wood, G.C., Richardson, J.A., Goode, J.S. 1978. Nucleation and growth of porous anodic films on aluminium. *Nature* 272:433–35.

Thompson, G.E., Wood, G.C. 1981. Porous anodic film formation on aluminium. *Nature* 290:230–32.

Thorne, N.A., Thompson, G.E., Furneaux, R.C., Wood, G.C. 1986. Electrolytic colouring of porous anodic films on aluminum. *Proceedings the Electrochemical Society* 86–11:274–90.

Trompoukis, C., Abass, A., Schttauf, J.-W. et al. 2018. Porous multi-junction thin-film silicon solar cells for scalable solar water splitting. *Solar Energy Materials and Solar Cells* 182:196–203.

Vilanova, A., Lopes, T., Spenke, C., Wullenkord, M., Mendes, A. 2018. Optimized photoelectrochemical tandem cell for solar water splitting. *Energy Storage Materials* 13:175–88.

Volpe, M., Inguanta, R., Piazza, S., Sunseri, C. 2006. Optimised bath for electroless deposition of palladium on amorphous alumina membranes. *Surface and Coatings Technology* 200:5800–806.

Wang, C.-M., Hsieh, C.-H., Chen, C.-Y., Liao, W.-S. 2018a. Low-voltage driven portable paper bipolar electrode-supported electrochemical sensing device. *Analytica Chimica Acta* 1015:1–7.

Wang, H.-F., Tang, C., Zhang, Q. 2018b. Template growth of nitrogen-doped mesoporous graphene on metal oxides and its use as a metal-free bifunctional electrocatalyst for oxygen reduction and evolution reactions. *Catalysis Today* 301:25–31.

Wang, M., Wu, Z., Yang, H., Liu, Y. 2018c. Growth orientation control of Co nanowires fabricated by electrochemical deposition using porous alumina templates. *Crystal Growth and Design* 18:479–87.

Wang, Z., Yang, H., Wang, M. et al. 2018d. SERS-based multiplex immunoassay of tumor markers using double SiO2@Ag immune probes and gold-film hemisphere array immune substrate. *Colloids and Surfaces A: Physicochemical and Engineering Aspects* 546: 48–58.

Wood, G.C., O'Sullivan, J.P. 1970. The anodizing of aluminium in sulphate solutions. *Electrochimica Acta* 15:1865–76.

Wu, Z., Jiang, L., Zhu, Y., Xu, C., Ye, Y., Wang, X. 2012. Synthesis of mesoporous NiO nanosheet and its application on mercury (II) sensor. *Journal of Solid State Electrochemistry* 16:3171–77.

Yao, L., Ao, J., Ming-JerJeng, M.-J. et al. 2017. A CZTSe solar cell with 8.2% power conversion efficiency fabricated using electrodeposited Cu/Sn/Zn precursor and a three-step selenization process at low Se pressure. *Solar Energy Materials & Solar Cells* 159:318–24.

Yu, C., Li, X., Liu, Z., et al. 2016. Synthesis of hierarchically porous TiO_2 nanomaterials using alginate as soft templates. *Materials Research Bulletin* 83:609–14.

Yu, K., Carter, E.A. 2016. Determining and controlling the stoichiometry of $Cu_2 ZnSnS_4$ Photovoltaics: The physics and its implications. *Chemistry of Materials* 28: 4415–20.

New Trends in Nanoscale Electrodes and Electrochemistry: The Role of Dimensionality

P. R. Bandaru
University of California

24.1 Introduction

The past few decades have witnessed considerable progress into exploring the scientific aspects of probing materials at the nanometer scale and the consequent impact on ushering new and improved technology. Considering that the harness and storage of energy have been considered *the* major problem of humanity (Nelson & Strano 2006), it would be natural to consider the role that nanostructures could play in solving the problem. At the very outset, the chief attributes of the nanoscale refer to size ranges of the order of 1–100 nm, with the relevant nano-object have been defined as a material with one, two, or three external dimensions in the nanoscale (Anon n.d.). Examples of such nano-objects include atom thick graphene sheets (Kastnelson 2012; Novoselov et al. 2007), cylindrical carbon nanotubes (CNTs) (Bandaru 2007; De Volder et al. 2013), and quantum dots (Jacak et al. 1998). Consequently, the (i) large surface area-to-volume ratio, as well as the possibility of (ii) the discretization/quantization of the energy levels, due to quantum confinement, are seen as two immediate consequences of the nanoscale. From the point of view of electrical charge based energy storage (as in batteries (Linden & Reddy 2010) and capacitors (Burke 2011; Meller et al. 2014)), it may be thought that while larger surface area/volume ratio may imply an easier access to charge storage (with the possibility of fast charging and discharging), quantization could imply the tuning and obtaining of a specific voltage. On the other hand, the reduced volume would also imply lower charge and energy capacity overall, and it is unclear whether carrier confinement and energy quantization would be useful for energy delivery. Moreover, several practical issues such as durable and reliable contacts to the nanoscale objects that tap the charge and energy to a three-dimensional world, and the possibility of contaminating influences from the ambient could be problematic. To truly understand the potential of the nanoscale objects and materials, it would then be relevant to further probe the attributes, and it is the purpose of this review to consider in detail universal and fundamental characteristics related to the rate constant of an electrochemical reaction, modulated by the underlying dimensionality of the nanostructure. Much of the materials' focus will be on graphene related electrodes, considering that the wonder material of early 21st century, has been extensively studied due to both the underlying science and potential technological applications. Due to their closeness to activated carbon (AC), of widespread use in the battery and capacitor technologies.

24.1.1 The Fundamental Electrokinetic Model: Arrhenius Theory

A critical understanding of the thermodynamics and kinetics inherent to electrochemical reactions is necessary for scientific insights into charge transfer (Kuznetrsov 1995) as well as in applications ranging from biochemical reactions (Savéant 2006) to charge storage in capacitors (Yamada & Bandaru 2013, 2014) and batteries (Linden & Reddy 2010). While the foundational attributes have almost always been reckoned in terms of one-electron-based charge transfer (Bard & Faulkner 2001; Compton & Banks 2011), much of the theoretical and experimental analysis has only obliquely referred to the considerations of dimensionality. Consequently, three-dimensional electrode characteristics and classical thermodynamics have been implicitly assumed in heterogeneous electron transfer kinetics, encompassing the widely used Butler–Volmer (BV) formulations and the subsequent Marcus (Marcus 1956; Sutin 1986) – Hush (Hush 1999) interpretations. In this regard, Arrhenius-based activation theory, leading to the BV approaches, has been used for over a century and extensively documented in standard electrochemistry textbooks (Bard & Faulkner 2001). In the BV case, the rate constant (K^{BV}), considering that for the forward reaction rate (K_F) and for the backward reaction (K_B), is:

$$K^{BV} = K_F + K_B = K^o \exp\left[\frac{\alpha e\eta}{k_B T}\right]$$
$$+ K^0 \exp\left[-\frac{(1-\alpha)e\eta}{k_B T}\right] \qquad (24.1)$$

In Eq. (24.1), α is the electron transfer coefficient and η refers to the overpotential $(= V - V^0)$, with V as the applied voltage and V^0 as the standard redox potential. The e is the elementary unit of electronic charge, k_B is the Boltzmann constant, and T is the temperature. While simple to use, in principle, such an approach does not yield substantial insight into the type and involvement of the specific constituents (redox species as well as the electrode)

and α is phenomenologically determined. The Marcus–Hush (MH) theory then seeks to better understand the rationale for K^0 and α through a more detailed consideration (Henstridge et al. 2012) of the reorganization dynamics of the solvent and the redox species *vis-à-vis* the electrochemical reactions and the electrolyte (through the macroscopic dielectric constant).

24.1.2 A Review of the Prevalent Electrochemical Kinetics Theories

Conventional electrokinetics models, employing, *e.g.*, the BV (Bard & Faulkner 2001), MH (Marcus 1956, 1992; Hush 1958), or the Marcus–Hush–Chidsey (MHC) kinetics (Chidsey 1991; Henstridge et al. 2012), employ phenomenological constants to relate to the underlying electrochemical processes. For instance, in MH/MHC models, the reorganization energy (λ) is taken as a measure of the configurational change related to the species undergoing the redox (*oxidation* or *reduction*) reaction, *i.e.*, either the λ_{ox} or the λ_{red}. While configurational changes, both internally – *e.g.*, related to bond stretching yielding a λ_{int}, as well as externally – *e.g.*, related to the rearrangement of the solvation shell, surrounding electrolyte, *etc.*, yielding a λ_{ext}, are accommodated in the models, the underlying mechanisms are yet based on classical mechanics and electrostatics. We first briefly review the salient features of the MH kinetics approach and their extension by Chidsey (1991). Consequently, we consider typical (Marcus 1956; Sutin 1986) free energy (G) – reaction coordinate (q) curves: Figure 24.1. Typically, the reaction coordinate has been broadly interpreted, and may refer to the distance, in a multidimensional *extensive* variable sense (*e.g.*, the change of bond length and electrical charge), between the oxidized (O) and reduced (R) species in an electrochemical redox reaction, of the type: $O + e^- \rightleftharpoons R$. While the progressive lowering of the minimum energy of the R parabola (*e.g.*, through increasing the η) always decreases the free energy of reaction ΔG^0, the *free energy of activation* ΔG^a initially decreases, reaching zero

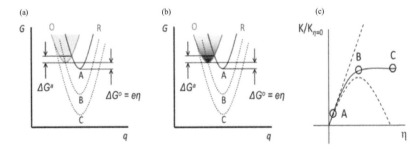

FIGURE 24.1 (a) Gibbs-free energy (G) – reaction coordinate (q) curves for the oxidized (O) and reduced (R) species in an electrochemical redox reaction, of the type: $O + e^- \rightleftharpoons R$. The free energy of reaction (ΔG^0) and the free energy of activation (ΔG^a) are indicated, for several applied voltages (η) and resulting R species configurations. (b) For low-dimensional structures, variation in the density of states (DOS) accessible for electron transfer, as in the lowering of the DOS for a one-dimensional system, would yield equivalent behavior. (c) The operating points corresponding to the various R curves in (a) and (b) represented in the normalized rate constant $(K/K_\eta = 0)$-η plot. (Reprinted with permission from Yamada, H. & Bandaru, P.R., *AIP Advances*, 6, 065325 (2016). Copyright @ The American Institute of Physics.)

when the R parabola passes through the minimum of the O parabola and subsequently increases, due to a shift of the R free energy curves to the left-hand side of the O parabola: Figure 24.1a. The concomitant increase and decrease of the electrochemical reaction rate constant K^{MH}, *i.e.*, as represented in Eq. (24.2), with v as the attempt frequency, reaches a maximum when $\Delta G^a = 0$.

$$K^{MH} = v \exp\left[-\frac{\Delta G^a}{k_B T}\right] = v exp\left[-\frac{(\lambda \mp e\eta)^2}{4\lambda k_B T}\right] \quad (24.2)$$

Such a non-intuitive increase and subsequent decrease of the reaction rate with increasing driving force (*i.e.*, η) constitutes the essence of the *inverted region*, particular to the MH theory. Such a notion on the maximum of a rate constant has been experimentally confirmed (Grampp 1993), *e.g.*, in intramolecular reactions, concerning molecules with bridged donor–acceptor units (Miller et al. 1984). It may also be derived that (Grampp 1993) $\Delta G^a = \frac{\lambda}{4}\left(1 + \frac{\Delta G^0}{\lambda}\right)^2$, with λ as the reorganization energy – which is related to the energy required for both the internal (*e.g.*, due to the bond configuration changes) and the external (*e.g.*, in the rearrangement of the solvation shell and surrounding electrolyte) configurational changes. Subsequently, it is evident that a zero ΔG^a would imply that the peak of the K^{MH} is at a value of $\lambda \sim -\Delta G^0$.

However, such a theory seemed to be incompatible with the notion of long-distance interfacial electron transfer where the rate constant decreases exponentially with increased donor–acceptor separation distances (Li & Weaver 1984) as considered through the seminal work of Chidsey (1991). Additionally, the experimental observation, in certain metal electrode-based electrochemical ensembles, of the saturation of the electrochemical current with increasing η, prompted the consideration of a continuum of energy level states. The consequently derived rate constant K^{MHC}, considering energy-level occupancy through the Fermi-Dirac distribution f_{FD}, and the explicit introduction of a constant metallic electronic density of states (DOS) ($= \rho$), was of the form (Chidsey 1991):

$$K^{MHC} = v(\rho k_B T) \int_{-\infty}^{\infty} \frac{1}{1 + \exp(x)}$$
$$\times \exp\left(-\frac{k_B T}{4\lambda}\left[x - \frac{\lambda \mp e\eta}{k_B T}\right]^2\right) dx \quad (24.3)$$

The variable, $x = \frac{E - E_F}{k_B T}$ refers to the normalized energy of a relevant participating level (E), *e.g.*, in the electrode, relative to the Fermi energy (E_F) and the negative sign is used for $\eta > 0$. The integration limits may be narrowed down to either the negative interval $(-\infty, 0)$ or the positive interval: $[0, \infty]$, if $f_{FD}(E)$ can be approximated by a step function, which would be applicable when the η exceeds 26 mV ($= k_B T/e$). The MHC relation, indicated in Eq. (24.3), may also be interpreted as related to the area of overlap between the f_{FD} and an Arrhenius-based rate law: Figure 24.2.

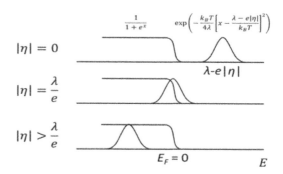

FIGURE 24.2 The Chidsey formulation for the rate constant in Eq. (24.3) can be interpreted as indicating the relative overlap between the Fermi-Dirac distribution function $f_{FD}(E)$ on the left, and the Gaussian curve on the right, corresponding to the MH models. (Reprinted with permission from Yamada, H. & Bandaru, P.R., *AIP Advances*, 6, 065325 (2016). Copyright @ The American Institute of Physics.)

The MHC relation seems to yield excellent agreement with the experimental observation of the rate constant, which tends to a constant value at larger electrode overpotential. It is to be noted that the rate constants, *e.g.*, the K^{MHC} are typically obtained through chronoamperometry (CA) experiments, through the electrical current I decay with time t (in response to a step-voltage change) of the form: $I = I_0 \exp\left(-K^{MHC}t\right)$. However, even in such molecular systems, the Chidsey modification to the MH theory adopts an intrinsically continuum point of view, through assuming a constant ρ.

24.1.3 The Influence of Dimensionality: The Role of the Density of States (*DOS*)

It is well known that the DOS, primarily depending on whether quantum confinement in a particular dimension is relevant. Through considering the *DOS* of the electrode to be a distribution over energy instead of a constant value, we broadly aim to extend the utility of the MHC kinetics to a larger class of materials and situations. While constant *DOS* can explain K-η trends for macroscale electrodes, the limited carrier density for nanoscale electrodes causes unexpected behavior for K at higher values of η. For instance, we observe in zero-dimensional (0-D) or one-dimensional (1-D) nanostructured electrodes, electrical current oscillations as a function of the η, corresponding to the gradual population (and de-population) of each successive sub-band. We posit that the consideration of a variable/non-constant *DOS* leads to a deeper appreciation of the MHC formulations and may yield tests of dimensional character and concomitant contribution to electrochemical systems. We specifically indicate the utility of our theory in explaining the electrochemical characteristics of single-layer graphene (SLG)-based electrodes – a unique prototype of a two-dimensional system.

24.2 Reinterpreting Electrochemical Kinetics through a Density of States (*DOS*) Perspective

First, we reinterpret the classical free energy – reaction coordinate curves depicted in Figure 24.1a in the context of lower dimensional structures. The initial decrease in ΔG^a followed by a subsequent increase can be related by analogy to the availability and subsequent lack in the number of energy levels (related to the *DOS*) accessible for electron transfer. Such a modulation is apparent in the *DOS* of one-dimensional nanostructures, with increasing carrier concentration and change of the E_F, and may be induced through appropriate η. We have then observed that such non-constant *DOS* yields novel electrical current–voltage response in related electrochemical systems. Figure 24.1b indicates the correspondence for lower dimensional systems where the decreasing *DOS* at higher energy may be taken analogous to the increasing λ. Indeed, saturation of the electrical current/rate constant curves may be indicative of the limit of a finite *DOS*.

The consequent plots of the respective normalized K value variation, through Eqs. (24.1)–(24.3), with the η (corresponding to BV, MH, or MHC kinetics) are indicated in Figure 24.3. From CA-related experiments and $I = I_0 \exp\left(-K^{\text{MHC}}t\right)$, such characteristics may be considered equivalent to electrical current I-η plots. The figure also indicates a re-plotting of experimental data previously obtained and were selected with the rationale that they represented three diverse situations, where nonlinear variation of the rate constant with η was observed, *i.e.*, (i) an inverted region of the rate constant, per MH kinetics: Miller et al. (1984), (ii) an initial linear variation followed by a saturation of the rate constant, marking the

MHC mechanisms: Chidsey (1991), and (iii) rate constant variation corresponding to electron transfer across solid–solid interfaces typical to carbon-coated Li$_x$FePO$_4$ crystals: Bai et al. (Bai & Bazant 2014). It is to be noted that while the BV kinetics indicate a linear variation (on the semi-log plot) with η, the MH model exhibits a peak as a function of the η. It is also relevant to note that the experimental curves were also fit through employing Poisson statistics assuming homogeneous charge transfer, with a net λ ($= 1.2$ eV) constituted from (i) an *external* solvation energy $\lambda_s = 0.75$ eV and (ii) an *internal* vibrational energy component $\lambda_v = 0.45$ eV. The incorporation of Poisson (*cf.*, Gaussian distribution) statistics also yields an inversion of the K, while avoiding the steeper drop-off of the MH curve and was considered (Miller et al. 1984) the best fit to certain CA data.

We now consider the influence of a variable *DOS* on the K variation with η. The number of electrons available for the redox reaction, $n = \int_{E_C}^{\infty} f_{FD}\left(E - E_F\right) DOS\left(E - E_C\right) dE$, where E_c is the energy at the bottom of the conduction band. We concomitantly introduce a new *DOS*-based reaction rate constant: $K^{\text{MHC}-DOS}$, considering the influence of the energy levels, through:

$$K^{\text{MHC}-DOS} = vk_BT \int_{-\infty}^{\infty} \frac{DOS\left(\left|x + \frac{E_F - E_C}{k_BT}\right|\right)}{1 + \exp(x)}$$
$$\times \exp\left(-\frac{k_BT}{4\lambda}\left[x - \frac{\lambda \mp e\eta}{k_BT}\right]^2\right) dx \quad (24.4)$$

The integration may again be either over the negative interval $(-\infty, 0)$ or the positive interval: $[0, \infty]$, as previously discussed. In a limiting case corresponding to Eq. (24.3), the *DOS* would be a constant (*e.g.*, ρ), reverting to the original Chidsey formulation (Chidsey 1991). In the subsequent treatment, the E_c was taken as reference energy and set to zero. Such a formulation involving the energy variation of the *DOS* as a function of the dimensionality, D (*e.g.*, $DOS_{3D} \sim$ constant or $\sim E^{1/2}$ – for a semiconductor, $DOS_{2D} \sim E^0$, $DOS_{1D} \sim E^{-1/2}$, $DOS_{0D} \sim$ Dirac delta function like) also allows for a variable height of the step function, related to the product of the *DOS* and the $f_{FD}(E)$, as implicitly assumed in the Chidsey modification to the MH theories. The resulting $K^{\text{MHC}-DOS}$-η curves, as a function of the dimensionality-dependent *DOS*, are indicated in Figure 24.4. In addition to the parabolic *E-k vector* dispersion, we have also incorporated a linear *E-k* dispersion as seems to be necessary to describe the characteristics of grapheme (Narayanan et al. 2015) and related 2D materials. The respective influences of the dimensionality and the dispersion are clearly evident.

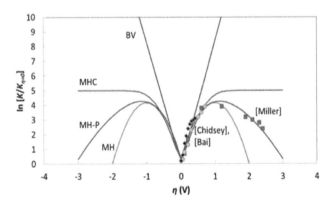

FIGURE 24.3 The variation of the normalized reaction rate constants, K, as a function of electrochemical kinetics, i.e., corresponding to BV, MH, MHP, or MHC, models – with the overpotential η. The K values have been normalized to their minimum values in each case. The experimental values were adapted from Miller et al. (1984), Chidsey (1991), and Bai and Bazant (2014). (Reprinted with permission from Yamada, H. & Bandaru, P.R., *AIP Advances*, 6, 065325 (2016). Copyright @ The American Institute of Physics.)

24.2.1 Novel Electrochemical Behavior in Two- and One-Dimensional Systems

While the traditional MHC-based formulations assumed a constant *DOS*, particular to bulk-like/three-dimensional (3D) *metallic* electrodes, the energy variation of the *DOS* in lower dimensional systems yields rich and involved behavior.

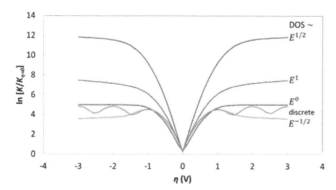

FIGURE 24.4 The variation of the normalized $K^{\text{MHC–DOS}}$ based reaction rate constants with η – obtained from Eq. (24.4), for electrodes with DOS $\sim E^a$. The exponent a, indicated in the figure, is a function of the electrode dimensionality (i.e., $a = \frac{1}{2}$ for a three-dimensional semiconductor; $a = 0$ or 1, for a two-dimensional system, $a = -\frac{1}{2}$ for a one-dimensional system) and is Delta-function like for zero-dimensional systems, such as quantum dots. The case of $a = \frac{1}{2}$ involves a bandgap, which causes the $K_{\eta} = 0$ to be smaller than that for the other cases. Generally, a reduction of the K corresponds to a decreasing DOS with energy. (Reprinted with permission from Yamada, H. & Bandaru, P.R., *AIP Advances*, 6, 065325 (2016). Copyright @ The American Institute of Physics.)

For instance, the behavior of a two-dimensional (2D) material with parabolic energy dispersion, *e.g.*, involving a quantum well, is seen to differ compared to one with linear energy dispersion, *e.g.*, graphene. In the latter case, an increasing DOS with electron kinetic energy is responsible for the observed variation. The situation for a one-dimensional (1D) material, *e.g.*, a CNT, constituted electrode – with parabolic energy dispersion along the long axis and quantization along the two perpendicular directions, with a decreasing DOS *vis-à-vis* energy, corresponding to inversion in the K-η curves at a sufficiently large η as posited in the original MH formulations. In one-dimensional

systems, the initial increase of the DOS upon the E_F reaching the band edge and the subsequent $E^{-1/2}$-induced decrease yields a corresponding modulation of the K and the electrical currents.

We then predict the occurrence of oscillations in the $K/K_{\eta} = 0$-η curves in one-dimensional nanostructures as a function of chirality in Figure 24.5. As is well known (Bandaru 2007), the specific nature of wrapping of a constituent graphene sheet, through the chirality index $[m, n]$, dictates whether the resulting CNT is metallic/semiconducting. We depict the corresponding DOS for a semiconducting [10,0] nanotube and a (ii) metallic [9,0] nanotube: Figure 24.5 b and c, respectively. While K oscillations are particularly pronounced in the former, they are less so in the latter case. The underlying reason may be related to the smaller (/larger) separation of the energy sub-bands, respectively. Moreover, the oscillations in the semiconducting and the metallic cases occur at different voltages, corresponding to the DOS variation.

It is to be noted that the width of the energy sub-bands (ΔE) in the electrode considered on the horizontal axis of DOS (E), with respect to λ – typically electrode adjacent (*e.g.*, electrolyte) side would be another important variable in interpretation of the K-η modulations. Generally, ΔE is indicative of the energy-level spacing and inversely related to the size of the nanoscale electrode, *e.g.*, in a one-dimensional CNT of diameter d, the equivalent ΔE would be proportional to $1/d^2$. The magnitude of the λ, as necessary to proceed from an O states to an R state (as in $O + e^- \rightleftharpoons R$), *cf.*, Figure 24.1, can be considered analogous to an energy-level width. When ΔE is larger (/smaller) compared to the λ, the interaction of the electrode energy levels (and relevant electron exchange/redox interactions) with respect to the electrolyte would be more (/less) sharply defined and yield an oscillatory (/smooth) K-η variation. A small λ implies that the nuclear reconfiguration and the coordinating solvent interactions (Fletcher 2010) accompanying

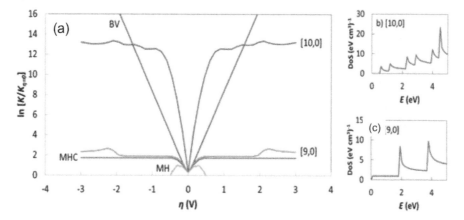

FIGURE 24.5 (a) The predicted variation of the reaction rate constants with η for a [10,0] and [9,0] single-walled CNT with $\lambda = 0.25$ eV. The respective DOS (E) variation, for the (b) [10,0], and (c) [9,0] single-walled CNT are indicated. The modulations in the rate constants are due to the successive population and de-population of sub-bands in the nanostructure. The E_F was taken to be zero in these simulations. (Reprinted with permission from Yamada, H. & Bandaru, P.R., *AIP Advances*, 6, 065325 (2016). Copyright @ The American Institute of Physics.)

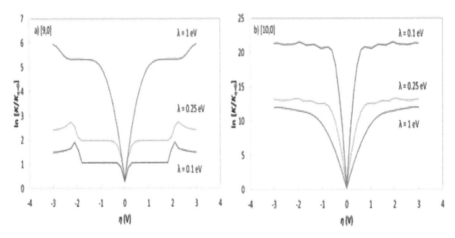

FIGURE 24.6   The $K/K_{\eta} = 0$ vs. η variations for a (a) [9,0] CNT, and (b) [10,0] CNT, at various λ, indicates the competing effects of the classical reorganization energy (λ) and the sub-band energy separation (ΔE), with respect to the influence of the DOS. The bandgap for the [10,0] CNT actually causes $K/K_{\eta} = 0$ to decrease with increasing λ, in contrast to the [9,0] CNT. (Reprinted with permission from Yamada, H. & Bandaru, P.R., *AIP Advances*, 6, 065325 (2016). Copyright @ The American Institute of Physics.)

the redox reaction are negligible. At a large enough $\lambda/\Delta E$, a continuous electronic distribution/*DOS* may be assumed yielding smooth MHC kinetics, with an increase of the K up to $\eta \sim \lambda/e$ and subsequent saturation of the K. The discussed K-η variation as related to the $\lambda/\Delta E$ ratio is indicated in Figure 24.6. As it was recently indicated that a λ of ~ 0.2 eV seemed to be effective for modeling MHC-based charge transfer kinetics at LiFePO$_4$battery electrode interfaces (Bai & Bazant 2014), such modulations could be experimentally probed. Additionally, the $K/K_{\eta=0}$ increases with λ for a [9,0] CNT, as was previously indicated (Henstridge et al. 2012) but shows the opposite variation in a [10,0] CNT. The bandgap in the semiconducting [10,0] CNT causes the $K_{\eta = 0}$ value to be smaller than that for the metallic [9,0] CNT; such an effect is stronger for smaller λ, *cf.* Figure 24.2. We further note that the *net DOS* relevant to the carriers participating in the electrochemical/redox reaction should be considered. Consequently, if the charge carriers from the surface states (*e.g.*, with energy levels inside the bandgap) are involved in the kinetics, then the local *DOS* may need to be considered. The rate constant may be increased at low η due to a sufficient number of surface states.

24.3 Experimental Validation of Lower Dimensional Electrokinetics

The *all-surface* characteristic of SLG offers a remarkable paradigm for the exploration of novel physical and chemical phenomena. However, many of the related observations, *e.g.*, the half-integer quantum hall effect (Zhang et al. 2005), enhanced electrical and thermal conductivity due to reduced phase space (Das Sharma et al. 2011), wrinkling, *etc.* are mostly concerned with electromechanical attributes, while the chemical aspects unique to graphene are less distinctive. For instance, in many cases, it is unclear as to how

a graphene sheet manifests clearly different electrochemical characters compared to any other surface (Ambrosi et al. 2014). Here, we report on an electrochemical characteristic unique to the two-dimensional nature of SLG, adding another facet to interpreting reaction rates in terms of dependence on the dimensionality of charge transport and positing the *necessity* for considering an alternate viewpoint.

SLG was synthesized via chemical vapor deposition (CVD) on copper foil and subsequently transferred to a SiO$_2$/p-Si substrate. The materials' preparation-related details have been previously reported (Zaretski et al. 2015). The transferred SLG films were first characterized through Raman spectroscopy (*inset* to Figure 24.7), from which the single-layer aspect was verified through the ratio of the intensities of the 2*D*- and the *G*-peak, *i.e.*, from an I_{2D}/I_Gratio of >2. The small *D*-peak intensity, *i.e.*, with I_D/I_Gratio of ~ 0.05 is indicative (Narayanan et al. 2015) of the relatively high quality of the SLG. Surface structure characterization through atomic force microscopy reveals distinct wrinkles: Figure 24.7a, that have characteristic modulations of the order of 10 nm. Electrochemical characterization of the SLG was performed using a three-electrode setup, under potentiostatic control. An annular electrical contact using silver epoxy was used to surround the SLG surface, that served as the working electrode (WE), while a platinum wire and a saturated calomel electrode were used as the counter electrode (CE) and a standard calomel electrode-based reference electrode (REF), respectively: Figure 24.7b. The supporting electrolyte was an aqueous solution of 1M KCl and was comprised of a 1 mM concentration ($= C$) redox couple: potassium ferri-/ferrocyanide: $K_3\mathrm{Fe}\,(\mathrm{CN})_6^{3-}$/$K_4\mathrm{Fe}\,(\mathrm{CN})_6^{4-}$, to study the specific reduction and oxidation reactions, respectively. A droplet (~ 100 µL) of electrolyte was cast onto the SLG, over the area enclosed by the annular contact (silver epoxy). The contact was covered with Kapton® tape to prevent exposure to electrolyte.

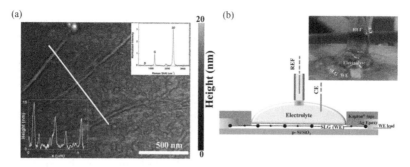

FIGURE 24.7 (a) Atomic force microscopy image of a section of the SLG sample transferred onto a p-Si/SiO$_2$ substrate. The wrinkles on the sample surface corresponding to the line scan (white line) are displayed in the lower left inset. The Raman spectrum of the transferred SLG is indicated in the top right inset. (b) Schematic of the three-electrode droplet electrochemical cell (actual experimental arrangement shown in the top right inset). The SLG WE, Pt wire CE, and a reference (REF) saturated calomel electrode are indicated. (Reprinted with permission from Narayanan, R., et al., *Nano Letters*, 15, 3067 (2015). Copyright @ The American Chemical Society.)

24.3.1 Electrochemical Characterization

The reduction and oxidation of the electroactive moieties at the SLG electrode/electrolyte interface, considering the oxidizing species, O: K$_3$Fe(CN)$_6^{3-}$, and the reducing species, R: K$_4$Fe(CN)$_6^{4-}$, is represented through:

$$O + e^- \underset{k_b}{\overset{k_f}{\rightleftharpoons}} R \qquad (24.5)$$

Here, k_f and k_b are the forward and backward heterogeneous reaction rate constants, respectively. Generally, k_f (/k_b) is proportional to the magnitude of the overpotential η (= $V - V^{0\prime}$) with the applied potential: V being negative (/positive) with respect to an equilibrium (formal) potential $V^{0\prime}$. Since at equilibrium, the reaction in Eq. (24.5) must obey the Nernst relation, we employ the relation: k_f/k_b = θ (=exp [$F\eta/RT$]), with F (= 96,487 C/mole) as the Faraday constant, R (= 8.3 J/mol·K), as the gas constant, at a given temperature: T. At any given V, a *net* reaction rate: $k = k_f + k_b$, may also be defined. At equilibrium (*i.e.*, at potential $V^{0\prime}$), $k_f = k_b = k_0$.

Typically, cyclic voltammetry (CV) or the related Nicholson method (Bard & Faulkner 2001; Velicky et al. 2014) cannot be used to determine k_f (/k_b) since the current response beyond V_{p-r} (/V_{p-o}) (*i.e.*, far from equilibrium) is mostly due to mass transport control. Consequently, CA (involving potential step experiments) was used to determine k_f and k_b at potentials far from equilibrium. The details of the arrangement and the experimental analysis are published elsewhere (Yamada & Bandaru 2016). The observed $i(t)$ was parameterized (Bard & Faulkner 2001) as in terms of the involved rate constants: k_f and k_b.

The *net* rate constant: k (= $k_f + k_b$) normalized to k_0 is plotted as a function of η in Figure 24.8. It was immediately obvious that the observed ln (k/k_0) − η variation differs markedly from both BV kinetics – where a linear relationship is expected on the semi-log plot, as well as the MHC kinetics– which predicts a saturation of the rate constants with increased η Alternatively, we observed for the SLG samples, that while the slope monotonically decreases with

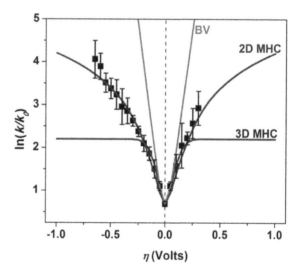

FIGURE 24.8 A plot of the extracted rate constants ($k = k_f + k_b$, sum of the forward and the backward rate constants, respectively) normalized to the $k_{\eta} = 0$ V as a function of η. The experimental data are a poor fit with the theoretical fits expected from conventional BV kinetics, three-dimensional MHC kinetics (MHC) assuming a constant DOS but could be modeled well through dimensionality-dependent electrochemical kinetics (**2D DOS model**), following Eq. (24.3) in the text. (Reprinted with permission from Narayanan, R., et al., *Nano Letters*, 15, 3067 (2015). Copyright @ The American Chemical Society.)

increasing $|\eta|$ it never reaches zero within the stability limits of the electrolyte. We have also noted that the observed ln (k/k_0) − η variation, cannot be fit through varying the series/film resistance.

24.3.2 Charge Transfer between a Nanoelectrode and Liquid Electrolyte

A schematic representation of the implications of Eq. (24.4) for (i) η = 0, (ii) η < 0, and (iii) η > 0 is indicated in Figure 24.9. Electron transfer from electrode to

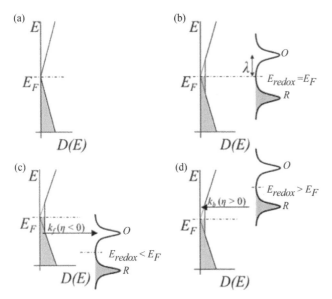

FIGURE 24.9 A schematic of (a) the energy (E)-dependent DOS (E), of an ideal SLG sheet. The DOS (E) for a nonideal SLG (say, with wrinkles, electron, and hole puddles, etc.) vis-à-vis the distribution functions of the O and R species in the redox couple, with respect to the overpotential (η), for (b) $\eta = 0$, (c) $\eta < 0$, and (d) $\eta > 0$, respectively. The E_F and the $E_{redox}(= eV^{0\prime})$ are the corresponding Fermi energy of the electrodes and the energy related to the formal/equilibrium potential: $V^{0\prime}$. The nonzero value of the DOS at the Dirac point is responsible for a finite electrochemical rate constant at $\eta = 0$: k_0. (Reprinted with permission from Narayanan, R., et al., *Nano Letters*, 15, 3067 (2015). Copyright @ The American Chemical Society.)

electrolyte occurs from the filled carrier states in the electrode (following the Fermi-Dirac distribution), to the empty states in the electrolyte (which may be distributed in energy, following for instance a Gaussian distribution (Gerischer 1997). With varying η, the electrolyte states effectively scan over the *DOS* distribution of the electrode, implying that the k/k_0 variation *mimics* the electrode *DOS*. For a pristine graphene sheet (*e.g.*, defect or wrinkle free), $n = 0$ at the Dirac point, and it would be expected that $k_0 = 0$. However, a finite value of for k_0, with the implication of a non-zero *DOS* was observed. We then fit the experimental data (depicted through the "2D *DOS* model" in Figure 24.8) assuming a finite and constant *DOS* close to $\eta = 0$ and found that the energy width of such a constant *DOS* region could be at most 2λ. A best fit, with a width of 2λ where $\lambda = 4k_B T$ ($\sim$104 meV at room temperature), is indicated previously in Figure 24.8. Beyond $e\eta \sim \pm \lambda$, the linear variation of the *DOS* resumes yielding an increasing k/k_0 with η.

Physically, the source of this additional, nonideal *DOS* variation could be due to the formation of both (Martin et al. 2008) electron and hole puddles as well as inevitably induced wrinkles (Zhu et al. 2012) – as in Figure 24.7a. Up to a threshold $|\eta| = 4k_B T \sim 104$ meV, corresponding to the constant *DOS* near the Dirac point, increasing (/decreasing) η would only result in reducing electron (/hole) puddle concentrations; which seem to be present

in equal numbers as deduced from the symmetry of the data in Figure 24.8. Only beyond a threshold $|\eta|$ would the *DOS* increase linearly, concomitant with an increase of the carrier concentration and resulting Fermi energy: Figure 24.9. Assuming each electron/hole puddle has sufficient number of carriers to yield concomitant *quasi*-Fermi levels of ± 104 meV, we estimated an intrinsic puddle carrier concentration of $\sim 8 \times 10^{11}/cm^2$, which is plausible for undoped graphene on substrate; previous work on graphene in electrochemical cells has shown such intrinsic doping (Xia et al. 2009) of up to $10^{12}/cm^2$.

24.4 Electron Tunneling in Nanoscale Electrodes for Battery Applications

The use of nanostructured materials as electrodes in energy storage devices (Bandaru et al. 2015), such as batteries and electrochemical capacitors (Yamada & Bandaru 2014), is typically based on considerations related to a large ratio of the surface area to the volume. An optimal packing of the nanostructures, spanning scales and dimensionality, is necessary to ensure both a large mass-based as well as volume-based energy density, as well as the seamless passage of a large magnitude of concomitant electrical current. Considering the relevance of multidimensional current passage, careful attention needs to be paid to the relevant mode of charge transfer both within the electrode, as well as from the electrode to an electrolyte. Generally, electron/carrier transfer occurs when filled states on one side and empty states on the other are aligned (Gerischer 1997). Such an alignment is shifted by an applied voltage difference, *i.e.*, the overpotential, between the electrode and electrolyte. The relationship is reversible and may be observed experimentally through techniques such as chronoamperometry (Bard & Faulkner 2001; Chidsey 1991; Narayanan et al. 2017). While a theory of electron transfer has been posited based on perturbation theory (Fletcher 2010), such formulations typically consider a constant DOS (Yamada & Bandaru 2016) in the electrode and a continuum of energy states in the electrolyte (Gerischer 1997). However, in nanostructures, the *DOS* is highly variable, *e.g.*, as related to the quantum capacitance C_Q (Yamada & Bandaru 2013) in SLG (Narayanan et al. 2015), and non-smooth variations/oscillations of the electrochemical rate constant were predicted in one-dimensional CNTs, due to chirality-dependent *DOS* (Yamada & Bandaru 2016). It is the aim of this chapter to clarify such foundational aspects, probe alternate viewpoints, and interpret experimental results.

24.4.1 Modeling of Batteries through Electrokinetic Theory

Conventional models, *e.g.* BV (Bard & Faulkner 2001), MH (Marcus, 1956, 1992; Hush 1958), or MHC (Henstridge et al. 2012; Chidsey 1991) kinetics, express traditional electrode battery operation through chemical reaction dynamics, which is chiefly considered through the reaction rate constant k (1/s). These models employ phenomenological constants to relate to the underlying electrochemical processes. This approach is well suited for traditional electrode materials but may not be appropriate for nanomaterials. For instance, in the BV model, α and β are electron transfer coefficients for the chemical reaction (backward and forward, respectively), where $I = I_0 \, (e^{-\alpha\eta} + e^{\beta\eta})$ and $\alpha + \beta = 1$, where η is the overpotential. They represent the ease of reducing the potential barrier for the backward and forward chemical reaction, respectively. In the MH/MHC models, the reorganization energy λ (eV) is taken as a measure of the configurational change related to the species undergoing the redox (*oxidation* or *reduction*) reaction. The Chidsey formalism (Chidsey 1991) considers electron occupancy by invoking the Fermi-Dirac distribution, *i.e.*, with $f(E) = 1/(\exp[(E - E)_F /k_B T] + 1)$, with electron energy E, Fermi energy E_F, and thermal energy $k_B T$. For metallic electrodes, implicitly assuming constant DOS ρ (1/eV) is reasonable. Under this assumption, k saturates when $|\eta| > \lambda/e$, where e is the elementary charge.

24.4.2 Analytical Representation of Electrical Current in Batteries

However, such an assumption may be unsuitable for nanostructured electrodes. Recent experimental k-η characteristics for SLG electrodes (Henstridge et al. 2012; Narayanan et al. 2017) do not saturate and instead monotonically increase with η. A better understanding of such aspects is achieved below by introducing an energy-dependent *DOS*. The details of this new theory will be presented elsewhere (Yamada & Bandaru 2017).

We first consider the relation between the k and an observed electrical current I (A) to reconcile the experimental data with the energy-dependent *DOS* theory. k is the net chemical reaction rate for a redox reaction involving oxidized O and reduced R ion species as in $O + e^- \rightleftharpoons R$ and is often experimentally determined (Bard & Faulkner 2001; Chidsey 1991; Narayanan et al. 2017) as a time decay constant. For such a reaction,

$$I = FA \left[k_f C_O - k_b C_R \right], \tag{24.6a}$$

where C_O and C_R (mol/m^2) are the molar concentrations of the O and R species, respectively, and k_f and k_b (1/s) are the forward and backward reaction rate constants, respectively. F (C/mol) is the Faraday constant, and A (m^2) is the electrode–electrolyte contact surface area. For the determination of k_f or k_b alone, large negative or positive η is used, whereby Eq. (24.6a) reduces to

$$I = en_{2D} Ak, \tag{24.6b}$$

where n_{2D} (1/m^2) is the area density of the relevant ion species and $k = k_f$ or k_b. The nanostructured experimental data may thus be written equivalently in terms of k or I.

When considering charge transfer through electron transport, I is defined as

$$I = eAnv, \tag{24.7a}$$

where n (1/m^3) is the electron carrier density and v (m/s) is the electron velocity. The shift from k to v allows us to discuss spatial coordinates. However, such a specific form is relevant for current flow through a homogeneous material. I from the electrode to the electrolyte (or vice versa) may be broadly described through an equation of the form

$$I = eAnv\Theta, \tag{24.7b}$$

with Θ as the tunneling probability of the electrons participating in the redox reaction (Muller & Kamins 1977; Schrieffer 1964).

Considering Eq. (24.7b) in more depth, it is necessary to use both $f(E)$ and the DOS for both the electrode and the electrolyte to determine A, n, and v, as well as the coefficient from Fermi's Golden Rule to determine v and Θ (Bardeen 1961, 1962; Binnig et al. 1982, 1986). Consequently,

$$I_{lt} = e \int \left[f_{eld} \times A\ell D_{3D} \right] \frac{2\pi}{\hbar} M^2 \left[(1 - f_{rdx}) \times \rho \right] dE, \tag{24.8a}$$

$$I_{rt} = e \int \left[(1 - f_{eld}) \times A\ell D_{3D} \right] \frac{2\pi}{\hbar} M^2 \left[f_{rdx} \times \rho \right] dE. \tag{24.8b}$$

Here, I_{lt} represents left-going current (right-going electrons) and I_{rt} represents right-going current (left-going electrons). The equation explicitly considers the electrode *DOS* $D(E)$ (1/(eV m^3)) and occupancy probability $f_{eld}(E)$ with Fermi energy E_F, and the electrolyte *DOS* (integrated over volume) $\rho(E)$ (1/eV) and occupancy probability $f_{rdx}(E)$ with redox energy E_{rdx}. Depending on the dimensionality of D, a spatial normalization representing the active electrode region is multiplied, *e.g.*, a volume $A\ell$ (3D), an area A (2D), a length ℓ (1D), or nothing (0D), such that the normalization times D has units of 1/eV. $\hbar$ (eV s) is the Planck's constant, and the matrix element of interaction M (eV) couples the electrode and electrolyte energy levels, smaller in magnitude compared to the level broadening, and hence considered to be relatively energy independent (Harrison 1961). The electrolyte properties and matrix element together represent the Fermi's Golden Rule, which describes electron transfer rate. Moreover, for the occupancy probability distribution, the Fermi-Dirac distribution was considered over the Boltzmann distribution for both electrode and electrolyte to represent the occupancy of the states and implicitly considers electron–electrolyte interactions.

Typically, ρ may be considered through a Gaussian: $\rho(\Delta E) = \rho_0 \exp\left[-(\Delta E)^2/(4\lambda k_B T)\right]$. The ΔE is related to the deviation of a redox species electron energy from the most probable energy, corresponding to say, that for reduction (*red*) or oxidation (*ox*), *i.e.*, $\Delta E = \Delta E_{red} \left(= E_{red} - E_{red}^0\right)$ or $\Delta E = \Delta E_{ox} \left(= E_{ox} - E_{ox}^0\right)$. The electrolyte *DOS* is represented through two peaks for the *red* and *ox* levels, which may be correlated to electron affinity and ionization energy, with a range of energies and corresponding states. Moreover, the possibility of a solid electrolyte may be easily understood through considering the ρ of a solid as well as replacing E_{rdx} with another E_F, distinct from the electrode's E_F. It is then interesting to note that the electrolyte dimensionality may need to be considered in the performance of a related device, *e.g.,* a solid-state battery. For instance, a confined two-dimensional electrolyte may yield significantly enhanced power due to a step-like onset of the electrical current with increased voltage (Datta 2005).

It is to be noted that the use of a constant *DOS* for *D* is acceptable, so long as the applied voltage range is less than $|E_F - E_C|$: Figure 24.10. Here, the product of the respective f with the distribution of the O and the R species in the electrolyte – per Eq. (24.3) yields an electrical current variation with η of the MHC type: Figure 24.10d, with a saturation at $\eta > \lambda$. The electrical currents have been normalized to the equilibrium current value I_o (the current at $\eta = 0$). If a constant DOS is assumed, the result is a curve consistent with the MHC theory. Furthermore, if λ is taken to an infinite limit, the result is a line consistent with the BV theory.

24.4.3 Electrical Current in Low-Dimensional Electrodes

We investigate the implications of Eq. (24.10) for other low-dimensional electrodes, where there would be considerable variation in the corresponding *DOS*. For example, in the case of SLG with $DOS \sim E$ and in contact with an electrolyte, we obtain a $I(\eta)/I_0$ variation indicated in Figure 24.11a for the defect-free/pristine case ($E_F = 0$). The remarkable aspect here is that I does not saturate as in MHC kinetics but instead monotonically increases with η over a considerable range mimicking the *DOS* variation and would be dictated only by the electrolyte stability limits (Goodenough & Kim 2010). Higher η values result in more graphene electrode states being continually aligned with the electrolyte. It is again assumed that f can be approximated by a step function – applicable when η exceeds 26 mV ($= k_B T/e$). However, the inevitable presence of defects (*e.g.,* structural wrinkles (Zhu et al. 2012; Pereira et al. 2010) and electron-hole puddles (Gilbertini et al. 2012; Martin et al. 2008; Narayanan et al. 2015), *etc.*) on graphene has been well documented and for such an unintentionally doped material, $E_F > 0$ assuming n-type doping. In this case: Figure 24.11b, there would be an initial monotonic variation with η followed by a dip in the $I(\eta)/I_0$ vs. η plot, when the Dirac point eventually aligns with the electrolyte *DOS* peak, beyond which there would be an increase again.

While such interesting features in I need experimental confirmation, we have seen both in our own experiments and those in literature the need for explicitly considering the dimensionality relevant *DOS*. For instance, the oxidation and reduction kinetics of ferricyanide at SLG electrodes were fit to BV kinetics (Ritzert et al. 2013) – see inset to Figure 24.12a. However, a more detailed analysis indicates unequal

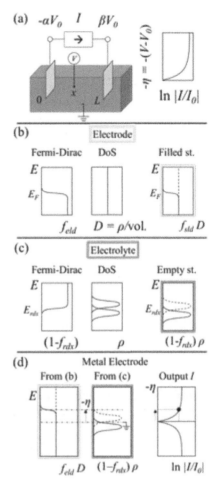

FIGURE 24.10 (a) The consideration of electrode dimensionality of nanoscale electrodes, together with such electrode–electrolyte interactions, in a typical electrochemical system (e.g., a battery or capacitor), yields novel electrokinetics. (b) The interaction of the electrode/electrolyte probability distribution function, e.g., the Fermi-Dirac function: $f_{FD}(E)$ and $f_{ely}(E)$, respectively left, with the relevant DOS and $W(E)$ – middle, yields the number of participating carriers as a function of energy (E) – right. The electrolyte oxidation (O) and reduction (R) levels are indicated. (c) The consequent electrical current (I) with respect to an overpotential: η (with the standard redox potential as a reference) and normalized to I_0 – the current at $\eta = 0$, is due to the mutual overlap of the carrier density from the electrode with the electrolyte. It is seen that beyond a certain overlap, a constant I was observed, yielding MHC kinetics. (Reprinted with permission from Yamada, H., et al., *Chemical Physics Letters*, 695, 24 (2018). Copyright @ Elsevier.)

forward and backward electron transfer coefficients of 0.08 and 0.12, respectively, with a sum less than unity. However, a sum of unity is required (Fletcher 2010) from fundamental thermodynamic considerations related to microscopic reversibility (Henstridge et al. 2011). Alternately, the use of Eq. (24.10) for the electrical current yields a much better fit: Figure 24.12a, compared to BV kinetics. Yet another independent confirmation of the necessity of

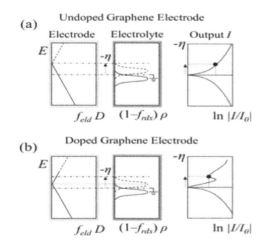

FIGURE 24.11 A non-monotonic $I(\eta)/I_0$ variation for (a) defect-free/pristine, and (b) defective/n-doped graphene. The variation closely follows the relevant DOS shape. A dip is observed in the latter case, when the Dirac point eventually aligns with the electrolyte DOS peak. (Reprinted with permission from Yamada, H., et al., *Chemical Physics Letters*, 695, 24 (2018). Copyright @ Elsevier.)

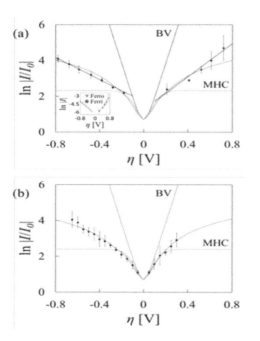

FIGURE 24.12 Modeling the experimentally obtained electro-chemical rate/current variation for (a) SLG, using the formalism relevant to Eq. (24.4) yields a better fit than conventional BV kinetics, as used by Ritzert et al. (2013) – see inset, (b) Another independent confirmation of the necessity of DOS considerations from our own experiments (Narayanan et al. 2017) with a finite DOS at the Dirac point. (Reprinted with permission from Yamada, H., et al., *Chemical Physics Letters*, 695, 24 (2018). Copyright @ Elsevier.)

DOS considerations is indicated by our own experiments (Narayanan et al. 2017) on defective SLG with a finite *DOS* at the Dirac point, with the data and fit plotted in Figure 24.12b.

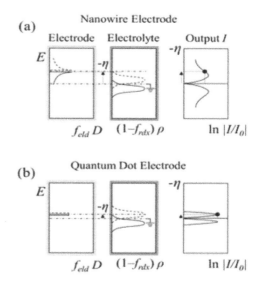

FIGURE 24.13 A decrease in the electrical current: $I(\eta)$, with increasing η is predicted in (a) one-dimensional structures, such as single-walled CNTs, and (b) in zero-dimensional quantum dots, due to a depletion of constituent electrical carriers. (Reprinted with permission from Yamada, H., et al., *Chemical Physics Letters*, 695, 24 (2018). Copyright @ Elsevier.)

The implications of considering the specific *DOS* of the electrode as well as the electrolyte were also extended to one-dimensional (*e.g.*, CNTs) or zero-dimensional (*e.g.*, quantum dot) based systems. From the aspect that I is an indicator of the integrated DOS variation, we see that for one-dimensional structures where $DOS \sim E^{-1/2}$, that there is actually a reduction in the current with increasing η: Figure 24.13a, while a delta-function like *DOS* in quantum dots, with a finite number of states yields a concomitant limited range of η over which a finite current may be obtained: Figure 24.13b.

24.5 Conclusions

In summary, we have shown that considering the specific nature of the *DOS*, as would be necessary in nanostructured materials, leads to a modification of the expected MHC electro-kinetics, considering graphene as a prototype example. While MHC-based models are seemingly appropriate for most macroscale metallic electrodes, they need correction (Bai & Bazant 2014) for nanoscale electrodes. We have also predicted, most notably, the occurrence of oscillations of the rate constant and the concomitant electrical current in semiconducting nanotubes, the experimental verification of which would be a significant test of the nature of electrical conductivity as well as dimensionality. Moreover, electron tunneling, considering energy-dependent *DOS* of the electrode and the electrolyte, may be used to describe current flow in electrochemical systems deploying nanostructured electrodes. We have indicated the possibilities of unique signatures of electrical current variation that may be obtained in electrochemical systems through specific consideration of the *DOS* variations of nanoscale electrodes. The application of such formulations to systems with solid electrolytes would enable estimates of the maximal current and expected efficiency. The study of nanoscale electrochemical kinetics bares open intriguing aspects related to the necessity for a deeper theoretical and experimental investigation (Park et al. 2007) of dimensionality-related effects in electrochemical kinetics. The implications of such study extend to properly modeling the electrical current that could be obtained in electrochemical and biological systems, with applications incorporating battery-related systems, wearable sensors, *etc.*, where such electrical current modulations would impact energy and power delivery.

Acknowledgments

The authors are grateful for support from the Defense Advanced Research Projects Agency (DARPA: W911NF-15-2-0122) and the National Science Foundation (NSF: CBET 1606192 and CMMI 1246800). The assistance of Prof. D. Lipomi and his research group as well as the discussions with Prof. P. Asbeck are appreciated.

References

Ambrosi, A. et al. 2014. Electrochemistry of graphene and related materials. *Chemical Reviews*, 114, pp. 7150–7188.

Anon, ISO/TS 27687:2008. Nanotechnologies—Terminology and definitions for nano-objects—Nanoparticle, nanofibre and nanoplate.

Bai, P. & Bazant, M.Z., 2014. Charge transfer kinetics at the solid-solid interface in porous electrodes. *Nature Communications*, 5, p. 3585.

Bandaru, P.R. et al., 2015. Charge transfer and storage in nanostructures. *Materials Science and Engineering: R: Reports*, 96, pp. 1–69.

Bandaru, P.R. 2007. Electrical properties and applications of carbon nanotube structures. *Journal of Nanoscience and Nanotechnology*, 7, pp. 1239–1267.

Bard, A.J. & Faulkner, L.R., 2001. *Electrochemical Methods: Fundamentals and Applications*, Hobeken, NJ: John Wiley & Sons, Inc.Bardeen, J., 1961. Tunnelling from a many-particle point of view. *Physical Review Letters*, 6(2), pp. 57–59.

Bardeen, J., 1962. Tunneling into superconductors. *Physical Review Letters*, 9(4), pp. 147–149.

Binnig, G. et al., 1982. Surface studies by scanning tunneling microscopy. *Physical Review Letters*, 49(1), pp. 57–61.

Binnig, G. Quate, C.F., & Gerber, C., 1986. Atomic force microscope. *Physical Review Letters*, 56(9), pp. 930–933.

Burke, A.F., 2011. Electrochemical capacitors. In T. B. Reddy & D. Linden, eds. *Linden's Handbook of Batteries*. New York, NY: McGraw Hill, p. 39.1.

Chidsey, C.E., 1991. Free energy and temperature dependence of electron transfer at the metal-electrolyte interface. *Science (New York, N.Y.)*, 251(4996), pp. 919–22.

Compton, R.G. & Banks, C.E., 2011. *Understanding Voltammetry*, London, UK: Imperial College Press.

Datta, S., 2005. *Quantum Transport: Atom to Transistor*, New York: Cambridge University Press.

Fletcher, S., 2010. The theory of electron transfer. *Journal of Solid State Electrochemistry*, 14, pp. 705–739.

Gerischer, H., 1997. Principles of electrochemistry. In P. J. Gellings & H. J. M. Bouwmeester, eds. *The CRC handbook of solid state electrochemistry*. Boca Raton, FL: CRC Press, pp. 9–74.

Gilbertini, M. et al., 2012. Electron-hole puddles in the absence of charged impurities. *Physical Review B*, 85, p. 201405(R).

Goodenough, J.B. & Kim, Y., 2010. Challenges for rechargeable Li batteries. *Chemistry of Materials*, 22(3), pp. 587–603.

Grampp, G., 1993. The Marcus inverted region from theory to experiment. *Angewandte Chemie International Edtion*, 32(5), pp. 691–693.

Harrison, W.A., 1961. Tunneling from an independent-particle point of view. *Physical Review*, 123(1), pp. 85–89.

Henstridge, M.C. et al., 2011. Marcus–Hush–Chidsey theory of electron transfer to and from species bound at a non-uniform electrode surface: Theory and experiment. *Chemical Physics Letters*, 517(1–3), pp. 108–112.

Henstridge, M.C. et al., 2012. Marcus–Hush–Chidsey theory of electron transfer applied to voltammetry: A review. *Electrochimica Acta*, 84, pp. 12–20.

Hush, N.S., 1958. Adiabatic rate processes at electrodes. I. Energy-charge relationships. *The Journal of Chemical Physics*, 28(5), p. 962.

Hush, N.S., 1999. Electron transfer in retrospect and prospect 1: Adiabatic electrode processes. *Journal of Electroanalytical Chemistry*, 470(2), pp. 170–195.

Jacak, L., Hawrylak, P., & Wojs, A., 1998. *Quantum Dots*, New York: Springer.

Kastnelson, M.I., 2012. *Graphene: Carbon in Two Dimensions*, Cambridge, UK: Cambridge University Press.

Kuznetrsov, 1995. *Charge Transfer in Physics, Chemistry and Biology: Physical Mechanisms of Elementary Processes and an Introduction to the Theory*, Boca Raton, FL: CRC Press.

Li, T.T.T. & Weaver, M.J., 1984. Intramolecular electron transfer at metal surfaces. 4. Dependence of tunneling probability upon donor-acceptor separation distance. *Journal of the American Chemical Society*, 106(20), pp. 6107–6108.

Linden, D. & Reddy, T.B., 2010. *Linden's Handbook of Batteries*, New York: McGraw Hill.Marcus, R.A., 1956. On the theory of oxidation-reduction reactions involving electron transfer. I. *The Journal of Chemical Physics*, 24(5), p. 966.

Marcus, R.A., 1992. Electron transfer reactions in chemistry: Theory and experiment. *Review of Modern Physics*, (2), p. 599.

Martin, J. et al., 2008. Observation of electron–hole puddles in graphene using a scanning single-electron transistor. *Nature Physics*, 4(2), pp. 144–148.

Meller, M. et al., 2014. Electrochemical capacitors as attractive power sources. *Solid State Ionics*, 265, pp. 61–67.

Miller, J.R., Calcaterra, L.T., & Closs, G.L., 1984. Intramolecular long-distance electron transfer in radical anions. The effects of free energy and solvent on the reaction rates. *Journal of the American Chemical Society*, 106(10), pp. 3047–3049.

Muller, R.S. & Kamins, T.I., 1977. *Device Electronics for Integrated Circuits*, New York: Wiley.Narayanan, R., et al., 2015. Modulation of the electrostatic and quantum capacitances of few layered graphenes through plasma processing. *Nano Letters*, 15(5), pp. 3067–3072.

Narayanan, R. et al., 2017. Dimensionality dependent electrochemical kinetics, at the single layer graphene-electrolyte interface. *The Journal of Physical Chemistry Letters*, 8, p. 4004.

Nelson, D.J. & Strano, M., 2006. Richard Smalley: Saving the world with nanotechnology. *Nature Nanotechnology*, 1(2), pp. 96–97.

Novoselov, K.S. et al., 2007. Room temperature quantum hall effect in graphene. *Science*, 315, p. 1379.

Park, B.Y. et al., 2007. A case for fractal electrodes in electrochemical applications. *Journal of the Electrochemical Society*, 154, pp. P1–P5.

Pereira, V.M. et al., 2010. Geometry, mechanics, and electronics of singular structures and wrinkles in graphene. *Physical Review Letters*, 105(15), p. 156603.

Ritzert, N.L. et al., 2013. Kinetics of interfacial electron transfer at single-layer graphene electrodes in aqueous and nonaqueous solutions. *Langmuir*, 29(5), pp. 1683–1694.

Savéant, J.-M., 2006. *Elements of Molecular and Biomolecular Electrochemistry: An Electrochemical Approach to Electron Transfer Chemistry*, New York: John Wiley & Sons, Inc.

Schrieffer, J.R., 1964. Theory of electron tunneling. *Reviews of Modern Physics*, 36(1), pp. 200–204.

Das Sharma, S. et al., 2011. Electronic transport in two-dimensional graphene. *Reviews of Modern Physics*, 83, p. 407.

Sutin, N., 1986. Marcus' contributions to electron-transfer theory. *The Journal of Physical Chemistry*, 90(16), pp. 3465–3466.

Velicky, M. et al., 2014. Electron transfer kinetics on mono-and multilayer graphene. *ACS Nano*, 8, pp. 10089–10100.

De Volder, M.F.L. et al., 2013. Carbon nanotubes: present and future commercial applications. *Science (New York, NY)*, 339(6119), pp. 535–539.

Xia, J. et al., 2009. Measurement of the quantum capacitance of graphene. *Nature Nanotechnology*, 4, pp. 505–509.

Yamada, H. & Bandaru, P.R., 2013. Limits to the magnitude of capacitance in carbon nanotube array electrode based electrochemical capacitors. *Applied Physics Letters*, 102(17), p. 173113.

Yamada, H. & Bandaru, P.R., 2014. Enhanced electrical current densities in electrochemical systems through the use of nanostructured electrodes. *Applied Physics Letters*, 104(21), p. 213901.

Yamada, H. & Bandaru, P.R., 2016. Electrochemical kinetics and dimensional considerations, at the nanoscale. *AIP Advances*, 6(6), p. 065325.

Zaretski, A.V. et al., 2015. Metal-assisted exfoliation (MAE): green, roll-to-roll compatible method for transferring graphene to flexible substrates. *Nanotechnology*, 26(4), p. 045301.

Zhang, Y. et al., 2005. Experimental observation of the quantum Hall effect and Berry's phase in graphene. *Nature*, 438, pp. 201–204.

Zhu, W. et al., 2012. Structure and electronic transport in graphene wrinkles. *Nano Letters*, 12(7), pp. 3431–3436.

Index

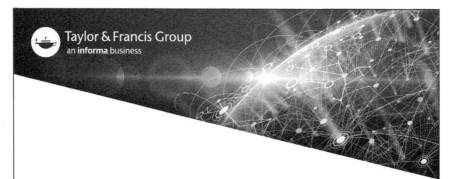

Printed and bound by CPI Group (UK) Ltd, Croydon, CR0 4YY

17/10/2024

01775672-0015